Principles of Thermal Ecology

Principles of Thermal Ecology

Temperature, Energy and Life

Andrew Clarke
British Antarctic Survey, Cambridge, UK
and
School of Environmental Sciences, University of East Anglia, Norwich, UK

With best wishes
Andrew Clarke

OXFORD
UNIVERSITY PRESS

Great Clarendon Street, Oxford, OX2 6DP,
United Kingdom

Oxford University Press is a department of the University of Oxford.
It furthers the University's objective of excellence in research, scholarship,
and education by publishing worldwide. Oxford is a registered trade mark of
Oxford University Press in the UK and in certain other countries

© Andrew Clarke 2017

The moral rights of the author have been asserted

First Edition published in 2017

All rights reserved. No part of this publication may be reproduced, stored in
a retrieval system, or transmitted, in any form or by any means, without the
prior permission in writing of Oxford University Press, or as expressly permitted
by law, by licence or under terms agreed with the appropriate reprographics
rights organization. Enquiries concerning reproduction outside the scope of the
above should be sent to the Rights Department, Oxford University Press, at the
address above

You must not circulate this work in any other form
and you must impose this same condition on any acquirer

Published in the United States of America by Oxford University Press
198 Madison Avenue, New York, NY 10016, United States of America

British Library Cataloguing in Publication Data
Data available

Library of Congress Control Number: 2017943726

ISBN 978–0–19–955166–8 (hbk.)
ISBN 978–0–19–955167–5 (pbk.)

DOI 10.1093/oso/9780199551668.001.0001

Printed and bound by
CPI Group (UK) Ltd, Croydon, CR0 4YY

Links to third party websites are provided by Oxford in good faith and
for information only. Oxford disclaims any responsibility for the materials
contained in any third party website referenced in this work.

Preface

This book has taken several years to write, but it has been much longer in preparation. In fact I can probably trace it right back to growing up in central London. As a young schoolchild I was a regular visitor to the local library where the staff were very tolerant of me sitting on the floor to read. I devoured anything I could find on birds and animals, and one day as I turned over the pages of a newly arrived book I came across a grainy but evocative image of a forbidding mountainous landscape. Little did I know that I would later spend almost five years of my life undertaking research at this very spot: South Georgia, an isolated island in the Southern Ocean and on the edge of Antarctica.

Fieldwork in polar regions inevitably focuses attention on temperature, and the more I thought about the links between temperature and ecology, the more interesting they became. They proved to be the focus of my entire research career, so it was perhaps inevitable that once I retired from active fieldwork, I would write a book on temperature.

Although I often referred to the classic review edited by Precht and colleagues (*Temperature and life*, Springer, 1973), for much of my research career the book I kept immediately to hand was the ground-breaking *Strategies of biochemical adaptation* by Peter Hochachka and George Somero (Saunders, 1973). With its regular revisions this has remained a definitive work for ecological physiologists. Apart from those general physiology texts that had chapters on temperature, a few highly specialised symposium volumes and *Temperature biology of animals* (Andrew Cossins and Ken Bowler, Chapman & Hall, 1987), there was nothing else available for those needing an overview of thermal ecology until Michael Angilletta's *Thermal adaptation* (OUP, 2009). This takes an innovative behavioural and evolutionary perspective, reflecting the approach of those ecologists who work with animals in the field rather than cells or proteins in the laboratory. It therefore complements the approach I have tried to take, which is a more traditional path through physiology to ecology.

In writing this book I have been able to present some topics in the way I would like to have been introduced to them as a student. To take one example, metabolism is so much easier to understand when it is built from its foundations through a single, simple idea: energy carried by electrons is used to generate a proton gradient which powers the synthesis of ATP. With this one concept, all of the bewildering detail of the metabolic chart and the enormous evolutionary variety of microbial metabolism have a common framework. I have also been able to indulge myself by exploring the history of some topics. It has been fun to track down some of the very early literature but salutary to find that some of this did not say what I (and many others) thought it said, and that some classic references had clearly not been read for many years as they are systematically mis-cited.

Like all scientists I have benefited from support, encouragement and the continual exchange of ideas with colleagues. One of the great unspoken benefits of fieldwork in distant places is the time spent travelling, or in bad weather, when there is an opportunity to talk science away from the everyday concerns of a modern research institute. These conversations are essential to scientific development; they are where ideas are explored, errors corrected and new projects planned.

All young scientists need guidance as they set out on a career, and I was lucky to receive this from a series of outstanding mentors. In particular I benefitted from support from two notable department heads, Dick Laws and Nigel Bonner, and also from John Lawton at the Natural Environment Research Council; in addition there was sound advice at important moments from Inigo Everson, John Croxall, John Gray and John Sargent. Throughout my career in Antarctic science I have enjoyed discussion with many colleagues, both within the institute and abroad, conversations which have done much to sharpen my thinking and eradicate errors, so thanks are due to Mike Angilletta, David Barnes, Chris Cheng, Steven Chown, Melody Clark, Pete Convey, Alistair Crame, Paul Dayton, Bill Detrich, Art DeVries, Hugh Ducklow, Brian Enquist, Inigo Everson, Ray Huey, Roger Hughes, David Jablonski, Tim Jickells, Ian Johnston, Adrian Friday, Kevin Gaston, Nick Lane, Simon Laughlin, Barry Lovegrove, Carlos Martínez del Rio, Mike Meredith, John Morris, Eugene Murphy, Lloyd Peck, Hans-Otto Pörtner, Langdon Quetin, Robin Ross, Bruce Sidell, Victor Smetacek, George Somero, Chris Todd and Paul Tyler.

Science involves teamwork, and I was particularly lucky to have worked with a series of wonderful research assistants, Lesley Holmes, Liz Prothero-Thomas and Nadine Johnston. It is also a pleasure to thank two librarians at the British Antarctic Survey, first Christine Phillips and then Andrew Gray, for their outstanding professional support in tracking down every obscure historical reference I could challenge them with. Scholarship is all about checking primary sources, and I could not have done this without them.

I am also grateful to those colleagues who agreed to read through draft chapters. All of them took care and time, and they all found things I had got wrong. Needless to say any remaining errors are down to me. My thanks to David Atkinson, Martin Baker, Chris Cheng, Melody Clark, Pete Convey, Paul Cziko, Richard Davies, Art DeVries, Jack Duman, Fritz Geiser, Michael Kearney, John King, Nick Lane, Barry Lovegrove, Andrew McKechnie, Ian Renfrew, Brent Sinclair, George Somero, Jim Staples, John Turner and Eric Wolff.

Enormous thanks are due to Ian Sherman, who tried for years to persuade me to write this book before finally succeeding, and his team at OUP, especially Lucy Nash and Bethany Kershaw, who saw it through to completion. During the writing of this book I have also received institutional support from the British Antarctic Survey, and the School of Environmental Sciences at the University of East Anglia, for which I am extremely grateful. And finally, because it is the most important, deep thanks to my patient wife, Gill, who has provided unstinting support and encouragement through my entire career.

Andrew Clarke
Norfolk, February 2017

Contents

1 Introduction — 1
    1.1 A note on units — 3
    1.2 General principles and the importance of natural history — 3

2 Energy and heat — 6
    2.1 Energy — 7
    2.2 The First Law of Thermodynamics — 7
    2.3 Energy at the atomic scale — 11
    2.4 Internal energy — 14
    2.5 Energy and heat — 15
    2.6 Thermodynamic systems — 16
    2.7 Work and energy revisited — 17
    2.8 Entropy — 17
    2.9 The Second Law of Thermodynamics — 22
    2.10 The Third Law of Thermodynamics — 23
    2.11 Gibbs (free) energy — 24
    2.12 Summary — 25

3 Temperature and its measurement — 29
    3.1 The Zeroth Law of Thermodynamics — 29
    3.2 What exactly is temperature? — 29
    3.3 Thermometry: the measurement of temperature — 32
    3.4 The gas thermometer and absolute zero — 34
    3.5 The thermodynamic temperature scale — 35
    3.6 A new definition of temperature? — 37
    3.7 Types of scale — 37
    3.8 Thermometry: the practical measurement of temperature — 38
    3.9 The International Temperature Scale — 39
    3.10 The measurement of temperature in ecology — 39
    3.11 Some general points about thermometry in ecology — 46
    3.12 Measuring past temperatures: palaeothermometry — 47
    3.13 Summary — 51

4 Energy flow in organisms — 56

- 4.1 Conduction — 57
- 4.2 Convection — 60
- 4.3 Thermal radiation — 62
- 4.4 Thermal relationships of organisms — 68
- 4.5 Evaporation — 73
- 4.6 The full biophysical treatment of fluxes — 73
- 4.7 Flows of chemical potential energy — 74
- 4.8 A complete energy budget? — 79
- 4.9 Summary — 79

5 Water — 82

- 5.1 Water and life on Earth — 82
- 5.2 Why water? — 83
- 5.3 The hydrogen bond — 85
- 5.4 The physical properties of water — 86
- 5.5 Temperature and water — 88
- 5.6 Solubility of gases (oxygen and carbon dioxide) — 92
- 5.7 Calcium carbonate solubility — 94
- 5.8 The hydrophobic interaction — 95
- 5.9 Water in cells — 98
- 5.10 Summary — 99

6 Freezing — 103

- 6.1 The freezing of water — 103
- 6.2 Ice in the natural environment — 104
- 6.3 Freezing and organisms: some general principles — 107
- 6.4 Cryobiology and ecology — 109
- 6.5 Single cells in the environment — 110
- 6.6 Freezing and multicellular organisms — 113
- 6.7 Freezing and fish — 113
- 6.8 Freezing in intertidal marine invertebrates — 117
- 6.9 Freezing in arthropods — 118
- 6.10 Freezing in terrestrial vertebrates — 123
- 6.11 Freezing in plants — 124
- 6.12 Concluding remarks: the wider context — 125
- 6.13 Summary — 127

7 Temperature and reaction rate — 131

- 7.1 Rate of reaction — 131
- 7.2 Temperature and reaction rate — 134
- 7.3 Transition state theory — 137
- 7.4 Enzyme-catalysed reactions — 138
- 7.5 Evolutionary adaptation of reaction rate to temperature — 140
- 7.6 Managing the cellular environment — 147

7.7	The paradox of adaptation	152
7.8	A subtle problem: describing the relationship between physiological rates and temperature	153
7.9	Summary	159

8 Metabolism — 163

8.1	Metabolism is simple	163
8.2	Gaining electrons	166
8.3	Photosynthesis	166
8.4	Where is the energy?	167
8.5	Using electrons: metabolism	168
8.6	Control of ATP synthesis	171
8.7	Metabolism in ecology	172
8.8	What processes comprise resting metabolic rate?	177
8.9	Temperature and metabolic processes at the molecular level	182
8.10	Temperature and whole-organism metabolism	182
8.11	Why does resting metabolism increase with temperature?	188
8.12	Metabolic cold adaptation	189
8.13	Summary	191

9 Temperature regulation — 196

9.1	Behavioural thermoregulation	200
9.2	How do animals sense temperature?	201
9.3	Generating heat	204
9.4	Avoiding overheating	209
9.5	Summary	211

10 Endothermy — 214

10.1	What is endothermy?	214
10.2	Body temperature in endotherms	215
10.3	Control of body temperature	218
10.4	Ecology and body temperature in endotherms	219
10.5	What sets the maximum body temperature an endotherm can have?	221
10.6	How do endotherms generate their heat?	222
10.7	A general picture of obligate heat generation in endotherms	227
10.8	Thermoregulation	228
10.9	Winter is coming: keeping warm below the TNZ	230
10.10	Keeping cool: endotherms above the TNZ	232
10.11	The evolution of endothermy	234
10.12	Summary	240

11 Torpor and hibernation — 245

11.1	Daily torpor	247
11.2	Hibernation	249
11.3	Ecology of torpor and hibernation	250
11.4	The physiology of torpor	257

11.5	Torpor, hibernation and size	260
11.6	A phylogenetic perspective	263
11.7	Summary	264

12 The Metabolic Theory of Ecology — 267

12.1	The influence of size: scaling	267
12.2	The scaling of metabolic rate	268
12.3	The West, Brown and Enquist model	271
12.4	The influence of temperature	276
12.5	The central equation of the Metabolic Theory of Ecology	277
12.6	The status of the MTE as a theory	278
12.7	MTE and mammals revisited	279
12.8	Does the Metabolic Theory of Ecology constitute a general theory of thermal biology?	280
12.9	Summary	280

13 Temperature, growth and size — 285

13.1	How should we measure growth rate?	285
13.2	The energetics of growth	286
13.3	Temperature and growth	289
13.4	Growth and temperature in the natural environment	293
13.5	Modelling growth	296
13.6	Temperature and size	299
13.7	Summary	304

14 Global temperature and life — 308

14.1	The global distribution of temperature on land	308
14.2	The global distribution of oceanic temperatures	314
14.3	Are there thermal limits to life?	316
14.4	Temperature thresholds for life	318
14.5	Low temperature limits	319
14.6	The special case of endotherms	321
14.7	High temperature limits to life	321
14.8	The thermal limits to life on Earth	324
14.9	Summary	325

15 Temperature and diversity — 329

15.1	What is biological diversity?	329
15.2	Some practical issues	330
15.3	Global diversity: land and ocean	331
15.4	Global patterns of diversity	333
15.5	Mechanism	336
15.6	Temperature and diversity	338
15.7	Diversity and energy	342
15.8	Confounding energy with temperature	345
15.9	Can energy explain everything?	346

15.10	A final comment: the role of time	346
15.11	Summary	349

16 Global climate change and its ecological consequences — 354

16.1	The Earth's atmosphere	354
16.2	Why is the Earth's temperature what it is?	355
16.3	The greenhouse effect	356
16.4	The temperature at the Earth's surface	359
16.5	Ice ages	361
16.6	Recent climate change	363
16.7	Organism response to climate change	371
16.8	Responses to recent climate change	374
16.9	Final comments	383
16.10	Summary	384

17 Ten principles of thermal ecology — 391

17.1	A few final thoughts	393

References — 395
Index — 455

CHAPTER 1

Introduction

> With the ratification of long tradition, the biologist goes forth, thermometer in hand, and measures the effects of temperature on every parameter of life. Lack of sophistication poses no barrier; heat storage and exchange may be ignored or Arrhenius abused; but temperature is, after time, our favourite abscissa. One doesn't have to be a card-carrying thermodynamicist to wield a thermometer.
>
> **Steven Vogel**[1]

Temperature affects everything. It influences all aspects of the physical environment and governs any process that involves a flow of energy, setting boundaries on what an organism can or cannot do. Temperature is at once the most pervasive aspect of the environment in relation to the physiology and ecology of organisms, and also probably the most frequently measured. But, as Steven Vogel expressed so memorably (above), ecologists are not always rigorous in their approach to temperature. Even today it is possible to find newly published studies which confuse temperature with energy, or muddle temperature and heat.

This book is not intended to be a thorough review of the entire field of thermal ecology. It could not be, for the field is now so enormous that any such review would be massive and unreadable[2]. Instead my aim is to provide a simple path through the subject, hoping that the key principles emerge clearly.

When dealing with a subject as vast as the relationship between organisms and temperature, it is helpful to break the subject down into manageable blocks. The material in this book falls into three basic broad topic areas:

1. The flow of energy in and out of the organism.
2. The influence of temperature on what an organism can do.
3. How these affect the way an organism interacts with its environment, including other organisms around it.

This is a fairly traditional approach but it allows us to establish a rigorous framework for understanding how organisms perform before we move to the ecological consequences.

The first four chapters deal with physics. Why physics in an ecology book? The short answer is because it is important, and while the basics of thermal physics can be found in any standard textbook they are not always presented in a way that is helpful to an ecologist. The book therefore opens with a simple treatment of energy, work and heat (these are not the same things) and then relates these to temperature (which is different again); these topics are covered in Chapters 2 and 3. This involves the tackling some simple thermodynamics, but despite this subject's daunting reputation the basic concepts are straightforward and they lay important foundations for a treatment of how organisms exchange energy with the environment (Chapter 4).

The next two chapters (5 and 6) could be seen as forming what musicians would call a bridge passage, in that they link the thermal physics section with the thermal physiology that follows. The first of these chapters is concerned with water. Organisms are mostly water and water is a most unusual substance. So unusual in fact that its quirks and peculiarities are integral to why life on Earth exists at all. Moreover, the way water behaves as its temperature changes affects all of physiology and ecology. Without water there is no life and

Principles of Thermal Ecology. Andrew Clarke, Oxford University Press (2017).
© Andrew Clarke 2017. DOI 10.1093/oso/9780199551668.001.0001

without an understanding of water's eccentricities our knowledge of physiology is incomplete. The chapter that follows then deals with what happens when water freezes, a more widespread ecological challenge than is often appreciated.

The central section deals with the relationship between organism performance and temperature. The first chapter of this section (7) tackles the fundamentals of how reaction rate is affected by temperature, including the important issue of how best to capture this relationship mathematically. This leads naturally to a discussion of metabolism at the whole organism level (Chapter 8). The next two chapters deal with the regulation of body temperature, first in ectotherms (Chapter 9) and then in mammals and birds (Chapter 10). Endothermy allows mammals and birds to maintain a high and stable body temperature, but it is expensive, and so when food is short it can be relaxed either by short-term torpor or longer-term hibernation (Chapter 11). We follow this by examining a highly influential attempt to build a coherent integrated theory of how temperature affects whole-organism metabolism, the Metabolic Theory of Ecology (Chapter 12). Finally in this section we explore how temperature and energy influence the adult size of an organism, and how fast it grows to achieve this (Chapter 13).

In the final section we explore thermal ecology itself. Having examined how just one feature of the environment, temperature, affects organism performance, we now add the important extra dimension of interactions with other organisms. The first chapter (14) sets the scene by examining the broad-scale relationship between temperature and life on Earth. In it we explore what might be the thermal limits to life on Earth, asking whether there is anywhere on the planet with liquid water at temperatures that preclude life. This is also something of a bridge chapter, and it is followed (Chapter 15) by an examination of the potential links between temperature, climate and diversity, asking the question of how (if at all) purely physical factors influence the number and variety of species living in a particular place. This chapter also touches upon the important aspect of deep time.

We then tackle the important subject of how organisms are responding to climate change (Chapter 16). This is such an important topic that it requires a detailed treatment. We start by seeing how the Earth's climate is regulated and why it is now changing. This necessitates some atmospheric physics and a little oceanography, because a thorough grasp of the way climate is regulated is essential for understanding its effects on organisms. It also needs a strong historical perspective because it is only once we can distinguish long-term climate change caused by man from the natural variability that characterises the Earth's climate system that we can understand what is happening to the flora and fauna. This chapter may spring a surprise or two, for not all of the much-discussed examples of recent climate change are necessarily caused by man and his activities (but many others are). The book then concludes with a very brief survey of what I believe to be the principles of thermal ecology.

In many of the physiological and ecological chapters I have taken time to introduce the basics of the topic before moving on to explore the influence of temperature. For example in discussing the way temperature influences metabolism in Chapter 8 I first lay out some of the principal features of metabolism. While this might seem to be duplicating material to be found in any standard physiology or ecology text, it is important. 'Metabolism' can mean different things to physiologists and ecologists, so I feel it is essential to establish precisely what I mean by the term. I have taken a similar approach in discussing the influence of temperature on growth and its relationship with diversity[3].

Examples are important and I have used these extensively, mixing classic studies with more recent work and where possible selecting examples from all continents and across diverse taxa. I have also, quite deliberately, concentrated on referring back in history to the key studies that defined particular disciplines. These are two reasons for this (well, the third is that I find the history fascinating). The first is that, as the French philosopher Auguste Comte famously commented, to understand a science it is necessary to know its history. In some cases, such as thermodynamics, history provides a clear and logical way to introduce a potentially difficult topic; in others, such as temperature, it is simply a fascinating intellectual story.

The second reason is that modern science advances so rapidly that for students starting out on a research career there is often little time to look backwards at where and how their chosen topic of study arose. The requisite nod to the development of the topic typically consists of citation to a popular recent review article (often unread). While this is understandable, it is also unfortunate: the field of thermal ecology has several examples of old concepts being rediscovered, or inadvertently renamed, and presented as novel. It is a salutary lesson to explore history, especially as it can remind us that the pioneers of our science were smart.

1.1 A note on units

I have standardised on the use of SI units. These will be familiar to anyone with a scientific or engineering background. The SI (Système international d'unités) was developed in France and adopted at about the same time as the Greenwich meridian was taken to define zero longitude. Ironically, given the subject of this book, SI units are least strictly observed in the field of temperature. The SI unit of temperature is the Kelvin (K, not °K, as we shall see in Chapter 3), but ecologists rarely use this because it is inconvenient over the range of temperatures in which they are most interested. Instead we use Celsius, while in North America Fahrenheit is still widely used domestically. In this book I have opted for clarity and given temperatures in Celsius, but with the SI unit added where needed.

However for technical reasons that are discussed in Chapter 3, wherever temperatures are manipulated mathematically, I have used exclusively SI units. This leads to a slightly unfamiliar presentation where a particular temperature may be given in Celsius, as this is familiar, but temperature differences and rates are given in SI units (for example the difference between the melting and boiling point of water is 100 K, and rates of temperature change are given as K per unit time). This approach is not always intuitive, and can cause problems with some referees and journal editors, but it is rigorous and I remain unrepentant.

1.2 General principles and the importance of natural history

Ecologists are a varied bunch. Some are motivated by finding general principles, others are interested in why various species do things differently. I am fascinated by both. I want to know how and to what extent organisms are constrained by the laws of physics, and I have spent most of my professional life thinking about how temperature affects organisms. As an undergraduate I spent a summer field season in Svalbard and like many before me, and since, I became fascinated by the polar regions. For my doctoral thesis I worked for over two years in the Antarctic, at South Georgia. My animal (we always think of the species we work on as 'our' animal) was a small benthic shrimp, *Chorismus antarcticus*, and I became intrigued as to why this species showed such marked seasonality in its growth when the annual variation in seawater temperature was only a few degrees. I had been taught by ecologists who knew the North-East Atlantic where everything was (apparently) driven by seasonal temperature change, and yet this omnivorous animal remained highly seasonal in an environment that was only mildly seasonal in terms of temperature. I was perplexed. And why, despite the low temperature of the water, did it grow so fast in summer? Was its growth limited by food, or temperature or both? These questions have continued to fascinate, for as is so often the case in ecology, simple questions turn out to have complex answers.

As well as pursuing a career as a professional ecologist, I have been a birder for as long as I can remember. The process of learning to identify birds raised quite different questions. I grew up in England where we have three leaf-warblers (*Phylloscopus*), similar in appearance but with different songs. Careful work by the early naturalists had established the subtle differences in habitat that meant you tended not to find all three species together, but when I eventually made it to the Himalaya and far eastern Asia I was confronted with a bewildering array of species of leaf-warblers which defied immediate explanation[4].

These are the two contrasting features of ecology: the simple underlying rules and the bewildering detail. These two ways of looking at ecology are not, or should not be, mutually exclusive. Their interplay can be exemplified by a plot of basal metabolic rate as a function of size (Figure 1.1). I have chosen to plot data for fish, as these data came from a study of fish energetics I did with Nadine Johnston in the 1990s, but I could equally have used data for mammals, birds, reptiles, insects or any one of several groups of marine invertebrates.

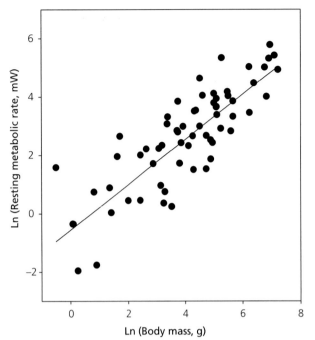

Figure 1.1 Standard metabolic rate in teleost fish as a function of body mass. Each data point represents the standard metabolic rate of a fish of the median body mass for the fish used in the original study. There is one data point per species, and the data have not been corrected for temperature. Metabolic rate data have been converted to SI units before plotting, and the line is fitted by least-squares linear regression.

This plot has three key features. The first is so obvious that it might easily be overlooked: the plot is clearly linear. To achieve this, however, both variables have been transformed (in this case to natural logs). The second feature is that the relationship has a clearly defined slope, and the third is that there is marked variance about the line.

The slope of a fitted least-squares regression line is less than unity (it is actually 0.78 for this data set). Since we might have expected the slope to be unity, so that a doubling of size would lead to a doubling of oxygen demand, a slope of 0.78 suggests the influence of one or more constraints of some type. Much effort has gone into elucidating what the constraint(s) might be, and why the slope is what it is (this is discussed in Chapters 8 and 12). The variance about the line reflects ecological variety and is a measure of the extent to which evolution has been able to work around these constraints. In simple terms, the slope reflects the constraint(s) of physics and the variance reflects evolution and ecology. To see ecology in the round we have to understand both: we need to recognise the constraints under which organisms operate, and the extent to which these can be circumvented.

In recent years there has been a dramatic revival of interest in broad patterns in ecology. A key factor in this has been the availability of electronic data storage and powerful computers for analysis, but the raw material for these broad-scale analyses stems from the dedicated work of thousands of fieldworkers and natural historians, professional and amateur, who have documented the details of the lives of myriads of organisms from all over the world. Sometimes satirised by the ignorant as 'stamp collecting' or 'pebble counting', this enormous body of work provides the basis without which ecological generalisations would not be possible.

The analysis of large ecological data sets to detect patterns over large spatial scales (typically continental or global) has become known as macroecology. More recently the analysis of patterns in the physiological performance of organisms over similarly large scales has been termed macrophysiology[5]. Both will feature throughout this book.

Macroecology and macrophysiology involve the statistical analysis of large data sets. These analyses range from the simple, concentrating on defining the dominant patterns, to the highly complex, examining the subtle ecological signals contained in the variance. Which approach is best depends on the question of interest. There is, however, an important point which is often missed in the strident arguments to be found in this literature: the nature of the individual.

The mathematics that describe the behaviour of atoms and provide our understanding of the effect of temperature on chemical reactions (discussed in Chapter 7) assume that all the atoms in the population are identical: one atom of oxygen or one molecule of glucose behaves exactly as another. On this assumption are built simple equations that describe the bulk behaviour of a large number of atoms or molecules to a very high degree of accuracy and precision.

Organisms are not like this: every species is different from all others, for each has a unique evolutionary history. This history is reflected in the chemical makeup of the species: indeed we use differences in nucleic acid chemistry to reconstruct that history. Furthermore, not all individuals within a species are identical, either in their history or their physiological performance. The statistical techniques used by ecologists were developed specifically to cope with this variability between individuals[6].

There is a really important issue here: the extra variability introduced by the process of evolution makes it difficult to predict the thermal behaviour of an organism from first principles. We cannot, for example, start from physics and build a mechanistic model of how insect metabolism will (or should) vary with habitat temperature; its physiology is just too complicated. Furthermore even if we could do this, the variability between individuals of a species means that the answer would be different for each and every individual. All we can do is observe how metabolism varies with temperature in the species of interest and describe the relationship statistically. The results of such an analysis may point to the existence of a constraint (as in Figure 1.1), and may even give hints as to what that constraint might be. We need to remember, however, that a description is all that it is. If we want to use the relationship to predict what might happen in different circumstances, say for an organism larger than any in the data set, or for a different temperature, we have to be careful. If we do not know the mechanism underpinning the pattern we describe, and typically we do not, we cannot know how reliable our prediction is.

This emphasises how important it is to have an interplay between ecologists interested in fundamental principles who aim for simple patterns and ecologists interested in the natural history underpinning the variability in the observed patterns. Neither is complete in itself, and both are essential to our understanding of Nature. We could not have a better example of this than in the famous final paragraph of Darwin's *Origin of species*, which captures both the profundity of Darwin's grasp of a simple fundamental process and his wonder at the outcome.

Notes

1. This quotation comes from the opening chapter of Steven Vogel's delightful book *Life in moving fluids* (Vogel 1981). Reproduced with permission of Princeton University Press.
2. The thorough review by Precht et al. (1953) runs to 779 pages, and the field has grown enormously since then.
3. Definitions are important for clarity of communication, in science as elsewhere. Otherwise we live in the looking-glass world, where as Humpty Dumpty said to Alice 'When I use a word . . . it means just what I choose it to mean — neither more nor less'. This comes from Lewis Carroll's *Through the looking-glass, and what Alice found there* (1871).
4. It was the English parson and naturalist Gilbert White (1720–1793) who first recognised the differences between *Phylloscopus collybita* (Chiffchaff), *Phylloscopus trochilus* (Willow Warbler) and *Phylloscopus sibilatrix* (Wood Warbler), and described these in his famous collection of letters to Thomas Pennant and other leading naturalists, *The natural history and antiquities of Selbourne* (1789). The Himalaya is now recognised as the centre of diversification for leaf-warblers (Family Phylloscopidae). The classic description is Ticehurst (1938), and a modern review is Price (2010).
5. The term 'macroecology' was coined by James Brown and Brian Maurer in a paper in *Science* in 1989. Both subsequently developed the idea in books (Brown, 1995; Maurer 1999). Macrophysiology was introduced by Chown et al. (2004) and Gaston et al. (2009).
6. The foundations of ecological statistics were laid by the English mathematician and biologist Sir Ronald Fisher (1890–1962). Of particular importance was his development of analysis of variance (ANOVA) and maximum likelihood techniques (see Fisher et al. 1943). He was one of the founders, along with J.B.S. Haldane and Sewall Wright, of population genetics.

CHAPTER 2

Energy and heat

> It frequently happens, that in the ordinary affairs and occupations of life, opportunities present themselves of contemplating some of the most curious operations of nature; and very interesting philosophical experiments might often be made, almost without trouble or expence (*sic*) by means of machinery contrived for the mere mechanical purposes of the arts and manufactures.
> **Count Rumford (Benjamin Thompson)**[1]

No ecologist wanting to understand how organisms interact with their environment can afford to ignore the fundamental aspects of energy and how it flows. Whether our concern is a bacterium in a hot spring, a lichen on a frozen mountain-top or a coral on a tropical reef, we need to know what heat is, why energy flows and how heat and temperature are related.

In this chapter we will examine the nature of energy and heat, laying the foundation for the discussion of temperature that follows. In doing so, we will introduce some key concepts from thermodynamics. This is a topic which frightens many a biologist, but it need not. The basic ideas are straightforward (well, maybe except for entropy) and a grasp of these is important for any ecologist interested in temperature, metabolism or energy flow. It will help avoid some of the mistakes which can still be found in the ecological literature, where it is not uncommon to see heat and temperature confused, entropy misunderstood and some fundamental concepts ignored.

Part of the reason for this may be that heat and temperature are such familiar features of everyday life that we feel intuitively we understand them. Like many animals we possess sensors which give us an indication of the temperature of both ourselves and our environment, sensors that are essential to our wellbeing. They tell us when we are too warm or too cool, and they warn us that fires are hot and ice is cold. And yet these very same sensory systems also make us aware that things may not be quite so straightforward. For example steel feels cold whereas wood feels warm, even when we know instinctively (or can show with a thermometer) they are actually at the same temperature. And once we had thermometers to hand, we could see that objects supplied with the same amount of heat change temperature at different rates. It is even possible for considerable quantities of heat to be supplied to an object while its temperature remains unmoved, as in the melting of ice.

Clearly temperature and heat are different things. While a grasp of this difference goes back, as so often in science, to the ancient Greeks, the modern science of thermodynamics only developed at the time of the Industrial Revolution in the late eighteenth and early nineteenth centuries. Although the two events are often linked causally, the improvement in the performance of steam engines actually proceeded quite independently of the intellectual formulation of concepts of heat and energy. Indeed the Industrial Revolution was well underway, with locomotives pulling trains and steam engines powering industry, decades before the concept of energy was developed[2].

Energy is central to the thermal ecology of organisms, and so we start by examining this most basic topic of all.

Principles of Thermal Ecology. Andrew Clarke, Oxford University Press (2017).
© Andrew Clarke 2017. DOI 10.1093/oso/9780199551668.001.0001

2.1 Energy

Energy is one of the fundamental concepts in all of science, and the flow of energy has been central to our view of how organisms and ecosystems function since the very start of ecology. We might therefore expect physicists to have a very clear idea of what energy is. Surprisingly, this is not so.

2.1.1 What is energy?

Ask any scientist what energy is and you are likely to get the reply that energy is *the capacity to do work*. This is the definition we were probably all taught at school[3]. The problem with this definition, however, is that it tells us what energy *does*, not what energy *is*. Moreover, it leaves us with the problem of then explaining the nature of work. One reason that energy is so difficult to define or describe, and thus remained elusive for so long, is that unlike physical quantities such as mass, volume or pressure, it is not directly measurable. We cannot place an object into an instrument and obtain a measure of its energy content; all we can do is infer its energy from other characteristics, such as its temperature or the ability to do work[4].

This difficulty was explained in typically idiosyncratic style by Richard Feynman in his famous *Lectures on physics*[5]:

There is a fact, or if you wish, a law, governing all natural phenomena that are known to date. There is no known exception to this law—it is exact as far as we know. The law is called the conservation of energy. It states that there is a certain quantity, which we call energy, that does not change in the manifold changes which nature undergoes. That is a most abstract idea, because it is a mathematical principle; it says that there is a numerical quantity which does not change when something happens. It is not a description of a mechanism, or anything concrete; it is just a strange fact that we can calculate some number and when we finish watching nature go through her tricks and calculate the number again, it is the same.

This might imply that the conservation of energy, which we now recognise as one of the most fundamental laws of physics, is little more than an accounting device. The power of the idea of the conservation of energy is that once it was realised that energy could be converted between its different forms, but could never be created or destroyed, then scientists had a window into the most basic operation of the universe, and a powerful rule that has never been found to be broken[6].

2.2 The First Law of Thermodynamics

The principle of the conservation of energy is now formalised as the First Law of Thermodynamics. A typical statement of the first law is:

Energy can neither be created nor destroyed, but only changed from one form to another.

No principle such as this ever springs fully formulated into the scientific consciousness, and the essential idea underpinning the first law was formulated by the Swiss-born Russian chemist Germain Hess in 1840. Hess recognised that the energy change for a reaction that takes place in a series of steps was identical to the sum of the energy changes at each individual step. Then in 1841, the German Julius Robert von Mayer enunciated one of the original statements of the conservation of energy, when he stated that *energy can be neither created nor destroyed*[7].

The First Law has important consequences for ecologists concerned with the flow of energy through organisms or ecosystems: it tells us that the inputs and outputs must balance, exactly. If the energy inputs are not accounted for by the sum of all outputs, then something has been missed. This may well be a result of failing to track energy that changes in form within the organism, particularly energy dissipated to the environment as heat.

We now recognise that all conservation laws in physics are a reflection of fundamental symmetries in nature, a relationship known as Noether's theorem[8]. The conservation of energy is an inevitable consequence of the symmetry (or invariance) of the laws of physics with time. An example of this symmetry would be the physicist's familiar model of colliding billiard balls. If we were to film one ball striking another and then replay the film in reverse, it would look perfectly natural. Indeed without ancillary information it would be impossible to decide which of the two versions of the film was the original. This is because the equations that describe how billiard balls interact are invariant with respect to time: Newton's laws of motion work just as well backwards as forwards. The same applies to motion such as the orbiting of the Moon about the Earth, or the planets around the Sun.

This is fine theoretically, but we are all familiar with processes in everyday life that are clearly irreversible and move only forwards in time, such as the burning of wood in a fire. Biological examples might be the development of an embryo into an adult organism, a cheetah eating a gazelle or the decay of a plant cut down and left to rot on the compost heap. These do not indicate, as was once thought, that there is something special about the energetics of living things. But they do exemplify what the physicist Arthur Eddington called the *arrow of time*, and they indicate that something else is involved[9]. Quite what that something is we will return to later.

2.2.1 Potential energy and kinetic energy

All forms of energy can be classified into one of two fundamental categories, potential energy and kinetic energy (Box 2.1). Potential energy is energy a body has by virtue of its position, for example a body's gravitational potential energy depends on that body's position in the gravitational field. Kinetic energy is energy a body has by virtue of its motion.

This classification is useful in clarifying different forms of energy, but it still does not tell us what energy is. Neither does it resolve the nature of the relationship between energy and work, or tell us whether all forms of energy are the same thing. These are deep and troubling questions which go to the very heart of nature, for as we now know there are circumstances under which energy and matter can be converted and energy is linked intimately with the nature of time.

2.2.2 Interconversion of energy and mass

Our intuition might suggest that mass and energy are different things: a given amount of matter, say a lump of copper, should have a particular mass (the amount of copper) and contain a defined amount of energy. However, since the work of Albert Einstein we have known that mass can be converted to energy, and vice versa[10]. The mathematical relationship between mass and energy is captured in perhaps the most famous equation in all of science:

$$E = mc^2$$

Box 2.1 Potential and kinetic energy

Physicists divide energy into *potential energy*, which is energy a body possesses by virtue of its position, and *kinetic energy*, which is energy a body possesses by virtue of its motion. All the familiar forms of energy can be assigned to one of these two fundamental categories.

Kinetic energy	Potential energy
Kinetic energy	Gravitational energy
Radiation energy	Electrodynamic (magnetic) energy
	Electrostatic energy
	Chemical (bond) energy
	Elastic energy

This simple classification encapsulates some surprising ideas. For example, that the mechanics of light (photons) can be treated identically to that of matter, and so radiation is a form of kinetic energy. Note that thermal energy is not listed explicitly, as this is a form of kinetic energy. Some physicists would also classify mass as a form of potential energy, but the interconversion of mass and energy is not relevant to ecology so we do not do so here.

In considering potential energy, it is also important to determine the limits to the system under consideration, if only to make the mathematics tractable. For example, in calculating the change in gravitational potential energy of a body taken to the top of a mountain, we focus on the gravitational field of the Earth, but ignore that of the Sun. This is because the change in potential energy from the Sun's gravitational field over this distance is very small. In contrast, for calculating the orbits of the planets or the trajectory of a spacecraft, then the gravitational field of the Sun is critical, as are those of the planets, but we can ignore that of the nearby stars.

where E is the energy, m is the mass and c is the velocity of light *in vacuo*[11]. One consequence of this equivalence is that mass is sometimes regarded as a form of potential energy (Box 2.1).

The very large value of c means that a tiny amount of mass is equivalent to an enormous amount of energy. For example, the difference in mass between a helium nucleus and the two protons and two neutrons from which it is forged is extremely small, but the energy released in nuclear fusion when helium is formed from hydrogen is what makes the Sun shine, and thereby is the source of energy of most biological energy on Earth.

It is worth noting in passing that there can be very small changes in mass during chemical reactions, caused by the difference in bond energies between the initial reactants and the final products. Strictly, what is conserved in a chemical reaction is thus mass plus energy. However the changes in mass are so small that while important theoretically, they are of no practical relevance to the flows of energy involved in physiology or ecology, and we will ignore them from now on.

2.2.3 Energy and work

If we are to understand the nature of energy in relation to organisms, it is necessary first to have a clear idea of the various forms of energy, the nature of work and how these are related.

A simple introduction to the concept of work is to consider what is involved in lifting a heavy object. Here work is being done against gravity, and the amount of work done is calculated simply by multiplying the weight of the object by the height through which it was lifted. Since weight is a force this relationship can be generalised: work is done when any kind of force is exerted through a distance, and the work can be calculated as the product of the force and the distance. For our example of a heavy object being lifted, the work done is given by:

$$W = mg\Delta h$$

where m is the mass of the object, g the acceleration of free fall and Δh the height through which the body was lifted. Note the mathematical convention here: Δ symbolises a finite, measurable, difference, so Δh is the difference between the height of the object before being lifted and its height afterwards[12].

Whenever a body does work on its surroundings, that body loses energy equivalent to the work done. Equally when a system does work on a body, that body gains energy equivalent to the work done. Work and energy thus have the same dimensions and units. In lifting a heavy object such as a suitcase against gravity, we have done some work. This work has been done on the suitcase, and the suitcase has thereby gained potential energy. Recall that potential energy is energy that an object has by virtue of its position in a field, in this case the Earth's gravitational field. So here the change in potential energy, ΔE_p, is a function of the gravitational field (mg) and the change of position within that field (Δh):

$$\Delta E_p = mg\Delta h$$

It is immediately obvious from these two equations that the gain in potential energy of the suitcase is identical to the work done in lifting it against gravity.

Let us change the example slightly and consider lifting a ball from the floor. As with the suitcase example above, here we have done work on the ball, lifting it against the force of gravity, and the ball has gained potential energy equivalent exactly to the work done. If we then allow the ball to fall to the ground, it gains energy of a much more obvious kind: it starts to accelerate and gains kinetic energy (Figure 2.1). As the ball falls it loses potential energy, because it is moving through the gravitational field of the Earth, but gains kinetic energy in exact proportion. Total energy is conserved (the First Law), but the nature of that energy has changed. The kinetic energy, E_k, is given by:

$$E_k = \frac{mv^2}{2}$$

where m is the mass of the ball and v its velocity.

When the ball hits the ground the kinetic energy is converted briefly to elastic potential energy, which is then immediately converted back to kinetic energy as the ball bounces. The ball slows as it rises, with kinetic energy being converted back to gravitational potential energy. Were this a perfectly elastic collision, the ball would return exactly to the

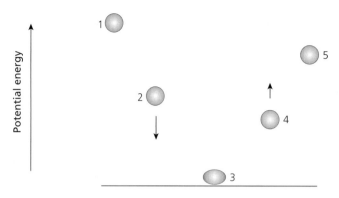

Figure 2.1 A bouncing ball shows the interconversion of potential energy and kinetic energy. At 1 the ball is stationary with a potential energy dictated by its height above the ground. On being released the ball gains kinetic energy as it accelerates down (2), in direct proportion to the decrease in potential energy. When the ball hits the ground (3) the kinetic energy is converted briefly to elastic potential energy, with some also going to other forms of energy such as sound and frictional heat. The ball then rebounds (4) as the elastic potential energy is converted to kinetic energy but the ball never reaches the original height (5) because of the conversion of some of the kinetic energy to sound and heat.

height from which it was released. In reality some of the kinetic energy is also converted to heat and sound, and so at each successive bounce the ball rises to a lower height. Although the conversion of kinetic energy to heat and sound is sometimes described loosely as a loss of energy, the First Law is not being broken. It is simply that in the real world the energy is partitioned into a wider range of forms than in an idealised thought experiment. The First Law remains intact.

This simple illustration contrasts very nicely the two fundamental types of energy, potential and kinetic energy, which together these make up the total energy of a body:

$$E = E_k + E_p$$

where E is the total energy, E_k the kinetic energy and E_p the potential energy.

Although we have established the basic principles with simple physical examples, it is worth using these principles to see what is happening in a more complex biological example, namely a bird such as a hawk taking off. Here the bird is doing work against gravity, and the energy for that work comes from chemical energy in the muscles. As the hawk rises its potential energy increases in proportion to the work done. If it spots a prey item, this potential energy is converted to kinetic energy as the hawk stoops. However, as with the bouncing ball, none of these conversions is perfectly efficient and some energy will be dissipated as sound and heat: many birds make noise in flight and in all organisms muscular activity generates heat. The First Law tells us that a budget calculated for the hawk taking off and stooping on prey will only balance if all of these forms of energy are accounted for. Some of these are very difficult to measure, and this is why it is so difficult to make energy budgets for organisms or ecosystems balance.

2.2.3 Work, energy and heat

For centuries scientists and philosophers had struggled to understand the nature of heat. Although Isaac Newton had suggested in the early eighteenth century that heat was comprised of small particles in motion, the idea that came to dominate science was that heat was a fluid, *caloric*, which was released when heat moved from one body to another. This was proposed by Antoine Lavoisier in 1789 as an alternative to the long-held idea that combustion involved the release of the hypothetical substance *phlogiston*. Lavoisier was an exceptionally careful experimenter, and he gave particular attention to weighing all of the reactants and products in the processes he was studying. Lavoisier became convinced that phlogiston did not exist, and instead he proposed that heat was an invisible, tasteless, odourless, weightless fluid, and he called this *calorific fluid*. He postulated, not unreasonably, that hot bodies contained more of this fluid than did cold ones, and that a given body contained a set amount of heat[13].

The first indication that there was something incomplete in these ideas came from the work of Benjamin Thompson (later Count Rumford, the name by which he is usually known). Rumford spent time in Germany, where he was given the task of supervising the improvement of munitions at the arsenal in Munich. At that time cannons were produced by casting a solid block of metal and then boring this out (hence the modern term bore for the size of a rifle or artillery piece). Rumford was intrigued by the immense amount of heat produced during the boring process, and he set about measuring this. He devised a deliberately blunt boring tool to maximise the amount of friction, and immersed the cannon in water to measure the heat released. He was surprised to find he could boil the water, and intrigued that the cannon appeared to be the source of an inexhaustible amount of heat: the longer the boring went on, the more heat was produced. Thompson quickly realised that Lavoisier's idea that the cannon contained only a fixed amount of heat, because it contained only so much caloric fluid, could not possibly be correct. He recognised that what was going on was the conversion of work into heat. He was even able to estimate the amount of heat produced by a given amount of work[14].

Very soon afterwards this was followed by the seminal work of James Joule. In a series of very careful experiments, Joule was able to make accurate and precise estimates of both the amount of work involved and the quantity of heat produced, and thereby estimate what he called the *mechanical equivalent of heat*. The apparatus consisted of a paddle wheel which was driven by weights which were allowed to fall through a set distance. The paddle was designed very carefully to move past a series of baffles so that the work was used to heat up the water and not simply move it around, and great care was taken to minimise the loss of heat to the surroundings. What Joule had measured was how much energy it takes to increase the temperature of water, what we now call the heat capacity of water[15].

This early work looked at the world macroscopically, that is at the everyday scale of cannons, billiard balls and hawks. This is the realm of classical thermodynamics and while this is appropriate for exploring the thermal ecology of organisms, it lacks any underlying mechanism. In essence, using the equations of classical thermodynamics we can describe what is going on, and do so in very precise mathematical terms, but we do not know why. To explain what is going on, and provide a mechanism, we need to move to the atomic scale.

2.3 Energy at the atomic scale

An understanding of the mechanisms that explained the properties of bodies described by classical thermodynamics only came with the recognition of the existence of atoms, coupled with the development of the kinetic theory of gases and statistical mechanics (Box 2.2).

Statistical mechanics explains how the concepts of classical thermodynamics, such as temperature or heat, arise from the behaviour of atoms when the number of atoms is so vast that we cannot know what each and every atom is doing individually. Instead we use probability theory to predict the bulk properties that emerge, an approach usually termed statistical thermodynamics.

In any body at a temperature above absolute zero the atoms and molecules that make up that body are in constant motion. For a monoatomic gas such as argon at room temperature, the energy of the gas is essentially all in translation (that is, movement of the atoms in space). For a molecule containing more than one atom then there is also energy of rotation (the molecules are spinning) and of vibration (the two atoms in the molecule are moving closer together and then further apart, as if on a spring, and there are also bending motions). Because of these complications, the basic theory was worked out for what physicists term an *ideal gas*. This is a purely theoretical ideal, where the atoms are point-like and of zero size, undergo perfectly elastic collisions and do not interact (that is, van der Waals forces are ignored).

Because the gas molecules are moving rapidly, and continually colliding with other gas molecules, it might be thought that given sufficient time the energy of the molecules in a gas would even out such that all the molecules have more or less the same energy. Perhaps surprisingly, this is not so. The distribution of molecular speeds in an ideal gas was determined in 1860 by James Clerk Maxwell[16]. The distribution is:

$$f(v) = 4\pi \left[\frac{m}{2\pi k_B T} \right]^{\frac{3}{2}} v^2 e^{-\frac{mv^2}{2k_B T}}$$

Here $f(v)$ is the fraction of molecules with speed v, m is the mass of the atom, T is the absolute

> **Box 2.2 Atomic theory and thermodynamics**
>
> Although a number of ancient civilisations appear to have speculated that everyday objects are composed of tiny discrete and indestructible units, the birth of atomism is traditionally ascribed to the Greek philosophers Leucippus and Democritus. Democritus is said to have considered the problem of whether an object could be cut into ever-smaller pieces, or whether there was a point beyond which no further subdivision was possible. He concluded there must be a limit, defined by the Greek adjective ατομος (atomos), meaning uncuttable, thereby giving us the word atom. This is captured in his famous statement (one of very few direct quotations from Democritus that survive):
>
> > by convention sweet and by convention bitter, by convention hot, by convention cold, by convention colour; but in reality atoms and void
>
> Here Democritus is offering a profound philosophical view of the world, arguing that there is a difference between what our senses tell us and the underlying reality (which remains an important message for any scientist working today). Although Democritus believed that the existence of atoms explained many features of the natural world that were otherwise difficult to understand (such as the ability of salt to dissolve in water, or of fish to swim through the sea), atomism largely disappeared as a view of the world for over two millennia[17].
>
> It was John Dalton, widely regarded as the father of atomic theory, who placed the existence of atoms on a firm theoretical basis with his analysis of the proportions in which elements and chemical compounds combine. His atomic theory comprised five key propositions:
>
> 1. Elements are made of extremely small particles (atoms).
> 2. Atoms of a given element are identical in size, mass and other properties; atoms of different elements differ in size, mass and other properties.
> 3. Atoms cannot be subdivided, created or destroyed.
> 4. Atoms of different elements combine in simple whole-number ratios to form chemical compounds.
> 5. In chemical reactions, atoms are combined, separated or rearranged.
>
> Apart from the slight complications introduced by the existence of isotopes and nuclear processes (notably radioactivity), Dalton's atomic theory remains a cornerstone of chemistry[18].
>
> While a few physicists remained unconvinced of the existence of atoms (most famously Ernst Mach), for the majority the crucial empirical evidence for their existence came from Albert Einstein's 1905 explanation of the Brownian motion of suspended pollen grains as being caused by the impact of rapidly moving water molecules.

temperature and k_B is what we now call Boltzmann's constant. The distribution is determined completely by m and T.

Although the distribution of molecular speeds in oxygen at 20°C looks fairly symmetrical (Figure 2.2), there is a long tail produced by a small fraction of molecules with high speeds. As a result the mean speed (440 m s^{-1}) is slightly higher than the most probable (modal) speed (390 m s^{-1}). However the high frequency of collisions and the consequent small mean free path length (~70 nm for dry air at sea-level) mean that the rate at which an individual molecule covers the linear distance between two fixed points is much slower. This is why it takes time to smell the food after an oven door has been opened[19].

In air the nitrogen molecules are moving slightly faster than the oxygen molecules, and the carbon dioxide molecules slightly slower. The slight differences in speed are caused solely by the small difference in mass; their kinetic energies are similar because they are continually exchanging energy as they collide.

Maxwell's distribution of molecular speeds was extended to consider the distribution of molecular energy by Ludwig Boltzmann[20]. Boltzmann's derivation was explicitly statistical and showed that the distribution of energy among identical particles at thermal equilibrium depends only on the mass of the molecule and the temperature of the body. Specifically $f(E)$, the probability that a particle will have energy E, is given by:

$$f(E) = Ae^{-\frac{E}{k_B T}}$$

where A is a normalisation factor (such that $f(E)$ varies between 0 and 1), k_B is the Boltzmann constant and T is the absolute temperature. What this equation tells us is that the higher the temperature,

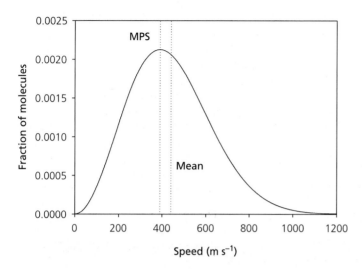

Figure 2.2 The Maxwell–Boltzmann distribution of molecular speeds for oxygen at 20°C. The plot shows the fraction of oxygen molecules that have a speed at, or very close to, a given speed. The dotted lines show the most probable speed (MPS) and the mean speed. Note that the mean speed is slightly higher than the most probable speed because of the long tail comprising a small fraction of molecules with high speeds.

the greater the probability that a particle will have energy E. The Maxwell–Boltzmann distribution of molecular energies is a result of profound importance to thermal ecology.

Statistical approaches are so much a part of the modern ecologist's toolkit that it is easy to underappreciate just how radically Maxwell and Boltzmann changed our view of the universe. The giants of classical physics, Johannes Kepler, Galileo Galilei and Isaac Newton, had built a view of the universe that was resolutely deterministic (A causes B), and the metaphor they used was that of clockwork. Maxwell and Boltzmann showed that the outcome of any physical process is purely statistical (that is we can only ascribe a likelihood that A will result in B). For the processes dealt with by classical thermodynamics the number of particles involved is so vast that the likelihood is overwhelming, and to all intents and purposes the outcome is deterministic. For some biological processes, however, the number of particles or entities can be sufficiently small that behaviour is far less predictable (an example here would be the number of free protons in the intracellular milieu of the smallest cells).

2.3.1 Characteristic energy, thermodynamic beta and the Boltzmann factor

One important aspect of the Maxwell–Boltzmann distribution is that the average energy of a particle is approximated by k_BT. This is sometimes called the *characteristic energy*; at 25°C its value is 4.11×10^{-21} J (= 0.026 eV), or in macroscopic systems 2.479 kJ mol^{-1}. This tells us something important about the nature of Boltzmann's constant: it is simply the conversion factor between energy and temperature. Its numerical value is set by the units of energy we select and the value we assign to the fixed point that defines the absolute temperature scale.

The inverse of the characteristic energy, β, is of profound significance to thermodynamics. It is sometimes referred to as *thermodynamic beta*, and it captures the relationship between temperature and energy (indeed as we shall see in Chapter 3, it is regarded by some physicists as a more useful indication of the temperature of a system than temperature itself):

$$\beta = \frac{1}{k_BT}$$

The exponential factor in the Boltzmann distribution is particularly important in thermal physics, and it is often referred to as the *Boltzmann factor*:

$$Boltzmann\ factor = e^{-\frac{E}{k_BT}}$$

A Boltzmann factor appears in any fundamental equation involving temperature and energy. We will encounter both thermodynamic beta and the Boltzmann factor again in Chapter 7 when we look at temperature and its effect on physiological reaction rates.

We now have enough background to start exploring in more detail the nature of energy in objects such as organisms. The first concept we need to explore is that of *internal energy*.

2.4 Internal energy

The internal energy of a body is the sum of all the kinetic energy and potential energy associated with the atoms and molecules that make up that body. It is usually designated U, and is one of the most important measures for keeping track of the energy changes in a system.

An organism, even an apparently simple one such as a bacterium, is an extremely complex object. So to get a handle on the concept of internal energy we should start with a much simpler example, such as oxygen or nitrogen gas. Here each molecule consists of two atoms linked by a covalent bond, and the molecules are moving around at great speed (Figure 2.2) so they have significant kinetic energy. However the molecules are also vibrating and tumbling around, so the internal energy of the gas is distributed among the component molecules in three types of movement: translation, vibration and rotation. For a more complex molecule, such as a cellular metabolite or a protein, there are a range of further bond motions and molecular flexions that contribute to internal energy.

By convention, internal energy comprises only those forms of energy that can be modified by heat, work or chemical reactions. That is, classical thermodynamics deals with only two of the four fundamental forces of nature, namely gravity and electromagnetism. The other two, the strong and weak nuclear forces, operate only over distances too small to be of relevance to the bulk flow of energy; internal energy thus does not include nuclear energy.

Internal energy is an *extensive* property; that is, its magnitude depends on the mass of the body: a larger animal or plant has more internal energy than a smaller one (see Box 2.3). Unfortunately, while we can define internal energy, we cannot measure it. There is no instrument known in which we could place our lump of copper (or indeed any other object) and obtain a measurement of its internal energy. Neither can we calculate the internal energy from first principles, it is just too complicated.

Box 2.3 Extensive and intensive properties

An *extensive* property of an object depends on its size (extent), which must therefore always be given. Examples of extensive properties include mass and internal energy. A 2 kg lump of pure copper has exactly twice the number of atoms, and twice the internal energy, as a 1 kg lump of pure copper at the same temperature and pressure. An *intensive* property is independent of the size of the object. Examples here would be temperature or molar heat capacity.

A simple way to visualise the difference is to imagine the block of copper as being divided into two halves. Each half-block contains exactly half the number of atoms and exactly half the internal energy (extensive properties), but each is at precisely the same temperature (an intensive property).

This might suggest that internal energy is not a very useful concept. The reason this is not so is that we can measure a *change* in the internal energy of a body very precisely indeed, and this makes the concept extremely valuable. So much so that internal energy is central to understanding the interaction between a body and its surroundings. The change in internal energy (ΔU) is the difference between the internal energy at the start (U_1) and the end (U_2) of the process:

$$\Delta U = U_2 - U_1$$

The change in internal energy of a body is given by:

$$\Delta U = \Delta Q + w$$

where ΔU is the change in internal energy, ΔQ is the change in energy content resulting from a gain or loss of energy and w is the work done on the body (for example increasing its potential energy by lifting it). By convention, energy flow into a body and work done on that body are regarded as positive. Equally, loss of energy and work done by the body on the environment are negative.

To take a biological example, a lizard basking in the sun is increasing its internal energy by absorbing radiant heat (Plate 1). This extra energy causes all the component molecules of the lizard to move around more rapidly, and we can measure this greater motion as an increase in temperature (we will explore the relationship between thermal energy and temperature in Chapter 3). If the lizard climbs a rock to bask, it also increases its potential energy.

2.4.1 Enthalpy

An important property of a system is its enthalpy, usually designated H. It is a state function (Box 2.4) and closely related to internal energy.

Enthalpy is a measure of the total energy content of a system; it includes the internal energy together with the amount of energy required to displace the environment and thereby establish the system's volume and pressure. Thus:

$$H = U + PV$$

where H is the enthalpy of the system, U its internal energy, V its volume, and P the pressure at the boundary between the system and its surroundings. One way to visualise the difference between enthalpy, H, and internal energy, U, is that U represents the energy required to create the system and PV represents the energy required to create room for the system in an environment of pressure P.

Enthalpy is sometimes referred to as the total heat content of the system. Although a useful shorthand, this statement is not strictly true, and things are not helped by the etymology of the term: enthalpy derives from the Greek ενθαλπος (enthalpos), which translates as 'put heat into'[21].

As with internal energy, the enthalpy of a system cannot be calculated from first principles, nor can it be measured directly. The change in the enthalpy of a system is, however, exactly equal to energy added to or taken from the system as heat, provided that the system is at constant pressure and the only work done involves a change in volume. Since under almost all ecological conditions pressure is effectively constant, but volume is allowed to change, enthalpy is an immensely useful state function for understanding heat flow and energetics in ecology.

2.5 Energy and heat

We can see that the list of state variables of importance to thermal ecology (Box 2.4) includes internal energy and enthalpy, but not heat. So what exactly is heat?

> **Box 2.4 State functions**
>
> A *state function* (also called a *state variable*) is a physical property of a system that depends only on the present state of that system, and is independent of the path by which that state was reached. The state functions of importance to thermal ecology are:
>
> | State function | Symbol | SI unit |
> |---|---|---|
> | Volume | V | litre, l (or sometimes L) |
> | Pressure | P | Pascal, Pa |
> | Temperature | T | Kelvin, K (not °K) |
> | Amount | n | mol |
> | Internal energy | U | Joule, J |
> | Enthalpy | H | Joule, J |
> | Entropy | S | J K^{-1} (occasionally called 'entropy units') |
> | Gibbs (free) energy | G | Joule, J |
>
> State functions are usually represented by a capital letter. The essential feature of a state function is that if a system is taken through a cyclic process and returned to its initial conditions, then the net change in all state functions is zero ($\Delta V = 0$, $\Delta T = 0$, $\Delta H = 0$ and so on). The opposite of a state function is a *path function*. Here the final state of the system depends on the route (path) by which that final state was achieved; an example of a path function would be work.

Perhaps surprisingly, it was only in the late nineteenth century that the Scottish physicist James Clerk Maxwell was finally able to make a clear statement of the nature of heat, developing the pioneering work of Rumford and Joule described in section 2.2.3, and based on the kinetic theory of gases (see Box 2.5).

A century on, and physicists now view heat in slightly more rigorous terms. Heat is the spontaneous transfer of energy through random molecular motion: energy is what is being transferred, and heat is the process. Energy can also be transferred through work, and a molecular view allows us to distinguish work and heat. The thermal motion of atoms is random and chaotic, and heat is the transfer of this energy through random atomic motions. In contrast, work is the transfer of energy through a coherent movement of atoms (Figure 2.3).

> **Box 2.5 James Clerk Maxwell and the nature of heat**
>
> In 1716 Jakob Hermann proposed that the heat content of a body was proportional to its density and *the square of the agitation of its particles*. Hermann spent some time in St Petersburg, where he undoubtedly encountered his compatriot Daniel Bernoulli who published his influential book *Hydrodynamica* in 1738. In this book Bernoulli ascribes the pressure of a gas to *very minute corpuscles, which are driven hither and thither with a very rapid motion*. Using this idea, Bernoulli was able to derive the relationship between pressure and volume of a gas (which we now know as Boyle's Law), and also proposed that as temperature is raised, the speed of the particles increased, and thereby the pressure[22].
>
> Although the prevailing view at the time was that heat was a form of fluid, Isaac Newton subscribed to the view that heat consists of the internal motion of the constituent particles, as did Henry Cavendish. In a recently discovered unpublished work on heat, probably written in 1787, Cavendish speculates that the heat content of a body consists of an active component (the term current at the time was *vis viva*) which affects a thermometer, and an inactive component which resulted from the relative positions of the particles and was a measure of the 'latent' heat of the body. This hints clearly at the then unknown concept of entropy and was a remarkably prescient remark for the late eighteenth century[23].
>
> It was James Clerk Maxwell who finally defined the nature of heat in rigorous terms, listing four key features[24]:
>
>> Heat is something that can be transferred from one body to another.
>> Heat is a measurable quantity, and hence can be treated mathematically.
>> Heat cannot be treated as a material substance.
>> Heat is one of the forms of energy.
>
> He also laid to rest the old idea of heat as a fluid with characteristically ruthless brevity:
>
>> Heat may be generated and destroyed by certain processes, and this shows that heat is not a substance.

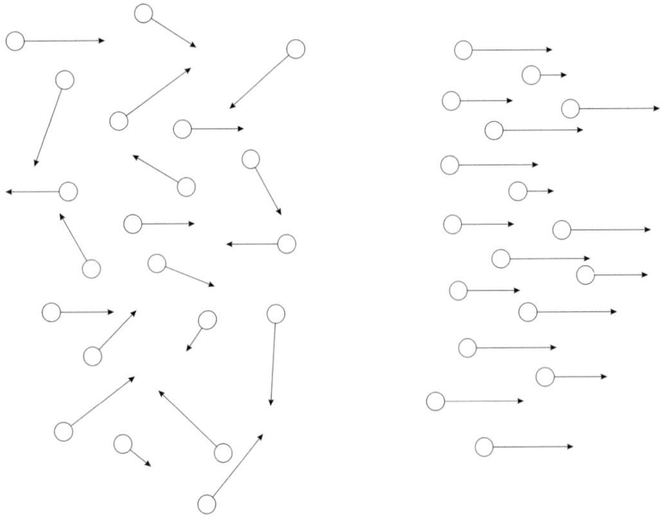

Figure 2.3 Molecular motion in thermal energy (left) and work (right).

Biologists are less rigorous in their terminology, and the thermal energy of a body is often referred to as its heat content. Moreover we often speak of heat being transferred between an animal and its environment, when strictly we should be saying that energy is being transferred through heat. In this book I will try to be rigorous, except when it would otherwise lead to convoluted or tortuous language, in which case I will slip into more informal language (though never without pointing out what I have done).

2.6 Thermodynamic systems

So far we have been considering objects that exchange energy with their environment, or have work done on them. At this point we need to introduce an important thermodynamic convention, defining the nature of the body under consideration in terms of precisely how that body interacts with the environment around it.

The first convention is that we divide the universe into two compartments: the system under consideration and everything else (its surroundings). We then consider the nature of the exchanges between the system and its surroundings:

An *open* system exchanges both matter and energy with the surroundings.
A *closed* system exchanges only energy with the surroundings.
An *isolated* system exchanges neither matter nor energy with the surroundings.

These are illustrated in Figure 2.4.

Since energy can be exchanged between the system and the surroundings by both heat and work, we can also define an *adiabatic* system (from the Greek ἀδιάβατος, impassable); this is a closed system where the exchange of energy with the surroundings is through work alone.

A moment's reflection should convince you that all of ecology deals exclusively with open systems: all living things have to exchange both energy and materials with their surroundings.

2.7 Work and energy revisited

Although the concepts of work and work were developed from mechanics, a molecular view of work shows us that in bioenergetics, work encompasses a wide range of physiological processes. Some examples of relevance to thermal physiology are shown in Table 2.1.

Table 2.1 Some examples of physiological work.

Type of work	Examples
Mechanical work	Movement of bulk material, such as in locomotion, or the movement of cilia or flagella, or of macromolecules within cells
Osmotic work	Movement of molecules and ions across a membrane against a concentration gradient
Electrical work	The directed movement of electrons or ions to create a difference in electric potential
Chemical work	Driving chemical reactions that would not happen spontaneously

Earlier in this chapter we saw that Count Rumford demonstrated that work could be converted completely to heat. The development of steam engines showed that heat could be converted to work, and Sadi Carnot had shown that the efficiency of this conversion was dependent on the temperature difference involved (see Box 2.6).

It was Rudolf Clausius who first recognised that this indicated a fundamental asymmetry in nature: while work can be converted completely to heat, the reverse is not true. There is some fraction of the energy of a body that can never be recovered as useful work. To explain why this is so, he introduced the concept of *entropy*[25].

2.8 Entropy

Entropy is one of the most important concepts in thermodynamics, but it is subtle and elusive and difficult to define precisely in simple language. Maybe because of this, or perhaps because many

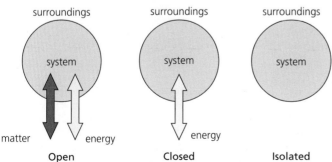

Figure 2.4 Thermodynamic systems.

> **Box 2.6 Carnot and efficiency**
>
> Carnot showed that for a heat engine taking energy from a hot reservoir at temperature and moving this to a cold sink at temperature to generate work, the maximum efficiency that could be achieved, η, is given by:
>
> $$\eta = 1 - \frac{T_h}{T_c}$$
>
> where T_h is the absolute temperature of the hot reservoir and T_c that of the cold sink. This formalises the engineering principle known to James Watt that for a steam engine, the hotter the steam the more efficient the conversion to work. Carnot recognised that this was a theoretical maximum, achievable only for a fully reversible process operating infinitely slowly (that is, doing no work!). For a heat engine where the processes of heat transfer are not reversible the maximum achievable efficiency, η^* is:
>
> $$\eta^* = 1 - \sqrt{\frac{T_h}{T_c}}$$

discussions of entropy are couched in terms unfamiliar to non-physicists, few ecologists are conversant with it. Indeed, despite the central importance of entropy to biological energetics, few physiology texts even mention it. When entropy is discussed, it is usually through analogies such as randomness or disorder which do little to help understanding and may even mislead.

Although the concept of entropy is sometimes difficult to grasp, it is a precisely measurable quantity, just like length or mass. It is also of profound importance to thermal ecology, though this importance is sometimes hidden. A useful indication of the importance of entropy is to see how much of physiology and ecology is influenced by it; some examples are given in Table 2.2. This list should be enough to convince you that entropy is of fundamental importance to physiology.

Perhaps the best way to introduce entropy is through a few examples. All of these involve spontaneous change, that is a change which has a natural tendency to occur and which happens without the need for work to bring it about.

Table 2.2 Some biological processes where entropy plays an important role.

Process
Physiological processes
Formation of membranes
Protein folding
Binding of RNA and proteins to DNA
Binding of cofactors to proteins
Enzyme function (indeed all cellular chemical reactions)
Heat generation in endotherms
Evolutionary processes
Adaptation of enzymes to temperature, pressure and pH
Ecological or environmental processes
Heat generation in a compost heap
Melting of snow and ice

Consider a volume of gas allowed to expand isothermally into a vacuum (Figure 2.5). As soon as the barrier is removed (say by opening a stopcock) the very rapid random movements of the gas molecules mean that the gas rapidly fills the vacuum and re-equilibrates. The change in state variables is described by the ideal gas equation (Box 2.7):

$$PV = nRT$$

Since there has been no change in energy (no heat has flowed, and no work done on or by the gas since the movement of the molecules has been entirely random), then T remains the same: the distribution of speeds among the gas molecules is exactly the same as it was before. Since the amount of gas (n moles) also remains the same, the increase in volume (V) is offset by a decrease in pressure (P).

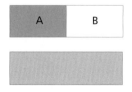

Figure 2.5 A spontaneous process: the expansion of a gas into a vacuum. In the upper diagram a volume of gas under pressure (A) is separated from an evacuated container (B) by an impermeable barrier. In the lower diagram the barrier has been removed and the gas now fills both spaces; the temperature remains the same but the pressure is lower and the entropy greater.

What else has changed is that the gas can now do less work. Consider using the gas to inflate a balloon, and it is easy to see that a larger balloon could be inflated by the gas before opening the stopcock than after. The energy content is the same, but for some reason the gas can now do less work.

A second example might be the release of a small amount of sugar or a highly coloured compound such as potassium permanganate, $KMnO_4$, into water: the sugar or $KMnO_4$ diffuses slowly until the distribution is uniform. A final example would be bringing together two identical blocks of metal, say one at 200°C and the other at 100°C. Once they are in thermal contact one block warms and the other cools until both are at 150°C.

In all of these cases the process always goes in one direction: the gas always fills the vacuum, the sugar or $KMnO_4$ always spreads out by diffusion and the two metal blocks always come to the same temperature. We have an individual explanation for each of these (pressure always equalises, substances always diffuse to equilibrium, heat always flows from hot to cold), and so it was a major conceptual leap to recognise that there was a single deeper principle behind all of these. That step was taken by Rudolf Clausius in his recognition of entropy.

At this point it is helpful to take a final example, this time one where energy is changing. Consider a mass of ice at subzero temperature, and track its temperature as heat is supplied at a constant rate (Figure 2.6). First the ice warms, and then there is a plateau during which heat is continually being supplied but there is no change in temperature, which remains at the melting point of 0°C. Once all the ice has melted, the water then warms (though more slowly because the thermal capacity of water is greater than that of ice) until once again a plateau is reached. Here again, heat continues to be supplied, but the temperature remains constant, this time at the boiling point of 100°C. Once all the water has been vapourised to steam, the steam then starts to warm (or as an engineer would say, to superheat).

What is happening at the two plateaux, when energy is being supplied but there is no change in temperature? The original explanation was that the energy was providing latent heat, that is heat which is in some way hidden, and to distinguish this from energy supplied when the temperature changed, which was sensible heat (because it could be sensed, such as with a thermometer)[28].

This is perfectly correct, but a better way to look at it is that during the phase change from ice to water, and then from water to water vapour (the plateaus in Figure 2.7), the energy supplied is being used entirely to meet the difference in entropy between ice and liquid water, or between liquid water and water vapour. This tells us something important: entropy is real, and the change in entropy is a

---

Box 2.7 The Ideal Gas Law

The Ideal Gas Law developed from a number of separate relationships established in the seventeenth century.

Robert Boyle established that the pressure, P, of an ideal gas is inversely proportional to its volume, V:

$$PV = constant$$

Boyle was a careful experimentalist, and recognised the need to control the temperature during measurements, and to allow for atmospheric pressure. This relationship is now known as Boyle's Law[26].

A short while later, in 1699, Guillaume Amontons showed that different volumes of air have their pressure increased equally by the same rise in temperature, and in a later paper showed that this relationship is independent of the initial pressure. Almost a century later, the French chemist Joseph Louis Gay-Lussac showed that when pressure is held constant, the relationship between volume and temperature, T, is similar in several gases:

$$\frac{V}{T} = constant$$

When he published his results, Gay-Lussac generously acknowledged that similar results had been established earlier by the French scientist and entrepreneur Jacques Charles, but were unpublished. This relationship is now known as Charles' Law, and Gay-Lussac is remembered today through his Law of Combining Volumes.

Boyle's Law and Charles' Law were combined by Émile Clapeyron, to produce the first formulation of the Ideal Gas Law[27]. The modern representation of this law is:

$$PV = nRT$$

The constant of proportionality, R, is known as the *universal gas constant*, and has a value of 8.314 J (mol K)$^{-1}$; n is the number of moles.

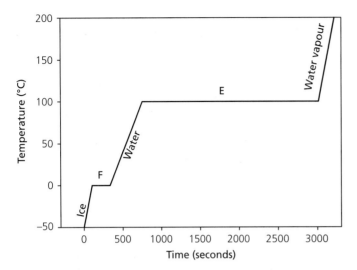

Figure 2.6 A heating curve for water. The plot shows the change in temperature of 1 kg of water, initially ice at −50°C, as energy is supplied at a rate of 1 kW, and the sample remains at atmospheric pressure. The two plateaux represent the entropy change at melting (F, the latent heat of fusion) and evaporation (E, the latent heat of evaporation).

precisely measurable property. Its definition is simple and explicit:

$$\Delta S = \frac{\Delta Q}{T}$$

where ΔS is the change in entropy, ΔQ is the energy supplied and T is the absolute temperature. This relationship was established by Clausius, and as always with equilibrium thermodynamics, it is assumed that the energy is supplied reversibly, that is very slowly and in very small amounts such that each step is, at least theoretically, perfectly reversible. It is important to note that entropy is not energy: its units are J K^{-1}.

It is only during phase changes (ice to liquid water, liquid water to water vapour and vice versa) that the energy supplied is utilised entirely by the change in entropy. At all other times, when the ice, liquid water or water vapour is also increasing in temperature, the energy input is increasing both the entropy and the temperature.

This is all very well, but we still have not said what entropy actually *is*. The explanation came from Ludwig Boltzmann, who considered the enormous number of ways that a given amount of energy could be divided within a system and used probability arguments to deduce the most probable distribution of energy across all the component entities (typically atoms or molecules). The details need not concern us here, but the outcome was that Boltzmann linked entropy to the number of available microstates, W (or sometimes Ω) in the system. The modern formulation of this relationship is:

$$S = k_B \ln W$$

where S is the entropy and the constant of proportionality and k_B is Boltzmann's constant. Since W (Ω) is a pure number, the dimensions and units of S and k_B are identical. This version of entropy is sometimes called statistical entropy (or Boltzmann entropy), but for an ideal gas it is identical to the entropy defined in terms of classical thermodynamics by Clausius[29]. This equation is carved on Boltzmann's memorial in Vienna, although Boltzmann never expressed his relationship in this way, and the inclusion of what we now refer to as Boltzmann's constant is down to Max Planck.

We now recognise that microstates are descriptions in terms of quantum mechanics of the different ways that molecules can differ in their energy distributions in a macroscopic system. When a macroscopic system is warmed, a great many additional microstates become available: its entropy increases. An element of this is captured by the change in molecular speed distribution of a gas when warmed; the broadening of the speed distribution indicates the greater range of microstates available at the higher temperature (Figure 2.7). There is a similar increase in available microstates when the volume of a gas is increased.

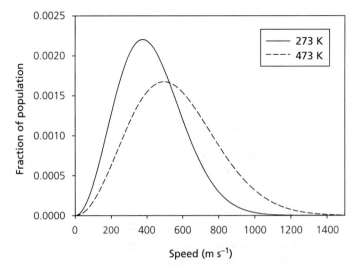

Figure 2.7 The distribution of molecular speeds in oxygen at 0°C (273 K) and 200°C (473 K). Note the broadening of the speed distribution at the higher temperature, reflecting the larger number of microstates available (greater entropy).

A simple analogy that captures the relationship between statistical likelihood and the number of microstates is the rolling of a pair of dice. Here the macrostate is the total score and the microstates are the faces of each die. A score of 2 or 12 is least likely because in each case there is only one way of achieving that score (1,1 and 6,6 respectively). In contrast a score of 7 is much more likely because this can be achieved in six different ways (1,6; 2,5; 3,4; 4,3; 5,2; 6,1). In this example of two dice there are 36 (6^2) possible microstates and the general result is that for n dice, the most probable number (represented by the greatest number of different microstates) is $3.5n$. If you pour 100 dice onto a table, the total face value will be close to 3500; the most probable total (macrostate) is that with the greatest number of microstates (that is, the highest entropy). It is just the same for the entropy of a gas (or any other system).

Most formal definitions of entropy are daunting for an ecologist or physiologist, so at this point I will suggest a working description as a basis for exploring the role of entropy in ecology:

Entropy is a measure of the way energy is distributed within a system that captures how much of the total energy is available to perform work.

This is not a formal definition of entropy; it is a broad-brush description that attempts to capture its essential features in a way that relates directly to thermal ecology. The important caveats are that the system must be in thermal equilibrium and the work isothermal, but this simple description conveys the key features important for understanding the role of entropy in ecology.

In some of our examples of spontaneous change discussed earlier, for example the diffusion of sugar in water, there has clearly been a change from a highly ordered state to a less ordered state and an increase in entropy is often described in terms of an increase in disorder. However in other processes, such as the expansion of a gas into a vacuum, it is difficult to explain what has happened in terms of order or disorder: the distribution of molecular speeds of the gas molecules is exactly the same before and after. A better general description is that the energy has become more spread out or dispersed in space. Although both of these analogies are common in the literature, neither is precise nor always easy to visualise. Perhaps the best simple description is that in the absence of any constraint the energy in a system always moves to the most probable distribution across the component atoms and molecules. This explanation leads directly to a simple statement which forms the basis of the Second Law of Thermodynamics:

In an isolated system, entropy always tends to the maximum.

Organisms are not, of course, an isolated system, so we need to examine the Second Law in a little more detail.

2.9 The Second Law of Thermodynamics

The Second Law of Thermodynamics has achieved an almost mythical status in science[30]. It is of fundamental importance to biological energetics, for it captures how and why physiological reactions proceed (or not).

As we have seen, an organism is a thermodynamically open system, continuously exchanging energy and materials (nutrients and waste products) with its environment. In this context, the isolated system of relevance to the Second Law is the universe; to understand how the Second Law influences physiology we need to consider the entropy of the organism and of its surroundings (the environment) separately. We can then see that the key factor is that for any reaction to be thermodynamically favoured it is the entropy of the universe that is critical. A reaction that involves a decrease in the entropy of the organism (for example growth) simply requires that there is an even greater increase in the entropy of the environment. In formal terms:

$$\Delta S_{uni} = \Delta S_{org} + \Delta S_{env}$$

where ΔS_{uni} is the change in entropy of the universe, ΔS_{org} is the change in entropy of the organism and ΔS_{env} is the change in entropy of the surroundings (the organism's environment). For a reaction to be thermodynamically favoured, the Second Law requires only that $\Delta S_{uni} > 0$.

It really is that simple.

The important thing to keep in mind is that the change in entropy is not some mysterious force, it is simply the inevitable tendency of a population of atoms and molecules to adjust to the most probable distribution of energies among the particles (which is the distribution with the highest entropy). A second important point is that the Second Law is statistical in nature; it not impossible that things may go in the opposite direction, just extremely unlikely. So unlikely in fact that in any macroscopic system (such as an organism), the number of atoms is generally so huge that the system always goes in one direction: the gas always fills the vacuum, the sugar always disperses and the heat always flows from hot to cold. It is the Second Law of Thermodynamics that provides the arrow of time.

We can therefore restate the Second Law of thermodynamics in a different way:

In any spontaneous reaction, the entropy of the universe always increases.

This simple but powerful principle underpins all energy flow in physiology and ecology. It explains why metabolic processes generate heat (energy is being dissipated to increase the entropy of the universe so that the process is energetically favoured), and it explains how organisms can grow and increase in complexity (a decrease in the entropy of the organisms simply requires a greater increase in entropy of the surroundings).

2.9.1 Thermodynamics and kinetics

A critical point is that thermodynamics only tells us in which direction a reaction will go; it says nothing about the rate. In the examples above, the gas equilibrated very rapidly, but the diffusion of sugar proceeded slowly. It is perfectly possible that a spontaneous reaction, allowed by the Second Law, does not proceed at all. A classic example here is the conversion of diamond to graphite. Entropy dictates that this reaction is favoured, but thankfully for anyone owning diamond jewellery, its speed at room temperature and pressure is effectively zero. The question of reaction rate is the subject of kinetics, and we will return to this in Chapter 7 when we look at physiological reactions.

2.9.2 Entropy and information

A formal similarity between the Boltzmann definition of entropy and the expression for the information content of a message points to a relationship between entropy and information, persuading some that the two concepts are essentially identical: entropy is information and vice versa (Box 2.8). At present it remains unclear whether there really is a deep relationship between thermodynamic entropy and information, and if so what its nature is. The debate continues, with a consensus yet to emerge[31].

> **Box 2.8 Entropy and information**
>
> In 1948 the American mathematician and electrical engineer Claude Shannon proposed a measure of the information content of a string of text[32]. Shannon was interested in how difficult it is to predict the next letter in the string, and it is easy to see that the more different letters are present, and the more equal the proportion of these letters, the more difficult such a prediction becomes. If there are only two letters (or, as in computing, just two states, 0 and 1), then determining the identity of a letter taken at random is far easier than if there are 26 letters, each present with different frequency, as in English. Shannon's formula is:
>
> $$H = -K \sum_i^\infty p_i (\log p_i)$$
>
> Here K is a constant that relates to the units of measurement, and p_i is the probability that letter i will appear next in the message. If K is set to 1 and the base of logarithms is 2, then the units are 'binary digits' or 'bits'. A similar approach had been taken slightly earlier by Norbert Wiener, and Shannon was explicit in his debt to Wiener in deriving his formula.
>
> Shannon was concerned with the problems of coding and decoding messages, and the efficiency of information transmission; he described his index H as '*a measure of information, choice or uncertainty*'. One way to think about how H represents the degree of uncertainty is to consider the number of binary (yes/no) questions that would need to be asked to determine the identity of the next letter in the sequence: the more symbols, the more questions that must be asked to determine the identity. Shannon's index is a measure of capacity; it quantifies the number of messages that could be sent with a given set of characters.
>
> Possibly as the result of a mischievous comment from John von Neumann, Shannon also referred to his index as an 'entropy' (and H is sometimes referred to as 'Shannon entropy'). There is actually a close mathematical similarity in the expression derived by Shannon to express the uncertainty in a message, and that derived by Ludwig Boltzmann to define thermodynamic entropy in terms of statistical mechanics.
>
> *Maximum (information) entropy in ecology*
> Maximum entropy has its origins in the work of Edwin T. Jaynes[33]. The idea is to arrive at the least-biased estimate of a relationship of interest (say, for example, the relationship between the number of species and area). This estimated relationship needs to be as smooth and flat as possible within the constraints of what we already know about the relationship, because any more complex relationship would mean assuming information that we do not have. Jaynes showed that the shape of the least-biased relationship can be determined by maximising the information entropy. This approach is proving to be very useful in ecology for determining possible relationships between complex variables in the absence of theory to predict the relationship.

2.10 The Third Law of Thermodynamics

Clausius established that as energy, Q, is added reversibly to a system at temperature, T, its entropy, S, increases in a precise way:

$$\Delta S = \frac{\Delta Q}{T}$$

This tells us that as a system cools it has less entropy, and suggests there must be a temperature when the entropy reaches zero. This is captured in the strong form of the Third Law[34]:

At absolute zero, the entropy of a perfect crystal is zero.

This statement of the Third Law was formulated by Max Planck. The real world, however, is non-ideal, and a weaker form of the Third Law is:

As a system approaches absolute zero, its entropy approaches zero.

In other words, even at absolute zero there will remain a small but finite residual entropy. This is because although particles would have zero kinetic energy (and hence zero thermodynamic temperature), they would continue to vibrate and rotate. In other words, although the atoms would not be moving anywhere, they would continue to jostle because of *zero-point energy*. This zero-point energy is the result of quantum mechanical fluctuations and a nice example of its consequences is that at atmospheric pressure helium remains liquid no matter how close it gets to absolute zero. This is because the zero-point energy exceeds the very small latent heat (entropy change) associated with the freezing of helium; the pressure needs to exceed ~ 2.5 MPa before the balance between the entropy change and zero-point energy shifts and helium freezes.

An implication of the Third Law is that absolute zero can never be achieved. We will return to the topic of absolute zero, and why we cannot achieve it, in the next chapter.

2.11 Gibbs (free) energy

The state variables we have discussed so far are the internal energy, U, the closely related enthalpy, H, and the entropy, S. The entropy dictates how much of the internal energy at temperature T is not available for work:

Energy unavailable for isothermal work $= ST$

The reason this energy is not available for isothermal work is that it can only be extracted by lowering the temperature (and the process is then not isothermal). For an organism this would be impossible, for as soon as any of this energy is extracted, the organism would cool and any further extraction would involve moving energy to a warmer environment (that is, uphill thermodynamically). To return to our example of a gas expanding into a vacuum, it is the increase in entropy that explains why the gas after expansion can do less work than it could before.

The difference between the total energy in the system and the energy unavailable for work is the energy the organism can use to do things (such as drive its physiology). Traditionally this has been called *free energy*, because it is free (in the sense of being available) to be used. For a system operating at constant temperature and constant volume, this is given by:

$$A = U - ST$$

where A is the Helmholtz free energy. This is an important measure for chemical reactions where pressure can change markedly, such as in explosions. Much more relevant to organisms, however, is the energy available under conditions of constant temperature and constant pressure, but where the volume is free to change; this is the Gibbs free energy, G:

$$G = H - ST$$

Strictly, G is a measure of the free enthalpy, and it is now recommended practice to drop the adjective 'free'; A is now referred to as Helmholtz energy and G as Gibbs energy[35]. The relationship between these various state functions is shown diagrammatically in Figure 2.8.

The importance of Gibbs energy is that it captures the thermodynamic driving force for a reaction at constant temperature. Essentially all cellular physiology takes place under isothermal conditions; while the specific temperature may differ between a polar lichen, a temperate insect or a tropical fish, that temperature generally does not change during a physiological reaction. It is Gibbs energy that organisms use to drive their ecology, and it is so important to thermal ecology that we will encounter it frequently in the rest of the book.

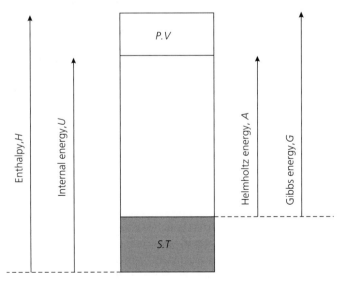

Figure 2.8 A diagrammatic representation of the various state functions of relevance to thermal ecology. The total energy of the system is represented by the height of the bar, and is divided into three components. The shaded area shows the energy unavailable for isothermal work ($S.T$, where S is the entropy and T the absolute temperature), and $P.V$ denotes the energy required to create space V for the system in an environment of pressure P.

2.12 Summary

1. Energy is the capacity to do work. It cannot be created or destroyed, only changed in form. The First Law of Thermodynamics states that the total energy of the universe never changes.
2. Heat is the spontaneous flow of energy from one body or system to another through the random movement of atoms or molecules. Work is the transfer of energy through the coordinated movement of atoms or molecules.
3. The entropy of a system determines how much of its internal energy is unavailable for work under isothermal conditions. The energy available for work under isothermal conditions and constant pressure is the Gibbs energy.
4. The Second Law of Thermodynamics states that for any reaction proceeding spontaneously the total entropy (system plus surroundings) must increase, which is why metabolic processes dissipate energy as heat.
5. All organisms are thermodynamically open systems, exchanging both energy and matter with their surroundings. Organisms can decrease their entropy through growth and development simply by ensuring a greater increase in the entropy of the environment.
6. Thermodynamics only tells us in which direction a reaction will go. It says nothing about the rate, which is the subject of kinetics.
7. For an ideal gas in thermal equilibrium the distribution of energy across the component atoms or molecules is described by the Maxwell–Boltzmann equation. This distribution is fixed by the temperature of the system.
8. The Third Law of Thermodynamics states that absolute zero can never be attained.

Notes

1. This quotation is the opening to his paper describing the work on the generation of heat by friction that prompted his recognition of the relationship between work and heat (Rumford 1798).
2. The harnessing of steam power and the Industrial Revolution were essentially a feature of central England, where coal and iron ore were available locally or could be delivered by canal. Here James Watt (1736–1819) and Matthew Boulton (1728–1809) oversaw the widespread adoption of ever more efficient steam engines in mines and factories. Elsewhere other engineers such as Thomas Savery, Thomas Newcomen, Richard Trevithick and George Stephenson developed the use of steam for locomotives, fuelling the construction of an extensive railway system. The theory of how these steam engines worked was developed by the Frenchman Nicolas Léonard Sadi Carnot (1796–1832). This parallel development of thermodynamic theory and engineering practice led the noted American historian of science Charles Gillespie to comment that *'the French . . . formulate things, and the English do them'* (Gillespie 1960, p. 205). A thorough history of the concept of energy is given by Jennifer Coopersmith (Coopersmith 2010).
3. This is the definition I was taught, and which remains in many textbooks. A slightly broader version, more common today, is: *the ability to do work or supply heat.*
4. An ecologist might query this statement, and point to the use of bomb calorimetry to determine the energy content of an organic compound or biological tissue. Calorimetry actually measures only part of the internal energy content of the sample (specifically the difference in chemical potential energy in the interatomic bonds of the molecules in the sample and those in the oxidised products).
5. This quotation is from Chapter 4 in *Six easy pieces* (Feynman 1998), which reproduces selected extracts from the celebrated *Lectures on physics* (Feynman et al. 1966) with an introduction by Paul Davies. Feynman clearly considers energy a difficult concept, and as Jennifer Coopersmith comments: *'if Feynman says the concept of energy is difficult then you know it really is difficult'* (Coopersmith 2010).
6. This is a wonderful example of how inductive science works. First comes a series of empirical observations on the conversion of work into energy, or from one form of energy to another. Then comes the inductive step (or intuitive leap) that suggests that these observations might reflect a general principle, namely that energy can be converted from one form to another but not lost or created. Finally this principle is strengthened into a fundamental law, which remains intact because no subsequent test of its predictions has ever failed. Should an exception to the conservation of energy ever be found, and experimental work in high-energy physics has subjected the law to very stringent examination, then scientists will have to rethink one of the foundation blocks of modern science.
7. Germain Hess (1802–1850) actually formulated his ideas in terms of enthalpy, because he was considering processes where no heat is exchanged with the surroundings. This statement became known as Hess' Law, which is important in determining

the overall energy required for a chemical reaction which can be divided into synthetic steps that are individually easier to characterise. It remains an important principle in the design of complex syntheses by chemists. Julius Robert von Mayer (1814–1878) is best known for enunciating in 1841 one of the original statements of the conservation of energy. However Mayer also made fundamental contributions to our understanding of the way organisms use energy. In 1842 he identified oxidation as the primary source of energy for any living creature, and he also proposed that plants convert light into chemical energy. Sadly his achievements have often been overlooked, and priority for the discovery of the *mechanical equivalent of heat* was attributed to James Joule in the following year.

8. Amalia ('Emmy') Noether (1882–1935) was one of the greatest of all mathematicians. She is most famous for her recognition that for any continuous symmetry in the laws of physics there exists a corresponding conservation law. Her life and work are described by Lederman & Hill (2004).

9. Arthur Stanley Eddington (1882–1944) was a British physicist famous for introducing Einstein's theory of relativity to scientists in the English-speaking world at a time when developments in German science were largely unknown. He organised an expedition to test the theory of relativity by determining the small change in the apparent position of stars caused by the bending of light passing close to the sun in the sky, as predicted by Einstein. To do this, it was necessary to make observations during a total eclipse of the sun, and Eddington made the observations on the island of Principe off west Africa, during the total eclipse of 29 May 1919. This proved to be a significant step in the acceptance of Einstein's ideas. Eddington coined the term *the arrow of time* in his book *The nature of the physical world* (Eddington 1928).

10. Albert Einstein (1879–1959) is probably the most famous name in twentieth century science, perhaps in all of science. Born in Ulm, Germany, he was an undistinguished student. After graduating and following a long and unsuccessful search for a teaching position, he eventually took a job in the Swiss Patent Office in Berne. In his spare time he worked on a series of remarkable papers, all published in 1905 (often referred to as Einstein's *annus mirabilis*). These papers were on the nature of Brownian motion, the photoelectric effect, the special theory of relativity and the equivalence of mass and energy. In 1908 he published his general theory of relativity, and in 1921 he received the Nobel Prize in Physics for his work on the photoelectric effect. In 1933 he moved to the Institute of Advanced Studies in Princeton, where he stayed for the rest of his life. Here he worked on the unification of gravity with electromagnetism, a goal that eluded him and which remains an unresolved problem in physics. Einstein's life and work are described beautifully in a biography written by his good friend and colleague, Abraham Pais (Pais 1982).

11. The stipulation that c is the velocity of light in a vacuum is important. Light travels more slowly when passing through a medium such as air, water or glass, by an amount that depends on the wavelength. It is this slowing that gives rise to phenomena such as rainbows and the ability of a glass (or protein) lens to focus an image. The speed of light *in vacuo* is a fundamental constant, and invariant. Its value is 299,792,458 m s^{-1}, or in round figures 300,000 km s^{-1}.

12. Although biologists often equate weight with mass, these are actually two quite different things. Mass is a measure of the amount of material in an object, whereas weight is a measure of the force exerted by that object in a gravitational field. If the mass is m, then the weight is mg, where g is the acceleration of free fall (often referred to as the acceleration due to gravity). The value of g varies from place to place, and decreases with altitude (because the object is then further from the centre of gravity of the Earth, and the force of gravity obeys an inverse-square rule). An object of a given mass thus has less weight up a mountain, much less weight on Mars or the Moon and no weight at all if it is neutrally buoyant in water because here the weight is balanced exactly by the opposing force of buoyancy.

13. Antoine-Laurent de Lavoisier (1743–1794) is probably the greatest scientist France ever produced. With exacting quantitative work, he showed for the first time that chemical reactions involve precise relationships, and he can truly be regarded as the founder of modern chemistry, and also incidentally of biological stoichiometry. Lavoisier was intensely interested in the nature of heat, which he believed to be a fluid. He suggested that the constituent particles of this calorific fluid repelled each other, thus providing a simple and intuitive explanation of why heat flows from a hot to a cold body and not the other way around. Lavoisier included heat (and light) on his list of the chemical elements alongside the 23 true elements known at the time. Lavoisier was an aristocrat, and also associated with the hated tax-collection system in France. Caught up in the French Revolution, he was accused of various crimes by Jean-Paul Marat, with whom he had crossed swords scientifically some years earlier. Lavoisier was tried, convicted and guillotined on 8 May 1794, a year after Marat himself had become a victim of the revolution.

14. Benjamin Thompson, later Count Rumford (1753–1814), was a physicist born in Woburn (formerly Rumford), Massachusetts, and one of the more colourful characters in the history of thermodynamics. Following service in the American War of Independence he moved to England, and then in 1785 to Bavaria. Here he undertook his seminal experiments on the conversion of work to heat and this work was published in the *Philosophical Transactions of the Royal Society* (Rumford 1798). In 1804 he married Lavoisier's widow, Anne-Marie, and although they separated after a year, Rumford lived out the rest of his life in Paris.
15. James Prescott Joule (1818–1889) was an English physicist most famous for his exceptionally careful work determining what was then known as the mechanical equivalent of heat (effectively the conversion of work to energy). He is recognised for this work in the use of his name for the SI unit of energy.
16. The Scottish mathematical physicist James Clerk Maxwell (1831–1879) was one of the most important figures in the development of modern science. Indeed, Einstein regarded Maxwell as the most significant figure in physics after Isaac Newton. Born and educated in Edinburgh, Maxwell studied mathematics at Cambridge and accepted the Chair of Natural Philosophy at Aberdeen in 1856. From 1860 to 1865 he was Professor of Natural Philosophy at King's College London and in 1871 he became the first Cavendish Professor of Physics in Cambridge, where he remained until his death. He is most famous for his derivation with Ludwig Boltzmann of the statistical distribution of molecular speeds, and his unification of electricity and magnetism, which unravelled the nature of light (Maxwell 1871). The physicist Ivan Tolstoy gives very readable account of Maxwell's life and work (Tolstoy 1981).
17. Democritus (ca 460–ca 370 BC) was an important pre-Socratic philosopher. Born in Abdera in Thrace, he formulated an early atomic theory of the universe. His work is difficult to disentangle from that of his mentor Leucippus, and only fragments of his work survive (Taylor 1999).
18. John Dalton (1766–1844) was an English chemist and meteorologist. He is best known for his work in the development of modern atomic theory, and is often regarded, together with Antoine Lavoisier and Robert Boyle, as one of the founders of modern chemistry. He also undertook important work on colour blindness (still sometimes referred to as Daltonism).
19. Estimates of the mean free path in air vary considerably (from ~35 to > 100 nm), depending on the assumptions made. This figure comes from Jennings (1988). The mean free path in air is an order of magnitude greater than the average separation of the molecules, and two orders of magnitude greater than the size of a molecule.
20. Jakob Hermann (1678–1733) was a Swiss mathematician who made important contributions to the development of classical mechanics. Daniel Bernoulli (1700 – 1782) was a Swiss mathematician and physicist, important in the early development of fluid mechanics (a modern translation of his major 1738 work is by Carmody & Kobus 1968).
21. Henry Cavendish (1731–1810) was a shy and reclusive English chemist, noted for his discovery of hydrogen, which he called 'inflammable air'; the modern name is down to Lavoisier. An extremely careful experimenter, he undertook pioneering work on the properties of gases, electricity and gravity (including a famous experiment to determine the density of the Earth). The recently discovered work on heat is described by Coopersmith (2010).
22. Ludwig Eduard Boltzmann (1844–1906) was a philosophically inclined Austrian physicist whose major achievement was his development of statistical mechanics, explaining the physical properties of matter in terms of the statistical behaviour of the component atoms. He committed suicide, possibly in part because of the difficulties of persuading some of his more conservative colleagues of the existence of atoms.
23. The origin of the term *enthalpy* is not entirely clear. Although often ascribed to Josiah Willard Gibbs, or jointly to Benoit Clapeyron and Rudolf Clausius, it is likely that the term was actually coined by the Dutch physicist Heike Kamerlingh Onnes (1853–1926), but first used in a scientific publication by J. P. Dalton (see Howard 2002 and Van Ness 2003).
24. See Maxwell (1871).
25. Rudolf Clausius (1822–1888) was a German physicist. In 1865 he gave a presentation (in German) to the Philosophical Society of Zurich, entitled *On several forms of the fundamental equations of the mechanical theory of heat*, which was published by the society later that year. In this he assigned the symbol S to a property, and named it entropy: 'I propose to call the magnitude S the entropy of the body, from the Greek word τροπη, transformation. I have intentionally formed the word entropy so as to be as similar as possible to the word energy'. This definition was incorporated into his book '*Mechanical theory of heat*' (the English translation of which is Clausius 1879).
26. Robert William Boyle (1627–1691) was an Irish chemist, most famous for his pioneering work on the nature of gases. His most important work is his book *The sceptical chymist* (Boyle 1661).
27. Clapeyron (1834).

28. Latent heat was discovered independently by at least three eighteenth century scientists. Jean André Deluc (1727–1817) noted in 1755 that a thermometer in a glass of ice did not change its reading as the ice was melting. At almost the same time the Scottish chemist Joseph Black (1728–1799) noted this phenomenon, and named it latent heat, though he did not publish his findings despite including them in his lectures. Slightly later, in 1772, the Swedish physicist Johan Carl Wilke (1732–1796) also discovered latent heat, and published his findings. The definitive study of this period (McKie & Heathcote 1935) affords priority to Black because of his more complete analysis. It may not be a coincidence that latent heat was discovered in countries that experience substantial winter snow, and the apparent reluctance of snow to melt on warm spring days was certainly a feature of the natural environment that intrigued Black.
29. A subtle difference is that the Clausius formulation deals with a change in entropy, whereas Boltzmann's statistical definition is an absolute entropy. Boltzmann probably chose W to symbolise the number of microstates from the German *Wahrscheinlichkeit*, meaning probability.
30. At a lecture delivered in Cambridge in 1959 the British chemist and novelist Charles Percy Snow (invariably referred to simply as C. P. Snow) famously equated knowledge of the Second Law of Thermodynamics with knowledge of Shakespeare. The lecture and subsequent book, *The two cultures* (Snow 1959), induced a vitriolic personal attack from the literary critic F. R. Leavis, and much debate. It is this debate, as much as the difficulty of the underlying concept of entropy, that has given the Second Law of Thermodynamics its iconic status.
31. An argument for entropy as information is advanced by Arieh Ben-Naim, who regards thermodynamic entropy as a subset of the subject matter covered by the Shannon information index (Ben-Naim 2012). A more traditional view of entropy is taken by Peter Atkins, and an especially readable introduction is Atkins (2007). Wicken (1986) and Brooks & Wiley (1986) take opposing views on the role of entropy and information in evolution. The role of information in biology is discussed by Maynard Smith (2000) with his customary elegance, clarity and style.
32. Shannon (1948); Shannon & Weaver (1949).
33. Jaynes (1957); Harte et al. (2008) provide an example of the application of maximum entropy to ecology.
34. The Third Law of Thermodynamics was developed by Walther Nernst between 1906 and 1912. It is sometimes referred to as *Nernst's theorem* or *Nernst's postulate*.
35. For a succinct and readable description of Gibbs energy and its relationship to entropy see Keeler & Wothers (2003). Josiah Willard Gibbs (1839–1903) was a reclusive American scientist who made important theoretical contributions to thermodynamics. Building on the work of James Clerk Maxwell and Ludwig Boltzmann, he created statistical mechanics (a term that he coined), explaining the laws of thermodynamics as consequences of the statistical properties of large ensembles of particles. The enormous impact of his work is in contrast to his quiet and solitary private life. Hermann Ludwig Ferdinand von Helmholtz (1821–1894) was a German physicist who made significant contributions to a wide range of scientific fields, but particularly the mechanical foundations of thermodynamics. He is also remembered for his philosophical work on the relation between the laws of perception and the laws of nature.

CHAPTER 3

Temperature and its measurement

'What's the good of Mercator's North Poles and Equators,
Tropics, Zones, and Meridian Lines?'
So the Bellman would cry: and the crew would reply
'They are merely conventional signs!'

Lewis Carroll (The hunting of the snark)

Understanding the role of temperature in ecology is impossible unless we have a clear picture of what temperature is and a practical method for measuring it. The former is an intellectual problem, the latter a technological one; both were recognised long before they were solved.

Temperature is a surprisingly elusive concept. The recognition that some things are hotter than others must go back to the dawn of human history and we can be certain that for as long as there have been bakers, smiths and potters, language will have had words for warm, hot and cold. This practical experience shaped early ideas of temperature, but it was not until the development of modern concepts of the structure and behaviour of matter that we came to understand what temperature is and how it differs from heat.

A typical everyday definition of temperature might be that it is a measure of how hot or cold an object is. This is clearly based on the experience of using our senses to detect whether an object is safe to handle, and it is related to some intuitive notion of heat. But even in everyday life things quickly get a little complicated for it is a common experience that metal feels colder to the touch than wood. Exploring why this is so sharpens our understanding of temperature, and shows that it is not as simple a thing as we might have thought.

A good place to start exploring the nature of temperature is the law of thermodynamics we left out of the previous chapter, the so-called zeroth law.

3.1 The Zeroth Law of Thermodynamics

The zeroth law of thermodynamics expresses in formal terms that way that energy flows between bodies in thermal contact (Figure 3.1)[1].

When the energy flow between different bodies is in exact balance, the net flow of energy is zero and these bodies are said to have attained the same temperature. A typical statement of the zeroth law is thus:

If two systems are in thermal equilibrium with a third system, then they are in thermal equilibrium with each other.

This provides a thermodynamic definition of temperature: it is that property of a body which determines whether it will gain or lose energy in particular circumstances. As we saw in the previous chapter, the spontaneous flow of energy in and out of a body is what we term heat. This thermodynamic definition is obviously related to our intuitive concept of temperature as the degree of hotness of a body. It does not, however, tell us what temperature actually is, and neither does it tell us how we might assign a value to that temperature.

3.2 What exactly is temperature?

The Zeroth Law of Thermodynamics provides us with a firm theoretical basis for defining temperature. As with energy and entropy discussed in the previous chapter, different insights into the nature

Principles of Thermal Ecology. Andrew Clarke, Oxford University Press (2017).
© Andrew Clarke 2017. DOI 10.1093/oso/9780199551668.001.0001

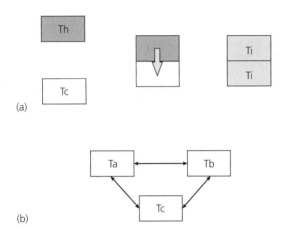

Figure 3.1 Temperature and thermal equilibrium. A (upper): if two bodies, one hot and the other cold, are brought into thermal contact then energy will flow (arrow) and they will come to thermodynamic (thermal) equilibrium at an intermediate temperature. If one of those bodies is a thermometer, then at equilibrium the temperature of the thermometer is identical to that of the other body. B (lower): the Zeroth Law of Thermodynamics states that if two bodies (T_a, T_b) are each in thermal equilibrium with a third body (T_c), then they are in thermal equilibrium with each other.

of temperature are provided by classical thermodynamics and statistical mechanics.

3.2.1 Temperature in classical thermodynamics

In classical thermodynamics, temperature is defined by the Clausius equality:

$$\Delta S = \frac{\Delta Q}{T}$$

In other words:

$$\frac{1}{T} = \frac{\Delta S}{\Delta Q}$$

The temperature of a system, T, is thus defined by the increase in entropy, ΔS, when a quantity of energy, ΔQ, is supplied to that system reversibly. Formally, the reciprocal of the temperature is equal to the partial derivative of the entropy with respect to the internal energy. While this definition is couched in terms unfamiliar to an ecologist, it is important because it leads directly to the thermodynamic temperature scale, use of which is essential for any physiologist or ecologist.

3.2.2 Temperature in statistical mechanics

The Maxwell–Boltzmann distribution of molecular speeds (see Chapter 2) tells us that, for an ideal gas, the mean kinetic energy of the gas molecules, E_k, is given by:

$$E_k = \frac{3}{2} k_B T$$

where k_B is Boltzmann's constant and T the absolute temperature (we will define absolute temperature shortly). In other words, the temperature of an ideal gas is a direct measure of its internal energy, and can be calculated from the mean kinetic energy of the constituent molecules. Note how Boltzmann's constant relates temperature to energy.

3.2.3 Thermodynamic beta

When two objects are placed in thermal contact (Figure 3.1) energy flows and the internal energies of the two bodies come to equilibrium. As we saw in Chapter 2, when two systems come to thermal equilibrium it is the entropy that is maximised. This can be formalised in terms of the quantity β (*thermodynamic beta*) that we met in Chapter 2, this being what equalises in systems at thermal equilibrium. Recall that:

$$\beta = \frac{1}{k_B T}$$

Intriguingly there are good thermodynamic arguments that β would be a better measure of temperature of a system, were it not for the historical evolution of our present concept of temperature[2]. Thermodynamic beta and absolute (thermodynamic) temperature are inversely related, though over the range of ecological temperatures the relationship is close to linear.

Thermodynamic beta is increasingly used as the abscissa by ecologists in studies of biological rate processes, where it is often labelled either 'temperature' (which it is not) or 'Boltzmann temperature' (ditto). Neither uses are helpful, and may be misleading if the underlying thermodynamic principles are not appreciated.

3.2.4 Temperature in biological systems

Various definitions of temperature are summarised in Table 3.1. For a biologist, the most useful definition is from classical thermodynamics (definition 2), because the exchange of energy with the environment is a key factor in the thermal balance of an organism. In terms of physiology, equally useful is the equivalence of thermodynamic temperature with the mean translational kinetic energy of constituent molecules. It is important to recognise, however, that an exact equivalence between thermodynamic temperature and the mean kinetic energy of the atoms applies only to an ideal gas, though it approximates well enough the behaviour of a monoatomic gas such as helium or neon at low pressures.

A monoatomic gas such as argon has only three degrees of freedom, which are the three spatial dimensions in which the atom can move. When two or more atoms are coupled together in a molecule, such as O_2, N_2 or CH_4, that molecule not only moves in space but also moves internally as each of the bonds between the atoms stretches, rotates or flexes. These bond movements represent extra degrees of freedom, and the total degrees of freedom for a molecule (and hence its internal entropy) increase rapidly with the number of atoms. These internal motions all utilise energy, and at equilibrium the internal energy is distributed equally across all the available degrees of freedom (a principle known as the *equipartition of energy*): as temperature increases the molecules not only move more rapidly, but have more intense internal motions. It is only the kinetic energy of translation that contributes to temperature, but the relationship between internal energy and temperature remains monotonic (that is temperature always increases as internal energy increases).

The temperature of a system can only be defined rigorously when the component particles (atoms, molecules, together with electrons and photons if these are also present) are in thermodynamic equilibrium. Under these conditions the distribution of their energies is described by the Maxwell–Boltzmann equation, and temperature is the parameter that defines that distribution (definition 5 in Table 3.1).

A consequence of this is that if a system is not in thermodynamic equilibrium, it has no meaningful temperature. If this seems odd, consider a fluorescent tube. Here the kinetic energy of the electrons within the discharge tube is equivalent to ~30000°C, the colour temperature of the radiation emitted by the phosphor coating the inner surface of the tube is ~5600°C and yet the glass is cool enough to touch. While the fluorescent tube is in steady state it is not in thermodynamic equilibrium and in a situation such as this, the concept of temperature is at the very least ambiguous, and quite possibly misleading[3].

Organisms are highly complex entities, and are typically not in thermodynamic or thermal equilibrium. To take a specific example, a mammal may have a warm core, but a cool surface and extremities; like the fluorescent tube, the mammal is in steady state but not in thermal equilibrium. A mammal is closer to thermodynamic equilibrium than a fluorescent tube, but we should always bear in mind that we are dealing with systems that do not adhere to the strict requirements for the rigorous application of classical thermodynamics or statistical mechanics. The thermal gradients within the mammal are, however, fairly small and so we can regard temperature as a spatially varying local property within the body. This allows us to use temperature in a meaningful way in ecology or physiology, as classical thermodynamics is still useful for near-equilibrium circumstances.

Table 3.1 Some definitions of temperature.

Definition
Intuitive definition
1 Degree of hotness
Definitions from classical thermodynamics
2 The property of a body that determines whether it gains or loses energy relative to its surroundings
3 The reciprocal of the rate of increase of entropy when energy is supplied reversibly
Definitions from statistical mechanics
4 A measure of the mean translational kinetic energy of constituent atoms or molecules
5 The parameter which defines the most probable equilibrium distribution of the energy of a population of atoms or molecules over all available states

3.3 Thermometry: the measurement of temperature

A thermometer is simply an instrument which provides a signal that is proportional to its own temperature. To measure the temperature of an object, we place the thermometer in direct thermal contact with that object and wait until the object and the thermometer have come to thermal equilibrium (Figure 3.1). When this has happened, the object and the thermometer are at the same temperature; the temperature of the thermometer (which we can read) is then the same as the temperature of the object (which is what we want to know). We have probably all had the experience of taking our own temperature with a clinical thermometer placed under the tongue, and being advised to wait while the thermometer comes to the correct temperature. At all times the thermometer is showing its own temperature but time is needed for the thermometer to achieve thermal equilibrium with the object whose temperature it is measuring.

This may seem an unduly pedantic description of an everyday activity, but the details are important. Indeed it is easy to take this familiar action for granted, but the development of accurate and reliable thermometers was a major intellectual and practical challenge. A brief history of how these challenges were met is a valuable way to introduce some key concepts.

3.3.1 The development of the thermometer

The invention of the thermometer is traditionally ascribed to Galileo (Box 3.1), but the origins of scientific thermometry go back considerably further and are lost in the mists of time[4].

The earliest devices for registering changes in temperature were probably very similar to that shown in Plate 2: a glass sphere containing air atop a narrow column containing water is immersed in a container of water. Changes in temperature cause the volume of air in the sphere to increase or decrease, and the length of the water column changes in response. Records of devices such as these go back to antiquity[5]. The main problems with these early instruments were twofold: the height of the water column would vary with air pressure as well as temperature and there was no scale. Without a numerical scale there was no way in which a number could be ascribed to the observed temperature, nor could different instruments be compared. These early instruments are best described as thermoscopes rather than thermometers[9].

3.3.2 The development of the temperature scale

The key step in developing a temperature scale is the selection of fixed points. A fixed point is a physical phenomenon that occurs repeatedly at the same temperature, such as the melting of ice or the boiling

---

Box 3.1 Galileo and the thermometer

The earliest written account of a true thermometer comes from the Italian physiologist Santorio, but credit for its invention is traditionally assigned to Galileo, who is thought to have developed his air thermometer around 1592. A letter to Galileo from his friend Giovanfrancesco Sagredo in 1613 refers to Galileo having invented the thermometer, but also indicates that he (Sagredo) had developed the instrument further. He used it to show that in winter the air can be colder than ice or snow, and that small bodies of water are colder than large ones. These observations indicate that Segredo had a means for determining that the height of the water column had changed in response to temperature, and it is likely that these were marks in the stem and a numerical scale[6].

The air thermometer was probably invented independently by the controversial English physician Robert Fludd, sometime between 1612 and 1617, and also by the flamboyant inventor and showman Cornelis Drebbel between 1598 and 1622[7].

The next significant advance was to use liquid rather than air to record changes in temperature, and the invention of what we would recognise today as a thermometer is believed to have been by Ferdinand II, Grand Duke of Tuscany, who constructed a sealed alcohol-in-glass thermometer around 1641. The final step was the development of a universal scale that would allow different thermometers to be compared. This was achieved by Daniel Fahrenheit[8], whose skills as a glassblower enabled him to make exquisitely sensitive mercury-in-glass thermometers.

of water at a particular atmospheric pressure. The problem for the early experimenters was, of course, how could they know that the melting of ice or the boiling of water always occur at the same temperature when there was no reliable thermometer to hand? The problem of fixed points was a major struggle in the history of thermometry, but eventually an understanding and consensus emerged[10].

The Danish astronomer Ole Rømer appears to have been the first person to make reproducible thermometers using the melting point of ice and the boiling point of water as fixed points, probably in 1702[11]. It seems that he first assigned a value of 60 to the upper fixed point (60 being a number familiar to astronomers), and then to ensure that all meteorological temperatures would have positive values, he divided the scale into eighths, with seven eighths above the melting point of ice and one below. This assigned a value of 7.5 to the freezing point of water.

Rømer used his thermometers to record the daily air temperature in Copenhagen during the exceptionally cold winter of 1708–1709 (Figure 3.2). It would appear from his original notebooks that following this severe winter Rømer assigned a rounded value of 8 to the melting point of ice, possibly because of the very low temperature recorded in January 1709. Rømer died the following year, and this modification to his original scale was never adopted.

In 1708, Rømer was visited by Daniel Fahrenheit. Fahrenheit accepted the principles behind Rømer's scale, but divided Rømer's degrees by four to obtain a greater resolution from his thermometers. Fahrenheit eventually settled on the temperature *under the armpit of a living man* (actually probably his wife) *in good health* as an upper reference point, the temperature of an ice-water bath as an intermediate reference point, and a mixture of ice, water and sal-ammoniac (ammonium chloride in its natural mineral form) or sea salt as the lowest. To these he assigned values of 96, 32 and 0 degrees respectively, although it seems clear that he only used the melting point of ice and the temperature of the human body as true fixed points. The ice and salt mixture was simply the lowest temperature that was easily achievable experimentally (and domestically) at the time.

Fahrenheit published his scale in 1724[13]. Soon after his death, the scale was modified slightly, by taking the boiling point of water as the upper fixed point and assigning values of 32 and 212 degrees respectively for the melting point of ice and boiling point of water. This had the advantage of making the interval between the two fixed points exactly 180 degrees, a highly convenient number as it can be divided into many integer fractions. The modified Fahrenheit scale is still used widely in America today, though elsewhere it has been superseded by the Celsius scale.

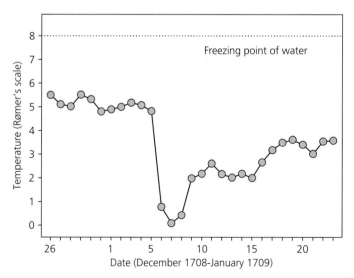

Figure 3.2 Air temperature measured once each day in Copenhagen during the exceptionally severe winter of 1708–09. The data are from the notebook of Ole Rømer, and temperature scale is Rømer's final modification of his original scale, in which the freezing point of pure water is assigned a value of 8 rather than the original 7.5[12].

The Swedish astronomer Anders Celsius devised his scale around 1742, setting the melting point of ice to 100 and the boiling point of water to 0. This is a centigrade scale, in that there are exactly 100 divisions (degrees) between the two fixed points, but it is the inverse of the scale we know today. Quite why Celsius did this is unclear, but it is sometimes supposed that because Celsius came from a country with long winters, he was more interested in colder temperatures than warmer, and his scale meant that no matter how cold the winter, or warm the summer, he would never have to deal with negative values of temperature. Whatever the reason, his scale was inverted soon after his death, probably by Carl von Linné (Carolus Linnaeus), to produce the Celsius scale used throughout the world to this day. A centigrade scale was also developed in by Jean Pierre Christin in Lyons around the same time[14].

What lessons can we learn from this brief outline of the early history of thermometry? The first is that the development of the field involved a close association between technical development and intellectual exploration. As so often, scientific advance and technical innovation went hand in hand.

The second, and very important, point is that the Celsius and Fahrenheit temperature scales are completely arbitrary. There is no particular meaning to the values of 0°C or 32°F; they are, as Lewis Carroll said through the mouth of the Bellman, merely conventions. This means that we cannot proceed very far with temperatures on these scales if we wish to explore the effects of temperature on biological processes. For example, a temperature of 50°C is not twice as hot as one of 25°C. If this seems odd, just consider the ratio of the temperatures of the two fixed points (boiling water, melting ice) in the different scales: in Fahrenheit the ratio is 6.6 (212/32), in Rømer it is 8 (60/7.5) and in Celsius it is infinity (100/0). The arbitrary nature of the Celsius and Fahrenheit scales limits their usefulness for thermodynamics, and also for ecology. We will return to this point shortly, but first we need to see how this problem was circumvented, through the derivation of the absolute, thermodynamic, temperature scale.

3.4 The gas thermometer and absolute zero

In parallel with the development of the liquid-in-glass thermometer, physicists were probing the fundamental properties of gases. In particular the work of Amontons and, almost a century later, Charles and Gay-Lussac showed that when pressure is held constant, the positive relationship between volume and temperature is similar in several gases (described in Chapter 2). They recognised that this relationship could be used to estimate temperature, and the result was the development of the gas thermometer.

A key concept to emerge from this work was that there must exist a lower limit to temperature. Amontons recognised that if the pressure of a gas is caused solely by its heat content, then extrapolation to zero pressure should represent a point of zero heat content. We would call this point the absolute zero of temperature but Amontons never used this phrase because at this time the distinction between heat and temperature was not understood. Others argued that a limit to low temperature would be achieved by reduction of volume at constant pressure to zero. Many of the leading experimentalists at the time measured the change in pressure or volume of gases between the ice point and boiling point of water to estimate the value of absolute zero, and their estimates were remarkably accurate (Figure 3.3). Perhaps the most accurate was that made by Henri-Victor Regnault, who estimated that the limit would be achieved at −272.75°C[15]. This is strikingly close to the modern value of −273.15°C.

Absolute zero can also be estimated from kinetics, quite independently from thermodynamics. This was first done by determining the temperature dependence of the reaction between hydrogen peroxide and hydrogen iodide, and estimating the temperature at which the reaction ceases. This turned out to be −276°C, remarkably close to the currently accepted value derived from thermodynamics[16].

The Third Law of Thermodynamics tells us that we can never achieve absolute zero, but experimenters have got remarkably close (Box 3.2).

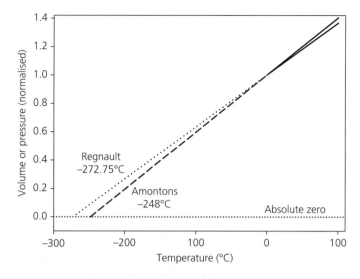

Figure 3.3 The discovery of Absolute Zero. The volume or pressure of fixed amount of gas was determined at the freezing point (0 °C) and the boiling point (then called the steam point, ~100 °C), and then extrapolated linearly to zero volume or pressure. Two early estimates are shown. The solid lines show the observed data, and the dotted/dashed lines the extrapolation to zero volume or pressure.

Box 3.2 Closing in on Absolute Zero

The coldest natural environments we know of are in the depths of the universe. Measurements of the cosmic microwave background radiation indicate that the average temperature of the universe is currently 2.725 K. The coldest object yet observed outside the laboratory is the Boomerang Nebula, where rapid expansion has cooled the nebula gas to ~1 K[17].

On Earth, conventional approaches to cooling gases allowed experimenters to liquefy helium over a century ago, and using this the Dutch physicist Onnes discovered superconductivity in mercury[18].

As atoms and molecules get very cold, quantum effects start to dominate and strange things happen. At temperatures close to absolute zero, all the atoms in a system can occupy the lowest quantum state simultaneously and the resultant Bose–Einstein condensate behaves as a single atomic entity. The first pure Bose–Einstein condensate was created in 1995 by cooling a dilute gas of about 2000 atoms of ^{87}Rb to a temperature below 170 nK by a combination of laser and magnetic evaporative cooling. In 2003 a temperature of 450 pK was reported for a Bose–Einstein condensate of 2500 atoms of ^{23}Na. In the late 1990s, researchers in Finland reported temperatures below 100 pK for atoms of rhodium, and even lower temperatures (< 50 pK) were reported in 2010. These are the lowest temperatures yet produced. They are, however, the temperature of a particular degree of freedom (nuclear spin, a quantum property), and not a conventional thermodynamic temperature.

A practical difficulty in achieving the lowest thermodynamic temperatures is that the very small samples involved may approach the minimum size for such a temperature to be meaningful. The lowest reported thermodynamic temperature for a directly cooled molecule is currently 2.5 mK, achieved by magnetic trapping of the diatomic molecule strontium monofluoride (SrF)[19].

3.5 The thermodynamic temperature scale

The recognition that there exists an absolute lower limit to temperature paved the way for a fundamentally different scale of temperature, based on thermodynamics and independent of the physical properties of any particular substance. It was William Thomson (Lord Kelvin) who developed the thermodynamic temperature scale. His starting point was Carnot's theory of ideal heat engines (Figure 3.4).

A heat engine works by using the difference in temperature between a high temperature source and a lower temperature sink. The heat engine uses

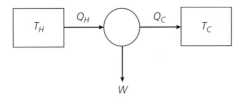

Figure 3.4 An ideal heat engine, as formalised by Sadi Carnot. The boxes represent a heat source and heat sink, and the circle represents the Carnot heat engine. Q_H is the energy supplied by the high temperature source T_H; some of this energy is used for work, W, with the remaining energy Q_C passed to the low temperature sink, T_C.

some of the energy that flows from the source to the sink to undertake work, and Carnot showed that the maximum efficiency that could be obtained from such a heat engine is:

$$1 - \frac{Q_C}{Q_H}$$

where Q_H is the energy supplied by the high temperature source and Q_C the energy passed to the low temperature sink. This follows directly from the First Law of Thermodynamics (this being an ideal engine, there are no losses to friction or other factors). Kelvin recognised that Carnot's theorem could be used to define a thermodynamic temperature by setting the ratio of the temperatures equal to the ratio of the energies:

$$\frac{T_C}{T_H} = \frac{Q_C}{Q_H}$$

Temperatures defined in this way are independent of material properties; in other words they are absolute temperatures[20]. The beauty of this absolute scale is that a natural zero emerges, which is the temperature at which a Carnot heat engine can do no work: absolute zero. Kelvin was also able to show that this approach to defining a temperature scale also leads to an equation for the behaviour of an ideal gas, identical to that which had emerged from the work of Amontons, Boyle, Charles and Gay-Lussac. The thermodynamic temperature scale could thus have been derived by consideration of the ideal gas law; it just so happened that Kelvin approached the problem through the theory of an ideal heat engine.

Kelvin's original suggestion was to bring his new scale into line with the existing Celsius scale by simply adding 273 to the temperature in Celsius. Because the thermodynamic temperature scale has a natural zero, absolute zero, it requires only a single fixed point to define it. The fixed point selected was the triple point of water (Figure 3.5), the unique temperature and pressure where the solid, liquid and vapour phases of water coexist in equilibrium. The triple point of water was chosen because it is very precise and highly reproducible, in contrast to

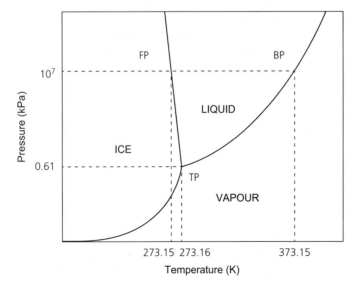

Figure 3.5 The phase diagram for water, showing the triple point (TP), as well as the melting point (FP) and boiling point (BP) of water at atmospheric pressure.

the historical fixed points (the ice and steam points) which are neither.

The Conférence Générale des Poids et Mesures (CGPM) in 1948 fixed the triple point of water at precisely 0.01°C, and assigned it a value of 273.16 K. This had two important consequences: it set absolute zero to 0 K and defined an interval of 1 K as being exactly the same as an interval of 1°C[21].

An absolute (thermodynamic) temperature scale was also proposed by the Scottish engineer and physicist William Rankine in 1859, in which he assigned a value of 491.688°R (or °Ra to distinguish this from the Réaumur or Rømer scales) to the triple point of water. This defined an interval of 1°R as identical to an interval of 1°F. This scale is largely obsolete, though it is still used in some engineering applications in the United States[22].

3.6 A new definition of temperature?

The Kelvin is one of the seven base units of the Système Internationale d'Unités (SI), known colloquially in many countries as the metric system. A long-standing problem with these base units is that some are defined in terms of the fundamental constants of nature, while others are not. The International Committee for Weights and Measures (CIPM: Comité International des Poids et Mesures) is considering a proposal to define all seven base units in terms of fundamental constants. These changes will not change the size of any SI unit, so there will be perfect continuity with present day science and engineering.

The proposed change is to define temperature in terms of Boltzmann's constant, which will itself be ascribed an agreed fixed value of 1.38065×10^{-23} J K^{-1}. This makes the definition of the Kelvin dependent on the definition of the second, metre and kilogram, but the triple point of water becomes a value to be determined experimentally. At the time of writing it seems likely that this definition will be adopted at the twentieth CGPM in 2018.

3.7 Types of scale

Many ecologists regard the Celsius and absolute (thermodynamic) temperature scales as equivalent, such that it does not matter which one is used. This is not so. The Celsius and absolute temperature scales are fundamentally different things and the difference matters a good deal. This is not just because many ecological situations involve the use of awkward negative numbers on the Celsius scale, but more importantly because only the absolute scale can (or should) be included in equations or models. To understand why, we need to distinguish the different types of scale used in science.

The theory of scales goes back to a seminal paper published by the American psychologist Stanley Smith Stevens[23]. In this he distinguished a number of types of scale, differing in their mathematical properties and the statistical operations that can be applied to them. While this approach has been criticised, and even subject to satirical attack, the classification of types of scale has proved both influential and valuable for clarifying what is actually involved in making a measurement of temperature.

The differing types of measurement scale are summarised in Table 3.2, with examples. A *Nominal Scale* (from the Latin *nominalis*, pertaining to a name) is essentially only a labelling and has little to do with measurement at all. Typical examples would be the names of plants or animals, or the stellar spectral types (O, B, A, F, G and so on) used by astronomers. These names can actually be numbers, such as those given to sports players. These numbers carry no quantitative meaning, and could just

Table 3.2 Types of measurement scale.

Scale	Description	Examples
Nominal	Names of objects	Colloquial names of plants and animals; formal taxonomic names; habitat types
Natural (counting)	Counts of objects or events (effectively an integer interval scale)	Individuals
Ordinal	Establishes order	Moh hardness; Beaufort wind scale
Interval	Establishes meaningful differences	Time; latitude; longitude; Celsius and Fahrenheit temperatures
Metric (Ratio)	Establishes meaningful ratios	All SI scales: length, mass, thermodynamic temperature

as easily be replaced by letters: a player in position 9 is not better than one in position 1 (they may be, of course, but this has nothing to do with the number assigned to their position on the field).

Ordinal Scales are those that establish simply the order (or ranking) by which things should be arranged. An example here would be the Moh hardness scale long used by geologists to order minerals in relation to their degree of hardness. In this scale each mineral standard is harder than all those below it on the scale. Thus gypsum (Moh hardness 2) is harder than talc (Moh hardness 1), and at the top of the scale, diamond (10) is harder than everything else. This does not mean, however, that topaz (8) is twice as hard as fluorspar (4); the Moh scale is purely a ranking. Another example of the ordinal scale would be the Beaufort wind scale familiar to mariners. Both the Moh hardness scale and the Beaufort wind scale have standards associated with each rank; for the Moh scale these are mineral standards, for the Beaufort wind scale they are the behaviour of the sails, or the state of the sea. While the numbers and standards can be compared to see which is higher or lower on the scale, they cannot be manipulated mathematically: we cannot multiply diamond by topaz, or subtract a calm sea from a storm.

Interval scales have no natural zero, and therefore have to be defined by two processes. The first is to define an arbitrary zero point, and the second is to assign a number to a different fixed point to define the size of the units. The Celsius temperature scale is a good example of an interval scale. The arbitrary zero was assigned to the melting point of ice, and the size of a degree Celsius set by assigning a value of 100 to the boiling point of water at a specified atmospheric pressure. Other familiar interval scales include latitude and longitude, and all the scales we use to tell time. The key features of an interval scale are that they tell us the order of data, and the size of the intervals between them. While interval scales are valuable, they have subtle but important limitations. It is possible to add or subtract Celsius temperatures or degrees of latitude, but as we saw above, we cannot meaningfully calculate a ratio. Neither can we multiply a value on an interval scale. This means that we cannot use temperature on the Celsius (or Fahrenheit) scale in a statistical or ecological model. If Celsius temperatures are all positive, then things will usually not go wrong, but the nature of an interval scale is such that this is risky and ill-advised.

Ratio (or Metric) Scales are scales that have a natural zero, and therefore require only a single point to define them. Examples of metric scales are all of the familiar SI scales we use for scientific measurement, such as mass, length and thermodynamic (absolute) temperature. In all of these scales, zero has a clear meaning: we all know what zero mass or zero length mean, and absolute zero provides the natural zero for thermodynamic temperature (in contrast to the arbitrary zero discussed above for Celsius and Fahrenheit). Furthermore we can add or subtract masses, lengths and absolute temperatures, and we can meaningfully divide masses, lengths or absolute temperatures into half or quarters, and also double or triple them.

Data expressed on ratio scales can be subjected to the complete suite of mathematical operations, and these are the only types of measurements that can (or should) be included in statistical or mathematical models. Because the Celsius and thermodynamic temperature scales have been constructed to establish identity between intervals of 1°C and 1 K, for many ecological situations the two scales are effectively interchangeable. Problems come, however, when negative Celsius temperatures are encountered. The frequent solution here is to rescale the Celsius temperature, say by adding 10, but this can itself lead to mathematical or statistical difficulties. It is far better to use thermodynamic temperature; that is what it was designed for.

3.8 Thermometry: the practical measurement of temperature

To measure temperature we use some physical property that varies in a precise and predictable way with temperature. Particularly useful have been the volume of a gas or liquid, and electrical resistance. In measuring temperature we need to distinguish between primary and secondary thermometers.

A *primary thermometer* is one whose equation of state does not involve any additional factors that are unknown. Temperature can thus be calculated directly from the physical process that forms the basis of the thermometer. Examples of primary thermometers include gas thermometers, acoustic thermometers and radiation thermometers, all of which are in widespread use in laboratories. They are, however, often inconvenient for practical thermometry in ecological situations.

All other thermometers are *secondary thermometers*, and these include almost all of the thermometers used by ecologists. An example of a widely used secondary thermometer is the platinum resistance thermometer. We cannot use this as a primary thermometer because we do not know enough about the temperature dependence of the electrical resistance of platinum to write down the equation of state. Instead, as with all secondary thermometers, we have to calibrate the platinum resistance thermometer against a primary thermometer. This calibration involves the erection of a practical temperature scale, so that any secondary thermometer we use in ecology or physiology provides a meaningful measurement of temperature that is as close as possible to the absolute (thermodynamic) temperature scale. This is the *International Temperature Scale* (ITS).

3.9 The International Temperature Scale

The ITS is a practical temperature scale that is sufficiently reproducible to allow us to assign an absolute (thermodynamic) temperature with reasonable accuracy and precision in everyday science, engineering and medicine. The ITS is revised at intervals, and the most recent version was devised in 1990 (referred to as ITS-90)[24].

The ITS approximates the absolute (thermodynamic) temperature scale through the use of a series of well-defined fixed points together with highly accurate secondary thermometers to interpolate between these. The fixed points used to define ITS-90 are predominantly phase changes (usually solid to liquid) or triple points. This is because these occur under precisely defined physical conditions and a thermodynamic temperature can be established with high precision using a primary thermometer. The main fixed points defining ITS-90 are shown in Table 3.3.

The secondary thermometers used to interpolate between the fixed points are helium vapour pressure or helium gas thermometers (temperatures

Table 3.3 Some fixed points defining ITS-90. TP: triple point, MP: melting point, FP: freezing point.

Element	Nature of fixed point	Thermodynamic temperature (K)
Ne	TP	24.5561
O (as O_2)	TP	54.3584
Ar	TP	83.8058
Hg	TP	234.3156
Ga	MP	302.9146
In	FP	429.7485
Sn	FP	505.078
Zn	FP	692.677
Al	FP	933.473
Ag	FP	1234.93
Au	FP	1337.33
Cu	FP	1357.77

below 24 K), platinum resistance thermometers over the range 24 K to 962 K and radiation thermometers at temperatures above 962 K. The thermodynamic temperature assigned to each of the fixed points is used to construct an equation relating the properties of the thermometer (for example the electrical resistance of platinum) to temperature. Using this equation, the thermometer can then be used to estimate the thermodynamic temperature at any point within its calibration range. For ecologists the relevant part of ITS-90 is covered by platinum resistance thermometers.

3.10 The measurement of temperature in ecology

Essentially all of the thermometers used by ecologists are secondary thermometers, and for these calibration is critical. Only then can we be sure that when a thermometer tells us it is 20°C, it really is 20°C and not 18°C or 22°C.

Thermometers for use by physiologists in the laboratory or by ecologists in the field come with calibrations that can be traced back to a primary thermometer in a measurement laboratory such as the National Institute of Standards and Technology (NIST) in the United States or the National Physical

Laboratory (NPL) in the United Kingdom. This traceability to a primary standard is essential to ensure accuracy, precision and comparability between different observers or laboratories. For example where large-scale climate analyses involve the pooling of data from many different sources, it is essential that an air temperature measured in Siberia is as accurate as one made in Australia, or any analysis of global temperature will be biased.

Ecologists use a range of types of secondary thermometers, depending on the scientific requirements and field conditions. These thermometers differ in their range, precision and suitability for particular tasks.

3.10.1 Liquid-in-glass thermometer

This is the oldest type of thermometer still in scientific use, and is probably what most people think of as a thermometer. Temperature is estimated from the length of a column of liquid within a glass capillary; as the temperature increases the molecules in the liquid move ever faster; as a result the spaces between them become larger and the liquid expands. The accuracy and precision of the measurement depend on the volume of liquid, the evenness of the capillary bore and the quality of the glass.

The choice of liquid depends on the temperature range of interest and the intended use of the thermometer. As with all metals, mercury expands (increases in volume) with temperature. Mercury expands more than most metals, and at room temperature does so similarly to water (Table 3.4). Although many early thermometers used water, this is not very suitable because the rate at which it expands is itself strongly temperature-sensitive (in other words, a water thermometer has a highly non-linear response).

Table 3.4 Volumetric expansion coefficients of liquids used in thermometry, and typical thermometer glass (all data for 20 °C).

Liquid	Coefficient (K^{-1}) × 10^6
Mercury	182
Ethanol	1090
Water	207
Glass	25.5

Mercury has long been the preferred liquid for routine thermometry, because of its high thermal expansion and the linearity of this expansion with temperature (though it is, of course, the difference between the thermal expansion of the liquid and that of the glass comprising the thermometer bulb that determines the overall linearity of the thermometer). Mercury also has the advantage of a high boiling point (356.7°C), which allows it to be used in a variety of high-temperature applications. A disadvantage is that mercury freezes at −38.8°C, a temperature well within the range of interest of meteorologists and ecologists. The lower range can be extended to −61°C by the use of a mercury–thallium alloy, but liquid-in-glass thermometers intended for lower temperatures typically use alcohol (ethanol). This remains liquid down to −114°C, but boils at 78.4°C.

From the very earliest days of the scientific investigation of temperature, the mercury-in-glass thermometer has offered a quick, simple and reliable measurement of temperature for medical, scientific, household and industrial requirements. Sadly, its days are numbered. This is because the mercury vapour released when a thermometer is broken is highly toxic and mercury is being phased out for liquid-in-glass thermometers, being replaced by galinstan (an alloy of gallium, indium and tin) or a variety of non-toxic organic liquids. Many countries in the European Union banned the use of mercury-in-glass thermometers in 2013, and in the United States official calibration services for these ceased in 2011.

3.10.2 The platinum resistance thermometer

A characteristic of metals is that the electrons of the outer shell are bound rather loosely. The interior of a solid metal thus has ionised atoms retained within the crystal lattice, and a large number of electrons free to move about. Normally the electrons move randomly, with no net movement in any one direction. However if an electric field is applied to the metal, the electrons move rapidly along this field towards the positive terminal; in other words, an electric current flows. As they move, the electrons are scattered by the metal ions they encounter; the electrons lose energy when scattered, and this energy is converted to heat (or in an incandescent bulb, also to light). The higher the temperature, the greater the thermal motion of the ions, and the more the electrons are scattered. The resistance to the flow of the current thus increases with temperature. This positive relationship between electrical resistance and temperature can be used to measure temperature very precisely[25].

While this relationship is a feature of all metals, the best resistance thermometers utilise platinum. This is because the relationship between electrical resistance and temperature in platinum is smooth and almost linear over a very wide range of temperatures (Figure 3.6).

Platinum resistance thermometers are produced in a wide variety of forms, depending on their intended use. The highest quality units, used to define ITS-90, utilise a strain-free wire coil supported within a sealed housing containing an inert gas. While these provide the highest accuracy and precision, their fragility makes them of limited use for ecologists. More robust units utilise thin films of platinum, platinum wires wound over an insulating support or a coiled element.

Platinum resistance thermometers have the advantage of high accuracy, a low drift and a wide operating range; typical instruments have a nominal resistance of 100 Ω at 0 °C, and sensitivities are typically 0.37–0.92 Ω K^{-1}. Those intended for ecological use typically have a precision of ±0.01 K.

3.10.3 Thermistors

Like resistance thermometers, thermistors utilise the thermal dependence of electrical resistance to estimate temperature, but instead of using a conductor such as platinum thermistors use semiconductors (often metallic oxides). These compounds conduct electricity only poorly because almost all the electrons are bound tightly to their parent atom. As temperature increases, the increased thermal motion of the atoms results in more electrons becoming free and available to flow. This leads to a decrease in electrical resistance as temperature increases (Figure 3.6).

The precise relationship between resistance and temperature depends on the material used to construct the thermistor. The relationship can be inverse, when the thermistor is said to have negative temperature coefficient. This contrasts with the positive temperature coefficient of a typical resistance thermometer, although some thermistors do have positive temperature coefficients.

The advantage of thermistors for ecology is that they generally have excellent accuracy (±0.001 K when carefully calibrated), and response times are generally faster than platinum resistance thermometers.

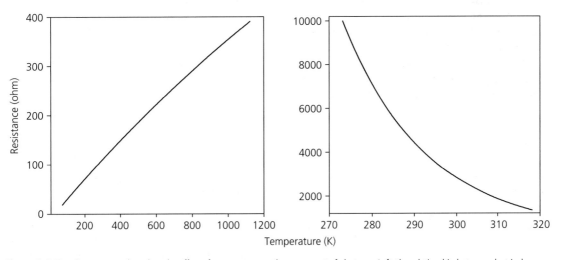

Figure 3.6 Two thermometers based on the effect of temperature on the movement of electrons. Left: the relationship between electrical resistance and temperature for platinum, normalised to a resistance of 100 Ω at 0 °C (273.15 K)[26]. Right: the relationship between electrical resistance and temperature for a typical thermistor. Note the very high electrical resistance of thermistors compared with platinum resistance thermometers, because thermistors are constructed from semiconductor materials.

3.10.4 Thermocouples

Thermocouples work by exploiting the thermoelectric effect, the direct conversion of a temperature difference to a voltage in a conductor. As with almost all the properties that distinguish metals, this effect is caused by the behaviour of the free electrons. As they move, the electrons carry both thermal energy and potential energy. The thermoelectric effect arises because of interactions between the free electrons and the metal ions in the lattice. These interactions differ between various elements, and there is a strong relationship between the electrical and thermal conductivity in metals (Figure 3.7).

There are several types of thermoelectric effect, all known by the name of their discoverer. The Peltier effect occurs when a current flows across a junction between different metals, causing the junction to either warm or cool. The Peltier effect occurs only at the junction and only when current flows; it is much used to cool microscope stages and cold plates. The Thomson effect is the heating or cooling of a current-carrying conductor subject to a temperature gradient; the metal then either absorbs or releases energy depending on the direction of current flow.

The important thermoelectric effect for thermometry is the Seebeck effect, which is the generation of a voltage in a conductor subject to a thermal gradient through the direct conversion of a temperature difference into an electromotive force[27]. The size of this effect is quantified by the Seebeck coefficient, which varies between metals (Table 3.5). When two different metals are joined in a loop, there is a net difference in the voltage generated. The size of this voltage depends on the magnitude of the difference between the Seebeck coefficient of the two metals and the temperature difference between the two junctions.

A thermocouple utilises the Seebeck effect to measure temperature. One junction (the sensor) is allowed to vary while the other (the reference) is maintained at a known temperature. As the reference junction temperature is known, the voltage

Table 3.5 Seebeck coefficients of selected pure metals, relative to platinum. The absolute Seebeck coefficient for platinum is −5 µV K⁻¹ at room temperature.

Metal	Seebeck coefficient (µV K⁻¹), relative to platinum
Selenium	900
Germanium	330
Antimony	47
Molybdenum	10
Rhodium	6
Platinum	0 (by definition)
Nickel	−15
Bismuth	−72

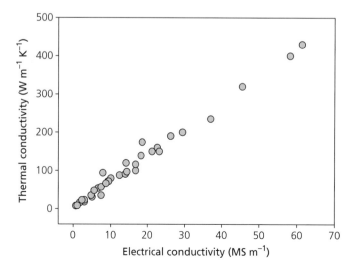

Figure 3.7 The correlation between thermal conductivity and electrical conductivity across different metals. Each symbol represents a different metallic element. This relationship exists because both electrical and thermal conductivity in metals involves the movement of free electrons.

difference can be used to estimate temperature. It is not the junction that generates the signal; the Seebeck effect is a feature of the thermal gradient along the entire wire between the junctions. However as long as the wires are homogeneous, and the reference junction is maintained at a constant (and known) temperature, the voltage difference depends only on the sensor temperature.

The thermal characteristics of thermocouples vary widely between the different types, depending on the conductors used (Table 3.6). These conductors are often alloys, of which the commonest used in thermocouples are chromel (90% nickel, 10% chromium), constantan (55% copper and 45% nickel) and alumel (95% nickel, with 2% manganese, 2% aluminium and 1% silicon).

Thermocouples are among the most widely used of temperature sensors, particularly in industry where their simplicity, reliability and wide temperature range are valuable characteristics. In ecology, however, their usefulness is limited by their relatively low accuracy (typically ±2 K).

3.10.5 Radiation thermometers

Radiation thermometers are fundamentally different from any of the thermometers described above. They work not by coming to thermal equilibrium with the object to be measured, but by measuring the radiation emitted by that object.

That hot objects emit light whose colour depends on temperature has been known since humans first mastered fire, and has provided language with vivid metaphors. We are all familiar with the dull red glow of the dying embers of a fire (about 700°C) and the fierce white glow of an incandescent bulb (above 1800°C). In simple terms, we know that the whiter and brighter the light from an object, the hotter it is. Although the human eye cannot detect radiation from an object cooler than about 800 K, every object above absolute zero radiates energy in the electromagnetic spectrum. This radiation stems from the conversion of the kinetic energy of the atoms into electromagnetic energy, brought about as random thermal motion of the atoms causes the electrons and protons of adjacent atoms to interact. The emission of this energy reduces the internal energy of the body and its temperature decreases. This continues until the loss of energy is balanced by energy absorbed from the environment. The loss of internal energy through thermal radiation is of immense importance to the energy balance of organisms, as we shall explore in Chapter 4.

The spectral distribution of the energy emitted by a perfect emitter (known as a black-body) is described by an equation derived by the German physicist Max Planck in 1890. The equation is:

$$u_\lambda(T) = \frac{2hc^2}{\lambda^5} \frac{1}{e^{\frac{hc}{\lambda T k_B}} - 1}$$

where $u_\lambda(T)$ is the wavelength-specific spectral radiance, h is Planck's constant, c is the velocity of light *in vacuo*, k_B is Boltzmann's constant, λ is the wavelength and T the absolute temperature. The Planck radiation law can be expressed in a number of ways, the equation above captures the power emitted from the surface of a black-body per steradian per unit wavelength. The Planck equation for black-body radiation marks a watershed in science, for it not only solved a number of theoretical difficulties but it also marked the start of modern physics (Box 3.3).

Figure 3.8 shows the spectral distribution of radiation emitted by a black-body at 5780 K, which is the temperature of the surface of the Sun. It can be seen that the peak emission is in the visible spectrum (the light we see) but that there is also considerable energy radiated in the near infrared wavelengths (the heat we feel).

Table 3.6 Thermoelectric characteristics of some common thermocouples; Seebeck coefficients are relative to platinum.

Type	Metals	Seebeck coefficient (µV K^{-1})	Useful range (°C)
E	Chromel–constantan	60	−100 to 1260
J	Iron–constantan	51	−190 to 760
T	Copper–constantan	40	−200 to 37
K	Chromel–alumel	40	−100 to 1260
S	Platinum–10% rhodium in platinum	11	0 to 1482
R	Platinum–13% rhodium in platinum	12	0 to 1482

> **Box 3.3 Thermal radiation and Max Planck**
>
> When Max Planck proposed his equation for black-body radiation in 1900, he was trying to solve what was known as the *infrared problem*. As spectroscopists probed ever further into the infrared, there had appeared small but important discrepancies between the spectral energy distribution predicted by existing theory and the new measurements. Planck was attempting to explain the observed distribution from first principles, basing his arguments on the statistical methods developed by Ludwig Boltzmann. He found he was only able to do this by assuming that the energy of the entities in the system existed in discrete amounts (quanta), in contrast to the continuous distributions assumed previously. This was an audacious leap and, as Planck himself acknowledged, without any theoretical foundation. It did, however, produce an equation that described observations to a high degree of accuracy.
>
> This step is often regarded as the point at which physics broke with its classical foundations, and also as the origin of quantum mechanics. Perhaps ironically, given his life-long struggle with the consequences of quantum mechanics, it was Einstein who first realised the significance of what Planck had done. Initially Planck regarded the quantisation of energy as a hypothesis, which would probably be abandoned when the real mechanism was discovered. It was only later that he himself realised the importance of the step he had taken[28].
>
> *The Ultraviolet Catastrophe*
>
> One of the other successes of Planck's expression was to predict correctly black-body radiation at short wavelengths. All attempts to model black-body radiation based on classical physics had predicted that power should climb to infinity at short wavelengths, a prediction that became known as the *ultraviolet catastrophe*. Because of this, a scientific myth has arisen that Planck solved the ultraviolet catastrophe, a story repeated in many books and articles. In fact the term was only coined many years after Planck had first presented his formula, and careful historical research has established the correct sequence of events. Nevertheless the story that Planck solved the ultraviolet catastrophe with an audacious flash of inspiration is so entrenched in the popular science literature that, like Archimedes in his bath, Galileo and the leaning tower of Pisa, and Newton's apple, it will probably never be erased[29].

From his radiation law, Planck was able to derive a number of previously established relationships, including the Stefan–Boltzmann law. This law was derived initially by the physicist Jožef Stefan, on the basis of measurements made by John Tyndall, and later derived from first principles by Ludwig Boltzmann. The law relates the energy radiated, summed across all wavelengths, from a body of area A to its absolute temperature, T, as:

$$P = A\varepsilon\sigma T^4$$

where P is the radiative power, ε the emissivity of the surface and the constant of proportionality and σ is the Stefan–Boltzmann constant. The emissivity

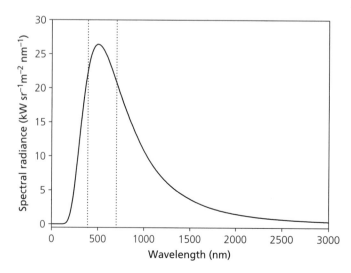

Figure 3.8 The spectral energy distribution for a black-body at 5780 K derived from the Planck radiation law. The dotted lines show the approximate upper and lower limits to visible light. Wavelengths below these are the near ultraviolet, wavelengths above them are the near infrared.

of a perfect black-body is 1, and the value of the Stefan–Boltzmann constant is 5.6704×10^{-8} W m^{-2} K^{-4}. This tells us that the total power radiated by an object increases rapidly with temperature[30].

Radiation thermometers do not measure the total power output, but utilise the spectral distribution of radiation to determine temperature. The first radiation thermometer was the bolometer invented by the American astronomer Samuel Langley in 1880[31]. The principle of the bolometer is simple: infrared light falling on a metal wire warms the wire, which changes its electrical resistance. The design of the bolometer is such that only a relatively narrow range of wavelengths is allowed to fall onto the sensor, and the change in resistance is measured by a bridge circuit.

Most radiation thermometers measure radiance over a narrow band of wavelengths, typically within the range 0.5–25 μm. Whereas the total energy radiated by an object varies with T^4 (Stefan's law), the shape of the spectral distribution of radiant energy means that within this range of infrared wavelengths, radiance increases with $\sim T^{12}$. This marked sensitivity of radiance to temperature means that although radiation detectors have low precision (typically ±1%), temperature can be measured very accurately.

All objects above absolute zero emit radiation. Because of this a radiation thermometer actually measures the radiance of the target object minus the radiance of the thermometer itself. For measurement of typical ecological temperatures correction for thermometer temperature is important; for objects hotter than ~150°C, the correction is very small because of the dependence of radiance on T^{12}.

A powerful advantage of radiation thermometers is that they be used to measure the temperature of distant objects remotely, for example the temperature of stars from the Earth, or the temperature of the Earth from a satellite. A disadvantage is that they can only measure the temperature of the surface; in measuring oceanic temperatures from space, for example, the sensors measure the temperature of the top few microns and can say nothing about the temperature of the underlying water (which might be 4 km deep). Indeed when winds are light, the skin temperature can differ significantly from the surface waters immediately below.

The most commonly used instrument to measure sea-surface temperature from space is the Advanced Very High Resolution Radiometer (AVHRR) which has been carried on all polar-orbiting meteorological satellites since 1978. The radiometer measures infrared radiation emitted from the surface of the Earth in five wavelength bands, three in the infrared (3.55–3.93 μm, 10.30–11.30 μm and 11.50–12.50 μm) one in the near-infrared band (0.725–1.00 μm) and a visible-light band (0.58–0.68 μm). Because each of these wavelength bands contain signals from a variety of sources, error correction is an important feature of all satellite observations. The most important errors are caused by unresolved or undetected clouds, water vapour (which reduces the apparent temperature of the sea surface), aerosols and dust. These errors can be reduced by careful cross-comparison of the signals in different channels and the measurement of surface temperatures from space has driven major advances in our understanding global patterns of temperature on Earth.

3.10.6 Thermal imaging cameras

Although the fundamental technique was developed almost a century ago, it is only recently that the technology has advanced sufficiently for thermal imaging cameras to be used routinely by ecologists. High-performance thermal imaging devices use sensors cooled to 60–100 K, though some are cooled as low as 4 K with liquid helium. This reduces noise in the sensor and thermal input from the camera itself, but these devices are typically too large and cumbersome for ecological use. Thermal imaging cameras designed for field use have imaging systems that operate at ambient temperature, although these may be thermally stabilised to reduce noise. The sensors used in these cameras comprise units (pixels) with a highly temperature-sensitive property, the nature of which varies from sensor to sensor. The increase in temperature when infrared radiation falls on the sensor is converted to an electrical signal and the image is then typically displayed in false colour (Plate 3). A significant problem is that the accuracy of the conversion of the sensor signal to temperature is dependent on a number of factors, and in many current models is only ±2%.

Thermal imaging cameras are proving to be of value in thermal ecology because of their ability to capture the variation in surface temperature over a wide spatial area, and also show how this varies in time. They have also proved valuable for population census (for example of marine mammals hauled out on land or on ice floes)[32].

3.11 Some general points about thermometry in ecology

The single most important factor in the use of any thermometer, except for the radiation thermometers just discussed, is that a thermometer is reporting its own temperature. This only corresponds to the measurement of interest when the thermometer and what it is measuring are in thermal equilibrium. Anything that influences the temperature of the thermometer can thus affect the accuracy of the result.

If the thermometer has a large thermal mass, for example a large cold thermometer placed in a small volume of warm water, it can change the temperature of the liquid as it equilibrates. The small thermal mass of thermistors and thermocouples make these particularly useful for ecology and the small volumes of liquid that typify physiology. A disadvantage of a small thermal mass, however, is that the temperature of the thermometer is easily affected by factors such as direct radiation or fluid flow. Ecologists have taken advantage of the latter effect by using thermistors to measure flow rate.

Small sensors such as thermistors can be immersed completely in the environment of interest, but larger instruments may have only part of the thermometer immersed in a liquid with the rest exposed. Most high-quality liquid-in-glass thermometers have a mark on the stem to show the level of immersion for which they are calibrated. If the thermometer is immersed above or below this mark then the thermal gradient along the thermometer stem may introduce an error into the reading.

For many ecological applications a fast response, in other words a low thermal mass, is a useful characteristic. However this can also be a problem as such a thermometer can change its reading very quickly after being removed from thermal equilibrium with the object whose temperature is to be measured. For example a small medical thermometer will start to cool down as soon as it is removed from under the tongue to be read. To overcome this problem, clinical thermometers often have a small constriction in the capillary which slows down the contraction of the liquid in the column. This is so effective that old-fashioned clinical thermometers had to be shaken to aid re-equilibration of the reading to ambient air temperature.

A particular problem is set by the measurement of deep-ocean temperatures. Below about 200 m ocean temperatures are almost universally < 4 °C. The practical problem is that the temperature registered by a thermometer being retrieved from depth will be affected as it passes from the cold deeper waters through the warmer (or in polar regions, colder) surface waters. The solution was the development of the reversing thermometer, the key feature of which is a constriction in the mercury capillary that causes the thread of mercury to break at a precisely determined point when the thermometer is turned upside down. To measure a deep-sea temperature the thermometer is lowered in the normal (upright) position to the required depth and allowed to come to thermal equilibrium with the water. The thermometer is then flipped upside down; the mercury column breaks and falls to the far end of the thermometer bore (which is now at the bottom). The thermometer is raised to the surface while still in the reversed position, and the length of the mercury in the capillary of the reversed thermometer is read. This broken column of mercury records the temperature of the water at the time the thermometer was reversed, but the reading has to be corrected slightly for the fact that the thermometer is now warmer (or in polar regions, colder) than it was when it was reversed. The reversing thermometer is carried inside a glass tube which protects the thermometer from the pressure at depth because high pressure can squeeze additional mercury into the capillary.

This was a very clever solution to the problem and reversing thermometers were the primary source of measurements of oceanic temperature from around 1900 to 1970. The mercury-in-glass reversing thermometer has now been superseded, first by the digital reversing thermometer (a platinum resistance thermometer which recorded the temperature at the time of reversal) and then by platinum resistance thermometers measuring and logging temperature continuously as they descend and ascend through the water column. This is typically done as part of a combined unit for measuring conductivity (from which seawater salinity is calculated), temperature and depth (a CTD unit, the mainstay of modern oceanographic research).

Contact thermometers such as liquid-in-glass and resistance thermometers, thermistors and thermocouples are all highly susceptible to thermal radiation. That is, their readings can be in error if sunlight falls on the sensor. This is a particular problem for thermometers used in meteorology, where the measurement of interest is the temperature of the air. If the thermometer is placed in the sun its reading will be inaccurate because of heating by thermal radiation from the sun (hence the usual stipulation *air temperature in the shade*).

The solution to this problem was the Stevenson screen, which is now standard for the measurement of air temperature all over the world (Box 3.4).

Box 3.4 The Stevenson screen

The Stevenson screen is an enclosure that shields meteorological instruments from precipitation and direct solar radiation, while still allowing air to circulate. It also prevents interference from drifting leaves and wandering animals. It was designed by Thomas Stevenson in 1864, and, although the design was modified later, it is still referred to as a Stevenson screen through much of the world[33].

The instrument box is louvred to allow the free flow of air, and painted white to reflect solar radiation (Plate 4). This ensures that what is measured by the thermometer inside is the temperature of the air in the shade.

The positioning of the screen is governed by criteria recommended by the World Meteorological Organisation in 2010. It is placed 1.25 m above ground to be above the strong temperature gradients close to the ground, with the opening away from the sun (in the northern hemisphere the opening faces north so that when the screen is opened to read the thermometer, no sunlight can fall on the thermometer and distort the reading). The screen must also be placed such that the thermometer reading is not affected by the influence of vegetation or buildings. The general rule is that the screen be sited at twice the distance of the height of any nearby object, for example at least 20 m away from a tree or building of height 10 m.

3.12 Measuring past temperatures: palaeothermometry

Reliable records of temperature measured by thermometer go back only to the early eighteenth century. A climatologist wanting to know the temperature earlier in history, a geologist needing to determine the Earth's temperature through deep time or a palaeobiologist interested in the body temperature of an extinct dinosaur must all rely on inferring temperature from a proxy.

For any temperature proxy to be useful, it must conform to a basic set of principles:

1. It must have a clearly defined relationship with temperature, and the magnitude of the effect must be sufficient to allow precise estimation of palaeotemperature.
2. Temperature must be the main control on the proxy, or if there are two controls it must be possible to distinguish their effects (usually with independent ancillary data).
3. The relationship with temperature determined today (the calibration of the proxy) must reflect that in the past.
4. The proxy must not have been affected subsequently (for example by ecological or geological processes); in other words the temperature signal must be frozen in time.

These criteria are stringent, but critically important. Palaeothermometry is technically difficult, and a wide range of proxies has been used (Table 3.7). Of these, by far the most important have proved to be those based on isotopes.

3.12.1 Isotope palaeothermometry

Both oxygen and hydrogen have three naturally occurring isotopes, and in each case the lightest isotope is by far the most abundant (Table 3.8). ^{17}O and the naturally radioactive tritium (^{3}H) are so rare that they play no practical role in palaeothermometry.

The different isotopes fractionate (that is, they change in their relative proportions) whenever the molecules containing them change phase (solid to liquid, liquid to vapour and the reverse) or participate in a chemical reaction. This fractionation results simply from the differences in mass, which affect

Table 3.7 Some important palaeothermometers.

Type	Proxy
Physical proxies	
	Fractionation of hydrogen isotopes
	Fractionation of oxygen isotopes
	Clumped isotope geochemistry
	Noble gas solubility
Chemical proxies	
	Concentration of Mg^{2+} in calcite or Sr^{2+} in aragonite
	Alkenone unsaturation
Ecological proxies	
	Leaf morphology
	Nearest living relative analogy
	Community analysis

Table 3.8 The isotopes of hydrogen and oxygen. Mass is relative molecular mass, and abundance on Earth is % by atom. Tritium is produced naturally by cosmic rays, but undergoes radioactive decay and exists only in trace amounts (< 10–15‰).

	Hydrogen		Oxygen	
Mass	Abundance	Mass	Abundance	
1	99.985	16	99.76	
2 (deuterium, D)	0.015	17	0.038	
3 (tritium, T)	unstable	18	0.200	

their thermodynamic equilibrium and kinetic behaviour. The critical factor for palaeothermometry is that this fractionation is affected by temperature.

In 1947 Harold Urey suggested that variation in the temperature of water would lead to measurable differences in the ratio of ^{18}O to ^{16}O in calcium carbonate precipitated from that water. The development of precise mass spectrometers in the 1950s allowed this idea to be tested and by 1953 the first palaeothermometer calibration had been established (Figure 3.9). Isotope palaeothermometry is now a powerful established technique, central to our understanding of the thermal history of the Earth[34].

The differences in thermodynamic and kinetic behaviour are more marked the lighter the atom: deuterium (^2H or D) is twice as heavy as ^1H, whereas ^{18}O is only 12.5% heavier than ^{16}O and ^{208}Pb less than 2% heavier than ^{204}Pb. Because of this, hydrogen and oxygen are extremely useful as palaeothermometers, but heavier elements such as lead are of no use at all. Isotopes of heavy elements have, however, been invaluable in helping unravel the geological history of the Earth[35].

The temperature signal is not the $\delta^{18}O$ or δ^2H of the sample; it is the difference between the sample and the source from which it was fractionated (Box 3.5), and herein lies the greatest challenge in palaeothermometry. For example, the δ^2H of the

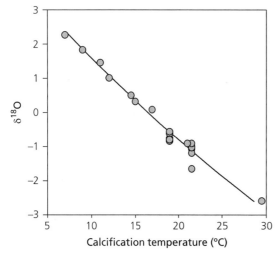
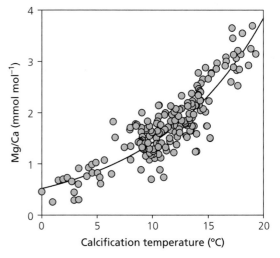

Figure 3.9 Two palaeothermometry calibrations. A (left panel): relationship between calcification temperature and $\delta^{18}O$ in a range of molluscs. B (right panel): relationship between Mg/Ca ratio in (six) species of foraminifera[36].

> **Box 3.5 Isotope ratios**
>
> The heavy stable isotopes of water (^{18}O and ^{2}H) are very much less abundant than the lighter isotopes (^{16}O and ^{1}H) and the relative abundances can be expressed as a ratio, for example:
>
> $$^{18}R = \frac{[H_2{}^{18}O]}{[H_2{}^{16}O]}$$
>
> Here ^{18}R is the ratio of the abundance of ^{18}O in water to ^{16}O in water. Because the proportion of the abundant isotopes (^{16}O, ^{1}H) are ~1, the statistical problems associated with calculating ratios of proportions can be ignored.
>
> The numerical value of isotope ratios such as these, although typically small, can vary widely in the natural environment, so it is useful to express them relative to the isotopic composition of a defined standard. Again using ^{18}O as the example:
>
> $$\delta^{18}O = \left[\left(\frac{^{18}R_{sample}}{^{18}R_{standard}}\right) - 1\right]10^3$$
>
> The factor of 10^3 is included because the abundance of the rarer heavy isotope is typically very small. This has become known as delta (δ) notation, and given the units 'per mil' (‰). Ratios such as ^{18}R and $\delta^{18}O$ are actually dimensionless and hence do not have units; the use of 'per mil' and the symbol ‰ have, however, become standard practice to indicate that a factor of 10^3 has been applied to make the numbers tractable. This example uses the ratio of ^{18}O to ^{16}O; the approach is identical for ^{2}H and ^{1}H (where the measure is designated δ^2H).
>
> Stable isotope geochemistry typically considers the concentration of molecules containing only a single heavy isotope: $H_2{}^{18}O$ versus $H_2{}^{16}O$ for $\delta^{18}O$, or $^{1}H^{2}H^{16}O$ versus $^{1}H_2{}^{16}O$ for δ^2H. Molecules containing two heavy isotopes (for example $^{1}H^{2}H^{18}O$) are so rare that ignoring them causes little error.
>
> For isotopic studies of water, the standard is purified seawater of standard isotopic composition prepared by the International Atomic Energy Authority in Vienna, and known as Vienna Standard Mean Ocean Water (VSMOW). For studies of calcium carbonate or organic carbon the standard was the isotopic composition of internal calcite from *Belemnitella americana*, a fossil belemnite from the Cretaceous Pee Dee Formation in South Carolina (and hence known as Pee Dee Belemnite, PDB). The original standard has now all been used, and has been replaced by Vienna PDB (VPDB), of identical isotopic composition to the original. Other standards are used occasionally for very depleted samples.

water in polar ice depends strongly on the condensation temperature at which the rain or snow was formed and this relationship can be seen in time-series of snow or rain for a particular site. Other important complications come from variations in the temperature of the source water, seasonality in temperature and atmospheric factors such as the strength of the inversion. To use δ^2H in polar ice cores to determine past temperatures, we must assume that these many and complex relationships were the same in the past as they are now, or else we must be able to deduce how these have changed. On long time scales, there are also variations in the isotopic composition of the oceanic source water.

A second example is that the $\delta^{18}O$ of skeletal aragonite or calcite from marine organisms such as foraminifera or molluscs depends both on the temperature and the isotopic composition of the seawater in which the organism is living. In some species the relationship between $\delta^{18}O$ and temperature may be complicated further by metabolic fractionation, often referred to as 'vital effects' (which is shorthand for fractionation during metabolism that we cannot explain or model). A particular difficulty with $\delta^{18}O$ of foraminifera to determine the Cenozoic thermal history of the oceans over the past 60 million years is that the isotopic composition of the seawater depends on the fraction of the Earth's water that is bound up in polar ice. This must be estimated and allowed for before the temperature can be inferred. Despite these difficulties, analysis of the $\delta^{18}O$ of carbonates from foraminifera has provided our best window into the thermal history of the oceans[37].

3.12.2 Clumped isotope palaeothermometry

An important recent development in palaeothermometry is based on the slight thermodynamic tendency for the heavy isotopes ^{13}C and ^{18}O to bond with each other, leading to a slight excess of ^{13}C–^{18}O bonds in skeletal carbonates compared with the number predicted on the assumption of random bonding[38]. This excess is greater at lower temperatures, and this allows for a determination of the temperature at which the carbonate formed, a technique known as *clumped isotope thermometry*. For measurement the carbonate is converted to carbon dioxide with phosphoric acid, and the concentrations of $^{13}C^{18}O^{16}O$ (relative molecular mass 47) and $^{12}C^{16}O_2$ (relative molecular mass 44) are measured. The slight increase in concentration of the heavier form compared with that expected assuming random bonding is termed Δ_{47}. Quantum mechanical calculations and measurement of natural samples with known crystallisation temperatures have shown that Δ_{47} varies with the inverse square of temperature. The value of this measure is that because it determines the extent of non-random binding within the carbonate, it does not depend on knowing (or estimating) the ^{18}O concentration of the source water.

This powerful new technique has been used to reconstruct oceanic temperatures as far back as the Silurian, and to estimate the body temperature of extinct vertebrates[39].

3.12.3 Mg/Ca/Sr palaeothermometry

When skeletal carbonate is synthesised by organisms, the Ca^{2+} may be replaced by small amounts of the Group 2 elements immediately above (Mg^{2+}) or below (Sr^{2+}) calcium in the periodic table. The differences in ionic size mean that Mg^{2+} substitutes for Ca^{2+} in calcite, whereas Sr^{2+} substitutes for Ca^{2+} in aragonite. Inclusion of Mg^{2+} in calcite is endothermic, and hence Mg/Ca ratio increases with environmental temperature; in contrast, inclusion of Sr^{2+} in aragonite is exothermic and Sr/Ca ratio declines with temperature.

The residence time of these elements in seawater is long, leading to fairly stable Mg/Ca/Sr ratios in seawater over periods of ~10^6 years. This reduces (but does not eliminate) the problem of variation in source composition and makes the Mg/Ca and Sr/Ca palaeothermometers particularly valuable. The Mg/Ca palaeothermometer has been much used in foraminifera, though care has to be taken to allow for differences in temperature sensitivity between different taxa, and for the effects of size and ecological variables. Calibrations have been achieved by growing foraminifera at different temperatures, analysing sediment trap samples in regions with a strong seasonal variation in temperature (Figure 3.9) or examining samples from sediment surfaces in regions with differing temperatures. The Sr/Ca palaeothermometer has been particularly important in reconstructing Holocene temperatures from corals[40].

3.12.4 Alkenone palaeothermometry

Alkenones are long-chain ketones produced by haptophyte algae, the most important of which is *Emiliania huxleyi*. These compounds typically contain between 35 and 41 carbon atoms, and differ in the number of carbon–carbon double bonds (unsaturation). The alkenones fall to the seabed in phytodetritus (marine snow), and can be recovered from sediment cores. The relative proportions of the different alkenones, and specifically the degree of unsaturation, have been shown to vary with the temperature of the surface waters where the algae grow. The unsaturation is expressed by calculating the ratio of C_{37} ketones containing 2 or 3 double bonds:

$$U_{37}^{K'} = \frac{[C_{37:2}]}{([C_{37:2}]+[C_{37:3}])}$$

The relationship between this measure ($U_{37}^{K'}$) and surface oceanic temperature has been determined from the alkenone composition of surface sediments from areas with differing surface temperatures[41].

3.12.5 Ecological proxies

The most important ecological palaeotemperature proxies are fossils or subfossils of species whose climatic niches are known from the present day. This approach has proved valuable in reconstructing Late Pleistocene and Holocene climates, and particularly valuable insights have come from beetles, diatoms and pollen.

There are a number of criteria for a particular species to be valuable as a climate proxy. It must be sensitive to environmental change, it must be highly mobile so that its distribution can shift rapidly

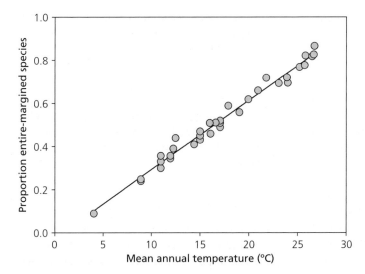

Figure 3.10 The first data showing a relationship between mean annual temperature and the proportion of woody dicotyledon species in a given assemblage whose leaf margins are entire (that is, not toothed or serrated)[44].

as the environment changes, it must have a well-defined climatic niche, and it must fossilise. Beetles (Coleoptera) have proved ideal in meeting these requirements, and they also have the benefit of being a highly diverse group with an abundant fossil record. Climate reconstructions using fossil beetles were the first to show the very rapid shifts in climatic zones that accompanied the shift from glacial to interglacial conditions in northern latitudes after the Last Glacial Maximum, rapid changes that were only confirmed much later by isotope studies from ice cores in Greenland[42].

In a similar way, diatom assemblages preserved in lake sediments have been valuable in reconstructing past lake environments, and fossil pollen has been enormously important in reconstructing past terrestrial plant assemblages (from which climate can be inferred)[43].

Where fossil species are extinct, we may be able to infer temperature from the climate niche of a close living relative. This approach carries the implicit assumption that climate niche tends to be conserved within lineages, which while generally true cannot necessarily be established for a particular case. Organisms can adapt their climate niche as the environment changes, and so the further back in time we look, the less reliable this approach is.

Where extinct species have no close living relative, we may be able to infer some aspects of their ecology from their morphology. This has been a particularly useful avenue with plant leaves, whose morphology is correlated with climate. The rationale underpinning this correlation is that selection will tend to converge on an optimal solution for leaves having to capture sunlight, conserve water and regulate their temperature. The first correlation between leaf shape and mean annual temperature was produced by Jack Wolfe in 1979 (Figure 3.10).

The method has been refined with the use of a wider range of measures of leaf size and morphology (nowadays often termed leaf physiognomy), and multivariate statistics. While relationships are established for living floras with a minimum of 20 species, fossil assemblages may force palaeobotanists to work with far fewer, with consequent loss of resolution. Nevertheless leaf size and morphology has proved an extremely valuable tool for reconstructing past terrestrial climates[45].

3.13 Summary

1. The concept of temperature is defined by the Zeroth Law of Thermodynamics. Temperature is that property of a body which determines whether it gains or loses energy in a particular environment.
2. In classical thermodynamics temperature is defined by the relationship between energy and entropy. In an ideal gas, temperature is a measure of the mean kinetic energy of the molecules.

3. Temperature can be defined only for a body that is in thermodynamic and thermal equilibrium. Organisms do not conform to these criteria, but the errors in assuming that they do are generally small.
4. The Celsius and Fahrenheit temperature scales are arbitrary. They require two fixed points, one to define the zero and the other to set the scale. The thermodynamic (absolute) scale of temperature has a natural zero (absolute zero) and is defined by a single fixed point, the triple point of water. The thermodynamic unit of temperature is the Kelvin. The Celsius scale is convenient for much ecological and physiological work, but where temperature is included in statistical or deterministic models, only thermodynamic temperature should be used.
5. Thermometers used in ecology and physiology are secondary thermometers, which require calibration against a primary thermometer. Over the temperature range of interest to ecologists, the International Temperature Scale (ITS-90) is defined by the platinum resistance thermometer.
6. Past temperatures can only be reconstructed with the use of proxies, the most important of which are based on isotope fractionation.

Notes

1. The zeroth law was named by Sir Ralph Fowler (Fowler & Guggenheim 1939). It was so-called because it was formalised after the first, second and third laws, but being regarded as fundamental was designated the zeroth law rather than the fourth. Although the basic idea goes back at least to William Rankine in the nineteenth century (see Truesdell 1979) and was also stated by James Clerk Maxwell, it was Fowler who formalised and named it.
2. This is described nicely by Atkins (2007).
3. This neat example comes from Nicholas & White (1994).
4. An excellent history of the thermometer is Middleton (1966). Other useful sources are Chang (2004) and Sherry (2011).
5. Recognition that a change in the temperature of air resulted in a change of its volume goes back at least to the third century BC. The Greek engineer Philo of Byzantium (around 280–220 BC), who probably lived most of his life in Alexandria, demonstrated that air expanded when warmed. Both he and the much later Greek mathematician and experimentalist Hero of Alexandria (around 10–70 AD), are believed to devised instruments that could show this. As with so much early science, these works were lost to the west during the Medieval Period (also referred to rather disparagingly as the Dark Ages) and only became available in the west again after being retranslated from Arabic. The role of Arab scholars in preserving early scientific knowledge, and extending this, is described by Al-Khalili (2010), and the modern recognition that that the Dark Ages were not so dark after all is described by Hannam (2009) and Freely (2012a).
6. The description of the early thermometer comes from Santorio Santorii (Sanctorius, 1561–1636) of Padua, in his commentary on Galen, published in Venice in 1612. At this time Padua was at the centre of European intellectual life, and the location of many significant advances in anatomy and botany. The letter to Galileo is quoted extensively by Middleton (1966).
7. Robert Fludd (1574–1637) was an astrologer, mathematician and cosmologist, who entered into a celebrated correspondence with Kepler. He published diagrams of an air thermometer (a thermoscope with a numerical scale) in 1626 and again (posthumously) in 1638, although by this time air thermometers were well known in England (see Middleton 1966). Cornelis Jacobszoon Drebbel (1572–1633) was one of the more colourful characters in the history of science. Born in the Netherlands, he travelled widely, eventually settling in England around 1604. He is believed to have refrigerated the Great Hall of Westminster Abbey in the summer of 1620 to impress King James 1 of England and Scotland, and to have sailed a submarine carrying 16 passengers (including himself) at a depth of 4–5 metres from Westminster to Greenwich in 5 hours. This required a way to replenish the air, and a note written by Robert Boyle in 1662 suggests that Drebbel may have discovered oxygen over a century before the Swedish pharmacist Carl Scheele and the English clergyman Joseph Priestley (see Shachtman 1999). The problem for historians is that Drebbel earned his living as an inventor and showman (some would say charlatan) and so he kept few records of his discoveries, presumably to keep them from falling into the hands of his rivals.
8. Daniel Gabriel Fahrenheit (1686–1736) was born in Danzig (Gdansk), then in the Polish-Lithuanian Commonwealth, but lived most of his life in the Dutch Republic. He visited England in 1724, the year he was elected as Fellow of the Royal Society of London. That

same year he published a series of five papers, all in Latin, that represent his entire written scientific output.

9. This distinction between thermoscopes and thermometers was proposed by Middleton (1966), and is now generally accepted.

10. The struggle to decide fixed points for calibrating thermometers is described in detail by Chang (2004).

11. Ole Rømer was a Danish astronomer, most famous for establishing a finite value for the speed of light by resolving discrepancies in the observed timing of the orbits of the Galilean satellites of Jupiter. The history of Rømer's scale was described (in Danish) by Kirstine Meyer; it is summarised by Middleton (1966). See also Meyer (1910).

12. Plotted using tabulated data from Rømer's original notebook, reproduced in Middleton (1966).

13. Fahrenheit (1724). For a full discussion of Fahrenheit's scale see Middleton (1966).

14. Anders Celsius (1701–1744), was born in Uppsala in Sweden, and became professor of astronomy at Uppsala University in 1730. He made important contributions to our understanding of the aurora (the Northern Lights), and died of tuberculosis when only 43. It is likely that the inversion of the original Celsius scale was undertaken independently by several people, including Pehr Elvius, Secretary to the Royal Swedish Academy of Sciences, the instrument maker Daniel Ekstrøm and Mårten Strømer, who studied astronomy under Celsius. Middleton (1966) argues that the key person was probably Linnaeus (Carl von Linné, 1707–1778), whereas Shachtman (1999) suggests that the reversed scale was introduced by Celsius' successor as professor of astronomy, Mårten Strømer, in 1750. Middleton (1966) also discusses the centigrade scale developed by Jean Pierre Christin.

15. The French physicist Henri-Victor Regnault (1810–1878) was one of the great experimentalists. He did not provide science with innovative breakthroughs or lasting insights, and hence is largely ignored in most histories of thermodynamics. What he did do, however, was to undertake exceptionally careful measurements of the thermal properties of gases, and many of the constants he determined are still in wide use today by scientists and engineers. He reported his estimate of absolute zero in his monumental work of 1847.

16. Harcourt & Esson (1895).

17. The universe is not in thermal equilibrium; some objects are extremely hot, whereas most of the universe is very cold. The universe as a whole thus does not have a defined thermodynamic temperature. It is, however, bathed in the cosmic background microwave radiation which can be regarded as providing an overall temperature for the universe averaged by volume. The best estimate is that the energy spectrum of the this radiation is equivalent to that of a black body at 2.725 K (Fixsen 2009). The cosmic background microwave radiation was discovered serendipitously in 1964 by Arno Penzias and Robert Wilson, who received the 1978 Nobel Prize in Physics. The temperature of the Boomerang Nebula was measured by Sahai & Nyman (1997).

18. Helium was first liquefied by Heike Kamerlingh Onnes (1853–1926), on 10 July 1908. He used this to explore the relationship between electrical resistance and temperature, first with platinum and gold, and he discovered superconductivity in mercury on 8 April 1911, in Leiden. Onnes received the Nobel Prize in Physics in 1913 for this work.

19. The form of matter we now call a Bose–Einstein condensate was predicted in 1924 by Albert Einstein on the basis of calculations by the Indian physicist Satyendra Nath Bose. Temperatures low enough to test Einstein's prediction were, however, not achievable until the 1990s, and the first Bose–Einstein condensate was created by Eric Cornell, Carl Wieman and co-workers on 5 June 1995. About four months later, an independent effort led by Wolfgang Ketterle created a larger condensate made of ^{23}Na. Cornell, Wieman and Ketterle won the 2001 Nobel Prize in Physics for their achievements. For recent measurements of very low temperatures in the laboratory see Tuoriniemi & Knuuttila (2000), Leanhardt et al. (2003), Medley et al. (2011) and Barry et al. (2014). The minimum size for thermodynamic temperature to be meaningful is discussed by Hartmann (2006).

20. Thomson (Kelvin) developed his temperature scale in a series of papers between 1848 and 1851. Kelvin's route to his conclusions is explained beautifully by Chang (2004).

21. The 9th CGPM (Conférence Général des Poids et Mesures, 1948) fixed the triple point of water at 0.01 °C. The 10th CGPM (1954) confirmed the triple point as the defining fixed point of the thermodynamic temperature scale and assigned it a value of 273.16 K precisely. The 13th CGPM (1967/8) renamed the unit of thermodynamic temperature the Kelvin (symbol K), thereby rendering the previous unit of degree Kelvin (°K) obsolete. Finally, in 2005 the Comité International des Poids et Mesures agreed that for the purposes of defining the triple point, the thermodynamic temperature scale would refer to water having an isotopic signature identical to Vienna Standard Mean Ocean Water (VSMOW). Ironically, one consequence of the newly defined absolute scale was that the boiling point of water at standard atmospheric pressure was now 99.974 °C (Nicholas &

White 1994); this meant that the Celsius scale was no longer a true centigrade scale because the difference between the two fixed points that define it was no longer precisely 100.

22. William John Macquorn Rankine (1820–1872) was a Scottish engineer, physicist and mathematician who made significant contributions to the theory of heat engines and thermodynamics in general. His absolute temperature scale is now little used outside the United States, and the US National Institute of Standards and Technology recommends against its use. Nowadays, reference to absolute or thermodynamic temperature (almost) always means the Kelvin scale.

23. The original paper is Stevens (1946). The approach was criticised by Velleman & Wilkinson (1993) as being unduly strict. The nominal scale was also satirised by Lord (1953), using football numbers as the focus for his humour; this paper does, however, also make some important points about the statistics of large numbers.

24. Previous versions were in 1927, 1948 and 1968. The 1948 version was amended slightly in 1960 when the scientific community adopted the Systéme International (SI) for units, and renamed the International Practical Temperature Scale (IPTS-48). The IPTS-68 was itself modified in 1975 and 1976.

25. The principle of using the thermal dependence of electrical resistance to measure temperature was probably first enunciated by William Siemens at a Bakerian Lecture to the Royal Society in London in 1871. However the invention of the platinum resistance thermometer is usually credited to Hugh Longbourne Callendar, who described this to the Royal Society in 1886 and published a description the following year (Callendar 1887).

26. The plotted relationship is derived from the equation of Callender & van Dusen. The ITS-90 uses a more complex relationship, combining a 12th-order polynomial between 13.80 K and 273.16 K with a 9th-order polynomial between 273.15 K and 1234.93 K.

27. The Seebeck effect was discovered by the German-Estonian physicist Thomas Johann Seebeck (1770–1831) in 1821, when he discovered that a closed loop formed by two different metals disturbed a compass needle when there was a temperature difference between the junctions. He called this the thermomagnetic effect, and it was the Danish physicist Hans Christian Ørsted (1777–1851) who recognised that it was an electric current that was generating the magnetic field that disturbed the compass needle, and renamed it the thermoelectric effect. The Peltier effect was discovered by the French physicist Jean Charles Athanase Peltier (1785–1845) in 1834. In the 1850s, William Thomson (Lord Kelvin) laid out the principles of using the thermoelectric effect to measure temperature, explained the relationship between the Seebeck and Peltier effects, and predicted what we now know as the Thomson effect.

28. Max Karl Ernst Ludwig Planck (1858–1947) was a German theoretical physicist, widely regarded as the founder of quantum theory. He received the Nobel Prize in Physics in 1918, having been twice nominated by Einstein. Planck referred to the statistical step he introduced as *an act of desperation*, and later wrote that *experience will prove whether this hypothesis is realised in nature* (Pais 1982, p. 370–371). The historian of science Max Jammer wrote in 1966 that *Never in the history of physics was there such an inconspicuous mathematical interpolation with such far-reaching physical and philosophical consequences* (cited in Kragh 1999, p. 61).

29. The phrase 'ultraviolet catastrophe' was coined by the Austrian physicist Paul Ehrenfest in 1911. An excellent account of the history of the Planck Radiation Law, and its importance to the evolution of modern physics, can be found in Abraham Pais' vivid account of the life and work of Albert Einstein (Pais 1982), and in Helge Kragh's history of twentieth century physics (Kragh 1999). This episode is also the focus of a detailed study by the philosopher of science Thomas Kuhn (Kuhn 1978).

30. Jožef Stefan (1835–1893) was Austrian by nationality, but a member of a minority ethnic group, the Carinthian Slovenes. He spent most of his scientific life in Vienna, making major contributions to thermal physics. Using his radiation law, he made the earliest meaningful estimate of the surface temperature of the Sun (5760 K, as against the modern value of 5780 K).

31. Samuel Pierpont Langley (1834–1906) was a largely self-taught astronomer, who developed the first radiation thermometer which he used to undertake pioneering work on the solar constant and the infrared part of the solar spectrum. His invention is recognised memorably in a limerick of unknown origin:

Oh Langley devised the bolometer,
It's really a kind of thermometer,
That can measure the heat,
From a polar bear's feet,
At a distance of half a kilometer.

Sadly for fans of the limerick, Langley actually measured the temperature of a cow from a quarter of a mile, but perhaps that was difficult to scan or rhyme.

32. For a review of thermal imagery in ecology see McCafferty (2007), and an example of its use in biophysical modelling see McCafferty et al. (2011).

McCafferty et al. (2013) report the thermal imaging of emperor penguins in Antarctica.

33. Sir Thomas Stevenson (1818–1887) was a British civil engineer, who designed many lighthouses around Scotland; he published his design for a thermometer screen in 1864 (Stevenson 1864). This was not the first design for such a screen, but it appears to be the first to incorporate double louvres (see Middleton 1966). The design was updated by Bilham (1937). Thomas Stevenson was the father of the novelist Robert Louis Stevenson.

34. Harold Urey (1893–1981) was an American geochemist who won the 1934 Nobel Prize in Chemistry for the discovery of deuterium. As well as initiating the field of isotope palaeothermometry, together with his graduate student Stanley Miller he undertook the seminal experiment in which they synthesised organic molecules by applying an electric discharge to a mixture of gases believed to represent the early atmosphere of Earth. The basic principles of isotope palaeothermometry were described by Urey (1947), and the first calibrations presented by Epstein et al. (1951, 1953).

35. While hydrogen fractionates more strongly than oxygen, analytical difficulties in determining the hydrogen/deuterium ratio mean that it is similar in precision to using oxygen isotopes.

36. The molluscan data are from Epstein et al. (1953), which corrects a few errors in the original report (Epstein et al. 1951); note that here they are plotted with temperature as the independent variable, because it is temperature that is believed to be driving the isotope fractionation. However a calibration curve has the isotope fractionation as the independent variable, as we wish to use $\delta^{18}O$ to estimate temperature. Given the variability in the data, these two relationships will not be the same. The foraminiferal Mg/Ca data are replotted from Anand et al. (2003).

37. Both benthic (bottom-living) and surface (planktonic) foraminifera are preserved in sediment cores, and as these can be distinguished, isotope fractionation can provide data on both surface and deep water temperatures. This is critical for reconstruction of past atmospheric climate and water circulation.

38. This excess arises through slight differences in the zero-point energy of carbonate ions with different C and O isotopes (isotopologues).

39. The clumped isotope technique is described by Eiler (2007, 2011) and its use in estimating vertebrate body temperatures discussed by Eagle et al. (2010, 2011).

40. The Mg/Ca palaeothermometer is discussed by Lear et al. (2002), Barker et al. (2005) and Hertzberg & Schmidt (2013) and results compared with alkenone palaeotemperatures by Timmermann et al. (2014). Useful introductions to the Sr/Ca palaeothermometer are Marshall & McCulloch (2002) and DeLong et al. (2007).

41. A valuable general background to alkenones is provided by Marlowe et al. (1984), and calibration of the unsaturation index is described by Prahl & Wakeham (1987). For more recent applications see Prahl et al. (2006, 2010) and Rosell-Mele & Prahl (2013).

42. The use of fossil beetle assemblages to reconstruct past climates is described by Elias (1994, 2013), and some nice examples of the work are from the founder of the field, Russell Coope (Coope et al. 1997, 1998).

43. The analysis of diatom assemblages and fossil pollen is well described by Roberts (1998).

44. The data plotted here are from the original study of the forests of Eastern Asia by Wolfe (1979).

45. The first demonstration that leaf morphology was correlated with mean annual temperature was Wolfe (1979). Useful recent reviews are Wolfe (1995), Royer et al. (2005) and Peppe et al. (2011). Guerin et al. (2012) have shown that leaf morphology has responded to recent climate change in South Australia.

CHAPTER 4

Energy flow in organisms

> Ecology is done poorly if either the biotic or abiotic aspects of the subject are not treated in a fully correct and rigorous scientific manner
>
> David M. Gates[1]

We now have enough background to bring physics to bear on ecology and examine how energy flows in and out of organisms. All organisms exchange energy and materials with their environment and, as we saw in Chapter 2, are thus open thermodynamic systems[2].

The basic flows of energy can be summarised simply (Figure 4.1).

Organisms gain or lose energy in three ways: exchange of kinetic energy (light and heat), uptake and discharge of chemical potential energy and work. These all change the internal energy of the organism, and they are connected intimately; for example metabolism uses chemical potential energy from nutrients and dissipates much of this as kinetic energy (heat). In endotherms (mammals and birds) this dissipated heat is used to maintain the internal body temperature (see Chapter 10).

We saw in Chapter 2 that while we cannot measure or calculate the total internal energy of a body, we can quantify changes in its internal energy very precisely. We do so by capturing the movement of energy in a simple balance equation:

$$\Delta U = \Delta E + w$$

Here ΔU is the change in internal energy, ΔE is the difference between energy taken up and energy passed to the environment and w is work. The w term is positive when work is done on the organism (because it thereby gains energy, for example an albatross riding the wind increasing its gravitational potential energy as it soars in the air) and negative when the organism does work on the environment and thereby loses energy (for example by an animal walking about, or a plant extending its height against gravity). This equation underpins all movements of energy in ecology, from the individual level to whole ecosystems[3].

Internal energy increases when energy enters the organism either as potential energy (E_p, mainly chemical potential energy in food and nutrients) or as kinetic energy (E_k, mainly as heat but in plants and autotrophic bacteria also as light). Internal energy decreases through the loss of potential energy in waste products, loss of kinetic energy as heat dissipated to the environment and by any work done by the organism. We can therefore expand the balance equation:

$$\Delta U = \left[\left(E_p + E_k \right)_{in} - \left(E_p + E_k \right)_{out} \right] + w$$

This equation captures all of the energy flows through the organism and because energy and work have the same dimensions, it is balanced. In this form it uses terms unfamiliar to ecologists working with animals or plants but we can rewrite the equation in ecological terms, while maintaining its structure and balance:

$$\Delta U = \left[\begin{array}{l} (nutrients + heat + light)_{in} - \\ (waste + heat)_{out} \end{array} \right] + work$$

It is, of course, essential that all the elements of this equation be expressed in the same units, even though the individual components may be measured in different ways. The unit of choice is energy (J), or if we are interested in rates the units are

Principles of Thermal Ecology. Andrew Clarke, Oxford University Press (2017).
© Andrew Clarke 2017. DOI 10.1093/oso/9780199551668.001.0001

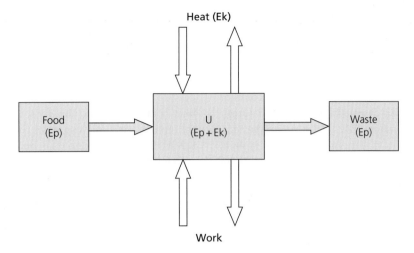

Figure 4.1 The major pathways of energy flow in an organism. Transfers of potential energy (E_p) are shown in grey, and transfers of kinetic energy (E_k) in white; U is the internal energy. This basic structure applies to all organisms, whether autotrophic or heterotrophic, bacterium, archaean or eukaryote.

power (W). Recall from Chapter 2 that in classical thermodynamics heat is the spontaneous transfer of energy between a body and its surroundings. In everyday language we describe this as a flow of heat (just as we did above), and for convenience we will follow this informal usage here.

The modelling of energy flow through organisms has traditionally taken one of two approaches. One is to model the flow of chemical potential energy while ignoring movement of energy as heat, and the other is to concentrate on the thermal relationship between the organism and its environment, while ignoring changes in chemical potential energy. A few workers have attempted to combine the two into a full energy budget, but these more complex budgets have yet to enter the ecological mainstream. This is not as bad as it might sound; the aim of a model is to capture the essence of a problem and formalise it in as simple a way as possible so that the mathematics remain tractable. In the case of modelling the fate of chemical potential energy taken in as food, for example, the assumption is made that the temperature of the organism remains constant over the period of interest. Similarly, in modelling the flow of energy in and out of the organism as heat, it is often assumed that energy flows can be simplified by assuming that the heat contributed by metabolism can be ignored. This is a more questionable assumption, especially for endotherms where a high body temperature is maintained by (temporarily) retaining metabolic heat.

Organisms exchange energy with their environment through four physical processes: conduction, convection, radiation and evaporation. In all of these, it is the surface of the animal that is the site of energy exchange. The relative balance of these processes varies with the organism and its situation, but all four can be important. We will examine each in turn, before exploring the total energy exchange between an organism and its environment.

4.1 Conduction

Energy is transferred by conduction when a system is not in thermal equilibrium, and some places are hotter than others. Energy flows from the higher temperature to the lower, which means that conduction has both magnitude and direction and is thus a vector quantity. The flow of energy under these circumstances is described by an equation derived by Joseph Fourier in 1822[4]:

$$\frac{dQ}{dt} = -kA\frac{\partial T}{\partial n}$$

This describes the rate of flow of heat, Q, through a small plane of area A within the medium, driven by the temperature gradient normal to the plane. This equation defines k, which is the *thermal conductivity* of the medium. This has units of W m^{-1} K^{-1}, and its value has to be determined experimentally (that is, it cannot be derived from theory). The negative sign indicates that heat flows down the temperature gradient. One complication is that k varies with temperature, but this is not generally a problem for biological materials at physiological temperatures.

It is generally impossible to determine the temperature gradient at a point within the system, and so Fourier's equation is usually used in its integrated form. To take the classic example of heat flowing along a metal bar, familiar to many from school physics, the integrated equation is:

$$\dot{Q} = \bar{k} A \left(\frac{T_2 - T_1}{L} \right)$$

Here $\dot{Q}$ is the rate of flow of energy as heat, A is the cross-sectional area of the bar, T_2 and T_1 are the temperatures at either end of a section of bar of length L and $\bar{k}$ is the mean thermal conductivity over the temperature range T_2 to T_1 (Figure 4.2).

A metal bar is not a useful approximation to an organism. A slightly more realistic model, though still imperfect, would be a tube. If we consider a sort of tubular mammal, perhaps a stoat or weasel, heat is generated internally to maintain a core body temperature of ~37 °C. Because its surface is cooler than this, the stoat is not in thermal equilibrium and heat flows outwards. For radial conduction within a cylinder, the rate of heat flow is again given by the Fourier equation:

$$\dot{Q} = \bar{k} 2\pi L \left(\frac{T_2 - T_1}{\ln\left(\frac{r_2}{r_1}\right)} \right)$$

Here T_2 and T_1 are the temperatures at distances (radii) r_2 and r_1 from the centre of the tube.

This equation tells us that the steady-state distribution of temperature through the body (from core to skin) is logarithmic, in contrast to the bar where the gradient is linear. In a real stoat, of course, the actual distribution of temperature within the body is not a logarithmic decline from core to skin, because heat is moved around the body in the blood, and the stoat exerts a fine degree of control over where this heat does and does not go. But it is important to start with the fundamental principles to understand the framework that the stoat's anatomy and physiology allows it to modify. This is seen very clearly in the distribution of temperature through the body of another animal that might be viewed as roughly tubular, a seal (Figure 4.3). Here the temperature in the inner core of the body is fairly uniform, because of conduction within the core tissue and the movement of heat in the circulating blood, but there is a steep gradient through the insulating fat layer (blubber) to the skin, which is at the temperature of the water. The outer parts of this gradient are roughly exponential, indicating that towards the outside of the blubber layer, heat transfer is primarily by conduction.

If we wish to apply the equations of heat conduction to an organism, we need to know the area of its surface and the thermal conductivity of its body tissue. Of these two, the area of the body is often much the more difficult to determine accurately and precisely. The thermal conductivity of tissue is similar to that of water, which is perhaps not surprising given that tissue is predominantly water. Typical values of thermal conductivity for some substances are shown in Table 4.1.

The data in Table 4.1 show us that air does not conduct heat very well. It is the air in snow that makes this such a good insulator, allowing small mammals to remain active beneath the winter

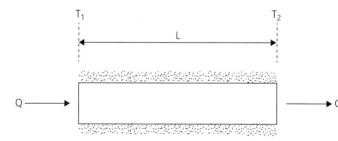

Figure 4.2 Conduction of heat through a metal bar. Temperature decreases linearly from T_2 to T_1 along length L of the bar.

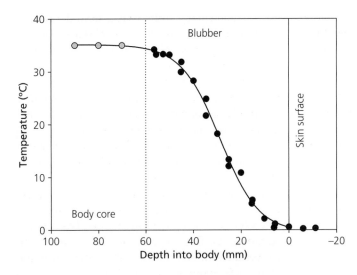

Figure 4.3 Radial heat conduction through a roughly tubular animal, the harbour seal, *Phoca vitulina*. Temperature decreases through the insulating blubber layer, from the core temperature (assumed to be 35 °C, a typical value for a seal), to water temperature of ~0 °C. The observed data are shown in black, the assumed core temperature is shown in grey and the fitted curve is sigmoidal[5].

Table 4.1 Thermal conductivity of some materials; data are provided for room temperature (25°C, 298 K) unless otherwise indicated.

Substance	Thermal conductivity (W m^{-1} K^{-1})
Materials of biological importance	
Air (1 bar pressure)	0.026
Water	0.58 (293 K)
Ice	2.22 (273 K)
Fresh snow (dry)	0.03–0.25 (273 K)
Old snow (wet)	0.4
Soil (dry)	0.3
Peat (dry)	0.08
Sand (dry)	2–4
Solid rock	2–7
Wood	0.04–0.17
Fur	0.038
Animal fat	0.16–0.23
Tissue (human)	0.46
Other materials for comparison	
Polystyrene foam	0.01
Cork	0.07
Asbestos	0.08
Glass	1.5
Mercury	8.3
Aluminium	205
Copper	385
Silver	406
Diamond	>2000

snow cover in northern latitudes (and for igloos to work). It is also why fur is so good at preventing the transfer of heat; the air retained in the fur provides the insulation. Fat is also a good insulator, hence the value of a blubber layer, but this is greatly exceeded by the insulating properties of fur. On the other hand, many of the surfaces on which organisms stand conduct heat quite well, and this can be an important pathway for heat loss under some circumstances.

4.1.1 The mechanism of heat conduction

In thermal conduction energy is transferred by two mechanisms, the movement of free electrons and the transfer of vibrational energy directly between molecules. The energy is quantised and vibrational energy transfer is by high-frequency elastic waves called *phonons*[6].

Conduction by phonons occurs in all solids and is often the dominant mechanism of energy transfer. In metals, however, the phonons are scattered by the free electrons and the dominant mechanism of energy transfer is by movement of the electrons themselves. Because these free electrons also carry electrical energy, there is a strong correlation between electrical conductivity and thermal conductivity in metals (see Chapter 3).

Conduction is the main mechanism for the movement of heat in soil, and is also important in the transfer of heat within the body of an animal or plant: wherever metabolism generates a local thermal disequilibrium, conduction transfers energy from warmer to cooler parts. Similarly when the surface of the body is warmed, then heat is transferred by conduction into the body. A nice example of this is when a reptile or mammal lies down on a sun-bathed rock to warm their body by the conduction of heat inwards, or the opposite, when a desert insect or reptile keeps its body well away from the hot sand to prevent the conduction of heat into the body. Overall, however, direct conduction of energy between an organism and the ground is usually a minor factor in its thermal balance, though it is extremely important in the movement of heat across the body surface where it is carried away by convection.

4.2 Convection

Convection is the transfer of heat by the bulk flow of material. *Free convection* arises when differences in temperature lead to differences in density through thermal expansion. These spatial variations in density produce instability, which leads to movement of the fluid; free convection thus arises through buoyancy forces. *Forced convection* arises when some external agent drives the fluid past the organism; stream flow, tidal currents and wind are typical examples of forced convection. Unfortunately for thermal ecologists, understanding the movement of fluids is one of the most difficult of problems in all science[7].

It is possible, however, to describe what happens in simplified situations fairly well in qualitative terms. To take the example of a warm mammal in still cool air, the temperature gradient decreases quasi-exponentially with distance from the surface but the velocity gradient varies from zero at the surface (the boundary layer resulting from friction) to a peak and then it decays again (Figure 4.4). The profiles for a cool surface in a warm fluid (such as ice on a pond during a cold night) are, however, not a mirror image of those for a warm surface in a cool liquid; fluid dynamics are complicated.

Even in relatively simple situations the flows in involved in convection are complex. For a warm mammal in still cool air heat is transferred from the skin to the adjacent still air of the boundary layer by conduction. The warmer air rises, and is replaced by cooler air flowing in from the side. The detailed pattern of this flow depends on factors such as the temperature difference, viscosity and density of the fluid, surface roughness and so on.

Specific equations for convection do exist, but you will rarely find them in physiological textbooks. This is because most equations describing convection are relevant only to the specific circumstances of a particular situation. Where an equation is needed to capture convective heat transfer between an organism and its environment, physiologists typically fall back on very general equations, where the complexities of heat transfer from the surface to the adjacent fluid are captured by a simple equation:

$$Q = h_c A \Delta T$$

where A is the area of the body surface, ΔT the difference in temperature between the body surface and the surrounding fluid (air or water) and h_c the convective constant or *surface heat transfer coefficient*.

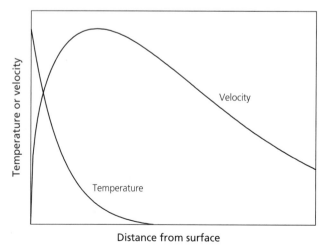

Figure 4.4 Schematic diagram showing representative profiles of temperature and velocity as a function of distance from a warm surface under free convection in a cooler fluid. Shape and magnitude will vary depending on temperatures and the nature of the fluid.

The value of h_c is a complex function of the wind (or water) velocity profile, the temperature profile and surface roughness.

For free convection h_c varies as:

$$h_c = f(\Delta T)^n$$

where n typically lies in the range 0.6 to 0.8. Free convection does occur in ecological situations, but most of the time organisms are exposed to moving air or water and convection is forced. Here the critical factor is the thickness of the boundary layer, which has been shown experimentally to depend on the dimension, D, of the organism (typically its diameter) and wind speed, V, as:

$$h_c = f\left(\frac{V^{k_1}}{D^{k_2}}\right)$$

where the exponents $k_1 \sim 1/3$ and $k_2 \sim 2/3$. Three physical properties of a fluid are important in convection: its thermal conductivity, k, its thermal capacity, C_p, and its thermal expansion coefficient. Equations to calculate the value of h_c from these properties are available, but many of the variables involved are difficult to estimate. Instead, physiologists generally have to determine a value for h_c experimentally.

Experimental determination of h_c for an organism is technically demanding, and so is not often done. However h_c was determined experimentally in a classic study of the thermal characteristics of the desert iguana *Dipsosaurus dorsalis* by Warren Porter and colleagues in the 1970s. They made an accurate model of the lizard and using a wind tunnel they showed that when the models were oriented across the airflow, heat transfer was about 25% higher than when parallel to the airflow because of the shorter flow lengths and thinner boundary layers (Figure 4.5). This important study shows us that while convection is dauntingly complex physically, it is of enormous importance ecologically.

The values of h_c obtained for the *Dipsosaurus* models were specific to those models under the particular circumstances of the wind-tunnel experiments. Luckily, dimensional analysis can help generalise the results and thereby reduce the requirement for experimental work. This is because convectional heat transport can be described by a suite of dimensionless numbers, the most common of which are the Reynold's, Prandtl, Nusselt and Grashof numbers.

The Prandtl number, Pr, is the ratio between the kinematic viscosity and the thermal diffusivity, and it relates the velocity distribution to the temperature

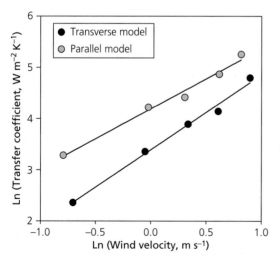
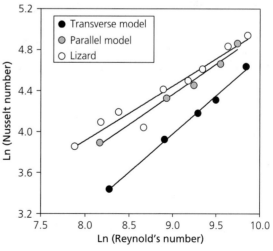

Figure 4.5 Convective heat transfer in the lizard *Dipsosaurus dorsalis*. A (left panel): heat transfer coefficients as a function of wind speed (both variables natural log transformed) for model lizards in a wind tunnel, with model either parallel to wind flow (grey symbols) or across the wind flow (black symbols). B (right panel): relationship between the dimensionless Nusselt number and Reynold's number (both natural log transformed) for model lizards (symbols as for left panel) and lizards (white symbols)[8].

distribution and thereby gives an indication of the convective heat transfer ability of the fluid. It is calculated as:

$$Pr = \frac{C_p \mu}{k}$$

where C_p is the thermal capacity, μ the dynamic viscosity and k the thermal conductivity. Air has a Prandtl number ~0.72 and water ~6.

The Nusselt number, Nu, captures the relationship between the length of an object L and the thickness of the boundary layer:

$$Nu = \dot{Q} \frac{L}{kA\Delta T} = \frac{h_c . L}{k}$$

Here $\dot{Q}$ is the rate of heat transfer by convection. An important feature of the Nusselt number is that if Nu were constant, heat transfer by convection would be proportional to the temperature difference (which is Newton's Law of Cooling, which we discuss in section 4.4.3 later in the chapter). Sadly for thermal ecologists seeking simple relationships, Nu is not constant.

The Grashof number, Gr, is the ratio of buoyant force times inertial force to the square of the viscous force. It gives an indication of the tendency for free convection to occur in a particular system. It looks the most complex, but is in some ways the easiest to use.

$$Gr = \frac{\rho^2 g L^3 \Delta T \beta_v}{\mu^2}$$

Here ρ is the density of the fluid, g the acceleration due to gravity and μ the dynamic viscosity. These dimensionless numbers can be combined to define two others: the Rayleigh number, Ra, is the product of Pr and Gr and the Peclet number, Pe, is the product of Pr and the Reynold's number, Re (which we will meet in Chapter 5).

These dimensionless numbers do not allow us to calculate the specifics of convection in a particular case, but they do provide valuable ball-park figures to allow us to determine the general situation. They are particularly useful when used in conjunction, for example plotting Nu as a function of Re, for a given value of Pr under forced convection, as shown by data for *Dipsosaurus* in Figure 4.5, where the relationship between Nu and Re for the transverse model showed excellent agreement with data calculated for live lizards[9].

Convection can dominate the energy budget of plant leaves, and is important in the thermal relationships of terrestrial animals. The movement of heat around the body in circulating blood or haemolymph is a form of forced convection (or better, *convective mass transport*). Where fluids circulate to distribute oxygen or remove waste products, then heat is transferred incidentally. This means, for example, that it is very difficult for an aquatic organism breathing with gills to retain metabolic heat. All the heat generated by metabolism is carried to the gills by the circulating blood or haemolymph, where it equilibrates rapidly with the water passing over the gills. This is essentially the coupling of two forced convections, circulating blood passing through the gills and water passing over them. The only way this loss of heat can be avoided is by a counter-current heat exchange structure (a *rete mirabile*), such as is found in tuna and related fishes, which removes heat from the blood before it reaches the gills (we will discuss this in more detail in Chapter 9)[10].

4.3 Thermal radiation

All bodies above absolute zero lose energy by radiation. This happens because the constituent atoms contain charged elements (electrons and protons) which interact as the molecules vibrate or move about. These interactions force electrons to higher energy levels, and photons are released as the electrons drop back to lower energy levels. The production of thermal radiation is thus a consequence of quantum mechanical processes at the molecular level. The details of this are complex but the key point is that the energy (and thus the wavelength) of the photons produced depends on the energy of the interactions. Because the molecules have a wide range of energies, photons are emitted over a wide range of wavelengths. The distribution of photon energies is the continuous black-body spectrum (see Chapter 3), with the shape and the intensity being a function of temperature.

Examples of black-body radiation spectra for objects at the temperature of the Sun and a lizard are shown in Figure 4.6. Two features are important here: the Sun emits far more radiation than the

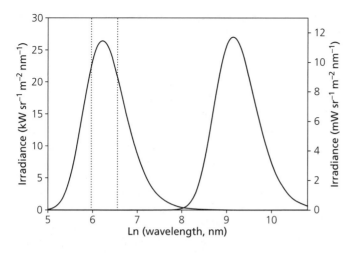

Figure 4.6 Radiation spectra for perfect black-bodies at the temperature of the surface of the Sun (5780 K, left-hand curve and ordinate) and a basking lizard (310 K, 37 °C, right-hand curve and ordinate). The dotted lines mark the upper and lower limits for visible radiation, showing how all of the thermal radiation emitted by the basking lizard is invisible, being well into the infrared. Note the large difference in peak irradiance for the two curves (a factor of ~ 10⁶)[11].

lizard, because it is very much hotter, and the peak wavelength (colour) of the emitted radiation is at a shorter wavelength.

The relationship between the peak wavelength (colour) of radiation emitted by a body and its temperature has long been known as Wien's Displacement Law. It was discovered empirically before the theoretical foundations were built by Max Planck. It is:

$$\lambda_{max} = \frac{b}{T}$$

where λ_{max} is the wavelength of peak emission, T is the absolute temperature and the constant of proportionality, b, is *Wien's displacement constant* (2.898 × 10⁻³ m K). Over the physiological temperature range this effect is small, with the wavelength of peak emission decreasing from 10.62 µm at 0 °C to 9.26 µm at 40 °C; the radiative power, however, changes significantly (Figure 4.7).

The total power radiated by an object depends on its temperature, as described by the Stefan–Boltzmann Law we met in Chapter 3. The energy radiated, summed across all wavelengths, from a body of area A and absolute temperature T is given by:

$$P = A\varepsilon\sigma T^4$$

where P is the radiative power, ε the emissivity and σ the Stefan–Boltzmann constant. The emissivity is a dimensionless number that varies from 1 (perfect emissivity, as in a black-body) to 0. Emissivity is wavelength dependent, but for most biological objects radiating in the infrared, ε is ~ 0.98, no matter what their colour in the visible wavelengths might be. We can use this relationship to see how much energy is lost from organisms at different temperatures (Figure 4.8). The energy radiated per unit area differs between an insect overwintering in polar regions and a desert plant by a factor of ~ 3.5. The great difference in size of organisms, however, means that the absolute power radiated may vary by a far larger factor.

We can see from this graph that thermal radiation from a typical adult human is about 513 W m⁻²; assuming a body area of 1.8 m² an unclothed human will thus be radiating ~920 W (that is, almost 1 kW). Given this rate of heat loss you might ask why a human does not radiate away all of its internal energy and freeze to death? The answer is that all bodies not only emit radiation, they also receive it from the surroundings and over time these two will come to balance. The key measure is the difference between radiation emitted and radiation received, and the net loss of energy as heat is therefore given by:

$$\text{Net radiation} = A\varepsilon\sigma\left(T_o^4 - T_e^4\right)$$

where T_o is the surface temperature of the human and T_e the temperature of the environment; this gives a net radiative loss of only ~190 W for an unclothed human in a room at 20 °C (the loss of 920 W is offset by a gain of 730 W). The influence of this balance between gain and loss of heat by radiation is easily appreciated if you imagine standing unclothed outside under a clear winter sky or next to

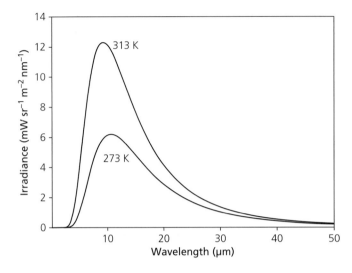

Figure 4.7 Radiation spectra for black-bodies over the physiological temperature range, from 0 °C (273 K) to 40 °C (313 K). At the higher temperature, the wavelength of peak power shifts only slightly but the power at the peak wavelength increases by a factor of 1.98.

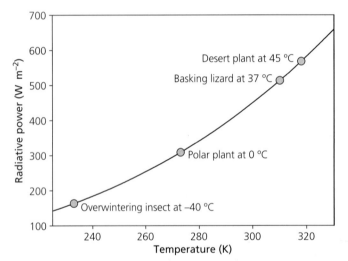

Figure 4.8 Radiative power (W m^{-2}) of objects at different temperatures, assuming an emissivity of 0.98. The line shows the Stefan–Boltzmann law, with four indicative organisms marked by grey symbols.

a fierce bonfire. In the former case heat loss is so powerful that death can be rapid, and in the latter you will overheat and cook very quickly[12].

An important point here is that the critical factor is not the core temperature but the surface temperature, for it is this that dictates the radiative balance. The lower the surface temperature, the lower the energy loss from radiation. For many polar animals, heat loss from radiation is lowered greatly by keeping the outer surface cool (as can be shown easily with thermal imaging: see the Emperor Penguins in Plate 3). This does, however, increase the thermal disequilibrium within the body and drives a flow of energy by conduction from the core to the skin, but this flow can be reduced by the presence of an insulating layer (blubber, fur or feathers). A cool skin also increases the net energy balance in favour of energy from the environment, although the incident energy still has to be transferred through the insulating fur or feathers to the body core.

The radiative balance of an organism is complicated by two factors, its shape and its colour. The Stefan–Boltzmann equation discussed above applies only for an object in a uniform environment. Under these circumstances, an organism with a complicated shape can be regarded as effectively having a spherical surface visible to the incoming radiation. Where the environment contains a non-uniform source of radiation (such as the Sun) then the calculations become very much more complicated. As with convection, the mathematics become tractable only with the use of simplifying assumptions.

4.3.1 Colour

Not all the visible radiation that falls on a body is absorbed. Some is reflected, and it is this that gives objects their colour. A leaf, for example, absorbs red and blue light strongly, but the wavelengths in between are absorbed only a little, which is why a leaf looks green. Colour is an important attribute of an organism (Box 4.1), but perhaps not in the way often portrayed, because the wavelengths that dominate the thermal balance of organisms are all invisible to the human eye (and hence play no part in the perceived colour).

Box 4.1 Colour

Ask anyone to describe an elephant and you are likely to receive answers along the line 'large and grey'; similarly a lady beetle would probably be described as 'small and red with black spots'. In other words, the two most obvious features of an animal are its size and its colour. Colour is such a key feature that it becomes part of the vernacular name of many organisms. While size continues to attract considerable attention from ecologists, these days colour often goes unremarked. Once the subject of whole books[13], animal colour is now more or less taken for granted.

William Hamilton listed three key functions for colour in an animal[14]:

1. Optimisation of heat exchange and radiation relationships
2. Camouflage
3. Communication

He also suggested a fourth, incidental, category in which the colour has no function in itself but simply reflects underlying physical or physiological characteristics.

While colour has significant implications for energy exchange, an organism's colour will be the result of trade-offs with other ecological factors. It is also influenced by the ability of the organism to obtain the raw materials to manufacture the pigments required, which in turn may indicate fitness.

The radiation that falls on an organism is either absorbed or reflected. The *absorptance* (or the *absorption factor*), α, is then:

$$\alpha = \frac{Q_a}{Q_i}$$

where Q_i is the energy incident on the surface and Q_a the energy that is absorbed; α clearly varies from 1 (all incident energy is absorbed) to 0 (none is absorbed)[15]. Similarly, a *reflectance*, ρ, can be defined:

$$\rho = \frac{Q_r}{Q_i}$$

where Q_r is the energy reflected. For an animal exposed to incident radiation, absorptance and reflectance are the only possibilities, and so $\alpha + \rho = 1$. For a liquid or transparent solid, then a third possibility exists, which is that the radiant energy enters the object but is not absorbed by a molecule. This is transmitted energy, for which there is an analogous factor, *transmittance*, τ. Transmittance is not generally relevant to animals (for exceptions see Box 4.2), though it may apply to plant leaves, and is an important factor in bodies of water. The complete relationship is then $\alpha + \rho + \tau = 1$, though for most organisms $\tau \sim 0$ and so the simpler equation applies.

Sunlight received on the surface of the Earth has 52–55% of its energy in the infrared (>700 nm) and 42–43% in the visible (400–700 nm), with the rest being in the ultraviolet (<400 nm). Organisms generally have absorptances (and emissivities) in the infrared ~0.98, regardless of skin, plumage or pelage colour. In other words, organisms are effectively perfect absorbers and perfect emitters of infrared radiation, which couples them to the radiation environment very tightly[16]. In the visible range, however, absorption depends strongly on colour. Because absorbed visible radiation is converted to heat, colour exerts a strong influence on thermal balance. Put simply, colour influences body temperature through its control of the absorbance of visible light, but is irrelevant to the absorption and emission of thermal radiation in the infrared (where all organisms are effectively black).

A consequence of this is that darker organisms should warm up faster and achieve a higher equilibrium temperature than paler ones[17]. This idea was

> **Box 4.2 Transparent animals**
>
> Almost every animal and plant we are familiar with is opaque. While there are a few terrestrial organisms that are translucent, it is only in the sea that truly transparent animals can be found. These range from the more or less familiar, such as the glass shrimp, to the less well known deep-sea organisms. Many animals that live near the ocean surface are highly transparent, as this provides them with valuable camouflage. It can also make them extremely difficult to pick out in a sample! They include gelatinous forms (ctenophores, and the diverse assemblage of organisms we tend to refer to collectively as jellyfish), crustaceans and planktonic polychaete worms. The most familiar, however, are probably cranchiid squid (family Cranchiidae), as they often appear in natural history documentaries. These are found at surface and midwater depths, and they often combine great transparency with luminescent organs. There are also transparent octopus, such as the glass octopus, *Vitreledonella richardi* (the sole member of its genus and family). The most familiar terrestrial examples are the glass frogs (family Centrolenidae) of Central and South America, where the abdominal skin is highly translucent. There are also a few fish that are transparent, such as the glass catfish, *Kryptopterus vitreolus*, and the deep-water barreleyes (or spook-fish) of the family Opisthoproctidae.
>
> Transparency is very difficult to achieve in bodies that have a different refractive index from the surrounding medium, which is why it is easier in water than in air (and why glass frogs are translucent at best). The majority of organic molecules do not absorb visible light; the opaque nature of tissue comes about because of scattering at the many discontinuities in refractive index at the junction between different organelles within cells, and between tissues. Maintaining transparency requires energy, which is why animals that are transparent in life are opaque in death.
>
> In many organisms that maintain transparency in the bulk of their tissue, the gut and gonads can be highly coloured and visible through the body. For organisms living in the twilight zone of the mid-waters, luminescent organs are used to offset shadow effects, or silhouette when viewed from beneath by a potential predator[18].

first tested by comparing the temperatures of insects of different colours under identical illumination, and darker individuals were found to warm more rapidly and to achieve a higher equilibrium temperature. For example in natural sunlight the dark morph of the desert grasshopper *Calliptamus coelesyriensis* is ~4 K warmer than the pale morph, and similar results are found with locusts (*Schistocerca gregaria*) under controlled conditions of artificial sunlight[19].

In lady beetles (Coccinellidae) melanic forms are well known, and in the Netherlands the proportion of these in populations of *Adalia bipunctata* varies with the level of spring sunshine (Figure 4.9). Under controlled conditions in the laboratory melanic forms warm faster and achieve a slightly higher equilibrium temperature than the usual red colour morph. These colour differences influence activity patterns and the timing of reproduction, indicating that they have real ecological consequences. Similarly, in *Colias* butterflies the extent of wing melanism influences thermoregulation, and through this reproductive behaviour and fitness (we will discuss the energetics of *Colias* in more detail in Chapter 7)[20].

4.3.2 The thermal melanism hypothesis

The biophysical principles underpinning the relationship between surface colour and body temperature were laid out in the 1960s by Warren Porter and David Gates. Subsequent ecological work discussed in section 4.3.1 showed that differences in equilibrium body temperature between individuals with high and low surface reflectance carried through to physiological performance and fitness, and the concept that surface reflectance has adaptive significance for thermoregulation has become known as the *thermal melanism hypothesis*[21].

This hypothesis was established largely with detailed studies of individual species. More recently the hypothesis was tested by Susanna Clusella-Trullas and colleagues in a broad comparative (macrophysiological) study of the surface reflectance of 68 species of heliothermic lizards (lizards that rely on basking in sunshine to increase their body temperature). They found that surface reflectance varied from 0.02 to 0.34, with half of the values lying within the range 0.1 to 0.2. The thermal melanism hypothesis predicts that surface reflectance should

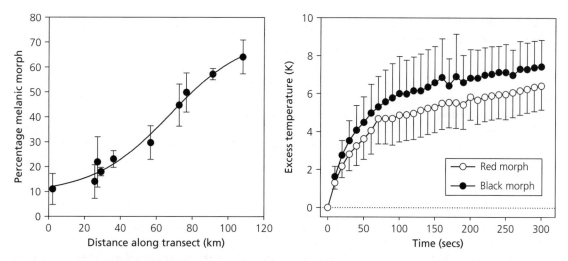

Figure 4.9 Thermal melanism in the lady beetle *Adalia bipunctata*. A (left panel): variation in the frequency of melanic morph in populations along a cline from N to S in the Netherlands, with hours of sunshine decreasing along transect. B (right panel): heating curves for red (white symbols) and melanic (black symbols) morphs in still air. Data are presented as mean and SD, expressed as difference between body temperature and body temperature at the start[22].

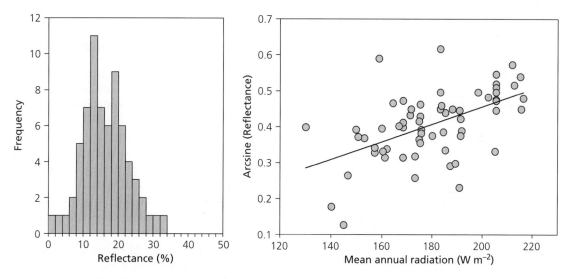

Figure 4.10 Lizard reflectance. A (left panel): frequency histogram of reflectance (here presented as a percentage, which is 100.ρ) for 68 species of lizard. B (right panel): relationship between reflectance (arcsine-square root transformed) and mean annual radiation for 66 species of lizard. The fitted line is a least-squares regression ($F = 24.1$, $p < 0.001$)[23].

be low (that is, the capacity to absorb visible radiation should be high) in species living where incident radiation is less intense or available for less time. This proved to be the case (Figure 4.10), demonstrating a broad ecophysiological pattern that matches the individual studies.

4.3.3 Why are so many polar animals white?

The results from insects and lizards point clearly to an advantage in having a darker surface where incident radiation is lower. This would suggest that animals living at high latitudes should be dark, and yet the opposite is true: many (but by no means all) of the well-known Arctic animals are white. These include the polar bear (*Ursus maritimus*), Arctic fox (*Vulpes lagopus*) and stoat (*Mustela erminea*, often referred to as ermine in its white winter coat) among the mammals, and the rock ptarmigan (*Lagopus muta*), snowy owl (*Bubo scandiacus*) and gyrfalcon (*Falco rusticolus*) among the birds.

The explanation is that in these species the benefits of camouflage outweigh thermal considerations. Camouflage may be important because species are preyed upon (such as ptarmigan) or are predators that use stealth to approach their prey (such as the polar bear, or snowy owl). The value of camouflage is shown by those species that are white in winter, but change to match the snow-free surroundings in summer, or like the gyrfalcon show a cline in whiteness of plumage towards the Arctic. Another nice example of the importance of camouflage is the difference in pelage colour in the pups of polar seals. In the Arctic, where there are many land or ice-based predators, the pups have white coats. In the Antarctic, where land-based predators are lacking (though there are marine predators waiting for the pups when they go to sea), they have black coats[24].

Another example of the interplay between the thermal consequences of colour and an animal's ecology comes from the black beetles of hot deserts.

4.3.4 The black beetle paradox

The pioneer of desert ecology Patrick Buxton noted that in general desert animals are coloured to resemble the ground on which they live, thereby affording camouflage against predators. He also remarked that any desert animal not so camouflaged tended to be black. Because a black colour enhances heat uptake, and water for evaporative cooling is scarce in deserts, a black surface would seem to be maladaptive for a desert existence. The prevalence of black among desert beetles was termed the *black beetle puzzle* by the controversial desert explorer Richard Meinertzhagen[25].

This apparent paradox was resolved by William Hamilton in the late 1960s, in his study of the tenebrionid beetles of the Namib Desert. While black beetles are a feature of many deserts, the Namib is unusual in also having white species (for example the spectacular *Onymacris bicolor*). Hamilton showed that while the thermal tolerances of black and white species were similar (lethal temperatures for both were 48–49 °C), black species warmed more rapidly and achieved a higher equilibrium temperature in sunlight and shade, and lost water less rapidly than white species. These beetles are small, so their temperature responds rapidly to the local environment, and also flightless, so they must live in the narrow band of very hot air close to the desert surface. Like all deserts, the Namib gets very cold by night under a clear sky, but in the middle of the day surface temperatures can exceed 60 °C, well above the thermal tolerance of the beetles. This means that the beetles are active only in the early morning before the ground heats up or in the early evening as it is cooling down. Black colouration is thus beneficial because it allows the beetles to warm up rapidly and become active even when the ground and air temperatures are low[26].

4.4 Thermal relationships of organisms

To understand the thermal relationships of an organism we need to combine the various mechanisms discussed above into an overall picture of the energy balance. This is not straightforward, because the pathways of energy exchange between an organism and its environment are many and various (Figure 4.11). These pathways are affected by a suite of environmental variables, including the radiation field (radiant heat from direct sunlight, scattered skylight, reflected light from the surroundings, diffuse radiation from a cloudy sky and so on), air temperature, wind speed and humidity. Modelling this is a highly complex task, but necessary if we are to define the thermal balance of an organism in fully quantitative terms. This might be needed, for example, to define its thermal niche precisely, or to predict how a particular species might react to a change in climate.

If we assume a steady state, and that over the period we are considering the organism does no net work, then the rate at which the organism gains energy as heat is balanced by the rate at which it is lost:

$$\dot{Q}_{in} = \dot{Q}_{out}$$

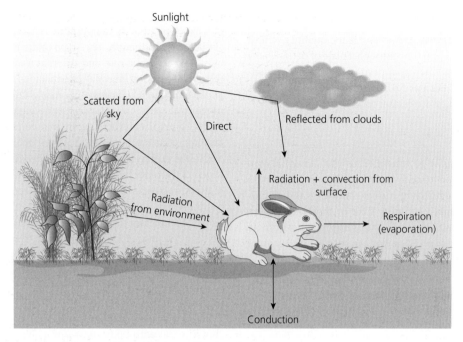

Figure 4.11 Energy flows in and out of an organism.

Here $\dot{Q}$ is the rate of change of internal energy resulting from a flow of energy as heat and the units are W. The energy input comes principally from the dissipation of chemical potential energy as heat during metabolism, and the absorption of radiant energy; there may be a contribution via conduction from the ground, but this is generally small. The losses are principally from convection, radiation and the latent heat required to evaporate sweat or water in the respiratory tract. Thus, in steady state:

$$\dot{Q}_{metab} + \left(\dot{Q}_{cond} + \dot{Q}_{rad}\right)_{in} = \left(\dot{Q}_{cond} + \dot{Q}_{rad}\right)_{out} + \dot{Q}_{conv} + \dot{Q}_{evap}$$

We have seen in previous sections in this chapter how each of these terms is itself complex. For many ecologists this complexity is too daunting, and it has prompted a debate over the best way to model heat flow in real animals. For some a complex heat-transfer model is essential, for others a simplified approach is the only practical way to make progress. As always with modelling, the answer is not that only one particular way is valid, but that the relevant approach is used for the question being tackled and all simplifying assumptions are kept in mind[27].

We can rearrange the above equation to balance heat from metabolism with net radiation, together with losses from convection, evaporation and conduction, and assuming that heat transfer into the organism by conduction is negligible:

$$\dot{Q}_{metab} = \left(net\ radiation\right) + \dot{Q}_{cond} + \dot{Q}_{conv} + \dot{Q}_{evap}$$

If we further assume that the net radiation term represents a loss of energy (the organism is not warming up from radiant heat) and that heat transfer is dominated by convective processes, then with some mathematical approximations because $(T_o - T_e)$ is small, this equation reduces to what is effectively Newton's Law of Cooling:

$$\dot{Q}_{metab} = C(T_o - T_e)$$

The coefficient C is usually termed *thermal conductance*, and has units of W K^{-1}. This equation is a gross simplification of the thermal exchanges of an organism, but it does capture the heat loss from a warm animal to a cool environment reasonably well. It is only of practical relevance to an endotherm under conditions where evaporative losses are minimal, when in practice the relationship between resting metabolic rate and the temperature difference $(T_o - T_e)$ is often observed to be linear.

4.4.1 A note on terminology

In physics, *thermal conductance* is the quantity of heat that passes in unit time through a plate of defined area and thickness when its opposite faces differ in temperature by 1 K. For a plate of thermal conductivity k, area A and thickness L, thermal conductance, C, is thus:

$$C = k\frac{A}{L}$$

C has units W K^{-1} and k W m^{-1} K^{-1}. Thermal conductivity and conductance are thus analogous to electrical conductivity (units A m^{-1}V^{-1}) and electrical conductance (units A V^{-1}). In thermal biology, the term conductance is analogous in concept to its use in thermal physics, but cannot be applied quite as rigorously.

4.4.2 Thermal conductance of endotherms

The thermal conductance of endotherms is determined empirically from the relationship between metabolic rate and ambient temperature, when the organism is held below its thermoneutral zone (TNZ, this being the range of ambient temperatures over which a resting endotherm can maintain a constant internal body temperature). The logic here is that within the TNZ the bird or mammal can regulate its body temperature by altering insulation (and hence thermal conductance; this is discussed further in Chapter 10), but once ambient temperature falls below the TNZ, the only way the organism can maintain its set-point body temperature is to increase metabolic heat production. The slope of the relationship between metabolic rate and external temperature is then a measure of thermal conductance (or strictly, *minimal thermal conductance*, because within the thermoneutral zone conductance may be higher) (Figure 4.12). Thermal conductance in this context is thus an integrated measure of the flux of energy as heat across the surface of the animal to the environment. Much of the older literature gives the measured thermal conductance in units of oxygen consumption, but these can be converted to energy with standard conversion factors. Important in this context is that holding the animal in a respirometer to measure its resting oxygen consumption incidentally means that air movement around the animal, and hence convective heat loss, is not typical of that experienced by the organism in the real world.

We cannot predict from theory how conductance should scale with body mass, for as we saw earlier in section 4.4 the biophysical processes that remove energy from an animal's surface are complex. A simple plot of conductance as a function of body mass reveals a curvilinear relationship with plenty of scatter. Transformation (natural logs) of both variables produces a relationship that is linear, indicating that a power law is a reasonable statistical description of the relationship. In mammals the slope of the scaling relationship is ~0.57 (Figure 4.13)[28].

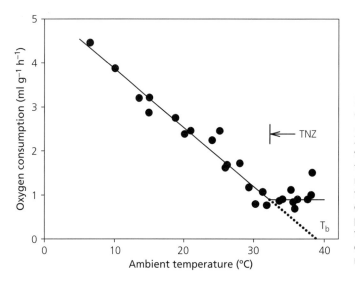

Figure 4.12 The definition of minimal thermal conductance. Data for the pygmy possum *Cercartetus nanus*, plotted in the original units. The Thermoneutral Zone (TNZ) represents the range of temperatures over which metabolic rate is constant, and internal body temperature (T_b) is maintained by thermoregulatory mechanisms including variation in conductance. Below the TNZ thermal conductance is at its minimum and T_b can only be maintained by increasing metabolic heat production. The slope of the line is thus a measure of the minimum total thermal conductance. Extrapolation of this line to zero metabolic rate gives a measure of resting T_b in the TNZ (here 38.9 °C)[29].

Early studies almost always expressed conductance on a mass-specific basis (that is, C/M_b where M_b is body mass), reflecting the convention at the time of expressing metabolic rate on a mass-specific basis. The scaling relationship of this derived variable was then examined by plotting it as a function of M_b. This is not a sensible procedure statistically (Box 4.3).

A striking feature of the scaling of minimal thermal conductance is that the slope is far lower (shallower) than that of resting metabolic rate (BMR), which for mammals >150 g is ~0.75 (we discuss the scaling of metabolic rate in more detail in Chapter 8). This indicates that the balance between conductance and BMR varies with size. Calculation of the ratio between heat loss (conductance) and heat production (BMR) indicates that on the basis of the BMR in larger mammals extrapolated to small sizes (<150 g), heat loss becomes an increasingly difficult problem for small mammals (Figure 4.13). Biophysical considerations indicate that small mammals are less able to adjust insulation of their pelage than larger ones, and a study of ten species of mammal ranging from 16 g to 70 kg suggested that conductance per unit area was invariant with size (the mean value was 21.8 W m^{-2} K^{-1}). This may explain why small mammals have a relatively high BMR: they need to generate more heat metabolically to maintain their body temperature in the face of the higher absolute conductance that follows from their relatively greater surface area[30].

4.4.3 Newton's Law of Cooling

We saw above that the complex equation for energy flow from an organism to its environment can, with some assumptions, be reduced to a form analogous to what is often referred to as Newton's Law of Cooling. The 'law' itself is fairly straightforward, being a simple treatment of convective cooling[31]. It is:

$$\frac{dQ}{dt} = \dot{Q} = h_c A \left(T_{b(t)} - T_e \right)$$

where $\dot{Q}$ is the rate of energy transfer as heat, h_c the heat transfer coefficient (W m^{-2} K^{-1}), A the surface area of the body, $T_{b(t)}$ the temperature of the body at time t and T_e the environmental temperature (this is actually Fourier's relationship for heat transfer by conduction we met earlier in section 4.1). Where h_c is independent of temperature, then cooling objects follow Newton's Law well. Newton's Law of Cooling can also be expressed in terms of temperature:

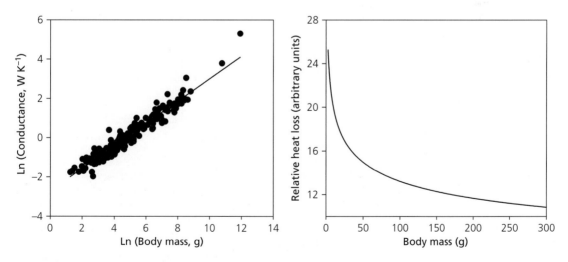

Figure 4.13 Thermal conductance in mammals. A (left panel): scaling relationship between minimal thermal conductance and body mass in mammals; the line is a least-squares fit. B (right panel): ratio between heat loss (as estimated by minimal thermal conductance) and heat generation (as estimated by BMR for mammals >150g in body mass, extrapolated to smaller body masses). This shows the increasing difficulty in maintaining body temperature at small sizes, because of the greater surface area to body volume ratio[32].

Box 4.3 Mass-specific variables in physiology and ecology

In both physiology and ecology it has long been common practice to 'correct' for the effect of body mass on a variable by expressing it in a mass-specific form. For example oxygen consumption, VO_2, is often expressed as VO_2/M_b, where M_b is body mass. This is inadvisable on a number of grounds. Firstly this only removes the effect of mass in the special case where the variable scales directly (isometrically) with mass. In all other cases, the mass-specific variable still varies with mass; nothing has been gained and the error associated with the measure has been increased. Secondly, the statistical behaviour of a ratio can be affected by the change in the frequency distribution.

A particular problem arises with the scaling of mass-specific variables. For example, a plot of VO_2/M_b as a function of M_b (not uncommon in the literature) is effectively a plot of M^{-1} against M, with variation added by VO_2. A plot such as this will yield a significant negative relationship, even when the two variables are random numbers (Figure 4.14).

The example here is simple and obvious; the problem is more difficult to spot when the same variable is hidden within complex derived variables.

The lesson is clear: avoid mass-specific variables, and work only with absolute rates. Where the effect of body mass needs to be allowed for (as it very often does in ecology and physiology) use generalised linear models or related techniques[33].

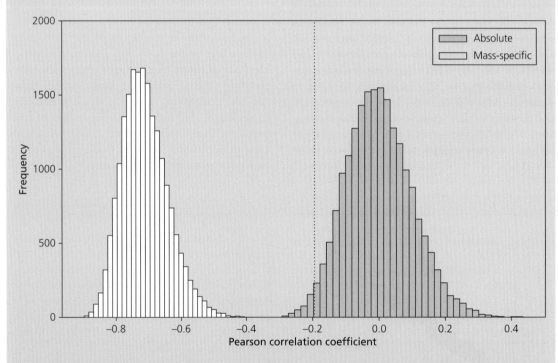

Figure 4.14 A null model of mass-specific respiration. Frequency histograms of the Pearson correlation coefficient between respiration rate and body mass (grey bars), and mass-specific respiration rate and body mass (white bars) where both variables are random numbers. For each simulation, 100 data points for respiration rate were sampled randomly from the range 1–300 mW, and 100 data points for body mass were sampled randomly from the range 1–1000 g. This was repeated 20 000 times. Both variables were transformed (natural logs) before calculating a correlation coefficient. The dotted line shows the two-tailed probability for $p = 0.05$ and 100 degrees of freedom. The absolute rates have a mean correlation coefficient of 0, with ~2.5% of data in each tail (only lower tail shown), thus conforming to statistical theory. For simulated mass-specific data all 20 000 simulations produced a highly significant correlation, despite respiration rate and body mass variables being random numbers.

$$\frac{dT_{b(t)}}{dt} = -r\left(T_{b(t)} - T_e\right)$$

Here the coefficient r is a positive constant characteristic of the system and with units of inverse time:

$$r = \frac{h_c A}{C_p}$$

where C_p is the thermal capacity (at constant pressure) of the body. This equation can be integrated:

$$T_{b(t)} = T_e - \left(T_{(t=0)} - T_e\right)e^{-rt}$$

In other words, the temperature of the body decays exponentially. The rate at which the temperature decays depends on the value of h_c (since this is a component of r). The value of h_c is higher in moving air than in still air, and much higher in moving water (which is why water is a better coolant than air, and why aquatic ectotherms generally cannot maintain a higher temperature than the water in which they swim).

In reality an organism is rarely in a situation where the flow of energy as heat to the environment follows Newton's Law, although for endotherms in a cool, still environment it is a reasonable first-order approximation. For ectotherms, or endotherms in a warm or hot environment, or when the wind is blowing, it is usually necessary to use a more detailed biophysical treatment.

4.5 Evaporation

We saw in Chapter 2 that it takes considerable energy to convert liquid water to water vapour. This high latent heat of evaporation (2.26 kJ g^{-1}) means that terrestrial organisms can use considerable amounts of energy in evaporating water from their respiratory tract or body surface. While this is an effective way of cooling under thermal stress, many hot areas are also very dry and here the thermoregulatory benefits have to be offset against water economy. The danger of dehydration in terrestrial environments is such that only two groups of animals have been able to diversify extensively on land (arthropods and vertebrates).

Table 4.2 Evaporation of water from the body surface of various animals at room temperature. These data are indicative only, to demonstrate the order of magnitude variation; exact rates will depend on environmental conditions[34].

	Evaporation (µg cm^{-2} h^{-1} mm^{-1} Hg saturation deficit)
Earthworm (*Lumbricus*)	400
Frog (*Rana*)	300
Salamander	600
Garden snail (active)	870
Garden snail (inactive)	39
Man (not sweating)	48
Rat	46
Lizard	10
Mealworm (*Tenebrio molitor* larva)	6

As with convection, evaporative water loss is difficult to model, as it depends on a large number of factors; particularly important are wind speed, temperature and atmospheric humidity. The rate of water loss also depends critically on the nature of the organism's integument: arthropods have a hard cuticle which is typically fairly impervious to water, whereas many other invertebrates have an exposed surface that loses water easily. In mammals and birds the presence of fur or feathers restricts water loss from the surface (though does not eliminate it), the scaly skin of many reptiles is more permeable than might be thought and amphibians can lose water rapidly through their skin (Table 4.2)[35].

General relationships linking water loss to size are available for some broad groups of animals, but for a full biophysical treatment of energy flow, energy loss through evaporation has to be determined experimentally for the species of interest.

4.6 The full biophysical treatment of fluxes

In all organisms heat is generated in the cell by metabolic processes ($\dot{Q}_{met}$). In terrestrial vertebrates some of this metabolic heat is used to evaporate water in the respiratory tract ($\dot{Q}_{evap}$) and the rest passes to the body surface through an insulating layer. Energy may also added to the body by solar radiation ($\dot{Q}_{rad(in)}$), so for a terrestrial organism the energy input is:

$$\dot{Q}_{rad(in)} + \dot{Q}_{met} - \dot{Q}_{evap}$$

In steady state this must be balanced by loss processes at the body surface. These are:

$$\dot{Q}_{rad(out)} + \dot{Q}_{conv} + \dot{Q}_{cond} + \dot{Q}_{evap}$$

We can express the various components using relationships explored earlier in the chapter:

$$Loss = \varepsilon\sigma A_1 \left(T_o^4 - T_e^4\right) + h_c \left(\frac{V^{k_1}}{D^{k_2}}\right) A_2 \left(T_o - T_e\right)$$
$$+ kA_3 \left(T_o - T_e\right) + E_\lambda A_4 \frac{dM_{wat}}{dt}$$

where A_n is the area (likely to be different for each term), T_o and T_e are the surface temperature of the organism and the ambient temperature respectively, E_λ is the latent heat of evaporation and M_{wat} is the mass of water evaporated. This equation is difficult (but far from impossible) to apply to a real organism, because the various areas involved differ between the terms and are mostly unknown, and the complexities of the convective constant h_c.

This equation is presented here in terms of energy. Assuming knowledge of the thermal capacity of the organism, it can be recast in terms of temperature. In the latter form it has been used successfully to model the thermal relationships of lizards and other reptiles, and also estimate those of extinct dinosaurs[36].

4.7 Flows of chemical potential energy

The second major thread within energetics has been to consider the flow of chemical potential energy from food that is either directed to growth and reproduction or dissipated in metabolism. The groundwork for this was laid in the middle of the twentieth century, with the development of the balanced energy budget by the Russian G. G. Winberg, and this approach was developed in particular by fish biologists, ecologists exploring the energetics of small mammals and agriculturalists interested in the energetics of domesticated animals. Much important work consolidating the balanced energy budget as an important tool for understanding the ecology and energetics of organisms was undertaken during the International Biological Programme (IBP); for this reason it is often referred to as the IBP balanced energy budget[37].

The balanced energy budget is based on a thorough understanding of how organisms use energy. Developments in our knowledge of physiology and biochemistry have resulted in some elaborations, but the basic equation has provided the backbone to our understanding of the flow of potential energy through organisms for well over half a century. The basis of the budget is a discrete-time balance equation. Taking the chemical potential energy, $E_{p(t)}$, as the state variable, this is:

$$E_{p(t+\Delta t)} = E_{p(t)} + (inputs) - (losses)$$

In other words, when integrated over a suitable time interval Δt, there is an exact balance between inputs and outputs. Expressing this in terms of physiological processes determining these inputs and outputs, and rearranging, the basic budget is:

$$C - F = \Sigma P + \Sigma R + E$$

where C is the consumption (the energy taken in as food), F the unassimilated food voided as faeces, ΣP the sum of all production processes, ΣR the sum of all respiration losses and E the energy excreted. For this equation to be balanced, and thus conform to the conservation of energy, we must assume that over the period for which the budget is calculated there is no net change in heat content, nor any work done on the environment.

The body tissue of an organism comprises three very different functional components, namely tissue, energy reserves and skeleton. The production term ΣP includes all the energy contained in newly synthesised material in all three of these compartments (or four if somatic tissue and new reproductive tissue are considered separately). Depending on the organism it may also be necessary to include other terms, for example mucus production in molluscs, but these three are generally considered to be the most important. Traditionally, reserve has been considered explicitly only when it is large and long-lasting (for example wax storage in marine copepods or fat in mammals and birds). It is, however, important to distinguish reserve as a distinct functional compartment, even when its mass is small and turnover fast[38]. This is because it decouples the assimilation of nutrients from their subsequent utilisation, and hence it can unbalance the energy budget in the short term. Also the conversion of nutrients to storage and reconversion to precursor monomers during growth or vitellogenesis are a key factor in determining growth efficiencies. The various components thus need to be handled independently in any model that deals with both energy and material flow. The production term ΣP thus breaks down:

$$\Sigma P = \Delta P_{som} + \Delta P_{gon} + \Delta P_{res} + \Delta P_{ske} + \Delta P_{muc}$$

Here the various ΔP terms represent the energy sequestered into new somatic tissue (ΔP_{som}), reproductive tissue (ΔP_{gon}), storage reserve (ΔP_{res}), skeleton (ΔP_{ske}), and mucus (ΔP_{muc}).

The respiration term ΣR is at once the simplest and most complex term in the budget. It is the simplest because it represents the energy dissipated in metabolism, and it is complex because the physiological processes driving respiration are many and varied. The principal drivers of respiratory metabolism are maintenance, locomotor activity and the thermodynamic costs of synthesis but there are others and in truth we do not know the full range or relative importance of the various cellular and organismal processes requiring the ATP generated by intermediary metabolism (this is discussed in more detail in Chapter 8). The basic components of ΣR are:

1. The cost of maintenance processes, R_m, usually estimated by resting oxygen consumption of an inactive, post-absorptive animal. These comprise cellular maintenance processes as well as whole organism support processes.
2. The costs associated with normal day-to-day locomotor activity, R_a, usually estimated by the day energy expenditure minus the costs of maintenance. These costs include the extra organismal costs that are required to support activity (for example enhanced cardiovascular and nervous activity).
3. The costs associated with the processing of food. These include the costs associated with the digestion and absorption of food in the gut, R_d, and the metabolic (thermodynamic) costs associated with the consequent use of these nutrients in growth, R_{som} (or R_{gon} if the nutrients are directed

to gonad synthesis). These are difficult to distinguish as they are combined in the pulse of metabolism that accompanies a meal in episodically feeding organisms (called the *specific dynamic action*, SDA, in the older literature, but more commonly now referred to as the *heat increment of feeding*, HIF), or they lead to a general elevation of metabolism over maintenance in herbivores that feed more or less continuously. These costs are discussed in more detail in Chapter 13.

The respiration term thus breaks down:

$$\Sigma R = R_m + R_a + \left(R_d + R_{som} + R_{gon}\right)$$

where the brackets group the three components measured in SDA/HIF.

The basic form of the IBP balanced energy budget implies that ΣP and ΣR are independent sinks for energy, whereas in reality they are partially linked: R_{som} and R_{gon} are coupled tightly to P_{som} and P_{som} through the thermodynamic cost of synthesis (we discuss this in more detail in Chapter 13), and the cost of maintenance, R_m, and the cost of locomotor activity, R_a, are both functions of body mass. These relationships mean that when fitting empirical data for ΣR and ΣP to the model is straightforward, but parameterising the model to predict energetics is more complex.

The different components of the energy budget have to be measured in different ways. Growth is usually assessed by the change in mass over a set period of time, which may need to be long for a slow-growing organism. Respiration and excretion are usually measured in individuals held in respirometers, typically for a few hours to ensure that a precise measurement can be obtained. Respiration measurements may involve simultaneous estimates of oxygen uptake and carbon dioxide production, which is useful at indicating the metabolic substrates being utilised (see Chapter 8). Excretion is usually estimated from uric acid (in terrestrial organisms) or urea and ammonia (in aquatic organisms). In marine invertebrates, nitrogen is excreted mainly as ammonia, although it typically also involves a number of other nitrogen compounds; for example, in the limpet *Nacella concinna* excretion is dominated by ammonia (50–90% of total N excreted) and whilst urea (8–10%) and other nitrogenous compounds such as amines (~3%) are generally less important, at times they may each comprise up to 40% of the excreted N. The excreted amines typically include arginine, allantoin, purines and pyrimidines[39].

Any formulation of the energy budget must be consistent in units, and balanced in dimensions. Ecologists traditionally track energy and work in terms of power, whereas biological oceanographers usually track materials (typically carbon or nitrogen). The choice of units will depend principally on the purpose to which the energy budget is put, but as the various components are necessarily measured in different units some interconversion will always be needed.

The IBP balanced energy budget makes no explicit allowance for temperature, although this will affect every component of the budget. Often all components of the budget are measured at the same temperature, and so while the resultant budget applies only to an organism at a given temperature, thermal effects can be ignored. They do become important, however, in modelling the energetics of ectotherms experiencing strong diurnal variations in body temperature, or which live in a seasonal environment. Most components are also affected by size, and this can be allowed for by expressing all physiological processes as a function of body mass. Size is usually integrated into the IBP balanced energy budget by assuming a power law relationship between physiological rates and body mass[40].

A subtle problem arises because the different components of an energy budget are typically measured on very different timescales. Thus physiological measures such as respiration or excretion are usually laboratory based and average rates over a period of hours, whereas growth rate may require much longer periods of integration. These very different timescales can result in unbalanced budgets, especially where an organism utilises short-term storage. The pragmatic approach to this problem is to select some intermediate timescale and scale all measured variables up or down to this. This does, however, introduce problems in the scaling of error (multiplying up a variable measured over a short timescale to a long integration period may result in very wide confidence intervals).

Constructing a full balanced energy budget is a time-consuming task, which may be why, even considering the insight they can provide, there are relatively few complete budgets in the literature. Those that are available, however, share a number of general features. The most important is that, even when constructed with the greatest care and attention to detail, budgets often fail to balance. Since organisms necessarily follow the First Law of Thermodynamics, their energy budget must balance, exactly. The lack of balance arises from two main sources. The first is measurement error, and the second is the failure to include some processes that utilise energy. One common source of error is the inability to accurately capture the costs of activity.

The important ecological feature common to all budgets is that the major sink for assimilated energy is respiration. For example, in fifteen budgets for young, rapidly growing fish, in carnivorous species respiration accounted for on average 55% of assimilated energy, and for herbivorous species it averaged 63%. For eight species of gastropod, respiration utilised 53% of assimilated energy. A slightly higher value would be expected in fish if they utilise more energy for swimming than gastropods do for crawling[41]. That respiration represents the largest sink for assimilated energy remains a general result; it is also the least understood.

The IBP balanced energy budget remained the mainstay of energetics until the late twentieth century, when two new approaches were developed. One was the *Metabolic Theory of Ecology* (MTE), which was concerned initially with a mechanistic explanation for the scaling of metabolic rate rather than tracking all the energy flows within an organism. We will examine the MTE in more detail in Chapter 12.

The most important development was *Dynamic Energy Budget* (DEB) theory developed by Bas Kooijman, which is designed to track the flow of energy and materials through an organism over its entire life cycle. Whereas the IBP formulation provides a framework for integrating empirical observations of respiration and growth, DEB theory takes a fundamentally different approach in that it predicts these from input variables such as temperature and food availability. It is also fully dynamic, in contrast to the IBP discrete-time balance equation. As such,

DEB theory represents a pioneering attempt to predict an organism's life history from a set of simple physiological and biophysical principles[42].

4.7.1 The dynamic energy budget

A key feature of DEB theory is the partitioning of body mass into two abstract quantities, namely *structural volume* and *reserve*. The structural volume represents the permanent tissue, which utilises energy for its maintenance; this maintenance cost is assumed to be directly proportional to structural volume. The reserve constitutes a temporary storage pool of energy and materials, between assimilation and mobilisation, in the form of lipid, protein or carbohydrate; in DEB theory this fraction is turned over (as it is in the cell) but requires no maintenance. Energy uptake is related explicitly to food availability through a functional response curve, and the energy mobilised from the reserve is partitioned between growth (κ) and reproduction ($1-\kappa$) in a defined proportion (in DEB notation this is the κ-allocation rule, and it operates throughout the complete life cycle). Integral to the structure of any DEB model is the interplay between surfaces and volumes and when food availability is constant, the rate of growth equation in DEB theory is mathematically equivalent to the von Bertalanffy version of the Pütter equation (though its derivation is quite different). The onset of reproduction is determined by a threshold denoting sexual maturity; above this threshold reproductive material accumulates and is either allocated directly to eggs or sperm (income breeding) or earmarked for future reproduction (capital breeding)[43].

All processes in the model are temperature-sensitive and this sensitivity is captured with an Arrhenius relationship. The form of the Arrhenius relationship used in the DEB model is:

$$k_{T_2} = k_{T_1} \cdot \exp\left(\frac{T_A}{T_2} - \frac{T_A}{T_1}\right)$$

Here $k_{(T)}$ is the rate of the reaction at a particular temperature, T_2 and T_1 are the absolute (thermodynamic) temperatures and T_A is the Arrhenius temperature. It is T_A that dictates the temperature sensitivity of the reaction and typical values for growth lie in the range 5000 to 14,500. The value of T_A cannot be derived theoretically, it can only be determined empirically for a given system. The logic for applying a single temperature correction across all reactions is that for the cell or organism to maintain its integrity across a range of temperatures, the temperature sensitivity of the various physiological reactions must be broadly similar to prevent potentially lethal metabolic imbalances. It is also possible that the physical processes that determine the temperature sensitivity of enzyme-catalysed reactions are similar across all enzymes, and hence temperature sensitivity is constrained within a narrow range. We do not know whether either of these possibilities is correct, but it remains a widespread observation that the temperate sensitivity of biological processes tends to lie within narrow bounds, and so the assumption of a single temperature sensitivity for an organism's physiology is reasonable. (This is an important topic in thermal biology, which we explore in more detail in Chapter 7.)

DEB theory is based on a rigorous picture of how energy and materials flow through organisms (Figure 4.15). It is different from previous models in that respiration is not parameterised in itself, but emerges from the estimated costs of diverse processes within the model. This approach does, however, assume that all the processes that dissipate energy through respiration are represented correctly in the DEB model. Given that respiration dominates the energy budget of organisms, accurate prediction of the energy dissipated in metabolism is a strong test of the model.

Given the comprehensiveness and rigour of DEB theory, one might ask why it has not been adopted more widely by ecologists. It is more abstract than the IBP balanced energy budget, and the relatively slow adoption of DEB theory is probably related to the non-intuitive relationship between core variables in the theory and the measurements that ecologists and physiologists actually make in the field or laboratory: it is not immediately clear, for example, what measurement is needed to estimate an abstract quantity such as 'structural volume' or 'maturity rate coefficient'. Interface programs are available to estimate the core variables required by a DEB model from physiological and ecological measurements, and it may be that as more examples of DEB models appear in the literature more ecologists will be persuaded to use it[44].

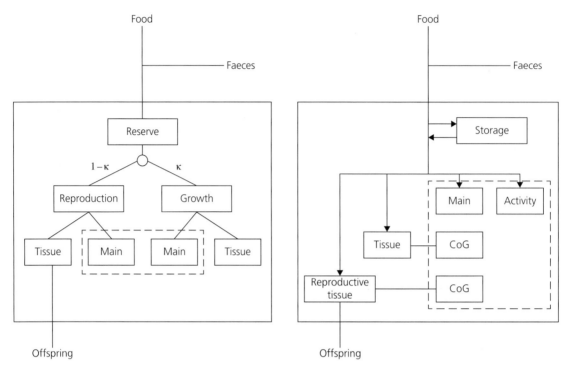

Figure 4.15 Energy flow in two energy budget models. A (left panel): energy flow in a DEB model. B (right panel): energy flow in the IBP balanced energy budget model. Main: maintenance costs; CoG: cost of growth. Dashed lines group components of respiration (oxygen consumption).

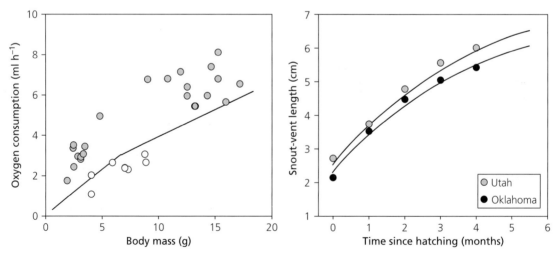

Figure 4.16 A DEB model of the lizard *Sceloporus undulatus*. A (left panel): oxygen consumption of *S. undulatus* (white symbols) and the related *S. occidentalis* (grey symbols) and the DEB prediction (line). B (right panel): observed growth in laboratory raised lizards from Utah (grey symbols) and Oklahoma (black symbols) with growth trajectories predicted from DEB theory[45].

4.8 A complete energy budget?

Biophysical models allow us to predict the thermal (kinetic) energy content of organisms and energy budgets track the flow of chemical potential energy, including its dissipation as heat. Very few attempts have, however, been made to link these two approaches to generate a complete picture of the energy flows within an organism. A notable exception has been Michael Kearney's model of the widespread North American lizard *Sceloporus undulatus*.

In this study a biophysical model was used to predict the hour-by-hour steady-state body temperature of lizards, based on real climate data. The output from this was then coupled to a DEB model to predict the growth, metabolism and reproduction of lizards throughout North America. The predicted growth trajectories matched field data closely, and the modelled reproductive output explains the distribution very well (Figure 4.16). Modelled respiration rates are generally lower than those observed, but overall the match between model and field data is highly impressive. Coupling a DEB model to a biophysical model requires fewer parameters than previous attempts to link an energy budget to the biophysics, and this combination clearly has enormous potential[46].

4.9 Summary

1. An organism is an open thermodynamic system exchanging both energy and materials with its environment.
2. Although organisms are not in thermal or thermodynamic equilibrium, flows of energy in and out must balance exactly with changes in internal energy.
3. Organisms exchange energy with their environment by radiation, conduction, convection and evaporation of water. The relative importance of these varies with the organism and its situation.
4. Energy loss to the environment can be approximated by Newton's Law of Cooling and quantified as minimal thermal conductance, but this simplification is only useful for warm endotherms in a still, cool environment. For all other circumstances a full biophysical treatment is necessary.
5. Flows of chemical potential energy can be described by the balanced energy budget. Energy flows through an individual organism can be predicted by the Dynamic Energy Budget model.
6. A full description of the energy balance of an organism requires the coupling of a biophysical model of heat flow with an energy budget model. This combination provides a powerful tool to model the thermal and energetic niches of organisms, and to predict how these might change in the future.

Notes

1. This quotation comes from David Gates' classic book *Biophysical ecology* (Gates 1980). Although now over 35 years old this book remains an essential text for anyone concerned with the energy balance of organisms. Reproduced with permission of Springer-Verlag.
2. Organisms are not, however, in either thermal or thermodynamic equilibrium, which makes the rigorous application of thermodynamics a little trickier than in the idealised systems beloved of theoreticians.
3. In physiological modelling, we assume we are working at constant temperature and pressure, so the change in internal energy is identical to the change in enthalpy. For organisms the assumption of constant pressure is reasonable, but the assumption of constant temperature is a simplification that may not always be valid. The classic study of energy flow within ecosystems is Lindemann (1942).
4. Jean-Baptiste Joseph Fourier (1768–1830) was a French mathematician and physicist, best known for his analysis of heat transfer. His name is remembered in mathematics for the Fourier series and Fourier transform, and in physics for his law of heat transfer. He is also often credited with the discovery of the greenhouse effect (see Chapter 16). The treatment of heat flow here follows Sprackling (1991).
5. Modified with permission of Cambridge University Press from Schmidt-Nielsen (1975), with data from Irving & Hart (1957) and assumed core temperature from Clarke & Rothery (2008).
6. A *phonon* is a quantum of vibrational mechanical energy, just as a photon is a quantum of electromagnetic (light) energy. Phonons and electrons both redistribute energy in any solid not in thermal equilibrium. Because the movement of energy down a thermal gradient by conduction is an irreversible process, there is an associated increase in entropy.

7. This complexity is captured in an apocryphal story attributed to the German physicist Werner Heisenberg (1901–1976). He is reputed to have said that when he eventually met God, he was going to ask two questions: why relativity and why turbulence, and that he was expecting an answer for the first. A similar witticism has been attributed to the British physicist (and expert on hydrodynamics) Horace Lamb (1849–1934).
8. Redrawn, with permission of Springer-Verlag, from Porter et al. (1973), following conversion to SI units. Lizard data are from Weathers (1970). Heat transfer coefficients measured under field conditions are typically 50% to 100% higher than those measured under idealised conditions in wind tunnels (Pearman et al. 1971, Porter et al. 1973).
9. This section is based on the discussion of convection in Steven Vogel's marvellous book *Life in moving fluids* (Vogel 1981). Although concerned with fluid flow, heat transfer by forced convection is so important to thermal ecology that this book is essential reading for any ecologist or physiologist involved with the transfer of heat in or out of an organism. Sprackling (1991) provides a rigorous physical treatment of convection.
10. Vogel (2012) provides a wonderful introduction to the energy balance of plant leaves. Leigh et al. (2012) use biophysical arguments to explain how plants living in hot climates and still air benefit from thicker leaves. The *rete mirabile* of tuna fish is described by Carey & Teal (1966).
11. Note that these irradiance curves are plotted in terms of wavelength (which is how they are usually plotted in biophysics). In this presentation peak power is in the green part of the visible spectrum (560 nm). However if they are plotted in terms of frequency, then peak power lies in the infrared (880 nm). This difference arises because the Planck energy function is a density distribution function and is defined in terms of differential calculus (Soffer & Lynch 1999).
12. The loss of heat to the night sky is so effective that it allowed the ancient Persians in the fourth century BC to make ice in the desert. This ice was then stored in carefully designed buildings that were covered with straw during the day (to limit radiant heating from the sun) and removed by night (to cool the building), thus pre-dating the development of ice-houses in Europe by many centuries.
13. For example Thayer (1909) and Cott (1940). Thayer's ideas were roundly criticised by big-game hunter and 26th President of the United States, Theodore Roosevelt (Roosevelt 1911).
14. Hamilton (1973).
15. Strictly absorptance needs to be defined in terms of radiant flux (energy per unit time). Direction is also important, as the incident radiation may be isotropic (of equal intensity in all directions), although in ecological situations it will almost always be strongly directional (typically from the Sun above). The absorptance also varies with wavelength, and so may need to be specified in terms of spectral radiance (energy per steradian per square metre per unit wavelength or frequency).
16. Smith et al. (2016) have recently reported an interesting example of the adjustment of infrared emissivity allowing for subtle control of thermal balance in the Australian lizard *Pogona vitticeps*.
17. The higher body temperature arises because at equilibrium energy absorbed must balance energy radiated. A darker species absorbs more energy in the visible range than a paler species, and thermal equilibrium is achieved only when this is matched by increased energy radiated away in the infrared, which requires a higher surface temperature. Because the intensity of radiation varies with the fourth power of absolute temperature, the effect of colour on equilibrium body temperature is quite small, but it is real.
18. For a valuable review of transparency in organisms see Johnsen (2001).
19. The early studies of orthopterans are Buxton (1924) and Hill & Taylor (1933), and a more recent study is Harris et al. (2013).
20. The relationship between colour, thermal characteristics and ecology in lady beetles is described by Brakefield (1984a, b) and de Jong et al. (1996), and in *Colias* butterflies by Watt (1968, 1977, 1983), Watt et al. (1983), Kingsolver (1987) and Ellers & Boggs (2004).
21. Porter & Gates (1969). See also Clusella-Trullas et al. (2007) for a review of thermal melanism.
22. Diagram modified from Brakefield (1984a, b).
23. Replotted, with permission of John Wiley and Sons, from Clusella-Trullas et al. (2008). The arcsine-square root transformation is necessary because the reflectance data are percentages.
24. Dawson et al. (2014) undertake an interesting comparison of camouflage and insulation in the polar bear and the marsupial koala.
25. See Buxton (1923) and Meinertzhagen (1954).
26. This work is described in Hamilton (1973).
27. See McNab (1970, 1973) for arguments in favour of a simplified approach, and Porter & Gates (1969) for the detailed biophysics. McNab (1980) discussed practical and statistical issues in the measurement of thermal conductance.

28. The nature of the scaling relationship to be expected for conductance is discussed by Turner (1988). Analyses of empirical data for birds and mammals are Herreid & Kessel (1967), Bradley & Deavers (1980), Aschoff (1981) and Schleucher & Withers (2001). Boyles & Bakken (2007) demonstrated the effect of wind speed on the measured conductance in two small mammals.
29. Plotted with permission of Cambridge University Press from Schmidt-Nielsen (1975); original study Bartholomew & Hudson (1962).
30. The relationship between conductance per unit area and size is Roberts et al. (2010). Naya et al. (2013) show how BMR and conductance act synergistically to regulate body temperature in small mammals (127 species of rodent).
31. Isaac Newton published his experiments on cooling anonymously and in Latin in 1700 (often quoted as 1701, because the relevant volume of *Philosophical Transactions* was published across both years). He did not regard his generalisation about the rate at which bodies lose heat as an empirical law, but we know very little about this aspect of his scientific endeavour. This work provides the first mathematical formulation of heat transfer, and was incorporated by Fourier in his later mathematical treatment of heat flow. See Ruffner (1963) and Cheng & Fuji (1998). Sir Isaac Newton (1643–1727) was one of the greatest of all scientists. He made fundamental contributions to our understanding of mechanics (Newton's Laws of Motion), gravity and optics. From 1665 he also developed the calculus, which was developed independently by Gottfried von Leibniz (1646–1716) around 1675. Newton's great work is *Philosophiæ Naturalis Principia Mathematica* (Principles of Natural Philosophy), a three-volume work published in 1687, and usually referred to simply as the *Principia*.
32. Thermal conductance data are from Tables 1 to 6 in Bradley & Deavers (1980) and Table 1 in Aschoff (1981b).
33. Useful discussions are Atchley et al. (1976) on the statistical behaviour of ratios, Beaupre (2005) on mass-specific measures and Freckleton (2002, 2009) on comparative studies in general.
34. Data from Schmidt-Nielsen (1969).
35. Useful summaries of evaporative water loss are Schmidt-Nielsen (1969, 1975) and McNab (2002).
36. See Porter et al. (1973), Spotila et al. (1973) and O'Connor & Dodson (1999).
37. Winberg (1960), this being an English transpation of the 1956 Russian original. See also Petrusewicz & Macfadyen (1970). The International Biological Programme (IBP) was organised by Charles Waddington and ran from 1964 to 1974, initially with Canadian and European scientists, and from 1968 also including scientists from the USA. It was modelled on the successful International Geophysical Year (1967–1968), and was built around a series of detailed studies of selected biomes with a focus on productivity. Its history and influence is discussed by Hagen (1992).
38. The concept of reserve as a separate and important energetic compartment was introduced by Bas Kooijman in his development of the Dynamic Energy Budget (Kooijman 1993).
39. Nitrogen excretion data come from Clarke et al. (1994), Clarke & Prothero-Thomas (1997) and Fraser et al. (2002). Nitrogen excretion in marine organisms is a remarkably understudied topic, so few generalisations are possible but Nicol (1960), Barrington (1967) and Vanni & McIntyre (2016) provide valuable summaries. McNab (2002) reviews nitrogen excretion in marine and terrestrial vertebrates.
40. Blaxter (1989) uses a power law function to allow for body mass in the energy budgets of mammals. Clarke (2013) incorporated scaling and a temperature function into the IBP balanced energy budget to explore possible energy budgets for dinosaurs.
41. The energy budgets are reported by Brett & Groves (1979) and Priede (1985) for fish, and Hughes (1986) for gastropods.
42. The theory is presented in detail in a series of books: Kooijman (1993, 2000, 2010).
43. The κ-allocation rule directs a defined fraction of mobilised reserves to reproduction throughout the life cycle. For immature stages this is explained as providing energy for the *increasing complexity or information content within the organism*, during which no biomass accumulates and the energy is dissipated as heat. This is a clear indication that entropy is involved (see Sousa et al. 2006 for a formal discussion of entropy in the context of DEB theory).
44. Sousa et al. (2010) list six state variables, eight other variables and 19 parameters for a DEB model, though these can be reduced for specific models covering particular stages of the life cycle, or restricted ecological circumstances. Recent reviews of DEB theory and its applications are Maino et al. (2014) and van der Meer et al. (2014).
45. Plotted with data kindly supplied by Michael Kearney. Observed oxygen consumption data are from Angilletta (2001a, b) and Dawson & Bartholomew (1956).
46. Kearney (2012). Previous studies coupling biophysical and energy budget models include Grant & Porter (1992) and Buckley (2008).

CHAPTER 5

Water

Biology has forgotten water, or never discovered it.

Albert Szent-Györgyi[1]

One of the most profound images of the last century was that of the Earth floating in space, taken by Apollo astronauts on their way to the Moon. Many aspects of this image emphasise features which can escape the Earth-bound spectator: the swirls of cloud capture the importance of lower atmosphere processes in redistributing energy and water, and the lights of cities and slash-and-burn agriculture reveal the extent to which man has modified the terrestrial environment. But most powerful of all, the predominantly blue nature of the globe reminds us that ours is a planet of water.

Water has held a symbolic place in human culture since the earliest western civilisations, possibly because so many of these arose in arid environments. It was the ancient Greeks who appear to have been the first to recognise the central importance of water to life. It is, however, easy to misinterpret this as prescient of modern ideas, especially as the ancient Greek philosophers were not scientists in the way we understand the term today. They observed the world around them but did no experiments to test their ideas; rather their aim was to build convincing abstract pictures of the way the universe might be constructed. Nevertheless it is fair to say that the ancient Greek philosophers did focus intellectual attention on the fundamental importance of water[2].

We now recognise that life on Earth depends absolutely on liquid water, and the existence of the liquid state sets the boundary conditions for life itself. Although some forms of life are capable of surviving almost complete dehydration, they are unable to grow or reproduce when dry and can only complete their life cycle once they have returned to the hydrated state. Clearly, if we wish to understand how temperature affects animals and plants, we need to understand the properties and thermal behaviour of water.

5.1 Water and life on Earth

Organisms are mostly water. In the large complex organisms we are most familiar with, the overall body water content is governed by the relative proportions of cells, extracellular body fluids, adipose tissue, skeleton and so on. To understand the importance of water to life we therefore need to look deeper, into the water content of the living cell.

It is actually quite difficult to determine the water content of an individual cell; the few measurements have been made indicate that a typical animal or microbial cell contains 65–80% water (Table 5.1). These measurements come from an evolutionarily diverse range of organisms, suggesting that a cellular water content of 65–80% is probably a general feature of life on Earth. In some specialised cells the water content can be very different; for example, in adipose (fat storage) cells of vertebrates the huge lipid deposit within the cell reduces cell water content to ~ 10%. There are no data for plant cells in Table 5.1, because the presence of a vacuole makes their water content extremely labile (and also raises the question of whether calculating a mean water content for a plant cell has any physiological meaning)[3].

Principles of Thermal Ecology. Andrew Clarke, Oxford University Press (2017).
© Andrew Clarke 2017. DOI 10.1093/oso/9780199551668.001.0001

Table 5.1 Water content of cells[4].

Species	Water (% fresh mass)
Bacteria	
Escherichia coli	70
Amoebozoa (Mycetozoa: slime moulds)	
Dictyostelium discoideum	70
Ascomycota (Saccharomycetes: yeasts)	
Saccharomyces cerevisiae	65
Vertebrata	
Rattus norvegicus (heart)	76
Erythrocytes (nine species of mammal)	59–64
Lithobates pipiens (seven visceral tissues)	68–82

Table 5.2 Whole body water content of some organisms[5].

Species	Water content (% fresh mass)
Plants	
Aquatic plants	>90
Terrestrial plants	40–90
Xerophilous plants	<10
Seeds	50
Animals: invertebrates	
Gelatinous zooplankton	94–96
Planktonic marine invertebrates	76–92
Benthic marine invertebrates	78–92
Terrestrial insects	54–76
Terrestrial molluscs	80–90
Resting stages	<10
Animals: vertebrates	
Generalised vertebrates	~70

The whole body water content of complex multicellular animals and plants is variable because of the varying proportions of tissues with differing water contents. Aquatic plants and animals can have very high water contents (Table 5.2), partly because water is present in abundance and partly because of the hydrostatic support the environment provides. In gelatinous zooplankton (jellyfish and sea-gooseberries) water content may be as high as 96%. In contrast, terrestrial organisms need to invest in skeletal components such as bone, chitin or lignin to maintain their structure and shape, and these reduce the proportion of water in the organism as a whole. In some cases, such as worms, water provides the mechanism for maintaining body shape (hydrostatic skeleton) and is essential for movement.

Complex organisms can also have significant quantities of water outside the cells, and this extracellular water also provides a transport medium for moving nutrients and waste products about the body. In a typical vertebrate water comprises about 70% of the overall body mass, of which most (45%) is within the cells, with the remainder being either located in tissue spaces (the interstitial fluid, about 20%) or in the blood (plasma fluid, about 5%).

Where water is extremely scarce, as in deserts, some xerophilous plants have adapted by reducing their water content while others have mechanisms for taking up considerable quantities of water during the short periods it is available and storing this for times of drought. This water may be stored in the plant stem (such as in cacti and the baobab tree *Adansonia*) or in the leaves (such as in *Aloe* species and stonecrops, *Sedum*), leading to a high water content in the plant overall.

Where metabolic processes are largely or completely suspended in resting stages, water content may be reduced to almost zero. Examples of these are plant seeds, tardigrade tuns, dehydrated nematodes and also bacterial and fungal spores. In these cases, however, the tissue must be rehydrated before metabolism can continue. We will explore this further in later chapters.

5.2 Why water?

Water is widely recognised as an excellent solvent; indeed it dissolves more substances than any other known liquid and is sometimes referred to as the 'universal solvent'. The influence of water on life, however, is all-pervasive and extends far beyond its role as solvent.

Water is integral to the chemical reactions of physiology, not just as a solvent but also as a factor influencing the structure of the interacting molecules, and as a reactant. Of the four main classes of chemical reaction involved in physiology, namely oxidation, reduction, condensation and hydrolysis, water participates as a reactant in all four. Of equal importance to cellular integrity are the substances that dissolve only sparingly in water. In particular the interplay between lipids and water provides much of the important architecture of the cell, and waxy outer coatings are essential in regulating the passage of water in and out of many terrestrial organisms.

The importance of liquid water to life on Earth is related to two fundamental factors. The first is that water is very common; it is the second most common molecule, behind hydrogen, in the universe. The second is that the physical properties of water are highly unusual. We rather take water for granted as a fairly ordinary and unremarkable compound. It is the simplest compound of the two most common reactive elements on Earth; being odourless and tasteless it does not announce its presence in the way that ammonia or hydrogen sulphide do, and it is ubiquitous. Water is, however, a most extraordinary substance.

5.2.1 The anomalous nature of water

Many elements form compounds with hydrogen, and generally the melting points and boiling points of these compounds increase with molecular mass. This increase comes about because as the molecules get larger they have more electrons. In consequence the Van der Waals interactions between the molecules become stronger, and it takes more energy to move a molecule from the liquid to the gaseous state.

However, something unusual happens with the hydrogen compounds of elements in Groups 15, 16 and 17 of the periodic table. In all three groups the hydrides of the lightest members (NH_3, H_2O and HF respectively) have anomalously high boiling points. This is because in each case the hydrogen is bonded to a strongly electronegative element (N, O, F). The result is the formation of hydrogen bonds between the molecules, and these additional intermolecular forces, absent in the other members of the groups, lead to a significant change in physical properties.

We can illustrate these anomalous properties with water, the physical properties of which are very different from those that would be predicted on the basis of the properties of the dihydrides of the other Group 16 elements (Figure 5.1). Were we to predict the behaviour of water on the basis of the analogous compounds of the other group 16 elements, it would melt around −89 °C and boil at around −73 °C. An important ecological consequence of this anomalous behaviour is that water occurs naturally in all three phases (ice, water and vapour) at the surface of the Earth, often simultaneously (Plate 5). The reason for this anomalous behaviour is the hydrogen bond.

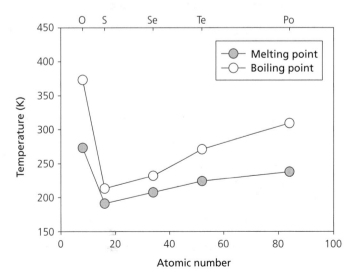

Figure 5.1 Melting and boiling points of the dihydrides of Group 16 elements, showing how the presence of hydrogen bonds in water elevates its melting and boiling point well above those to be expected on the basis of the properties of the dihydrides of the other elements in the group.

5.3 The hydrogen bond

The hydrogen bond in water results from the asymmetrical charge distribution on the water molecule. The oxygen atom is strongly electronegative and thus tends to attract the electron cloud from around the hydrogen nucleus. This decentralised electron cloud gives the water molecule its polar nature (Figure 5.2). The hydrogen bond is essentially an electrostatic attraction between partial charges on adjacent molecules. It is not a chemical bond in the sense of the covalent bonds linking the hydrogen atoms to the oxygen atom in the water molecule itself, but rather a particularly strong electrostatic attraction between molecules. This hydrogen bonding holds water molecules closer together (by about 15%) than if water was a simple liquid with van der Waals interactions as the dominant intermolecular force.

In general, hydrogen bonds form when a hydrogen atom lies between two strongly electronegative atoms. The bond is usually represented as:

$$X-H\cdots Y$$

where X and Y are the electronegative atoms N, O or F. Conventionally the hydrogen atom is termed the hydrogen bond donor, and the electronegative atom the acceptor. Although predominantly an electrostatic interaction, the hydrogen bond does have some covalent features: it is directional, stronger than Van der Waals interactions and involves a limited number of atoms. Hydrogen bonds vary in strength (Table 5.3) and in the case of HF can exceed 150 kJ mol^{-1}. Typical values for water are ~5–25 kJ mol^{-1}, depending on the influence of other atoms nearby.

Because the hydrogen bond is predominantly electrostatic in nature, it is described reasonably well by Coulomb's Law:

$$E_p = \frac{Q_1 Q_2}{4\pi(\varepsilon_r \varepsilon_0) r}$$

Here E_p is the potential energy (that is, the bond energy), Q_1 and Q_2 are the partial charges, r is the distance separating the partial charges, ε_0 is the vacuum permittivity and ε_r is the relative permittivity (previously known as the dielectric constant) of the medium. The strength of a hydrogen bond thus depends linearly on its length, which is the distance between the partial charges, r. This means that a hydrogen bond does not break in the all-or-nothing manner that characterises a covalent bond; it simply gets weaker the further apart the partial charges are. A hydrogen bond is thus not a bond in the sense of a conventional covalent bond which has a well-defined length and energy.

The strength of a hydrogen bond depends on distance, angle and the local environment, so not all hydrogen bonds in water will have the same energy. This dependence of the strength of a hydrogen bond on its local environment is termed cooperativity. Cooperativity has significant consequences for structural biology because it exerts a strong influence on how macromolecules fold and interact. In particular, once one hydrogen bond has formed it is easier to attach subsequent hydrogen bonds, and conversely once one hydrogen bond breaks, the local structure becomes more likely to disintegrate.

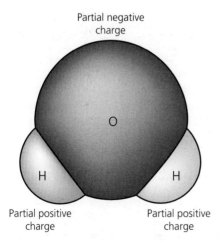

Figure 5.2 Water molecule showing distribution of partial charges that gives the molecule its polar character.

Table 5.3 Properties of some hydrogen bonds of biological importance[6].

Bond	Mean bond length (nm)	Typical bond energy (kJ mol^{-1})
O—H....O	0.270	−22
O—H....O$^-$	0.263	−15
O—H....N	0.288	−15 to −20
N$^+$—H....O	0.293	−25 to −30
N—H....O	0.304	−15 to −25
N—H....N	0.31	−17
HS—H....SH$_2$		−7

Hydrogen bonds form between relatively few types of atom, but since these include some of the commonest atoms to be found in biological molecules, hydrogen bonds are of enormous importance in physiology. Not only do they give to water many of the properties that make it so central to life as we know it, but they also are important in influencing the solubility of compounds in water, determining the secondary and tertiary structure of proteins and nucleic acids and the binding of substrates and co-factors to enzymes. Of particular importance is the way that hydrogen bonds influence the structure of liquid water.

The partial positive charges in the water molecule are associated with the two hydrogen atoms and the partial negative charges with two lone pairs of electrons in the oxygen atom. Possessing two donor sites and two acceptor sites, the water molecule has the ability to form up to four hydrogen bonds, the spatial distribution of which gives rise to a three-dimensional network. X-ray diffraction studies of hexagonal ice have shown that each oxygen atom in water is surrounded by four other oxygen atoms, with a hydrogen atom and hydrogen bond along each O–O axis (Figure 5.3).

The structure of liquid water is less well understood, but the increased molecular motion is believed to result in an average of fewer than four hydrogen bonds per water molecule and a currently accepted value is ~3.4. This is a time-averaged number because in liquid water the hydrogen bonds break and reform rapidly as the water molecules move around. Furthermore the hydrogen bonds in water are at slightly different angles and lengths, and hence are of different strengths, compared with ice. Perhaps surprisingly, physicists still lack a complete picture of water structure, but a useful conceptual model of bulk water has water molecules linked by a constantly shifting network of hydrogen bonds of differing lengths and orientations leading to local variations in structure and density[7].

Calculation indicates that in liquid water a small but significant fraction of collisions between molecules have sufficient energy to disrupt a hydrogen bond. In other words, at typical physiological temperatures, hydrogen bonds in water, and in biological molecules, will be breaking and reforming very rapidly (on a timescale of 1–100 picoseconds).

5.3.1 Temperature and the hydrogen bond

The strongest experimental evidence for the temperature dependence of the length (and hence strength) of the hydrogen bond in water comes from studies of hexagonal ice. These show that the cube root of the specific volume has a quadratic relationship with temperature (Figure 5.4). The ice lattice is maintained solely by hydrogen bonds, so variation in the specific volume of ice can only come about by changes in hydrogen bond length (and hence bond strength). In liquid water the effect of temperature on the strength of the hydrogen bond can be seen from the temperature dependence of the enthalpy of vapourisation (Figure 5.4)[8].

One of the consequences of the variation in hydrogen bonding structure with temperature in water is that ice has a lower density than liquid water at the freezing point (Box 5.1).

5.4 The physical properties of water

The high density of hydrogen bonds in liquid water influences all of its physical properties. Some properties of particular importance to organism and their ecology are summarised in Table 5.4.

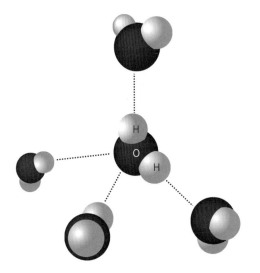

Figure 5.3 Structure of water molecules and hydrogen bonding in ice, as revealed by X-ray diffraction. Each water molecule is hydrogen bonded to four other water molecules, arranged as a regular tetrahedron.

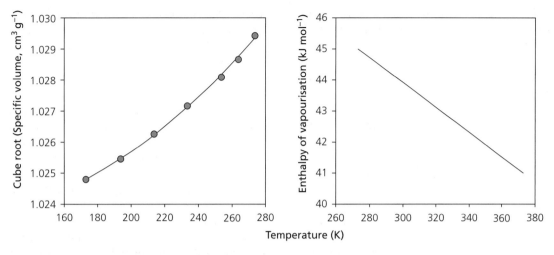

Figure 5.4 Two manifestations of the temperature dependence of hydrogen bond strength in water. A (left panel): cube root of the specific volume of hexagonal ice as a function of temperature. The experimental data are shown as solid symbols, and the line is the least-squares quadratic fit. B (right panel): a decrease in the enthalpy of vapourisation of water with increasing temperature[9].

Box 5.1 Why ice floats on water (and why pipes burst in winter)

One of the more extraordinary properties of water, and one that has the most profound consequence for life on Earth, is that ice floats. We take this everyday experience for granted, but it is actually a most unusual property. Were water to behave as most other compounds, the solid phase (ice) would be heavier than the liquid phase (water); water bodies would then freeze solid in winter rather than having an insulating layer of surface ice, killing all life in any lake, stream or pool. In the sea, ice forming in seawater would sink to the seabed rather than floating and the seabed of the continental shelves at high latitudes would be permanently frozen and lifeless. Instead it forms an important habitat for life (Plate 5).

The reason for the lower density of ice is that the rigid crystal lattice maintained by the four hydrogen bonds produces more space between the water molecules than does the more fluid and dynamic arrangement of molecules in liquid water. This difference in spacing, following directly from the pattern of hydrogen bonding, results in ice having a density ~9% lower than liquid water.

As the temperature of liquid water approaches the freezing point, an increasing fraction of the water molecules start to assume a more ice-like arrangement in which there is increased space between molecules. At the same time the space between the other water molecules decreases because of the lower thermal motion as temperature decreases. The former process decreases density, and the latter increases it. At the temperature where the effect of a more ice-like molecular arrangement balances the lower thermal motion water exhibits a maximum density. This temperature is 3.98 °C; above and below this temperature water density decreases. Seawater does not exhibit a similar maximum density above because the presence of ionic solutes affects the pattern of hydrogen bonding.

Since ice has a lower density than water, a given amount of water occupies more space as ice than it does as water. This expansion on freezing is what causes pipes to burst in winter. It can also cause extensive damage to tissues, especially if ice forms within the cell.

The latent heats of both fusion (melting/solidification) and vapourisation of water are high. It takes considerable energy to melt snow or ice, and also to evaporate water, and these effects act to ameliorate potentially rapid changes in climate. They also exert a powerful influence over the thermal physiology of organisms: the evaporation of water from the surface of an animal or plant can remove a great deal of its internal thermal energy. The high thermal conductivity of water, coupled with its relatively

Table 5.4 Some physical properties of liquid water and their ecological consequences. Where these vary with temperature, data are provided generally for 0 °C (~273 K).

Property	Value	Comment
Equilibrium freezing point	273.15 K	At 1 atm pressure (0.1 MPa); anomalously high
Boiling point	373.12 K	At 1 atm pressure (0.1 MPa); anomalously high
Specific heat capacity	4.75 kJ (g K)$^{-1}$ 75.3 J (mol K)$^{-1}$	Unusually high; varies with temperature with minimum value at 35 °C
Relative permittivity (dielectric constant)	78.5 (at 298 K)	One of the highest values known for a liquid; renders ionic compounds very soluble
Latent heat of vapourisation	2257 J g^{-1} (at 373 K)	Very high; allows for rapid cooling by evaporation, and thereby regulates upper bulk temperature of aquatic habitats
Latent heat of fusion	334 J g^{-1}	High, exerting a strong moderating effect on climate
Thermal conductivity	0.561 W m^{-1} K^{-1} (at 273 K)	High; allows rapid equalisation of temperature within cells and tissues without damage from convection
Thermal diffusivity	0.143 × 10^{-6} m^2 s^{-1} (at 298 K)	High
Surface tension	75.7 mN m^{-1} (at 273 K)	High; important for fluid motion in plants, and for droplet formation in clouds.
Dynamic viscosity	1.792 mPa s (at 273 K)	Relatively low, enabling use of water as an efficient transport medium
Density	999.97 kg m^{-3} at 277 K (water, maximum density); 916.7 kg m^{-3} (ice)	Relatively high; density varies with temperature and pressure, and with concentration of dissolved solutes.
Compressibility	0.46 GPa^{-1} (at 298 K)	Very low, so density changes only very slowly with pressure (and hence depth)

low viscosity, also mean that any change in temperature can equilibrate rapidly throughout the organism. Desert mammals and birds can thus use evaporation to cool down, though the same process can also be highly damaging to structures such as leaves which can cool so rapidly that they suffer thermal damage.

5.5 Temperature and water

The temperature sensitivity of hydrogen bonding means that many of the physical properties of water vary with temperature. This variation can have important ecological consequences, in that water at 0 °C and water at 30 °C are quite different environments for organisms living there. It also means that the properties of the organism's internal environment, both intracellular and extracellular, vary with body temperature.

5.5.1 Ionisation and pH

One of the most important properties of water from a physiological perspective is its ionisation. This ionisation is often represented as:

$$H_2O \rightleftharpoons H^+ + OH^-$$

The equilibrium constant for this reaction, K_W, is calculated as:

$$K_W = \frac{[H^+].[OH^-]}{[H_2O]}$$

where, by convention, square brackets represent the concentrations of the relevant ions or molecules. However water is a very weak electrolyte and the number of water molecules ionised at any one time is so incredibly small that the proportion of unionised water molecules [H_2O] is effectively unchanged at ~1. The expression thus reduces to:

$$K_W = [H^+].[OH^-]$$

Here K_W is also referred to as the ionic product of water, because it is calculated from the product of the concentrations of the hydrogen and hydroxide ions.

This description is something of a simplification, for the ionisation of water is more complicated than the above equations imply. The process actually involves two water molecules reacting to form a hydroxonium ion and a hydroxide ion:

$$2H_2O \rightleftharpoons H_3O^+ + OH^-$$

This happens because the electronic structure of water allows it to function as both an acid (by donating a hydrogen ion) and as a base (by accepting a hydrogen ion) (Figure 5.5).

Because the resulting hydroxonium ion is itself a very strong acid and the hydroxide ion a very strong base, they react to form water again almost as soon as they appear and at any given time there are only a very small number of hydroxonium and hydroxide ions present in water. The prevalence of hydrogen bonding means that the hydrogen ion tends to exist in even more complex ions, notably $H_5O_2^+$ and $H_9O_4^+$. This physical chemistry might seem a little esoteric for biologists, but it is not: much of physiology depends on the movement of hydrogen ions (protons) and other ions, all of which actually exist in hydrated forms. These hydrated ions are significantly larger than the naked ion, which can exist only fleetingly in the free form under physiological conditions in the cell, and this has important consequences for how cells handle them.

Because so few water molecules ionise, K_W has a very low value: ~10^{-14} at ~25 °C. This is a very small number and so for convenience the ionisation of water is usually expressed as pH, where:

$$pH = -\log_{10}[H^+]$$

Since at ~25 °C $[H^+]$ is 10^{-7} mol dm^{-3}, pH = 7. Similarly, at 25 °C, pOH = 7.

The ionisation of liquid water is temperature sensitive, and hence its pH varies with temperature (Figure 5.6). It is important to recognise that although the pH gets higher as water cools, the water has not become alkaline. This is because the increase in pH (which means less ionisation and hence fewer hydrogen ions) is matched exactly by an increase in pOH; over the entire temperature range pH = pOH, and so water remains neutral. Water is acid only when pH > pOH, and alkaline only when pOH > pH.

The relationship between pH and temperature for pure water is curvilinear (Figure 5.6). Our lack of a complete understanding of the structure of liquid water means that this relationship cannot be deduced from theory, and can only be described by empirical equations (that is, a statistical fit to observed data). Over the range of temperature of interest to physiologists (0–40 °C), the variation of pH with temperature is ~ −0.0171 K^{-1}.

Organisms are mostly water, and the pH of their body fluids, both intracellular and extracellular, also varies with temperature. This pH is actively regulated by the organism, and the form of the relationship ($\Delta pH/\Delta T$) is often strikingly similar to that of pure water (Figure 5.6). This preserves an approximately constant alkalinity relative to pure water; in other words the ratio $[OH^-]/[H^+]$ is

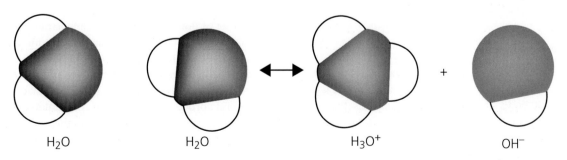

Figure 5.5 Space-filling diagram showing the process of ionisation in water, whereby one water molecule donates a hydrogen ion to another, resulting in the formation of a hydroxonium ion (H_3O^+) and a hydroxide ion (OH^-).

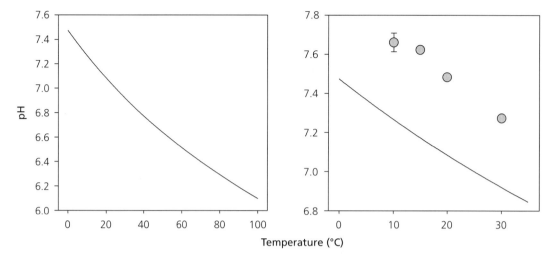

Figure 5.6 Temperature and ionisation of water. A (left panel): variation of pH of pure water with temperature between the equilibrium melting point and boiling point. B (right panel): variation of pH of pure water with temperature over the physiological temperature range (line) with pH of the haemolymph of *Limulus polyphemus* (mean and standard error); error bars fall within symbol in some cases (n = 8–19)[10].

maintained despite the change in pH. The physiological significance of this is that about 80% of the titratable residues of proteins come from the imidazole group of histidine residues. The temperature sensitivity of the ionisation of imidazole is closely similar to that of water, and so the maintenance of relative alkalinity maintains the charge on proteins in the region required for their activity. This maintenance of a constant relative alkalinity is an important component of cellular homeostasis, which we explore further in Chapter 7.

In seawater the situation is complicated by dissolved gases, electrolytes and buffering. Diurnal variation in surface pH of seawater are related mainly to changes in CO_2 content associated with photosynthesis and respiration, and these mask changes associated with temperature.

5.5.2 Viscosity

The viscosity of water is of profound significance for organisms that must move water around within their body, and for aquatic organisms that must move through it in the course of their daily lives.

Dynamic viscosity, usually given the symbol μ or η, is a measure of how easily water moves past itself. Kinematic viscosity (usual symbol ν) is the ratio of dynamic viscosity to density, and is a particularly useful derived measure when considering, for example, the dynamics and energetics of filter-feeding organisms. The significance of dynamic viscosity to movement through water depends critically on the organism's size. The influence of size is captured by the dimensionless Reynold's number, which expresses the ratio of inertial to viscous forces acting on a body moving through a fluid. The Reynold's number, R_e, is given by:

$$R_e = \frac{\rho u a}{\mu}$$

where u is the velocity of the organism relative to fluid of density ρ and dynamic viscosity μ, and a is some linear dimension of the object relevant to the direction of movement. R_e quantifies the relative importance of inertial and viscous forces for a given set of flow conditions and can characterise the flow regime that the organism will encounter. Laminar flow occurs at low R_e where viscous forces are dominant, whereas at larger values of R_e flow is turbulent and the inertial forces produce chaotic eddies and vortices. For very small zooplankton, R_e is low and water is viscous; for the fish that prey on them, R_e is large and water flow is turbulent[11].

The dynamic viscosity of water is sensitive to temperature, increasing as temperature decreases (Figure 5.7). This increase becomes very steep below

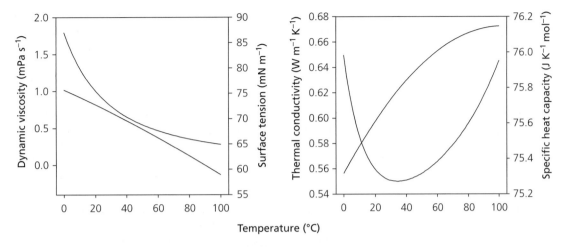

Figure 5.7 Variation in properties of water with temperature. A (left panel): dynamic viscosity (upper curve) and surface tension (lower curve). B (right panel): thermal conductivity (upper curve) and specific heat capacity at constant pressure (lower curve); note the minimum in heat capacity at ~36 °C[12].

0 °C in undercooled water. Within the physiological temperature range, dynamic viscosity increases by a factor of 2.75 between 40 °C and 0 °C.

The viscosity of seawater is slightly higher than that of freshwater (1.077 mPa s at 20 °C, ~7% higher than pure water). The variation with temperature is similar, though it rises a little more steeply at temperatures below ~ 5 °C. The variation of dynamic viscosity with temperature means that those forces operating in fluids to oppose motion will increase as temperature decreases. Examples of ecological processes that will be affected by these forces include the sinking of diatom frustules, the circulation of blood or haemolymph and filter feeding by suspension feeders. Despite the significance of water viscosity to filter feeders, relatively little work appears to have been undertaken exploring the effect of temperature on viscosity in relation to the anatomy, physiology or ecology of aquatic filter feeders living at different temperatures.

The viscous drag, F, on a small sphere of radius r falling through water of dynamic viscosity μ at velocity u is given by Stokes' formula:

$$F = 6\pi\mu r u$$

Thus for a sphere, the viscous drag will be about 2.5 times stronger in a polar sea at 0 °C than in a tropical sea at 30 °C, though the net accelerating force will not be the same as water density also differs. The change in viscosity exceeds the change in density and the net outcome is that it is easier for a diatom or other planktonic organism to remain suspended in the water column in cold waters than in warm waters, though the direct impact on ecology appears to be small as these differences are far outweighed by movement of the water itself[13].

5.5.3 Surface tension

The surface tension of liquids arises because of the cohesive forces between the molecules of the liquid, which in the case of water are predominantly hydrogen bonds. In the interior of the liquid, a water molecule is attracted equally in all directions, as it has other water molecules all around. At the surface the molecules only have other water molecules on one side (we can ignore the small number of water molecules in the gas phase, even when the air is saturated with water vapour). These water molecules are therefore attracted more strongly in towards the liquid than towards the overlying air, creating an internal pressure and forcing the water surface to contract to its minimal area. The result is a surface film with an internal tension (surface tension) which makes it more difficult to move an object through the surface than to move it through the liquid when it is completely submersed.

Surface tension is usually designated γ, and has units of N m^{-1}. At 25 °C the surface tension of water is 72 mN m^{-1} (often quoted in non-SI units as 72 dynes cm^{-1}). As would be expected from the temperature dependence of hydrogen bond strength, the surface tension of water decreases with increasing temperature (Figure 5.7).

The surface tension of water leads to a number of effects, such as the beading of water on the waxy surface of plant leaves or the feathers of aquatic birds. Surface tension also allows small animals to walk on water. The variation of water surface tension with temperature is small over the physiological range, and hence unlikely to have a significant effect on the ecology of insects and other organisms whose habitat is the water surface (γ is 71.2 mN m^{-1} at 30 °C, and 75.0 at 5 °C, a difference of only ~5%).

Surface tension is also of critical importance to the mechanism used by plants to move water and sap around the plant body. Evaporation of water from the surface of mesophyll cells causes the water–air interface to contract into the cellulose matrix of the plant cell wall. This effect of surface tension places the bulk water in the plant under tension (negative pressure), which is transmitted throughout the continuous water column within the plant because the strong hydrogen bonding within the water allows this tension to be transmitted wherever the water is continuous[14].

5.6 Solubility of gases (oxygen and carbon dioxide)

All gases dissolve in water to some extent. The two most important gases from a biological perspective are oxygen and carbon dioxide.

5.6.1 Oxygen

Dissolved oxygen is essential for all aerobic organisms living in water. Its concentration in seawater also governs the availability of many physiologically important elements, such as iron and phosphorus.

The concentration of oxygen in water, [O_2], establishes an equilibrium with the oxygen in the air according to Henry's Law:

$$pO_2 \rightleftharpoons k_H [O_2]$$

The constant of proportionality is the solubility coefficient, usually designated k_H in recognition of William Henry who formulated his law in 1803[15]. Here pO_2 is the partial pressure of oxygen in the atmosphere and Henry's constant k_H has dimensions of pressure divided by concentration. Rather confusingly for biologists attempting to work their way through the literature, k_H comes in a variety of forms with different dimensions, depending on how the atmospheric pressure and gas concentration in solution are expressed. When the concentration of gas in both air and water are both expressed as mole fractions, Henry's constant is dimensionless.

The dependence of the concentration of a gas in solution on the partial pressure of the gas in the overlying atmosphere has some important ecological consequences. Since atmospheric pressure varies with altitude, lakes at high altitude will hold less oxygen that at lower altitudes. On average, solubility decreases by ~1.4% for each 100m increase in altitude. The availability of oxygen in aquatic habitats will also have varied through evolutionary history as the partial pressure of oxygen in the atmosphere has varied.

The value of k_H depends on salinity, meaning that a given volume of freshwater contains more oxygen than the same volume of seawater at an equivalent temperature and partial pressure. At 0 °C the difference is ~24% and at 30 °C it is ~22% (Figure 5.8). It is perfectly possible for water to contain more oxygen, for example by entrainment of air as water passes over a streambed riffle or a spillway, or when aquatic plants are photosynthesising rapidly and releasing oxygen into the water. Under these conditions the water becomes supersaturated with oxygen and this can be so great that the oxygen forms bubbles, as can frequently be seen on sunny days with underwater plants in ponds or from microbial mats in streams.

The increased solubility of oxygen at low temperatures has a number of interesting ecological consequences. It has allowed one group of marine fish to exist without the use of haemoglobin or myoglobin as oxygen carriers. These are the icefish, a family of 15 species found almost exclusively in the cold waters of the Southern Ocean (Box 5.2).

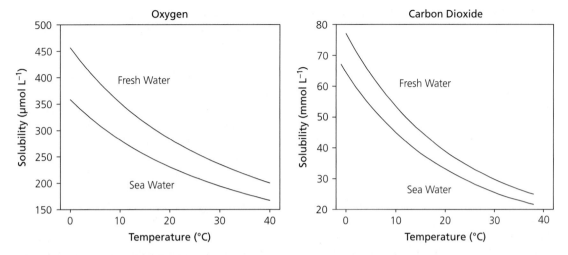

Figure 5.8 Solubility of oxygen (left panel) and carbon dioxide (right panel) in fresh water and seawater. Both gases are more soluble in cold water, and also more soluble in freshwater than in seawater[16].

Box 5.2 The bloodless icefish of Antarctica

In December 1927 the biologist Ditlef Rustad arrived at remote Bouvet Island in the Southern Ocean as a member of the 1927/28 Norwegian Antarctic Expedition. Within weeks he had caught a very strange fish. This fish lacked scales, had large eyes and a long protruding jaw, and, astonishingly, gills that were white. Rustad photographed the fish, which he called a 'white crocodile fish', and wrote in his notebook 'blod farvelöst' (colourless blood). Rustad had collected the first known example of an icefish (Plate 6).

As is often the way with large expeditions, it was many years before the description was published (by Orvar Nybelin, in 1947). In the interim, in 1929, another Norwegian biologist, Steinar Olsen, was visiting South Georgia (then the focus of a large whaling industry), where he was told by the local whalers that there were three local species of 'blodlaus-fisk' (bloodless fish). This was also recorded by Harrison Matthews in his classic account of the island. Olsen collected samples of the fish, and their blood, and discovered a fourth species, this time outside the Southern Ocean in the harbour at Punta Arenas in Chile. These specimens were examined back in Norway by Johan Ruud, who established the lack of erythrocytes in their blood. Ruud finally got his hands on living specimens when he visited South Georgia in 1953 (and there is an apocryphal story that his first specimen was eaten by the ship's cat before he could sample it!). He was able to undertake measurements of the oxygen-carrying capacity of the blood and other physiological measurements; he also was instrumental in drawing the attention of the general scientific community to these remarkable fish.

Icefish (Channichthyidae) comprise 15 species and are part of the dominant superfamily of fish in continental shelf waters of the Southern Ocean (Notothenioidea). All species lack haemoglobin, but there is low and variable expression of myoglobin across the family. It would appear that the high oxygen solubility of cold water (and blood fluid) allows for sufficient oxygen to be carried in physical solution to meet metabolic demands. The lack of scales allows uptake of oxygen through the skin in addition to the gills and the blood volume is large, circulated by a large heart[17].

5.6.2 Carbon dioxide

When carbon dioxide dissolves in pure water it forms a weak acid, carbonic acid. The ionisation of carbonic acid in pure water is low, and hence may be neglected for many purposes. The further fate of CO_2 in fresh waters depends on the pH, and the presence of carbonate (leached from rocks) and ions such as Mg^{2+}. As with oxygen, the solubility of CO_2 in pure water increases at low temperature (Figure 5.7).

In seawater the situation is more complex, because of the presence of a wide range of ionic species. Just as in pure water, some of the dissolved CO_2 reacts with water to form carbonic acid:

$$CO_2 + H_2O \rightleftharpoons H_2CO_3$$

This process is slow (seconds to minutes at seawater pH) but once formed the carbonic acid dissociates rapidly, releasing a proton to form bicarbonate, some of which loses a further proton to form a carbonate ion:

$$H_2CO_3 \rightleftharpoons HCO_3^- + H^+$$

$$HCO_3^- \rightleftharpoons CO_3^{2-} + H^+$$

The various equilibrium constants for these reactions differ somewhat and are all pH-sensitive. At a typical surface seawater pH of 8.2, the relative proportions of $[CO_2]$, $[HCO_3^-]$ and $[CO_3^{2-}]$ are respectively 0.5% (of which only a trace is in the form of carbonic acid), 89% and 10.5%. In other words, for every 100 molecules of CO_2 dissolved in seawater, 89 are in the form of bicarbonate, 10 have formed carbonate ions and only one remains as CO_2. The sum of the dissolved carbonate species is the total dissolved inorganic carbon, ΣCO_2 (though it is also variously designated DIC, TIC and TCO_2 in the oceanographic literature). ΣCO_2 is measured experimentally by adding excess acid to convert all the carbonate species to CO_2, and then measuring this.

An important physiological aspect of this complex equilibrium is that, being charged, HCO_3^- cannot enter or leave the cell interior. For inorganic carbon to enter or leave the cell it must be in the form of CO_2, and the enzyme carbonic anhydrase is of fundamental importance in catalysing the interconversion of CO_2 and HCO_3^-.

5.7 Calcium carbonate solubility

The carbonate system discussed in the previous section is of profound ecological significance because precipitated calcium carbonate is used to provide skeletal support in a wide variety of aquatic organisms. Most well known are probably the reef-building scleractinian corals of tropical seas, but many other groups of aquatic organisms use calcium carbonate to construct an internal skeleton or an external shell. Calcification has probably evolved many times independently and in geological history there have been many periods of extensive reef-building, by a wide variety of organisms[18].

Natural $CaCO_3$ exists in a number of crystalline forms, of which only calcite and aragonite are of widespread biological importance. Aragonite is thermodynamically metastable at atmospheric pressure and physiological temperatures (Figure 5.9), though the time scale of conversion to calcite in the ocean is of the order 10^7 to 10^8 years. (Recall from Chapter 2 that while thermodynamics tells us where the reaction will go, it is the kinetics that determine how fast it will get there.) The slow rate at which aragonite changes to calcite under physiological conditions means that this is not relevant physiologically, and both calcite and aragonite are utilised by marine organisms to build skeletons: calcite in echinoderms and some foraminiferans, aragonite in most bivalves and gastropods, with many organisms using a mixture. Aragonite precipitation generally predominates in warmer seawater, and scleractinian corals (whose skeleton is exclusively aragonite) only form reefs where the seawater temperature is above ~18°C. In contrast, brachiopods (whose skeletons are calcitic) are today a cold-water group, though this was not always so[19].

$CaCO_3$ precipitates according to the following overall reaction:

$$Ca^{2+}(aq) + CO_3^{2-}(aq) \rightleftharpoons CaCO_3(s)$$

$CaCO_3$ precipitates as a solid, so by definition its thermodynamic activity is 1 and the equilibrium constant is thus the solubility product, K_{sp}:

$$K_{sp} = [Ca^{2+}].[CO_3^{2-}]$$

The value of K_{sp} for calcite and aragonite in pure water is very low. In seawater K_{sp} is higher by several orders of magnitude, because of the presence of ions such as PO_4^{3+}, SO_4^{2+} and Mg^{2+}. Absolute solubility products are very difficult to measure in seawater because of the presence of these ions, but the apparent solubility products (K'_{sp}) of both calcite and aragonite increase with decreasing temperature (Figure 5.9) and increasing pressure. At all temperatures aragonite is more soluble than calcite.

The precipitation of carbonate to build a skeleton by aquatic invertebrates requires metabolic energy. The temperature dependence of carbonate solubility

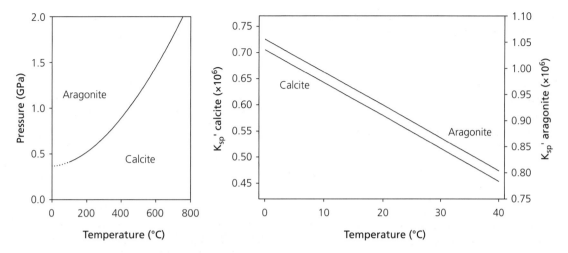

Figure 5.9 Calcite and aragonite solubility. A (left panel): phase diagram showing phase space where either calcite or aragonite is thermodynamically stable. Solid line shows best fit to experimental data, and dotted line an extrapolation to modern oceanic temperatures. B (right panel): apparent solubility coefficients for calcite (lower line, left axis) and aragonite (upper line, right axis) in seawater. Note the different scales for calcite and aragonite; over the physiological temperature range aragonite is more soluble than calcite[20].

indicates that more energy must be provided at lower temperatures (where carbonate is more soluble) than at higher temperatures: it is therefore energetically more expensive for an organism to build a carbonate skeleton in polar waters than it is in tropical waters. It does not matter what physiological mechanisms the organism uses to control the kinetics of precipitation, Hess' Law (see Chapter 2) tells us the energy required is dictated solely by the difference prevailing between the thermodynamic conditions in the surrounding seawater and those in the fluid precipitating the carbonate.

The increasing energetic cost of secreting a carbonate skeleton at lower temperatures means that in some calcifying organisms the mass and composition of the skeletal carbonate varies geographically. The tendency for molluscs from polar and deep-sea environments to be small and thin-shelled has long been remarked upon, and a classic quantitative study was an examination of shell size in shallow-water gastropods from the western north Atlantic (Figure 5.10). This study demonstrated a marked increase in the relative mass of the shell towards warmer waters. The mass of shell was quantified by a calcification index, CI, calculated as:

$$CI = \frac{\text{Dry mass of the shell}}{\text{Internal volume of the shell}}$$

This index provides a modicum of control for the fact that gastropods vary greatly in size between species. More recent work has confirmed this trend for gastropods, and also for echinoderms where the skeleton is internal. In addition there are geographical differences in the balance between the organic and inorganic constituents of the shell[21].

The precise metabolic cost of calcification in living organisms is exceptionally difficult to measure, but the available evidence suggests strongly that the variation in carbonate solubility with temperature is of considerable energetic significance. This in turn has ecological and evolutionary importance. For example it is generally only in tropical waters that we can find gastropods and bivalves with large or highly ornamented shells (Plate 8).

5.8 The hydrophobic interaction

While the physiology of the cell is dominated by the role of water as solvent and reactant, cell structure is determined by compounds that are almost insoluble in water, namely lipids. Not only do lipids form the core of the outer membrane that defines the cell's size and shape, but they also provide an internal architecture (the endoplasmic and sarcoplasmic reticulum) and delimit organelles such as mitochondria, chloroplasts and the nucleus.

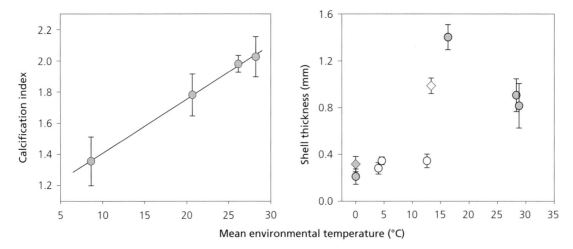

Figure 5.10 Calcification and temperature in marine invertebrates. A (left panel): relationship between calcification index and mean annual surface water temperature for 105 species of shallow-water marine gastropods from the western Atlantic Ocean. Data show mean and 95% confidence intervals for mean. B (right panel): relationship between shell thickness (mm) for an individual of standard size and mean environmental temperature in gastropods (family Buccinidae, circles) and echinoids (family Echinidae, diamonds). Data show mean and 95% confidence intervals, with northern hemisphere species in white and southern hemisphere species in grey[22].

The driving force in the construction of lipid membranes is the *hydrophobic interaction* (also known as the *hydrophobic effect*), which is the tendency for nonpolar substances to aggregate in water. The hydrophobic interaction is what drives the separation of oil and water, and it is also important in the folding of newly synthesised proteins.

The Gibbs energy change for a nonpolar compound dissolving in water, ΔG_S is given by:

$$\Delta G_S = \Delta H_S - T\Delta S_S$$

where ΔH_S is the enthalpic contribution to the Gibbs energy change, ΔS_S the change in entropy and T absolute temperature (we will discuss the derivation of this important equation in Chapter 7). For hydrocarbons of moderate to large size, ΔH_S ~0. Under these circumstances, ΔS_S dominates the Gibbs energy change ($\Delta G_S \sim -T\Delta S_S$) and the hydrophobic interaction is an entropically driven process. This change in entropy arises because the arrival of a hydrophobic molecule in water disturbs the conformational entropy of the water surrounding that molecule very significantly.

The precise molecular details underpinning the change in entropy are subtle. The simplest way to picture the hydrophobic interaction is that it minimises the Gibbs energy of the system (that is, it maximises the entropy) by reducing the surface area of contact between hydrophobic molecules and the surrounding water. The assembly of lipid molecules in water is thus not driven by an attraction between the lipid molecules, or by repulsion of the lipids by the water (as is sometimes portrayed); it is simply an entropy-driven process that minimises Gibbs energy.

5.8.1 The hydrophobic interaction builds membranes

The phospholipid molecules that provide the primary structure of biological membranes combine a large hydrophobic portion formed by two fatty acid chains, with a hydrophilic nitrogen base attached to a glycerol core through a phosphate group (Figure 5.11).

In a typical biological membrane the base is generally choline, ethanolamine or serine. All of these are electrically charged and because they contain nitrogen they form hydrogen bonds with the surrounding water. The hydrophobic interaction means that these molecules will align themselves with the hydrophobic tails forming the inner core of the membrane, and the charged bases facing outwards towards the surrounding water[23].

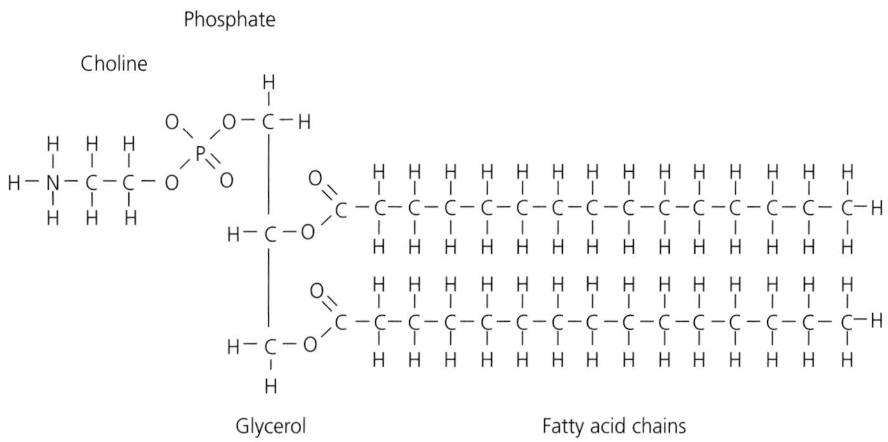

Figure 5.11 A typical membrane phospholipid, phosphatidyl choline, showing the hydrophobic fatty acid chains and the polar choline head-group attached to the glycerol backbone.

5.8.2 Folding of proteins

The hydrophobic interaction is also of enormous importance in the folding of newly synthesised proteins. The correct three-dimensional structure (the *native structure*) is essential to a protein's function. A misfolded protein is generally inactive, and in some cases may even be toxic (as in a number of neurodegenerative diseases). A newly synthesised protein leaves the ribosome as an unfolded polypeptide or a random coil. This polypeptide lacks any stable three-dimensional structure and folds spontaneously either during, or immediately after, its synthesis. The native structure of a protein is determined by its amino acid sequence and during folding the amino acids interact with each other. This process depends on the local environment (pH, presence of ions and salts, temperature) and is aided by chaperone proteins.

The most stable structure for a protein is that with the lowest Gibbs energy. For this to be the most likely outcome under the conditions under which the protein folds the native structure needs to be unique, stable and kinetically accessible. This is known as Anfinsen's dogma (or the thermodynamic hypothesis), and appears to be true for many small globular proteins[24].

The hydrophobic interaction is crucial to folding, because the lowest Gibbs energy is achieved by minimising the number of hydrophobic side-chains exposed to water (Figure 5.12). The folded structure of the native protein is thus dictated by the nature of the amino acid residues in the primary structure. Some amino acids (for example arginine or glutamate) are charged and hence form strong electrostatic interactions with the surrounding water molecules, whereas a larger number (for example histidine, asparagine or tyrosine) are polar and thus form hydrogen bonds with nearby water molecules. A significant contribution to protein stability includes the formation of hydrogen bonds within the protein molecule and the five hydrophobic amino acids (alanine, isoleucine, leucine, phenylalanine and valine) are therefore typically buried in the core of the protein.

For proteins that are embedded in membranes, the principles of folding are similar but the distribution of residues is different. In proteins spanning the membrane, that part of the protein embedded in the membrane tends to be highly hydrophobic, so that it is held within the membrane, whereas the portions of the protein that are exposed to the aqueous environment on either side of the membrane tend to be more hydrophilic.

5.8.3 Temperature and the hydrophobic interaction

We have seen above that at typical ecological temperatures, the enthalpy of transfer of a non-polar compound into aqueous solution is negligible but

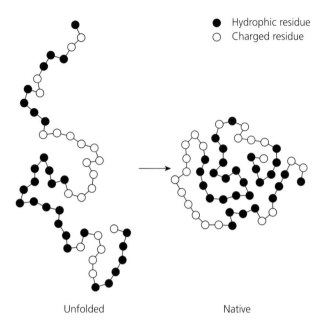

Figure 5.12 A schematic illustration of the role of the hydrophobic interaction in the folding of a newly synthesised polypeptide into the protein native state. Hydrophobic amino acid residues are shown in black, and hydrophilic (charged or polar) residues in white. The state of minimum free energy is reached when the hydrophobic residues are buried in the protein core.

the entropy change is large and negative. As temperature increases, the entropic contribution also increases, and the hydrophobic interaction effect therefore gets stronger. Conversely, as temperature decreases the hydrophobic interaction also decreases, and this decrease in the strength of the hydrophobic interaction is a major cause of cold-denaturation in proteins. It may also influence the likelihood of successful folding in proteins synthesised at different temperatures, a topic we will return to in Chapter 13[25].

5.9 Water in cells

Biochemists typically work in the realm of the very small or the utterly disintegrated[26]. We study how cells function by smashing them up, extracting and purifying the component of interest and then redissolving this in water or a suitable dilute aqueous buffer. This is convenient, and sometimes the only possible approach. Indeed much of what we know about how macromolecules function in cells comes from data obtained in this way. For example, the foundations of our understanding of how temperature affects enzyme catalysis stems almost entirely from studies of purified enzymes in dilute buffers.

However the cell is not like this. Although water is the major component of living cells the cytoplasm is very far from being a dilute aqueous solution of ions, metabolites and proteins. The cell is crowded with molecules and the space between these is often small[27]. The cytoplasm is much more like a colloid than the dilute solution of the physiologist's test tube (Plate 9) and this colloid-like nature has significant consequences for cell survival at sub-zero temperatures. There are thus two features of water in cells that are important in understanding how cells respond to a temperature. The first is the interaction between water and the other molecules in the cell and the second is the crowded nature of the cell.

The ions, small metabolites, macromolecules and membrane surfaces in cells are all electrically charged. Because of this they interact with the polar water molecules that surround them to form a hydration layer of highly structured water (Figure 5.13).

In the example shown here, the hydration shell around a sodium ion arises because of an electrostatic attraction between the positively charged ion and the partial negative charges from the lone pairs of electrons on the oxygen of the water molecules. A hydration shell also forms around the negatively charged chloride ion, although here the electrostatic attraction is with the partial charges on the hydrogen atoms of water.

The key biological ions (Na^+, K^+, Ca^{2+}, Cl^-) all have six water molecules in their inner hydration

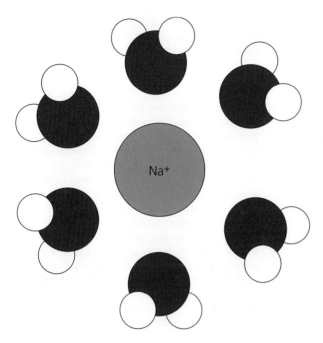

Figure 5.13 The first hydration shell of water around a sodium ion.

shell. The influence of the ion extends beyond this inner hydration shell, though decreasing in intensity with distance. Weakly hydrated ions may be regarded as having two or three hydration shells, whereas in more strongly hydrated cations the effects may extend out for as many as seven to nine hydration shells.

The behaviour of the highly structured water in these hydration shells differs from that of bulk water, and because of this it is sometimes referred to as vicinal water (although this term has somewhat gone out of fashion recently). The extent of vicinal water around ions or charged surfaces is not easy to assess, but recent estimates suggest that it may extend as much as 0.02–0.05 µm from a surface such as a membrane or protein. Since nowhere in a typical eukaryote cell is likely to be more than 0.005–0.01 µm from a surface, be it a membrane or a macromolecule, this suggests that all of the water in the cell must be structured by proximity to other cellular components. Or putting it the other way around, none of the cell's water can be regarded as behaving like bulk water[28].

Not all physiologists would agree with such an extreme view of the cell's water. Most biologists adopt the default view which dominates the textbooks, which is that cell water simply provides the environment for physiological processes, the dynamics of which are determined primarily by the properties of the macromolecules involved. There is, however, growing evidence that many biological components in the cell modify the structure and behaviour of the cell's water such that it differs in important ways from that of bulk water. To take one example, the osmotic response of cells indicates that ~20% of the cellular water is osmotically inactive, suggesting that a significant fraction of the cell's water does not behave as bulk water. This is clearly an area in need of resolution, for it has great significance for our understanding of how cells react to a change in temperature.

5.10 Summary

1. Liquid water is essential for life, and a metabolically active cell is ~70% water.
2. Water contains the highest density of hydrogen bonds of any known liquid. The physical properties of liquid water, and their temperature dependence, are dictated to a significant extent by the properties and dynamics of these hydrogen bonds.

3. From an ecological perspective, the important properties of liquid water are its high latent heats of fusion and vapourisation, high specific heat, low dynamic viscosity, high surface tension and its ionisation.
4. The solubility in water of oxygen, carbon dioxide and the calcium carbonate used to build skeletons in many invertebrates groups all increase with decreasing temperature.
5. The hydrophobic interaction is important in the formation of cellular membranes and the folding of proteins. Its strength increases with temperature, and this may be a factor in the cold-denaturation of cellular macromolecules.
6. The cell is extremely crowded with macromolecules. Coupled with the highly structured water close to membranes or protein surfaces and the hydration shells around ions, this means that the behaviour of water in cells differs significantly from that of bulk water.
7. The thermal behaviour of isolated cellular components studied in dilute aqueous buffers may not reflect accurately their behaviour in the intact cell or tissue.

It is worth leaving the final word to Philip Ball:

That the only solvent with the refinement needed for nature's most intimate machinations happens to be the one that covers two-thirds of our planet is surely something to take away and marvel at[29].

Notes

1. This quotation comes from Szent-Györgyi (1971); reproduced with permission of Johns Hopkins Press.
2. The key figures here are the pre-Socratic philosophers from the school at Miletus (a Greek colony in Asia Minor, in what is now modern Turkey): Thales (~624–~546 BC), who was regarded by Socrates as the first true philosopher in the Greek tradition, Anaximander (~610–~546 BC) and Anaximenes (~585–~526 BC). Thales viewed water as the origin of everything, and water was included as one of the four fundamental elements by the later philosophers Empedocles and Aristotle. This identification of water as fundamental to life was not simply a lucky chance, but based on careful observation of the world coupled with everyday practical knowledge of the necessity of water for plants to grow and animals to survive. The school of Miletus is usually regarded as the origin of the western tradition of philosophical exploration of the world. The word scientist was only coined in 1834 by the Cambridge philosopher of science William Whewell. An excellent introduction to ancient Greek speculation on the nature of the universe is Toulmin & Goodfield (1962), and Freely (2012b) gives a highly readable account of the flourishing of early Greek philosophy in Asia Minor. A more wide-ranging analysis of the symbolic place held by water in early civilisations is given by Marc Henry (2005).
3. Expressing water content as a proportion or percentage of body mass is not very sensible from a statistical perspective, as the mass of water occurs as a variable in both numerator and denominator. Furthermore, because this measure is insensitive to small changes in water content, it can obscure subtle changes in water content that have profound physiological and ecological consequences. But it is how such data are almost universally presented.
4. The data on cell water content in Table 5.1 are from Beilin et al. (1966), Albe et al. (1990), Aliev et al. (2002) and Ling (2004).
5. The data in Table 5.2 came from Öpik & Rolfe (2005) for plants, Clarke et al. (1992) and Kiørboe (2013) for gelatinous zooplankton (60 species from three phyla), Clarke (2008) for benthic marine invertebrates (six species from four phyla), Studier & Sevick (1991) for insects (137 species from 11 orders), Lyth (1982) for terrestrial molluscs (12 species of land slug, Pulmonata) and Schmidt-Nielsen (1975) for vertebrates.
6. The data in Table 5.4 are from Haynie (2001).
7. See Finney (1982) and Chaplin (2000, 2006) for technical discussions of water structure, and Ball (1999) for an excellent general introduction to water and its properties.
8. Modelling the hydrogen bond involves quantum mechanics, and theory predicts a quadratic temperature dependence of bond strength on temperature, as observed. Assuming the transfer of one water molecule from the liquid to the gaseous state involves the breaking of two hydrogen bonds (water molecules at the surface can only form hydrogen bonds with other water molecules on the liquid side) then the bond energy at 25 °C (298 K) is 22 kJ mol^{-1}.
9. Both diagrams from Dougherty (1998), reproduced with permission of the American Institute of Physics.
10. The variation of the pH of pure water plotted from the equation of Hepler & Woolley (1973). The data for *Limulus* haemolymph are from Table 2 in Howell et al. (1973).
11. The concept of a dimensionless number capturing the balance between viscous and inertial forces was

introduced by George Gabriel Stokes (1819–1903) in 1851, but the number is named after Osborne Reynolds (1842–1912), who popularised its use in 1883. For a lucid introduction to the relationship between aquatic organisms and their fluid environment see Steven Vogel's *Life in moving fluids* (Vogel 1981), and for excellent discussions of the relationship between animals and their physical environment in general see Alexander (1968) and (Denny 1993).

12. The plots in Figure 5.9 were from drawn equations or tabulated data in Vargaftik et al. (1983), Sengers & Watson (1986) and Ramires et al. (1995).

13. For a review of suspension in planktonic organisms see Smayda (1970).

14. This is the cohesion-tension theory for the ascent of xylem sap in plants, proposed independently by John Joly and Henry Dixon in 1894 and by Eugen Askenasy in 1895, and still the dominant explanation today. For a modern description see Öpik & Rolfe (2005).

15. This equation captures the relationship as expressed in the usual formulation of Henry's Law: *At a constant temperature, the amount of a given gas that dissolves in a given type and volume of liquid is directly proportional to the partial pressure of that gas in equilibrium with that liquid.*

16. The oxygen and carbon dioxide data plotted in Figure 5.12 come from Riley & Skirrow (1975). Carroll et al. (1991) provide the most recent definitive data for carbon dioxide solubility, though as with most literature data in this field, the data are for a partial pressure of CO_2 of 1 atmosphere. This results in much more CO_2 being present in solution than is the case when pure water is equilibrated with a normal atmosphere, where the partial pressure of CO_2 is only 39 Pa (because CO_2 is a trace gas in the atmosphere, with a mole fraction of 0.00039, or 0.039% by volume).

17. The definitive publication is Ruud (1954), though the original description is Nybelin (1947), which probably makes Matthews (1931) the first published record of Antarctic icefish. The physiology of *Chaenocephalus aceratus* was studied by Ralph & Everson (1968) and Holeton (1970). Recent studies of the evolutionary path to the haemoglobinless condition are Sidell et al. (1997), O'Brien & Sidell (2000), Moylan & Sidell (2000) and Near et al. (2006).

18. Geologists recognise anywhere between five and eight periods of reef-building in geological history, depending on how a reef is defined (some geologists consider that structures built by microbial organisms such as cyanobacteria do not qualify as reefs). Rachel Wood (1999) gives a thorough account of reef structure and biology, and Andrew Knoll (2003) reviews the evolution of biological calcification.

19. Calcite is trigonal in crystal form, and may contain varying amounts of Mg^{2+}. Conventionally low-Mg calcite contains <4 mole% $MgCO_3$, and high-Mg calcite contains >12 mole%. Aragonite is orthorhombic. The hexagonal form vaterite is of minor importance in biology, as is amorphous $CaCO_3$. The hexahydrate ikaite is monoclinic and is stable only in near-freezing seawater, where it can form spectacular underwater structures (for example in Ikka Fjord in SW Greenland). The mineralogy of bivalve shells is discussed by Lowenstam (1954) and Carter (1980). For a general discussion of the distribution of calcite and aragonite in marine organisms in relation to temperature see Vermeij (1978). The evolutionary history of calcification through geological time is discussed by Knoll (2003). The standard Gibbs energy change for the conversion of aragonite to calcite is small, −1.05 kJ mol^{-1}, and the entropy increase is 3.7 J (mol K)$^{-1}$.

20. Calcite and aragonite solubility in seawater plotted from equations given by Edmund & Gieskes (1970), solved for a chlorinity of 19.73 (equivalent to a salinity of 35: Lyman 1969).

21. Latitudinal variations in the balance between the organic and inorganic components of marine invertebrate shells are discussed by Watson et al. (2012). Richard Palmer (1992) made a careful assessment of the cost of calcification in molluscs. The thin shells of cold-water molluscs were documented by David Nicol (1967), and the influence of predation on molluscan shell morphology is discussed thoroughly by Geerat Vermeij (1978).

22. Calcification index data are from Graus (1974). The data are for, in order of increasing mean annual seawater temperature, Woods Hole, Massachusetts (15 species), Beaufort, North Carolina (28 species), southern Florida (69 species) and Puerto Rico (25 species). The shell thickness data are replotted from Watson et al. (2012). Both plots reproduced with permission from John Wiley and Sons.

23. This is actually the structure of membranes in bacteria and eukaryotes (Singer & Nicolson 1972). Although the membranes of archaeans are also built from lipids, these are quite different lipids. Originally it was thought that these different lipids reflected the extreme environments in which some (but by no means all) archaeans live, but now it is recognised as a result of a deep evolutionary divide between bacteria and archaeans. Nick Lane gives an excellent account of modern ideas on the significance of these differences (Lane 2015).

24. The postulate comes from the Nobel Laureate Christian B. Anfinsen, from his work on the folding of ribonuclease A. The uniqueness requires that the amino acid sequence does not have any other configuration with a comparable Gibbs energy. If this were not so, a significant number of newly synthe-

sised proteins would fold into non-functional configurations. The stability requires that small changes in the local environment do not drive changes in the configuration or the protein's function would be unduly sensitive to small fluctuations in the cellular environment. The requirement for kinetic accessibility implies that the path from the unfolded to the native structure must be reasonably smooth. This last point is related to another postulate, Levinthal's paradox, which states that the number of possible conformations available for a given protein is so great that it would take more time than the universe has existed to explore all possibilities for even a relatively small protein (say 1000 amino acid residues). This tells us that the process of folding is such that many of these possible configurations cannot be achieved, and that thermodynamics drives the folding in a very particular direction. Prions and other amyloid diseases (such as Alzheimer's disease and Parkinson's disease) are an exception to Anfinsen's dogma, in that they are stable conformations of proteins that differ from the native state.

25. The temperature dependence of the hydrophobic interaction in protein folding was explored by Robert Baldwin. By examining the thermodynamics of the transfer of six different hydrocarbons to water he was able to parameterise a model capturing the temperature dependence of the hydrophobic interaction in protein folding. He was able to show that the entropy of transfer reaches zero at ~113 °C (Baldwin 1986). John Schellman suggested that the hydrophobic interaction was the dominant factor in protein folding (Schellman 1997), and Dias et al. (2010) explored the role of the hydrophobic effect in the cold-denaturing of proteins. Tanford (1980) gives an overview of the hydrophobic interaction.

26. Coulson et al. (1977).

27. For vivid images that bring the crowded nature of the cell to life, see David Goodsell's wonderful illustrations (one of which is reproduced here as Plate 9) in his book *The machinery of life* (Goodsell 2009).

28. For a discussion of vicinal water, see Drost-Hansen (1982). The biologist Jim Clegg has long argued that the traditional physiologist's view of cell water as being essentially bulk water is misleading. Taylor (1987) and Clegg & Drost-Hansen (1991) provide useful discussions of this contentious topic.

29. Ball (1999).

CHAPTER 6

Freezing

> The wave, over the wave, a wierd thing I saw,
> Through-wrought, and wonderfully ornate:
> A wonder on the wave—water become bone.
>
> The Exeter Book[1]

Life began in the sea and living organisms carry a reminder of this in their body fluids which are often similar to seawater in concentration, if not always in composition. Like seawater, these body fluids will freeze a little below 0 °C, and the avoidance of freezing is one of the greatest physiological challenges faced by living organisms. If a cell freezes, it dies[2].

While freezing temperatures are an obvious feature of many high-latitude or high-altitude habitats, freezing is a more widespread environmental challenge than might be expected. Even in subtropical areas, there may be occasional freezes: in Florida two days of unusually cold weather in January 1985 wiped out over 90% of the citrus fruit crop.

The global land surface has been divided by the United States Department of Agriculture into 13 plant hardiness zones based on the average minimum winter temperature. Ten of these zones have minimum temperatures below 0 °C, and these cover over half the terrestrial land surface. Freezing is indeed a widespread physiological challenge[3].

We will start by exploring some fundamental aspects of the freezing of water, and how these affect the ability of unicellular organisms to survive when their environment freezes. We will then see how these principles apply to multicellular organisms.

6.1 The freezing of water

As water freezes, the change in entropy releases latent heat, the volume increases and solutes are excluded from the ice. All of these have important consequences for organisms.

The freezing point is the temperature at which the solid phase (ice) and liquid phase (water) are in equilibrium (hence the more formal name of *equilibrium freezing point*). For pure water at atmospheric pressure the freezing point is 0 °C. Above the freezing point an ice crystal in water tends to melt, below the freezing point it tends to grow.

The freezing of pure water starts with an ice nucleus. In *homogeneous nucleation* the random motion of water molecules causes localised fluctuations in the density and orientation of the water molecules, and occasionally by chance a few molecules form a small cluster with the configuration of ice. The surface of this cluster, the ice nucleus, is effectively an interface between solid and liquid phases, and other water molecules can bind to this surface, allowing the cluster to grow. This only happens if the cluster exists for long enough for other water molecules to bind before the same random fluctuations cause it to disintegrate. The phase change from liquid water to solid ice, crystallisation, thus consists of two processes, nucleation and crystal growth.

Principles of Thermal Ecology. Andrew Clarke, Oxford University Press (2017).
© Andrew Clarke 2017. DOI 10.1093/oso/9780199551668.001.0001

Because it is based on the random fluctuations of thermal motion, homogeneous nucleation is a statistical phenomenon. As a result the onset of crystallisation is unpredictable and water can remain liquid below its equilibrium freezing point of 0 °C. This is a highly unstable state, almost universally referred to as *supercooled water*, though the better term is *undercooled water*[4]. Its discovery is traditionally ascribed to Gabriel Fahrenheit, who stumbled when carrying a flask of undercooled water and observed the spontaneous appearance of flakes of ice[5].

At 0 °C the critical size of an ice nucleus is 1–2 nm and it contains ~ 350 water molecules; this critical size decreases rapidly as the temperature is lowered, increasing the likelihood of nucleation. At temperatures above 0 °C, these nuclei are disrupted before they can grow further, but below 0 °C the thermodynamics favour continual growth. The nucleation of water is complex; for ecologists the key factors are that water can remain liquid well below its equilibrium freezing point, and the chances of spontaneous nucleation are lower the smaller the volume. The absolute limit at which the liquid state can exist is the *homogeneous nucleation temperature*; for pure water at atmospheric pressure this is −38.5 °C[6].

Nucleation can also be initiated by particulate matter present in the water. The particle has to be larger than ~ 10 nm, and bind water molecules in a configuration that other water molecules recognise as ice. This is *heterogeneous nucleation*, and it is the reason that most natural water bodies start to freeze at ~ 0 °C, since almost all will contain sufficient nucleators to initiate freezing at high subzero temperatures.

6.1.1 Ice phases

Ice that forms at atmospheric pressure and at high subzero temperatures is hexagonal ice. There are about eighteen different crystalline phases of ice and three amorphous (that is, non-crystalline) phases. Some of these are denser than liquid water, and hence would not float, but they are all only stable at temperatures and pressures that are far outside those normally experienced by living organisms, and hence need concern us no further[7].

6.1.2 The influence of solutes

It is only pure water that has an equilibrium freezing point of 0 °C. The presence of any dissolved substances (solutes) will lower the freezing point because of the increase in entropy[8]. For an ideal solution, the depression of the freezing point, ΔT_F, depends only on the solute concentration, in a simple linear relationship:

$$\Delta T_F = K_F b i$$

where b is the molality of the solute, i is the van't Hoff factor (the number of ions per individual molecule of solute) and the cryoscopic constant, K_F, is a characteristic of the solvent. For water, K_F is 1.86. The van't Hoff factor is necessary to allow for the fact that many salts ionise when dissolved in water; for example 1 molecule of NaCl produces 2 ions, Na$^+$ and Cl$^-$, and here the van't Hoff factor is 2. For BaCl$_2$ the van't Hoff factor is 3, and for sucrose (which does not ionise) it is 1[9].

This effect means that seawater has an equilibrium freezing point of between −1.86 and −1.92 °C at the surface, depending on its salinity. Because the freezing point is also influenced by pressure, the freezing point of seawater decreases with increasing depth (Box 6.1).

6.2 Ice in the natural environment

All ice that forms in the environment has a hexagonal crystal lattice, but the macroscopic form of that ice differs depending on the environment in which it forms.

6.2.1 Ice in water bodies

When freshwater lakes, streams or rivers start to freeze, the ice forms initially at the surface, mainly because in winter the air is often colder than the water. This ice typically has three layers, reflecting different modes of formation. The primary ice layer is the initial complete layer at the surface. The secondary layer comprises ice that grows downward from the primary layer, driven by the temperature gradient between the primary ice layer (itself cooled by the air) and the water below. As this secondary ice grows, it generally does so as long

Box 6.1 The freezing point of oceanic water

The presence of salts in seawater lowers its freezing point. The freezing point is also affected by pressure, and so it decreases with increasing depth in the oceans (Figure 6.1).

The mean depth of the ocean is about 3686 m, and here the freezing point is −4.7 °C. Over the abyssal plain, assuming a typical depth of 5000 m, the freezing point is −5.7 °C and at the deepest point yet known, the Challenger Deep in the Mariana Trench, at 10 990 m, it would freeze at −10.2 °C. Seawater at these depths never freezes, because the temperature is always above 0 °C.

Because the temperature of seawater below the surface mixed layer (typically the top 100 m of the ocean) is fairly constant and above 0 °C, in most of the ocean ice does not form and freezing is not a problem that organisms ever have to face. The exceptions come in those areas of the world where sea-ice is forming, as this can produce large volumes of cold (~ −2 °C) bottom water and also small volumes of highly concentrated brines which are very cold (possibly as cold as −20 °C).

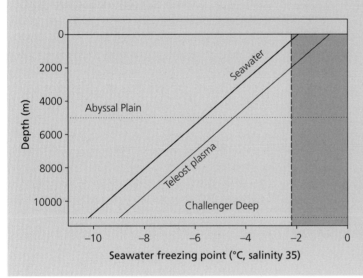

Figure 6.1 Variation in equilibrium freezing point of seawater (salinity 35) and teleost blood plasma with depth. The solid line shows the surface and the two dotted lines represent the mean depth of the abyssal plain and the deepest point known in the ocean, the Challenger Deep. The grey area shows the range of subzero temperatures recorded in surface oceanic waters[10].

columnar crystals oriented vertically. This layer can constitute the bulk of the ice growing on lakes and rivers. Finally there may be a third layer, on top of the primary layer, resulting from the freezing of precipitation or water that has flooded onto the ice.

In seawater, ice nucleates at the surface when the overlying air is sufficiently cold. This early stage is called grease ice, reflecting the characteristic appearance of the sea surface as freezing starts. The formation of a complete covering of ice on seawater is a highly complex process which varies depending on factors such as air temperature, wind and wave action. Once there is a complete cover, ice grows downwards as columnar crystals similar to that in freshwater. A superficial layer of ice can form on top of the primary layer, from frozen precipitation or meltwater.

As ice forms, it excludes from the growing crystal any particles or solutes that are present in the bulk water. In seawater, this process produces brine and under equilibrium circumstances the composition of this brine depends only on its temperature: at −6 °C the brine will have a salinity of 100, at −10 °C it will be 145 and at −21 °C it will be 215 (about six times the concentration of seawater). This happens because as temperature falls, more and more of the seawater is in the form of ice, leaving the salts dissolved in an ever-smaller volume of water.

The brine formed by exclusion of salts from the ice is confined to pores and channels between the ice crystals, and this gives oceanic secondary ice a very different structure from that in freshwater. These brine channels and pockets are an important habitat for organisms (Box 6.2). At times this very

Box 6.2 Life in ice and snow

In the sea-ice fields of polar regions, the most obvious signs of life are the marine mammals and seabirds that use sea-ice as platforms to rest or breed. But there is also life within the ice, as the brine channels that form as ice freezes out from seawater provide a habitat for a wide variety of organisms. These include primary producers such as diatoms and other phytoplankton, as well as bacteria that utilise the carbon compounds these exude, viruses and a range of predators.

The increase in temperature through sea-ice, from the cold surface to the warmer interface with the seawater beneath, leads to gradients in the size of brine channels and salinity of the brine they contain (Figure 6.2).

Although organisms can be found through the ice, most life is found close to the seawater. This can easily be seen when floes are turned over by the passage of a ship and the underside of the ice is revealed as coloured deep green or brown; the brine is less concentrated here, and hence osmotic stresses are less intense. Towards the upper surface the brine channels are very narrow, restricting movement of organisms and nutrients, and brine concentrations can be very high. Here the fluids may also be extremely viscous because of high concentrations of extracellular polymers excreted by the phytoplankton.

Generally we can only study these organisms by melting the ice, and this changes their environment drastically. At present we do not know to what extent organisms in the most concentrated brines are growing and dividing, or simply waiting for the ice to melt before resuming their life cycle[11].

The underside of sea-ice is a surprisingly rich and active system. The abundance of phytoplankton makes this a valuable resource for grazers able to utilise this unusual habitat (such as euphausiids) and predators of these grazers, such as fish. The architectural complexity of the habitat also provides a refuge for crustaceans and fish from their own predators.

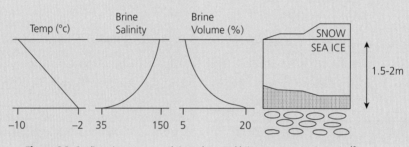

Figure 6.2 Gradients in temperature, brine volume and brine concentration in sea ice[12].

cold and highly saline brine leaks from the ice into the seawater below. Because this brine is very cold, the seawater freezes around it and slowly a column of very cold brine encased in ice extends downwards, a so-called brinicle. Where this comes into contact with marine organisms, it causes them to freeze (because the brine has a temperature below the freezing point of the body fluids).

Ice can also form occasionally within oceanic water. For this to happen, the temperature of the water has to drop below the freezing point of seawater at the particular depth. Typically this is a feature of shallow water (where it forms anchor ice, often encasing and killing benthic organisms) or very cold seawater flowing beneath ice shelves at depths of ~ 200 m or more, where ice forms as platelets up to 15 cm in diameter. These platelets rise through the water column, and they can form dense layers as they accumulate beneath surface ice. Here they can freeze into the ice and increase its thickness significantly[13].

6.2.2 Ice in the atmosphere: snow

Ice also forms in the atmosphere, and may eventually fall as snow. Freezing of water in the atmosphere is most commonly the result of heterogeneous nucleation triggered by particles also present in the air. The ice crystals can grow by direct transfer of water molecules from the vapour phase, which is why snowflakes often display the six-fold symmetry of the hexagonal ice lattice.

Freshly fallen snow can have a density as low as 50 kg m^{-3} because of the air trapped within, and a layer of such snow can provide an insulating blanket allowing many terrestrial organisms to continue with their lives on the ground beneath. Where snow accumulates year on year it gradually packs down under its own weight. It is classified as ice once it exceeds a threshold density of 830 kg m^{-3}. Some of the air remains trapped as bubbles within the ice, and this trapped air has proved to be important in studies of climate by providing direct evidence of the atmosphere at the time the snow was deposited.

6.2.3 Ice in the terrestrial environment

Nucleators are so widespread within soil that ice will start to form as soon as the temperature drops to the freezing point. This only happens at the shallow depths to which cold can penetrate, and so many organisms can avoid the risk of freezing by retreating deeper within the ground. Above ground water vapour will condense from the air as frost onto any surface that cools below the dew point of the surrounding air. This is particularly frequent on clear cold nights when exposed objects such as plant leaves lose heat by radiation faster than it can be supplied by convection (wind) or conduction from the rest of the plant, and hoar frost forms[14].

6.3 Freezing and organisms: some general principles

Many tropical and temperate organisms are injured or killed when cooled, even at temperatures well above the freezing point. This is *chilling injury* and it affects any organism exposed to a non-freezing temperature below that to which its physiology is adapted. The causes of chilling injury can be many and varied, and differ from organism to organism[15]. Physiological adaptations that help minimise chilling injury, such as changes to proteins and cellular membranes, are not necessarily the same as those needed to counter the challenge of freezing, and we discuss them in Chapter 7.

Once ice is present in the environment, organisms are faced with a whole new suite of challenges. Complex multicellular plants and animals are able to ameliorate these challenges to some extent, as we shall see, so it is easier to discern the fundamental principles in simpler systems such as bacteria, archaea and unicellular eukaryotes which are exposed to the environment in a way that the cells within complex organisms are not.

As an example we can see what happens when the unicellular alga *Chlamydomonas nivalis* is exposed to freezing temperatures. This organism is well known to those who walk in the mountains, as it grows on snow banks which it colours red or green, and is commonly referred to as the snow alga. If snow algae are frozen in liquid nitrogen (−196 °C), their survival is found to depend on the rate at which the cells were cooled (Figure 6.3).

In this experiment peak recovery was observed at a cooling rate of 3 K min^{-1}; survival was lower

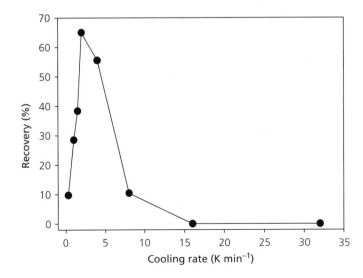

Figure 6.3 Recovery following thawing of snow alga, *Chlamydomonas nivalis*, cooled to liquid nitrogen temperature (−196 °C) at different rates[16].

at faster rates of cooling and, perhaps surprisingly, also at slower rates of cooling. This pattern, often referred to as the 'inverted-U', is typical of many unicellular organisms, including bacteria, protozoa and also plant protoplasts (isolated individual cells of higher plants that have had their cell walls removed). While the general pattern is observed in many cell types, the optimal cooling rate for survival is enormously variable: it is 1 K min^{-1} or less for mouse stem cells and embryos, ~10 K min^{-1} for the yeast *Saccharomyces cerevisiae* and > 1000 K min^{-1} for human erythrocytes.

Snow algae in the natural world would never be exposed to liquid nitrogen temperatures, nor would they ever likely experience the rates of cooling used in this study, so these results are not immediately relevant to ecology. The general pattern observed in cryobiological experiments does, however, point to some fundamental principles that are relevant to ecology.

The most influential explanation for the general pattern exemplified by the *Chlamydomonas* study came from Peter Mazur, who proposed a *two-factor hypothesis* of cooling rate injury[17]. In cryopreservation experiments the cells are contained in a small volume of fluid in which ice nucleates and hence the remaining fluid increases in osmotic strength as solutes are excluded from the ice. Mazur's hypothesis was that at slow rates of cooling the high osmotic strength of the bathing medium pulls water from the cells, causing damage by shrinkage and an increase in the concentration of the cytosol, whereas at high rates of cooling cells were killed by the formation of intracellular ice. Survival was thus highest at intermediate rates of cooling (Figure 6.4).

This hypothesis has two important implications. Firstly it is not the low temperature of storage itself that is important, but rather damage caused at intermediate temperatures (typically in the range −15 °C to −60 °C), a zone that cells must cross twice, first as they are cooled and then again as they are warmed. Secondly an important factor is the cell membrane and its permeability to water. The likelihood of intracellular ice formation in survival can be modelled in terms of the rate of cooling, water permeability of the cell and the extent of undercooling of intracellular water, and the predicted incidence of intracellular ice matches the observed survival data quite well[18]. Despite this, many unanswered questions remain about exactly how intracellular ice is formed.

Our understanding of what precisely causes cellular damage at slow rates of cooling is even less complete. For very slow rates of cooling we can assume that the intracellular environment is in equilibrium with the bathing medium, and as cooling proceeds the internal cell environment also becomes more concentrated. It is thought that under these circumstances the cell is damaged by

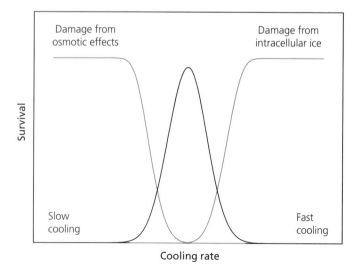

Figure 6.4 A conceptual model of Mazur's two-factor hypothesis of why the optimum survival of isolated cells from exposure to liquid nitrogen temperatures is often at intermediate rates of cooling. Here the 'inverted-U' survival frequency is represented by a Gaussian curve (black), and the frequency of damage by osmotic effects and intracellular ice by sigmoidal responses (grey); both are arbitrary choices to illustrate the principle.

the toxicity of the increasingly concentrated solutes (notably electrolytes), unwanted interactions between proteins and perhaps also by physical changes within the cell as water is withdrawn. Other suggestions have included damage to intracellular components caused by packing, and destabilisation of the cell membrane on shrinkage. These are very general statements and in none of these cases are we yet able to point to a more specific cause of damage[19].

6.4 Cryobiology and ecology

Cryobiology is not ecology, but it has provided important clues as to what is important in adaptation to a freezing environment in the natural world. In particular it has demonstrated the key role played by the cell membrane and the importance of the movement of water.

An important aspect of the natural environment is that organisms living in polar or alpine environments frequently have to content with multiple freeze/thaw cycles. The significance of repeated freeze/thaw cycles was identified in the eighteenth century by Gilbert White, who commented:

. . . I would infer that it is the repeated melting and freezing of the snow that is so fatal to vegetation, rather than the severity of the cold[20].

In polar and alpine habitats, as well as many temperate areas in winter, freeze and thaw can be a daily event. To take just one example, microclimate recording at the surface of a moss bank in Antarctica detected 30 freeze/thaw events in a single month alone, essentially one each day (Figure 6.5). These temperature fluctuations will be experienced by any organisms living on the moss surface, such as tardigrades, nematodes and microarthropods, as well as the microbial fauna. All of these must be resilient to frequent freeze/thaw: even if each event resulted in only 5% mortality, < 10% of the population would remain after 50 freeze/thaw events. This implies a resilience to freezing in the natural world that exceeds the survival typically observed in cryopreservation.

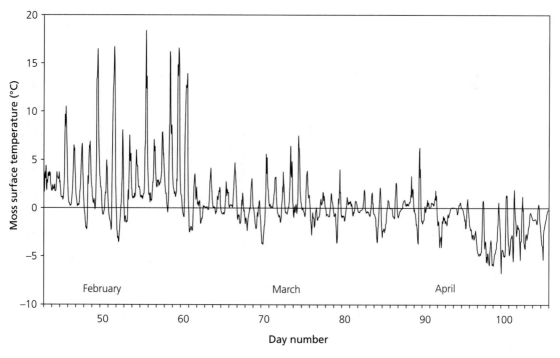

Figure 6.5 Temperature recorded from the surface of a moss bank at Anchorage Island, Antarctica (67° 36′ S, 68° 13′W) over three months in 2006[21].

Freezing-tolerant organisms can experience significant removal of water from cells and many organisms that tolerate extensive dehydration are incidentally also tolerant of freezing, even though they may never naturally experience freezing temperatures. Examples are organisms living in saline habitats, and many microscopic invertebrates such as tardigrades, rotifers and nematodes that are characteristic of terrestrial habitats that may dry out[22]. Tolerance of dehydration may be a very ancient feature of organisms, as the earliest photosynthetic organisms and their associated heterotrophs likely lived in shallow habitats subject to intermittent drying out. The acquisition of tolerance to freezing may simply be building on existing ancient capabilities, rather than something special in itself.

6.5 Single cells in the environment

Much of the diversity of life exists as single cells, unobserved and unremarked. In addition to all bacteria and archaeans, a great many eukaryotes also exist as unicells. In many cases there is a cell wall or other protective structure between the environment and the cell membrane itself, but because the cell has to exchange nutrients and waste products with its surroundings, these protective structures must be permeable. All single-celled organisms are thus affected directly by changes in their external environment. If the external environment cools sufficiently for ice to form, unicellular organisms are exposed to two potential threats: ice formation within the cells, and damage associated with osmotic stress (dehydration and cell shrinkage). In understanding how organisms respond to these stresses ecology has much to learn from the work of cryobiologists attempting to preserve cells and tissues.

6.5.1 Intracellular ice formation

Intracellular ice can form when the cooling rate is sufficiently high that the cell cannot maintain osmotic equilibrium with its surroundings, but this generally requires much faster cooling than is typical of the natural environment. For example in bacteria, intracellular ice only forms when cells are cooled rapidly (>100 K min^{-1}) in dilute aqueous media, and in the yeast *Saccharomyces cerevisiae* intracellular ice formation requires cooling rates faster than 20 K min^{-1}. Cooling rates in the environment rarely exceed 1 K min^{-1}, and most aquatic habitats change temperature even more slowly. While some specialised habitats such as rock or leaf surfaces can change temperature more rapidly it would seem that intracellular ice formation is rarely, if ever, a problem for single-celled organisms outside the laboratory[23].

An exception to these generalisations is cloud droplets, which because of their small size can undercool to the homogeneous nucleation temperature. Cloud droplets thus offer an environment, possibly the only such environment on Earth, where microbial cells may continue to metabolise and grow, albeit slowly, at temperatures as low as ~−40 °C. The ecological importance of this is that viable microbial cells within clouds can be dispersed globally[24].

Cells in undercooled cloud droplets are, however, extremely susceptible to being killed by intracellular ice formation should the droplet nucleate, when they are subject to extremely fast cooling. This fate can be avoided, however, if the droplet nucleates at a high subzero temperature when it is only slightly supercooled (that is closer to 0 °C than to −40 °C). This may even be induced by the bacterial cells themselves, mediated through ice-nucleating proteins on the external surface[25].

6.5.2 Ice-nucleating proteins

Bacteria that promote ice nucleation were discovered by Steven Lindow and colleagues when exploring the causes of frost injury in agricultural plants. In frost-sensitive plants, ice forms in supercooled extracellular water between −2 °C and −5 °C and then propagates through the plant. This was shown to be related to the presence of the bacteria *Pseudomonas syringae* and *Erwinia herbicola* on the leaf surface. Where these bacteria were absent, plants did not suffer damage until about −8 °C. The bacteria express one or more proteins on the outer membrane which act as a heterogeneous nucleation site. These proteins are anchored in the membrane, and appear to act by exposing a central core of a highly repetitive domain which hydrogen bonds to water molecules and aligns these in an ice-like arrangement, thereby providing a seed for heterogeneous nucleation.

At about the same time as the discovery of the activity of *Pseudomonas syringae*, ice-nucleating agents were also identified in the haemolymph of beetles from the mountains of southern California. Subsequently ice nucleation has also been reported more widely from insects and also from plants, though relatively few of these nucleators have been isolated and characterised. Ice nucleation has also been reported from some intertidal molluscs, though here it is not always clear whether the cause is an incidental bacterium or a deliberately expressed nucleator[26].

6.5.3 Cellular dehydration

In habitats such as running streams, extended water bodies or snow banks, cells will be exposed to moderately low temperatures but not to dehydration. Where the water body is restricted, such as within sea ice or in soil, then as ice forms the remaining water will increase in osmotic strength, pulling water from the cells. This movement of water is passive, that is the cells can do little about it and they act more or less as perfect osmometers. This dehydration can damage the cell, and it is now clear that cells respond to this challenge in a number of ways.

The first is the production of small molecular weight compatible solutes. The essential features of these are that they increase the intracellular osmotic strength, thereby reducing the tendency for water to leave the cell, and that they do not interfere with cellular structure or processes. The latter is critical because macromolecules such as proteins interact not only with cell water but also with the many small molecules in their environment. Compatible solutes include polyols, sugars, free amino acids, methylamines and urea. These do not act simply as colligative osmolytes, they also interact with macromolecules and cellular structures to stabilise these in the face of dehydration[27].

More recently it has been discovered that when faced with dehydration stress many cells synthesise a number of intrinsically disordered proteins known as LEA proteins (Box 6.3). The function of these proteins appears to be to protect macromolecules against irreversible aggregation as cell water content drops. Although first investigated in relation to a reduction in cell water driven by drought, it is now becoming clear that they may also be important in protecting cells against the dehydration induced by freezing of the extracellular environment.

Box 6.3 LEA proteins

In the early 1980s, Leon Dure and colleagues identified a number of proteins that accumulate in large amounts during the maturation phase of embryogenesis in cotton seeds, which they termed *Late Embryogenesis Abundant (LEA) proteins*. These proteins were characterised by their hydrophilic nature and high levels of particular amino acid residues (notably glycine, alanine, glutamic acid, lysine, arginine and threonine). Unusually for proteins, they proved to be highly unstructured in aqueous solution. These proteins were shown to be a subset of *dehydrins*, a group of proteins associated with dehydration stress in plants. LEA proteins were subsequently identified in the vegetative tissues of a range of plants exposed to dehydration, osmotic or low temperature stresses. They are now recognised as being ubiquitous in the plant kingdom, and have also been identified in bacteria, yeasts and terrestrial invertebrates such as nematodes, tardigrades and insects. It is quite possible that they are universal.

LEA proteins were long recognised as being associated with low tissue water content, and it was assumed they must have a protective function. Recently it has been shown that they have a chaperone-like role, protecting proteins and cellular structures against irreversible aggregation at low water contents. Furthermore they appear to operate synergistically with organic compounds that also accumulate as cell water decreases, such as trehalose.

LEA proteins are a field of active research at present, and they may prove to be a ubiquitous response to low cell water content, whether this is induced by dehydration or the osmotic challenge of freezing of extracellular water[28].

6.5.4 Vitrification

Vitrification (also known as the glass transition) occurs when a liquid begins to behave as a solid during cooling, but without any substantial change in molecular arrangement or thermodynamic state variables (pressure, volume, internal energy, entropy) (Figure 6.6). In bulk liquids, as temperature decreases all molecular motions (translational and internal) become progressively slower until a critical temperature is reached where there is insufficient energy for significant translational molecular motion to take place over a meaningful timescale. This is the vitrification or glass transition temperature, T_G, and it is defined operationally as the temperature at which viscosity exceeds 10^{12} Pa·s.

Free-living cells exposed to osmotic dehydration at the slow cooling rates typical of the natural environment can often maintain osmotic equilibrium with the surrounding fluid. As cell water decreases, the viscosity of the cytoplasm increases and organisms such as bacteria, archaeans or unicellular eukaryotes may then vitrify.

Intracellular vitrification is more complex than in simple bulk liquids. This is principally because, as we saw in Chapter 5, the interior of the cell is extremely crowded, approximating a colloid in physical structure. As colloids dehydrate they exhibit a sharp increase in viscosity and undergo a colloid glass transition. The vitrification of the interior of a free-living microbial cell exposed to ice in the external environment is primarily the result of dehydration and is closer in nature to the vitrification of a colloid than it is to the glass transition of bulk water[29].

In free-living unicells, and in the absence of cryoprotectants, dehydration driven by freeze-concentration of the external environment may trigger vitrification at temperatures between −10 and −25 °C. The vitrification temperature of a cell will vary with the precise composition of the cytosol, and this may allow the intracellular vitrification temperature to be adjusted by natural selection to match ecological circumstances (for example by varying the level of small cryoprotectant molecules).

The very high viscosity of the vitrified cell means that movement of oxygen and metabolites effectively stops. Under these conditions, metabolism ceases. The cell does, however, maintain its internal integrity, and metabolism can start again once the cell warms and rehydrates. This process carries the danger that once water molecules can move within the cell, any small ice nuclei present may grow rapidly (a process usually termed devitrification) and the consequent mechanical damage may be lethal. The presence of chaperone proteins such as dehydrins or LEAs, as well as cryoprotectants such as polyols may be important in ensuring a safe transition from the vitrified to the normal fluid state, though this is an area where more research is needed.

The possibility of vitrification of tissues at low temperatures in the natural world appears to have been first proposed by Basile Luyet in the 1930s, but this possibility was never followed up by

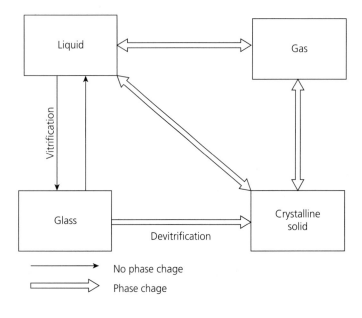

Figure 6.6 The relationship between vitrification (glass transition) and primary phase changes. Note that devitrification is not the reverse of vitrification (as might be thought) but rather a phase change to the crystalline solid[30].

ecologists. While vitrification has been explored extensively in the seeds and tissues of plants and more recently in bacteria, until recently it has not been considered in discussions of the low-temperature behaviour of isolated cells or multicellular animals. It may, however, be a far more widespread and important mechanism in low-temperature ecology than recognised so far[31].

6.6 Freezing and multicellular organisms

We now have a framework for exploring how more complex multicellular organisms respond to the challenge of ice in the environment. In these organisms most of the cells are exposed not to the external environment with its ice, but to the extracellular fluids. The critical factor is the composition of these extracellular fluids, and what the organism can do to influence the presence of ice there. These two factors suggest that more complex organisms can be divided into a small number of broad categories in terms of the nature of the freezing stress they experience (Table 6.1).

Table 6.1 Major categories of organism in relation to freezing stress.

Category	Comments
Unicellular organisms	
Bacteria	Cell membrane exposed to the external environment.
Archaea	Cell membrane exposed to the external environment.
Unicellular eukaryotes	Cell membrane exposed to the external environment.
Multicellular organisms	
Marine invertebrates	Extracellular fluids typically isosmotic with seawater; rarely exposed to temperatures below −2 °C (intertidal forms excepted).
Terrestrial invertebrates	May be exposed to very low temperatures (below −60 °C).
Teleost fish	Dilute blood, may be exposed to temperatures significantly below plasma freezing point (which is ~ −0.7 °C).
Amphibians, reptiles	May be exposed to very low temperatures (below −40 °C).
Endothermic vertebrates	Freezing danger for extremities; body core kept warm.
Plants	May be exposed to very low temperatures (below −60 °C).

We will look first at fish, partly because they are subject to a relatively small challenge from freezing temperatures, but also because it was work on polar fish that first established the existence of protein antifreezes.

6.7 Freezing and fish

Marine teleosts (bony fish) have a low blood osmolality, typically 350–450 mOsm kg^{-1}. This is much lower than in other marine fishes such as hagfishes, elasmobranchs and also the coelacanth, where values are usually ~ 1000 mOsm kg^{-1}. The reason is generally believed to be that the early evolutionary history of teleosts was in freshwater, where a low blood osmolarity confers a physiological advantage in reducing the metabolic costs of osmoregulation. While the earliest fossils of jawed fishes (the heavily armoured ostracoderms) are usually found in freshwater or brackish water deposits, the protochordate ancestors of vertebrates were almost certainly marine and the evolutionary reason for the dilute blood of teleosts is still a subject of debate[32].

The low concentration of ions in the blood of teleosts means that the equilibrium freezing point of their plasma is ~ −0.7 °C, and temperate or tropical fish cooled to −1 °C do indeed freeze. Despite many apocryphal stories of goldfish in garden ponds, no fish survives being frozen. This raises the obvious question of how polar fish avoid freezing when swimming in water at −1.9 °C.

The first clues came from work by Per Scholander and colleagues in Arctic Canada, who showed that in deep water, where there was no ice present because of the pressure, fish were able to survive in a permanently supercooled state[33]. In contrast, the freezing point of blood plasma of a shallow-water sculpin (probably *Myoxocephalus scorpius*) was sufficiently low to allow the fish to live in these polar waters without freezing. They were, however, unable to identify the cause of the lowered freezing point. This was done by Art DeVries, working with fish from Antarctica which live in intimate contact with ice (Plate 9). In a simple and elegant experiment he showed that the polar fish contained one or more proteins that acted as an antifreeze (Figure 6.7).

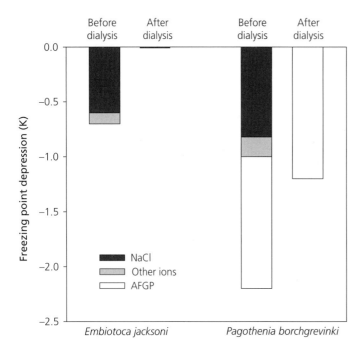

Figure 6.7 The experiment which showed that the antifreeze of Antarctic fish is a protein. In the temperate black perch (*Embiotoca jacksoni*) the blood plasma freezes at ~−0.7 °C, and at ~0 °C after dialysis. By comparison, in the Antarctic *Pagothenia borchgrevinki* the blood plasma freezes at −2.2 °C, but after dialysis it still freezes at −1.2 °C, because of the presence of an antifreeze glycoprotein (AFGP)[34].

DeVries compared the thermal behaviour of the blood plasma of a temperate (the black perch *Embiotoca jacksoni* from California) and a polar fish (*Pagothenia borchgrevinki*), before and after dialysis. In the temperate *Embiotoca* the blood plasma froze at −0.7 °C, and most of the freezing point depression was caused by inorganic ions (principally Na^+ and Cl^-). After dialysis the freezing point was −0.01 °C, indicating the dialysis removed essentially all of the components responsible for depression of the freezing point. In the Antarctic *Pagothenia borchgrevinki*, the freezing point was much lower, and the causative agent remained after dialysis; this indicated clearly that the antifreeze was a large molecule, and moreover one that remained after TCA had precipitated all of the previously known blood proteins. The antifreeze agent was isolated and shown to be a series of at least eight distinct glycoproteins comprised of a repeating sequence of the tripeptide alanine-alanine-threonine, with a disaccharide attached to the threonine through a glycoside linkage (Figure 6.8). The molecular weight varied from 2600 to 33 700 daltons, depending on the number of repeating units[35].

The fish fauna of the Antarctic continental shelf is highly unusual in being dominated in terms of both biomass and species by a single group, the notothenioids[36]. Antifreeze glycopeptides (AFGPs) have been found in all shallow-water notothenioids examined, and the chemical structure is highly conserved across the species (though in the smaller AFGPs proline can replace one of the alanines).

Attention then turned again to the Arctic, and in the two decades that followed essentially identical AFGPs were found in northern cods (Gadidae), and a number of antifreeze proteins (AFPs) without attached sugars were found in a diverse range of fish (Table 6.2). Antifreeze proteins have also been found in Antarctic fish other than notothenioids. These broad classes of antifreeze peptide are very different in chemical composition and in three-dimensional structure. What they have in common is the ability to prevent freezing within the narrow range of temperatures that these fish experience. Current research is directed at determining the three-dimensional structure of these antifreeze peptides, to understand how they interact with ice[37].

The obvious questions are how do these antifreezes work, and do all the different antifreeze molecules act in the same way? The first clue was that the effect of concentration on freezing point depression was far stronger than a simple colligative effect; that is, the

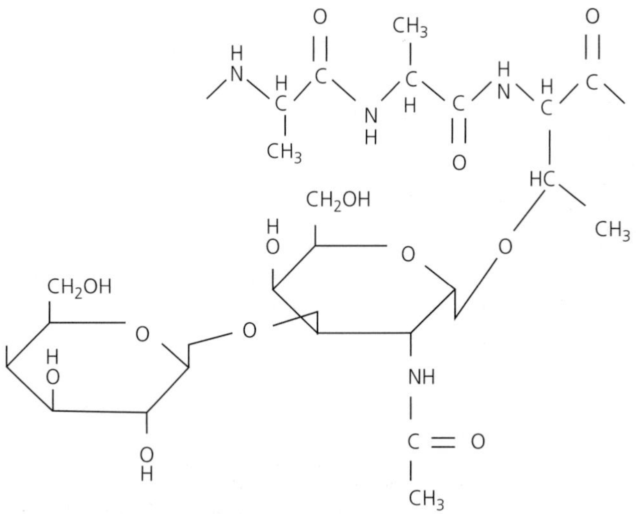

Figure 6.8 The basic repeating tripeptide of the antifreeze glycoprotein isolated from Antarctic fish, showing the disaccharide attached to the threonine moiety[38].

Table 6.2 Major classes of antifreeze molecules in teleost fish.

Type	Comments and some representative species (order)
Glycoprotein antifreezes	
AFGP	Nototheniods (five families) (Perciformes); Antarctic.
AFGP	*Gadus morhua, Boreogadus saida* (Gadiformes); Arctic.
Protein antifreezes	
Type I AFP	Small (4–7 kDa), alanine-rich, forming α-helix. *Pseudopleuronectes americanus* (Pleuronectiformes), *Myoxocephalus scorpius, Liparis gibbus* (Scorpaeniformes); Arctic.
Type II AFP	11–24 kDa, cysteine-rich, β-structured protein with a few disulphide bridges. *Clupea harengus* (Clupeiformes), *Hemitripterus americanus* (Scorpaeniformes), *Osmerus mordax* (Salmoniformes); Arctic.
Type III AFP	7 and 14 kDa globular proteins. *Macrozoarces americanus, Pachycara brachycephalum, Anarhichas lupus* (Perciformes); Arctic and Antarctic.
Type IV AFP	12.3 kDa, with a bundled helix structure. *Myoxocephalus octodecemspinosus* (Scorpaeniformes); Arctic. The functional status of this protein as an antifreeze is unresolved.

depression of the freezing point was far greater than would be expected simply on the basis of the number of molecules involved. It was also evident that the effectiveness depended on size of the molecule.

Careful observations of the growth of ice showed that when a seed ice crystal is present in an aqueous solution of antifreeze, ice does not form until about $-1.2\,°C$, but once formed these crystals melted closer to $0\,°C$. This difference between the freezing point and the melting point is termed *thermal hysteresis*, and is frequently used to quantify the extent of antifreeze action in biological systems. Indeed the presence of thermal hysteresis >0.05 K is sometimes taken as proof of the existence of an antifreeze, even without further characterisation of the agent(s) involved[39].

The generally accepted mechanism of action for all antifreezes is adsorption-inhibition. The hypothesis is that the antifreeze molecule binds to ice when hydroxyl groups on the outside of the antifreeze molecule form hydrogen bonds with the periodic water molecules in the ice crystal lattice. The binding of the antifreeze to the ice then inhibits further growth of the crystal. The mechanism is believed to be that water molecules can only join the lattice between the adsorbed antifreeze molecules, and this produces strongly curved growth fronts. This enhanced curvature increases the surface energy, and makes further growth thermodynamically unfavourable at temperatures normally experienced by the fish[40].

For this mechanism to work, the three-dimensional structure of the antifreeze is critical, since this must provide hydrogen-bonding moieties in a spatial arrangement that binds to the ice. This single requirement explains why a variety of very different molecules can have antifreeze activity, since the underlying composition can vary as long as suitable hydrogen-bonding moieties are arranged on the outside of the molecule at the correct orientation and spacing to bind to ice. This binding is reasonably well understood for notothenioid AFGP and Type I AFP from winter flounder (*Pseudopleuronectes americanus*), though the ice crystal planes to which they bind are different. The binding of other fish AFPs to ice is not yet well understood, but is believed to similar.

The adsorption-inhibition hypothesis implies that ice is already present within the fish for an antifreeze molecule to bind to. This ice is there because in shallow waters ice is present in the environment and is ingested as the fish feed or drink, or may enter through small skin lesions. While the antifreeze is effective at stopping these ice crystals from growing further, it does not remove or melt them. For fish in much of the Arctic the ocean warms enough in summer to clear the fish of ice. For Antarctic fish that may pass their entire life history in cold water, the ice remains. Recent studies have suggested that ice crystals with adsorbed AFGP may be recognised and ingested by phagocytes in the spleen, thereby locking the ice away. Although the shallow waters warm in summer, this ice may not melt because the adsorption of AFGP inhibits melting of the ice crystal and in many years the melting temperature may never be reached[41].

6.7.1 Where, when and how much?

Because the freezing point of seawater varies with pressure (Box 6.1) ice is only present in shallow waters. In deeper waters where there is no external ice, fish can live permanently supercooled (as Scholander first identified in his early work on Arctic fish). Antifreezes are needed only in shallow waters, and here the concentration varies with the likelihood of encountering ice, being highest in species such as *Pagothenia borchgrevinki* which lives much of its life just beneath or within the platelet ice layer. Where the environment is seasonal, AFP production is also seasonal, such as in the winter flounder *Pseudopleuronectes americanus* (Figure 6.9).

The key points of entry for ice into the fish are through the integument, the gills and the gut. These are the important sites for protection and studies of Antarctic fish have shown that these tissues tend to have high concentrations of antifreeze. Antifreeze is also found in almost all other fluid compartments in the fish, exceptions being the endolymph and urine. The AFGP of notothenioid fish is a small enough molecule to be filtered into the urine, and this would result in the steady loss of antifreeze. This does not happen because their kidneys either lack glomeruli or these are non-functional, and the urine is produced by secretion. Some fish with

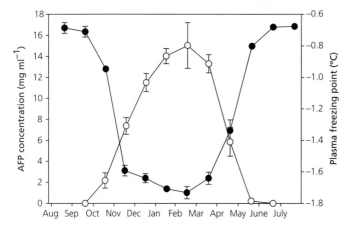

Figure 6.9 Seasonal variation in antifreeze protein concentration (white symbols) and freezing point of the blood plasma (black symbols) in winter flounder (*Pseudopleuronectes americanus*)[42].

AFPs do, however, have glomerular kidneys, and in these species the AFP is believed to be retained by a charge-repulsion mechanism, although some AFP does appear in the urine[43].

6.7.2 Evolutionary aspects

The polar oceans have not always been cold. It is likely that notothenioid fish evolved antifreeze between 22 and 42 million years ago, a window which encompasses the early Miocene cooling of the Southern Ocean. Genomic studies have shown that the notothenioid AFGP evolved from the control regions of a pancreatic trypsinogen-like protease and, remarkably, a chimaeric gene and protein can be detected. The gene for AFGP is present in multiple copies, and each gene encodes a large polyprotein. Post-translational modification of the polyprotein produces many copies of AFGP per gene, and this allows the fish to maintain a high level of antifreeze in the blood (up to ~ 3.5 mg ml^{-1}).

The AFGP of northern cods (gadoids) is also encoded by a family of polyprotein genes, but the gadoid cod AFGP shares no sequence identity with the pancreatic trypsinogen gene and its origin is currently unknown. The difference in evolutionary origin, together with differences in the coding and the very different climatic history of the Arctic, indicates that these two AFGP molecules evolved independently, and are thus a nice example of convergent evolution at the molecular level[44].

The Arctic seas have been cold for far less time than the Southern Ocean, with freezing temperatures for only about the past 3 million years. Type I AFPs appear to have evolved independently in four lineages, though the parent genes are currently unidentified. Type II AFPs are derived from a Ca^{2+}-dependent C-type lectin, and the highly conserved intron and exon structure in divergent lineages of fish has led to the suggestion of lateral gene transfer. Type III AFPs are found only in a single lineage (zoarcids or eelpouts), a lineage with both Arctic and Antarctic representatives. This AFP appears to have originated from the C-terminal domain of a sialic acid synthase gene[45].

The diversity of antifreezes with a wide variation of parent molecules might suggest that the acquisition of antifreeze capability in fish is not an especially difficult evolutionary challenge. On the other hand, the cooling of the Southern Ocean is associated with the loss of most of the warm-water fish fauna, and the dramatic radiation of the notothenioids, which might suggest that many lineages were unable to evolve the antifreeze necessary to survive in the cold shallow waters of Antarctica. The two polar regions thus offer biologists contrasting patterns of evolutionary response to freezing temperatures in the sea.

6.8 Freezing in intertidal marine invertebrates

The intertidal is a tough place to live. Although mobile organisms can avoid the harshest conditions by moving beneath macroalgal fronds, sessile invertebrates such as barnacles and mussels must withstand dehydration in strong sun, flooding with freshwater during rain and, in winter, scouring by ice and a regular switch between freezing in air and thawing in water at each tide. In temperate regions unusually cold periods can kill large numbers of intertidal organisms, such as in the exceptionally cold winters of 1946–47 and 1962–63 in North-west Europe.

Working at Woods Hole on the eastern seaboard of North America, John Kanwisher showed that a number of intertidal marine invertebrates (three bivalves and two gastropods) tolerated freezing of their extracellular water. Some of these species undercooled down to −5 °C, but below this temperature all started to freeze. At −15 °C between 54% and 67% of the total body water was frozen. Measurements of oxygen consumption showed that the tissues were still metabolising, although rates were very low at −15 °C. The barnacle *Semibalanus balanoides* has also been shown to tolerate freezing of extracellular water, and this appears to be a general feature of intertidal animals.

A thermal hysteresis glycoprotein has been isolated from the mussel *Mytilus edulis*, and an antifreeze agent has also been reported from the pedal mucus of the Antarctic limpet *Nacella concinna*, although the highly viscous nature of this mucus may be the critical factor in slowing nucleation of mantle or tissue water. Despite the obvious relevance of freezing to intertidal invertebrates, we still lack a coherent and comprehensive picture of how these organisms cope with the stress of freezing temperatures[46].

6.9 Freezing in arthropods

In many parts of the world terrestrial invertebrates must survive much lower temperatures than are ever encountered by fish or intertidal invertebrates. By far the most studied such group are arthropods, and especially insects.

The response of arthropods exposed to sub-zero temperatures fall into four broad categories (Table 6.3). The first is *chilling injury* which is essentially an inability to cope with the decrease in temperature, and is independent of the presence of

Table 6.3 Some terminology in ecological studies of freezing. SCP: supercooling point (the temperature at which the organism freezes)[47].

Term	Comments
General terms	
Cold-hardiness	A general term covering everything that an organism does to allow it to survive periods of low temperature; the process of acquiring this is *cold-hardening*.
Non-freezing injury	
Chilling injury	Damage at low temperatures, independent of the presence of ice. Can be sub-lethal.
Responses to freezing stress	
Behavioural avoidance	Overwinter in microhabitats where freezing can be avoided.
Freeze tolerance	A term applied to those organisms that survive the freezing of extracellular water. The sub-categories below are somewhat arbitrary, but useful.
Partial freeze tolerance	Can survive some formation of extracellular ice, at high subzero temperatures. Die if ice formation equilibrates at, or above the SCP.
Moderate freeze tolerance	Freeze at a high subzero temperature, but die if cooled to more than ~10 K below the SCP.
Strong freeze tolerance	Freeze at a relatively high temperature, but can survive much lower temperatures (> 10 K below their SCP).
Freeze tolerant with low SCP	Species with a very low SCP (below −25 °C), but which can survive freezing to temperatures a few degrees below their SCP.
Freeze intolerance	A term applied to those organisms that are killed by the freezing of extracellular water. Also referred to as *freezing-susceptible* or *freeze-avoiding*. These organisms generally survive freezing temperatures by undercooling, but die when they freeze.

ice. It is temperature alone that is damaging, and chilling injury can affect tropical and temperate species at temperatures well above 0 °C. Chilling injury arises from the failure of one or more physiological systems, and the effects can either be fatal immediately or sub-lethal.

If air temperatures fall below freezing, terrestrial invertebrates can avoid freezing stress by selecting an overwintering site where temperatures remain above 0 °C. Typically this involves burrowing deep into litter or soil, where the microclimate removes any danger of ice formation[48]. If ice is present in the environment, it presents a serious physiological challenge to which organisms respond with a variety of mechanisms broadly termed *cold-hardiness*. Traditionally these have been divided into two broad categories: freezing tolerance and freezing intolerance. Although this binary classification has been useful, more recent work suggests that the overall range of responses is more nuanced.

Experimental work on the response of organisms to freezing extends back over two centuries, but the key historical work on insect cold-hardiness was undertaken by Reginald Salt, who first clarified the distinction between insects that survive freezing and those that do not, and identified the presence of high levels of glycerol in the haemolymph of overwintering insects[49].

Working principally with an important agricultural pest, the wheat stem sawfly *Cephus cinctus*, Salt showed that the temperature at which the overwintering larva froze (which insect physiologists usually call the *supercooling point*, SCP) was influenced by cooling rate: doubling the cooling rate lowered the mean supercooling point by 0.24 K. He also showed that whereas slow cooling rates allowed the larva to be in thermal equilibrium with its surroundings, faster rates produced marked temperature gradients within the body. On the basis of this work Salt recommended a standard cooling rate of 1 K min^{-1} for studies of insect cold-hardiness, a standard which has largely remained to this day.

Salt also showed that when larvae were cooled and then held at constant temperature, the mean time to freezing depended strongly on the temperature (Figure 6.10). The very high temperature sensitivity ($Q_{10} > 10^5$) indicates that this is not a

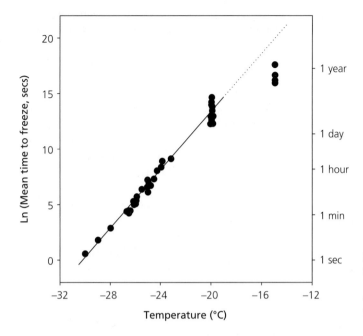

Figure 6.10 The effect of temperature on the mean time to freezing in larvae of the wheat stem sawfly, *Cephus cinctus*[50]. The warmest data do not match the overall trend and so these have been excluded from the least-squares linear regression (solid line).

biological process, and suggests that freezing is a stochastic physical event. A striking feature of this relationship is that the lowest temperature likely to be experienced in the wild in a normal winter allows the bulk of the population to survive in an supercooled state (but note that because each data point is a mean value, half the insects exposed to that temperature did not survive this long). This also implies that an unusually severe winter can result in many larvae freezing and dying. Horticulturalists and farmers have long recognised the value of a cold winter for reducing the size of pest populations; Salt showed why[51].

6.9.1 Supercooling in freezing-intolerant arthropods

Arthropods that are intolerant of freezing must survive low temperatures by avoiding nucleation of their body fluid, and the extent to which a given species can remain unfrozen below the melting point is termed supercooling capacity. Since the nucleation of ice is a stochastic event, the supercooling capacity needs to be sufficient for a high proportion of the population to survive a typical winter.

Supercooling capacity often varies with the stage of the life cycle. For example the eggs of some Lepidoptera can remain unfrozen to −50 °C. The small size of the eggs and the absence of ice nucleators are important factors here. Where arthropods overwinter as larvae, these can remain unfrozen down to below −30 °C. With a careful choice of overwintering site, this is sufficient to ensure survival. The supercooling capacity of adult insects varies extensively, with no consistent pattern across phylogeny, though few are able to remain unfrozen below ~ −30 °C.

Where supercooling capacity has been measured throughout the year, a strong seasonal pattern typically emerges: insects increase their capacity to avoid freezing in winter (Figure 6.11). This suggests that the insects have to prepare physiologically for winter, and implies that increasing supercooling capacity carries a cost.

A significant metabolic cost to overwintering is the need to synthesise cryoprotectants, which are typically low molecular mass polyhydroxy alcohols (polyols) and sugars. Glycerol is the commonest and most abundant such cryoprotectant, but sorbitol, mannitol, *myo*-inositol, trehalose and fructose are also important in some species. These are often present in multicomponent mixtures, possibly because each component brings a different physiological benefit. For example glycerol concentrations in

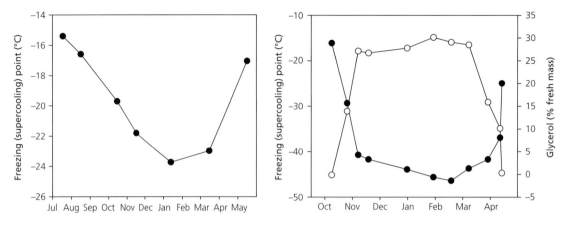

Figure 6.11 Seasonal variation in the freezing point (supercooling point) of two insects. A (left panel): seasonal variation in the freezing point of unfed adults of the beech leaf mining weevil *Rhynchaenus fagi*. B (right panel): seasonal variation in freezing point (black symbols) and whole insect glycerol content (% fresh mass, open symbols) in larvae of the pine shoot tortricid moth *Retinia (Petrova) resinella*[52].

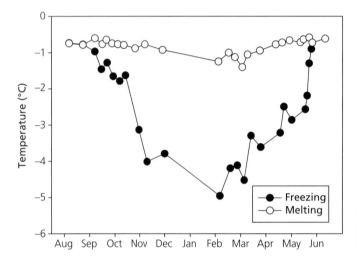

Figure 6.12 Seasonal variation in thermal hysteresis of haemolymph of larvae of *Meracantha contracta*, indicating the seasonal production of an antifreeze[54].

the haemolymph are often high enough to provide a significant colligative protection, whereas studies of organisms that tolerate extensive dehydration have shown that trehalose is important in maintaining membrane integrity. While the association between supercooling capacity and cryoprotectant presence is clear, there are many details yet to be worked out.

Arthropods that are intolerant of freezing also synthesise antifreeze proteins (AFPs). Although the presence of thermal hysteresis was actually discovered in an insect before its demonstration in polar fish, its significance was not immediately recognised. The first recognition of the existence of AFPs in arthropods was a seasonal study of the larvae of the tenebrionid beetle *Meracantha contracta* by Jack Duman (who had started out working with Art DeVries on antifreeze in Arctic fish) (Figure 6.12).

In arthropods the proteins that produce thermal hysteresis are referred to as either AFPs or thermal hysteresis proteins (THPs)[53]. They are now known from over 50 species of arthropod, and have been particularly well studied in beetles (Coleoptera). Most have been reported from North America and Europe, but they are also known from Antarctica. Unlike fish AFPs, those from arthropods bind to both the basal and prism planes of ice crystals, which may account for their greater hysteresis effect. As in fish, arthropod AFPs may comprise a suite of related molecules which are expressed differentially in various tissues, and they may act synergistically with small molecular mass solutes to enhance their effect[55].

The critical factor for freezing-intolerant arthropods is to avoid nucleation. In addition to the synthesis of cryoprotectants and antifreeze proteins arthropods eliminate potential nucleators from the gut (which can include ice-nucleating bacteria ingested with the food). Although we do not know precisely where nucleation takes place, there is a clear association between SCP and feeding in a range of arthropods. They also reduce the level of body water, which has two principal benefits: it lowers the chances of spontaneous nucleation (since this is volume-dependent) and increases the concentration of cryoprotectants[56].

6.9.2 Freeze tolerance in arthropods

Arthropods tolerant of the freezing of extracellular water typically freeze at a high subzero temperature (generally above −10 °C). The extent to which they can survive lower temperatures varies extensively, but some species can tolerate subsequent exposure to −70 °C (for example adults of the beetle *Pterostichus brevicornis* which overwinter in rotting wood in Alaska). This indicates that the freezing point and lower lethal temperature have evolved independently (Figure 6.13).

Arthropods that freeze at high subzero temperatures often initiate freezing with ice-nucleating proteins. These were first identified in beetles from the mountains of southern California, and the only insect ice-nucleating protein to be characterised so far comes from the crane-fly *Tipula trivittata*. This proved to be a lipophorin, suggesting that as with antifreeze proteins, the ice activity is a modification of the function of a pre-existing protein. As the ice forms, the remaining fluid haemolymph increases in concentration, which pulls water from the cells. The dehydration of the cells can be controlled by production of colligative cryoprotectants in the cell, and also modification of the water permeability of the cell membranes (for example by expression of aquaporin proteins which increase water flux).

In frozen arthropods there is also a danger from recrystallisation, the process by which small ice crystals become larger (this process is why ice cream gets crunchy when held in the freezer for a long time). If this happens, the larger crystals could cause extensive tissue damage, which is why many freeze-tolerant species also express antifreeze proteins. This is not as incongruous as it seems, because an important function of AFPs is to prevent ice crystals from growing in size. Quite how arthropods orchestrate the balance between ice nucleators promoting nucleation, and antifreezes preventing crystal growth is unclear.

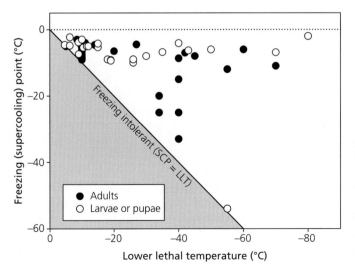

Figure 6.13 Variation in the freezing point (supercooling point, SCP) and lower lethal temperature (LLT) in freeze-tolerant insects. Data for adults are shown as black symbols and data for larvae or pupae as white symbols. The grey area marks where the organisms die at temperatures above their freezing point, and the line thus shows the limiting relationship for freezing-intolerant species (where freezing and death occur at the same temperature)[57].

6.9.3 Vitrification in arthropods

As temperature drops, ice forms in the haemolymph of freezing-tolerant insects and the concentration of sugars and other cryoprotectants in the remaining fluid increases sharply. Mixtures of sugar and water can vitrify at relatively high temperatures, and recently vitrification has been reported in the haemolymph of larvae of the freeze-avoiding beetle *Cucujus clavipes*. When cooled, most larvae froze at temperatures of −35 °C to −42 °C (below the temperatures normally experienced in winter), but some remained unfrozen down to −80 °C. DSC suggested that the haemolymph of these larvae underwent a glass transition at ∼ −58 °C, and could then avoid freezing as cold as −150 °C, and the vitrified larvae could survive exposure to −100 °C. These are the only reports of vitrification in overwintering arthropods, though the phenomenon may be more widespread than so far recognised[58].

6.9.4 Relating the laboratory to the natural world

We now have cold-hardiness data for almost 300 species of arthropod, and these indicate a wide range of responses to freezing temperatures. Mapping these data onto an insect phylogeny shows no clear association between lineage and nature of cold-hardiness, although sampling across the full diversity of arthropods remains patchy[59].

An interesting nuance here is that at least two insects have been noted as changing the nature of their cold-hardiness, both at locations in northern Indiana (41 °N). In the winters of 1977–78 and 1978–79 *Dendroides canadensis* was freezing-tolerant, freezing between −8 °C and −12 °C and with a lower lethal temperature of −28 °C. However beetles studied in the 1979–80 and 1980–81 winters were intolerant of freezing; they exhibited strong undercooling and did not freeze until ∼−26 °C, which was also their lower lethal temperature. The interesting aspect of this change is that the lower lethal temperature remained the same, but the freezing point did not, associated with an absence of ice-nucleating proteins. A similar switch has also been reported from another beetle, *Cucujus clavipes*[60].

There is a marked asymmetry in the pattern of cold-hardiness between northern and southern hemispheres (although there are far fewer data from the southern hemisphere at present) (Figure 6.14). The obvious question is what

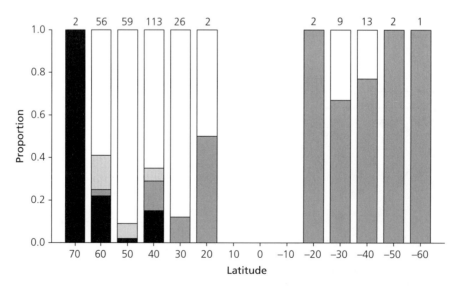

Figure 6.14 Latitudinal variation in the proportion of insects that are freezing intolerant (white bars) or freezing tolerant. Data are analysed in latitude bins of 0°–10°, 10°–20° and so on, plotted on the lower bin boundary. Freezing tolerant species are classified as species that are strongly freezing tolerant (black), moderately freezing tolerant (dark grey) or slightly freeze tolerant but with low freezing points (light grey). Southern latitudes are shown as negative, and the number of species in each latitude bin is shown at the top[61].

ecological factors drive this diversity of responses? Clearly lower lethal temperature must be related to the lowest temperature a given species is likely to experience over winter. That all species do not have a lower lethal temperature that ensures complete survival under any circumstances indicates that there are costs to having a low lethal temperature, though as yet we do not know what these are. The interplay between energetics, life history and ecology is clearly central to the evolution of cold-hardiness that matches the environment, but as yet we have only a general picture of how this plays out. Although much derided in some quarters, and out of fashion with many funding agencies, further detailed comparative studies of physiology and ecology will be integral to our understanding here[62].

6.10 Freezing in terrestrial vertebrates

For active endotherms in polar regions freezing is not a problem, except possibly for extremities such as ears, nose or limbs. Ectothermic vertebrates such as amphibians and reptiles are, however, faced with the same thermal challenges as invertebrates. While most amphibians and reptiles are creatures of warm or hot environments, frogs are common in cold northern forests and I have watched lizards foraging on the Tibetan plateau, where winter temperatures routinely drop below −40 °C. Selection of a suitable overwintering site can ameliorate temperatures to some extent, but it is now clear that a few amphibians and reptiles survive winter frozen.

The freezing of frogs in winter was first reported from the boreal forests of North America by William Schmid, who showed that three species of terrestrial frog survived freezing of ~35% of body water, whereas two species of aquatic frog died when frozen[63]. Subsequently a number of other frogs, as well as a few reptiles, have been shown to tolerate freezing. The best studied vertebrate freezer is undoubtedly the wood frog *Lithobates sylvaticus* (previously *Rana sylvatica*), and the general features established for this species appear to be fairly general.

As winter approaches frogs seek out hibernacula beneath the forest leaf litter, but fairly close to a pond. Wood frogs monitored in natural hibernacula in Alaska experienced an average of 193 days of subzero temperature, with a winter mean temperature of −6.2 °C, and an average seasonal minimum of −14.6 °C. All 18 frogs monitored survived the winter, which contrasts with a long-term study of the freezing-intolerant *Rana lessonae* and *Rana esculenta* in Switzerland. Over a six-year period winter survival varied from 20% to 80% for both species, with the main determinant of survival being weather: survival was lowest in winters with a low minimum temperature[64].

The forest floor provides a wide range of natural nucleators, so ice forms in the environment once air temperature drops below zero and this ice inoculates freezing in the frogs. Ice forms in the extracellular fluids (blood, lymph) but does not penetrate the cells. The freezing point of frogs is higher than in many insects (that is they have a lower supercooling capacity), but does not differ significantly between frogs that can tolerate freezing and those that cannot; what does differ is the lower lethal temperature[65].

As they start to freeze, frogs accumulate large quantities of cryoprotectants in their blood and tissues; these include glycerol, glucose and urea. The concentration of these is enhanced because the frogs also dehydrate. These cryoprotectants can be circulated throughout the body because the release of latent heat as ice forms allows the heart to continue pumping while freezing proceeds; it then stops. The cryoprotectants also enter the cells, where they help to prevent damaging dehydration of the cytosol by their colligative effect on osmotic pressure. There is also evidence for the presence of glycolipid antifreezes within some tissues[66].

At present physiological studies of winter freezing are limited to a small number of amphibians and reptiles. With this caveat, it nevertheless appears that the pattern of response is broadly similar across vertebrates that freeze for more than a transient period of a few days. Current evidence suggests that the ability to tolerate freezing has evolved several times independently in different lineages of anurans (wood frogs, tree frogs) and at least one salamander (the Siberian salamander *Salamandrella keyserlingii*), but not in toads. It has also evolved independently in several reptilian clades (snakes, lizards and possibly turtles)[67].

6.11 Freezing in plants

Insects and frogs can move to overwintering sites that ease the direct impact of cold; plants have to take whatever the environment throws at them. In the boreal forests (taiga) of Siberia or North America, for example, this can mean withstanding continuous winter temperatures below −40 °C and minima below −60 °C (see Chapter 14). In contrast, plants growing on mountains have to contend with freezing temperatures overnight coupled with warm or hot days, and those growing in temperate regions may be faced with periods of freezing temperatures that are unpredictable in their timing, intensity and duration.

Plants subject to frequent but light frosts can use structural features to minimise the chances of ice propagation into their tissues. One unusual adaptation is used by giant lobelias (*Lobelia telekii*) on the upper slopes of Mt. Kenya. Here the tall inflorescence is covered by long thin bracts which reduce radiative heat loss by night, but the delicate internal parts of the plant are surrounded by water. This water freezes by night, prompted by a nucleating agent, and the release of the latent heat of fusion is sufficient to keep the inner parts of the inflorescence close to 0 °C throughout the night until the warmth of the sun melts the water by day[68].

The classic early work on cold tolerance and freezing resistance of plants was done by Akira Sakai, working both in his native Japan and in North America. He showed that the cold tolerance of plants is related closely to the environments from which they come. This points clearly to a genetic underpinning, which has been confirmed recently by common-garden experiments[69].

In regions that experience long periods of intense cold, woody plants can exhibit deep supercooling in selected tissues (notably xylem, ray parenchyma and overwintering floral buds). In these, the extracellular fluids do not freeze down to −40 °C, or even below. Elsewhere the plant extracellular water freezes at a relatively high subzero temperature. This is almost inevitable because the presence of abundant nucleators (including the ice-nucleating bacteria discussed in section 6.5.2) on the surface of plants means that ice forms readily. This ice propagates easily into the plant through stomata, lenticels and surface lesions, although ice also nucleates within the plant. Plants appear to be reasonably tolerant of ice forming in extracellular (apoplast) fluids, and the main cause of injury is from osmotic dehydration of the cells. Damage can arise from the increased concentration of cellular electrolytes, macromolecular crowding and disruption to membranes; indeed the release of intracellular electrolytes is often used as a measure of low-temperature damage in plants. As ice forms it excludes dissolved gases, which can form small bubbles. In xylem water these bubbles cause embolisms and if these persist they can damage or kill shoots by preventing xylem flow[70].

Plants acquire cold-hardiness with similar physiological adjustments to those shown in animals. There are increases in the concentration of small molecular mass carbohydrates and compatible solutes, which act colligatively to reduce the osmotic stress on cells, and antifreeze proteins have been found in the extracellular fluids of some, but not all, freezing-tolerant plants. These AFPs appear to be important in preventing damaging recrystallisation of extracellular ice. Thermal hysteresis activity has been found in a wide variety of Antarctic plants, although interestingly of the two higher plants (angiosperms) found on mainland Antarctica one (the grass *Deschampia antarctica*) shows activity indicative of an antifreeze protein and the other (the pearlwort *Colobanthus quitensis*) does not. The two species extend equally far south, and *Deschampsia* is the first plant in which antifreeze activity appears to be expressed constitutively (that is, it is present year-round). Dehydrins have also been identified extensively in freezing-tolerant plants, and it appears that these function to prevent deleterious interaction between membranes[71].

The dehydration and very low temperature experienced by many plant cells in winter means that they are likely to undergo a colloid glass transition, although their different ionic composition together with the presence of chloroplasts, vacuoles and starch granules mean that the temperature(s) at which this occurs will differ from animal and bacterial cells. Vitrification has been demonstrated in cortical cells of winter hardened *Populus balsamifera*, and suggested to be widespread among northern trees and cold-hardy woody plants. Interestingly, plants can also show multiple vitrification events, suggesting that different fluid compartments undergo a glass transition at different temperatures, depending on their water content[72].

6.12 Concluding remarks: the wider context

Cold conditions have existed somewhere on Earth for a very long time. Adaptation to freezing temperatures is therefore probably also very old, and it has many features that are common across organisms. Ice within cells is generally lethal but when ice forms externally to the cells the main stress is dehydration as water is pulled from the cell osmotically. This causes the cell to shrink, intracellular concentration to increase and membranes to fold and buckle. A common response across many organisms is to synthesise compatible solutes, often sugars or polyols, to offset the osmotic imbalance. These small organic osmolytes may also act to protect membranes and macromolecules, and the particular osmolytes used by different organisms may be related to function (for example trehalose) and ease of production (for example glucose in frogs) in addition to compatibility. Protective proteins such as dehydrins or LEA proteins may also be involved.

All proteins interact with their aqueous environment, but two classes of proteins are important in managing ice: ice-nucleating proteins promote freezing at high subzero temperatures and thermal hysteresis proteins (antifreezes) prevent the growth of small ice crystals or recrystallisation of bulk ice. Antifreeze proteins are many and varied, indicating that a wide variety of parent molecules have been co-opted for ice management. It would appear that producing an antifreeze is not a difficult evolutionary task. The key structural feature of these diverse proteins is the presence of an array of hydroxyl groups in a configuration that allows binding to hexagonal ice.

That the principal cellular stress of the presence of external ice is dehydration indicates a strong physiological and evolutionary link between freezing and drought as ecological challenges. It is highly likely that adaptations to dehydration are extremely ancient, probably extending as far back as the initial conquest of the land. Tolerance of extensive or complete dehydration is, however, confined to unicells or very small multicellular organisms such as tardigrades or nematodes. Larger organisms cannot dehydrate completely, although many can tolerate some water loss. Selection of overwintering sites is central to water management. For example many amphibians move into ponds to overwinter, where freezing is not a problem. Others, however, select a moist terrestrial site where freezing is highly likely.

Vitrification has not attracted much attention by ecologists, but may prove to be important in organisms faced with very low winter temperatures. The lack of attention may be because vitrification is very difficult to demonstrate experimentally and, though long known from plants (both in seeds and overwintering woody plants), it has recently been demonstrated in insects. Together with the role of LEA and other chaperone proteins, this may prove to be more widespread than recognised to date.

6.12.1 The curious case of bacterial antifreezes

A microbial protein with thermal hysteresis activity has been reported from the widely distributed soil bacterium *Pseudomonas putida*. This proved to be a glycolipoprotein, and intriguingly exhibited both antifreeze and ice-nucleating activity. The first reports from polar regions were six bacterial strains isolated from Ross Island, Antarctica. One, a species of *Moraxella*, produced a lipoprotein with an N-terminal sequence suggesting it was anchored in the outer membrane. Subsequently thermal hysteresis activity was reported in 187 of 860 bacterial isolates from eight lakes in the Larsemann Hills, Antarctica, that ranged from freshwater to hypersaline. The thermal hysteresis activity was determined on whole-cell lysates, and one protein, from *Marinomonas primoryensis*, was studied in detail. It proved to be a Ca^{2+}-dependent protein and to induce >2 K of thermal hysteresis, though unlike fish or insect AFPs this protein did not produce the characteristic faceting on ice crystals grown below the hysteresis freezing point. Proteins with thermal hysteresis activity have now been reported in bacteria from sea-ice and cryoconite holes on glaciers, as well as in sea-ice microalgae and fungi[73].

Experimental work has established that bacteria only freeze when exposed to extremely fast cooling rates. The only natural environment where this is likely to happen is when a water droplet in a cloud nucleates. Bacteria on the surface of leaves or rocks can be subject to rapid changes in temperature but as far as we know these are never fast enough to freeze the bacterial cytosol. Bacteria in large water bodies such as lakes and streams will only cool or warm very slowly. This makes the discovery of proteins with thermal hysteresis activity in bacteria rather perplexing: why do they have them?

Bacteria exposed to very low temperatures may vitrify, so one possibility is that a protein with thermal hysteresis activity reduces the chances of damaging ice nucleation (devitrification) during warming as the cytosol returns to the fluid state. Another possibility is that they are not antifreezes at all: at least one bacterial protein with thermal hysteresis properties has been shown to have a role in binding the bacterial cell to ice surfaces[74].

A final possibility is that the thermal hysteresis activity is an incidental feature of proteins with other unrelated functions. And this highlights a subtle bias in the way we do science: the mass screening of bacterial strains for thermal hysteresis activity is done with isolates from cold habitats, since this is where we would expect to find an antifreeze. What if we repeated the screening in bacteria from temperate or tropical lakes?

6.12.2 A new classification

The discovery of an ever-widening range of ice-active proteins has prompted a more refined view of the diversity of ice-binding proteins in nature, based on their physiological or ecological function (Table 6.4).

6.12.3 Future prospects

Simple questions in ecology have a habit of becoming more complicated, but also more interesting. And this has certainly happened with our understanding of the ecology and physiology of freezing resistance. Antifreeze molecules have proved to be far more diverse than originally anticipated, with

Table 6.4 A recent classification of ice-binding proteins[75].

Function	Comments
Freeze avoidance	Proteins that produce thermal hysteresis and prevent ice crystals from growing. Found in marine teleost fish, arthropods and plants.
Freeze tolerance	Proteins that have only a small thermal hysteresis effect, but which prevent recrystallisation and thus control exiting ice. Reported from insects and plants.
Structuring of external ice	Proteins isolated from Antarctic marine algae. Stabilise brine pockets and help maintain a liquid habitat for the algae.
Ice adhesion	A protein isolated from the marine bacterium *Marinomonas primoryensis* is expressed externally, and couples the bacterium to ice surfaces.
Ice-nucleating proteins	Proteins that promote ice nucleation at relatively high subzero temperatures. Reported from plants, insects and bacteria.

entirely new classes still being identified, such as the glycolipid antifreezes now known from insects, two species of frog and one plant[76].

Integral to progress will be the burgeoning fields of genomics, proteomics and metabolomics. There is a long history of the use of these techniques in the field of freezing resistance, and they have proved especially valuable in unravelling the evolutionary history of antifreeze molecules, the physiological adjustments associated with freezing in frogs and the induction of cold-hardiness in the model plant *Arabidopsis thaliana* (thale cress). We are starting to see what changes are common across organisms, to distinguish adjustments to low temperature from those specific to freezing resistance and to unravel what is common between freezing resistance and dehydration resistance at the most fundamental level[77].

6.13 Summary

1. Freezing is a widespread ecological challenge, affecting organisms in over half the terrestrial environment as well as both polar seas. The challenge ranges from unexpected frosts in temperate regions, to the predictable long, deep winters of northern forests.

2. Seawater at atmospheric pressure freezes at about −1.9 °C. The only marine organisms liable to freeze in such cold water are teleost fish in shallow waters. These survive by synthesising a range of protein or glycoprotein antifreezes. Terrestrial organisms are faced with a far greater thermal challenge, and exhibit a more complex array of responses.
3. With very few exceptions, if a cell freezes internally, it dies. Unicellular organisms survive freezing temperatures by preventing ice nucleating within the cytosol, and tolerating the cellular dehydration and membrane disruption that follows from ice forming in the external environment. Some bacteria express ice-nucleating proteins on their surface to promote external ice formation at high subzero temperatures.
4. Multicellular organisms survive freezing temperatures by manipulating the composition of the extracellular body fluids. Terrestrial organisms may freeze at high subzero temperatures, often promoted by ice-nucleating proteins, and small molecular mass cryoprotectants (often sugars and polyols) moderate the osmotic stress on cells. When extracellular water freezes, control of the water content of cells is vital to their survival. A range of chaperone proteins (dehydrins, LEA proteins) help maintain the integrity of membranes and macromolecules. Thermal hysteresis (antifreeze) proteins prevent damaging recrystallisation of ice.
5. In some cases arthropods and higher plants prevent freezing in their extracellular fluids and survive by supercooling. This requires rigorous control of potential nucleators (especially in the gut), synthesis of antifreezes and cryoprotectants and often partial dehydration of the cells. This links freezing resistance and drought resistance at the cellular level.
6. Vitrification of extracellular water, or of the cell cytosol, may be a more widespread response to very cold temperatures than recognised to date.

Notes

1. An Anglo-Saxon riddle, to which the answer is ice. It comes from The Exeter Book, which is the largest extant collection of Old English literature, and this translation is by Michael Alexander (*The earliest English poems*, 1966, reissued as *The first poems in English* in 2008); the spelling of wierd is deliberate. Reproduced with permission.
2. The only known exceptions to this are two species of nematode, the Antarctic free-living *Panagrolaimus davidi* (Wharton & Ferns 1995) and the widely distributed pathogenic *Steinernema feltiae* (Ali & Wharton 2014). Quite how these nematodes survive a process which kills all other known organisms is unclear. Survival from intracellular ice has been described for the highly specialised fat body cells in several insects (Salt 1959, 1962; Davis & Lee 2001) and also other tissues (Worland et al. 2004). The impact of this at the organism level is unknown.
3. Long-term satellite data indicate that ~57% of the Earth's terrestrial surface has an annual minimum temperature at or below 0 °C, and 62% has recorded subzero temperatures at least once.
4. The terminology is discussed by Franks (1982).
5. Fahrenheit reported his discovery to the Royal Society in 1724; see Middleton (1966).
6. Taylor (1987) provides a nice summary for biologists, and Franks (1985) gives a more detailed treatment. It is difficult to determine the critical size of an ice nucleus experimentally, and modelling the nucleation of water is fiendishly complex. See Matsumoto et al. (2002), Liu et al. (2007) and Pereyra et al. (2011). The homogeneous nucleation temperature is from Hobbs (1974).
7. Kurt Vonnegut invented a fictitious form of ice-nine for his 1963 satirical novel *Cat's cradle*. This had the outrageous properties of being stable at normal temperatures and also catalysing the freezing of all the water in the world. Fortunately this ice-nine does not exist.
8. The addition of solute increases the entropy and hence changes the temperature at which entropy favours phase change. A substance which ionises changes entropy to a greater extent than one that does not because it introduces more particles to the solution. The addition of a solute also increases the boiling point, lowers the vapour pressure and reduces the chemical potential of the water. All of these are manifestations of the change in entropy.
9. See Atkins & de Paula (2011) for a more complete description. The simple formula works well for dilute solutions whose behaviour approximates that of an ideal solution. For more concentrated solutions, the relationship becomes far more complex (Ge & Wang 2009). The van't Hoff factor can be non-integral where dissociation is partial or ions associate.
10. Note that salinity, being expressed as the ratio of the electrical conductivity of the sample to that of

a reference standard, is dimensionless and has no units. The practice of adding the fictitious unit 'psu' (practical salinity unit) is officially discouraged.
11. For excellent introductions to the biology of sea ice see Thomas & Dieckmann (2002) and Thomas (2004, 2012).
12. Modified from Thomas & Dieckmann (2002); reproduced with permission of the American Association for the Advancement of Science.
13. The process of ice formation is described by Lock (1990) and Petrenko & Whitworth (1999). Anchor ice and its ecological effects are discussed by Dayton et al. (1969).
14. This term originates from the Old English word for someone (or something) grey with age.
15. Chilling injury is discussed by Morris (1987) and Wilson (1987) for plants and MacMillan & Sinclair (2011) for insects.
16. Modified from Morris et al. (1979).
17. The original presentation of the hypothesis was Mazur et al. (1972). A more recent comprehensive treatment of the ideas is Mazur (2004).
18. Mazur (1963, 2004).
19. Damage from increased solute concentration within the cell was first proposed by Jim Lovelock, based on work with human erythrocytes (Lovelock 1953). Meryman (1968) first suggested that damage may result when osmotic shrinkage exceeds a critical cell volume, and the membrane destabilisation hypothesis is presented by Steponkus & Lynch (1989). For reviews see Mazur (2004) and Muldrew et al. (2004).
20. Letter LXI to the Honourable Daines Barrington. Gilbert White (1720–1793) was an English naturalist who lived in Selborne, Hampshire. He wrote a series of letters to Thomas Pennant, a leading zoologist of the day, and Daines Barrington, both Fellows of the Royal Society, in which he documented many acute and detailed observations of the local flora and fauna. This included, compiled jointly with Robert Markwick, one of the earliest compilations of the phenology of plants and animals. Gilbert White is rightly regarded as the first ecologist, and the father of the field. His letters were published in 1789 as *The natural history and antiquities of Selborne*, and this book has remained continuously in publication ever since.
21. Data courtesy British Antarctic Survey, Cambridge.
22. Siminovitch & Cloutier (1983) provide a nice review of the early ideas here.
23. See Rapatz et al. (1966), Fonseca et al. (2006) and Seki et al. (2009). Clarke et al. (2013) discuss natural rates of temperature change in the environment, and Strimbeck et al. (1993) report rapid changes in temperature of pine needles.
24. The ecology of microbial life in cloud water droplets is discussed by Dimmick et al. (1975), Sattler et al. (2001), Hill et al. (2007) and Womack et al. (2010)
25. Temperature changes in nucleating water droplets in clouds are discussed by Murray et al. (2012) and Clarke et al. (2013).
26. The early work on the role of ice-nucleating bacteria in frost damage is described by Lindow et al. (1978a, b, 1982); Gurian-Sherman & Lindow (1993) describe the structure of the protein. Ice-nucleation activity in insects was first described by Zachiarassen & Hammel (1976) and Duman et al. (2010) provide a recent review. Aarset (1982) reviews freezing in intertidal marine invertebrates, and Madison et al. (1991) report the partial characterisation of an ice-nucleating protein isolated from an intertidal gastropod.
27. Yancey (2005) and Yancey & Siebenaller (2015) give valuable overviews of the nature and role of osmolytes and compatible solutes.
28. The original discovery is reported by Dure & Chlan (1981). Useful reviews of the distribution and function of LEA proteins are Garay-Arrogo et al. (2000), Battaglia et al. (2008), Tompa & Kovacs (2010) and Hand et al. (2011). Goyal et al. (2005) describe experimental work on the chaperone function of LEA proteins isolated from a nematode and wheat.
29. This apparently poor terminology has its roots in practical experience. Glass is perhaps the most familiar vitrified material (*vitrum* is Latin for glass), and it has been known since the earliest glassmakers of Mesopotamia or ancient Egypt that glass can change slowly to an opaque crystalline form, and this loss of glass-like quality was termed devitrification.
30. The vitrification of bulk fluids is described by Debenedetti (1996) and Wowk (2010). The crowded nature of the cell and its consequences are described by Ellis (2001) and the colloid glass transition by Zhou et al. (2009). The ecological aspects of the vitrification of microbial cells are discussed by Clarke et al. (2013) and Clarke (2014).
31. The earliest discussions of the vitrification of cells are probably Luyet (1937) and Luyet & Gehenio (1940).
32. This explanation is not universally accepted, and may be over-simplified. Brian McNab (2002) gives a thorough and balanced overview of the evidence and debate.
33. Scholander et al. (1957). See also Duman (2014) who revisits this classic early study in a modern context.
34. Modified from Eastman (1993), and reproduced with permission of Elsevier; the original study is DeVries et al. (1971).
35. DeVries & Wohlschlag (1969), DeVries et al. (1970, 1971).

36. Notothenioids are a suborder of Perciformes, the largest fish order, and are confined to the southern hemisphere. Eastman & Eakin (2000) list 122 species, of which 96 (78%) are found in the Southern Ocean, where they dominate the shallow-water fish fauna. There are eight families, one of which is the icefishes (Channichthyidae) which lack functional haemoglobin. The radiation of notothenioids in the Southern Ocean is discussed by Eastman (1993), Clarke & Johnston (1996), Eastman & Clarke (1998) and Eastman (2005).
37. Type I AFPs were first reported by Duman & DeVries (1974, 1976), Type II AFPs by Slaughter et al. (1981) and Type III AFPs by Hew et al. (1984) and Li et al. (1985) in Arctic zoarcids and Schrag et al. (1987) in Antarctic zoarcids.
38. Modified from Eastman (1993), and reproduced with permission of Elsevier; the original study is DeVries et al. (1971).
39. The advent of more sensitive instrumentation has revealed that there is also a small melting hysteresis. That is, ice crystals with antifreeze molecules attached have their melting point raised above the equilibrium freezing point of pure ice, caused by the presence of the antifreeze molecules. The total thermal hysteresis thus comprises two portions, a depression of the temperature at which ice forms and an elevation of the temperature at which that ice melts, relative to the equilibrium freezing point. In most discussions of thermal hysteresis it is just the depression that is considered.
40. This mechanism was proposed by Raymond & DeVries (1977). The growth of ice from a pre-existing nucleus is sometimes referred to as secondary nucleation (the primary nucleation being the formation of the nucleus itself).
41. See Evans et al. (2011) and Cziko et al. (2014).
42. Redrawn, with permission of John Wiley and Sons, from Petzel et al. (1980).
43. The excretion of AFP in Arctic fish is discussed by Boyd & DeVries (1983), Eastman et al. (1987) and Fletcher et al. (1989).
44. The evolution of antifreeze glycoproteins is discussed by Hsaio et al. (1990), Chen et al. (1997a, b), Cheng (1998), Cheng & Chen (1999) and Near et al. (2012). The loss of AFGP expresssion in temperate notothenioids is described by Cheng et al. (2003).
45. The evolution of Arctic AFPs is discussed by Baardsnes & Davies (2001), Deng et al. (2010) and Graham et al. (2008, 2012, 2013).
46. The early work is Kanwisher (1955, 1959). Dennis Crisp edited a series of papers reviewing the impact of the 1962–63 winter in the United Kingdom (*Journal of Animal Ecology*, 1964). Other key references are Crisp et al. (1977), Theede et al. (1976), Hargens & Shabica (1973) and Hawes et al. (2010). The most recent reviews are Aarset (1982) and Davenport (1992), and the field has advanced little since then.
47. This classification is based largely on Bale (1993, 1996), as modified by Sinclair (1999).
48. See Bennett et al. (2003).
49. The English physician Henry Power reported in 1663 that he had used a mixture of snow and salt to freeze nematodes, and that they were active when thawed some hours later. The French biologist René-Antoine de Réaumur also reported of freezing of insects in his classic memoire (de Réaumur 1736). Sømme (2000) provides a nice history of cold-hardiness work in terrestrial arthropods.
50. Modified from Salt (1966a, b, c); reproduced with permission from NRC Research Press.
51. The identification of thermal hysteresis in the darkling beetle *Tenebrio molitor* (mealworm) was by Ramsay (1964) and Grimstone et al. (1968). The important early work on insect cold-hardiness is Salt (1936, 1950, 1957, 1961, 1966a, b, c). The presence of glycerol in insect haemolymph was identified independently by Wyatt & Kalf (1957).
52. Modified from Bale (1980) and Hansen (1973); reproduced with permission of CryoLetters.
53. And in plants they are also referred to as ice recrystalisation inhibition proteins (IRIPs), another case of the unhelpful proliferation of terms for the same thing.
54. Modified from Duman (1977); reproduced with permission from Springer-Verlag.
55. For reviews of insect antifreeze proteins see Duman (2001, 2015) and Duman et al. (2010).
56. Worland & Block (1999) report the presence of ice-nucleating bacteria in the guts of two beetles from Antarctica.
57. This diagram has been modified from Sinclair (1999).
58. See Wasylyk et al. (1988) and Sformo et al. (2010).
59. Chown & Sinclair (2010).
60. These changes are reported by Horwath & Duman (1984) and Kukal & Duman (1989).
61. Redrawn from Sinclair & Chown (2005); reproduced with permission from John Wiley and Sons.
62. Sinclair (2015) provides a nice overview.
63. Schmid (1982).
64. The long-term study of *Rana* is Anholt et al. 2003. Recent field studies of freezing in frogs that provide a valuable ecological context are Larson et al. (2014), Costanzo & Lee (2013) and Costanzo et al. (2015).
65. See discussion in Hillman et al. (2009).
66. Storey & Storey (2004) give a valuable overview of freezing in the wood frog, while Costanzo & Lee (2013) consider vertebrate freezing in general. The

freezing of the Siberian salamander is reported by Berman et al. (1984).
67. The evolution of freeze tolerance in vertebrates is discussed by Voituron et al. (2002, 2009).
68. See Krog et al. (1979).
69. See Sakai (1960, 1970, 1983), Sakai & Okuda (1971) and Sakai & Weiser (1973). The common-garden work is Strimbeck et al. (2007) and Kreyling et al. (2015).
70. Sperry & Sullivan (1992), Sperry et al. (1994). See also Feild & Brodribb (2001).
71. The early reports of antifreeze (thermal hysteresis) proteins in plants are Urrutia et al. (1992), Duman (1993) and Duman & Olsen (1993). Valuable reviews are Atici & Nalbantoğlu (2003) and Griffith & Yaish (2004). The Antarctic work is Doucet et al. (2000) and Bravo & Griffith (2005).
72. Vitrification in woody plants is discussed by Hirsh et al. (1985), Hirsh (1987), Wisniewski et al. (2004) and Pearce (2004).
73. Xu et al. (1998), Yamashita et al. (2002), Gilbert et al. (2004, 2005), Raymond et al. (2007), Kawahara et al. (2007), Singh et al. (2014) and Hanada et al. (2014) describe proteins with thermal hysteresis activity from bacteria. Park et al. (2012), Jung et al. (2014) and Gwak et al. (2014) describe these from unicellular eukaryote taxa in polar regions.
74. See Janech et al. (2006), Guo et al. (2012) and Vance et al. (2014).
75. This table is based on Bar Dolev et al. (2016). See also Gilbert et al. (2005), Raymond et al. (2009), Raymond & Morgan-Kiss (2013).
76. See Walters et al. (2011) and Duman (2015).
77. Storey & Storey (2004) provide a nice overview of what genomics and proteomics have told us about the process of preparing for freezing in frogs, and the work on *Arabidopsis* cold acclimation is described by Tomashow (2001), Haake et al. (2002), Cook et al. (2004) and Van Buskirk et al. (2006).

CHAPTER 7

Temperature and reaction rate

Hotter is better[1]

In the 1920s, the eminent astronomer Harlow Shapley was working at the Mount Wilson observatory in Pasadena, California. While relaxing outside the observatory, he noticed how a local ant species always used the same trails and decided to time how long it took the ants to cover a 30 cm stretch of ground. He saw that the ants moved faster in the hot sun than in the shade and found that, as he had suspected, their speed was correlated strongly with temperature (Figure 7.1).

Here I have plotted speed as a function of temperature, which is the way a physiologist would view the data. Shapley actually plotted temperature as a function of speed, because he was interested in using the ants as a thermometer. He then went on to investigate two other ant species, this time experimentally, and revealed similar relationships (Figure 7.1). This work revealed two important features which are relevant to any discussion of biological rates in relation to temperature. Firstly the relationship between walking speed and temperature was non-linear, and secondly the precise relationship varied with species: not all ants moved at the same speed for a given temperature[2].

Shapley speculated that the relationships he observed reflected the temperature sensitivity of one or more of the chemical reactions involved in walking. Locomotion is, however, extremely complex at the molecular level, comprising many separate chemical processes underpinning muscle activity and its nervous coordination. This complexity makes it very difficult to tease out general principles, and so we need to start by considering a much simpler system. We will look first at the thermal behaviour of an elementary chemical reaction.

7.1 Rate of reaction

For two molecules to react they must meet. This simple requirement suggests immediately that the rate of reaction will depend on two factors:

1. Concentration: the more molecules are present, the higher the frequency of collisions.
2. Temperature: the faster the molecules are moving, the more frequently they will collide.

The theoretical principles of how concentration and temperature influence reaction rate were first worked out for the idealised system of a dilute perfect gas. It turns out that these principles can be applied without too many problems to reactions in dilute solution, and also extended to physiology. But we must never lose sight of the fact that, as we saw in Chapter 5, the complex intracellular environment in which physiological reactions takes place is very different from a dilute aqueous solution of the laboratory or the theoretical abstraction of a perfect gas.

7.1.1 Concentration

Consider a simple reaction where a single reactant forms a single product: A → B.

This is an *elementary reaction*, which is one that proceeds exactly as the chemical equation is written.

Principles of Thermal Ecology. Andrew Clarke, Oxford University Press (2017).
© Andrew Clarke 2017. DOI 10.1093/oso/9780199551668.001.0001

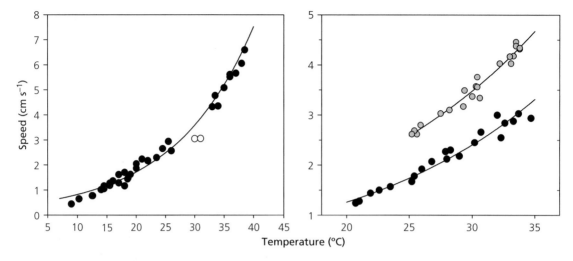

Figure 7.1 Walking speed as a function of temperature in ants. A (left panel): observational data for *Liometopum apiculatum*. Each data point represents the mean of 15–42 measurements, and the line is fitted assuming an exponential relationship, excluding two data points (white symbols) where the temperature measurement may have been unreliable. B (right panel): experimental data for *Tapinoma sessile* (black, mean of 7–20 observations per symbol) and *Iridomyrmex humilis* (grey, mean of 18–24 observations per symbol) with fitted exponential relationships[3].

An example might be the decomposition of dinitrogen tetroxide N_2O_4 to nitrogen dioxide NO_2:

$$N_2O_4 \rightarrow 2NO_2$$

This reaction will be familiar to many from school chemistry classes because it takes place at room temperature and involves a dramatic change in colour (N_2O_4 is colourless, whereas NO_2 is a deep reddish brown). The rate of the reaction depends on the concentration of N_2O_4 (note the convention whereby square brackets denote concentrations):

$$\text{Rate of reaction} = k_1[N_2O_4]$$

Where k_1 is the rate constant for the reaction. This reaction does not go to completion (that is, N_2O_4 is not converted completely to NO_2) because there is also a reverse reaction, where NO_2 forms N_2O_4. The rate constant here is k_2:

$$\text{Rate of reverse reaction} = k_2[NO_2]^2$$

Here the rate constant is determined by the square of the concentration; this is because of the stoichiometry (two molecules of NO_2 form one of N_2O_4). When these two reactions are exactly balanced, there will be an equilibrium mixture of NO_2 and N_2O_4, and the ratio of the concentrations is the equilibrium constant, K_{eq}:

$$K_{eq} = \frac{[NO_2]^2}{[N_2O_4]}$$

The position of this equilibrium is determined simply by the difference in the molar Gibbs energies of the reactant(s) and product(s), here $G_m(N_2O_4)$ and $G_m(NO_2)$:

$$\Delta G = G_m(N_2O_4) - G_m(NO_2) = -RT \ln K_{eq}$$

where R is the gas constant and T the absolute (thermodynamic) temperature. This relationship means that the greater the difference in molar Gibbs energy between the reactant and product, the greater the proportion of product in the equilibrium mixture (Figure 7.2). Note, however, that the position of equilibrium is also influenced by temperature: for our example, at −78.4 °C (the temperature of dry ice, solid carbon dioxide) the mixture is essentially pure N_2O_4, and this is also solid. At room temperature the mixture contains about 70% N_2O_4, and this percentage falls further as the temperature is raised. The effect of temperature has two components: the direct effect captured in the above equation and the temperature dependence of the molar entropies themselves. Even for the simplest of reactions, the effect of temperature is complicated.

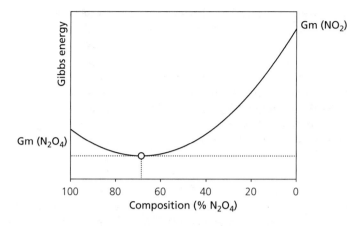

Figure 7.2 Diagrammatic representation of the relationship between Gibbs energy and equilibrium. The equilibrium for this reaction, shown by the dot, is at 68% reactants and 32% products, which is representative of the equilibrium between N_2O_4 and NO_2 at room temperature and pressure. At this point the Gibbs energy of the mixture is at its minimum. $G_m(N_2O_4)$ and $G_m(NO_2)$ denote the molar Gibbs energy for pure reactant (N_2O_4) and product (NO_2) respectively.

The change in Gibbs energy when 1 mole of reactant is converted to 1 mole of product under standard conditions is the standard Gibbs energy change, ΔG°. Standard conditions are very different from the conditions within the cell (Box 7.1), and so ΔG° is not an immediately helpful measure for physiologists. The change under cellular conditions, ΔG, can, however, be calculated relatively easily as:

$$\Delta G = \Delta G^\circ + RT \ln Q$$

where Q is the reaction quotient (Box 7.2).

As we saw in Chapter 2, a reaction proceeds spontaneously in the direction for which ΔG is negative ($\Delta G < 0$). For the example in Figure 7.2, this means that for any mixture to the left of the equilibrium, the direction of spontaneous change will be towards the right (towards the equilibrium). As equilibrium is approached the slope of the curve, which is a measure of the local value of ΔG, gets less and reaches zero at equilibrium. To the right of the equilibrium, ΔG is >0 and so the direction of spontaneous change is opposite; this means that again it is towards the equilibrium.

As we also saw in Chapter 2, the Gibbs energy captures the balance between the changes in internal energy and entropy, and thus expresses the maximum amount of work that can be achieved with a particular chemical reaction. At equilibrium the entropy is at its maximum and no useful work can be achieved; this is because the rate of change from reactant to product (yielding energy) is balanced by the reverse reaction (consuming energy). The

---

Box 7.1 The standard state and physiology

In thermodynamics it is essential to define a set of conditions that allow reactions can be compared in a meaningful way. This is known as the *standard state*, and is defined as:
  Gases: the pure gas at a pressure of 10^5 N m^{-2}
  Solids: the pure solid at a pressure of 10^5 N m^{-2}
  Liquids: the pure liquid at a pressure of 10^5 N m^{-2}
  Solutions: the solution at unit concentration (1 mol kg^{-1})
  The standard state applies only to an ideal solution where there are no solute–solvent interactions. Note that the standard state does not specify a temperature, although this is often taken as 25 °C (298 K).

The standard state is very different from the conditions within a typical cell, where solutions are far from ideal and most solutes are at concentrations very different from 1 mol kg^{-1}. Moreover most physiological reactions involve protons and water as reactants and products. Pure water is ~ 56 mol L^{-1}, and cells typically have pH ~ 7 (so protons are ~ 10^{-7} mol L^{-1}); both of these concentrations are thus far removed from standard conditions. By convention, however, for calculations of Gibbs energy change during a physiological reaction, where the concentrations of water and protons in the cell milieu effectively do not change, these two concentrations are set arbitrarily to 1.

> **Box 7.2 Law of Mass Action**
>
> Consider a simple chemical reaction where A and B react to form the products C and D[4]:
>
> $$a.A + b.B \rightleftharpoons c.C + d.D$$
>
> where a, b, c and d are the *stoichiometric coefficients* for the reaction (that is, they tell us how many molecules of each reactant and product are needed for the equation to balance). In most cases there is also a backward reaction, where C and D react to form A and B. When the system is at equilibrium, the forward reaction and backward reaction are in exact balance, and the Law of Mass Action states that the ratio of the reactants and products, the *mass action ratio*, is a constant:
>
> $$\frac{[C]^c [D]^d}{[A]^a [B]^b} = K_{eq}$$
>
> Note that the concentrations of each reactant and product are raised to the power of their stoichiometric coefficient. The constant, K_{eq}, is the *equilibrium constant*. When the system is not at equilibrium, the mass action ratio is called the *reaction quotient*, Q:
>
> $$\frac{[C]^c [D]^d}{[A]^a [B]^b} = Q$$
>
> Obviously, as equilibrium is approached, $Q \to K_{eq}$. Note that K_{eq} and Q are simple numbers, without units or dimensions.

energy changes involved in the making and breaking of chemical bonds in the conversion from reactant to product (or vice versa) dictate the change in enthalpy. The change in entropy arises through two processes. The first is the change in concentration, as some A is converted to B; since A and B have different molar entropies, the change in the proportion of A and B leads to a change in the overall entropy of the system. There is also a change in entropy because of the mixing of increasing amounts of B within the decreasing amount of A. The change of entropy with mixing usually dominates the entropy change.

In this brief summary I have emphasised the important role played by entropy in cellular physiology. Some of the topics have been treated very lightly so that the principles emerge clearly. More detailed treatments can be found in any standard textbook of physical chemistry or bioenergetics[5].

7.2 Temperature and reaction rate

Molecules must collide to react; temperature will therefore affect reaction rate simply because hotter molecules are moving faster and hence colliding more frequently. This effect is, however, small. Kinetic theory tells us that the frequency of two-molecule collisions in an ideal gas is proportional to the square root of the temperature. For an increase in temperature from 20 °C to 30 °C (293 K to 303 K), the frequency of collisions thus increases by a factor of 1.017 (that is, by less than 2%). A typical chemical reaction, however, at least doubles in rate for a 10 K rise in temperature. Clearly the increase in collision frequency explains only a very small fraction of the increase in reaction rate with temperature. Something else is going on.

The critical factor is that not all collisions are equal and only a very small fraction lead to a reaction. In a gas at room temperature and pressure there will be ~10^{33} collisions per second in a volume of 1 ml; if all that was needed was for molecules to collide, any reaction would go to completion in a fraction of a second. What dictates that only some collisions lead to a reaction is that the collisions need to be sufficiently energetic and the colliding molecules must be in the correct orientation. Collisions between molecules cause their covalent bonds to stretch and bend; this weakens the bonds temporarily but the bonds only break and reform if the collision delivers sufficient energy to the right places. The reaction mechanism involves the thermal excitation of bond vibrational levels, interactions between electron clouds and the redistribution of energy across the molecule but these details need not concern us

> **Box 7.3 The units of activation energy**
>
> Activation energy, E (or often E_a), is expressed in two ways, eV and kJ mol^{-1}.
>
> When activation energy is expressed as the energy per amount of material, the derived SI unit is J mol^{-1}, or for convenience kJ mol^{-1}. This is the usual unit for Gibbs energy, enthalpy and also frequently activation energy.
>
> Activation energy can also be expressed on a per particle basis in electronvolts (eV). Since 1 mol contains 6.02214×10^{23} particles (Avogadro's number), then the electronvolt is a very small number. 1 eV = 1.602×10^{-19} J, and an activation energy of 1 kJ mol^{-1} is equivalent to 1.04×10^{-2} eV.

here; the key point is that for a reaction to proceed, the energy of the collision must exceed a threshold *activation energy* (Box 7.3).

Although Leopold Pfaundler had developed qualitative arguments that only molecules exceeding a threshold energy could react, and Jacobus van't Hoff used a thermodynamic approach to explore the effect of temperature on the equilibrium, it was the Swedish physical chemist Svante Arrhenius who provided a physical interpretation of the temperature sensitivity of reactions in terms of the activation energy (Box 7.4)[6].

The implication of the activation energy for reaction rate is shown in Figure 7.3. Only those molecules with a translational kinetic energy greater than the threshold activation energy are able to react, and these are those molecules to the right (higher energies) of the threshold. In Figure 7.3a the distribution of translational kinetic energies is shown for diatomic oxygen at 273 K and 473 K, with a purely arbitrary threshold energy of 0.45 eV. It can easily be seen that the increase in temperature (200 K, quite a big increase) leads to a much greater proportion of oxygen molecules exceeding the threshold.

The proportion of molecules exceeding a given threshold energy, E, can be calculated easily, as it is given by the Boltzmann factor:

$$e^{\frac{-E}{k_B T}}$$

> **Box 7.4 The Arrhenius equation**
>
> The relationship between temperature and reaction rate perplexed physicists and chemists for over half a century. By the early twentieth century sufficient data had been accumulated for Jacobus van't Hoff to evaluate the various proposed relationships; he concluded that it was impossible to choose between them on the basis of empirical data. The two most favoured relationships were those proposed by Daniel Berthelot:
>
> $$k(T) = Ae^{DT}$$
>
> and van't Hoff himself[7]:
>
> $$k(T) = Ae^{\frac{-E}{RT}}$$
>
> Here $k(T)$ is the rate constant at absolute (thermodynamic) temperature T, A is a pre-exponential factor, R is the gas constant and D and E are constants. These two equations were based on theoretical treatments of the effect of temperature on the equilibrium constant, and hence the rate constants for the forward and reverse reactions. They assume that the changes in enthalpy and entropy for the reaction do not themselves vary with temperature. E was identified by van't Hoff as an energy term specific to a particular reaction, but there was no physical interpretation of this; van't Hoff's E and Berthelot's D were purely empirical fits that described the available data well.
>
> In deriving his equation, van't Hoff acknowledged the previous work of Leopold Pfaundler, who had argued from the Maxwell–Boltzmann distribution of energy that only molecules having energy greater than some critical value could undergo chemical change. Although Pfaundler's arguments were largely qualitative, his insights were profound. However it was Svante Arrhenius who put the relationship on a firm quantitative basis. Starting from the equation derived by van't Hoff, Arrhenius provided a clear mechanistic explanation in terms of the kinetic equilibrium between normal and reactive molecules, and recognised the role of a critical activation energy. For this reason nowadays we refer to van't Hoff's equation as the Arrhenius equation, and by 1910 it was generally accepted as capturing the important features of the effect of temperature on kinetics. The activation energy, E, is nowadays often denoted E_a[8].

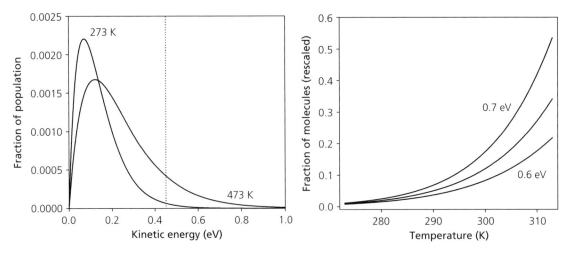

Figure 7.3 Threshold energy and temperature sensitivity. A (left panel): translational kinetic energy of oxygen molecules at 273 K and 473 K, with an arbitrary threshold shown at 0.45 eV. A much greater proportion of molecules exceed the threshold energy at 473 K than at 273 K. B (right panel): relationship between temperature and the fraction of molecules with translational kinetic energy above a threshold energy for three different values of the threshold energy: 0.6 eV (57.9 kJ mol^{-1}, lower), 0.65 eV (62.7 kJ mol^{-1}, middle) and 0.7 eV (67.5 kJ mol^{-1}, upper). Note that for these threshold energies, the fraction of molecules exceeding the threshold is very small and the numbers have been rescaled by factors of 10^9 (0.6 eV), 10^{10} (0.65 eV) and 10^{11} (0.7 eV) to render them visible on the same plot. Data plotted only for the physiological temperature range (0 °C–40 °C, 273 K–373 K).

where k_B is Boltzmann's constant and T is the absolute (thermodynamic) temperature. Typical activation energies for physiological reactions lie in the range 0.6–0.7 eV (Box 7.3), and activation energies in this range lead to temperature sensitivities whereby reaction rate doubles or trebles for an increase in temperature of 10 K (Figure 7.3b).

The importance of the activation energy to reaction rate can be seen by tracking the energy of the reactants and products as the reaction proceeds (Figure 7.4). For the reaction to proceed the potential energy of the reactants must increase to reach (or exceed) that of the energy of activation, E_a. This activation energy is the potential energy of the activated complex, where the reactants have combined but not yet yielded products. The size of the activation energy, E_a, thus dictates the kinetics: a larger E_a leads to a slower reaction because fewer collisions achieve the required energy. The energy required to achieve or exceed E_a comes largely from the translational kinetic energy of collision being converted to thermal excitation within the molecules forming the activated complex.

We saw above that collision frequency increases with temperature. While this effect is small, it does mean that even for simple reactions in the gas phase, the pre-exponential factor of the Arrhenius equation, A, is temperature dependent. The Arrhenius equation thus needs to be modified when being applied over a large temperature range. While these considerations are important theoretically, over the range of temperatures of interest to ecologists and physiologists, the temperature dependence of A is so weak compared with the large effect of the exponential Boltzmann factor that it can be ignored.

Because it offered a physical explanation of the activation energy, the Arrhenius equation largely superseded the others discussed by van't Hoff. An important point, however, is that even for the simplest of elementary reactions in the gas phase, the Arrhenius equation is a low-precision approximation that applies only over a narrow range of temperatures. While the value of the Boltzmann factor can be derived empirically from the temperature sensitivity of the reaction, attempting to derive the pre-exponential factor from kinetic theory can lead to estimates of the reaction rate that are out by several orders of magnitude. These limitations of the Arrhenius equation led to the development of *transition state theory* and a more complex rate equation.

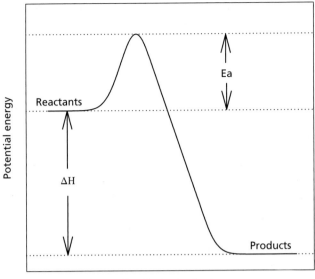

Figure 7.4 Energy change during the progress of a simple chemical reaction. The size of the activation energy, E_a, determines the kinetics, that is how fast the reaction will proceed under the given circumstances. The difference in potential energy between reactants and products (ΔH) determines whether the reaction is thermodynamically favoured and is unaffected by the activation energy. The reaction coordinate denotes the progress of the reaction.

7.3 Transition state theory

Transition state theory was developed independently in the 1930s by Henry Eyring at Princeton University, and Meredith Evans and Michael Polanyi at the University of Manchester[9]. Transition state theory can be applied to reactions taking place in solution, and is therefore directly applicable to cellular physiology. The theory is extremely successful at explaining the reaction rates of elementary chemical reactions where the reaction passes through a transition state:

$$A + BC \rightleftharpoons (ABC)^{\ddagger} \rightarrow AB + C$$

Here $(ABC)^{\ddagger}$ represents the transition state (Box 7.5). Typically this is extremely short-lived, with a lifetime of the order of that of molecular vibration (10^{-15}–10^{-12} seconds). The theory is based on three precepts:

1. The transition state lies at the peak of the potential energy surface.
2. The transition state is in quasi-equilibrium with the reactants.
3. The transition state converts into products, and kinetic theory can be used to calculate the rate constant for this conversion.

Box 7.5 The transition state

The *transition state* is the configuration of reacting molecules at the local energy maximum which defines the activation energy. It is a highly unstable intermediate state between reactants and products. It has partial bonds and cannot be isolated as an individual molecule, but its presence can be detected by extremely fast infrared laser spectroscopy[10].

The *activated complex* is a more general term for molecular configurations close to the energy maximum. It refers to the assembly of atoms around the energy of the transition state, incorporating not only the transition state itself but also structures at intermediate energies. The distinction is that a collision between reactant molecules may or may not result in a successful reaction. The outcome depends on factors such as the combined kinetic energy, relative orientation and the internal energy of the reactant molecules. Given suitable conditions, the colliding molecules may form an activated complex, but they do not necessarily go on to form products; the activated complex may instead fall apart and revert to reactant molecules. On the other hand if they achieve the transition state, then the only available pathway is to products and the necessary impetus for this is provided by internal molecular vibrations.

Transition state theory differs from classical chemical kinetics in that the reactants are only in quasi-equilibrium with the transition state. We can, however, still use classical equilibrium thermodynamics to express this equilibrium in terms of the difference in Gibbs energy between the transition state and the reactants. The molecules in the transition state do not exhibit a Boltzmann distribution of energies; the rate of formation of product follows instead from the concentration of the transition state and the frequency with which this is converted to products. This frequency can be obtained from the equivalence of the energy, E, of an excited oscillator derived from quantum theory:

$$E = h\nu$$

where ν is the frequency and h is Planck's constant, with the energy derived from classical physics:

$$E = k_B T$$

The frequency of conversion of the transition state to products, F, is thus given by:

$$F = \frac{k_B T}{h}$$

indicating a temperature dependency to the formation of product. A rigorous derivation also involves a *transmission coefficient*, κ, which is the probability that the transition state will decompose to form the products (and which includes the likelihood of quantum tunnelling). This coefficient is usually taken to be 1, and hence left out of the simpler formulation.

The first-order rate constant for the formation of product from the transition state is given by:

$$k = \frac{k_B T}{h} e^{\frac{\Delta G^{\ddagger}}{RT}}$$

where $\Delta G^{\ddagger}$ is the difference in standard Gibbs energy between the reactants and the transition state and other symbols are as before. The Gibbs energy can be separated into its enthalpic and entropic terms to yield the more explicit form of the Eyring–Polanyi equation:

$$k = \frac{k_B T}{h} e^{\frac{\Delta S^{\ddagger}}{R}} e^{-\frac{\Delta H^{\ddagger}}{RT}}$$

where $\Delta H^{\ddagger}$ is the enthalpy of activation and $\Delta S^{\ddagger}$ the entropy of activation. This equation shows clearly that for a given activation enthalpy ($\Delta H^{\ddagger}$), the reaction proceeds faster for a higher entropy of activation ($\Delta S^{\ddagger}$). This can be achieved, for example, by release of water molecules from the hydration sphere of the reactants as they form the transition state.

A consequence of the Eyring–Polanyi equation is that a plot of $\ln(k/T)$ against T^{-1} yields a straight line of slope $-\Delta H^{\ddagger}/R$, from which the enthalpy of activation can be calculated, and an intercept $\ln(k_B/h) + \Delta S^{\ddagger}/R$, from which the entropy of activation can be estimated. Estimating both $\Delta H^{\ddagger}$ and $\Delta S^{\ddagger}$ from the same plot is not, however, statistically sensible, as the two parameters are not estimated independently and the plot contains the same variable (T) on both axes. It is much better practice to estimate the two parameters independently.

The finer details of transition state theory and the derivation of the equation need not concern us here[11]. The important aspect is that when we consider the more complex chemical changes of physiology we have to abandon the Arrhenius equation and move to the more complex Eyring–Polanyi formulation based on quantum mechanics.

7.4 Enzyme-catalysed reactions

So far we have considered simple elementary reactions that typically follow from a single encounter between atoms or molecules. Complex reactions are very different because the overall reaction takes place through a series of elementary reactions, with the generation of intermediates that do not appear in the summary stoichiometric chemical equation.

In the cell, the complex chemical reactions that drive physiology are catalysed by enzymes. These enzymes increase the rate of reaction by a large factor, which can be as great as $\times 10^{19}$ relative to the uncatalysed rate. They do so by binding the transition state more strongly than the substrate, thereby lowering the activation energy for the reaction[12].

The kinetics of enzyme reactions were first studied using invertase, the enzyme that converts the disaccharide sucrose to its monosaccharide components glucose and fructose. This work revealed three principal features of enzyme-catalysed reactions:

1. For an initial concentration of substrate, $[S]_0$, the initial rate of product formation is proportional to the initial concentration of enzyme, $[E]_0$.
2. For a given value of $[E]_0$ and a low value of $[S]$, the rate of product formation is proportional to $[S]$.
3. For a given value of $[E]_0$ and high values of $[S]_0$, the rate of product formation becomes independent of $[S]$, reaching a maximum value, V_{max}.

These features were captured by Leonor Michaelis and Maud Menten in a simple mechanism. The substrate first binds to the enzyme to form an enzyme–substrate complex:

$$E + S \rightarrow ES$$

The rate of this binding step is given by

$$\text{Rate of binding} = k_a [E][S]$$

where k_a is the rate constant. The enzyme–substrate complex may then either dissociate, releasing the substrate unchanged:

$$ES \rightarrow E + S$$

where the rate is:

$$\text{Rate of dissociation} = k_a' [ES]$$

or the enzyme–substrate complex may move to the transition state, $ES^\ddagger$, which yields the products which are then released:

$$ES \rightarrow ES^\ddagger \rightarrow E + P$$

$$\text{Rate} = k_{cat} [ES]$$

These kinetics are described by the Michaelis–Menten equation, where the rate of conversion of substrate to product is given by:

$$\text{Rate} = V_{max} \left(\frac{[S]}{(K_M + [S])} \right)$$

Here K_M is the Michaelis constant, defined as:

$$K_M = \frac{(k_a' + k_{cat})}{k_a}$$

K_M is the substrate concentration at which the rate of reaction is exactly half of V_{max} (Figure 7.5). It is an inverse measure of the affinity of the substrate for the enzyme: a small value of K_M means that V_{max} is approached at relatively low values of $[S]$[13].

The kinetic parameters for enzymes vary widely, but it is often found that the substrate concentration in the cell is ~ K_M. This allows the cell to exert a fine degree of control over the rate of a reaction, and hence flux through a metabolic pathway, through small adjustments in substrate concentration.

Having established the broad features of the kinetics of enzyme-catalysed reactions, we can now examine the effects of temperature on reaction rate.

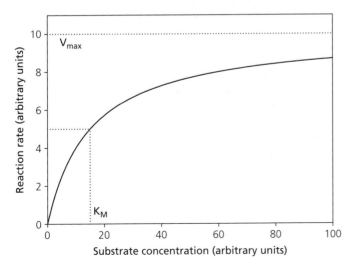

Figure 7.5 Michaelis–Menten kinetics for a simple enzyme-catalysed reaction, showing reaction rate as a function of substrate concentration. The dotted line shows the maximum reaction rate (V_{max}); also shown is the Michaelis constant (K_M), the substrate concentration at which reaction rate is precisely half V_{max}.

7.5 Evolutionary adaptation of reaction rate to temperature

Organisms live all over the Earth, in habitats ranging from the cold northern forests to boiling thermal springs. The thermodynamic effects of temperature on reaction rate discussed above might suggest that organisms from the extremes of this temperature range would differ enormously in their physiology. For example, all other things being equal, the rate of a typical reaction would increase by a factor of between 40 and 80 when comparing an organism with a body temperature of 0 °C with one whose body temperature was 40 °C. Comparing microbes living at –20 °C and 100 °C the factorial difference in reaction rate would be a staggering 60 000.

This does not happen, so clearly all things are not equal and there is compelling evidence that evolution has managed to ameliorate this direct effect of temperature on reaction rate. A key part of this amelioration involves changes in proteins: proteins isolated from organisms adapted to live in cold environments cease to function at temperatures well below those tolerated by organisms adapted to live at warmer temperatures. Conversely, proteins from tropical organisms fall apart or stop working at temperatures well above those at which polar organisms live. No enzyme has been found that is able to operate effectively across the entire temperature range of life. That is, no enzyme has been found that can maintain its kinetic properties over the full range of physiological temperatures. The challenge for the physiologist is to determine in what way proteins from various organisms differ, and how these differences affect function.

The important advances in our understanding of evolutionary adjustment to protein structure and function have come from comparative studies of related organisms living in different thermal environments. Marine organisms, and particularly fish, have proved important in unravelling how protein function is adapted to temperature. The advantage of using fish for these studies is that they generally exchange heat rapidly with the surrounding water (principally through their gills) and most fish are thus close to thermal equilibrium with their environment. The high thermal capacity of water means that its temperature changes relatively slowly and the enzymes of many aquatic organisms therefore operate at a well-defined and relatively constant temperature. This makes it fairly straightforward to undertake comparative studies of thermal adaptation, by selecting species living at different temperatures.

In contrast, intertidal and terrestrial organisms are generally subject to a highly variable thermal environment. This makes it difficult to decide on a representative temperature for a particular species in a given habitat. More importantly, it may also raises the possibility that the nature of the thermal physiology in terrestrial and aquatic organisms is different.

A nice introduction to the evolutionary adaptation of enzymes to temperature is the classic early study of muscle lactate dehydrogenase (A_4–LDH) in four species of barracuda (*Sphyraena*) from the eastern Pacific by John Graves and George Somero. The different species are similar morphologically, and all are schooling, pelagic, predatory fish. If we assume that the A_4-LDH in the various species should have similar activity at their normal environmental temperature (which seems a reasonable assumption ecologically and evolutionarily), then we would expect that two features in particular will be similar across the species: the affinity for substrate (K_M) and the rate at which the enzyme operates (k_{cat}). If these two characteristics are compared at the same experimental temperature, then the enzymes isolated from the different species differ in their properties (Table 7.1).

These data show clearly that the kinetics of enzymes from fish living at different temperatures

Table 7.1 Kinetic parameters for A_4-lactate dehydrogenase isolated from three species of barracuda (*Sphyraena*) living at different seawater temperatures in the eastern Pacific Ocean. Kinetic parameters for a fourth species, the temperate *S. idiastes*, were identical to those of *S. argentea*. K_M is measured for pyruvate[14].

Species	Mid-range temperature, TM (°C)	Measured at 25 °C		Measured at TM	
		K_M (mM)	k_{cat} (sec^{-1})	K_M (mM)	k_{cat} (sec^{-1})
S. argentea	18	0.34	893	0.24	667
S. lucasana	23	0.26	730	0.24	682
S. ensis	26	0.20	658	0.23	700

are themselves different. The critical observation, however, is that when these properties are compared *at the temperatures at which each species lives*, then the kinetic properties are remarkably similar (Table 7.1). Over evolutionary time, the A_4-LDH of these closely related species have become modified so that their kinetic properties match the thermal environment in which they live.

What we cannot do for *Sphyraena* is link the kinetics of the enzyme variants to the physiological performance of the fish, although it seems a reasonable assumption that selection has operated to optimise muscle performance in relation to habitat temperature. A study that did establish the links between enzyme kinetics and individual performance was the work by Ward Watt and colleagues on phosphoglucose isomerase (PGI) in butterflies of the genus *Colias*. These colourful and familiar butterflies are known as sulphurs in the USA, and clouded yellows in UK and Europe.

Flight is a key facet of butterfly ecology. Female *Colias* butterflies lay eggs singly and must fly between host plants and as adult lifespan is only a few days, the interaction between weather and flight performance may limit the reproductive output and hence population size. In *Colias* vigorous flight requires a body temperature of 35–39 °C, and this is also the temperature range over which wingbeat frequencies are highest. Because of this, the butterflies only fly in bright sunlight and at low wind speeds, and in consequence in non-ideal weather flight may be restricted to a few hours in the day.

PGI is an important enzyme in glycolysis, and is critical to the provision of energy during flight. PGI in *Colias* exists in a number of variants (allozymes: see Box 7.6), typically between four and six in natural populations but with only two or three present at a frequency >10%. Analysis of the kinetics of the different PGI allozymes isolated from homozygous individuals showed that while V_{max} did not vary much, there was significant variation in K_M. An enzyme with a lower K_M can work effectively at a lower substrate concentration, but it turns out that there is an inverse relationship between activity (V_{max}/K_M) and thermal stability: the more active allozymes are less stable (Table 7.2). Interestingly, the substrate utilisation efficiency for some heterozygotes is higher than for either homozygote, a nice example of heterozygous advantage (heterosis).

> **Box 7.6 Some definitions**
>
> The terminology surrounding enzyme variants and the genes that code for them can be confusing, but the terms are rigorously defined and the subtle differences are important[15].
>
> *Enzyme variants*
> **Isozymes** (also known as isoenzymes) are enzyme variants that catalyse the same reaction but which differ in amino acid sequence. Isozymes usually have different kinetic parameters (e.g. different K_M values), or different regulatory properties, and they are coded by genes at different loci.
>
> **Allozymes** (also known as alloenzymes) are enzyme variants that are coded by different alleles at the same gene locus.
>
> *Gene variants*
> **Homologous genes** (also known as homologues in the UK and homologs in the USA) are genes inherited by two species from a common ancestor. While homologous genes can be similar in sequence, similar sequences are not necessarily homologous. Orthologues are homologous genes where a gene is found in two different species, but the origin of the gene is a common ancestor.
>
> **Orthologous genes** (also known as orthologues or orthologs) are genes in different species that evolved from a common ancestral gene by speciation and typically retain the same function in the course of evolution. Over long periods of evolutionary time, the functions of orthologous genes may diverge; indeed this is an important process in the evolution of enzyme families.
>
> **Paralogous genes** (also known as paralogues or paralogs) are genes that arose by gene duplication.
>
> Note that orthologues and paralogues are both homologues. Orthologues are homologous genes that arose from a speciation event, whereas paralogues are homologous genes that arose from a duplication event.

Table 7.2 Kinetic parameters and thermal stability of phosphoglucose isomerase (PGI) allozymes in *Colias* butterflies. Kinetic parameters measured at 30 °C and pH 8.75; thermal stability is activity remaining at 30 °C after 60 min incubation at 60 °C.

Allozyme	V_{max}/K_M	Stability (%)
2	3.06	23 ± 2
3	1.87	72 ± 3
4	1.83	90 ± 4

This trade-off between activity and stability leads to a polymorphism that is maintained by the thermally variable environment. The more stable allozymes are favoured in warm conditions, but the more active allozymes allow the butterflies to fly over a wider temperature range and enhance survival in cooler or more variable environments. The distribution of allozymes also varies across species of *Colias* depending on the thermal environment they inhabit[16]. Interestingly, these patterns are not seen in two other metabolic enzymes, phosphoglucomutase (PGM) and glucose-6-phosphate dehydrogenase (G6PD).

Genetic variation in enzyme variants related to temperature was also shown by Dennis Powers and colleagues for *Fundulus heteroclitus* (the common killifish or mummichog). This small fish lives in the estuaries, bays and coastal waters of the eastern seaboard of the United States, from Newfoundland to Florida. The locus encoding the enzyme LDH-B has two alleles, LDH-Ba and LDH-Bb. In northern latitudes, where the mean water temperature is ~6 °C the populations express exclusively LDH-Bb, whereas in southern latitudes where the water temperature is ~ 21 °C, the population is fixed for allele LDH-Ba. In between, there is a cline in the proportion of the two alleles (Figure 7.6). This cline remains essentially identical some 40 years later, despite a significant shift in the distribution of temperatures along the coast.

The enzymes coded for by the two alleles are distinguished by a single nucleotide polymorphism and they differ in substrate affinity (K_M), reaction rate (k_{cat}) and thermal stability. The northern fish also have a greater concentration of enzyme, and they can swim faster at 10 °C than fish with the southern allozyme[17].

These three studies, all rightly now established as classics of ecological physiology, used slightly different approaches, but together they established a clear framework for understanding how organisms maintain metabolic activity in the face of differences in temperature. They showed how organisms in different thermal environments express different variants of key metabolic enzymes, and that the kinetic differences between these variants carry through to the performance and fitness of the individuals. The question they pose is what has changed to modify the kinetic behaviour of the LDH or PGI enzyme variants in the different populations or species?

7.5.1 Molecular mechanisms of adaptation to temperature

In theory the activity of an enzyme within the cell could be modified by changes in enzyme or substrate concentration, or changes in the cellular environment such as pH or the concentration of cofactors and inhibitors. While we cannot ignore changes in the cellular environment, in *Sphyraena*, *Colias* and *Fundulus* the changes were evident in the isolated enzyme studied *in vitro*; furthermore, differences in electrophoretic behaviour revealed differences in molecular charge indicating that the LDH molecules from the various species of *Sphyraena* were different in some way.

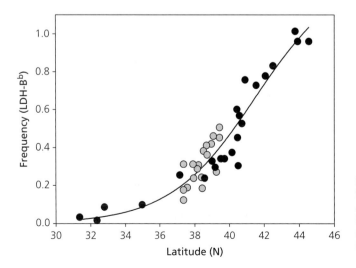

Figure 7.6 A latitudinal cline in the frequency of the allele for LDH-Bb in the coastal killifish *Fundulus hetreroclitus* along the Eastern seaboard of the USA. Data for coastal locations (black symbols) and estuaries (grey symbols) are distinguished, but the line (sigmoidal model) is fitted to all the data[18].

Insight into the nature of these changes came from comparative studies of a different enzyme, the Ca^{2+}-Mg^{2+}-activated ATPase involved in the action of muscle myosin. In a series of studies Ian Johnston and colleagues measured the activity of the myosin-ATPase isolated from fish living at very different temperatures. Because no single species, or even genus, of fish lives across the entire thermal spectrum, these studies necessarily involved fish from different families. Data for a tropical damselfish (probably *Chrysiptera cyanea*, though described as *Pomatocentrus uniocellatus*) and the polar icefish *Champsocephalus aceratus* are shown in Figure 7.7.

At any given temperature, the absolute activity of the polar enzyme is greater than the tropical enzyme, and at 20 °C, the difference is × 4.8. This might suggest that the enzymes from the warmer water fish are somehow deficient, and indeed cold-adapted enzymes are often described as being 'more efficient' (in that they have greater catalytic rate at any given temperature). This is somewhat misleading, in that it could imply that millions of years of selection have somehow failed to drive the evolution of an A_4-LDH that is efficient in warm water. The more sensible comparison evolutionarily is to compare activity at temperatures typical of those at which the fish actually live (0 °C for *Champsocephalus* and 25 °C for *Chrysiptera*); when this is done, activity is similar. As with *Sphyraena*, natural selection has driven the evolution of enzyme variants that conserve kinetic parameters in the face of the effect of temperature.

The slope of the Eyring–Polanyi plot is greater for the tropical *Chrysiptera* than it is for the polar *Champsocephalus*, indicating a greater enthalpy of activation, $\Delta H^{\ddagger}$. For the entire group of fish analysed, the result was very clear: over a temperature range of −2 °C to 30 °C the enthalpy of activation increased with temperature (Figure 7.7b). The activation energy, E_a, is related to the enthalpy of activation, $\Delta H^{\ddagger}$, as:

$$\Delta H^{\ddagger} = E_a - RT$$

where R is the gas constant and T the absolute (thermodynamic) temperature. For typical cellular conditions the value of $R.T$ is small relative to E_a, and so $\Delta H^{\ddagger} \sim E_a$.

Transition state theory tells us that the rate of conversion of reactants to products is dependent not on the enthalpy of activation alone but on the Gibbs energy of activation, which combines the enthalpic and entropic contributions:

$$\Delta G^{\ddagger} = \Delta H^{\ddagger} - T\Delta S^{\ddagger}$$

If the equilibrium constant between reactants and the transition state (and hence the Gibbs energy of activation $\Delta G^{\ddagger}$) is to remain broadly consistent across the range of temperatures, then the decrease

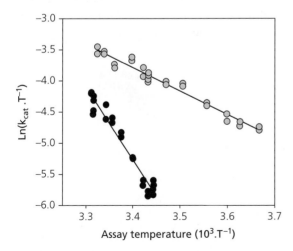 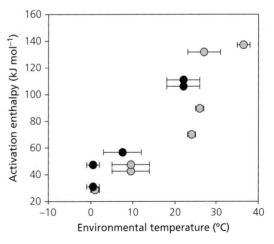

Figure 7.7 Activity as a function of temperature in Ca^{2+}-Mg^{2+}-activated myosin-ATPase from fish muscle. A (left panel): Eyring–Polanyi plots for *Chrysiptera cyanea* (tropical, black symbols) and *Champsocephalus aceratus* (polar, grey symbols). B (right panel): enthalpy of activation ($\Delta H^{\ddagger}$) for myosin-ATPase as a function of mean habitat temperature in 12 species of fish. Error bars show annual range of temperature experienced by each species, with the data symbol plotted at the midpoint. Grey and black symbols distinguish separate data sets[19].

in the enthalpic contribution to the Gibbs energy has to be offset by a change in the entropic contribution[20]. This relationship between the enthalpic and entropic contributions to free energy is termed enthalpy–entropy compensation, and has proved a highly contentious subject in physical chemistry (Box 7.7).

Since physiological constraints confine the Gibbs energy of activation to a narrow range of values, changes in the enthalpic contribution have to be associated with changes in the entropic contribution. Something in the myosin-ATPase molecule has changed to reduce the enthalpy of activation, and at the same time something must also have changed

Box 7.7 Entropy–enthalpy compensation

Entropy–enthalpy compensation is the name given to the strong correlation often observed between changes in enthalpy and entropy in biological processes. A typical example is shown by fish myosin ATPase (Figure 7.8a).

This is a striking relationship. It can arise when the equilibrium constant for the formation of the transition state, and hence the Gibbs energy of activation, is relatively constant but the enthalpy of activation changes. Under these circumstances an inverse relationship between $\Delta H^\ddagger$ and $\Delta S^\ddagger$ is inevitable and mathematically trivial.

A strong correlation can also arise deceptively when $\Delta H^\ddagger$ and $\Delta S^\ddagger$ are estimated from the same Eyring–Polanyi plot ($\Delta H^\ddagger$ from the slope and $\Delta S^\ddagger$ from the intercept), through a correlation in their respective errors. The slope of the line (which has dimensions of temperature) is then close to the geometric mean temperature at which the data points were obtained; in the example above it is 294 K. When this effect is dominant, it can lead to very tight correlations (as evident in the example below) without the scatter that normally characterises biological data.

There is, however, increasing independent evidence that variation in $\Delta S^\ddagger$ across enzyme orthologues is real. This comes from studies where $\Delta G^\ddagger$ and $\Delta H^\ddagger$ are estimated independently, as for example in isothermal titration calorimetry (ITC). Figure 7.8b shows the data for 171 studies of ligand binding by 32 different proteins as determined by ITC. While the limits to sensitivity of ITC set boundaries to observable data, statistical analysis indicates that the observed relationship between $\Delta H^\ddagger$ and $\Delta S^\ddagger$ is real.

While the statistical problems in demonstrating entropy–enthalpy compensation are genuine and serious, current evidence indicates that the phenomenon itself is also real and reflects the structural changes that underpin evolutionary adjustments of protein function to temperature and particularly the properties of the weak bonds that are critical in determining protein structure and function[21].

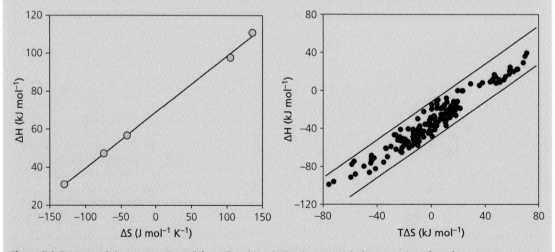

Figure 7.8 Entropy–enthalpy compensation. A (left panel): enthalpy (ΔH) and entropy (ΔS) of activation for Ca^{2+}-Mg^{2+}-activated myosin-ATPase from fish living at different temperatures. Enthalpy and entropy were estimated from the same Arrhenius plot and the fitted line is a least-squares fit. B(right panel): enthalpy and entropy (here expressed as energy, $T\Delta S$) for 171 protein–ligand binding reactions. Measurements, with thermodynamic parameters estimated independently using isothermal titration calorimetry (ITC). The two lines show the limits of detection for ITC, corresponding to Gibbs energy of -12 kJ mol^{-1} (upper) and -52 kJ mol^{-1} (lower)[22].

to alter the entropic contribution. These two somethings may, or may not, be the same.

The statistical problems of estimating $\Delta H^\ddagger$ and $\Delta S^\ddagger$ from the same data set clouded this issue for some time, but there is now clear physical evidence for an increased entropic contribution to the Gibbs energy of activation from enzyme variants in organisms living at low temperatures. The most powerful such evidence is that associated with the differences in enthalpy of activation, the different myosin-ATPase orthologues also exhibit striking differences in thermal stability. When measured at a common assay temperature of 37 °C, orthologues from three species of Antarctic fish had a half-life of thermal denaturation of less than 5 minutes, whereas those from eight species from the tropic Indian Ocean had half-lives of 40–200 minutes.

This matches the pattern seen in PGI from *Colias* butterflies (Table 7.2). A lower enthalpy of activation and a lower temperature of denaturation are now recognised as general features of enzyme orthologues from organisms adapted to living at different body temperatures (Figure 7.9).

The increase in the temperature of thermal denaturation in proteins adapted to higher temperatures is logical, but the slope of the relationship is typically much less than unity. In other words, the margin of safety is greater at lower temperatures than at higher temperatures.

All molecules are in a continual state of internal movement, with the covalent bonds vibrating, twisting and bending. These movements are considerably slower than the atomic vibrations, but it is these slower movements that dictate the rate at which a given enzyme molecule can process substrate. This is because the rate-limiting step in catalysis is usually the binding of the substrate to the enzyme; this step involves the reorganisation of many weak bonds, and typically has a time constant in the range 10^{-3} to 10^{-6} secs. In contrast, the reactions within the active site, where substrates are converted to products through covalent chemistry and quantum effects are extremely fast (~10^{-12} secs).

These internal movements mean that any given protein molecule is moving continually between microstates that differ in their three-dimensional configuration. The conformational entropy of the molecule is a direct measure of the number of these microstates (Chapter 2). For a protein, the number of available microstates is enormous, but only some of them are compatible with enzyme function. The population of enzyme molecules within the cell

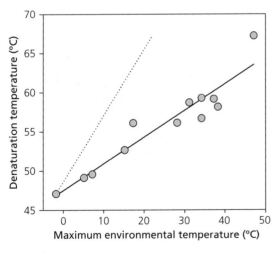
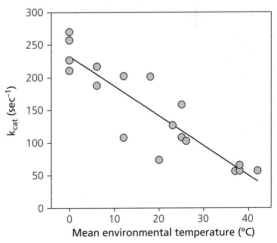

Figure 7.9 Enzyme stability, activity and temperature. A (left panel): denaturation temperature (the temperature at which half of the secondary structure has been lost, as measured by circular dichroism spectroscopy) of eye lens crystalline from a range of vertebrate species as a function of the average maximum environmental temperature experienced by that species. The solid line shows a least-squares fit, and the dotted line shows the 1:1 relationship. B (right panel): catalytic function (k_{cat}, sec^{-1}), measured at a common temperature of 0 °C, of A_4-LDH orthologues as a function of mean environmental temperature[23].

may thus be regarded as being a dynamic equilibrium between enzymes that are capable of activity and those that are, temporarily but reversibly, incapable of catalysis. The frequency distribution of microstates is temperature dependent and at higher temperatures a greater fraction of the population of enzyme molecules will be in a configuration which does not allow for ligand binding and recognition.

This reversible inactive state is quite different from the irreversible denaturation at higher temperatures. While activity or activation energy may vary across individual enzyme molecules, thermal denaturation of an enzyme molecule is catastrophic and irreversible. The variation in the temperature of irreversible thermal denaturation does, however, relate to the innate flexibility of the enzyme molecule. The greater thermal sensitivity of low-temperature orthologues is because they have a more flexible structure at any given temperature, which allows them to retain the necessary flexibility at their working temperature. Current evidence suggests that significant changes in flexibility, and hence activity, can be achieved with changes to a small number of amino acid residues, and sometimes only one[24].

The change in enthalpy when a ligand binds to the enzyme reflect the specificity of interactions between the two. In simple terms, the stronger the binding, the greater the change in enthalpy, and hence the greater the activation energy for the reaction. The strength of binding can be altered by changes in the amino acid residues influencing the binding. It is interesting, but not surprising, that the regions of the LDH molecule that appear to be the site of evolutionary adaptation to temperature are those that undergo large changes in conformation during the binding of ligands.

The changes in entropy, as always, are more complex. A major contribution comes from the loss of translational and rotational degrees of freedom when the ligand binds. There will also be significant contributions from solvation effects, notably the release of tightly bound water, and because of this changes at the surface of the enzyme molecule (where it interacts with the water of the cell) will be particularly important. A third contribution to the overall change in entropy comes from the conformational entropy of the enzyme, and the nature and magnitude of the contribution of this to the overall change in entropy on binding will be influenced by the amino acid composition of the enzyme. Some amino acids make a large contribution to conformational entropy and others a smaller contribution. For example, glycine makes the largest contribution of any amino acid because it has the greatest number of conformational states, whereas proline contributes the least. A switch from a glycine residue to a proline residue would thus exert a powerful effect on the overall conformational entropy of the enzyme molecule.

This delicate balance between non-covalent stabilising factors and entropic factors is necessary to allow the enzyme to function. While a given enzyme molecule will have many hundreds of weak bonds, the counterbalancing entropic factors mean that net stability is equivalent to only a few hydrogen bonds. Because of this, the thermal input from the cellular environment plays a significant role in driving the conformational changes that underpin catalytic activity. It is also why a single enzyme molecule cannot function over a wide temperature range, and certainly not over the full temperature range for life on Earth.

The recent explosive increase in the number of genome sequences for homologous enzymes and also in the number of high-resolution crystal structures for individual enzymes allows us to explore whether there are any general features of enzyme adaptation to temperature. In particular, comparative studies of proteins from microbial extremophiles (Bacteria and Archaea) living in polar regions or hydrothermal vents with those found in temperate regions have strengthened the general patterns of adaptation of enzyme function to temperature, and we are now able to draw a broad picture of what is involved in the adaptation of enzymes to temperature (Table 7.3).

This brief outline leads us to a critical point in interpreting comparative studies of enzyme adaptation to temperature, which is that there is a three-way relationship between structure, function and temperature (Figure 7.10). Comparison of properties such as kinetics or thermal stability at a common temperature will indicate whether orthologous enzymes from different organisms differ functionally. But it is only by comparing properties at the temperature to which the enzyme is adapted to operate that the physiological effectiveness of those evolutionary changes can be assessed.

Table 7.3 Some general features of evolutionary enzyme adaptation to temperature[25].

Feature
General features of enzymes adapted to a lower temperature
A more flexible structure
Higher activity at any given temperature
Denature at a lower temperature
Lower enthalpy of activation
Lower entropy of activation
Broad classes of change in amino acid residues in enzymes adapted to a lower temperature
Fewer ionic interactions
Fewer internal hydrogen bonds
Lower hydrophobicity
More polar (charged) residues

7.6 Managing the cellular environment

Laboratory studies of enzyme function have necessarily tended to examine isolated enzymes under controlled conditions in dilute solution *in vitro*. No molecule in the cell exists in isolation, however, and enzyme function is affected strongly by the cellular environment.

7.6.1 Managing concentrations

One way that the cell may offset the rate-limiting effect of a reduced temperature is to increase the concentration of enzyme. This was shown by Bruce Sidell for cytochrome c in green sunfish, *Lepomis cyanellus*; this protein increased by a factor of 1.5 following laboratory acclimation from 25 °C to 5 °C, largely because degradation was slowed more than synthesis when the fish were transferred to the lower temperature.

Examples of increased enzyme number at low temperature remain rare. This is partly because for a long time these were technically demanding measurements to make, requiring careful isolation and purification followed by assay with highly specific monoclonal antibodies[26]. There is also the physiological problem that there is only so much room in the cell and it is difficult to envisage how a cell could increase the number of all enzyme types by 50% and still function.

The advent of modern genomic techniques has allowed analysis of expression profiles in response to variations in temperature (both natural and in the laboratory). Microarray studies of *Fundulus* have revealed substantial variation (factors of 1.5 to 2) in natural expression of ~ 20% of > 900 genes. There are too few such studies to draw general conclusions, although there are indications that a change in temperature induces significant changes in expression of genes associated with growth and metabolism[27].

Cytochrome c is involved in the production of ATP by mitochondria, a process central to everything that the cell does; compensation for temperature in ATP regeneration is therefore critical. Some fish living in cool or cold climates have increased numbers of mitochondria in the

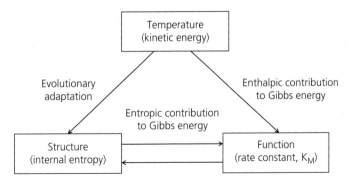

Figure 7.10 The three-way relationship between protein structure, protein function and temperature.

red muscle and the density of cristae within the mitochondria also increases, presumably as a mechanism to compensate for the fact that at the rate at which an individual mitochondrion can regenerate ATP is reduced at these low temperatures. The denser packing of mitochondria about the muscle fibres also reduces the diffusion distance between the sites where ATP is produced and where it is used. Similar changes in red and white muscle can be induced by acclimation in the laboratory in some species, and seasonal variations in mitochondrial density have been reported for the infaunal polychaete *Arenicola marina* (the lugworm). However studies of some Antarctic marine invertebrates suggest that this is not a universal response[28].

Control over reaction rate, and hence flux through a metabolic pathway, can also be exerted by managing the concentration of substrate. For many reactions, substrate concentration in the cell is close to K_M, which means that reaction rate is exquisitely sensitive to small changes in that concentration. This is both a problem (the intracellular environment needs careful management) and a benefit (it allows for subtle control)[29].

7.6.2 Managing pH

For many enzymes activity depends critically on pH. This is because pH controls the charge state of the active site, and thereby the binding of ligands. For example, in LDH the binding of ligands to the active site involves coordination with the guanidine group of an arginine residue and the imidazole group of a histidine residue. It has been known since the first half of the twentieth century that in ectotherms such as reptiles and fish there is an inverse relationship between the pH of body fluids and temperature. Two different but related theories were developed to explain these observations.

The first was that organisms manipulated their acid–base status to maintain a constant difference between plasma pH and the pH of water, that is to maintain a constant ratio of $[OH^-]/[H^+]$. This was termed *constant relative alkalinity*[30].

The ionisation of water is, of course, temperature dependent (Chapter 5): the pH of neutral water (usually designated pN) is 7.47 at 0 °C but 6.77 at 40 °C. The water remains neutral, because the change in pH is balanced exactly by a change in pOH. The relationship is not linear and so the rate of change of pH with temperature ($\Delta pN/\Delta T$) also varies. The oft-quoted value of -0.0171 pH units K^{-1} is true only at 25 °C; at 0 °C it is -0.0210 and at 40 °C it is -0.0145. Organisms maintaining a constant relative alkalinity will thus have a pH-temperature curve that tracks, but is offset from, pN (Figure 7.11).

A related hypothesis is that the pH of body fluids is regulated to maintain a constant ionisation of the imidazole moiety of histidine residues. At physiological pH only about half the imidazole groups are protonated (Im$^+$) and about half are not (Im). That is the dissociation ratio, α_{imid}, ~ 0.5, where:

$$\alpha_{imid} = \frac{[Im]}{([Im]+[Im^+])}$$

LDH illustrates the importance of this: when the imidazole group is protonated, pyruvate binds, whereas when it is deprotonated, lactate binds. Maintaining α_{imid} ~ 0.5 allows for reversibility and delicate control of LDH activity.

Because the intracellular and extracellular non-bicarbonate buffering is dominated by imidazole groups, predominantly from histidine residues in proteins, this buffering will vary with temperature. This prompted Robert Reeves to propose his *alphastat hypothesis*, which is that organisms actively regulate their intracellular pH so as to maintain α_{imid} ~ 0.5, and hence maintain net protein charge independent of temperature[31]. An example of the regulation of intracellular pH is shown in Figure 7.11b.

How strong is the evidence of the alphastat hypothesis? The pH of muscle and blood plasma in the turtle *Pseudemys scripta* acclimated to a range of temperatures suggested that the imidazole dissociation ratio (α_{imid}) maintained a constant value over a range of body temperatures (Figure 7.12). The value in muscle was considerably higher than 0.5, but both tissues maintained a constant dissociation ratio. This suggests that, at least in this species, alphastat regulation is an important factor in acclimation to temperature.

A compilation of data for a wide range of ectothermic vertebrates, however, suggests that while

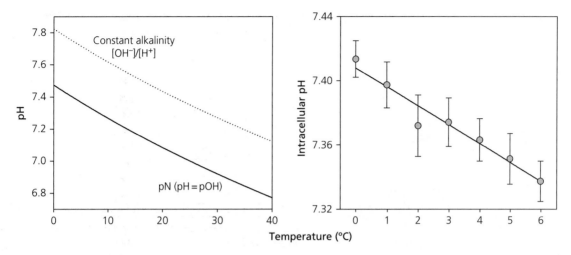

Figure 7.11 Constant alkalinity. A (left panel): variation in the pH of neutral water (pH = pOH, often designated pN: solid line) and a line of constant alkalinity (dotted line), where the concentration ratio [OH⁻]/[H⁺] is constant. B (right panel): variation in the intracellular pH of a teleost fish, the Antarctic eelpout *Pachycara brachycephalum*, with temperature. Intracellular pH was measured by ^{31}P-NMR. Note the narrow temperature range; the slope of the fitted least-squares regression line is −0.012[32].

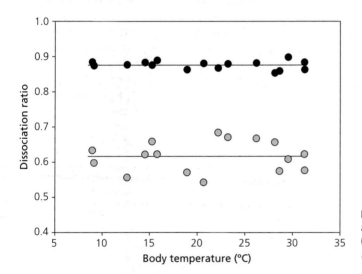

Figure 7.12 Imidazole dissociation ratio (α_{imid}) as a function of body temperature for skeletal muscle (black symbols) and plasma (grey symbols) in the turtle *Pseudemys scripta*[33].

the pH of extracellular and intracellular fluids does vary with temperature, the regulation of acid–base status does not conform to alphastat: the slope of the relationship (ΔpH/ΔT) is generally less than the ΔpK/ΔT of histidine imidazole (Figure 7.13).

A difficulty here is establishing a reference value for the ΔpK/ΔT of histidine imidazole in biological systems. Histidine exists as a free amino acid in some invertebrates, but is mainly found in proteins. Estimates of ΔpK/ΔT in these systems range from −0.018 to −0.024. Data for the temperature dependence of extracellular and intracellular pH are typically lower than this, though there is a small degree of overlap (Table 7.4).

The regulation of intracellular pH allows the organism to maintain enzyme integrity binding capacity (as captured by the apparent Michaelis–Menten constant, K_M), the ability to respond effectively to changes in substrate concentration, and also to reverse reaction direction. It also affects

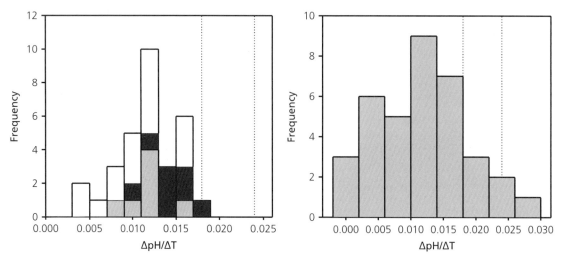

Figure 7.13 Frequency histogram of observed variation in pH with temperature ($\Delta pH/\Delta T$), compared with the range of values for $\Delta pK/\Delta T$ for imidazole (−0.018 to −0.024, dotted lines). A (left panel): extracellular pH in fish (grey bars), amphibians (black bars) and reptiles (grey bars). B (right panel): intracellular pH (6 tissue types from 12 species of fish, amphibian and reptile). Note that in the frequency histograms the negative signs for $\Delta pH/\Delta T$ and $\Delta pK/\Delta T$ have been omitted for clarity[34].

other important processes such as subunit assembly and the reversible binding of enzymes to structural elements in the cell. Intracellular pH regulation provides the organism with a very effective first line of defence in the face of temperature change, as acid–base balance can be adjusted far more rapidly than, for example, the expression of different enzyme variants.

The experimental investigation of alphastat regulation has been confined almost exclusively to laboratory acclimation studies of eurythermal vertebrates. There are very few data for organisms adapted evolutionarily to low temperatures, although the (tissue) pH of several Antarctic fish (*Dissostichus mawsoni*, *Pagothenia borchgrevinki* and *Trematomus bernachii*) and the isopod *Glyptonotus antarcticus* is high[36].

What can we conclude? Experimental data show clearly that ectotherms regulate their extracellular and intracellular pH in the direction predicted. The magnitude of the change is, however, variable and is typically lower than predicted from a strict adherence to either constant alkalinity or alphastat regulation.

7.6.3 Osmolytes, cofactors and chaperones

The stability and function of proteins are also affected by the presence of other compounds in the cell. Particularly important are small molecular mass osmolytes and chaperone proteins. The small organic molecules used by cells for this purpose are often called compatible solutes because, unlike inorganic ions, their concentration can vary without deleterious effects on the cell. Typical examples include polyols (glycerol, sorbitol, diglycerol phosphate), *myo*-inositol and methylamines such as glycine betaine and trimethylamine *N*-oxide (TMAO).

Table 7.4 Variation of pH with temperature in ectotherms[35].

Group	$\Delta pH/\Delta T$		
	Mean	Range	n
Extracellular fluid (plasma) pH			
Fish	−0.012	−0.007 to −0.014	7
Amphibians	−0.015	−0.011 to −0.018	8
Reptiles	−0.011	−0.005 to −0.018	16
Intracellular pH			
White muscle	−0.013	−0.005 to −0.019	11
Red muscle	−0.020	−0.003 to −0.028	5
Heart muscle	−0.010	−0.003 to −0.020	12
Liver	−0.016	−0.005 to −0.023	4

These compatible solutes have a range of functions, but an important one is maintaining the integrity of proteins. The continual internal movement of proteins runs the risk of entering a state which is both non-functional and not reversible, for example if the hydrophobic core is exposed. This danger appears to be minimised by a combination of osmolytes and chaperone proteins[37].

Organic osmolytes and chaperone proteins are always present in the cell, but the concentrations of both can be increased if the cell is exposed to an environmental challenge. Most attention has been directed at the damage induced by high temperatures, and the chaperone proteins expressed in response are often referred to as heat-shock proteins. Hyperthermophiles (organisms with a maximum growth rate at > 80 °C) have a particular problem with thermal denaturation of their proteins and nucleic acids, and a range of compatible solutes have now been identified that confer conformational integrity to their macromolecules. Some of these solutes appear to be restricted to hyperthermophiles, such as cyclic-2,3-bisphosphoglycerate, di-*myo*-inositol phosphate (DIP) and 2-α-O-mannosylglycerate (MG). Unlike compatible solutes from mesophilic and psychrophilic organisms, which typically are uncharged at physiological pH, many of those from hyperthermophiles are negatively charged[38].

While we are gaining important insights into the role of organic osmolytes and chaperone proteins in adaptation to high temperature, we do not yet have a clear picture of the extent to which these are involved in adaptation to very low temperatures, or indeed to intermediate (mesophilic) temperatures.

7.6.4 Membrane homeostasis

Not all proteins are based in the cytosol; many are located in membranes. For these enzymes it is the physical state of the membrane that is critical to their function. In particular the fluidity of the membrane is important for the lateral movement of proteins, and diffusion of substrates or cofactors.

Membrane lipids vary in their fluidity, depending on their composition and the temperature. Fluidity is a consequence of packing density, which is affected by the shape of the phospholipid molecules aligned in the bilayer (see Chapter 5). Where the fatty acids are fully saturated the chains are effectively straight, and the molecules can pack closely. Where the fatty acid molecules contain one or more double bonds this affects their shape and internal motion, and the strength of this effect depends on the number, position and type of double bond. In all cases, however, unsaturated fatty acid chains pack less densely, and the bilayer is then more fluid. A mixture of fluid membrane with embedded proteins and small areas ('islands') of more gel-like behaviour comprises the *fluid-mosaic model* of membrane structure, which remains the working model of membrane structure to this day[39].

The packing density, and hence membrane fluidity, is influenced by temperature because this determines the magnitude of the internal motions within the membrane. At higher temperatures or with more unsaturated lipid molecules, the membrane is fluid; at lower temperatures or with fewer unsaturated fatty acids, the membrane switches to a more gel-like state. This is shown well in an early study of fluidity and composition in membrane phospholipids from brain synaptosomes (Figure 7.14).

Here the fluidity of brain synaptosome membranes was estimated by measuring the depolarisation of the hydrophobic fluorescence probe 1,6-diphenyl-1,3,5-hexatriene (DPH). An increase in the fluorescence reading indicates a reduced rate of probe motion and by inference a more restrictive environment in the membrane within which the probe sits. This study shows two features that have proved to be general: the first is that higher body temperatures are associated with a reduced content of unsaturated fatty acids in the membranes, and secondly that there are strong differences between different phospholipids in the relationship. The relationship between membrane composition and fluidity is also affected by head-group composition and the presence of other molecules which can moderate the effect, the most important of which in vertebrates is cholesterol.

It is very difficult to study the effect of membrane fluidity on enzyme activity *in vivo*, so most work has involved enzymes reconstituted into phospholipid micelles or liposomes. These studies have shown that the state of the membrane exerts a powerful influence on enzyme activity. Most membrane-associated enzymes show a dramatic

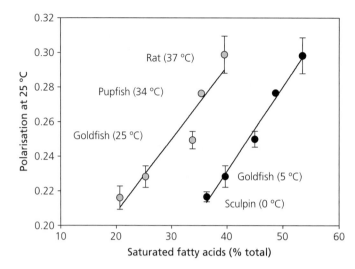

Figure 7.14 Relationship between fluidity of brain synaptosome membranes (assayed fluorescence polarisation of the membrane probe DPH normalised to 25 °C) and percentage of saturated fatty acids in the two major phospholipid classes, phosphatidylcholine (grey symbols) and phosphatidylethanolamine (black symbols). Data plotted as mean and 95% confidence intervals[41].

reduction in activity, or even become inactive, when the surrounding lipids are in the gel phase. There are a few examples of where enzyme activity is correlated with membrane fluidity, but even here the physical mechanism is far from clear[40].

It is now accepted that adjustment of phospholipid composition to conserve membrane physical state is a general response to a change in temperature across all classes of organism. When an organism is exposed to a new temperature, the change in membrane composition can happen rapidly (hours). It is also a feature of evolutionary adaptation to temperature.

Much attention has been directed at the unique membrane architecture of Archaea in respect of the ability of some taxa to live at very high temperatures. Archaeal membranes consist of isoprenoid hydrocarbon chains which are attached to the glycerol moiety by an ether link. Furthermore, the glycerol moiety has a different stereochemistry from that in bacteria and eukaryotes, and in some archaeans the isoprenoid chains are fused to form a single layer. This is a very different composition and architecture from the acyl-linked fatty acids in bilayers that are found in all other organisms. While this unique membrane architecture is found in hyperthermophilic archaeans, it is also found in archaea that live at low temperatures. This suggests that it may not necessarily be an adaptation to high temperature but simply a feature of archaeans in general. Its greater stability at high temperatures does, however, allow archaea to live at higher temperatures than bacteria or eukaryotes.

7.7 The paradox of adaptation

Our discussion of reaction rate has left us with something of a paradox, though one that is not often commented upon. Studies of isolated enzymes have frequently indicated a high degree of compensation for temperature in their catalytic capacity (as in the examples from fish and insects discussed in section 7.5). Since many of these studies have involved enzymes central to energy metabolism, we might expect that much of an organism's physiology and ecology would also show a marked independence from temperature. And yet, as exemplified by the ants studied by Howard Shapley, they do not. Many, indeed most, of the higher-level processes in organisms appear to be highly constrained by temperature and some examples are given in Table 7.5.

This is the paradox: studies of isolated enzymes *in vitro* typically indicate a high degree of compensation for temperature, but many whole-organism processes do not. Why is this?

The first consideration is that enzymes in the cell operate in a very different environment from the dilute solution of the laboratory. Diffusion rates in particular are much slower in the cytosol, mainly because of its greater viscosity. Diffusion rates will

Table 7.5 Examples of higher-level organism processes where maximum rate is constrained by temperature[42].

Process	Temperature-sensitive process that may constrain overall rate
Maximum muscle power output	Muscle fibre relaxation rate
Maximum mitochondrial ATP regeneration rate	Unknown (possibly ATP-synthase activity)
Nervous conduction rate	Possibly relative refractory period
Protein synthesis	Unknown (many possible rate-limiting steps)

slow as the temperature falls, partly because of the lower thermal motion but mostly because of the increase in viscosity. Some of the changes in cell architecture following adaptation to low temperature, or acclimation in the laboratory, can be interpreted as shortening diffusion distances to counteract slower movement of metabolites in the cell, especially in the number and distribution of mitochondria. It is not clear, however, to what extent this diffusion limitation keeps enzymes from operating at the maximum capacity that evolutionary modifications to the enzyme might otherwise allow[43].

This brings us to the important distinction between capacity and actual/realised rate. Physiologists probing evolutionary adjustments have typically explored the maximum capacity of the systems they study: what is the most power that a muscle can generate, what is the peak rate of ATP regeneration that a mitochondrion can deliver or what is the fastest that an organism can grow? This is because it is only the maximum capacity that can tell us to what extent there are thermodynamic or other constraints on what an organism can do. These studies have shown us that, within the limitations inherent in the design of organisms, temperature constrains many physiological processes that are important to an organism's ecology. All other things being equal, organisms with low body temperatures simply cannot move, grow or process information as fast as those with warmer bodies.

While maximum capacity is important, in that this sets the upper bound on performance, organisms do not operate routinely at full throttle, but at the rate required at the time. While evolution and thermodynamics set the maximum capacity, ecology and circumstances dictate the rate.

Present evidence would suggest that faced with a change in temperature, adjustments to acid–base status and membrane fluidity act to maintain metabolic integrity in the short term. For intermediary metabolism to be resilient to a change in temperature, both flux and control need to be maintained. These requirements underpin the evolutionary adaptations we see in many of the enzymes involved. It has long been assumed that there is also a requirement for the kinetics of all steps in a pathway such as intermediary metabolism to be matched. In essence this means a similar enthalpy of activation for each reaction, since this dictates the sensitivity to temperature. While this has long been something of an article of faith among physiologists, there have been surprisingly few tests of the idea, although model studies and experimental evidence point to such a matching in the yeast *Saccharomyces cerevisiae*[44].

Acclimatisaton to seasonal temperature change may involve changes in the relative expression levels of enzyme variants with different thermal characteristics, or expression of enzyme variants with different kinetic characteristics, and adaptation over evolutionary timescales involves the evolution of variants with kinetics matched to the new environment.

7.8 A subtle problem: describing the relationship between physiological rates and temperature

The recognition that the world's climate is warming has focussed attention on the thermal performance of organisms, and how this may influence their response. If we wish to compare the temperature sensitivity of different systems or model how these systems might respond, we need to be able to describe them mathematically. Otherwise we are left with simple verbal descriptions or arm-waving. This brings us to the somewhat contentious issue of how best to describe the temperature sensitivity of a system, be it an enzyme reaction, a physiological process or a suite of different organisms. Over a century ago, van't Hoff commented that it was impossible to choose between various mathematical descriptions of the relationship between biological rates and temperature, and the debate has continued ever since.

The problem is actually more complex than it might at first seem. This is because in examining the effects of temperature on an individual enzyme reaction, or a complex physiological process such as locomotion, or when comparing the effects of temperature across species, we are actually asking subtly different evolutionary questions. Because of this, the answers differ.

7.8.1 Enzyme activity

A typical relationship between enzyme activity and temperature is shown for α-amylase in Figure 7.15. Here activity increases non-linearly with temperature up to a peak (optimum) temperature, and then declines. The two enzymes illustrated were isolated from a psychrophilic and a thermophilic bacterium, and they differ markedly in their temperature sensitivity and optimal temperature.

Historically, physiologists fitting models to such data have concentrated on the rising part of the relationship, assuming that the decrease on the warmer side of the optimum temperature represents some form of denaturation. For the rising part of the curve, two models have been preferred, Arrhenius and Eyring–Polanyi. Both of these models describe a section of the rising part of the relationship reasonably well, but neither captures the overall relationship (Figure 7.16). In the region where the two models can be fitted, the slopes of the relationships are closely similar in the two species (−7.09 and −7.31 for the Arrhenius model, −19.01 and −19.22 for the Eyring–Polanyi model).

These results have proved to be fairly general: the Arrhenius and Eyring–Polanyi models both provide good statistical descriptions for part of the thermal behaviour of enzymes, but do not capture the entire pattern. The classical explanation for the overall relationship between activity and temperature is that it combines a quasi-exponential increase in activity with temperature, well explained by the Eyring–Polanyi model of enzyme kinetics, with an increasing loss of active enzyme through irreversible thermal denaturation. This explanation was questioned as long ago as 1998 by Theresa Thomas and Robert Scopes, who compared the activity of the highly conserved monomeric enzyme 3-phosphoglycerate kinase (PGK) isolated from a thermophilic and two mesophilic bacteria. They showed that k_{cat} values reached a maximum and then declined long before irreversible thermal denaturation exerted any detectable influence on activity. They concluded that the decline in activity at temperatures above the optimum must be caused by factors other than denaturation. Recently two models have been proposed to capture the full thermal behaviour of enzyme activity.

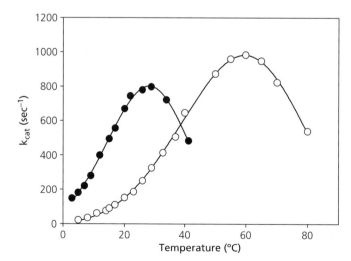

Figure 7.15 Effect of temperature on the activity of α-amylase orthologues isolated from the marine psychrophilic bacterium *Pseudoalteromonas haloplanktis* (black symbols) and the thermophilic soil bacterium *Bacillus amyloliquefaciens* (white symbols). The lines are Weibull curves[45].

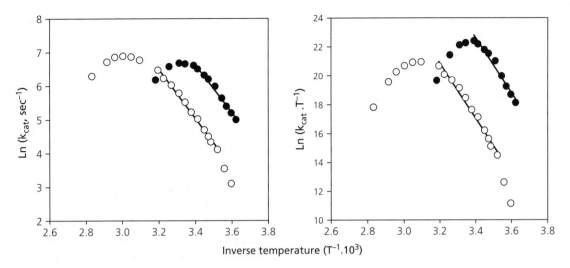

Figure 7.16 Enzyme activity data from Figure 7.10 with fitted least-squares regression models. A (left panel): Arrhenius model. B (right panel): Eyring–Polanyi model. T is absolute (thermodynamic) temperature (K).

The first posits an equilibrium between the active state of the enzyme, P_{act}, and an inactive but not denatured state, P_{inact}, and also includes an irreversible transition to the denatured state, P_{denat}:

$$P_{act} \rightleftharpoons P_{inact} \rightarrow P_{denat}$$

The key feature of this model is that the equilibrium between the active and inactive states is extremely rapid, and that the position of this equilibrium is affected by temperature. This equilibrium is underpinned by the rapid changes between active and inactive conformations discussed above. This *equilibrium model* has proved successful at explaining the full pattern of the thermal sensitivity of activity in a number of enzymes.

The second model is based on changes in heat capacity during binding of substrates and release of products, and the influence of these changes on the Gibbs energy of activation ($\Delta G^{\ddagger}$). An interesting implication of this *heat capacity model* is that for enzymes measured at their optimum temperature, the enthalpic contribution to enzyme catalysis is similar for all enzymes, and hence it is the entropic contribution to the Gibbs energy of activation that is paramount in determining the differences between reaction rates of different enzymes[46].

Despite their inability to capture the entire thermal behaviour of enzyme activity, the Arrhenius and Eyring–Polanyi models have remained popular descriptors of the thermal sensitivity of enzyme activity. This is partly because many (indeed most) studies are confined to the temperature range over which the relationship between reaction rate and temperature is more or less exponential (and hence linear in these models). Although we have made major strides in our understanding of the mechanism of enzyme reaction in the past century, in terms of selecting a model for describing the data, in some ways we have yet to progress beyond van't Hoff.

7.8.2 Describing physiological processes: the reaction norm

All physiological processes are affected by temperature. The problem for the ecologist or physiologist comes in deciding what the relationship with temperature might be. If we cannot define the relationship from knowledge of the underlying mechanism(s), all we can do is describe it statistically.

A nice example of an important physiological process that varies with temperature is the speed of locomotion in ants studied by Harlow Shapley (Figure 7.1). Here the relationship is roughly exponential, and Shapley himself speculated that the shape of this relationship might reflect the influence of temperature on the chemical processes involved in locomotion. If correct, this would allow us to apply an Arrhenius or Eyring–Polanyi model to this more complex system.

Locomotion is, however, a complicated series of closely integrated chemical and mechanical events. This poses the question of whether the overall relationship between locomotor performance and temperature is an emergent property of the individual relationships for components of the system, or simply a reflection of the thermal behaviour of a rate-limiting step? At present we do not know, and so in the absence of a viable theory or mechanistic model that allows us to predict the thermal response of the locomotor system from first principles, ecologists have been forced to rely on statistical descriptions.

A nice example of this is a classic early study of the locomotor performance of the iguanid lizard *Dipsosaurus dorsalis* by Ray Huey and Joel Kingsolver (Figure 7.17). In this species the temperature dependence of two measures of locomotor ability, burst speed and endurance, are both qualitatively similar to the relationship between temperature and enzyme activity: both increase with temperature up to a maximum (optimum), and then either plateau or tail off. Peak performance coincides with the preferred temperature of lizards allowed to choose their thermal environment in the lab, and also with the modal body temperature recorded from free-ranging lizards in the wild.

This type of response has been explored through the concept of the *thermal reaction norm* (also called more formally, the *norm of reaction*). The reaction norm describes how the phenotypic trait value (here locomotor performance) of a particular genotype varies as a function of some continuous environmental variable (here body temperature). Because a given reaction norm applies only to a specific genotype, the reaction norm of another genotype (say from a second individual of the same species) may well be different. However we can also capture the average relationship for all the genotypes in a species or population; this average relationship is the *thermal performance curve* for the species or population.

Thermal reaction norms and thermal performance curves typically have a characteristic shape in which performance increases with temperature, reaches a maximum (optimum) and then declines. Typically the decline at higher temperatures is steeper than the increase towards the optimum, which gives the reaction norm a characteristic shape (Figure 7.18). This shape implies that performance is relatively insensitive to a change in temperature when this is close to the optimum, whereas it is strongly dependent on temperature on the flanks where the gradient is steepest.

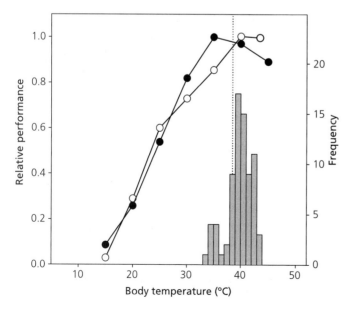

Figure 7.17 Thermal sensitivity of locomotor performance in the iguanid lizard *Dipsosaurus dorsalis*. The curves show the thermal sensitivity of burst speed (white symbols) and endurance (black symbols). To aid comparison, both measures have been scaled to unity at their maximum value. The frequency histogram shows the distribution of body temperature values in free-ranging lizards in the wild, and the dotted line shows the preferred temperature selected by lizards in the laboratory[47].

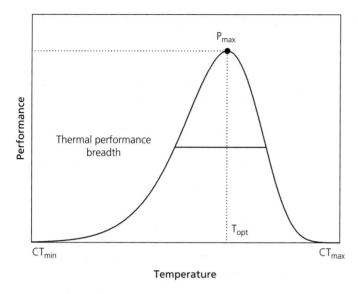

Figure 7.18 The thermal reaction norm. Maximum performance (P_{max}) is at the optimum temperature (T_{opt}), and the thermal performance breadth is measured at half P_{max}. CT_{max} and CT_{min} denote the upper and lower critical temperatures respectively, and thereby define the overall thermal performance range.

The key variables that capture the shape of the thermal reaction norm are the performance (P_{max}) at the optimum temperature (T_{opt}), the *performance breadth* and the overall thermal tolerance range. Performance breadth is a useful concept, but we need to stipulate at what performance level it is measured (here it is represented, arbitrarily, as the breadth in performance at half T_{opt}). If this is not done, the measure is meaningless.

A number of mathematical models can be used to describe the overall relationship between performance and temperature, though all are empirical and none have a strong theoretical foundation. Michael Angilletta explored a number of models that could be used to capture the key features of the performance curve, and concluded that none are ideal[48]. The reaction norm is, however, critical to understanding how thermal performance responds to selection (for example as environmental temperature changes, a topic we will explore in a later chapter). To do this, we need to understand the mechanism behind the shape of the performance curve, but at present we are far from being able to do this (Box 7.8).

7.8.3 Evolutionary comparisons across species

Comparison of physiological performance in different species has long been a mainstay of evolutionary physiology, and has provided important insights into adaptation. Comparative data are often presented and analysed using exactly the same statistical models as are used for studies of single enzymes or individual species. This may seem intuitively sensible, but it ignores some subtle difficulties.

A particular difficulty comes with the mixing of data from within-species and between- (across-) species studies. It is not uncommon to see plots where a relationship has been fitted to a data set where there is more than one data point for each species. While it might seem intuitive that the more data points are used the better the fitted relationship, there are important evolutionary differences between within-species and across-species data that can lead to misleading or erroneous conclusions when these are mixed.

Two different species will have different evolutionary histories. If they are ectotherms that have evolved to live at different temperatures then, as we have seen earlier, they will likely have physiologies with different thermal characteristics. The consequence of evolutionary adjustment is thus that the within-species and across-species relationships are fundamentally different in kind, though they may or may not be different in magnitude. Where there is a strong selective advantage to enhanced activity (for example in muscle power output), the two relationships may be similar; where there is a strong selective advantage to maintaining activity

> **Box 7.8 Explaining the thermal performance curve**
>
> The thermal reaction norm and the thermal performance curve are descriptive and not derived from theory. Can we explain their shape?
>
> The visual similarity of the left-hand side (below the optimum) to the thermal sensitivity of muscle function and mitochondrial ATP generation suggests that either or both of these may dictate the shape of the rising section of the curve.
>
> The traditional explanation for the decrease in performance at temperatures above the optimum has been an increase in enzyme degradation, but performance starts to decline before there is any detectable increase in irreversible protein denaturation. Recently Hans Pörtner has suggested that the drop in performance at higher temperatures is caused by an inability to deliver oxygen to the tissues fast enough to meet the increasing demand. Studies with marine invertebrates have shown that in this region haemolymph oxygen concentration drops and there is an increase in the concentration of anaerobic end-products such as succinate or octopine (Figure 7.19)[49].
>
> This model also includes the concept of the *pejus temperature*, which marks the temperature at which performance starts to decrease from the maximum observed at the optimal temperature (derived from the Latin *peius*, meaning worse). These temperatures could be identified clearly by eye in the haemolymph oxygen concentration of the spider crab *Maja squinado*, but assignment is far more subjective in many other physiological measures and other organisms. They could perhaps be determined more objectively by fitting a suitable statistical model and determining a critical change in gradient of the curve. But as we have no theory on which to base the choice of descriptive model, subjectivity remains.
>
> A cautionary note is that while the reaction norm and performance curve are useful descriptors, we cannot be sure that the same physiological processes underpins the shape of these curves in all organisms. The development of a theoretically based reaction norm or performance curve remains a significant challenge for thermal physiology.
>
>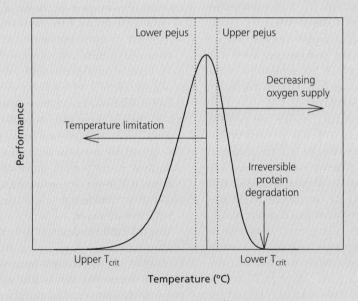
>
> **Figure 7.19** A mechanistic hypothesis for the shape of the thermal reaction norm.

independent of temperature (for example in some cellular homeostatic processes) the across-species evolutionary relationship may be significantly less steep than the within-species thermodynamic relationship. For any process that shows complete evolutionary compensation for temperature, the across-species relationship will have zero slope. In simple terms, the acute within-species relationship between a physiological rate and temperature expresses the thermodynamic challenge that

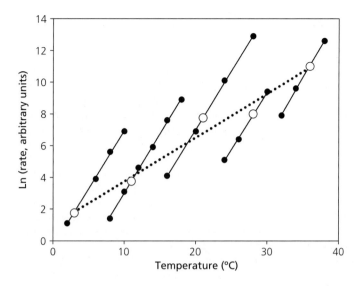

Figure 7.20 Comparison of within-species and across-species relationships between physiological rates and temperature. The black symbols represent the relationship between rate and temperature for five hypothetical species from different thermal regimes. The solid lines represent the acute (thermodynamic) response to temperature for each species as determined experimentally in the laboratory; for many physiological processes these relationships are typically exponential, and hence linear after logarithmic transformation. The white symbols represent the rate for each of the five species at their normal environmental temperature, and the dotted line is a regression fitted to these data. Note that in this model the difference in the slopes of the acute within-species (quasi-mechanistic) and across-species (statistical) relationships have been exaggerated for clarity; in reality they are often more similar in magnitude[51].

an organism faces from a change in temperature, whereas the across-species relationship indicates what evolution has (or has not) been able to do about this. This important difference between within-species and between- (across-)species relationships is shown schematically in Figure 7.20.

Obviously the temperature sensitivity of a physiological rate in a given species is itself the outcome of past evolutionary processes; the critical distinction here is that there are only small evolutionary differences between data points in the within-species relationship, but there are strong evolutionary differences between the data points in any across-species study[50]. Failure to distinguish these two fundamentally different relationships has led to some confusion in the ecological literature.

So far we have assumed that the main driver for a given physiological trait is adaptation to the environment. However, time is also a key factor here, in that the more recently two species have diverged, the more similar they are likely to be in ecology and physiology. We touched on this important consideration above, in that in the comparative study of A_4-LDH from fish we confined ourselves to comparisons within a genus (*Sphyraena*). A consequence of this is that in broad-based comparative studies, the influence of phylogeny may be an important confounding factor that must be allowed for.

7.9 Summary

1. All other things being equal, reaction rate increases roughly exponentially with temperature. The critical factor that determines the temperature sensitivity is the activation energy.
2. Organisms can minimise the acute effect of a change in temperature by adjustments to intracellular pH and membrane fluidity. The timescale here is typically hours.
3. Organisms that have adapted over evolutionary time to live at different temperatures can have enzyme variants that exhibit similar kinetics at the temperatures to which they have adapted to operate. Within species whose distribution covers a range of temperatures, there may be differential expression of enzyme variants with different kinetics across the distribution.
4. Enzymes adapted to different optimum temperatures differ in their amino acid sequence and thermal stability. The Gibbs energy of activation tends to be slightly lower in enzyme variants adapted to lower temperatures, but the big change is a decrease in the enthalpy of activation, with a corresponding change in the entropy of activation, both associated with a more open, flexible structure.
5. Despite evolutionary adjustments to individual enzymes involved in intermediary metabol-

ism (ATP regeneration), many whole-organism processes operate faster in tropical ectotherms compared with temperate or polar ectotherms. Examples include locomotion (muscle power output), ATP regeneration (mitochondrial function), nervous conduction and growth.

6. The across-species relationship between reaction rate and temperature is purely statistical. Comparison of performance between different species needs careful framing of hypotheses, and often requires allowance for the effects of phylogeny (more recently diverged taxa are more likely to share physiological or ecological traits).

Notes

1. The precise origin of this aphorism is unclear. The basic idea that physiological processes operate better (that is, faster) at higher body temperatures goes back at least to Hamilton (1973) and almost certainly much further. This succinct expression of the idea may well originate with the title of a paper by the American physiologist Al Bennett, where he asked *'is warmer better?'* (Bennett 1987). More recently the phrase has evolved to *'hotter is better'* (see for example Kingsolver & Huey 2008).
2. *Liometopum* relationship plotted with original data from Shapley (1920). *Tapinoma* and *Iridomyrmex* relationships plotted with original data from Shapley (1924), excluding outliers (standardised residuals > 2) as data for lowest and highest temperatures do not match overall relationships.
3. Harlow Shapley (1885–1972) was an American astronomer best known for using variable stars to estimate the size of the Milky Way. I first came across Shapley's work on ants through George Johnson's biography of Henrietta Leavitt, the computational assistant who discovered the correlation between period and luminosity in variable stars that Shapley used in his work (Johnson 2005). Shapley describes the ant study in his autobiography (Shapley 1969).
4. The Law of Mass Action was developed by the Norwegian chemists Cato Maximilian Guldberg (1836–1902) and his brother-in-law Peter Wage (1833–1900) in 1864, but as the original papers were in Norwegian (with a later, 1867, publication in French), they made little impact. The formulation was arrived at independently by the Dutch physical chemist Jacobus van't Hoff in 1877, but when van't Hoff became aware of the previous work, he acknowledged its priority.
5. Good discussions can be found in Wrigglesworth (1997), Haynie (2001), Keeler & Wothers (2003) and Atkins & de Paula (2011).
6. Leopold Pfaundler (1839–1920) was an Austrian physical chemist who formulated important ideas concerning the kinetics and thermodynamics of gases. Although his contributions were widely recognised at the time, he is now largely forgotten (but see Laidler 1984). Jacobus Henricus van't Hoff (1852–1911) was a Dutch physical chemist who laid the foundations for our understanding of how temperature affects reaction rate through its effects on kinetics and equilibria. He also made major contributions to the study of osmotic pressure and stereochemistry, and he was awarded the first Nobel Prize in Chemistry in 1901. The key works are van't Hoff (1884, translated 1896) and van't Hoff (1898).
7. Daniel Berthelot (1865–1927) was a French physicist who made fundamental contributions to the development of the kinetic theory of gases. His legacy is described by Wisniak (2010).
8. See Arrhenius (1889a, 1889b). Svante Arrhenius (1859–1927) was a Swedish physical chemist who made important contributions to theory of electrolytes, thermodynamics and cosmology. He was also important in recognising the importance of carbon dioxide in governing lower atmospheric temperatures, and in initiating ice ages. He was awarded the 1903 Nobel Prize in Chemistry. A valuable history of the Arrhenius equation is given by Laidler (1984).
9. Mihály (Michael) Polanyi (1891–1976) was a Hungarian physical chemist and philosopher, Henry Eyring (1901–1981) was a Mexican-born American physical chemist and Meredith Gwynne Evans (1904–1952) was a British physical chemist. Eyring and Polanyi initially worked together on quantum-mechanical approaches to calculating activation energy (Polanyi & Eyring 1932), and eventually published parallel papers describing the basis of what we now know as transition state theory (Evans & Polanyi 1935; Eyring 1935). The history of the development of transition state theory is described by Laidler & King (1983) and Keszei (2003).
10. This was pioneered by John Charles Polanyi (son of Michael Polanyi), for which he received the 1986 Nobel Prize in Chemistry.
11. For a succinct description of transition state theory aimed at a biological audience, see Haynie (2001).
12. This mechanism was first proposed by Pauling (1946). Linus Pauling (1901–1994) was an American chemist who made pioneering discoveries in physical chemistry, and is rightly regarded as the 'father of the chemical bond'. His seminal work is his book *The nature of the chemical bond* (1939). He was awarded

the Nobel Prize in Chemistry in 1954, and the Nobel Peace Prize in 1962, both unshared (he remains the only person to have received two unshared Nobel prizes in different fields).
13. The pioneering work on invertase was done by the French chemist Victor Henri (1872–1940). The kinetic theory that we use today was developed, also initially working with invertase, through a collaboration between the German chemist Leonor Michaelis (1875–1949) and the Canadian physician-chemist Maud Menten (1879–1960).
14. Data from Graves & Somero (1982).
15. See Jensen (2001).
16. The work on *Colias* is described by Watt (1977, 1983), Watt et al. (1983), Carter & Watt (1988) and Wheat et al. (2006).
17. Plotted from data in Place & Powers (1979), with permission from Spinger-Verlag.
18. See Powers & Place (1978), Place & Powers (1979) and Crawford & Powers (1989). The cline has recently been restudied by Bell et al. (2014).
19. Data for *Champsocephalus* and *Pomatocentrus* are from Arrhenius plots (Figures 3 & 4) in Johnston & Walesby (1977), converted to SI units and replotted as an Eyring–Polanyi plot. Note that *Notothenia neglecta* in the original paper has been renamed *N. coriiceps*. Data for enthalpy of activation and environmental temperature range are from data in Table 1 of Johnston & Goldspink (1975) (grey symbols) and Table 1 of Johnston & Walesby (1977) (black symbols); data converted to SI units before plotting.
20. Note that because the entropic contribution to the Gibbs energy of activation is subtracted from the enthalpic contribution, the correlation between $\Delta H^{\ddagger}$ and $\Delta S^{\ddagger}$ is positive.
21. The statistical problems were first pointed out by Exner (1964a, b, 1970), later by Krug et al. (1976a, b, c) and most recently argued forcibly by Cornish-Bowden (2002) (see also Chapter 9 for a discussion of the importance of error structure in least-squares regression). Lumry & Rajender (1970) discuss the role of water, Sharp (2001) discusses the various forms of compensation and Dunitz (1995) explores the role of weak bonds, principally hydrogen bonds, in enthalpy–entropy compensation.
22. Data for fish Ca^{2+}-Mg^{2+}-ATPase are from Table 1 in Johnston & Walesby (1977), converted to SI units. The ITC data for protein-ligand binding came from Olsson et al. (2011), reproduced with permission of John Wiley and Sons.
23. Plotted from data in McFall-Ngai & Horwitz (1990) and Fields (2001), and reproduced with permission of Elsevier.
24. Valuable reviews of the adaptation of enzyme molecules to temperature are Fields (2001), Hochachka & Somero (2002) and Fields et al. (2015). Much attention has been directed at how enzymes adapt to operate at low temperatures (because of their potential industrial applicability), and useful reviews here are Gerday et al. (2000), Cavicchioli et al. (2000, 2002), Georlette et al. (2004), D'Amico et al. (2006), Feller (2007, 2010) and Gu & Hilser (2009). For reviews of proteins adapted to high temperature see Jaenicke & Böhm (1998), Mukaiyama et al. (2008) and Sawle & Ghosh (2011).
25. Summarised from reviews by Sawle & Ghosh (2011), Gerday (2013a, b) and Fields et al. (2015).
26. See Sidell (1977, 1983) and Shaklee et al. (1977). A slightly earlier study of the increase in cytochrome oxidase in cold-acclimated goldfish, *Carassius auratus* was reported by Sidell et al. (1973).
27. See for example Oleksiak et al. (2002, 2005) and Podrabsky & Somero (2004).
28. The studies of evolutionarily adapted fish are Johnston et al. (1998), and the acclimation studies are Egginton & Sidell (1989) and Egginton et al. (2000). The invertebrate studies are Sommer & Pörtner (2002) and Morley et al. (2009).
29. For example Yancey & Somero (1978) showed that in LDH the K_M for pyruvate was conserved across 8 vertebrate species with body temperatures ranging from −2 °C to 38 °C.
30. That organisms maintain a constant relative alkalinity between plasma pH and the pH of neutral water (pN) was proposed independently by Winterstein (1954) and Rahn (1966). See also Rahn & Baumgardner (1972) and Howell et al. (1973).
31. Modified from Mark et al. (2002), using data only up to 6 °C.
32. The alphastat hypothesis was first presented by Reeves (1972, see also Reeves 1969). Later useful discussions are Reeves (1977, 1985), Malan et al. (1976) and Somero & White (1985), and a comprehensive review of the field is Burton (2002).
33. Plotted with data from Malan (1980). Dissociation ratio calculated from pH.
34. Histograms plotted with data from Table 4 (extracellular data) and 5 (intracellular data) of Heisler (1986).
35. Data taken from compilation by Heisler (1986).
36. See Qvist et al. (1977) and Jokumsen et al. (1981).
37. Interestingly, many of these compatible solutes are also used by cells as a response to freezing or dehydration stress (Chapter 6).
38. The role of small osmolytes in hyperthermophiles is discussed by Santos et al. (2008). See also Faria et al. (2008) and Mukaiyama et al. (2008). A useful general discussion of small osmolytes in the cell is Yancey (2005).

39. The fluid mosaic model was proposed by Seymour Singer and Garth Nicolson (Singer & Nicolson 1972)
40. Plotted with data from Cossins & Prosser (1978). The species were *Myoxocephalus verrucosus* (sculpin), *Carassius auratus* (goldfish), *Cyprinodon nevadensis* (desert pupfish) and *Rattus norvegicus* (laboratory rat).
41. For overviews of the relationship between membrane composition, fluidity and function, see Cossins et al. (1981), Gennis (1989) and Hazel (1995).
42. See Kiernan et al. (2001). Dell et al. (2011) provide a recent compilation of the thermal sensitivity of a wide range of physiological and ecological processes.
43. See Sidell & Hazel (1987) and Hubley at al. (1996). Despite its potential importance, this subject does not appear to have attracted much recent attention from physiologists, other than in relation to the extreme case of vitrification (see Chapter 6).
44. See Cruz et al. (2012). A recent survey by Elias et al. (2014) suggested that the thermal sensitivity of a wide range of enzymes was broadly similar.
45. Plotted with data from Georlette et al. (2004).
46. The early indications that thermal denaturation were insufficient to explain thermal behaviour of enzyme reaction rates came from Thomas & Scopes (1998). The equilibrium model is described by Daniel et al. (2007), Lee et al. (2007), Peterson et al. (2007) and Daniel & Danson (2010), and the heat capacity model by Hobbs et al. (2013).
47. Replotted from data in Huey & Kingsolver (1989), and reproduced with permission of Elsevier. A very similar result was shown for another lizard, *Sceloporus undulatus* by Angilletta et al. (2002).
48. See Angilletta (2006).
49. Pörtner (2001) introduces the key ideas with data from marine invertebrates and fish. The model is explored in relation to climate change by Pörtner (2002), Pörtner et al. (2006) and Pörtner & Knust (2007).
50. Modified from Clarke & Pörtner (2010).
51. All populations, of course, contain significant within-species variability in physiological performance; this variability may be sustained by competing selection and it provides the raw material for evolution when selection pressure changes. This means that the differences between within-species and across-species relationships is more of a continuum rather than as discrete as portrayed here, and the contrast is more one of degree than kind but the statistical point remains both valid and important.

CHAPTER 8

Metabolism

La réspiration est donc une combustion.

Antoine Lavoisier[1]

Metabolism defines the living state. It provides the energy an organism needs to power everything it does, and thereby underpins all of its ecology. Because metabolic processes involve the movement of energy, they are sensitive to temperature. This sensitivity is central to the ecological impact of temperature change, but before we can explore this, we need to establish what metabolism is.

8.1 Metabolism is simple

There can be few students who have not experienced a sinking feeling when first confronted with the bewildering complexity of the chart showing all the metabolic reactions of the cell[2]. And while the cell really is complicated structurally, its physiological complexity belies an underlying simplicity. Essentially all of metabolism is variation about a simple basic theme: an electron passes from a donor to an acceptor, losing energy as it does so. In this process the donor is oxidised (it loses one or more electrons) and the acceptor is reduced (it gains one or more electrons). The combination of reduction and oxidation is termed redox for brevity.

What evolution has done is to develop mechanisms for capturing the energy released by redox reactions and making it do work (Figure 8.1). The essence of metabolism was captured memorably by Albert Szent-Györgyi: *'life is nothing but an electron looking for a place to rest'*[3]. The secret of life is the electron.

8.1.1 Redox reactions in Nature

A considerable degree of the metabolic variety in nature comes from the array of different atoms and molecules that are used as electron donors and acceptors. Whether a given combination of donor and acceptor is suitable to drive metabolism depends on two factors: how much energy can be derived from the reaction and how common the donors and acceptors are in the environment.

The principles involved in a redox reaction are illustrated nicely by a simple experiment that many will remember from school chemistry classes. When a piece of zinc is placed into a solution of copper sulphate, the surface of the zinc rapidly becomes coated in gleaming fresh copper. Two things are going on here: zinc atoms are each releasing two electrons with the resultant zinc ions moving into solution, and copper ions in solution are each gaining two electrons and precipitating onto the zinc surface:

$$Zn \rightarrow Zn^{2+} + 2e^-$$
$$Cu^{2+} + 2e^- \rightarrow Cu$$

The complete reaction thus involves two separate half-reactions. These half-reactions are useful theoretical constructs, but they are slightly misleading in that the electrons are never free; they move directly from the zinc atom to the copper ion when thermal motion brings these together. The reducing species with its corresponding oxidised form is termed a *redox couple*. Here the redox couples are Zn/Zn^{2+} and Cu/Cu^{2+}.

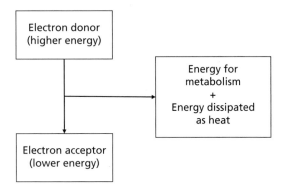

Figure 8.1 The essence of metabolism: an electron passes from a donor to an acceptor, losing energy as it does so. The difference in energy between the donor and acceptor is used by the organism to drive its metabolism.

The factor that determines that copper will deposit onto zinc, but not vice versa, is the standard reduction potential, E_h. This is a measure of the tendency of a chemical species to acquire electrons and thereby be reduced. By convention, standard reduction potentials are measured at pH 7, 25 °C, a concentration of 1 M for each ion participating in the reaction, a partial pressure of 1 atm for any gas that is part of the reaction, and with metals in their pure state. E_h is defined relative to a standard hydrogen electrode which is assigned (arbitrarily) the value of 0 V exactly[4]. The value of E_h for some redox couples of ecological importance is given in Table 8.1.

The conditions for the determination of standard reduction potential are valuable for comparative studies but decidedly non-physiological. The actual values of E_h in the cell, and hence the energy available from a given redox reaction, will depend on the concentrations of the redox couples and factors such as the pH of the intracellular environment. Some of these redox couples in Table 8.1 might seem somewhat esoteric, but many are ecologically important (see for example Box 8.1).

It is bacteria that make most use of the range of electron donors and acceptors available in the environment, and these are central to life on Earth for they are integral to the biogeochemical cycling of all important elements. In hydrothermal and anaerobic marine environments, for example, H and S are

Table 8.1 Standard reduction potentials of some inorganic redox couples of ecological importance.

Redox couple	Standard Reduction Potential, E_h (mV)
CO_2/CO	−540
SO_4^-/HSO_3^-	−516
$2H^+/H_2$	−420
CO_2/CH_4	−240
HSO_3^-/HS^-	−110
SO_3^{2-}/S^{2-}	−116
NO_3^-/NO_2^-	+430
MnO_2/Mn^{2+}	+380
NO_2^-/NH_4^+	+440
Fe^{3+}/Fe^{2+}	+772
O_2/H_2O	+815
$2NO/N_2O$	+1175
N_2O/N_2	+1355

Box 8.1 Redox in action: the nitrogen cycle in soils

When oxygen is unavailable, some bacteria in soils and anaerobic aquatic sediments switch to using nitrate, NO_3^-, as an electron acceptor. The NO_3^- is thereby reduced to NO_2^-, and eventually to N_2 gas, a process known as denitrification[5]. This process removes nitrogen from the soil, and since plants needs nitrogen for growth, in agricultural systems this nitrogen has to be supplied as fertiliser. Removal of nitrogen from soils by denitrification means that those higher plants which have entered into a symbiotic relationship with bacteria that can fix nitrogen (that is, reduce atmospheric nitrogen to NH_3, which can then be used by the plants) are of enormous agricultural importance. These are mostly legumes (peas, beans and clovers).

The microbial processes of nitrification and denitrification dictate much of traditional agricultural practice: tilling aerates the soil, thereby introducing oxygen so that the use of NO_3^- as electron acceptor, with consequent loss of N_2 to the atmosphere, is avoided. In addition the growing of legumes in a crop rotation adds nitrogen to the soil when these are ploughed in and break down in the soil. Where this is not done the soil nitrogen quickly becomes depleted and crop growth is reduced (as in slash-and-burn agriculture on nutrient-poor tropical soils, where the preferred solution is to move to less depleted soil).

the major electron donors and almost half the total electron flux may be through H in methanogenic bacteria.

While bacteria utilise a diverse array of electron donors and acceptors, both inorganic and organic, many of these redox pairs release relatively little energy, even under cellular conditions, and the growth rates of microbes using them can be quite slow. As can be seen from Table 8.1, one process that does release a large amount of energy is the reduction of oxygen to water. This is the basis of aerobic respiration, and thus the principal source of energy for most the organisms of the familiar terrestrial and aquatic systems we see around us. It does, however, require the presence of oxygen in the environment and it is now clear from geochemical evidence that the early environment of the Earth lacked oxygen, suggesting that many of the anaerobic metabolisms of microbes are extremely ancient, having evolved before there was significant oxygen in the environment[6].

The source of atmospheric oxygen is water. The oxygen is released as a by-product by organisms that had evolved the trick of pulling the electrons from water in the first place. Although we are most familiar with this process (photosynthesis) in green plants, a number of groups of microbes, some of them very ancient, have evolved various mechanisms for using the energy of sunlight to extract electrons from water (photolysis), and it is likely that the rise of oxygen in the atmosphere was connected with the evolution of photosynthesis by cyanobacteria.

Oxygen first arrived in the atmosphere 2.3–2.5 Ga ago, an event known as the *Great Oxygenation Event*, although atmospheric oxygen levels probably remained low until the later *Neoproterozoic Oxygenation Event* almost 2 billion years later. As aerobic beings that depend utterly on the presence of oxygen for us to breathe, we see oxygen as a life-giving element. This is only partly true, for oxygen can be extremely reactive and is highly toxic to many physiological processes. Organisms have to be careful to keep oxygen confined to places where it is needed (such as mitochondria) and away from places where it might wreak havoc. The pollution of the ancient anaerobic environment of Earth by oxygen must have been one of the most profound environmental changes that life on Earth has ever had to face.

There is thus a simple cycle of water, oxygen and energy through the biosphere. Photosynthesis extracts electrons from water, traps these in the reduced bonds of organic molecules; metabolism utilises this energy by recombining the electrons with oxygen to reform water (Figure 8.2). The importance of trapping the energy in organic compounds is that life requires a steady supply of energy whereas sunlight is intermittent; storing energy allows it to be used when needed. Other

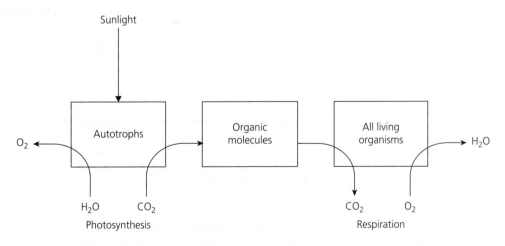

Figure 8.2 Photosynthesis and respiration as the two key processes cycling biological carbon on Earth.

important biogeochemical cycles involve P, S and N, but it is the photosynthesis/respiration cycle that dominates the flow of energy in the biosphere today.

8.2 Gaining electrons

All living things can be classified into a small number of trophic groups depending on the source of electrons used for metabolism, and the carbon used for building organic matter (Table 8.2). There is overlap between some of these terms, for example chemotroph and lithotroph. Also these terms may be combined: a bacterium or plant using light energy to extract an electron from water, and using CO_2 as carbon source is often termed a photoautotroph, whereas a bacterium in a hydrothermal vent using an inorganic electron donor may be termed a chemolithotroph, or even a chemolithoautotroph if the carbon source is CO_2.

While the diversity of microbial metabolism is of considerable interest in terms of the history of life on Earth, and in giving hints as to possible forms of life elsewhere, many of these are not well studied in terms of their thermal behaviour. The principles of how temperature affects the gaining of energy from the environment, and its use in metabolism, are best illustrated in the systems most familiar to us. These are photosynthesis, which uses light energy to drive the synthesis of organic compounds, and respiratory metabolism, which uses the energy retained in those organic compounds to fuel growth, reproduction and activity. We will look at these two key processes in turn, before examining how they are affected by temperature.

8.3 Photosynthesis

It takes a great deal of energy to extract electrons from water (often referred to colloquially as 'splitting water' as the products are electrons, protons and oxygen). This process is not favoured thermodynamically under physiological conditions, but plants can achieve it by using the energy of light photons, first to extract the electrons from water, and then to increase their potential energy sufficiently for them to be able to drive the synthesis of carbohydrate.

Photosynthesis in algae and higher plants takes place in the thylakoid membranes of the chloroplasts. The energy of sunlight is captured by the photosystems, each of which contains several hundred chlorophyll molecules, associated pigments (notably carotenoids) and proteins that together comprise the light-harvesting complex. Most of the chlorophyll and pigment molecules function simply to absorb photons and transfer the energy to a pair of chlorophyll molecules held in a very specific position; these are the *reaction centres* and they are where the transfer of light energy to chemical energy takes place. The structure of the light-harvesting complex increases the efficiency of light capture, which would otherwise be very low if photosynthesis had to wait for a direct hit by a photon on the reaction centre. The capture of a photon is a quantum-mechanical process, and hence completely insensitive to temperature.

The energy from the captured photons elevates the potential energy of electrons in the chlorophyll of photosystem II, which then pass along an electron transport chain to photosystem I, generating ATP in the process. These electrons are replaced immediately by others extracted from water by the *water-splitting complex*[7]. The potential energy of the electrons in photosystem I is then boosted by absorption of more photons and they now have enough energy to reduce NADP to NADPH, a process which involves a second electron transfer chain. This sequence of events comprising the extraction of electrons from water, their transfer to

Table 8.2 A classification of organisms based on their sources of energy and carbon.

Source of energy or carbon	Term
Energy source	
Sunlight (photons)	Phototroph
Atoms or molecules (electrons)	Chemotroph
Electron donor	
Inorganic (atoms or molecules)	Lithotroph
Organic molecules	Organotroph
Carbon source	
Inorganic molecules (typically CO_2)	Autotroph
Organic molecules	Heterotroph

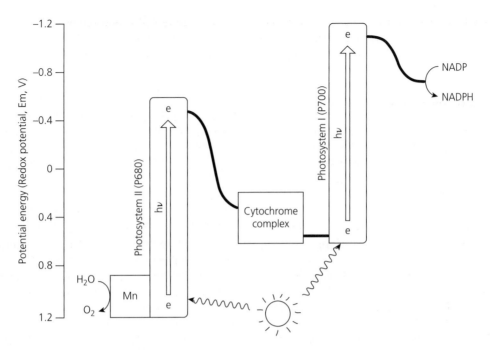

Figure 8.3 The flow of electrons in photosynthesis. There are two sites where the potential energy of the electrons is increased. In Photosystem II the P680 reaction centre absorbs light around 680 nm to extract electrons from water and boost their energy. After some of this energy has been used to synthesise ATP in the first electron transport chain (ETC1), the electron is boosted again by photon energy captured by Photosystem I. Here slightly longer wavelength (and hence lower energy) light is captured by reaction centre P700, and this boosts the potential energy sufficiently for the electron to drive the reduction of NAPH as it passes down the second electron transport chain (ETC2).

NADPH and the synthesis of ATP is often termed the *light reactions*.

The NAPDH and ATP from the light reactions are used by the plant to reduce CO_2 to glucose[8]. The essential component of this process is the Calvin–Benson cycle[9]. The key step in this cycle, the condensation of Ribulose 1,5-bisphosphate with CO_2, is catalysed by the enzyme ribulose-1,5-bisphosphate carboxylase/oxygenase. This enzyme is usually referred to as RuBisCO, and it is quite possibly the most abundant protein on Earth.

The complex series of reactions comprising photosynthesis can be summarised as the extraction of electrons from water, the use of light energy to increase the potential energy of these electrons and converting this to chemical potential energy in the reduced bonds of glucose (Figure 8.3). In most higher plants the energy is stored either as starch (which is a polymer of glucose) or as sucrose (a disaccharide combining one molecule of glucose with one of fructose). Some plants accumulate large quantities of these carbohydrates, either as starch in tubers (for example potatoes), rhizomes (for example irises and some grasses) and seeds, or as simple sugars (such as in sugar cane and sugar beet).

8.4 Where is the energy?

The energy in biological systems is carried as electrons. These electrons, however, are never free in the way they are in a conducting metal, and hence they do not carry kinetic energy as they do in a fluorescent tube or a particle accelerator. The energy used by organisms is a feature of the molecule of which the electrons are part: the electrons carry potential energy that depends on their energy level and the orbitals within the atom that binds them. As the electrons move from one molecule to another, their potential energy varies according to their electronic environment.

The energy derived from light by photosynthesis is stored in the electrons of the reduced bonds (C–C, C–H, C–N and N–H) in carbohydrates and lipids. The potential energy of these storage molecules comes from the addition of high-energy electrons to CO_2 as this is reduced to carbohydrate and lipid. This potential energy (specifically the enthalpy of combustion) can be estimated reasonably accurately by simply counting the number of these reduced bonds in the molecule:

$$H = 220k_C + 105k_N$$

where H is the enthalpy of combustion (kJ mol^{-1}), K_c is the number of C–C and C–H bonds in the molecule and K_N is the number of C–N and N–H bonds[10].

8.5 Using electrons: metabolism

Although living organisms utilise a wide diversity of electron donors and acceptors to provide the energy they convert to the reduced bonds of organic compounds, all organisms use the same metabolic pathways for making use of the chemical potential energy in organic compounds. These pathways form the link between the organic compounds in which the energy is stored, and ATP (adenosine triphosphate), which is the molecule all cells use to deliver energy to where it is needed. For this reason the sequence of reactions is often referred to as 'intermediary metabolism'.

8.5.1 Intermediary metabolism

The reactions comprising intermediary metabolism are common across all life, and are absolutely central to the use of energy by the cell[11]. They provide a range of important small molecules for use in synthesis, and they regenerate ATP from ADP. Intermediary metabolism consists of three sequential processes: glycolysis, the tricarboxylic acid (TCA) cycle and oxidative phosphorylation (Figure 8.4).

Glycolysis is anaerobic and takes place in the cell cytosol. It converts glucose to pyruvate, generating NADH plus some ATP from substrate-level phosphorylation in the process[12]. There are ten steps in glycolysis; most are more or less at equilibrium under cellular conditions, but three have large changes of Gibbs energy. This makes them effectively irreversible, and they are key points for regulation of glycolysis; the enzymes involved are hexokinase, phosphofructokinase and pyruvate kinase. It is important for the cell to have several points at which overall flux can be regulated, because although glycolysis is usually portrayed in textbooks as a linear sequence starting from glucose and ending with pyruvate, intermediate metabolites can leave the glycolytic pathway and other metabolic substrates can enter it at various locations[13].

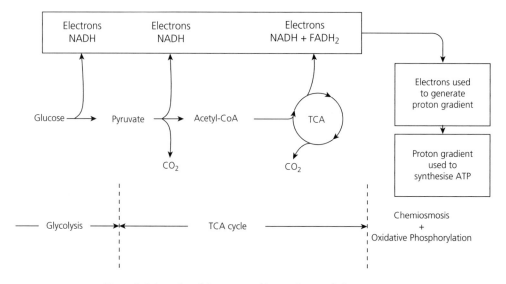

Figure 8.4 An outline of the structure of intermediary metabolism.

The pyruvate produced by glycolysis is transported into the mitochondrion where it is converted to acetyl-CoA and enters the citric acid or tricarboxylic acid (TCA) cycle[14]. The conversion of pyruvate to acetyl-CoA is sometimes referred to as the link reaction because of its position between the end of glycolysis and entry of acetyl-CoA into the TCA cycle. The TCA cycle generates NADH and $FADH_2$, each molecule of which conveys two electrons to the electron transport chain (ETC), where ATP is synthesised by oxidative phosphorylation.

The energy released by the electrons as they pass along the ETC is used to pump protons across the inner mitochondrial membrane, into the intramitochondrial space (Figure 8.5). This happens because the proteins of the ETC can bind protons from one side of the membrane as they accept electrons, and then release them to the other side of the membrane as they pass the electrons to the next carrier in the chain. Each time this happens the electrons lose potential energy.

The ETC contains about 20 different electron carriers, but the main sequence of events is as follows. Electrons carried on NADH pass first to complex I (NADH dehydrogenase), whereas those carried on $FADH_2$ pass to complex II (succinate dehydrogenase). Both then pass the electrons to CoQ_{10} (coenzyme Q_{10}, also known as ubiquinone when in its fully oxidised form, and ubiquinol when fully reduced). CoQ_{10} is free to move through the lipid bilayer by diffusion, and it passes electrons to complex III (cytochromes b and c1). From here electrons pass to cytochrome c, thence to cytochrome oxidase in complex IV and finally to oxygen, which is reduced to water[15]. For every pair of electrons that pass from NADH to water, 10 protons are pumped across the inner mitochondrial membrane.

The high concentration of protons in the intramitochondrial space leads to a pH and electrical charge difference across the inner mitochondrial membrane which provide the driving force (the *protonmotiveforce*, PMF) for the synthesis of ATP[16]. The electrical gradient across the membrane is about 150 mV over a distance of about 5 nm. These numbers are small, but they are equivalent to an electrical gradient of $\sim 3 \times 10^7$ V m^{-1}; this is an enormous electrical force, similar to that which generates a bolt of lightning[17].

Oxidation of a strong reducing agent such as NADH by a strong oxidising agent such as oxygen yields a large amount of energy. Under conditions in the mitochondrion the Gibbs energy change is ~ 225 kJ mol^{-1}. The organisation of the ETC means that this energy is not released in a single step, which would be both wasteful and dangerous, but in a series of smaller steps, each yielding ~ 25 kJ mol^{-1}.

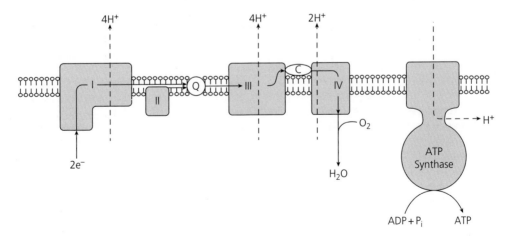

Figure 8.5 The basic structure of the mitochondrial electron transport chain. The solid line shows the path of electron flow, and the dashed lines show the associated movement of protons across the inner mitochondrial membrane.

8.5.2 Energy yield of intermediary metabolism

In theory the complete oxidation of 1 mol of glucose to CO_2 and water yields sufficient energy to synthesise 38 mols of ATP from ADP, but in the cell this is offset by the need to use ATP to move pyruvate, phosphate and ADP into the mitochondrion. The precise stoichiometry is difficult to establish because there is no direct coupling between electrons delivered to the ETC by NADH and $FADH_2$, and ATP generation. The best estimate is that in the cell each mol of glucose yields 30–32 mols ATP.

For glucose the standard Gibbs energy change on oxidation of glucose by gaseous O_2 to gaseous CO_2 and liquid water is -2879.3 kJ mol^{-1} [18]. If we assume the oxidation of 1 mol of glucose yields 32 mols of ATP, and use the standard Gibbs energy change for the formation of ATP from ADP and P_i (+ 30.5 kJ mol^{-1}), the fraction of Gibbs energy captured in ATP is:

$$(32 \times 30.5)/2879.3 = 0.34 \text{ (or 34\%)}$$

Under conditions in the cell, however, the Gibbs energy conserved in the regeneration of ATP from ADP and P_i is higher, $\sim +50$ kJ mol^{-1} (Box 8.2) and this means that a larger fraction of the energy liberated by the oxidation of glucose is retained within ATP[19].

These numbers are sometimes discussed in terms of the apparent inefficiency of energy capture by ATP, an implication being that evolution somehow has not managed to optimise metabolism. Given the 3 billion years or more that evolution has had to work on the core metabolism of cells, and the powerful selective constraint of efficiency in an energy-limited world, this would be surprising indeed were it true. In fact the dissipation of part of the potential energy of glucose as heat is not inefficiency at all; it ensures the overall change in entropy required by the Second Law of Thermodynamics to drive the reactions of metabolism forward. The dissipation of energy as heat ensures that intermediary metabolism is energetically downhill, and thus maintains a flux of energy through the pathway from storage compounds to ATP. Without this dissipation of energy as heat there could be no metabolism at all.

Box 8.2 ATP and energy

ATP (adenosine 5′-triphosphate) is the most important molecule that captures the energy released from storage compounds such as glucose or lipid by intermediary metabolism and delivers this to wherever it is needed in the cell. It is sometimes referred to as the *universal energy carrier*, but this is not strictly true as the cell uses other molecules as energy carriers for certain specialised tasks. ATP is, however, by far the commonest energy carrier in the cell and is typically present at concentrations of 1–10 mM.

The phosphoanhydride bonds that couple the terminal phosphate groups to adenosine are sometimes referred to as '*high energy bonds*'. This is also slightly misleading as the bonds themselves are relatively weak and hence easily broken. The key point is that the structure of the ATP molecule as a whole is high energy, related to electrostatic repulsions between the oxygens of the terminal phosphates, and resonance. In consequence the difference in potential energy between ATP and ADP + P_i is quite high, about 30 kJ mol^{-1} under standard conditions.

The Gibbs energy available from the hydrolysis of ATP depends on pH, on the relative concentrations of ATP, ADP and P_i, and also on the concentration of Mg^{2+}. Under cellular conditions the Gibbs energy from ATP hydrolysis may be as high as 70 kJ mol^{-1} (estimated for resting skeletal muscle) or as low as 50 kJ mol^{-1} (for exercising muscle or rapidly growing bacterial cells). The usual working number is a somewhat conservative 50 kJ mol^{-1}. These high values arise because the ATP/ADP ratio in the cell is many orders of magnitude away from equilibrium.

The cell uses other nucleotide phosphates to carry energy. GDP is important in protein synthesis and gluconeogenesis, CTP is involved in the synthesis of glycerophospholipids and UTP in the synthesis of glycogen. Other molecules are also used for specific energy requirements (for example creatine phosphate in muscle). This poses the interesting evolutionary question of why ATP is the dominant energy carrier and not GTP, CTP or UTP? There is no consensus answer to this question. The energetics of hydrolysis for the various nucleotide phosphates are similar, and many of the advantages proposed for ATP reflect conditions in living cells today, which are not necessarily the drivers for the original evolutionary choice. There may have been subtle structural or chemical factors that dictated the predominance of ATP, or it may simply be a frozen evolutionary accident. At present we do not know.

8.6 Control of ATP synthesis

One might expect a process as ancient and fundamental as intermediary metabolism to be very carefully managed by the cell, and so it proves. If intermediary metabolism simply proceeded regardless, it would use up valuable metabolic substrates to no good purpose. Instead intermediary metabolism is subject to a range of subtle controls, controls that are important because demand for ATP can vary over several orders of magnitude. Central to this control is the protonmotive force, PMF (Figure 8.6).

This simple diagram captures the central role performed by the PMF in regulating intermediary metabolism. Oxidation of metabolic substrates continues until the activity of the electron transport chain is halted by thermodynamic back-pressure from the PMF (that is, it becomes more and more difficult to pump protons across the inner mitochondrial membrane as the PMF increases). In the presence of ADP and P_i the PMF drives the synthesis of ATP: the greater the PMF, the greater the driving force for ATP synthesis. The flux of protons driving ATP synthesis reduces the PMF; this stimulates electron transport, and hence the production of NADH and $FADH_2$ from glycolysis and the TCA cycle, to restore the PMF.

Electron transport activity never goes to zero because there is always a small leak of protons back across the membrane. This leak is voltage-dependent, and may have evolved as a safety valve to limit the trans-membrane voltage gradient: the PMF cannot be allowed to increase to a level which causes electrical resistance to break down.

The relationship between PMF and ATP synthesis broadly follows Michaelis–Menten kinetics, and the greatest sensitivity of ATP synthesis to variations in PMF comes at intermediate levels of the PMF. The PMF thus provides an important mechanism for controlling the rate of ATP synthesis, and hence oxygen utilisation. An important ecological consequence of this control mechanism is that it is demand-driven: the oxidation of metabolic substrates and the associated oxygen consumption are driven by ATP utilisation.

While the use of ATP is the primary driver for the flow of electrons along the ETC, the cell also needs to regulate the supply of electrons provided by the NADH and $FADH_2$ from glycolysis and the TCA cycle. An important element of this control is feedback inhibition: high concentrations of ATP, NADH and $FADH_2$ inhibit the enzymes that catalyse key control reactions. Important among these are phosphofructokinase (PFK) in glycolysis, the link reaction catalysed by pyruvate dehydrogenase, and citrate synthase, isocitrate dehydrogenase and alpha-ketoglutarate dehydrogenase in the TCA cycle[20].

These feedbacks are subtle and effective, allowing intermediary metabolism to respond rapidly and sensitively to variations in the demand for ATP. They are cellular level controls, but it has been known for a long time that organisms exert a strong level of control on the overall level of metabolism, and also on the level of metabolism in different tissues. These can be increased or decreased as required, with the key regulation being through the concentration of Ca^{2+} within the cell. Ca^{2+} effects strong control by binding to key

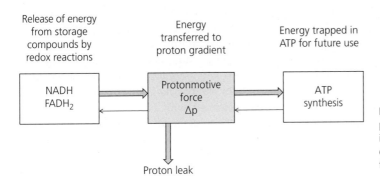

Figure 8.6 The central role played by the protonmotive force (PMF: Δp) in regulating intermediary metabolism. The block arrows show energy flow, and the line arrows regulatory feedbacks.

enzymes involved in intermediary metabolism, notably pyruvate dehydrogenase in glycolysis and isocitrate dehydrogenase and 2-oxoglutarate dehydrogenase in the TCA cycle. Control of enzyme activity is also exerted through phosphorylation by kinases, for example the pyruvate dehydrogenase complex, with this process also being regulated by Ca^{2+}. All of these are influenced themselves by systemic-level control through hormones such as insulin or the thyroid hormones T3 and T4. The organism can thus effectively set the overall level of metabolism by higher level controls, but also effect fine control of the short-term variation in intermediary metabolism set by ATP demand. The process of ATP synthesis is thus subject to complex and subtle controls at a variety of levels of integration (mitochondrial, cellular and organismal). These allow the organism to match ATP supply to the demand at the cellular level, while also providing for higher-level control.

The fundamental ecological point that emerges here is that intermediary metabolism is driven principally by demand for ATP. ATP is only synthesised, and hence oxygen used, because the organism needs to regenerate ATP that has been used for some task. Metabolism is demand-driven, and it is a cost to the organism, for it uses up resources that have to be replaced. A question any ecologist making a measure of metabolism should always ask is thus, what is the organism doing that required the ATP it is regenerating?

8.7 Metabolism in ecology

While knowledge of the mechanism of electron flow from metabolic substrates to oxygen and the associated synthesis of ATP is critical in understanding how intermediary metabolism works, what it does not do is allow us to predict how much ATP an organism requires. We cannot predict the level of metabolism of any whole organism from first principles; biological systems are just too complex. All we can do is measure the metabolic rate of the organism, and try to explain retrospectively why it is what it is. For anyone brought up on the statistical certainties of physics this might well be very frustrating, but such is the nature of physiology.

8.7.1 What do we mean by metabolism in a whole organism context?

We have seen above that at a molecular level the energy for metabolism is provided by the decrease in the potential energy of electrons as they pass from substrates to oxygen. This is, however, of little practical use to an ecologist: we cannot place an organism in a piece of laboratory equipment, or point an instrument at an animal in the wild, and measure its internal electron flow. Neither can we measure the production of ATP directly. What we can do, however, is measure the rate at which oxygen, the final electron acceptor in the ETC, is being used. So common is the measurement of oxygen uptake as a proxy for ATP synthesis that oxygen uptake itself is often referred to as 'metabolic rate'.

To convert a measure of oxygen consumption to a measure of energy flux energy we need to know the relationship between the two. This conversion factor is usually referred to as the *oxycalorific coefficient*, the name reflecting the early history of physiology when energy was measured in calories rather than joules[21]. Calculation of the oxycalorific coefficient is fairly straightforward: the energy released by the complete oxidation of a given metabolic substrate is divided by the amount of oxygen required. For glucose, the stoichiometry is:

$$C_6H_{12}O_6 + 6O_2 \rightarrow 6CO_2 + 6H_2O$$

The oxycalorific equivalent for glucose is thus the energy released divided by six, and the units are kJ mol^{-1}. Similar calculations are undertaken easily for other pure substrates (Table 8.3). Under normal circumstances most organisms actually use a mixture of metabolic substrates, but it is straightforward to calculate an overall coefficient by simple averaging. A typical substrate mix is usually assumed to be 70% carbohydrate, 20% fat and 10% protein.

Using an oxycalorific coefficient to convert measured oxygen consumption to estimate energy flux makes two key assumptions, neither of which is met fully in reality. The first is that that all oxygen use is associated with heat dissipation, and vice versa. In fact, all organisms generate some ATP anaerobically from glycolysis, and this dissipates heat that is not tied to oxygen consumption. Where the pyruvate generated by glycolysis is converted

Table 8.3 Oxycalorific coefficients for metabolic substrates. Data are expressed as energy dissipated per mol of diatomic oxygen gas[22].

Substrate	Oxycalorific coefficient (kJ mol^{-1})
Carbohydrates	
Glucose	467.5
Fructose	468.4
Sucrose	470.3
Lipids	
Palmitic acid	433.8
Generalised lipid	439.0
Generalised proteins	
N excreted as ammonia	427.5
N excreted as urea or uric `acid	435.2
Generalised mixed substrates	
Typical literature values	435–463

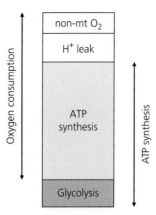

Figure 8.7 The relationship between oxygen consumption and ATP synthesis in organisms. ATP synthesis is shown in grey, and oxygen utilisation not associated with ATP synthesis in white. Note how some ATP synthesis does not involve oxygen, and also that some oxygen use is not associated with ATP synthesis. The sizes of the boxes are arbitrary as the relative contributions of these four processes are not known with certainty for any organism.

to acetyl-CoA and enters the mitochondrion to be fed into the TCA cycle, the overall stoichiometry of glucose oxidation, and hence the validity of the oxycalorific coefficient, is not affected. Where the pyruvate is converted to lactate or other fermentation product, then heat is generated but no oxygen utilised, and it is this process that affects the validity of the oxycalorific coefficient. The fraction of total ATP regeneration that is anaerobic is usually assumed to be small, but is actually known only poorly[23]. In addition, the cell uses oxygen for tasks other than intermediary metabolism. These are not fully characterised and are usually grouped as 'non-mitochondrial oxygen consumption', though one important role for oxygen is in the final stage of cholesterol synthesis.

The accuracy of the oxycalorific coefficient depends on the relative importance of anaerobic ATP generation and non-mitochondrial oxygen consumption compared with acting as the electron acceptor in mitochondria (Figure 8.7).

The second assumption is that all of the free energy released by substrate oxidation is dissipated immediately as heat. This cannot be entirely true, for if it were the cell could not build new tissue with the energy released from metabolic substrates. In protein synthesis for example, the formation of a peptide bond between two amino acids requires hydrolysis of the equivalent of four ATP molecules. Under cellular conditions this releases ~ 200 kJ mol^{-1} of energy, of which ~ 8–16 kJ mol^{-1} is incorporated as chemical potential energy in the bond, with the remainder being dissipated as heat. Simple calculation shows that if the ATP used in protein synthesis is provided by glucose oxidation, then only ~ 4% of the energy released by glucose is captured in the peptide bonds of proteins; the remainder (~ 96%) is dissipated as heat. A second major consumer of ATP in many organisms is muscular activity where the energy of ATP is used to effect work, but the energy here is eventually dissipated as heat through friction and drag.

The end result is that when averaged over a period of time, most (but importantly not all) of the energy released by the oxidation of metabolic substrates is indeed dissipated as heat, and so the assumption that we can equate oxygen consumption with energy dissipation through metabolism is reasonable. The relationship between metabolism, oxygen and heat had long perplexed scientists, and it was Antoine Lavoisier who paved the way for our modern views with a classic experiment on the metabolism of a guinea pig (Box 8.3). We now understand why Lavoisier's experiment worked, but we need to bear in mind that Lavoisier was working with a mammal. It is quite possible that the equivalence

> **Box 8.3 Unravelling the relationship between metabolism, oxygen and heat**
>
> The relationship between metabolism and heat production in animals (or least the endotherms most familiar to us) has been known since animals were first hunted or domesticated. It had also been known since ancient times that no animal could live where a fire would not burn, but experimental investigation only started in the eighteenth century. Joseph Priestley made a significant advance by comparing the behaviour of a flame and a mouse when held in a confined space. Priestley concluded that the flame and the mouse were altering the air in the same way[24]. We now recognise that both the flame and the mouse were utilising oxygen, leaving behind a mixture of nitrogen and carbon dioxide.
>
> The most important insights were, however, provided by the great French chemist Antoine-Laurent de Lavoisier. His most significant contribution to this field was made in the winter of 1782/1783 in association with the mathematician Pierre-Simon Laplace. Critical to this work was the design of an ice calorimeter for measuring the amount of heat given off during combustion or respiration. Lavoisier and Laplace placed a guinea pig in the calorimeter and measured the quantity of carbon dioxide and heat it produced (Plate 11). They then compared this with the amount of heat produced when sufficient carbon was burned in the ice calorimeter to produce the same amount of carbon dioxide as that exhaled by the guinea pig, and concluded that respiration was essentially a slow combustion process[25].
>
> This was confirmed a century later in 1889 by Max Rubner, who developed the first accurate respiration calorimeter and used this to show clearly that the First Law of Thermodynamics applied to living things. He kept a dog in the calorimeter for 45 days, measuring heat output and respiration simultaneously, and showed that the heat produced by the dog matched the energy content of its food, corrected for nitrogen excretion, to a remarkable degree of accuracy (0.3% over 45 days)[26].

between oxygen consumption, ATP generation and heat dissipation is less exact in organisms such as intertidal bivalves, or invertebrates living within sediments, where oxygen may be less freely available and in consequence there is a greater reliance on anaerobic processes for generating ATP.

8.7.2 A definition of metabolism for ecologists

Ecologists differ in what they mean by 'metabolism' and the term is often used without definition at all. A popular definition of late has been *'the flux of energy and materials through an organism'*[27]. This is clearly fully compatible with the usual definition used by biochemists, which is *'the sum total of all the chemical reactions in the body'*, but with an emphasis on flows and rates. These are entirely accurate definitions, but they are of limited practical use in that there is no known apparatus into which an organism can be placed that would deliver a measure of metabolism as so defined[28].

Organisms use ATP for almost everything that they do, and so a narrower definition of metabolism might be the rate of ATP regeneration. However, as we have seen, this too cannot be measured directly, only approximated by a measure of oxygen consumption. This can be expanded to include simultaneous measures of CO_2 production or nitrogen excretion, and while increasing the complexity of measurement this has the advantage of defining the balance of metabolic substrates being used. For most ecological applications, however, a simple measure of oxygen consumption is sufficient, but we always need to bear in mind that this is an indirect and incomplete measure of metabolic rate.

8.7.3 Levels of metabolism

The metabolic rate of an animal varies markedly, depending on what it is doing and the environment in which it finds itself. A mammal running to flee a predator may consume more than ten times the oxygen it does when resting beneath a tree, and a foraging honey bee may increase its metabolic rate in flight by a factor of ~150. In order to make meaningful comparisons between different individuals, or between species, it is therefore necessary to establish a set of standard conditions for measurement.

This was a key problem for the early pioneers of human physiology, and it was the American physiologist Francis Benedict who established the criteria for comparing the basal metabolic rate of different individuals[29]. The conditions he proposed were rigorous indeed, requiring metabolic rate to be measured at rest, with the individual held under maintenance rations but in a post-absorptive state, and within the thermoneutral zone. Specific guidelines were that the subject should be in complete muscular repose during the measurement (and before), which should be made with the subject lying down but awake. Because this approximates the cost of maintaining the body in a viable state, without the extra costs involved in digestion, reproduction or activity, it was termed *basal metabolism*. This measure of basal metabolism was recognised not to reflect the absolute minimal metabolism that might be observed, for it was well known by then that sleep lowers metabolic rate in humans by about 10%.

Interest soon spread to measuring the basal metabolic rate of mammals other than humans, particularly those of agricultural interest. These are, however, not as cooperative as humans and the early literature contains entertaining asides on the difficulties associated with such measurements. Samuel Brody, an important pioneer of metabolic studies, once commented wryly that *'the high values for the boar, bull and elephant reflect the technical difficulties in measuring metabolism of these temperamental animals'*[30]. One can only imagine the problems of encouraging a recalcitrant elephant, or even worse a cantankerous rhinoceros, into a respirometer. The early workers were well aware of the practical and conceptual difficulties of comparing, for example, ruminants with non-ruminants, or large slow-moving or sedentary animals with small active ones. It was also recognised that species take very different times to clear their gut, depending on diet and size and that some species ferment food in their gut, while others do not.

While it is impossible to meet some of the criteria established for humans when measuring metabolism of animals, comparative studies do require an agreed set of conditions for measurement. Nowadays these are taken to be an animal whose metabolic rate is not elevated by activity, digestion, lactation or the need to generate or lose heat. The metabolic rate measured under these agreed conditions became known as *resting metabolic rate*. The various terms that have been used to describe minimal metabolism are summarised in Table 8.4.

The term *basal metabolic rate* (BMR) was originally confined largely to human physiology. The current convention is to restrict the use of BMR to endotherms (where it includes the cost of generating heat to maintain internal body temperature), and use *standard metabolic rate* (SMR) for ectothermic vertebrates and invertebrates[31]. Very real difficulties arise, however, when we try and apply these terms more widely.

What should we do, for example, with fish, many of which swim continuously and thus for whom being at rest is a completely artificial situation? One solution has been to examine the relationship between metabolic rate and swimming activity, and then extrapolate the relationship back to estimate a theoretical value for the metabolic rate at zero activity. While a sound approach theoretically, this is extremely time-consuming and in consequence such studies are few in number[32].

The pragmatic approach for comparative studies of fish metabolism has been to select those data approaching closest to resting metabolism, and trust that the errors are not too great. When this is done, a clear relationship emerges between standard metabolic rate and lifestyle: more active species have a higher SMR than demersal, bentho-pelagic or benthic species (Figure 8.8), suggesting that it costs more to maintain a body selected for a pelagic lifestyle than it does one selected for a bottom-living lifestyle. Should we conclude that there are real differences in the cost of living for different lifestyles in fish, or is it simply that it is only in benthic or bathyal species that we can measure a true standard metabolic rate, and that in more active species the measurement contains an unavoidable contribution from the cost of swimming activity?

Things get more difficult still when we turn to invertebrates. Insects exchange respiratory gases through a system of tracheae. Oxygen enters through the spiracle, which can be closed off, and is then delivered deep into the tissues through a series of ever-finer tracheoles. Although simple diffusion is important, flow can be enhanced by abdominal pumping. This pumping can be very obvious, as anyone who has watched a dragonfly resting following a sortie after prey will know: here the abdominal pumping looks curiously as through the insect is panting (which in a sense it is). Many insects at rest

Table 8.4 The levels of metabolism[33].

Measure	Comments
Measure of minimal metabolism	
Basal metabolic rate (BMR)	This was defined by Francis Benedict as the metabolic rate of an awake but inactive, post-absorptive, non-growing, non-lactating mammal held in its thermoneutral zone.
Resting metabolic rate	The metabolic rate of a resting, post-absorptive animal; endotherms must be within their thermoneutral zone.
Standard metabolic rate (SMR)	A term introduced by August Krogh and used extensively by fish biologists in the 1940s, later extended to terrestrial ectotherms. Originally defined as the metabolic rate of a post-absorptive, minimally active fish in still water.
Least observed metabolic rate	A term suggested by Kenneth Blaxter for the lowest metabolic rate that is observed for short periods of time during a continuous record of metabolic rate (not the metabolic rate that is observed least often).
Maintenance metabolic rate	This was originally defined for mammals by Samuel Brody as the energy needed to fuel maintenance life processes, and is equivalent to the metabolic rate of an animal fed sufficient food to maintain constant body mass. A similar definition is now used for plants, where maintenance metabolism is defined the respiration rate when growth is zero.
Measures of routine metabolism	
Routine metabolism	The metabolic rate of an organism undertaking normal day-to-day activity.
Field metabolic rate (FMR)	The metabolic rate of a free-ranging animal in the wild, undertaking normal activity, usually averaged over a period of one or more days.
Daily energy expenditure (DEE)	The average energy expended each day in normal existence.
Sustained metabolic rate	The maximum metabolic rate that can be sustained over a period of time (days) while remaining in metabolic balance such that energy intake equals energy dissipation, and hence growth is zero.
Measures of peak metabolism	
Maximum metabolic rate (MMR)	The maximum aerobic metabolic rate that can be induced for a defined (short) period of time by running on a treadmill.
Summit metabolic rate	The maximum aerobic metabolic rate elicited by cold exposure in endotherms, but in the absence of activity or exercise.

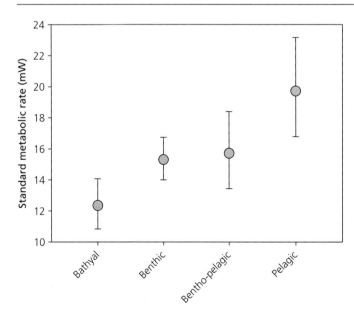

Figure 8.8 Standard metabolic rate (mean ± standard error) of 69 species of fish with different lifestyles, plotted with increasingly active lifestyles from left to right. Data are for a fish of body mass 50 g and a temperature of 15 °C, calculated using a General Linear Model[34].

do not, however, breathe continuously but instead undergo discontinuous gas exchange, in which periods of reduced oxygen uptake or CO_2 release are interspersed with periods of normal exchange. The pattern varies with species, and the range covers continuous but variable gas exchange, through cyclical patterns to the extreme where gas exchange is completely shut down between spikes of high flow. These patterns are driven by the spiracles and abdominal pumping, and are under central nervous control.

The question this poses is what is the correct measure of resting metabolic rate in a resting insect? Is it the peak rate observed at rest, or is it an average calculated over some period of time? This problem of interpretation set by a resting metabolic rate that varies in time is not confined to insects: how, for example, should we characterise the resting metabolic rate of an intertidal marine invertebrate which exhibits very different rates of oxygen uptake when immersed in water and when exposed at low tide? These are not the trivial questions they might seem at first glance, and they need resolving if we wish to tackle broad evolutionary questions such as whether the cost of existence is the same for all animal tissue.

8.8 What processes comprise resting metabolic rate?

The usual definitions of resting, standard or basal metabolic rate explicitly exclude many aspects of routine existence, such as locomotor activity, growth or reproduction, leaving just those processes that are necessary to keep the organism alive[35]. These processes have been investigated most thoroughly in humans and domesticated mammals, in which the energy used for maintaining organism viability is conventionally divided into the energy used for whole organism service costs, and that used for maintaining the existence of individual cells and tissues[36].

The service costs include the work involved in cardiovascular circulation, the immune system, information processing by the central nervous system and nitrogen excretion by the kidneys. In many vertebrates the cost of basal neural activity appear to be fairly constant at 2–8% of resting metabolic rate, the majority of this being the costs of the Na^+-K^+-ATPase activity essential for maintaining resting potential, or restoring this after the passage of an action potential[37]. In terrestrial vertebrates there is also a contribution from the muscular work involved in maintaining posture (that is, the cost of standing rather than simply lying down). Current data suggest that the costs of maintaining posture are small in larger grazing mammals (< 1% resting metabolic rate in cattle, for example), but are probably a little higher in man or sheep (perhaps 5% or more). Since most mammals will spend some time standing and some time lying down, we cannot realistically define a standard for measurement that requires either standing or lying down. Overall, whole-organism service costs in mammals may comprise 35–50% of resting metabolic rate[38]. We have no idea what these costs are for any invertebrate.

Teasing apart the many and varied processes involved in keeping a cell alive is not easy. The usual approach is to compare oxygen consumption before and after the addition of inhibitors that block specific processes. For example ouabain inhibits Na^+-K^+-ATPase, oligomycin prevents ATP synthesis and cycloheximide blocks protein synthesis. These studies are most easily performed with cell cultures, and most progress has been made with hepatocytes isolated from liver tissue. One major difficulty with this approach is that the isolation of cells from the parent tissue, such as hepatocytes from liver, is a highly disruptive process and we cannot guarantee that what we measure in the resulting cell culture is truly representative of what happens in the intact tissue within the organism. In particular the isolation process disrupts all higher-level control of physiological rates by hormones, and also changes the nutrient environment the cells experience. Nevertheless these are the best data we have, and they do point to some important general principles[39].

A study of hepatocytes isolated from the rat suggests that the key cellular processes under resting conditions are protein turnover, the maintenance of transmembrane ion gradients and the proton leak across the inner mitochondrial membrane (Table 8.5). It is likely that the balance will shift if metabolic rate increases. For example if there is increased demand for ATP, then as ATP regeneration

Table 8.5 Estimate of contribution of different consuming processes to whole animal resting oxygen consumption in the adult laboratory rat (*Rattus norvegicus*)[40].

Process	% BMR
Oxygen utilisation not involving ATP synthesis	
Non-mitochondrial O_2 utilisation	~10
Mitochondrial proton leak	~20
Oxygen utilisation associated with ATP turnover	
Protein synthesis	~20–25
Transmembrane Na^+ gradients	~20–25
Transmembrane Ca^{2+} gradients	~5
Gluconeogenesis	~7
Ureagenesis	~2.5
Actinomyosin ATPase	~5
Nucleic acid synthesis	Unknown (< 5)
Membrane lipid turnover	Unknown (< 5)
Total	~95

increases, the protonmotive force will decrease and the rate of proton leakage across the membrane will decrease as the flow of protons through ATP synthase increases.

Compared with many cell types, hepatocytes are relatively easy to isolate from their parent tissues, and this has meant that they are one of the most studied isolated cell lines *in vitro*. We do know, however, that different tissues in the body have a different balance across the various essential housekeeping cellular processes. For example liver and gut cells tend to have high rates of protein synthesis, whereas this comprises a lower fraction of resting metabolism in kidney, heart or brain. In contrast, ion pump activity is high in gut, kidney and brain, but lower in muscle and liver. These differences have a clear underlying rationale: for example, neural activity involves the rapid movement of ions and the liver is an important site for protein synthesis.

We must be careful not to generalise too far, because not all animals are rats. There appear to be no comparable data for invertebrates but we can surmise that in these the dominant consumers of oxygen are likely to be muscle, gut and those tissues that perform an analogous role to the vertebrate liver, with neural activity generally representing a small proportion of resting energy use.

8.8.1 How constrained is resting metabolic rate?

Given the highly conserved nature of intermediary metabolism and millions of years of selection for efficient use of resources, we might expect the cost of living, as reflected in the rate of resting metabolism per unit tissue, to be a fairly well-defined feature of physiology and similar across organisms. Perhaps surprisingly, this idea had not been tested until relatively recently. In 2009 Anastassia Makarieva and colleagues showed that the energetic cost of maintaining a unit mass of tissue (mass-specific resting metabolism) is fairly well conserved across organisms: despite a variation in body mass of 20 orders of magnitude, mass-specific metabolic rate varied only 30-fold (Figure 8.9).

This analysis raises some intriguing questions: what is it that sets the level of resting metabolism, which is typically in the range 0.3–9 W kg^{-1}? And what factors account for the variation within this range? We know that lifestyle plays a role in this variation (Figure 8.8), but body size is also important, as we shall see in later chapters.

8.8.2 Routine and active metabolism

While resting metabolic rate is beset with difficulties of interpretation and measurement, maximum metabolic rate is relatively straightforward. In mammals and reptiles the usual approach is to use a treadmill and measure oxygen consumption while the animal is running as fast as it can for a defined period of time. Flying organisms such as insects and birds are usually flown in a wind-tunnel while measuring gas exchange, and aquatic organisms are typically swum in a flume. These are technically demanding measurements to make, and so we have far fewer of them than we do for resting metabolic rate.

We do, however, have measurements for mammals, birds, reptiles, some fish and a range of insects. Almost all of the organisms studied tend to have a lifestyle characterised by periods of intense activity. In contrast we know very little of the maximum metabolic activity in sedentary species, or in whole classes of organisms whose activity is slow, such as molluscs or echinoderms. Our view of

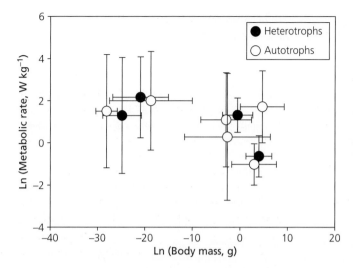

Figure 8.9 The mean mass-specific metabolic rate (W kg^{-1}) as a function of mean body mass (g) for selected groups of organisms. The heterotrophic groups are (left to right) bacteria, protozoa, insects, aquatic invertebrates, ectothermic vertebrates and endothermic vertebrates. The autotrophic groups are (again left to right) cyanobacteria, microalgae, tree seedlings and tree saplings[41].

maximum metabolic rate is thus coloured by concentration on those species which exhibit periods of intense activity.

The extent to which metabolic rate can be elevated above the resting rate varies widely between species. A human athlete in peak condition can achieve a maximum metabolic rate about 20 times their resting rate. In dogs this increase can be 30-fold, and the record for a mammal is held by the North American pronghorn, *Antilocapra americana*, which can uprate its metabolism about 65-fold[42]; a similar increase can be achieved by trained racehorses. A median figure across all mammals for which we have data is ~12. Reptiles typically achieve a lower increase, whereas insects can increase their metabolic rate by ~150-fold.

Whereas resting metabolic rate captures the utilisation of oxygen by all cells in the body, maximum metabolic rate is dominated by the oxygen demand of muscle. In a mammal exercising at its peak, over 90% of the blood flow is directed at muscles. While maximum metabolic rate is of interest in capturing the maximum rate at which an organism can operate, no individual ever operates at this intensity for anything other than a short period of time: any animal that spent all its time running about at breakneck speed would have no time to feed or find a mate.

Organisms actually function on a day-to-day basis with a metabolic rate that is above resting (which by definition excludes all routine activity) but well below the maximum possible for short bursts. This routine metabolic rate has perhaps been the hardest to pin down. As with other aspects there have been a number of approaches, of which the most important have been isotope turnover and heart-rate loggers (Box 8.4). Isotope turnover provides a time-averaged estimate of routine metabolic rate; this is usually termed the field metabolic rate (FMR) or the daily energy expenditure (DEE).

Animals going about their daily business, however, do not operate continuously at their FMR or DEE. Routine metabolic rate varies significantly from moment to moment as organisms move around, pause to rest, feed and may reach maximum rates for short periods of prey capture or predator avoidance. A typical pattern is shown by a widely foraging seabird, the Australasian gannet, *Morus serrator* (Figure 8.10). The continuous record of metabolic rate provided by a heart-rate logger can be divided into periods when metabolic rate drops to resting levels and periods of routine activity when metabolic rates are considerably higher. In the example shown here, both resting metabolic rate and routine metabolic rate show a slight but significant increase through the summer season. Most striking, however, is the difference in variability: resting metabolic rate varies little around its mean rate, whereas field metabolic rate is highly variable.

Box 8.4 Measuring metabolic rate in free-living organisms

Two techniques have been especially important for estimating the metabolic rate of free-living animals in the wild.

Isotope turnover

Although various isotope clearance techniques have been tried as ways of estimating metabolic rate, it was the introduction of doubly labelled water (DLW) by Nathan Lifson and colleagues in the 1950s that revolutionised the field. The key factor underpinning the value of DLW is that there is an isotopic equilibrium in the body between the oxygen in carbon dioxide and that in water, mediated through the activity of the enzyme carbonic anhydrase. Labelling the oxygen of water with ^{18}O results in half the label being present in body water and half in CO_2. Labelling the water with deuterium, ^2H, allows the elimination of ^{18}O in the water to be determined, and hence the loss of ^{18}O through CO_2, and thereby metabolic rate, to be estimated[43].

DLW offers a powerful technique for the measurement of the routine metabolic rate of a free-living animal in the wild, averaged over the period between the initial and final measurements. What it cannot do is tell us is how metabolic rate varies from moment to moment; for this we need other techniques.

Heart-rate loggers

Short-term variations in the metabolic rate of free-ranging animals can be tracked using heart-rate loggers. Comparison of the heart rate with directly measured oxygen consumption serves to calibrate the heart-rate monitor. Because these loggers record heart-rate continuously, they provide a valuable insight into the way that metabolic rate varies in response to short-term variations in activity, variations that measurement with DLW averages out.

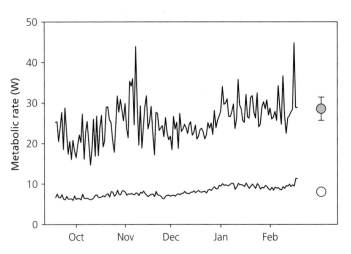

Figure 8.10 Field metabolic rate (FMR, upper trace and grey symbol) and basal metabolic rate (BMR, lower trace and white symbol) in a free-living Australasian gannet, *Morus serrator*, estimated from an implanted heart-rate logger. The symbols to the right show the mean and SE for all the data[44].

8.8.3 What physiological processes contribute to field metabolic rate?

While resting metabolism approximates the cost of keeping the organism alive, it is processes additional to BMR that power the organism's ecology. These everyday activities include moving around the landscape, feeding, growth and reproduction. Although these processes are all additional to BMR, in truth we do not know whether those processes that comprise BMR continue at the same intensity when the organism shifts from resting to activity. It is perfectly possible that when an organism is going about its daily business, basal processes are downgraded to allow energy to be diverted to ecologically important activities: FMR/DEE and BMR/SMR may not be strictly additive.

The major components of FMR/DEE are not fully characterised, but important uses of ATP are believed to include locomotion and the cost of synthesising new tissue, with the balance being made up by increased service costs (enhanced cardiovascular work, elevated neural function and so on). The cost of locomotion is fairly well characterised for organisms that run, swim or fly, but not for the quite different modes of locomotion employed by worms or snails. The energy used in locomotion involves both the inherent cost of a given mode of locomotion and the time spent moving. Estimates of these costs as a percentage of DEE for various vertebrates range from 5% in the least weasel *Mustela nivalis*, 3–36% in 27 species of lizard and as much as 80% in sockeye salmon, *Oncorhynchus nerka* during their spawning migration. This variation indicates a powerful influence of life history on how organisms use their energy but also precludes any broad generalisations[45].

8.8.4 The three levels of metabolic rate

We can therefore identify three ecologically relevant levels of metabolic rate. As we can see from Table 8.4, the terminology here is complex and confusing, as different authors use different terms that each have a subtly different basis. For the lower two levels of metabolic intensity a number of terms have been proposed, that have different conceptual bases but which in ecological terms are broadly synonymous:

- Level 1: Basal metabolic rate (BMR); Resting metabolic rate; Standard metabolic rate (SMR); Minimal metabolic rate; Least observed metabolic rate
- Level 2: Routine metabolic rate; Field metabolic rate (FMR); Sustained metabolic rate; Daily energy expenditure (DEE)
- Level 3: Maximum metabolic rate; Summit metabolic rate

The only solution to almost a century of the use of many different and sometimes mutually inconsistent terms is for any discussion of metabolic rate to make explicit precisely what level of metabolism is being discussed, and how it was measured.

8.8.5 Aerobic scope

The obvious questions for an ecologist or physiologist is: why are the levels where they are and what constraint(s), if any, prevent them from being different? For comparison across different species or different classes of organism, a useful measure is metabolic scope. The metabolic scope captures the extent to which FMR exceeds BMR/SMR, and can be calculated in two ways (another source of confusion in the literature):

$$\text{Factorial routine scope} = \text{FMR}/\text{BMR}$$
$$\text{Absolute routine scope} = \text{FMR} - \text{BMR}$$

Factorial scope is a ratio, and hence is dimensionless with no units. Absolute aerobic scope is the arithmetic difference between two rates and has dimensions and units of power.

Typical mean factorial scopes in vertebrates (mammals, birds, reptiles) lie in the range 3–6 (Table 8.6). These numbers are intriguing for they suggest that in vertebrates something constrains these scopes within a fairly narrow range.

Maximum aerobic scopes in mammals are much more variable; for moderately active species they lie in the range 2–20, but for highly active species they can be as high as ~65. For insects, however, the average maximum aerobic scope, based on the mean values for resting and active MR is ~150. Radically different body plans (insects versus vertebrates) can lead to very different factorial scopes for maximum performance[46].

Table 8.6 Factorial aerobic scope (the ratio of DEE or FMR to BMR) in selected groups of vertebrates. Aerobic scope has been calculated only for those species where both BMR and FMR/DEE have been measured.

Class/Order	n	Mean	SE	Min	Max
Birds	13	3.44	0.42	1.3	6.7
Mammals	18	3.53	0.34	1.9	6.9
Placentals	10	3.03	0.29	2.0	5.0
Marsupials	8	4.15	0.63	1.9	6.9
Reptiles	33	2.37	0.18	1.1	5.1
Lizards	28	2.35	0.19	1.1	5.1
Snakes	5	2.46	0.56	1.6	4.6

8.9 Temperature and metabolic processes at the molecular level

We have discussed metabolism in some detail, and the importance of this will become clear in later chapters, but we have not yet mentioned temperature. Does our understanding of the molecular processes involved in intermediary metabolism allow us to predict how sensitive it might be to a change in temperature?

In all organisms, intermediary metabolism involves a complex series of processes that includes diffusion, enzyme catalysis and quantum tunnelling, all of which differ in their inherent temperature sensitivities (Table 8.7).

To take the electron transport chain as an example, the transfer of electrons within complexes I to IV takes place through a chain of reaction centres containing transition metals (Fe-S centres in complexes I, II & III, and Cu in complex IV). These centres are not close to each other, as each is embedded in a protein (haem) molecule. The electrons pass between centres not along pathways in the protein molecules, but by quantum tunnelling. Quantum processes are completely unaffected by temperature, even close to absolute zero, but its likelihood is extremely sensitive to the distance between the reaction centres. The protein molecule itself is important in providing electrical insulation (that is, in preventing the electron from transferring to the wrong place), but thermal motions within the protein molecules are critical in bringing the reaction centres close enough for tunnelling to occur. As a result the likelihood of tunnelling becomes temperature dependent, for higher temperatures result in more rapid internal motions and a higher frequency of reaction centres coming close enough to allow tunnelling.

The two components of the ETC that transfer electrons between the respiratory complexes (CoQ10 and cytochrome c) do so by diffusing within the mitochondrial membrane. Being small, these components can move more rapidly within the membrane than can the larger protein molecules that form the respiratory complexes. Diffusion rates depend on the mean speed of molecules, which varies only slightly over the physiological temperature range.

Intermediary metabolism thus involves a complex series of closely linked processes, each with its own inherent temperature sensitivity. This makes it impossible to predict an overall temperature sensitivity, unless we can identify a particular step where temperature sensitivity is likely to dictate that of the process as a whole. It might be possible to explore this with a model of intermediary metabolism, but at present we lack data on the temperature sensitivity of many steps in the sequence (not least ATP-synthase itself) and so this approach is not yet feasible.

8.9.1 An aside: temperature and photosynthesis

Like electron transport, photosynthesis comprises a complex series of closely linked reactions, involving quantum effects, diffusion and enzyme catalysis. The relationship between photosynthesis and temperature is more complex than intermediary metabolism, because the rate of photosynthesis may be limited by factors such as the availability of light or carbon dioxide, thus complicating the effects of temperature.

8.10 Temperature and whole-organism metabolism

Many key concepts in science can be traced back to a single observation or experiment. For metabolism it is Lavoisier's experiment with a guinea pig in his ice calorimeter; for the relationship between metabolism and temperature it is an experiment with a goldfish (Box 8.5).

Table 8.7 Rate determining processes in intermediary metabolism.

Rate-determining process or factor	Temperature sensitivity	Examples and comments
Quantum tunnelling	Zero (independent of temperature)	Electron tunnelling within respiratory complexes in ETC; proton tunnelling in enzyme catalysis. Likelihood influenced by internal molecular motion of protein complexes.
Mean molecular speed	Low (7% increase from 0 °C to 40 °C in an ideal gas)	Diffusion of metabolites within the cytosol; diffusion of ubiquinone within the mitochondrial membrane.
Threshold activation energy	Moderate (rates double or treble for a 10 K rise in temperature)	Enzyme-catalysed reactions.

Box 8.5 August Krogh and the goldfish

A little over a century ago, the Danish physiologist August Krogh took a single goldfish, Carassius auratus, and measured its resting metabolism at a range of temperatures. The data showed a smooth monotonic change in resting respiration with temperature (Figure 8.11).

Krogh concluded that because there was no visual difference between data for a quiescent fish and an anaesthetised fish, the measurements were truly those of resting metabolism. Krogh also measured the resting respiration of a frog, a moth chrysalis and a dog. When expressed on a mass-specific basis, and after arbitrary adjustment of the dog data, these data matched those of the goldfish. The overall relationship became known as *Krogh's normal curve*, and for many years it was the canonical relationship between resting metabolism and temperature[47]. Subsequent investigations produced qualitatively similar, though not identical, results[48].

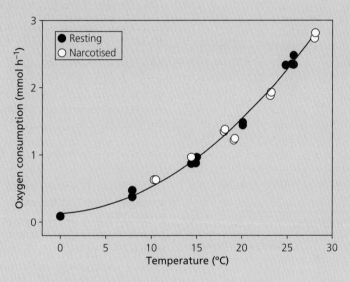

Figure 8.11 The seminal experiment of August Krogh: oxygen consumption of a single goldfish measured at different temperatures, both resting (black symbols) and narcotised (white symbols). The fitted line is a quadratic.

In the century since Krogh's experiment we have accumulated a vast array of data on the oxygen consumption of a wide variety of organisms. The effect of temperature has long been a common theme in these studies, with the temperature sensitivity typically expressed as a Q_{10} coefficient (Box 8.6). In exploring these data it is important, as we saw in the previous chapter, to distinguish between studies of temperature sensitivity within individuals or species, from studies across species. The former gives insight into the immediate (acute) response of metabolic processes to a change in temperature, and provides insight into the underlying thermodynamics. The latter tells us to what extent evolution has been able to modify this immediate effect, as each data point the pattern across species is the result of a long period of evolutionary adjustment to living at a particular temperature. In determining the temperature sensitivity of standard metabolism (or indeed any physiological rate) within a species and between species, we are measuring very different things. The two types of study should not be mixed, but often are.

8.10.1 Within-species relationships

The available data suggest that the temperature sensitivity of standard metabolism (as measured by oxygen consumption) is remarkably similar across a wide variety of organisms (Table 8.8).

> **Box 8.6 Capturing temperature sensitivity: Q_{10}**
>
> In Chapter 7 we saw that although the Arrhenius relationship has proved popular for describing the relationship between reaction rate and temperature, it has many limitations for capturing the thermal sensitivity of even simple chemical reactions.
>
> For whole-organism studies, a popular alternative for capturing thermal sensitivity in a simple way has long been the Q_{10} coefficient. This is calculated simply as the ratio of rates measured at two temperatures differing by 10 K:
>
> $$Q_{10} = \frac{\text{Rate at } T+10}{\text{Rate at } T}$$
>
> A Q_{10} coefficient can be calculated for any two temperatures, T_1 and T_2, as long as these are sufficiently far apart that the difference in rates can be determined reliably:
>
> $$Q_{10} = \left(\frac{\text{Rate at } T_2}{\text{Rate at } T_1}\right)^{\frac{10}{(T_2-T_1)}}$$
>
> For experimental data covering a wide range of temperature, an overall Q_{10} can be calculated from an exponential model relating the rate to temperature, T as:
>
> $$\text{Rate} = e^{bT}$$
>
> which can be linearised and fit by an ordinary least-squares regression model:
>
> $$\ln(\text{Rate}) = bT$$
>
> The overall Q_{10} for the data is then given by:
>
> $$Q_{10} = e^{(10.b)}$$
>
> Both the exponential (Q_{10}) and Arrhenius models are simply statistical fits that describe the central tendency in a data set with variability, and both have their limitations. For example, Krogh's goldfish data (Box 8.5) are well fitted by a quadratic, but neither an exponential nor Arrhenius relationship describes the data adequately.

Table 8.8 Within-species temperature sensitivity of standard metabolism[49].

Group	n (species)	Q_{10} Median	Q_{10} Range
Insects	29	2.60	1.30–6.60
Crustacean zooplankton	6	2.65	1.92–5.41
Echinoderms	14	2.21	1.22–2.82
Fish	5	2.50	1.90–3.41
Reptiles	60	2.34	1.15–3.66

Although individual studies can suggest high Q_{10} values (sometimes > 10), the median values for invertebrates and vertebrates fall in the range 2.2–2.7. In the two groups for which we have sufficient data, insects and reptiles, there is a slight but significant tendency for Q_{10} to increase towards lower temperatures. In insects there is also a slight tendency for Q_{10} to increase above ~ 25 °C.

This tells us something important, which is that a simple exponential relationship does not capture the full nature of the relationship between metabolic rate and temperature. An example which shows this very well is the springtail (collembolan) *Onychiurus arcticus*.

Onychiurus is a tiny insect, though large for a springtail, found in the mossy scree slopes below bird colonies in the high Arctic. Here the meltwater run-off is enriched with nitrogen and phosphate from bird droppings, which stimulates a rich growth of plants. The ecology and physiology of *Onychiurus* was studied in Svalbard by Bill Block and colleagues. Typically adult *Onychiurus* are 100–150 µg dry mass, they are active down to −4 °C but are able to withstand 25 °C. This broad temperature tolerance is important because although the air temperature stays low at the high latitude of Svalbard, sunlight can warm the ground significantly on a still day.

The metabolic rate of *Onychiurus*, as measured by oxygen consumption, shows a complex relationship with temperature (Figure 8.12). At intermediate temperatures, between about 10 and 30 °C, the relationship is well described by an exponential (Q_{10}) model. The Q_{10} over this intermediate range (1.66) is fairly low for an insect. At lower temperatures, where the springtails remain active, the temperature sensitivity is much greater, and is not captured well by the Q_{10} fitted to intermediate temperatures. In recent years the Arrhenius relationship has largely replaced the exponential (Q_{10}) as the favoured statistical model, but it too only captures the central portion of the

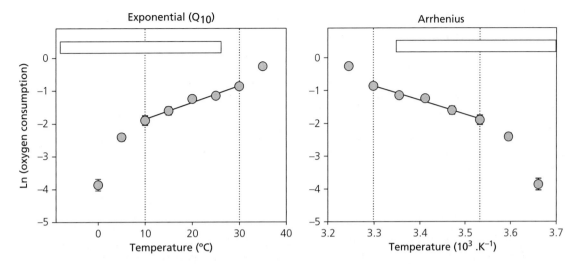

Figure 8.12 Relationship between standard oxygen consumption (mm³ h⁻¹) and temperature in the Arctic springtail *Onychiurus arcticus*. Data are plotted as mean and SE (n = 8–14 determinations per temperature). Linear fits for temperature sensitivity cover only the range between 10 and 30 °C (limits shown by dotted lines). The box shows the range of temperatures over which the springtails are active in the field, and do not show thermal stress in the laboratory[50].

overall relationship (Figure 8.12). At higher temperatures sensitivity again increases, though this is caused in part by thermal stress and associated activity. Clearly the value of any calculated Q_{10} or Arrhenius slope will be dependent on the range of temperatures over which it is calculated, and no single parameter captures the relationship in full.

Despite the limitations of simple statistical models, the data in Table 8.9 point to a general response, as a positive relationship between cell temperature and standard (resting) metabolic rate has been reported for practically every organism in which the relationship has been explored. This indicates a fundamental relationship within species: a rise in temperature leads to an increase in metabolic rate, and hence an increased use of resources. Because resources are often limiting, we might ask whether evolution has managed to mitigate this effect in any way. The way to explore this question is to look for the relationship between standard metabolic rate and temperature in organisms that have adapted over evolutionary time to live at different temperatures.

8.10.2 Across-species studies

Typical patterns for across-species studies are shown by data for teleost fish and bivalve molluscs (Figure 8.13). Because both the fish and the bivalves are ectotherms, body temperature is essentially identical to water temperature.

Two features are apparent from these plots: the maximum (and hence the mean) level of oxygen consumption increases with temperature and the spread of the data (its variance) also increases with temperature. The increase in variance comes about because at higher temperatures there are both species with high and low resting oxygen consumptions, whereas at the coldest temperatures we only see the lower rates.

The pattern of variance evident in these plots does, however, pose a problem for statistical analysis. The standard approach for dealing with this problem is to transform one or both of the variables; the usual transformations are logarithmic or inverse. The result of log-transformation is to both equalise the variance and linearise the relationship (Figure 8.14).

The log-linear model implies an underlying exponential relationship between metabolic rate and temperature, and provides an estimate of mean Q_{10} over the temperature range of the data. The Arrhenius model also estimates the thermal sensitivity over the entire data set. These two models explain essentially identical fractions of the variance (62.5% for the exponential model and 62.9% in the Arrhenius model). Since they fit the same number of parameters, neither model is preferred on purely statistical grounds.

186 PRINCIPLES OF THERMAL ECOLOGY

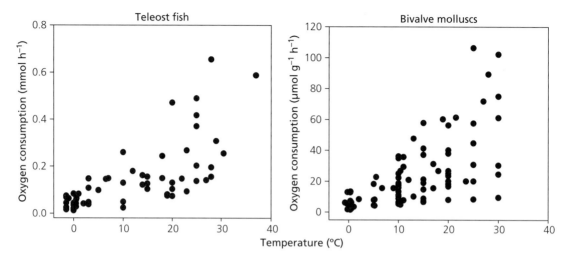

Figure 8.13 The relationship between standard (resting) oxygen consumption and temperature in two very different organisms, teleost fish (left panel) and bivalve molluscs (right panel). The fish data are plotted as oxygen consumption for a 50 g fish (close to the median size of fish in the data set), with one data point per species. The bivalve data are presented as mass-specific rates, and there are 86 data points covering 48 species, as some species are represented by data for both summer and winter[51].

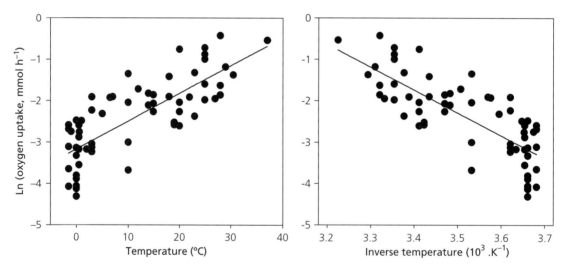

Figure 8.14 The relationship between standard (resting) oxygen consumption and temperature in teleost fish following transformation (natural logs) of the oxygen consumption. Two statistical models are fitted, exponential (Q_{10}) (left panel) and Arrhenius (right panel). Note that in the Arrhenius model, higher temperatures are to the left of the abscissa. Both lines were fitted by ordinary least-squares regression.

The similarity of the two models can be seen clearly if they are plotted in linear space (Figure 8.15). The two models are indistinguishable over most of the temperature range, and even above 30 °C the difference is trivial in comparison with the variance in the data.

Close examination of this plot shows that the two fitted lines do not lie in the middle of the data. This has lead some ecologists to suggest that fitting linear relationships to transformed data is misleading, and that the preferred technique is to fit a relationship directly to the untransformed

METABOLISM

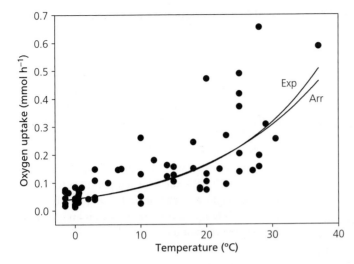

Figure 8.15 The relationship between standard (resting) oxygen consumption and temperature in teleost fish, with the two fitted models plotted (Exp: exponential, Arr: Arrhenius).

data using one of the non-linear fitting procedures that are now widely available in statistical software packages. This emphasises how important it is in ecology to have a clear question to be answered, and to understand precisely what is being done when any model is fitted to data. The critical factor in choosing the best model to fit to the data is in the nature of the distribution of error about the fitted line, and the extent of error in the independent variable (in this case, temperature)[52].

8.10.3 Do all species exhibit a similar temperature sensitivity of metabolism?

Since the mechanism of intermediary metabolism is common to all organisms, we might expect that the temperature sensitivity of metabolism might also be similar across the diversity of life. Resting metabolism has been particularly well studied in vertebrates, and a comparison of the temperature sensitivity of resting metabolism of the different classes shows that these are very similar (Figure 8.16). Statistical analysis indicates that the

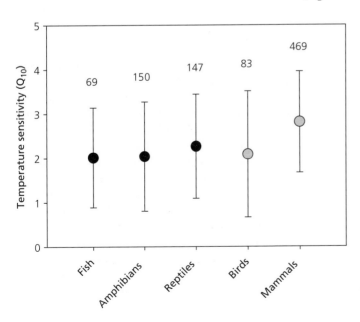

Figure 8.16 Temperature sensitivity (Q_{10}) of resting metabolism in different classes of vertebrates. Data are plotted as mean and 95% confidence intervals, with the number of species shown for each class. Ectotherms are shown in black, and endotherms in grey.

small differences are not significant, and suggest an overall Q_{10} for vertebrate metabolism of 2.19 across the temperature range –1.5 to 42 °C[53].

One important result is that the resting metabolism of the two endotherm classes (mammals and birds) has a similar temperature sensitivity to the ectotherms (fish, amphibians and reptiles). We are able to show this because, although any individual bird or mammal species has only a small variation in body temperature during normal daily life, mammals and birds as a whole have a sufficiently wide range of body temperatures for us to determine the pattern across species. Despite the significant extra metabolic costs of endothermy, the intermediary metabolism of mammals and birds is similar in its temperature sensitivity to that of all other organisms.

Data are also available for a number of invertebrate groups (Table 8.9), and these suggest a somewhat similar temperature sensitivity to vertebrates. The caveat here is that in all of these studies, many species were represented by more than one data point, and so the overall Q_{10} mixes within- and between-species effects.

Table 8.9 Across-species temperature sensitivity of invertebrate standard metabolism; nd: no data[54].

Group	n (species)	Temperature range	Q_{10}	Comments
Arthropoda				
Insects	20	0–42	2.34	Inactive when measured
Insects	346	nd	2.29	
Isopoda	15	–2–25	2.91	Marine benthic species
Crustacea	nd	0–30	2.26	Crustacean zooplankton
	35	–1.5–29	1.99	Epipelagic copepods
Echinodermata				
Echinoids	11	–2–28	1.41	
Mollusca				
Bivalves	48	–2–30	2.05	
Mixed taxa				
	143	1.4–30	1.63	Epipelagic zooplankton from 5 phyla

8.10.4 Has evolution modified the temperature sensitivity of metabolism?

If evolution has been able to ameliorate the acute effect of temperature to any extent, then we would expect to see the across-species relationship to have a lower temperature sensitivity (slope) than the within-species one. The extreme case, where evolution has achieved complete independence of metabolism from temperature, would result in the cost of living, as estimated by resting oxygen consumption, to be independent of temperature ($Q_{10} = 1$). Clearly we do not see this, indicating that having a higher cell temperature brings with it an inevitable higher cost of maintenance.

8.11 Why does resting metabolism increase with temperature?

Having established the general pattern that resting metabolic rate increases with body temperature, both within and across species, we need to explain it. There are two classes of explanation for the observed pattern:

1. Direct: resting metabolism is driven higher in a deterministic manner by an increased temperature.
2. Indirect: temperature affects aspects of cellular physiology such that there is an increase in ATP demand (and hence oxygen consumption) at higher cell temperatures.

We should also consider a third possibility, which is that the relationship is epiphenomenal; that is, it emerges fortuitously as the result of selection on other factors entirely. On the face of it, this does not seem likely, but it is very difficult to know how this possibility could be tested rigorously.

A direct relationship between temperature and resting metabolic rate is conceptually simple: an organism living at a higher temperature has no option but to synthesise more ATP and hence consume more oxygen. This relationship would apply equally to an individual organism subject to an acute temperature change, and to organisms adapted over evolutionary time to live at different temperatures, because the same physical mechanism underpins both.

Although perhaps intuitively appealing (after all, we all know that warmer temperatures speed things up), there is a problem with such a direct link between temperature and metabolism. This is that, as we have seen in Chapter 7, the outcome of evolutionary adaptation is to render activity of key metabolic enzymes relatively independent of temperature when organisms adapted to live in different thermal environments are compared at the temperature to which they are adapted to live. So if the activity of individual enzymes can compensate for temperature, why does the cost of living (basal metabolism) still vary with temperature?

The answer lies in the indirect effects of temperature on cellular metabolism and a return to the ecologist's question discussed in section 8.6: what is the organism using its ATP for? Part of the answer lies in the critical distinction between the maximum rate at which ATP can be regenerated if needed and the rate at which it is actually used.

The mechanism underpinning the relationship between temperature and resting metabolic rate is thus simple: temperature affects the rates at which basic cellular processes need to operate to maintain cellular viability, and these dictate ATP requirement and hence oxygen consumption (Figure 8.17).

Because the relationship between resting metabolic rate and temperature across species is the result of species-specific evolutionary optimisations, it cannot be predicted from first principles. It can only be described statistically. We can, however, ask what evolutionary pressures are likely to influence the resting metabolic rate of a particular species.

The cost of basal metabolic rate must be met from food or reserves, and so it might be expected that environments with chronic or seasonally acute food shortages would select for a reduced basal metabolic rate. Although a higher basal metabolic rate is costly, it also brings benefits. A higher basal metabolic rate allows a greater absolute aerobic scope, and hence more active lifestyles. An alternative way of looking at this is that an energetic lifestyle inevitably brings with it a high basal metabolic rate. A higher basal metabolic rate may also allow a more rapid response to an environmental challenge, through having more cellular machinery available. The basal metabolic rate of an organism thus depends on its ecology, with the level being set by an evolutionary trade-off between costs and benefits.

8.12 Metabolic cold adaptation

To conclude we need to touch upon a topic which once attracted considerable attention from physiologists, namely *metabolic cold adaptation*.

The origin of the idea goes back to August Krogh, who when discussing the metabolism and activity of organisms living at different temperatures in hot springs and polar seas, said:

It would be interesting to compare the respiratory exchange in such cases, because it would appear unlikely from a teleological point of view that it should differ so much as would ordinarily be implied from the temperature difference. One would expect that animals living at a very low temperature should show a relatively high standard metabolism at that temperature compared with others living normally at a high temperature[55].

It is important to quote what Krogh actually said, and to understand the context for his remarks, as both have been widely misunderstood since. Krogh was concerned specifically with the possibility of adaptation (Krogh actually called this acclimatisation) in organisms living at very low temperatures in polar regions. He was prompted to speculate about metabolism because in the course of his experiments with goldfish (Box 8.5), he noted that at

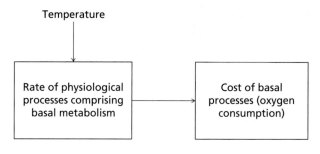

Figure 8.17 The relationship between temperature and resting oxygen consumption is indirect, mediated through the ATP demand for the processes comprising basal metabolism.

low temperatures the goldfish became very sluggish, in comparison with the active swimming exhibited by fish adapted to live at those same temperatures. The essence of Krogh's suggestion was that since the activity of polar fish was elevated compared with that of a goldfish at polar temperatures, the metabolism of polar fish would also be elevated relative to that of temperature or tropical fish cooled to polar temperatures.

Testing Krogh's prediction thus involves comparison of within-species measurements made on a temperate organism acclimated to a low temperature, with measurements from organisms adapted evolutionarily to that temperature. Physiologists have tended to avoid problems associated with laboratory acclimation by undertaking a slightly different test from that actually proposed by Krogh, namely to ask whether polar organisms show a metabolic rate higher than that predicted from the relationship between standard metabolic rate and temperature established for temperate and tropical organisms when extrapolated to polar temperatures.

Early studies of polar fish by Per Scholander and Donald Wohlschlag provided experimental support for Krogh's prediction, which Wohlschlag named *metabolic cold adaptation* (MCA). The diagrams showing the metabolism of Antarctic and non-polar fish were widely reproduced, and firmly established the concept in ecology and physiology textbooks. George Holeton was the first to query the existence of MCA in the 1970s, when he showed that polar fish were more sensitive to handling stress than had been previously recognised, and that the earlier studies had allowed insufficient time for the fish to recover from being caught and placed in the respirometer; the early studies were therefore not measuring true standard metabolism, and the elevated respiration results observed were caused by stress[56]. Work by John Steffensen also failed to find support for an elevated standard respiration in Arctic fish, and this work was valuable in broadening the range of lineages of fish in the comparison. The Antarctic fish fauna investigated by Scholander, Wohlschlag and Holeton is dominated by a single lineage, notothenioids, which are confined almost exclusively to the Southern Ocean and hence it was possible that the elevated metabolism observed in the early studies of Antarctic fish was related to phylogenetic differences rather than the evolutionary adjustment proposed by Krogh.

A comprehensive test can be made by taking the data for standard metabolic rate in tropical and temperate fish, extrapolating the across-species thermal sensitivity to polar temperatures and comparing this extrapolation with observed data for polar fish. This comparison shows clearly that MCA does not exist, at least in fish (Figure 8.18). Comparable studies of marine invertebrates also failed to provide any evidence for the existence of MCA[57].

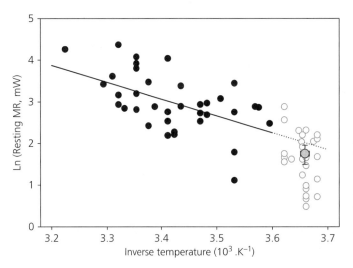

Figure 8.18 Testing metabolic cold adaptation (MCA) in fish. The relationship between temperature and resting oxygen consumption in teleost fish from tropical and temperature waters (black symbols and solid line), when extrapolated to polar temperatures (dotted line) falls within the spread of data for polar species (white symbols) and above the mean data for all polar fish combined (grey symbol, ±95% confidence intervals of the mean). If MCA existed in fish, the polar data should be above the extrapolated line, rather than below[58].

While it is easy to be critical of Krogh's suggestion now, it was a perfectly reasonable speculation at the time. We should recall that Krogh was working over a century ago, before we even knew that enzymes were proteins and well before the details of intermediary metabolism had been worked out.

The concept of metabolic cold adaptation has also been applied to studies of enzymes in polar organisms. While this is sensible in that any increase in metabolic rate has to be supported by the machinery of intermediary metabolism, it is not easy to distinguish a general evolutionary adaptation of enzyme machinery to low temperature, which can involve both changes to enzyme kinetics as well as quantitative changes in enzyme number, from any changes that might demonstrate MCA.

Krogh was proposing a general mechanism and the idea of metabolic cold adaptation was also taken up by physiological ecologists interested in the respiratory metabolism of terrestrial organisms. Here early studies also suggested that polar species did indeed have elevated metabolic rates, leading to an interesting division between marine ecologists who were largely abandoning the idea, and terrestrial ecologists where the idea remained firmly in the literature. In the most comprehensive analysis to date, data for insects suggest that, after allowing for the effects of body mass, body temperature and flight status, there remains a small but significant effect of ambient temperature: insects from cold climes had a small but significantly higher resting metabolic rate. This is the best evidence yet for a small but subtle effect which could represent metabolic cold adaptation in terrestrial organisms.

8.13 Summary

1. Metabolism is driven by redox reactions, in which part of the difference in potential energy between the electron donor and acceptor is used by the organism for its life processes (with the remainder being dissipated as heat).
2. Cellular metabolism involves a complex suite of chemical reactions. The key pathway is intermediary metabolism, by which the energy stored in reserves (glycogen, starch, lipid, protein) is transferred to ATP. In aerobic respiration the electrons released from reserves are passed to oxygen, which is thereby reduced to water.
3. Not all ATP regeneration involves oxygen as the final electron acceptor, and not all oxygen is used for ATP regeneration. Oxygen consumption is, however, often the simplest and most practical way to measure the rate of intermediary metabolism, and the errors in doing so are assumed to be small.
4. The costs of existence, as estimated by resting metabolism, represent only a part (~ 25%) of the daily energy expenditure of organisms. The costs of the organism's ecology (growth, reproduction, movement and so on) are additional to existence costs. Daily energy expenditure is typically estimated as the field metabolic rate.
5. Resting metabolic rate increases with cell temperature, indicating that it costs more energy to maintain a warm cell than it does a cool or cold cell. The temperature sensitivity of resting metabolism is highly conserved across organisms (Q_{10} ~ 2). We know essentially nothing of how the daily energy expenditure of organisms varies with body temperature, although this powers most of an organism's ecology and is thus likely to be a vital factor in fitness.

Notes

1. Although almost always ascribed to Lavoisier alone, this much-quoted aphorism comes from the famous joint report with Pierre-Simon Laplace, *Mémoire sur la chaleur*, read to the Académie des Sciences in Paris in 1783 and published in the following year (de Lavoisier & Laplace 1784).
2. Harold Morowitz (1999) has described the elucidation of the chemical reactions captured in this chart as *'one of the great intellectual achievements of mankind'*. He estimates that at least 19 Nobel Prizes have been awarded for the work done in elaborating this chart.
3. Albert Szent-Györgyi was a Hungarian biochemist who received the 1937 Nobel Prize in Physiology for the discovery of Vitamin C, and his work on the TCA cycle. This quotation comes from Trefil et al. (2009); a longer version appears in the introductory comments to a symposium on light and life (Szent-Györgyi 1961).
4. Reduction potential is different from electronegativity, which is a measure of the tendency of an atom to attract electrons to itself in a covalent bond. The two measures are related, both being influenced by

atomic and nuclear structure, and are loosely correlated across the elements. Some countries, notably the United States, have traditionally used standard oxidation potentials, which are simply the negative of standard reduction potentials. The reduction potential of a standard hydrogen electrode is defined as 0V for a pH of 0. Reduction potential depends on pH (varying by 59 mV per pH unit) which is why the $2H^+/H_2$ couple has a reduction potential of −420 mV at 25 °C and pH 7.
5. 'Nitrification' and 'denitrification' are nice examples of scientific terms that are confusing to the uninitiated and hence perhaps ill-chosen: after all nitrogen still exists following 'denitrification'. These terms arose because of their consequences for available nitrogen in soils. Denitrification converts NO_3^- to N_2 gas, thereby removing nitrogen from the soil. Nitrification is the reverse, and brings reduced nitrogen back into the system where it is once again available for plants.
6. The history of the oxygen content of Earth's atmosphere can be reconstructed from the mineral composition of ancient rocks, and the isotope composition of carbon and sulphur in them. See Rollinson (2007), Catling & Kasting (2007), Lenton & Watson (2011) and Och & Shields-Zhou (2012).
7. This is also known variously as the *water oxidising complex* or the *oxygen emitting centre*.
8. This used to be referred to as the *dark reactions*, but the term has fallen out of favour recently.
9. The cycle was discovered by Melvin Calvin (1911–1997), Andrew Benson (1917–2015) and James Bassham (1922–2012) at the University of California, Berkeley, using ^{14}C to trace the fate of CO_2 in cultured unicellular algae. The cycle is often referred to simply as the Calvin cycle, and occasionally as the Calvin–Benson–Bassham cycle. Somewhat controversially, Calvin alone was awarded the 1961 Nobel Prize in Chemistry for this work.
10. This formula comes from Kharasch (1929), and the rationale is described nicely by Stokes (1988).
11. A slight qualification is needed here, as anaerobic methanogenic Archaea lack the full TCA cycle (Goodchild et al. 2004).
12. Glycolysis is also known as the Embden–Meyerhof or Embden–Meyerhof–Parnas (EMP) pathway, after its discoverers Gustav Embden (1874–1933), Otto Meyerhof (1884–1951) and Jakub Karol Parnas (1884–1949). Meyerhof received the 1922 Nobel Prize in Medicine for his work on the metabolism of muscle, including glycolysis. Substrate-level phosphorylation is transfer of a phosphate group to ADP from another phosphate-carrying molecule by an enzyme-catalysed reaction.
13. The glycolysis pathway can be run in reverse, to make glucose (gluconeogenesis). For this to happen, different enzymes are used to run the irreversible steps in the reverse direction. Some bacteria can also run the TCA cycle in reverse, as a reductive synthesis cycle building reduced carbon compounds from CO_2 and water.
14. The tricarboxylic acid (TCA) or citric acid cycle is also known as the Krebs cycle, after its discoverer Sir Hans Krebs (1900–1981), who received the 1953 Nobel Prize in Physiology or Medicine for this work.
15. For details of the organisation, mechanism and energetic of the electron transport chain see Nicholls & Ferguson (2002).
16. The fundamental mechanism of a proton pump to generate a proton gradient which is then use to drive ATP synthesis is the *chemiosmotic hypothesis* of the English biochemist Peter Mitchell (1920–1992). Mitchell proposed his hypothesis (Mitchell 1961) at a time when the mechanism for ATP synthesis was unknown but widely believed to be substrate-level phosphorylation. Michell's chemiosmotic hypothesis was a radical alternative, but one that explained many previously puzzling features of ATP generation in mitochondria. The hypothesis prompted an extremely acrimonious debate but was eventually accepted, and Mitchell received the 1978 Nobel Prize in Chemistry.
17. I took this vivid metaphor from Nick Lane's marvellous book on mitochondria (Lane 2005). He also points out that on a per gram basis, a typical mammal generates heat 10 000 times faster than the Sun. Mitochondria are remarkable structures.
18. The change in enthalpy for the complete oxidation of glucose is −2803.1 kJ mol^{-1} but the energetics of biochemical reactions are better viewed in terms of changes in Gibbs energy, which combines the change in enthalpy with the change in entropy. The change in entropy depends on whether the water produced is in the vapour or liquid state.
19. The large fraction of the chemical potential energy of glucose not trapped in ATP was first pointed out by Prusiner & Poe (1968), who estimated dissipated heat as 74–81% of the enthalpy of combustion. These estimates, however, assumed an enthalpy change for ATP hydrolysis/synthesis of ~4.7 kcal mol^{-1} (19.7 kJ mol^{-1}). This is too low, and hence their estimates of dissipated heat were too high.
20. As with many physiology textbooks, the control is described here in terms of control by specific steps in the pathways. These are often referred to as rate-limiting steps, although for a pathway in steady state the flux of metabolites must be the same through all steps in the pathway. In the late 1960s and early

1970s, a number of physiologists developed an alternative approach, which they termed *metabolic control theory*. Nowadays it is more often referred to as *metabolic control analysis*, and it is closely related to sensitivity analysis in engineering. The key idea is that the control of flux through a pathway is a characteristic of the pathway as a whole, with each step in the pathway exerting some level of control on the flux, captured as the *flux control coefficient*. The flux control coefficients for a given pathway sum to 1, and the value for a given step is a system property (that is, it is affected by all other steps in the pathway). In practice the flux control coefficients for steps close to equilibrium tend to be smaller, and those for non-equilibrium steps tend to be larger. In other words, the non-equilibrium steps exert most control over flux through the pathway, but not in the all-or-nothing manner often portrayed. Metabolic control analysis has been very successful in providing a rigorous quantitative framework for modelling flux through metabolic pathways. The original descriptions of metabolic control analysis are Kacser & Burns (1973) and Heinrich & Rapoport (1974).
21. The use of the joule in place of the calorie was recommended by the International Union of Nutritional Sciences in 1969. This did not meet with unreserved enthusiasm among physiologists, as described by Max Kleiber in the introduction to the second (1975) edition of his influential book *The Fire of Life*. See Kleiber (1972) for the argument to retain the calorie as the unit of energy in physiology, and Hawkins (1972) for the counter argument. The counter argument won.
22. Literature values came from Brody (1945), Kleiber (1961), Schmidt-Nielsen (1975), Brafield & Llewellyn (1982), Gnaiger et al. (1989) and Lucas (1996). In the literature, the energy released is usually estimated by the standard enthalpy of combustion (that is, the heat released by complete oxidation under standard conditions at constant pressure). This takes no account of changes in entropy, and Gibbs energy would be a better measure.
23. To estimate the fraction of total ATP turnover that is anaerobic requires simultaneous measures of oxygen consumption and heat production, which is technically challenging. Early studies were concerned particularly with bivalves and other marine invertebrates that have well-developed abilities to withstand oxygen shortages, and hence might be expected to utilise glycolysis for ATP regeneration (Pamatmat 1978, 1983; Hammen 1979, 1980, 1984). Pough & Andrews (1985) used lactate accumulation to estimate the extent of anaerobic metabolism in lizards subduing prey, and Brafield (1985) reviewed the few studies where oxygen consumption and heat production were measured simultaneously in fish. For a more recent discussion see Scott (2005).
24. Priestley interpreted his observations in terms of the dominant theory of the time, which was that a burning substance released *phlogiston* to the air. Phlogiston was believed to be a fire-like substance contained within combustible bodies, and which was released to the air during combustion. It was also believed to be released during rusting, and thus was an attempt to explain the general suite of processes we now recognise as oxidation. The theory goes back to the seventeenth century alchemist Johann Joachim Belcher, but the name phlogiston was first proposed by the German chemist George Ernst Stahl in the early eighteenth century. Early quantitative work showed that metals such as magnesium gained weight when they burned, and this implied that phlogiston had negative weight. The critical work is usually credited to Lavoisier who showed that combustion required a gas, oxygen, which had mass and thus explained the increase in weight of metals as they were oxidised. Oxygen had actually been discovered independently by the Swedish chemist Carl Wilhelm Scheele in 1772 and by Joseph Priestley in 1774, but the name oxygen comes from Lavoisier, who also isolated the gas in 1775. Scheele was the first to extract the gas, and Priestley the first to announce its existence, but it was Lavoisier who understood what he had found and its significance.
25. The continuous slow 'combustion' that comprised metabolism thus enabled the guinea pig to maintain its body temperature above that of its surroundings, thus accounting for the puzzling phenomenon of animal heat. What Lavoisier and Laplace were unable to determine was precisely where this combustion took place, though they thought it was almost certainly in the lungs. Lavoisier continued his work in collaboration with the physiologist Armand Séguin, developing his ideas of respiration and combustion. This work was published in a series of exceptionally important papers, each following presentation to the Académie des Sciences in Paris (de Lavoisier & Laplace 1784; de Lavoiser & Séguin 1793, 1814). For excellent descriptions of Lavoisier's work see Mendelsohn (1964) and Holmes (1985). Everett Mendelsohn sets Lavoisier's work in the context of earlier work on combustion by Joseph Priestley and Adair Crawford in England, and Joseph Black and Patrick Dugud Leslie in Scotland, but also reveals its revolutionary nature. Frederik Holmes provides a thorough description of Lavoisier's wider work on the chemistry of life.

26. The German physiologist Max Rubner (1854–1932) was a major figure in the development of bioenergetics, and undertook a series of critical experiments which established many of the key aspects of the field. He was nominated (unsuccessfully) for a Nobel Prize on many occasions. For a brief biography of this important figure see Chambers (1952).
27. This, or variants of it, has been the definition used by the influential Metabolic Theory of Ecology (see Chapter 12). Jim Brown explores the nature and definition of metabolism very nicely in his 2002 MacArthur Award lecture (Brown et al. 2004)
28. The nearest we can get is a measure of heat dissipation. This is technically demanding, especially for aquatic organisms, and to interpret the result in terms of total metabolic fluxes we need to know the thermodynamics of every reaction taking place during the measurement. Measurement of heat dissipation is, however, the nearest we can come to a measure of total metabolism.
29. The key early studies were by August Krogh (Krogh 1914, 1916) and Eugene DuBois (DuBois 1927). The definitive study of the conditions affecting the measurement of resting or basal metabolism in vertebrates was the extensive work by Francis Benedict at the Boston Nutrition Laboratory (Benedict 1915, 1938). His proposals for the measurement of resting metabolism in endotherms have remained the definition of 'standard' metabolism ever since. In discussing his proposals for the conditions under which to measure resting metabolism, Benedict did, however, comment that *'These ideal, perhaps fantastic, conditions can not be met in all animals. It is impossible to dictate the precise conditions to a non-cooperative subject of research'* (Benedict 1938). Useful later discussions of the problems in measuring true basal metabolism are Kleiber (1961) and Blaxter (1989).
30. See Brody (1945).
31. Brian McNab (McNab 2002) defines BMR as including the costs of endothermy, and hence by inference the term is confined to mammals and birds; some text books also make this distinction explicit (see for example Hill et al. 2008). The concept of standard metabolism emanates largely from the work of August Krogh (see Krogh 1916), who viewed this is a measure for fish analogous to the concept of BMR as proposed originally for humans and then extended to all mammals.
32. The first such study was Spoor (1946) for goldfish. Later work was Smit (1965), also on goldfish, followed by the classic work of Roland Brett (Brett 1964, 1965, 1972), summarised in Brett & Groves (1979).
33. This table has been compiled from Krogh (1916), Benedict (1938), Brody (1945), Brett & Groves (1979), Priede (1985), Blaxter (1989), Peterson et al. (1990), Hammond & Diamond (1997) and McKechnie & Swanson (2010).

The concept of maintenance metabolism in plants is discussed by Amthor (2000) and Thornley (2011). Note that several of the acronyms used in studies of metabolism are ambiguous: RMR has been used to denote both resting and routine metabolic rate, SMR has been used for both standard and summit metabolic rate, and while FMR has been used most often to denote field metabolic rate, it has also been used as shorthand for flight metabolic rate in insects.
34. The data set is that analysed by Clarke & Johnston (1999), and the results are qualitatively identical to those of Killen et al. (2010) who used a larger data set (89 species as against 69 here), but expressed their data on a mass-specific basis and did not include body temperature in the statistical model.
35. Note that measures of basal or standard metabolism exclude all of the costs of the organism's ecology, except in the trivial sense that BMR/SMR is the cost of staying alive and a dead organism can have no ecology.
36. This categorisation goes back to August Krogh, who in his classic book (Krogh 1916) estimated that *'functional activities in the resting body . . . amounts to at least 25% of the total standard metabolism'*. Rigorous examinations of this distinction were made by Baldwin & Smith (1974), Baldwin et al. (1980), and the problem is discussed by Milligan & Summers (1986).
37. See Mink et al. (1981) and Niven & Laughlin (2008).
38. See Blaxter (1967) for a comprehensive review of this.
39. Coulson et al. (1977) provide an insightful and entertaining overview of these problems, which is still relevant today.
40. Modified from Rolfe & Brown (1997), Rolfe et al. (1999) and Hulbert & Else (2000).
41. Redrawn, with permission, from Makarieva et al. (2008).
42. This is similar to the aerobic scope of a trained racehorse. The physiological performance of the pronghorn is described by Lindstedt et al. (1991); see also Bishop (1999). Although often referred to as the pronghorn antelope, *Antilocapra americana* is not a true antelope. It is the only surviving member of its family (Antilocapridae) and the similarity to true antelopes is convergent. The pronghorn is famous for having a top speed second only to the cheetah. It can, however, sustain fast speeds for longer than the cheetah, suggesting that its athletic ability evolved in response to predators now long extinct.
43. The theory and practice of doubly labelled water (DLW) as a means of determining metabolic rate were developed by Nathan Lifson and colleagues at the University of Minnesota Medical School in the late 1940s and early 1950s (Lifson et al. 1949, 1955; Lifson & McClintock 1966). A valuable modern perspective is provided by John Speakman (Speakman 1997).

44. Modified from Green et al. (2013), with data kindly supplied by Jon Green.
45. The cost of locomotion is discussed by Alexander (1968), Vogel (1988), McNab (2002) and Biewener (2003); a useful model of locomotor costs in terrestrial vertebrates has been developed by Pontzer (2007). Christian et al. (1997) reviewed the energy used for locomotion in a range of lizards, Zub et al. (2009) quantified the energy used for locomotion in a captive population of the least weasel, and Brett (1986) studied the energetics of the anadromous sockeye salmon during their spawning run.
46. Hillman et al. (2013) present a meta-analysis confirming many previous suggestions that in vertebrates metabolic scope is constrained by the design of the cardiovascular system.
47. The goldfish data were reported by Ege & Krogh (1914) and the general relationship following the addition of the data from the frog and dog by Krogh (1914). August Krogh (1874–1949) was a Danish physiologist who received the 1920 Nobel Prize in Physiology for his work on the regulation of capillaries in muscle. He made major contributions to our understanding of respiratory physiology. He is also remembered for Krogh Principle, which is that for '. . . *a large number of problems there will be some animal of choice, or a few such animals, on which it can be most conveniently studied*' (Krogh 1929). This principle has guided comparative biological work for many decades. However it has also resulted in a concentration of work on a small number of representative organisms (think *Drosophila, Arabidopsis*), a trend which is entirely counter to what Krogh was actually arguing for. Later in the same essay Krogh elaborated a plea for broadly based comparative work that is usually omitted from any discussion of his principle: '*I want to emphasize that the route by which we can strive toward the ideal is by a study of the vital functions in all their aspects throughout the myriads of organisms.*' Krogh was actually arguing for comparative analyses to be based on the widest possible range of taxa, covering the full range of lineages.
48. For example Beamish & Mookerjii (1964) found a lower BMR for goldfish than that reported by Ege & Krogh. This emphasises that individual animals vary in their BMR, and that for a full understanding measurements of many individuals and appropriate statistical analysis is needed.
49. The insect data were taken from Hodkinson (2003), crustacean zooplankton data from Hirche (1984), echinoderm data from Hughes et al. (2011), fish data from Clarke & Johnston (1999) and the reptile data from White et al. (2006). In all cases where more than one Q_{10} value was available for a species, the median value was taken, so all species used to calculate the data in the table were represented by only a single estimate of Q_{10} (although the median value for a group was little different if it was calculated from all available data).
50. These plots were modified from Block et al. (1994).
51. The fish data are from Clarke & Johnston (1999) and the bivalve data are from Peck & Conway (2000).
52. For arguments against logarithmic transformation see Packard (2009), Packard & Birchard (2008) and Packard & Boardman (2008, 2009). For a succinct discussion of the problem see E. White et al. (2012). For the influence of error in the dependent variable on the choice of regression model see Xiao et al. (2011). See also Glazier (2013).
53. The data came from Clarke & Johnston (1999) for fish, White et al. (2006) for amphibians and reptiles, Clarke et al. (2010) for birds and mammals. In all cases each species was represented by a single data point; where data were available for a range of temperatures or masses, the median values were used. Data were fitted with a General Linear Model., with metabolic rate and body mass transformed to natural logs. The partial regression coefficient for temperature can then easily be converted to a Q_{10} value fitted over the range of body temperatures for which data are available. The GLM indicated that the differences in temperature sensitivity between vertebrate classes were not significant ($F_{4, 925} = 1.93$, $F = 0.103$).
54. The data in this table were compiled from analyses presented by Waters & Harrison (2012) and Addo-Bediako et al. (2002) for insects, Luxmoore (1984) for isopods, Ivleva (1980) for crustacean zooplankton, Ikeda (1985), Ikeda et al. (2001) for zooplankton, Brockington and Clarke (2001) for echinoderms and Peck & Conway (2000) for bivalves. In all cases many species were represented by more than one data point.
55. These key sentences come from *The respiratory exchange of animals and man* (Krogh 1916), being the opening to the section entitled *The possibility of acclimatization in cold-blooded animals* (p. 101—102).
56. See Holeton (1974); that handling stress or 'excitement' elevates metabolism in fish had been known since Smit (1965).
57. The key references here are Scholander et al. (1953), Wohlschlag (1960, 1964), Holeton (1970, 1972, 1973, 1974), Clarke (1980, 1991), Wells (1987), Steffensen et al. (1994), Clarke & Johnston (1999), Peck & Conway (2000) and Steffensen (2002). A more recent discussion (White et al. 2012) does not test Krogh's prediction directly, but uses a common-garden approach by correcting metabolic rate data to a common temperature using a Q_{10} of 2.4 (rather high for fish).
58. Modified from Clarke & Johnston (1999).

CHAPTER 9

Temperature regulation

> The behaviour of lizards obviously cannot be understood solely from physiological considerations, and thus lizard thermoregulation must be more complex than is generally believed
>
> **Raymond Huey and Montgomery Slatkin**[1]

In the previous chapter we saw how body temperature affects maintenance costs and hence resting metabolic rate. In this chapter we examine how, and why, organisms regulate that body temperature.

The why is fairly straightforward: the ability to generate metabolic power (ATP) and muscle performance both increase with temperature, and there is now widespread and compelling evidence that a warmer body brings increased fitness[2]. While there is a clear performance benefit to having a warm body, this option is not open to all. Many organisms do not, or cannot, regulate their body temperatures. These include almost all marine and freshwater ectotherms, because they exchange heat with the surrounding water so effectively that it is impossible to maintain a body temperature above that of the water in which they live (we will discuss some exceptions below). Where water bodies have spatial differences in temperature, then aquatic organisms can regulate their temperature to some extent by moving between these.

Thermoregulation is also difficult for organisms living in environments such as soil or sediments; here organisms have little choice but to track the diurnal or seasonal variations in habitat temperature. The only option available is to move up or down in soil to avoid daily or seasonal extremes. Sessile organisms also have little opportunity to thermoregulate, and must cope with whatever the environment throws at them. Plants can use evaporation to prevent excessive heating in daytime, and this may be an option for some intertidal organisms, but there is little opportunity to mitigate very low temperatures (an exception here are the large montane lobelias which retain water within the inflorescence, as we saw in Chapter 6).

For organisms that do have the opportunity to influence their own body temperature, there are a variety of approaches for thermoregulation, and these are summarised in Table 9.1.

The fundamental basis of thermoregulation is the balance between energy gains and losses discussed in Chapter 4. The flows of energy important to the regulation of body temperature are shown in Figure 9.1.

In all organisms the key inputs of thermal energy are from metabolism and the external environment. For almost all organisms it is energy from the environment that is most important in determining body temperature. The exceptions are endotherms (mammals and birds), and we will discuss these in the next chapter. The main routes for energy loss are through the outer surface and, in many terrestrial organisms, evaporation of water from the respiratory tract.

The importance of energy exchange with the environment means that for many organisms body temperature is coupled tightly to environmental temperature. As we saw earlier, this is the case for almost all aquatic organisms, but also for many terrestrial organisms. Two examples of tight coupling between environmental temperature and body temperature are shown in Figure 9.2.

Geckos of the genus *Nephrurus* are nocturnal and are widely distributed in the arid and semi-arid

Principles of Thermal Ecology. Andrew Clarke, Oxford University Press (2017).
© Andrew Clarke 2017. DOI 10.1093/oso/9780199551668.001.0001

Table 9.1 Categories of thermoregulation.

Category	Examples and comments
Little or no temperature regulation	
Habitat restriction	Many aquatic organisms; most organisms living within soil or aquatic sediments.
Lifestyle restriction	Sessile organisms (benthic invertebrates, plants); nocturnal organisms.
Behavioural thermoregulation	
Basking	Many reptiles, amphibians and insects; also endotherms. Regulation involves shuttling between sun and shade.
Migration	Diurnal vertical migration in the sea and lakes allows organisms to move between water at different temperatures.
Physiological mechanisms	
Retaining metabolic heat	Some fish (billfish, tuna, lamnid sharks).
Local generation of heat	Thermogenic plants; pre-flight warming in insects; some snakes incubating eggs; some fish that generate heat to warm the brain or eyes.
Full endothermy	Mammals and birds.

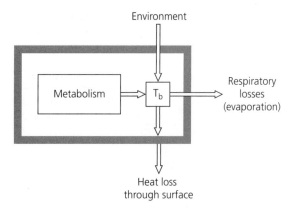

Figure 9.1 A conceptual model of the processes determining body temperature, T_b, in a terrestrial organism. This is a modification of the general model of energy flow in organisms (Figure 4.1), to emphasise the major sources of energy for maintaining T_b. Solid arrows indicate the main pathways for energy flow, and the grey symbolises an insulating layer (fat, fur or feathers). The model applies equally to ectotherms and endotherms, as these differ only in the relative balance of the various pathways for energy flow. In aquatic organisms the gills act as a very important extra route for dissipation of metabolic heat to the environment[3].

areas of Australia. They shelter in burrows during the day and forage by night, especially after rain when prey is more plentiful. Their nocturnal lifestyle means that they do not have the opportunity to bask, and so they operate with a body temperature identical to air temperature. On cool nights the geckos are cool, whereas on warmer nights they too are warm, improving their ability to move and feed.

The second example is the naked mole-rat, *Heterocephalus glaber*. This is a most unusual animal, being a mammal that has secondarily abandoned endothermy. It is physiologically an ectotherm, for its body temperature tracks ambient temperature between 12 °C and 42 °C (Figure 9.2). This is undoubtedly related to its subterranean lifestyle; temperatures recorded from their burrows suggest there is only a small daily and seasonal fluctuation and the mole-rats live more or less permanently at temperatures between 29 °C and 32 °C with no need to expend metabolic energy to maintain a warm body temperature (albeit one that is not particularly high for a mammal).

These two species are examples of thermoconformers, organisms whose body temperature matches that of the environment. Many reptiles and amphibians are thermoconformers, as are almost all terrestrial invertebrates because their small size means that they equilibrate rapidly with the environment. Some lizards, however, have a well-developed ability to regulate their body temperature, combining anatomical, physiological and behavioural mechanisms which together allow them to achieve high body temperatures during the daytime. An organism whose body temperature is independent of air temperature is termed a thermoregulator, and there are a range of possibilities in between (Figure 9.3).

The effectiveness (or intensity) of thermoregulation can be captured with a simple thermoregulatory coefficient, k, expressed in terms of the preferred temperature, T^*, of a particular species, the body temperature T_b and ambient temperature T_a:

$$T_b = T^* + k(T_a - T^*)$$

The coefficient, k, which is equivalent to the slope of the relationship between T_b and T_a (as in Figure 9.3), varies between 0 for a perfect thermoregulator and 1 for a perfect thermoconformer. Intuitively it might be better expressed as $1-k$, so that 0 is a thermoconformer (that is no thermoregulation)

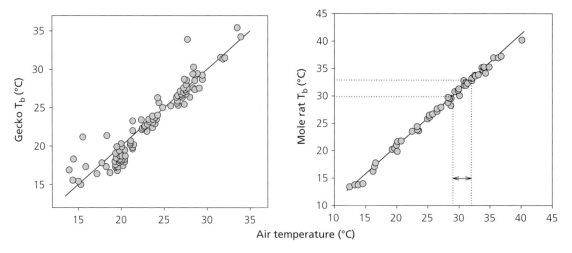

Figure 9.2 Two thermoconformers. A (left panel): body temperature as a function of air temperature in active individuals of the nocturnal Australian gecko *Nephrurus laevissimus*. The line shows equality (gecko temperature = air temperature). B (right panel): body temperature as a function of ambient temperature in the naked mole-rat *Heterocephalus glaber*. The line shows a least-squares linear regression (slope = 1.01, not significantly different from unity, $p > 0.1$). The arrow shows the range of burrow temperatures in the wild[4].

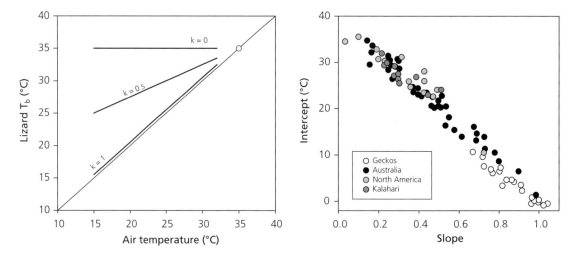

Figure 9.3 Lizard thermoregulation. A (left panel): conceptual model of lizard thermoregulation. The thin line shows equivalence (lizard temperature = air temperature). The symbol represents the preferred body temperature (35 °C) of a hypothetical lizard, and the three lines show the relationship between body temperature and air temperature for hypothetical lizards with thermoregulatory coefficients of 1 (perfect thermoconformer), 0 (perfect thermoregulator) and 0.5 (partial thermoregulator). B (right panel): range of thermoregulatory coefficients observed in a range of terrestrial reptiles from three continents[5].

and 1 a perfect thermoregulator, but the original formulation has largely stuck[6].

A body temperature above ambient air temperature does not necessarily indicate active thermoregulation. In a classic experiment, beer cans (filled with water, not beer) placed in the sun were found to be up to 8 K warmer than the air, with differences in the excess temperature related to factors such as orientation to the sun[7].

A beer can is clearly not a useful approximation to an active organism like a lizard, but the lesson that air temperature does not necessarily tell us the temperature of an organism is an important one. It led to the development of increasingly sophisticated physical models of lizards that were used to determine the *operative temperature*. Operative temperature is defined as the temperature of an inanimate object of zero heat capacity with the same size, shape and radiative properties as the animal exposed to the same environment[8]. In theory the operative temperature can be calculated using the complex thermal balance equations we explored in Chapter 4, but in practice it tends to be determined with physical models placed in the environment. These have varied from the simple (silicon tubing to approximate a snake) to the highly sophisticated (accurate copper casts of the lizard species of interest).

The operative temperature was conceived as a relatively simple means of determining whether an organism is likely to gain or lose energy by radiation and convection in particular circumstances. It provides a valuable conceptual and practical tool for comparative studies across different thermal environments, as long as the physical models have been fully calibrated against the real animals they are mimicking, under a range of radiation and air-flow conditions. What they do not do is tell us the actual body temperature of the organism, because they explicitly assume a thermal capacity of zero (biological tissue has a thermal capacity of ~3.4–3.9 kJ kg^{-1} K^{-1}, close to that of water, which is 4.18 kJ kg^{-1} K^{-1}). It is the thermal capacity of an organism that determines how a given intake of energy converts to temperature. It is only when a model or data logger has the correct thermal capacity that its temperature will track that of the organism of interest. This is shown nicely by the temperature loggers used to monitor the temperature of intertidal organisms, where careful matching of the size and thermal capacity of the logger to the organism of interest (the mussel *Mytilus californianus*) was essential to avoid significant errors[9].

Lizards as a group exhibit the complete range of thermoregulatory coefficients, with nocturnal species tending to be thermoconformers (Figure 9.3). It is tempting to interpret the relationship in Figure 9.3b as saying something about lizard physiology, but it actually reflects the nature of a least-squares regression fit: under many circumstances the slope and intercept are correlated. The relationship thus simply confirms statistical theory, but it does convey the wide range of thermoregulatory capacity in lizards[10].

This thermoregulatory coefficient requires experimental work in the laboratory, and it captures what an organism *can* do. In the real world where the organism lives, the thermal environment is spatially patchy and also varies through the day, and the organism has to juggle the demands of thermoregulation with the need to feed, defend a territory or find a mate. Paul Hertz and colleagues (including Ray Huey) therefore developed an improved thermoregulatory index to capture what the organism actually *does*. The first step is to determine the preferred temperature for the species, which involves presenting the organism with a thermal gradient and seeing which temperature it selects. Usually a range of preferred temperatures is found and the central portion (typically 50%) of these define the *set-point range*. The accuracy of thermoregulation is then the mean absolute deviation of the body temperature, T_b, from the mean of the set-point range, ΔT_b. A subtle problem here is that we can calculate an accuracy of thermoregulation from the body temperatures exhibited by an organism moving through a varied habitat but which is not thermoregulating at all. So we need to compare this value with the range of environmental temperatures experienced; again we can express this as the mean deviation of environmental temperatures, T_a, from the set-point range, ΔT_a. The effectiveness of thermoregulation, E, is then:

$$E = 1 - \frac{\Delta T_b}{\Delta T_a}$$

Note that here E is normalised so that a value of 0 indicates no thermoregulation and a value of 1 perfect thermoregulation. Although more time-consuming to determine than the original simple measure, this is far more meaningful ecologically. An example for *Anolis* lizards is shown in Figure 9.4.

In this plot the frequency distributions of body temperature, T_b, and ambient temperature, T_a, are compared with the preferred temperature range selected by lizards presented with a thermal gradient in the laboratory (the set-point range). In *Anolis cristatellus* in January (winter) the median body temperature is lower than the set-point range but the mean deviation (the arrow marked ΔT_b in the plot)

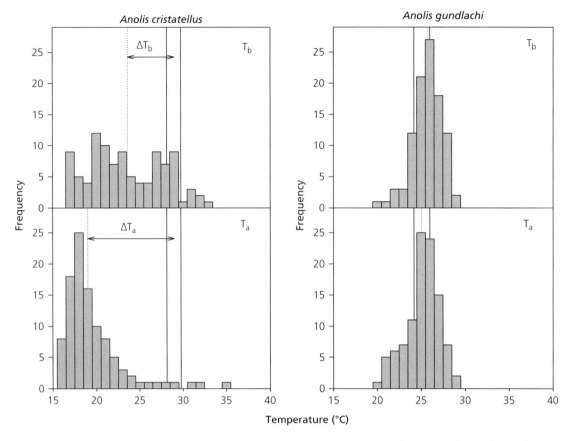

Figure 9.4 Effectiveness of thermoregulation in two species of *Anolis* lizard. The histograms show the frequency distribution of observed body temperatures (T_b, upper panels) and ambient temperatures (T_a, lower panels), with the median values shown as dotted lines. The range of set-point values for each species is shown by the two vertical solid lines[11].

is less than that of environmental temperature (ΔT_a) This indicates active thermoregulation, though with the lizards unable to achieve their set-point range. In contrast *Anolis gundlachi* was able to match its set-point range very well, principally because this fell within the range of ambient temperature at this location. A similar variation in thermoregulation has also been described for varanid lizards in Australia[12].

The mechanism used by these lizards to achieve their set temperature is *behavioural thermoregulation*.

9.1 Behavioural thermoregulation

The basis of behavioural thermoregulation is that the organism absorbs energy by basking when it needs to warm up, and moves to shade to avoid overheating. Typically terrestrial organisms cool down overnight and bask once the sun rises to bring their body temperature up to a level where physiological performance is optimal (or at least sufficient to start daily activity). Much studied in reptiles and insects, basking is also used by endotherms either to warm up in the morning or to help speed emergence from overnight torpor.

The efficiency with which a basking organism captures solar radiation varies with its colour and orientation. For example butterflies (Lepidoptera) vary their posture when resting depending on their need for heat. In dorsal basking the butterfly spreads its wings flat, allowing sunlight to warm the thorax directly. In addition dark pigmentation from melanin on the dorsal surface of the wings also absorbs energy, which is transferred to the thorax by conduction. A butterfly using lateral basking will close its wings over its back and orient itself at right angles to the sun; here warming will be enhanced by dark pigmentation at the base of the underwing. The butterfly can reduce the rate at which it gains energy by changing its orientation relative to the sun. Butterflies such as skippers and lycaenids hold their wings at an intermediate angle, and here the wings contribute little to the energy gain, which is almost exclusively by direct warming of the thorax. In all cases, the butterflies are using a mixture of behaviour, pigmentation and physiology to absorb and distribute energy in the body[13].

Once a basking lizard or butterfly attains its operating temperature it can move off to feed or to seek a mate. If this involves movement into shaded areas, the organism may cool down and need to return to basking later. For organisms in patchy thermal environments, it also means that obtaining and defending a territory with suitable basking sites is critical to survival. Anyone walking through European woodland in summer will have seen speckled wood (*Pararge aegeria*) butterflies spiralling in the air as they defend patches of sunlight from rivals.

9.1.1 Behavioural thermoregulation in water

Behavioural thermoregulation in aquatic organisms has been studied mostly in fish. A fish placed in a thermal gradient will explore, then spend most of its time in a relatively narrow range of temperatures which typically reflect the temperature to which they are adapted to live, though this can be modified by recent experience. The icefish *Chaenocephalus aceratus* selects temperatures between 1.5 °C and 3 °C, avoiding anything warmer, whereas *Tilapia rendalli* from African lakes selects 36 °C (close to its upper lethal temperature), and the hot springs pupfish *Cyprinodon macularius* selects water between 38 °C and 40 °C[14].

The preferred temperature can also be influenced by factors such as feeding and a diurnal vertical migration into surface waters by night is a widespread feature of both lakes and the open ocean. The surface waters are typically warmer and richer in food, but they are also regions where visual predators hunt. Early discussions of vertical migration in the sea emphasised the thermal environment, but more recent models suggest that the main drivers are the need to feed and the risk of predation, with thermal aspects relatively minor[15].

9.2 How do animals sense temperature?

The observation that many animals seek preferred thermal environments and avoid thermal extremes indicates that they must have a means of sensing temperature, and the sophisticated behavioural thermoregulation of lizards and butterflies implies that this must be quite precise.

Recent work has shown that these temperature sensors are a suite of ion channel proteins that span the plasma membrane of many cell types. First identified from *Drosophila*, they have been characterised most extensively in mice and rats. They are referred to as *transient receptor potential* (TRP) channels, and are typically composed of six membrane-spanning helices with both the N-terminal and C-terminal within the cell. TRPs are many and varied, and are sensitive to a range of chemicals as well as temperature; for example the best characterised is TRPV (the V indicating that it is also sensitive to the phenolic aldehyde vanillin).

The basis of temperature sensitivity is the ability of these TRPs to initiate action potentials in neurons through non-selective cation influx. This influx is temperature sensitive, and while the temperature-sensitive domain appears to be associated with the outer pore, the precise mechanism of this sensitivity is incompletely understood. Some TRPs are sensitive to cold, for example mouse TRPM8 is triggered at 26 °C with the current increasing at lower temperatures, whereas in two species of the clawed frog *Xenopus* it is 14 °C. Others are sensitive to heat, such as TRPV1, which is sensitive to a temperature rise above 37 °C in mice but has an activation threshold of around 46 °C in chickens[16].

It is likely that vertebrates express a range of TRPs with different temperature thresholds, and this provides them with an exquisite sensitivity to changes in temperature. These TRPs are not thermometers (although sometimes referred to as such); the mouse or chicken does not know what its temperature is, it merely detects a change in temperature which then influences its behaviour or triggers physiological responses. While we humans can estimate temperature from our skin sensors, this is only because we have come to associate a particular sensation with a culturally transmitted knowledge of temperature as determined by an external thermometer.

Some heat-sensitive TRPs also respond to capsaicin from chillies or allyl isothiocyanate from mustards such as wasabi, and some cold-sensitive TRPs are also triggered by menthol. This is why we perceive some foods or chemicals as 'hot' and others as 'cool'. Our own skin sensors will detect small local changes in heat flow that raise or lower temperature; when we touch an object with high thermal conductance, such as a metal, we sense the small loss of heat and feel that the metal is cool. Conversely when we touch an insulator such as wood, the small temporary build-up of heat in our skin surface makes the wood feel warm.

The sensing of temperature is also widespread among invertebrates; for example medicinal leeches (*Hirudo verbana*) select warmer waters when digesting a meal. Invertebrate thermoreception has been most studied in arthropods, many of which have the ability to distinguish quite fine differences in temperature. Insect thermoreceptors appear to be located in the antennae and mouthparts (maxillary palps), and, in some species of blood-sucking bugs, in the legs.

The signals produced by the TRPs are processed in the brain; in vertebrates it is the hypothalamus that is the seat of temperature perception and this region also regulates the autonomic response; behavioural responses involve the upper brainstem and the cerebral cortex. Quite how the TRP signals are processed in many invertebrates is unclear. For example the much-studied nematode *Caenorhabditis elegans* has only 302 neurons in total and no organised brain but still exhibits a sophisticated response to temperature.

9.2.1 Heat detection by pit vipers and jewel beetles

The TRP sensors discussed above provide a signal determined by the temperature of the sensor itself (just as a true thermometer provides a signal proportional to its own temperature: see Chapter 3). A few organisms, however, have sensors that indicate the temperature of an object other than themselves. These include many blood-feeding arthropods that need to locate warm animals on which to feed. In most cases this involves temperature sensors in the antennae or forelimbs which respond to infrared radiation, sometimes coupled with chemical sensors

detecting CO_2 or chemicals in sweat. The most intensely investigated, however, are snakes that can detect warm prey and insects that respond to distant fires (Box 9.1).

9.2.2 Costs of behavioural thermoregulation

The benefits of behavioural thermoregulation are clear: warming up allows the organism to run faster or fly more strongly. But it also has costs. These costs are conventionally regarded as being of three kinds: the direct cost of the behaviour, the enhanced risk of predation and trade-offs.

The direct costs arise because energy expended in thermoregulation cannot be used for processes contributing directly to fitness, such as growth or reproduction. These costs are not easy to quantify. It might be thought that as basking largely involves sitting still, they are probably small, but this may not always be so. For example a recent experimental study of the lizard *Sceloporus jarrovi* showed that the locomotory costs of moving between patches of shade, coupled with the risk of overheating while doing so, influenced how the lizards moved through the landscape: the accuracy with which the lizards were able to thermoregulate depended critically on the spatial distribution of sunlight and shade[17]. The trade-off is thus between the metabolic costs of shuttling behaviour, the chance of being predated and the loss of time for feeding, territorial defence or mate-seeking when thermoregulating.

The costs and benefits of behavioural thermoregulation have been extensively modelled, principally with lizards in mind. Ray Huey and Montgomery Slatkin developed the most influential of these models, and showed that when the opportunity to bask was limited, the costs of thermoregulation could outweigh any benefits. In the case of a nocturnal species there is no possibility of thermoregulating and so these remain as thermoconformers (Figure 9.2). An intermediate example is provided by the Andean toad *Bufo spinulosus*, which lives at high altitude where temperatures can vary as much in a single day (> 40 K) as they do seasonally. On sunny days the toads bask, raising their body temperature from near freezing to ~ 23 °C at mid-day (up to 7 K above ambient). On cloudy days, however, basking was reduced and body temperature tracked ambient

> **Box 9.1 Infrared detection**
>
> Among snakes infrared receptors are found in pythons and boas, but are most highly developed in crotalids (rattlesnakes and relatives) where they are located within the *pit organs*, found in the loreal region of the head just below and ahead of the eyes (and which give crotalids their colloquial name of pit-vipers). The pit organ consists of a hollow chamber, within which is suspended a thin membrane. This membrane is rich in mitochondria, highly vascularised and densely innervated by primary afferent nerve fibres from the somatosensory system; it is highly sensitive to infrared radiation in the wavelength range 8–12 μm. The precise details of how this radiation is detected are not yet clear, but we do know that a transient receptor potential protein (TRPA1) is involved. It is likely that TRPA1 responds to direct radiant heating. The sensitivity of the pit organ is thus related to the narrow aperture to the pit organ, which acts analogously to a pinhole camera, and the ability of the snake to balance the air pressure on either side of the sensing membrane. Pythons and boas also detect infrared radiation with labial pit organs distributed over the snout but these are far less sensitive than those of pit-vipers. Infrared imaging indicates that the pythons keep the region of the pit organs cooler than the rest of the face, which improves their ability to detect prey by reducing the problem faced by all infrared sensing devices of confounding the information of interest with radiation from the sensor itself[18].
>
> Vampire bats also have the ability to detect infrared radiation from a warm body, using three facial pit organs in the complex nose leaf structure. Interestingly, infrared imaging suggests that the nose of the vampire bat *Desmodus rotundus* is also cooler than the rest of its body by about 9 K. Experiments suggest that vampire bats can detect warm objects at a distance of up to 16 cm, and is thus used to locate areas of the prey to attack, rather than locating the prey itself at distance[19].
>
> Several groups of insects are able to locate newly burned trees, but most attention has been directed at the buprestid beetle *Melanophila acuminata*. This species is attracted to forest fires in huge numbers, and lays its eggs under the bark of newly burned conifers. *Melanophila* is able to detect forest fires at great distance using pit organs located on the mesothoracic (middle) legs and containing 50–100 densely packed sensilla. These sensilla are cuticular spheres with a fluid core into which projects the dendritic tip of a mechanosensitive neuron. It is believed that this neuron is triggered by the increase in pressure as the fluid is warmed by incident infrared radiation. The strength of the outer layers of the sensor are thus essential to ensure the increase in temperature leads to an increase in pressure (which will be detected by the neuron) rather than an increase in volume (which will not). A similar sensor appears to be present in the fire-seeking flat bug genus *Aradus*[20].

temperature. There are also few opportunities for basking in the depths of forests. Sunlight is, however, available higher in the canopy and the body temperature of the arboreal snake *Hoplocephalus stephensii* is correlated with its height above ground. Experimental tests give strong support for the general features of the Huey & Slatkin model, though its applicability is limited by our lack of knowledge of the precise costs and benefits of thermoregulation[21].

The costs and benefits of thermoregulation are not constant and so behaviour varies with circumstances. A nice example is the arboreal lizard *Chlamydosaurus kingii*, which has a lower set-point T_b in the dry season than in the wet season. In both seasons lizards bask early and late in the day, but in the dry season the lizards cease basking at a lower T_b than they do in the summer, indicating that they are thermoregulating in both seasons but the preferred temperature varies seasonally[22]. In aquatic environments, the patterns of migration into and away from warmer water involve a balance between opportunities for feeding, a favourable thermal environment and the likelihood of predation.

It is a widespread (though not universal) observation that ectotherms which have recently fed will select warmer environments, as these provide a benefit in speeding digestion. For example the newt *Triturus dobrogicus* selects warmer water, when available, after eating. A higher temperature also influences embryonic development, and this may have been a key driver in the evolution of vivipary in high-latitude reptiles. Reptiles that lay eggs in nests can do little to influence the thermal environment of embryonic development once the eggs are laid. In contrast, viviparous reptiles can use thermoregulation to maintain a higher, and more stable, thermal environment for their developing eggs. This may explain the greater incidence of vivipary in reptiles from higher latitudes[23].

For organisms that do not have the opportunity to thermoregulate, an alternative is to retain metabolic heat.

9.2.3 Retaining metabolic heat

Among ectotherms the retention of the heat dissipated in metabolism is highly developed in tunas (*Thunnus*, previously *Thynnus*). This was first reported by John Davy, brother of the notable chemist Humphrey Davy, although (as he noted) the warm muscles of tuna had long been known to fishermen[24].

In teleost fish routine swimming is powered by aerobic muscle. This is well vascularised and rich in mitochondria, hence its colloquial name of red or dark muscle. Typically this is a small strip along the mid-line of the fish, and the bulk of the muscle is the anaerobic white muscle used for intermittent bouts of rapid swimming for prey capture or predator avoidance (so-called burst swimming). In active predatory fish such as tuna the red muscle comprises a much greater fraction of the total muscle, forming extensive plates running laterally either side of the vertebral column. This is supplied by four artery–vein pairs which run subcutaneously, delivering oxygenated blood into the muscle. This is a different arrangement from most teleosts, where there is a single dorsal aorta from which arteries radiate laterally. The unusual anatomy of the vascular system in tuna allows for the retention of the heat generated by the muscle: the outgoing venous blood warms the incoming arterial blood, and the metabolic heat is thereby retained rather than being lost to the environment when the venous blood passes through the gills. This heat exchange is very efficient because of extensive elaboration of arterioles and venules into a close network, the *rete mirabile*[25].

The retention of metabolic heat by the *rete mirabile* is very efficient (> 90%) and allows the red muscle to be considerably warmer than the seawater in which the tuna swims. Generally muscle temperature is a function of seawater temperature, indicating little capacity for true thermoregulation. Some species, however, can adjust circulation and muscle fibre recruitment to exert some thermoregulatory control and prevent overheating at top speeds in warm water.

Heat retention by a *rete mirabile* in the muscle is also known from lamnid sharks. This was first reported from porbeagle (*Lamna nasus*) and mako (*Isurus paucus*), and allows these to maintain their muscles up to 10 K above ambient seawater temperature. The evolution of mechanisms for retaining metabolic heat allows these fish to increase their muscle power output to capture fast-swimming prey, and also to hunt in cooler water below the thermocline and extend their range into colder waters. In addition, the retained heat warms the viscera, speeding digestion, and the brain, enhancing neural function. These benefits suggest that the independent evolution of heat retention in tunas and lamnid sharks was a complex integrated process and not driven by either niche expansion or locomotor performance alone[26]. Lamnid sharks also use heat produced by the activity of red muscle to warm their eyes, as the blood is carried to a large sinus that houses a sub-orbital *rete mirabile*. There is anatomical evidence that a similar system may be present in some larger rays, but as yet no physiological confirmation.

More recently, a counter-current heat exchanger was reported from the opah, *Lampris guttatus*. In this species it is associated not with the muscle or brain but with the gills, and allows for a more general elevation of body temperature. These adaptations of retaining metabolic heat have often been referred to as endothermy, or sometimes as regional endothermy. They are, however, quite distinct from the endothermy of mammals and birds, and are therefore better termed heterothermy (this distinction is discussed more fully in Chapter 10)[27].

9.3 Generating heat

All muscles generate heat when working. In a few fish, however, some muscles have been modified such that their sole function is to generate heat to warm the eyes and brain. This occurs in the billfishes (Xiphiidae and Istiophoridae) where a portion of the superior rectus muscle of the extraocular muscles is non-contractile and functions solely to generate heat. The modified muscle cells lack contractile proteins and are rich in mitochondria. The heat is generated by rapid release and re-uptake of Ca^{2+} by the sarcoplasmic reticulum, and the key enzyme is Ca^{2+}-ATPase (SERCA). It is possible that the release of Ca^{2+} may also stimulate enhanced heat production by the mitochondria, but this appears to be unresolved. The heat produced is retained in the brain area by a cranial *rete mirabile*.

The generation of heat has also long been known in incubating pythons. This was first reported for an Indian python, *Python molurus* in Bengal by the French naturalist Christophe-Augustin Lamare-Picquot in 1832, but dismissed by the French Académie des Sciences. A short while later, in 1841, Achille Valenciennes reported observations on a python incubating eggs in the Jardin des Plantes in Paris, and in 1862 Sclater undertook careful study of an incubating python at the Zoological Gardens in London. More recently incubating pythons have been studied at the New York Zoological Gardens in 1932 and again in 1960. The latter animal was found to maintain a temperature up to 4.7 K above ambient air temperature. This was achieved by rhythmical contractions of the body musculature, at a frequency of up 15 min^{-1}, which elevated its metabolic rate by ~9-fold. It is possible that this is typical of larger constrictor snakes during incubation[28].

9.3.1 Pre-flight warming in insects

It has been known since the early nineteenth century that some larger moths can have a body temperature above that of the surrounding air, but real progress in understanding this only came with the development of the thermocouple, a device small enough to provide an accurate and precise measure of insect temperature. An important early study is shown in Figure 9.5.

In this study a large tethered moth (*Samia cecropia*) was monitored as it underwent intermittent activity. During periods of rapid wing movement thoracic temperature increased rapidly, but the abdomen warmed only slightly. Both temperatures dropped to ambient once activity creased. Subsequent work established that many large moths could not fly until their thoracic muscles were above ~32 °C. Flight in cool conditions thus required the muscles to be warmed, and this pre-flight warming was achieved by 'shivering' or 'wing-whirring'. Recordings from the nerves that activate the flight muscles showed that whereas in flight the nerves controlling the upstroke and downstroke fired alternately, during warm-up they fire together. As a result the muscle activity generates heat but not mechanical power for flight[30].

This pre-flight warming relies entirely on the energy from ATP being dissipated by muscular work. Because the speed of muscle fibre contraction is itself strongly temperature dependent, the wing-beat

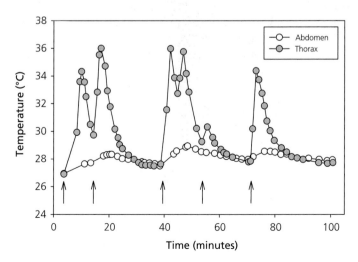

Figure 9.5 Thoracic and abdominal temperatures in an intermittently active *Samia cecropia* moth. Arrows show the onset of periods of continuous wing movements[29].

frequency increases as muscle temperature rises: when muscle temperature is low warming is slow, but as the thorax warms wing-beat frequency increases and the rate of warming itself accelerates.

Two features are of critical importance to pre-flight warming: size and hairiness (pile). In general insects are so small that convective cooling keeps their temperature close to ambient. A temperature above ambient can only be maintained in larger species, and the mean thoracic temperature during flight is strongly dependent on size (Figure 9.6).

Once the moth is warm enough it can generate sufficient power to fly. Temperature during flight is then a balance between heat generated by flight muscle activity and heat loss. The moth needs to ensure it does not overheat, and within different groups of moths flight temperature can be broadly independent of size, indicating evolutionary adjustments in factors such as wing-loading to allow regulation of temperature.

The above discussion might imply that a warm thoracic temperature is a prerequisite for flight in all moths, but this is far from the case. The sphingid moths on which most work has been done are large and heavy, and they fly continuously with very fast wing beats. In consequence they generate considerable amounts of heat and they operate close to their thermal limit. Although the thorax maintains a high temperature, the head is kept cool and the abdominal temperature is not regulated (compare the thorax, head and abdomen temperatures for *Manduca sexta* in Figure 9.6). While these large species have proved invaluable as model organisms for exploring insect thermoregulation they are far from typical of insects (or even moths) as a whole. Many moths fly at low temperatures, and some small northern species are active in winter at temperatures down to 0 °C. The advantage for these is that they avoid predation from bird and bats, which either hibernate or leave the northern forests in winter. The smallest moths generally cannot thermoregulate, but their low body mass and relatively large wing area can reduce flight costs. Some small noctuid moths, however, can begin pre-flight warming at temperatures close to 0 °C and despite being small (~200 mg) they achieve thorax temperatures of 30–35 °C in flight.

Much of what we know of insect thermoregulation comes from work on larger moths, but other insect groups have also been studied[32]. Another well-studied group of insects are butterflies. Flight is the centrepiece of butterfly ecology as it is necessary for feeding, finding a mate, searching for a host-plant on which to lay eggs and dispersal.

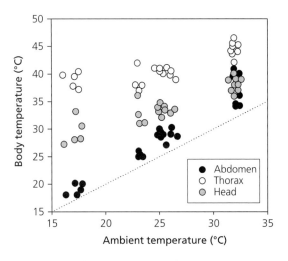
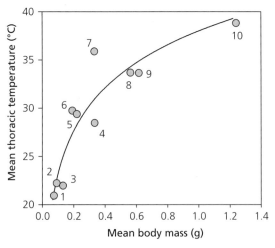

Figure 9.6 Moth flight temperatures. A (left panel): body temperature in the sphinx moth *Manduca sexta* in free flight, as a function of ambient temperature, with data for the thorax and abdomen plotted separately. The dotted line shows equality (body temperature = ambient temperature). B (right panel): relationship between mean body mass and mean temperature during flight for ten families of moth from Costa Rica (ambient temperature 15–17 °C). Families are 1: Pyralidae, 2: Ctenuchidae, 3: Geometridae, 4: Arctiidae, 5: Noctuidae, 6: Notodontidae, 7: Lasiocampidae, 8: Saturnidae, 9: Pericopidae, 10: Sphingidae. The line is an exponential model[31].

When active, all butterflies appear to maintain a relatively high thoracic temperature (between 20 °C and 40 °C). Very few butterflies are known to shiver to warm up, and so basking is essential to enable flight. Once flying, their leisurely wing beats generate less muscular heat than do the moths with their rapidly whirring wings. In my garden in England the day-flying hummingbird hawk-moth, *Macroglossum stellatarum*, flies continually, hovering to sup nectar before moving on to the next flower, all the time beating its wings so fast as for them to be invisible (hence its colloquial name). In contrast the butterflies flap slowly, and frequently rest to bask. There is more than one way to ensure warm muscles for flight.

Bumblebees are highly sophisticated thermoregulators, typically maintaining a thoracic temperature above 35 °C even at ambient temperatures down to 0 °C. They can be active at these low temperatures because they can initiate pre-flight warming at thoracic temperatures as low as 7 °C. A place with a warm microclimate is thus essential so that a resting bee does not cool down below this. In addition the presence of thoracic pile is essential to retain heat, and heat exchangers are used to retain heat when necessary. They also keep their heads cooler, and foraging on sugar-rich food sources provides the large amounts of energy needed for their expensive thermoregulation.

In contrast to the highly active bees, beetles are typically terrestrial, slow moving and reluctant to fly. Small beetles cannot thermoregulate at all, and hence like small moths must fly with cool muscles. Species with a body mass above ~1 g can generate and retain sufficient heat to allow pre-flight warming, and many very large species such as dung beetles activate thermogenesis to enable rapid running. The most powerful thermogenic beetles are the rain beetles (*Plecoma*) which can generate sufficient heat to be active in mountainous regions during winter[33].

Among the most characteristic insects of dry sunny places are the tiger beetles (carabids of the subfamily Cincindelinae). These brightly coloured beetles are active predators, relying on speed to capture their prey. To maximise their performance the beetles will bask to rise their body temperature, and they start hunting once they are warmer than ~ 35 °C (Figure 9.7). Between 35 °C and 40 °C their hunting speed increases markedly with body temperature (hunting speed is an average speed over the ground, as it combines periods of rapid running with short pauses to search visually). As their body temperature approaches ~40 °C, the beetles need to lose heat, which they do by stilting. That is, they elevate themselves as far above the ground as their legs will allow, to take their body as far as possible away from the hot air close to the ground.

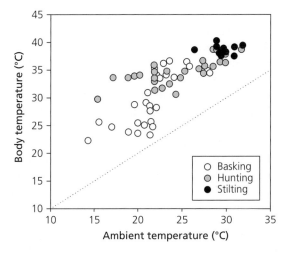 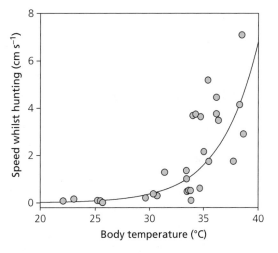

Figure 9.7 Temperature and locomotor performance in the tiger beetle *Cicindela tranquebarica*. A (left panel): Body temperature as a function of ambient temperature for beetles either basking, searching for prey or stilting to minimise warming. B (right panel): relationship between body temperature and searching speed. The fitted line is an exponential relationship, equivalent to a Q_{10} of 18.2[34].

Insects as a group thus use a wide variety of approaches to optimise muscle performance for flight, fight or locomotion. Effective thermoregulation is, however, mostly the preserve of larger species, but the enormous diversity and abundance of small (or very small) insects that cannot thermoregulate indicate that a warm body is not a prerequisite for success[35].

9.3.2 Thermogenic plants

The ability of certain plants of the genus *Arum* to generate heat was noted by Jean-Baptiste de Lamarck in 1778, and the role of respiration in this process established by the Swiss pioneer of photosynthesis Nicolas-Théodore de Saussure in the early nineteenth century[36]. Modern work on these plants follows from a visit by Roger Knutson to the forested edge of a frozen marsh in Iowa in February 1971, where he observed eastern skunk cabbage, *Symplocarpus foetidus*, emerging from the winter snow. Each plant was surrounded by a small depression in the snow suggesting that the skunk cabbage had melted its way through the snow. He showed that the temperature of the spadix (the characteristic swollen spike of flowers) was greater on warmer days, that the respiration rate was greater the colder the air temperature and that the fuel for metabolic heat production was starch stored in the tap-root. At the same time, Kenneth Nagy and colleagues were finding that another arum lily, *Philodendron selloum*, can warm its spadix to exceed air temperature by 40 K; importantly they also showed that the plant could regulate its heat production such that the flower temperature was somewhat independent of air temperature[37]. Heat production is now known from a number of plants from a range of families (Table 9.2).

The energy for this heat production comes from carbohydrate (starch) or lipid stores. As in normal mitochondrial respiration, oxygen acts as the final electron acceptor but, remarkably, the pathway is resistant to cyanide. This cyanide-resistance arises because the mitochondrial electron transport chain contains an alternative oxidase enzyme that accepts electrons from the ubiquinone pool and uses these to reduce oxygen to water. This essentially diverts electron flow at ubiquinone, by-passing the later stages of the electron transport chain (the cytochrome

Table 9.2 Thermogenic plants. Families in which at least one member has been demonstrated to produce heat[38].

Family	Order	Common name
Cycadaceae	Cycadales	Cycads
Nymphaceae	Nymphaeales	Water lilies
Magnoliaceae	Magnoliales	Magnolias
Annonaceae	Magnoliales	Custard apples
Aristolochiaceae	Piperales	Dutchman's pipes
Araceae	Alismatales	Arum lilies
Cyclanthaceae	Pandanales	Palms
Arecaceae	Arecales	Palms
Nelumbonaceae	Proteales	Lotuses
Rafflesiaceae	Malpighiales	Rafflesias and allies

oxidase to which cyanide binds irreversibly), and dissipates the electron energy as heat. As we saw in the previous chapter, under normal circumstances electrons from NADH enter the electron transport chain at complex 1. There are, however, two other dehydrogenase enzymes that can accept electrons from NADH (or NADPH) and pass these to the ubiquinone pool without pumping protons. Under these circumstances, all of the energy from the starch or lipid is dissipated as heat, as had been suggested by careful simultaneous measurement of heat production and respiration in thermogenic plants[39].

Those species with the most powerful thermogenic capacity are able to regulate the temperature of the spadix. For example the lotus *Nelumbo nucifera* can maintain a more-or-less constant spadix temperature of 30–36 °C while air temperature ranges from 10 °C to 45 °C. This means that not only must the plant generate heat at lower temperatures, it must lose it at higher temperatures (Figure 9.8).

Thermogenesis in plants appears to function principally as a means of increasing scent production and dispersion to attract pollinators, and thermogenesis and scent production are often tightly coupled. The heat can also allow plants to erupt through the last vestiges of winter snow, and may even provide a thermal reward to insects. The large scarab beetle *Cyclocephala colasi* that visits the floral chamber of *Philodendron solimoesense* in French Guiana can maintain a temperature ~4 K above ambient air at night. The beetle gains an energetic

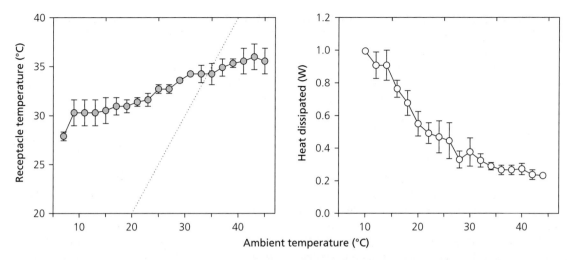

Figure 9.8 Thermogenesis in the sacred lotus *Nelumbo nucifera*. A (left panel): receptacle temperature as a function of ambient temperature, mean ± 95% confidence intervals for 13 individuals in free air, with data pooled into bins of 8–10 °C, 10–12 °C and so on. The dotted line shows parity (receptacle temperature = ambient temperature). B (right panel): heat production by the receptacle, calculated from respiration rate, as a function of ambient temperature for 1–19 plants; mean ± 95% confidence intervals calculated over similar temperature bins[40].

benefit from the flower by reducing its need for its own thermogenesis. Quiescent beetles can maintain a thoracic temperature of ~33 °C by thermoregulation over an ambient temperature range of 16 °C to 29 °C; if we assume an ambient temperature of 20 °C this thermoregulation represents an increase over resting non-thermogenic metabolic rate of 116-fold. The heat provided by the flower, reducing the beetle's requirement for thermogenesis, thus represents a significant energy saving, allowing them to feed and mate at a fraction of the cost required outside the flower[41].

9.4 Avoiding overheating

So far we have been concerned with warming up: getting sufficient heat into muscle, brain or eyes to optimise performance. Organisms living in hot environments, however, can face the opposite problem, which is to avoid overheating. Once again size is important, as small organisms heat up quickly whereas large animals like antelopes do so slowly.

Where shade is available many animals utilise this to avoid heating from direct solar radiation. Anyone who has visited the African savannah will have seen antelopes or large cats sheltering from the heat of the sun beneath spreading acacia trees.

This thermoregulatory behaviour carries some of the same fitness costs as does basking, in that time spent in the shade is time not feeding (although ruminants can use the time for digestion, as can predators who have fed during the night).

For small organisms where there is no shade, or which must hunt for food over open ground, overheating is a very real risk. Some reptiles will burrow into the desert sand to avoid direct sunlight (and hide from potential prey), others will minimise contact with the hot surface by regularly lifting their feet into the air. A common anatomical adaptation of desert organisms is long legs, so that the body is carried higher in the air, away from the very hot boundary layer close to the ground. Many insects will use any elevation they can to move into cooler air, including stones, dung piles or vegetation. For example the ant *Ocymyrmex barbiger* of the Namib desert forages over bare sand for insects that have succumbed to the heat. It spends most of its time on sand up to 51 °C; if it needs to forage over hotter ground, and sand in the Namib can exceed 60 °C, then the ants make frequent pauses to avoid the searing heat. They do so by climbing plant stems, stones or even faecal pellets to reach cooler air. Even a small elevation can bring the ants into air that is 15 K cooler than the surface, and ~7 K cooler than the air close to the surface. The frequency of this behaviour increases rapidly with sand temperature (Figure 9.9).

Many desert animals use evaporation of water to cool down, but this brings its own problems in an arid environment where water is in very short supply. For example in the White-browed Sparrow-weaver, *Plocepasser mahali*, of southern Africa, evaporative water loss is low up to an air temperature of ~39 °C, but then increases sharply (Figure 9.9).

Desert mammals and birds have been shown to have lower resting metabolic rates than related species from less arid environments, and this aids thermoregulation in two ways: less metabolic heat is being generated and respiratory water loss is reduced. Desert animals also reduce water loss by producing very dry faeces and highly concentrated urine. There must be costs to a lower resting metabolic rate (or else all mammals and birds would exhibit this) and these may include a reduced scope for activity or reproduction[42].

9.4.1 Retaining heat during the day

Large mammals may also allow their body to gain heat during the day, and then cool down again by night when radiative cooling is very effective under the clear sky. The classic study of this was by one of the pioneers of ecological physiology, Knut Schmidt-Nielsen, who showed that the body temperature of the camel, *Camelus dromedarius*, may rise from 34 °C to 41 °C during the day and estimated that this saves about 5 litres of water which would otherwise have been needed for evaporative cooling. Interestingly, camels given access to water reduce their daytime temperature rise, suggesting that allowing the body to heat up carries physiological costs. A recent study of the sand gazelle (the Arabian subspecies, *marica*, of the goitered or black-tailed gazelle *Gazella subgutturosa*) showed that in summer this allowed its body temperature to rise during the day from a minimum at dawn of ~38.2 °C to a maximum at dusk of ~40.5 °C. The extent of the diurnal variation was greatly reduced in winter, and the total energy stored during the day was related to ambient temperature (Figure 9.10). As with camels, gazelles allowed access to water reduced the extent of heat storage.

A potential risk from allowing body temperature to rise is impairment of brain function, which can start with only a modest rise in temperature. In some mammals there is a large sinus in the floor of the cranial cavity, which contains a capillary network (the cranial *rete mirabile*) whereby arterial

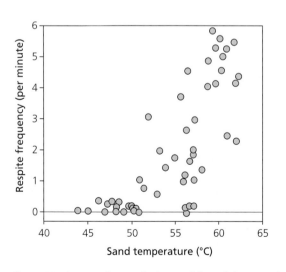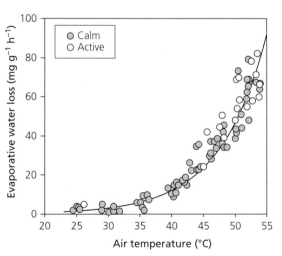

Figure 9.9 Thermoregulation in the desert. A (left panel): frequency of respite pauses during foraging in the Namib desert ant *Ocymyrmex barbiger* as a function of sand temperature. B (right panel): relationship between evaporative water loss and air temperature in the white-browed sparrow-weaver, *Plocepasser mahali*, in southern Africa. The fitted line is an exponential, equivalent to a Q_{10} of 3.87[43].

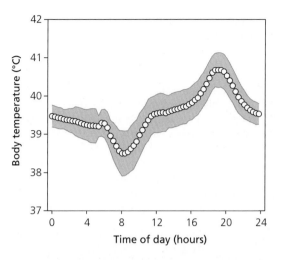 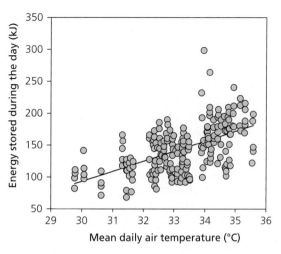

Figure 9.10 Heat storage during the day in the sand gazelle. A (left panel): daily variation in body temperature in summer. Data are mean (white symbols) and SE (grey); data are for 20-minute intervals, averaged over 6 animals for 56 days. B (right panel): energy stored during the day as a function of the mean air temperature for that day. The line shows a least-squares regression ($p < 0.0001$)[44].

blood flowing to the brain is cooled by venous blood returning from the nasal cavity where evaporative cooling has reduced its temperature[45].

9.5 Summary

1. Locomotor performance (running speed, flight ability) increases with muscle temperature. For many organisms there is a fitness advantage to being warm. Digestion and embryonic development are also faster at warmer temperatures.
2. Many organisms use behavioural thermoregulation to maintain a high body temperature during the day, basking in the sun to warm up and retreating to the shade to avoid overheating. This option is not open to most aquatic organisms, or those living in soil or sediment. It is also generally not possible for small or nocturnal organisms.
3. A small number of active predatory fish utilise a counter-current heat exchanger (*rete mirabile*) to retain metabolic heat and warm their muscles, brain or eyes. A few have modified optical muscles as heater organs, and a range of plants generate heat to aid dispersal of scent and attract pollinators.
4. A wide range of larger insects use rapid but unsynchronised muscle contraction to elevate their body temperature prior to flight, or other activity.
5. In hot climates organisms may need to dissipate heat to avoid overheating. The major behavioural mechanism is shade-seeking, or for small organisms stilting or climbing onto objects such as plants to move out of the hottest air next to the ground. Larger mammals may tolerate a limited degree of warming during the day, releasing this in the cool of the night. Evaporative cooling is very effective at losing heat, but because it loses valuable water it can only be used sparingly in arid areas.

Notes

1. This quotation comes from their influential review of lizard thermoregulation (Huey & Slatkin 1976); reproduced with the permission of Chicago University Press.
2. Angilletta (2009) provides a comprehensive list of comparative studies of thermal performance curves, showing that increased temperature generally leads to faster growth, improved locomotor ability, faster development and higher fitness.
3. Diagram modified and redrawn from Clarke & Rothery (2008).
4. *Nephrurus* data from Pianka (1978). The mole-rat data were from Buffenstein & Yahav (1991), with burrow temperature data from Bennett et al. (1988); redrawn with permission of Elsevier.

5. Conceptual model modified from Huey & Slatkin (1976); lizard thermoregulation data from Pianka (1985, 1986).
6. This coefficient was proposed by Huey & Slatkin (1976), as the simplest relationship that captures the nature of thermoregulation.
7. The beer can experiment is Heath (1964).
8. This definition of operative temperature comes from Bakken & Gates (1975).
9. For useful reviews of operative temperature see Walsberg & Wolf (1996), Dzialowski (2005) and Bakken & Angilletta (2014). Seebacher & Shine (2004) discuss the importance of thermal capacity. Helmuth & Hofmann (2001) and Helmuth (2002) discuss the accurate assessment of temperature in intertidal organisms.
10. The extent to which the slope and intercept of a least-squares regression are correlated depends on the error distribution, and specifically the relationship between the mean of x and the mean-square deviation of x. This problem is often unrecognised because it is not mentioned in most introductory statistical texts, or in the documentation accompanying commercial software packages.
11. Diagram modified from Hertz et al. (1993).
12. The *Anolis* study is Hertz et al. (1993) and the *Varanus* study is Christian & Weavers (1996).
13. This terminology for butterfly basking was introduced by Clench (1966) and important subsequent studies were Wasserthal (1975) and Rawlins (1980). The significance of angled wings was discussed by Kingsolver (1985a, b) and Heinrich (1990). A valuable review is Kingsolver (1985c).
14. See Lowe & Heath (1969), Crawshaw & Hammel (1971), Caulton (1978) and Elliott (1981).
15. McLaren (1963) proposed thermal benefits to vertical migration in the sea. A more recent model is Tarling et al. (2000).
16. Useful reviews of TRP sensors are Jordt et al. (2003), Vriens et al. (2014) and Bagrientsev & Gracheva (2015).
17. Sears et al. (2016).
18. The classic early papers are Newman & Hartline (1981) and de Cock Buning (1983). See also Campbell et al. (2002) and Gracheva et al. (2010).
19. Kürten & Schmidt (1982).
20. The key early papers on insect infrared detectors are Evans (1964, 1966a, b).
21. The model is described by Huey & Slatkin (1976). Herczeg et al. (2006, 2008) tested the model, Lambrinos & Kleier (2003) describe thermoregulation in *Bufo spinulosus* and Fitzgerald et al. (2003) in *Hoplocephalus stephensii*. Angilletta (2009) provides an excellent overview of optimality modelling of behavioural thermoregulation.
22. Christian & Bedford (1995).
23. The *Triturus* study is Gvoždík (2003). The evolution of vivipary in reptiles is discussed by Qualls & Andrews (1999) and Shine (2004).
24. Davy (1835) (this is the full report; short summaries of this work also appeared at least three other journals in the same year). John Davy (1790—1868) was a Cornish doctor and amateur chemist. He discovered and named phosgene in 1812, and also silicon tetrafluoride. He was elected a Fellow of the Royal Society in 1812, and edited nine volumes of the collected works of his more famous brother, Humphrey Davy.
25. The key historical references here are Carey & Teal (1966), Carey & Lawson (1973). Walli et al. (2010) report temperature data from free-swimming fish. See also Alexander (1996) and Bernal et al. (2001).
26. Heat retention in lamnid sharks was first reported by Carey & Teal (1969). A valuable up-to-date review is Block (2011). The evolution of heat retention in fishes is discussed by Block & Finnerty (1994) and Dickson & Graham (2004).
27. Wegner et al. (2015).
28. The original presentation (Lamare-Picquot 1835) is a short anonymous verbal report (recorded on p. 70, where his name is spelt Lamarrepicquot). The traveller and naturalist Christophe-Augustin is often confused with his more famous relative François-Victor, who was a surgeon. Benedict et al. (1932) review the early work, and Hutchinson et al. (1966) report the definitive study. Brashears & DeNardo (2013) revisit the problem.
29. Plotted with data from Oosthuizen (1939).
30. Dotterweich (1928), Kammer (1968). See also Krogh & Zeuthen (1941).
31. Redrawn from Heinrich (1993), with permission of Springer-Verlag and the author; original studies are Hegel & Casey (1982) and Bartholomew & Heinrich (1973).
32. For anyone interested in insect thermoregulation, two books by the doyen of the field, Bernd Heinrich, are essential reading. *Bumblebee economics* (1979) is a beautifully written account of thermal regulation in these important insects, and *The hot-blooded insects* (1993) provides a thorough and engaging survey of the wider field of insect thermoregulation. Both are illustrated with Heinrich's delightful drawings.
33. Morgan (1987) describes thermogenesis in rain beetles, Oertli (1989) covers the influence of size and other factors on thermogenesis in beetles and Chown et al. (1995) describe the thermal ecology of *Circellium*

bacchus, an entirely flightless and purely ectothermic dung beetle from Africa.

34. Redrawn from Heinrich (1993), with permission of Springer-Verlag and the author; original study is Morgan (1985).
35. *The race is not to the swift, nor the battle to the strong* (Ecclesiastes 11, King James translation). The author of Ecclesiastes was clearly an observant naturalist.
36. See de Lamarck (1778). Jean-Baptiste de Lamarck (1744–1829) was a French naturalist, who published major works on plants and invertebrates and developed the first coherent theory of evolution. While Lamarck is often portrayed only as the person who got evolution wrong, in France he is rightly revered as one of the greatest of biologists. Nicolas-Théodore de Saussure (1767–1845) was a Swiss geologist, chemist and botanist, who showed that plants acquire their carbon from atmospheric CO_2, and take up nutrients as well as water from the soil. His most important work is his book *Recherches chimiques sur la Végétation* published in 1804.
37. Knutson (1972, 1974, 1979), Nagy et al. (1972).
38. From Seymour (2001) and Patiño et al. (2000), where primary references can be found. Taxonomy from Angiosperm Phylogeny Website, version 13 (http://www.mobot.org/MOBOT/research/APweb), downloaded 17 Jan 2017.
39. Seymour (2001) provides a succinct review of thermogenesis in plants, Juszczuk & Rychter (2003) and Vanlerberghe (2013) review the function of alternative oxidase, and Kakizaki et al. (2012) describe the organisation and function of the respiratory chain in thermogenic plants.
40. Plotted with data taken from Seymour & Schultze-Motel (1998); reproduced with permission of the Royal Society of London.
41. Seymour et al. (2003, 2009). Seymour et al. (2010) describe thermal benefits to beetles in *Magnolia ovata* from Brazil.
42. Important early discussions of desert adaptations in general are Taylor (1970) and Baker (1982). McKechnie & Swanson (2010) review the metabolic adaptations of desert birds.
43. The *Ocymyrmex* data are redrawn from Heinrich (1993), with permission of Springer-Verlag and the author; original study is Marsh (1985). The *Plocepasser* data are from Whitfield et al. (2015), replotted with an exponential model fitted by the author.
44. Plotted with data from Ostrowski & Williams (2006); reproduced with the permission of Company of Biologists.
45. The camel study is Schmidt-Nielsen et al. (1956), the gazelle study is Ostrowski & Williams (2006).

CHAPTER 10

Endothermy

> The mammal is a highly-tuned physiological machine carrying out with superlative efficiency what lower animals are content to muddle through with.
>
> James Arthur Ramsay[1]

In the previous chapter we explored the behavioural and physiological mechanisms by which many organisms can maintain a high body temperature for at least part of the day. In this chapter we examine an alternative approach to maintaining a high body temperature, one that is very successful ecologically but also very expensive energetically: endothermy.

The ecological benefits of endothermy are clear. It has allowed mammals and birds to occupy a diverse range of niches and to live in a wide variety of habitats, including the air. They are more widely distributed on the Earth than any other terrestrial vertebrate group, and are found in all seas of the world. The costs of endothermy are also well understood: the high rate of metabolism needed to maintain a warm body requires a great deal of food. In this chapter we will examine how warm endotherms are, explore how that warm body is maintained and speculate on how endothermy might have evolved.

10.1 What is endothermy?

There are many definitions of endothermy in textbooks and journal articles. The most useful is probably:

The maintenance of a high and relatively constant internal body temperature, where the principal source of heat is a high metabolic rate at rest.

This is quite a rigorous definition, for it restricts endothermy among living organisms to mammals and birds. All other living organisms are ectotherms, and for these the principal source of heat is the environment[2].

Although ectothermy and endothermy are often cast as rigid alternatives, they simply represent a differing balance between the two major sources of heat for all organisms. Both endotherms and ectotherms generate heat metabolically and both exchange heat with their environment (Figure 10.1). The difference is in the principal source of energy used to maintain body temperature: in ectotherms it is the environment, in endotherms it is metabolism[3].

The language of thermal physiology has had a long and tortuous history, with a sizeable literature arguing about semantics without achieving either consensus or clarity. The older terms of *warm-blooded* (for birds and mammals) and *cold-blooded* (for reptiles, amphibians, fish and invertebrates) have long been recognised as inaccurate and misleading, and have largely been abandoned in the scientific literature. Nevertheless a number of overlapping terms continue to be used to differentiate between different types of body temperature regulation, and these are listed in Table 10.1.

Although some of these terms have been used more or less as synonyms, modern usage tends to distinguish between endothermy/ectothermy (which designate the principal source of heat for body temperature), homeothermy/poikilothermy (which indicate the extent to which body temperature, however generated, varies) and bradymetabolism/tachymetabolism (which describe the level of metabolic rate).

Principles of Thermal Ecology. Andrew Clarke, Oxford University Press (2017).
© Andrew Clarke 2017. DOI 10.1093/oso/9780199551668.001.0001

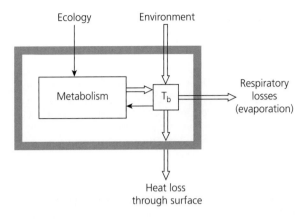

Figure 10.1 The conceptual model of the processes determining body temperature, T_b, in a terrestrial organism, here modified to show two important influences on resting metabolic rate. The first is ecology (lifestyle) which influences the intensity of resting metabolism and the second is the important feedback from body temperature itself. Solid arrows indicate the main pathways for energy flow and the line arrows the influence. The model applies equally to ectotherms and endotherms, as these differ only in the relative balance of the various pathways for energy flow.

Table 10.1 Some definitions[4].

Term	Definition and comment
Terms based on the primary source of heat	
Endotherm	An organism whose primary source of body heat is a high metabolic rate at rest.
Ectotherm	An organism whose primary source of body heat is the environment.
The use of the terms endothermy and ectothermy was first advocated by the American herpetologist Raymond Cowles.	
Terms based on stability of body temperature	
Homeotherm (Homoiotherm)	An organism that maintains a more or less stable body temperature, regardless of the primary source of heat. In some large ectotherms the loss of heat to the environment is sufficiently slow that internal body temperature remains relatively high and constant; these are often called *inertial homeotherms* (or, colloquially, *gigantotherms*).
Poikilotherm	An organism with a variable body temperature. This characterises most terrestrial ectotherms and also many mammals and birds when they are small. The boundary between homeothermy and poikilothermy has been defined as a seasonal or daily variation in body temperature > 2 K.
These terms were introduced by the German physiologist Carl Bergmann	
Terms based on intensity of metabolism	
Tachymetabolism	The high metabolic rate characteristic of mammals and birds.
Bradymetabolism	The lower metabolic rate characteristic of reptiles, amphibians and fish.
These terms have the advantage of distinguishing the mechanism (high or low resting metabolism) from the outcome (endothermy or ectothermy), but have not gained wide acceptance.	

The most confusing term to be found in the literature is *heterothermy*. As we saw in Chapter 9, some insects, fish and even a few plants can either retain or generate metabolic heat, but in all of these the warming is limited to specific tissues and is often confined to particular times. Although sometimes referred to as *facultative endotherms*, these organisms do not conform to our definition of endothermy and we need an alternative term to describe them (Box 10.1). They are perhaps better regarded as *heterotherms*, a term originally suggested for organisms that do not rely solely on either external or internal sources of heat. Although facultative endotherms match this definition of heterothermy, the term has mostly been used in a different context, which is to describe mammals and birds that allow extremities to cool in order to retain heat (*regional heterothermy*), or which go into torpor and hibernation (*temporal heterothermy*). These are quite different phenomena from the facultative endothermy of fish, reptiles, insects and plants, and it is unfortunate that the same term has been used for both[5].

Endotherms are characterised by a high, relatively stable body temperature. This poses two questions: how high is body temperature and how stable is it? We will explore the first in this chapter, and the second in the next.

10.2 Body temperature in endotherms

Although active mammals and birds maintain a high body temperature, its precise value varies between species and the overall range of temperatures (~15 K) is wider than often realised (Figure 10.2).

Box 10.1 Challenging the definition of endothermy

The rigorous definition of endothermy used in this book excludes many organisms often described as endothermic, such as some fish, many insects and a few plants. Although all of these do indeed generate heat internally (which is the etymological derivation of the term endothermy), they only warm parts of the body and often only when needed. For this reason they do not conform to the definition of endothermy used here.

Recently, however, it has been shown that a fish, the opah or moonfish, Lampris guttatus, maintains its entire body warmer than the seawater in which it swims, and does so continuously. It manages this by trapping the heat produced by its aerobic (red) muscles with counter-current heat exchangers within its gills. It also has a layer of fatty connective tissue which provides insulation. Because the opah swims by continual movement of its pectoral fins, rather than the more usual body undulation or tail movements used by most fish, heat production is also continuous. The opah thus maintains a high body temperature continuously, not by up-rating its general metabolism as mammals and birds do, but by a sophisticated anatomical mechanism for capturing and retaining the heat generated by routine muscular activity[6].

While the opah is a wonderful example of the parallel evolution of mechanisms for retaining metabolic heat to increase body temperature, and in the most challenging environment of all for retaining heat, it is not an endotherm.

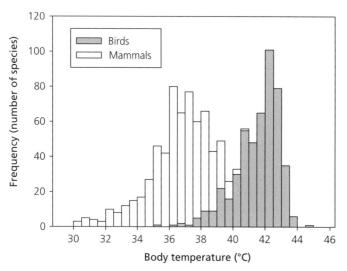

Figure 10.2 Frequency distribution of body temperature in mammals and birds (one value per species). Birds generally have warmer bodies than mammals, but the range of values in both groups is wide. The data for the two groups are stacked, not overlapped[7].

It has long been known that, on average, birds have a higher body temperature than mammals, although we do not really know why. Many older textbooks suggest that in mammals mean body temperature increases in the sequence monotremes through marsupials to placentals, with the implication that the more advanced (by which is usually meant the more recently evolved) forms have higher body temperatures. However as so often in ecology, the situation is more complicated than this.

The most important confounding factor is body size and the relationship between heat flow and body size is one of the oldest topics in thermal ecology. Since heat flow is integral to the maintenance of body temperature, and because the skin surface is a major route for heat transfer one might expect body temperature to vary with size. If we plot body temperature as a function of body mass, we see that in mammals body temperature increases slightly with size, but that in birds body temperature decreases with size (Figure 10.3)[8].

At the very broadest level there are statistically significant differences in the mean body temperature of the traditional major groups of mammals and birds (Table 10.2). The median and mean body temperature for mammals is lowest in monotremes, intermediate in marsupials and highest in placentals,

ENDOTHERMY 217

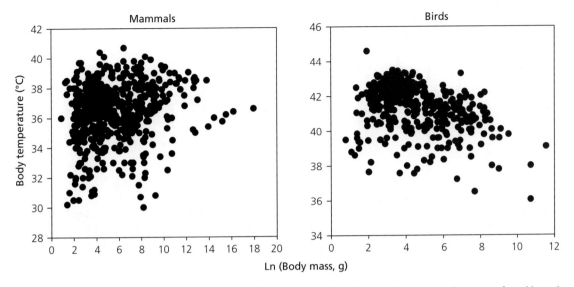

Figure 10.3 Scaling of body temperature (°C) with body mass (g) in mammals and birds. Note that the body mass data are transformed (natural logs) to improve the distribution of variance[9].

Table 10.2 Mean body temperature of the traditional major groups of mammals and birds[10].

Group	N	Body temperature (°C)			
		Median	Mean	SD	Range
Mammals					
Monotremes	4	30.8	30.9	0.88	30.0–32.0
Marsupials	80	35.7	35.5	1.32	30.8–39.9
Placentals	512	36.9	36.6	1.58	30.2–40.7
Birds					
Non-passerines	247	40.9	40.7	1.43	35.0–44.6
Passerines	236	42.2	42.1	0.82	39.3–43.5

but the number of species varies greatly between these groups and there is considerable overlap. In birds, although the traditional division into passerines (songbirds or perching birds) and non-passerines (everything else) is fairly balanced numerically, the non-passerine group lumps together a large number of birds of very different size and ecology (waterfowl, birds of prey, woodpeckers, parrots, hummingbirds and so on). Our understanding of the evolutionary relationships within both mammals and birds is in a state of considerable flux, and uncertainty surrounds the relationships of major lineages. At the level of order, however, the taxonomy of both birds and mammals remains relatively stable, and order thus makes a useful taxonomic level at which to examine variation.

In both mammals and birds, body temperature varies significantly across orders (Figure 10.4). In mammals mean body temperature varies from quite low values in monotremes, pangolins (Pholidota), anteaters and sloths (Pilosa) and armadillos (Cingulata) to high values in the odd-toed ungulates (Perissodactyla), carnivores (Carnivora), deer and other even-toed ungulates (Artiodactyla) and rabbits and hares (Lagomorpha). Some placental groups, such as anteaters, sloths and pangolins, have a lower mean body temperature than most marsupials. Across bird orders mean body temperature values are higher than in mammals, with the lowest values for ratites (emu, ostriches and relatives), and swifts and hummingbirds (Apodiformes), and high values for woodpeckers and allies (Piciformes) and songbirds (Passeriformes, by far the largest order of birds).

These patterns need to be interpreted with care, for some of the groups with a relatively low mean body temperature, such as bats in mammals and hummingbirds in birds, exhibit regular bouts of torpor, when their body temperature drops. It is always possible that some of the measured data for these species are not representative of the fully active state. Individual orders of mammals and birds also show differing patterns of the scaling of body temperature with body mass. Among mammals, in

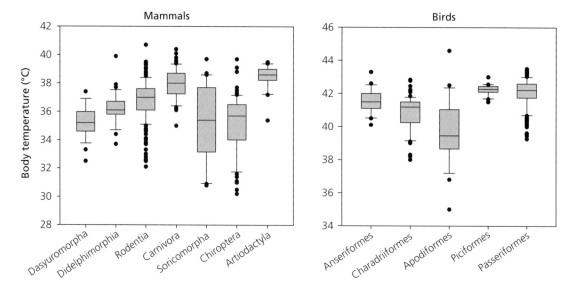

Figure 10.4 Boxplot showing the body temperature of different orders of endotherms. Data are plotted only for those orders with data for 20 or more species. The box defines the middle two quartiles (25th and 75th percentiles), with the mean value shown as a line. The whiskers show the 10th and 90th percentiles, and individual outliers are shown as data points[11].

dasyurid marsupials and bats body temperature increases with body mass, while in others such as carnivores and artiodactyls the relationship is the reverse. Among birds there is positive scaling of body temperature in tubenosed seabirds and passerines, and negative scaling in shorebirds.

The body temperature of individual orders of mammals and birds thus reveals a heterogeneous picture: mean levels vary between lineages, and patterns of scaling with body mass do the same. This suggests strongly that body temperature is not random, but is adjusted to match the particular ecology and physiology of each lineage. It also means that any overall relationship calculated for all mammals or all birds will depend critically on the relative representation of the different groups in the data set. Indeed when allowance is made for the taxonomic or phylogenetic structure in the data set, the scaling of body temperature with body mass becomes non-significant in both mammals and birds[12].

10.3 Control of body temperature

For a mammal or bird to maintain its body temperature within a narrow range, or to cycle its body temperature diurnally in a regulated manner, it must have a control system. The general structure of the temperature control system in mammals has been known for some time. The temperature regulating centre in the brain is the hypothalamus, which receives input from temperature sensors in the skin and mucous membranes (the peripheral thermoreceptors) and from the body core (central thermoreceptors), including the anterior portion (the preoptic area) of the hypothalamus itself.

As we saw in Chapter 9 the temperature-sensitive receptors are the *transient receptor potential* ion channels (TRPs) found across the animal kingdom. Some are sensitive to an increase in temperature, others to a decrease in temperature. The signals from these are conveyed to the posterior hypothalamus, which controls a complex thermoregulatory response involving other hypothalamic, autonomic and cerebral thermoregulatory centres. Some responses are involuntary, mediated through the sympathetic nervous system. These include pilo-erection (increasing the depth of fur to trap insulating air) and shivering, vasoconstriction of skin blood vessels, stimulated by norepinephrine released from sympathetic nerve fibres, or sweating, stimulated by the cholinergic sympathetic nervous system. There are also hormonal responses: the hypothalamus releases *thyroid stimulating hormone* (TSH), which induces the thyroid gland to liberate large amounts of thyroid hormones (T3 and T4) into the blood, prompting an increase in resting metabolic rate. The cerebral cortex may also induce voluntary behavioural responses, such as huddling to conserve heat or a move into the shade to reduce heat input from solar radiation. All of these will result in a change of skin and core temperature, which will be detected by the sensors, and the hypothalamus will adjust its output accordingly.

Mammals and birds do not know what their temperature is, they simply react to the information from the peripheral and central thermoreceptors. The adjustment of the set-point, diurnally, seasonally or evolutionarily, presumably involves changes in the hypothalamus, but as yet we do not know what these are. We cannot point to a particular feature of the hypothalamus, either anatomical or chemical, and say that this means that body temperature will be set relatively high or relatively low.

10.4 Ecology and body temperature in endotherms

The variation in mean body temperature and the difference in scaling patterns across different lineages of birds and mammals suggests strongly that body temperature is under selective control, and that the body temperature of a particular species has evolved to match its ecology. Two important aspects of ecology which appear to influence body temperature in endotherms are environmental temperature and diet.

10.4.1 Environmental temperature

It was long believed that the body temperature of humans varied with climate, and that this in turn influenced or even dictated the personality that could be ascribed to peoples living in particular places (for example the 'hot-blooded' peoples from the Mediterranean basin contrasting with the cold and unemotional peoples from the northern lands of the midnight Sun). These ideas have their origin in ancient views of physiology, and were so persuasive that they have given rise to vivid metaphors which remain in our everyday language[13].

While we now recognise that the body temperature of a healthy human varies little with climate or latitude, there are well recognised relationships when mammals as a whole are considered. Thus the body temperature of mammals is related to the mean environmental temperature of their habitat and the pattern suggests that the body temperature of a representative mammal increases from the tropics towards the poles (Figure 10.5).

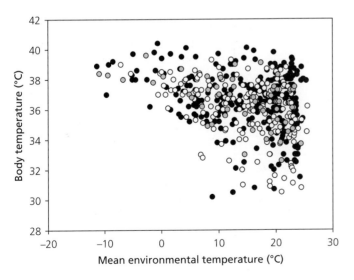

Figure 10.5 Relationship between body temperature and mean environmental temperature in 512 mammals. Data are coded by body mass: 0–50 g (white symbols), 51–100 g (grey symbols) and > 100 g (black symbols)[14].

A regression line fitted to these data might suggest that, all other things being equal, a polar mammal living at −10 °C has a body temperature ~2.7 K warmer than a tropical mammal of similar size living at 25 °C. This would, however, be slightly misleading because the pattern is more one of increasing variance towards the tropics. The maximum body temperature observed in mammals remains more or less constant across the environmental temperature range, whereas the minimum body temperature decreases systematically towards the tropics. There is no indication that this large-scale pattern varies with size, but the increasing variance and the greater data density in warmer environments do mean that the assumptions underpinning the use of least-squares regression are not met by these data.

This pattern is intriguing. It might be thought that the greater physiological challenge of maintaining a high body temperature in cold climates would result in a lower body temperature in polar regions. This is clearly not the case, for the data suggest a tighter regulation of a high body temperature towards the poles. In contrast there appears to be something about living in the tropics that allows a much greater range of body temperatures there. Since body temperature influences the level of basal metabolic rate, this suggests that a greater range of metabolic rates in tropical mammals. We return to this in Chapter 15 when we discuss energy and diversity.

10.4.2 Diet

The body temperature of endotherms is correlated with their diet (Figure 10.6). In herbivores mean T_b increases in the sequence nectarivores<frugivores< granivores<foliovores, suggesting that T_b is associated with the fraction of cellulose in the diet.

A relationship between diet and basal metabolic rate (BMR) has long been recognised in mammals, and a similar relationship has been reported for daily energy intake. Previous analyses, however, have not considered body temperature as a factor. When body temperature is included in the statistical model the relationship between diet and BMR vanishes because of a strong covariation between diet and body temperature[15]. This suggests (but does not prove, since it is only a correlation) that the fundamental relationship is between diet and body temperature, and BMR varies with diet simply because of its dependence on body temperature (see Chapter 8). The physiological interpretation of this pattern would be that a warmer body is better for utilising a diet rich in cellulose, but that this comes with the energetic cost of a higher BMR. For example a 150 g mammalian foliovore has a BMR, on average, 30% higher than a similarly sized mammal eating invertebrates. The mean body temperature associated with a particular diet must therefore represent a trade-off between energy intake from that diet and the energetic costs

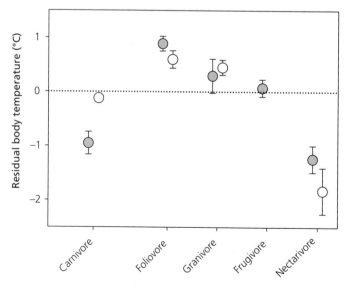

Figure 10.6 Variation in body temperature (°C) with diet in mammals (grey symbols) and birds (open symbols). Data are residuals from a general linear model (body mass, body temperature, with diet category as fixed factor), plotted as mean and SE; symbols plotted only where a diet category has data for five or more species (range 5 to 237)[16].

associated with processing it, as well as other lifestyle factors that affect body temperature and BMR. An example of the latter would be a higher BMR in a species having an active lifestyle compared with a more sedentary species (see Chapter 8).

An association between diet and body temperature has long been known in reptiles and fish, where herbivorous species tend to maintain relatively high body temperature through behavioural thermoregulation (lizards) or to be more numerous in warmer waters (fish). The intensity of insect herbivory also tends to be higher towards the tropics, and in warmer periods of history[17]. There is thus a clear association between diet and body temperature in ectotherms. It is intriguing that this is also seen in endotherms, but the mechanism is unclear. It is usually assumed that a higher body temperature improves digestion of plant tissues, but a warmer body will enhance the digestion of any dietary material. The unresolved physiological question is why the trade-off between energy intake and energetic costs should be optimised at different body temperatures for different diets.

Herbivory is widespread in mammals, greatly exceeding its incidence in reptiles, amphibians and fish. In birds foliovory is mainly a feature of two lineages (Galliformes and Anseriformes), and this may be in part be a consequence of the large gut necessary for digesting plant leaves, which in turn means that this is not viable as a diet for small passerine birds if they are to fly; granivory (seed-eating) is, however, very common in passerines.

While an increase in the extent of herbivory cannot have been the driver for the evolution of endothermy in mammals, it may have been one of its most important ecological consequences. The evolution of a permanently warm body allowed an increasing diversity of mammals to utilise a widespread and abundant food resource that was previously the preserve of insects, pelycosaurs during the Permian, and the large herbivorous dinosaurs until their demise at the Cretaceous–Palaeogene extinction event.

10.5 What sets the maximum body temperature an endotherm can have?

The maximum body temperatures achieved routinely by the warmest endotherms are similar to those of active desert lizards, being a little above 40 °C. This prompts the question of whether there is a maximum temperature at which a vertebrate can operate successfully, and if so what that is.

It is noted occasionally that typical endotherm body temperatures are close to the temperature at which the specific heat capacity of water, and hence energy transfer required to change temperature, is at a minimum (see Chapter 5)[18]. The slope of the relationship is, however, so shallow over the range 20 °C to 50 °C that it is difficult to see that the small changes in heat capacity can have any significant physiological effect. We must look elsewhere for a physical explanation.

One consequence of having a body temperature higher than the ambient temperature is that it allows for fine control of that body temperature. The net flow of energy (heat) is from the mammal or bird to the environment and the rate of this flow can be adjusted rapidly by control of peripheral blood flow and variation in plumage or pelage depth. This adjustment is far more rapid than can be achieved by increasing metabolic heat production. In some tropical regions, ambient temperature can exceed body temperature, presenting the animal with the opposite problem of energy (heat) tending to flow inwards. Perhaps maintaining a body temperature high enough to exceed ambient temperature in these tropical regions is just too costly.

An important energetic cost of a high body temperature is the cost of maintenance, a key component of which is the turnover of proteins (Chapter 9). All proteins suffer damage and the cell has sophisticated mechanisms for detecting damaged proteins and removing them. Some structural proteins have to last years and while a few proteins survive many days (for example haemoglobin in erythrocytes), most have lifetimes of only tens to hundreds of minutes. Sources of damage to proteins are many and varied but comparison of the temperature sensitivity of the rate of turnover of a typical mammalian protein with body temperatures of birds and mammals suggests an increased incidence of thermal damage to proteins might be a factor determining the upper thermal limit for mammals (Figure 10.7).

The data for both body temperature and protein turnover plotted here come from mammals, and so the pattern could also be interpreted as indicating no more than mammalian proteins have been adjusted by evolution to match the body temperatures

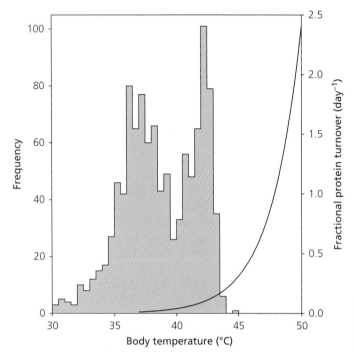

Figure 10.7 Turnover rate of an average mammalian protein (line) as a function of temperature, compared with the body temperature of mammals and birds (histogram)[19].

of mammals. It seems reasonable to suppose that there must be a temperature above which it is not possible for any protein to operate, although proteins from bacteria and archaeans living in hydrothermal vents continue to operate at temperatures > 100 °C. Vertebrates do not survive much above 45 °C, suggesting that there is something that sets an upper limit to their body temperature, but as yet we do not know what that something is.

10.6 How do endotherms generate their heat?

The body temperature of an organism is the result of a balance between heat gain and heat loss (Figure 10.1). The two major routes for heat gain are from the environment and from metabolism. The main routes for heat loss in terrestrial organisms are by evaporation during breathing and through the skin surface; for the latter the role of insulation is critical.

To a first approximation we might assume that the delivery of heat from the environment would be much the same for a lizard waiting to pounce on a passing insect and a similarly sized rodent foraging nearby. A crucial difference is that the lizard is using behavioural means to regulate its direct exposure to the Sun, while the rodent has insulating fur. This points to a major role for internal heat production in the mammal, and a significant difference in resting metabolism between endotherms and ectotherms has long been known[20].

A simple plot of resting metabolic rate as a function of body mass shows that the mean resting metabolism of mammals is about an order of magnitude higher than that of reptiles (Figure 10.8). A problem with this simple comparison is that the mean body temperature for the mammals and reptiles differs (in these data it is 37 °C for the mammals and 30 °C for the reptiles). We know that resting metabolic rate is influenced by body temperature and so this difference in temperature will exaggerate the difference in metabolic rate.

When the effect of temperature is allowed for in the fitted statistical model, then at larger sizes (between 150 g and 22 kg) mammalian resting metabolism exceeds that of reptiles at the same body temperature by a factor of ~5. At smaller sizes (below ~150 g) the relationship between resting metabolism and body mass in mammals is significantly curvilinear (though this is subtle and not

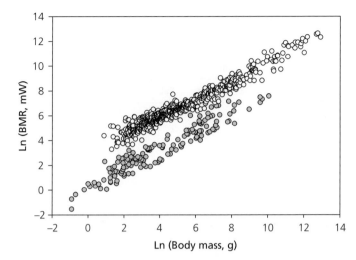

Figure 10.8 A comparison of the basal metabolic rate (BMR) of typical endotherms (mammals, open symbols) and typical ectotherms (reptiles, grey symbols)[21].

evident visually in a plot covering the full mass range), and this means that the difference increases sharply with decreasing size (Table 10.3). These complications are important when considering details of thermal ecology of small mammals and reptiles, but for the purposes of a broad comparison between mammals and reptiles, a simple important conclusion emerges. This is that for larger mammals, resting metabolism exceeds that of similar sized reptiles with the same body temperature by a factor of ~4–6. This is lower than the usually quoted values (a cursory scan of textbooks and papers will reveal estimates ranging from 10 to 30 for the ratio of mammalian to reptilian BMR), but more rigorous because it allows for the effect of both body mass and body temperature.

Table 10.3 Comparison of basal metabolic rate (BMR) in mammals, birds and reptiles. The BMR ratio is the factor by which the BMR of a mammal or bird exceeds that of a reptile of the same body mass and body temperature, T_b. Mammal data were calculated for a T_b of 35 °C, and bird data for a T_b of 40 °C. Comparison for the largest birds (*) was for a body mass of 5 kg, as this was close to the largest value in the bird data set.

Comparison	Ratio of BMR		
	10 g	150 g	22 kg
Comparison with mammals (T_b 35 °C)			
Reptiles (T_b 35 °C)	9.6	6.2	4.5
Comparison with birds (T_b 40 °C)			
Reptiles (T_b 40 °C)	9.8	6.9	4.5 *

When the analysis is extended to cover the full size range, however, it is clear that no single number captures the difference between the resting metabolism of a typical endotherm and that of a typical ectotherm: the uprating of endotherm metabolism is about ~5 for larger individuals and ~10 for smaller. The inclusion of body temperature in the analysis is critical because it allows us to distinguish the effects of body temperature on resting metabolism from that of endothermy itself. There is an important evolutionary point here, which is that the evolution of endothermy was not simply the acquisition of a warm body: a reptile with a body temperature identical to that of a living mammal or bird still has a significantly lower resting metabolic rate. Something else is clearly involved, something that leads to a significantly higher resting metabolism in mammals and birds.

We now recognise four main mechanisms for generating heat in endotherms: a general elevation of the metabolic rate at rest, the generation of heat from muscular contraction during daily activity, shivering and the activity of brown adipose tissue (BAT, a specialised heat-generating tissue) (Table 10.4). Shivering and BAT are not the most important sources of heat for endothermy, though they are sometimes the only ones discussed in textbooks. They are only switched on when needed, and hence are sometimes referred to as facultative thermogenesis, to distinguish them from the obligate thermogenesis which is a continuous process in all endotherms.

Table 10.4 Heat generation in endotherms.

Mechanism	Location
Obligate thermogenesis	
A general elevation of the metabolic rate at rest	Probably all tissues, but especially the visceral organs (liver, gut, kidney); muscle important in birds
Exercise (or activity) thermogenesis	
Routine daily activity	Muscle
Facultative thermogenesis	
Shivering	Muscle
Non-shivering thermogenesis	An increase in the level of thermogenesis by all tissues, plus the activity of brown adipose tissue (BAT) in those species that have this

Table 10.5 Contribution of the major oxygen-consuming organs of the body to body mass and resting metabolic rate[22].

Organ	Contribution to body mass (%)		Contribution to BMR (%)	
	Human	Rat	Human	Rat
Liver	2	5	17–27	20
Gut	2	5	10	5
Kidney	0.45	0.9	6–10	7
Lung	0.9	0.6	4	1
Heart	0.45	0.5	7–11	3
Brain	2.1	1.5	16–20	3
Skeletal muscle	41.5	42	15–20	30
Skin	7.8	nd	1.9	nd
Skeleton	10	10	nd	nd
Total	67.2	65.5	~90	69

10.6.1 Obligate thermogenesis

All metabolic processes dissipate heat. What endotherms have done is to trap some of this heat temporarily as it passes to the environment, and use it to maintain a high and stable body temperature.

The obvious question is where is this heat generated: do endotherms show a general elevation of metabolism throughout the body, or is it confined to specific organs or tissues? One way to examine this question is to estimate the contributions of different organs to the total resting metabolic rate from their size and measurements of tissue-specific metabolic rate from tissue slices *in vitro*. Unfortunately such estimates rarely add up to the measured whole organism metabolic rate. This mismatch undoubtedly reflects both the technical difficulties in making the measurements, and the uncoupling of tissue slices from regulatory factors such as nutrient supply and hormones. Organ metabolism can also be estimated by comparing the oxygen content of the arterial supply and venous outflow from organs, but again these rarely sum to whole organism values.

Despite these technical difficulties we can draw some robust general conclusions. Skeletal muscle is typically the largest tissue by mass in the mammalian body, and therefore contributes significantly to resting metabolism (Table 10.5). It is the visceral organs such as gut and liver, however, that provide the most important contribution. In a resting human, the visceral organs of the chest and abdomen comprise less than 10% of body mass, but contribute nearly 60% of the total heat. These data point to a key role for the visceral organs in the high metabolic rate at rest in mammals.

Our present understanding of the molecular mechanisms of routine heat production in endotherms comes principally from the work of Alan Hulbert and Paul Else in Australia and Martin Brand in England. An important clue came from a comparison of tissue mass and mitochondrial density in visceral organs in a typical mammal (the mouse *Mus musculus*) with a reptile of similar size (the lizard *Ctenophorus nuchalis*, previously *Amphibolurus nuchalis*): both the mass of the visceral organs and mitochondrial density within the tissues were higher in the mammal. When combined with data on the surface area of the inner mitochondrial membrane, there was a striking difference in the total mitochondrial surface area per tissue (Figure 10.9). The amount of mitochondrial inner membrane (which is where ATP synthase is located) is a powerful indicator of the cell's metabolic capacity, and these data thus indicate that the mouse has a considerably greater capacity for ATP generation than the lizard. While this comparison is based on just one mammal and one reptile, it confirms a range of more anecdotal analyses.

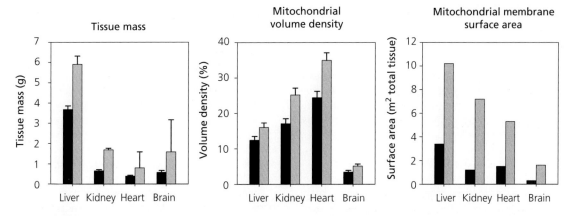

Figure 10.9 Comparison of tissue mass and mitochondrial characteristics in a mammal (*Mus musculus*, grey bars) and a reptile (the lizard *Ctenophorus nuchalis*, black bars) of similar mass and acclimated for 90 days to 37 °C. Error bars are standard errors[23].

10.6.2 The molecular basis of heat generation in endotherms

An important question is whether the higher metabolic rate at rest in the visceral organs of endotherms is the result of a general elevation of metabolism, or an increase in a particular process.

A comparison of the breakdown of resting metabolism in isolated hepatocytes from a rat and a similarly sized lizard with the same body temperature suggests that the mammal has uprated all of the major processes consuming oxygen in the cell (Figure 10.10). The fraction of total oxygen consumption used to drive the various processes is broadly the same in the rat and the lizard, suggesting that, at least in the liver, the evolution of endothermy has not involved the enhancement of any one particular process at the expense of another. Rather the extra oxygen consumption in endotherms results from a general uprating of metabolism.

A significant utiliser of oxygen at rest is the mitochondrial proton leak, in which protons return across the mitochondrial inner membrane into the matrix but without being coupled to ATP synthesis; the potential energy of the protonmotive force is dissipated entirely as heat. The movement of protons across the inner mitochondrial membrane is mediated by uncoupling proteins (Box 10.2).

The wider role of uncoupling proteins is not yet fully clear, but they appear to play an important role

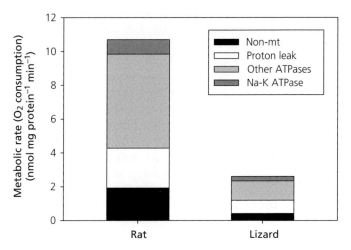

Figure 10.10 Comparison of the processes contributing to resting metabolic rate in hepatocytes isolated from a mammal (*Rattus norvegicus*) and a similarly sized lizard (*Pogona vitticeps*), with identical body temperatures (37 °C). Non-mt: non-mitochondrial oxygen use[24].

> **Box 10.2 Uncoupling proteins**
>
> Uncoupling proteins (UCPs) are proteins that span the inner mitochondrial membrane and allow protons that have been pumped into the intermembrane space by the electron transport chain to return to the mitochondrial matrix without being coupled to the synthesis of ATP. The potential energy of the proton gradient is thereby dissipated solely as heat.
>
> The first uncoupling protein to be described, UCP1 (known as *thermogenin* in the earlier literature because of its role in the generation of heat) is expressed only in brown adipose tissue (BAT). Other uncoupling proteins are found more widely: UCP2 is ubiquitous in mammalian tissues, whereas UCP3 is expressed mainly in skeletal muscle. Two other homologues are known (UCP4 and UCP5, the latter also known as BMCP1 or *brain mitochondrial carrier protein*, as it was discovered in mammalian nervous tissue). Uncoupling proteins are now known from birds, fish, invertebrates (insects) and plants. Comparison of vertebrate genomes indicates that the gene for UCP1 is present in fish and amphibians, but not in reptiles or birds. Furthermore, birds have also lost the gene for UCP2, emphasising the importance of UCP3 and muscle activity for their thermogenesis.
>
> Uncoupling proteins are closely related to anion mitochondrial carrier proteins, including the adenine nucleotide transporters that move ADP and ATP across the mitochondrial membrane from the matrix (where ATP is regenerated from ADP) to the cytosol (where ATP is used).
>
> Control of uncoupling protein activity has been studied mostly in UCP1. UCP1 is activated by fatty acids and inhibited by purine nucleotides (GDP, ADP). The fatty acids are released by triacylglycerol lipase, and this is the final stage of a signalling cascade initiated by release of norepinephrine by the sympathetic nervous system. This cascade provides a mechanism for central control of heat production in BAT. We might expect a similar control of the activity of UCPs in other tissues, but these are not so well studied as UCP1[25].

in regulating the level of reactive oxygen species (ROS), and perhaps also in maintaining the electrical gradient at levels below those which would cause the insulation of the membrane to break down. It may thus be that the use of uncoupling proteins for heat generation is essentially a modification of a system that has an ancient origin and has to do with regulating the mitochondrial protonmotive gradient. As so often in evolution, an apparently novel solution proves to be a modification of something that evolved originally to do something quite different.

The activity of uncoupling proteins appears to be linked to the unsaturation level of the membrane phospholipid fatty acids. Broad comparisons between ectotherms and endotherms point to important differences in phospholipid unsaturation which are correlated with strong differences in cellular resting metabolism. Phospholipid fatty acids from the rat (*Rattus norvegicus*) contained more double bonds than those from the lizard *Pogona vitticeps* (previously *Amphibolurus vitticeps*), caused mainly by an increase in the proportion of arachidonic acid (20:4ω6) and a decrease in that of linoleic acid (18:2ω6). A comparison within mammals indicated that variation in mass-specific resting metabolism with size was correlated with differences in the proportion of eicosapentaenoic acid (20:5ω3). Membrane lipid unsaturation has been proposed as the 'pacemaker of metabolism', but at present we have no clear physical mechanism to explain how this might work. The mechanism must presumably be through the link between the activity of membrane-bound uncoupling proteins and the unsaturation of the lipid environment in which they sit. While the relationship between membrane lipid unsaturation and resting metabolic rate is clear at a very broad level (for example comparing mammals with reptiles, or across all mammals), it is not always evident at the intraspecific level. This does not necessarily invalidate the idea that membrane lipid unsaturation is an important regulator of resting metabolism, but does indicate that the cell has a range of mechanisms for adjusting the level of resting metabolism[26].

So far we have concentrated on viscera, principally liver, and mammals. Muscle contributes considerable heat when active (which we discuss in the next section) but it is also a significant contributor to resting metabolism. Recently attention has been directed at resting muscle as a possible source of thermogenesis, particularly in birds. Compared with mammals, birds typically have a large muscle mass (especially pectoral muscle) and the demands of flight can lead to reduced visceral organ mass. It is now clear that vertebrate muscle can generate heat by the rapid cycling of Ca^{2+} across the sarcoplasmic reticulum, achieved by the uncoupling of Ca^{2+}-ATPase (SERCA) by a protein (sarcolipin). This is the same mechanism used by some fish in the modified muscle used as heater organs, suggesting that heat generation by resting muscle is an ancient facility, and may even have been the first mechanism for thermogenesis to have evolved[27].

10.6.3 Exercise or activity thermogenesis

Anyone who grew up in seasonal climes and complained about being cold in winter can probably recall being told by their parents to 'run about and keep warm'. This is because muscular activity generates considerable heat. Typically only 20–25% of the Gibbs energy in ATP is converted by the muscle into work and the rest is lost as heat[28]. This generation of heat by muscular activity is termed activity (or exercise) thermogenesis. Since many mammals and birds spend much of their time moving, we can surmise that activity thermogenesis routinely makes a significant contribution to their thermal balance. It is only when they are completely at rest that obligate thermogenesis peaks.

Under normal circumstances the heat required to maintain T_b in mammals and birds is thus provided by a combination of obligate thermogenesis (a high metabolic rate at rest) and activity thermogenesis (heat generated by routine activity). In human physiology this combination is referred to, rather confusingly, as *non-exercise activity thermogenesis* (NEAT). The stipulation 'non-exercise' is to distinguish heat generation by routine activity from that generated when exercise is specifically undertaken (such as playing sport or in the gym).

10.7 A general picture of obligate heat generation in endotherms

Accepting that the data emanate from a small number of studies of very few species, we can formulate a general picture of what is involved in endothermy. The key features of obligate thermogenesis would appear to be an increase in the size of some visceral organs, together with a proliferation of the mitochondria within them, resulting in a general uprating of their resting metabolic rate. This requires an increase in the proton conductance of the inner mitochondrial membrane in endotherms, as part of a general uprating of all metabolic processes. The enhanced rate of ATP production may be matched by an increased cycling of Na^+ and Ca^{2+} across the plasmalemma and sarcolemma membranes respectively. The heat generated from the routine muscular activity required by everyday activity is the second, and very important, contributor to obligate thermogenesis.

An important point here is that obligate thermogenesis mixes contributions from viscera (resting metabolism) and muscle (resting and active metabolism). These two sources of heat are independent, and hence may be subject to differing selective forces.

10.7.1 Variation in endotherm BMR

This general picture of heat generation in endotherms provides us with a clear rationale for explaining variations in their metabolic rate, be this seasonal within an individual, or when comparing individuals within a species or across species: differences in the relative size of organs, the mitochondrial density within the tissues or in the aerobic capacity of the mitochondria will all result in variations in resting (basal) and/or routine metabolic rate.

In recent years, an increasing number of studies have linked variations in BMR between different species of both mammals and birds to differences in the relative size of their organs (Table 10.6). Similarly, differences in BMR between individuals of a number of species of bird have been related to variations in organ size.

Many birds show strong seasonal or regional differences in BMR, and particular attention has been directed at passerine birds and migratory shorebirds. Small passerine birds wintering in cold northern forests typically have a higher BMR. This is presumed to be a response to the need for greater metabolic heat production in winter, and frequently is associated with larger organs, increased muscle mass and enhanced aerobic capacity. These variations are not confined to birds living in northern forests, for seasonal differences in metabolism have been found in African birds such as Knysna turaco (*Tauraco*

Table 10.6 Relationship between BMR and relative organ size in endotherms; in all cases the effects of body mass were allowed for in comparisons of metabolic rate and organ size[29].

Details of study

Differences between species

In 23 species of temperate European birds, resting metabolic rate is correlated with the relative mass of heart and kidney, but not muscle.

In 50 species of mammal, BMR is correlated positively with muscle mass.

Comparison of 32 species of tropical bird (Panama) with 17 species of temperate bird (Ohio), supported by literature data for 408 species from several continents, showed that the lower resting metabolic rate of tropical species is associated with a decreased size of the heart, flight muscle, liver, pancreas and kidneys.

Differences within species

Tree sparrows (*Passer montanus*) in China have higher resting metabolic rate in winter, associated with larger liver, heart and gut.

Red knots (*Calidris canutus*) in winter have a higher BMR and also maximum metabolic rate, associated with a greater body mass resulting from increased pectoral muscle.

In four passerine birds from China (*Aegithalos concinnus, Pycnonotus sinensis, Sturnus sericeus* and *Emberiza pusilla*), individual differences in resting metabolic rate are correlated with variations in the size brain, liver, kidney and gut.

Chinese bulbuls (*Pycnonotus sinensis*) have higher BMR in winter, associated with larger liver and gut, and also increased mitochondrial protein content and cytochrome c oxidase activity.

Cold acclimation-induced up-regulation of genes involved in muscle hypertrophy and angiogenesis in the North American passerine *Junco hyemalis*.

Differences in BMR in humans from different regions are associated with variations in the relative size of liver, spleen, heart, kidneys and brain.

corythaix) and southern red bishop (*Euplectes orix*), and in passerines living in different latitudes in China. A comparison of wintering populations of the migratory shore bird *Calidris canutus* (red knot) showed that birds wintering in colder temperate locations in Europe had a higher BMR and larger organ sizes than birds wintering in Africa[30].

10.8 Thermoregulation

The ability of endotherms to adjust their thermal conductance means that, within limits, body temperature can be maintained without the need to adjust metabolic rate; the range of ambient temperatures over which this can be achieved is the *thermoneutral zone*, TNZ (Figure 10.11).

Within the TNZ a constant body temperature is maintained solely by adjustments to heat loss, for which changes to peripheral circulation and insulation (fur in mammals, feathers in birds) are the most important mechanism[31]. At the cool end of the TNZ, the organism has reached the limit of what can be achieved by adjusting conductance and this point (the lower critical temperature, T_{LC}) marks the point of minimum conductance. Below the T_{LC}, body temperature can only be maintained by switching to new mechanisms of heat generation. Similarly T_{UC} marks the point above which the organism must employ extra mechanisms for losing heat.

Two examples of the TNZ, one each for a mammal and a bird, are shown in Figure 10.12. In the mouse, the TNZ is fairly narrow, from 22.5 °C to 27.5 °C.

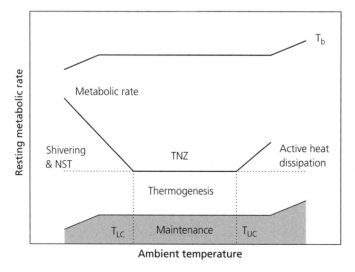

Figure 10.11 Conceptual model of the thermoneutral zone (TNZ) of an endotherm. The TNZ is the temperature range over which body temperature (T_b) remains constant and basal (resting) metabolic rate is independent of ambient temperature. The boundaries of the TNZ are the lower critical temperature (T_{LC}) and the upper critical temperature (T_{UC}). NST: non-shivering thermogenesis. Note that the cost of maintenance is related directly to body temperature.

Plate 1. A Namib day gecko (*Rhoptropus afer*) basking on the sunny side of a dark rock to warm up. (Image courtesy Barry Lovegrove). (see page 14)

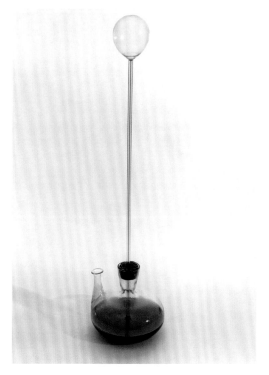

Plate 2. A reproduction of Galileo's thermoscope; the original was probably constructed in the late sixteenth century. (Image: Science Museum/Science & Society Picture Library). (see page 32)

Plate 3. Two emperor penguin, *Aptenodytes forsteri*, chicks photographed with a thermal imaging camera and with the image displayed in false colour. Note how the downy juvenile coat insulates the chicks so well that their surface temperature is similar to that of the ice and snow on which they sit (~−35 °C), whereas the face and flippers are warmer (~−10 °C). The insulating layer of downy feathers and the huddling behaviour allow the chicks to maintain an internal body temperature of ~38 °C. (Image courtesy of André Ancel, IPHC/CNRS/IPEV). (see pages 45 and 64)

Plate 4. The interior of a modern Stevenson screen housing a range of thermometers (dry bulb and wet bulb mounted vertically, and maximum and minimum mounted horizontally). Also present are a torch to take readings by night and a water bottle to keep the wet bulb wet (Image Wikimedia Commons). (see page 47)

Plate 5. Emperor penguins (Aptenodytes forsteri) on floating sea-ice in Antarctica. This would not be possible were it not for the unusual property of ice being less dense than liquid water. (Image courtesy Pat Cooper, British Antarctic Survey). (see pages 84 and 87)

Plate 6. An image of ice shelf and ocean, showing all three phases of water present simultaneously (although only the solid and liquid states are visible). (Image courtesy Chris Gilbert, British Antarctic Survey). (see page 93)

Plate 7. The icefish *Chaenocephalus aceratus*, photographed in shallow waters at Signy Island, South Orkney Islands, Antarctica; an example of an icefish, lacking functional haemoglobin and myoglobin. (Image courtesy Doug Allan, British Antarctic Survey)

Plate 8. An example of a highly ornamented tropical gastropod (*Murex altispira*). Such ornamentation is very rare among cold-water gastropods, where the production of a carbonate skeleton is energetically more expensive. (Image Wikimedia Commons). (see page 95)

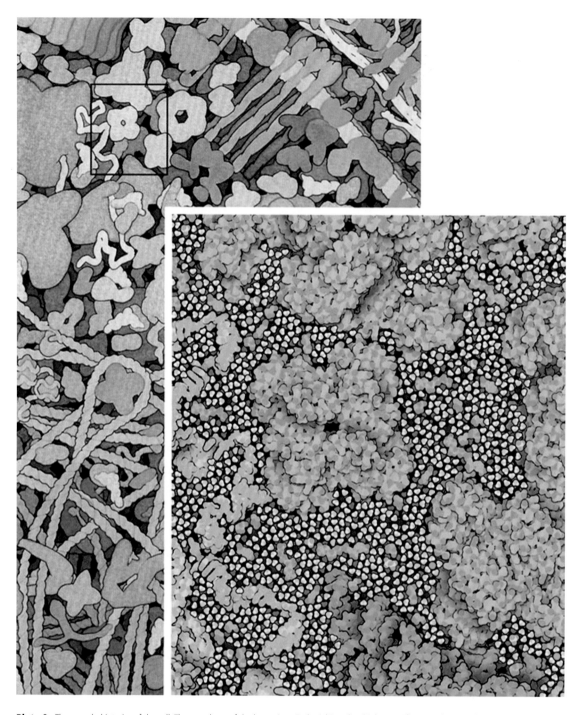

Plate 9. The crowded interior of the cell. The cytoplasm of the bacterium *Escherichia coli* at high magnification, showing the crowding of small molecules (water, ions, small metabolites) between the larger proteins and nucleic acids. The smallest molecules shown are water; note how few water molecules there may be between macromolecules. (Image courtesy David Goodsell, reproduced with permission of Springer-Verlag). (see pages 98, 102, and 114)

Plate 10. An immature *Trematomus bernachii* resting on ice in McMurdo Sound. The presence of antifreeze in this fish means that it does not freeze despite being in intimate contact with ice. (Image courtesy Paul Cziko)

Plate 11. An exact model of Lavoisier's Ice Calorimeter, with one side cut away to show the inner workings. The guinea pig was held in the mesh cage, surrounded by ice. The water produced from ice in the inner jacket by the dissipation of metabolic heat flowed down the tube to be captured and weighed. The outer jacket was also filled with ice to minimise heat transfer between the calorimeter and the laboratory. (Image: Science Museum/Science & Society Picture Library). (see page 174)

Plate 12. An example of a series of nunataks, the Behrendt Mountains in Antarctica. These isolated nunataks support microbial and microinvertebrate life, and may represent the coldest habitat with life on Earth. (Image courtesy Pete Convey, British Antarctic Survey). (see page 310)

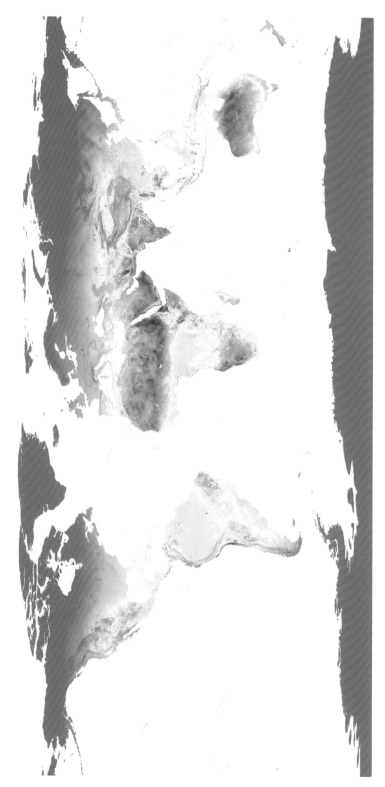

Plate 13. Map showing mean annual temperatures on land, 2002–2013. Data range from < −10 °C (green) to > 30 °C (red). Data from NASA MODIS Aqua satellite. Map courtesy Andrew Fleming, British Antarctic Survey. (see page 313)

Plate 14. Map showing seasonality of temperature (mean difference between January and June temperatures). Data range from < 10 K (green) to > 40 K (red). Data from NASA MODIS Aqua satellite. Map courtesy Andrew Fleming, British Antarctic Survey. (see page 313)

Plate 15. Map showing mean sea surface temperature, 2002–2013. Data range from < 0 °C (blue) to > 28 °C (red). Data from NASA MODIS Aqua satellite. Map courtesy Andrew Fleming, British Antarctic Survey. (see page 315)

Plate 16. Map showing seasonality of sea surface temperature (mean difference between January and June temperatures). Data range from < 5 K (blue) to > 20 K (red). Data from NASA MODIS Aqua satellite. Map courtesy Andrew Fleming, British Antarctic Survey. (see page 315)

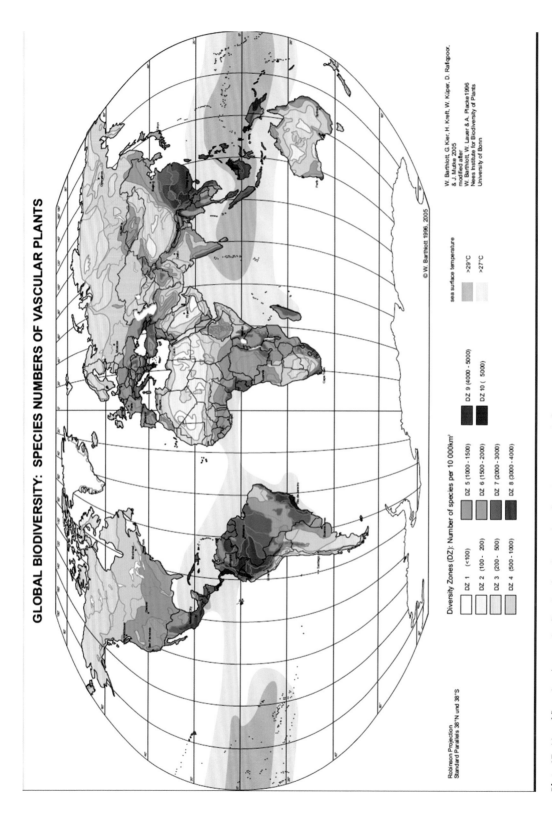

Plate 17. Map of flowering plant species diversity, plotted as number of species per 10 000 km². Map courtesy Wilhelm Barthlott, University of Bonn. (see pages 333 and 342)

Above the TNZ metabolic rate rises, partly because the body is warmer (and hence maintenance costs increase) and partly because the mouse is expending energy in attempting to keep its body temperature down. Below the TNZ metabolic rate again rises, this time because the mouse is generating extra metabolic heat in an attempt to prevent its core temperature from dropping. The TNZ of the mouse is fairly narrow, and in many small endotherms it can be difficult to define a TNZ at all. Instead there is often a smooth decline in BMR above and below some ambient temperature that marks the point of minimum BMR for that particular species.

In a classic study of body temperature and metabolism in the white-tailed ptarmigan (*Lagopus leucura*), a bird of the high mountain tundra of North America, George West showed that the T_{LC} shifted seasonally (Figure 10.12b). In winter the TNZ extended to lower temperatures, and the slope of the relationship between metabolic rate and temperature was shallower. This indicates that the ptarmigan have a lower thermal conductance in winter, achieved largely by an increase in the insulating properties of the plumage.

The seasonal change in the TNZ and thermal conductance of the ptarmigan shows that animals are able to adjust their thermal properties to match their environment. There are, however, powerful physical constraints on the thermal properties of endotherms which determine what they can and cannot do. The most pervasive of these is size. Smaller organisms have a greater surface area to volume ratio than larger organisms, and in endotherms this has significant consequences for metabolism. Since heat is lost principally through the body surface, all other things being equal, small mammals and birds lose their body heat far more easily than large ones. A consequence of this is that the breadth of the TNZ varies with size (Figure 10.13).

This variation of TNZ breadth with body mass indicates a fundamental constraint of size on thermal properties. Small endotherms have a very narrow TNZ (and as noted above, this may be so narrow as to be difficult to discern), but large endotherms can maintain a broad TNZ. Variation about the line indicates that evolution has been able to get around this constraint to some extent, and this is shown by the variation of T_{LC} and T_{UC} with ambient temperature (Figure 10.13). We can interpret these patterns as showing that while body size constrains the breadth of the TNZ, its position (that is, the absolute values of T_{LC} and T_{UC}) can be adjusted by selection. The strength of the relationship between T_{LC} and

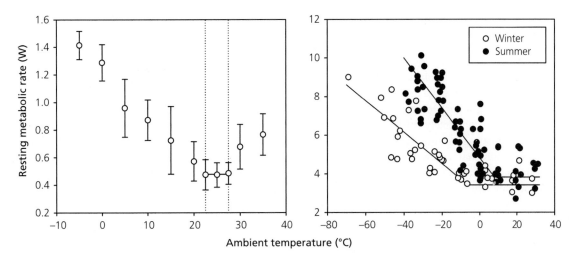

Figure 10.12 Two examples of the thermoneutral zone (TNZ). Left: TNZ in the mouse *Apodemus chevrieri* from the Hengduan mountains of China. The dotted lines show the upper (T_{UC}) and lower (T_{LC}) critical temperatures defining the breadth of the TNZ (solid line). Right: seasonal variation in the metabolic rate of the white-tailed ptarmigan, *Lagopus leucura*, from Alaska. The fitted lines show the resting metabolic rate within the TNZ for summer (higher) and winter (lower), the shift in the lower critical temperature (T_{LC}) from summer to winter and the lower thermal conductance in winter (shown by the shallower slope of the relationship between metabolic rate and temperature)[32].

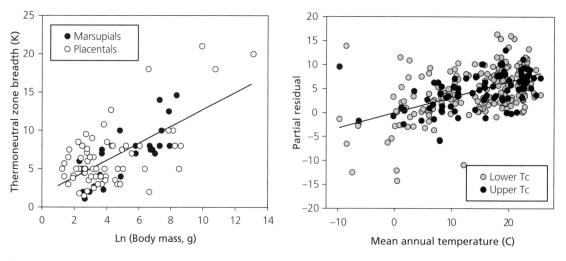

Figure 10.13 Variation of the thermoneutral zone with body size and environmental temperature. Left: variation of the breadth of the TNZ with body mass. A General Linear Model indicated no significant difference in the slope of the relationships for marsupials and placentals (p = 0.288). Right: variation of upper and lower critical temperatures with the mean annual temperature. Because both T_{LC} and T_{UC} vary with body mass, data are plotted as partial residuals from a General Linear Model. The line shown is for T_{UC}[33].

mean environmental temperature is similar to that for T_{UC} (the slopes are 0.32 and 0.30 respectively), suggesting that selection can shift these more or less in parallel, leaving the breadth of the TNZ essentially unaltered.

Within its TNZ a mammal or bird maintains its core temperature by controlling sensible heat loss and without the need to adjust heat supply from obligate thermogenesis and routine day-to-day activity. At temperatures above and below the TNZ, an organism has to bring additional processes into play. This is termed *thermoregulation*, which is somewhat unfortunate for it carries the implication that the organism is not regulating temperature when within its TNZ. This is completely misleading. An organism is still regulating its temperature when within its TNZ, it is simply that the mechanism is different[34].

The thermoregulatory processes used by a mammal or bird are different depending on whether it is above T_{UC} or below T_{LC}, so we will consider these separately.

10.9 Winter is coming: keeping warm below the TNZ

The challenge for an endotherm below its T_{LC} is to minimise or offset the loss of heat to the environment. Typically this challenge arises for endotherms that experience marked seasonal changes in ambient temperature, such as those living at high latitudes or high altitudes.

Although the physiological challenge of cold weather or climate can be severe, there are behavioural responses that go some way to alleviating this challenge. The huddling behaviour of emperor penguins in the depths of the Antarctic winter is well known, but many small birds will huddle at roost to reduce night-time heat loss. Selection of a suitably sheltered roost site is also important, and many species cache food for times of scarcity. Behavioural adjustments can be seen as an important first line of defence, but these are often insufficient on their own and a physiological response is also needed.

A critical aspect of heat loss is insulation and many endotherms living in cold environments have enhanced insulation. Small birds living at high latitudes have an increased density of feathering compared with related species from lower latitudes, and many polar mammals have longer, denser fur in winter. Associated with greater insulation is an increase in the rate of obligate thermogenesis. This is achieved by increasing the size and metabolic activity of visceral organs, and many birds and mammals show an increase in resting metabolic rate in winter (Table 10.6).

Sometimes, however, this increase in resting metabolism is not enough, and extra sources of heat need to be used. The most important of these are shivering and, in some mammals, brown adipose tissue.

10.9.1 Shivering

Shivering is a widespread response to the need to generate extra heat when a mammal or bird finds itself in a cold environment, and below its thermoneutral zone. In shivering the muscles contract in such a way that essentially all of the chemical energy of ATP is converted into heat. True shivering occurs only in mammals and birds (although the regular muscular contractions of incubating pythons might also be viewed as shivering).

Shivering is initiated by the same motor neurons that govern voluntary muscle contraction. However whereas in voluntary muscle contraction the control emanates from the motor nuclei of the central nervous system (CNS), in shivering the commands originate in thermosensitive and integrating parts of the CNS in the hypothalamus and cerebellum. In mammals, shivering is initiated when cooling is detected by one of several thermally sensitive parts of the body. This might be cooling of the skin, cooling of the internal organs, or of the hypothalamus, midbrain brain stem or the spinal cord. In birds the initiation is different, and skin receptors have the dominant role. The intensity of shivering is graded; it starts with increased muscle tone, through microvibrations to tremor of flexor and extensor muscles[35].

10.9.2 Substitution

Although birds living below their TNZ need to generate heat from shivering, they also need to move about to feed. Indeed anyone who has watched small birds in northern winter forests will know they spend much of their day searching ceaselessly for food.

One of the most studied species in this regard is the black-capped chickadee (*Poecile atricapillus*) of North America. This tiny bird is a year-round resident of the mixed deciduous and spruce taiga forest; while most of the species with which they share the forest in summer leave for warmer climes in winter, the chickadees and a few other species stay put. They feed during the day to build up fat reserves which are then utilised overnight. Studies in Alaska by Brina Kessel showed that the intensity of feeding activity is related to temperature: the colder the weather, the more food is needed to get through the night. This intense winter feeding activity leads to a daily energy expenditure which can exceed that in the breeding season[36].

The increased muscular activity as the chickadees search for food generates heat. Just as everyday activity in summer can reduce the need for obligate thermogenesis, so heat generated by activity in winter can reduce the requirement for shivering or enhanced heat production by the visceral organs. This has been termed '*activity-thermoregulatory heat substitution*' (or '*substitution*' for short). The principle is precisely the same whether heat generated by routine muscular activity is reducing the need for obligate thermogenesis when the organism is within its TNZ, or reducing the need for facultative thermogenesis when it is below its TNZ. However it is only the latter that is termed 'substitution'.

Substitution has been shown to be widespread in mammals and birds[37]. It is sometimes regarded as providing activity for free, in that the heat generated by this activity reduces the need for facultative thermogenesis. It is perhaps better to view this as a feature of the more general balancing of demands for energy that characterises all organisms at all times. For a chickadee in winter, the imperative is to gather sufficient food to see it safely through the night. It does not matter whether its body temperature during the day is maintained by shivering, an elevated rate of resting metabolism or as a by-product of foraging activity. The trade-off in winter is that time spent foraging increases the chances of finding food, decreases the demand for facultative thermogenesis, but also increases the chances of being predated. It is the balance between these trade-offs that dictates the chickadee's pattern of foraging activity, thermogenesis and diurnal variation in lipid storage; other factors that influence this balance include a diurnal variation in body temperature, caching of food for later use and selection of a suitable roosting site. Juggling energy in winter is more than a simple activity/thermogenesis trade-off, and this emphasises the need to view an organism's thermal physiology in the context of its ecology as a whole, not just the immediate physical environment[38].

10.9.3 Brown adipose tissue

Brown adipose tissue (BAT, also called brown fat) is a specialised tissue found only in placental mammals. It is known from a range of hibernators, small rodents living in cold habitats and in the new-born of many species. The range of species from which BAT has been described is growing, and it seems reasonable to suppose that it is very widely distributed in hibernators and small mammals living in polar or alpine regions. However we do know that BAT is absent from monotremes and marsupials, and it is also unknown from birds[39].

Brown adipose tissue is distinctive, and was once known as the 'hibernating gland'. We now know that BAT does not secrete hormones or other chemicals into the general circulation, and so this inappropriate term has been abandoned. BAT is typically found between the shoulder blades, around the neck and close to the heart, larger blood vessels and lungs, positions which enable BAT to pass heat rapidly into the general circulation and thereby warm the body core rapidly.

The colour that is so distinctive of BAT, and which gives it its name, comes from a combination of the cytochrome in the very high density of mitochondria, and the haemoglobin carried in the high level of vascularisation. In contrast to white adipose tissue (normal fat), the intracellular lipid droplets are smaller and dispersed through the cell, rather than being a single large globule. BAT generates heat without shivering, by a process often referred to as non-shivering thermogenesis (NST). This is something of a misnomer, since the obligate background generation of heat in endotherms is also a form of non-shivering thermogenesis[40].

The key feature of BAT is the expression of the uncoupling protein UCP1. UCP1 provides a channel in the inner mitochondrial membrane which allows protons to leak back without being coupled to ATP synthesis, and their potential energy is dissipated as heat (see Box 10.1). UCP1 is found only in brown adipose tissue, whereas other uncoupling proteins are found more widely in tissue. The generation of heat by BAT is under tight regulatory control. This and shivering are the two key forms of facultative thermogenesis, and they do not form part of the mechanisms of routine (obligate) thermogenesis.

10.10 Keeping cool: endotherms above the TNZ

Once an organism is above its TNZ and can no longer maintain its body temperature within the boundaries of viability, it must lose heat or die.

For endotherms living in deserts or other hot environments behavioural adjustments are the first line of defence. For small mammals the heat of the day can be avoided by retreating to a burrow and emerging to feed at night. Larger mammals utilise shade, where this is available, or adjust their posture to minimise heating by solar radiation. While shade is invaluable in reducing heat load, it usually does not provide food. A mammal in a hot environment must therefore shuttle back and forth between areas for feeding and shade. As with the lizards discussed in Chapter 9 we see that the physiological response to a thermal challenge has to be understood within a wider behavioural context.

The only physical mechanism an organism has to achieve heat loss is evaporation of water. The energy (latent heat) needed for evaporation is supplied by the organism, whose internal energy and body temperature thus fall. For this to be effective, the skin surface from which water is evaporating must be well supplied with blood vessels, so that the blood will be cooled before being returned to the body core.

The latent heat of vapourisation of water is high (Chapter 2), so evaporation is a very effective means of dissipating heat. It does, however, involve a loss of body water. In deserts with the twin challenges of high temperature and low water availability, the loss of water through evaporation can be a significant problem. For this reason, desert organisms use evaporative heat loss for alleviating acute short-term heat stress only when behavioural solutions such as changing posture or seeking shade are longer sufficient, or are not possible.

The driving force for evaporation is the difference in vapour pressure between the skin surface and the adjacent air. This means that a key factor in evaporative cooling is movement of air across the skin surface so that the air humidified by evaporation is moved away to allow evaporation to continue.

The mechanism of evaporative heat loss varies with species. Some sweat, some pant and others spread saliva over their body surface.

10.10.1 Sweating

Sweating involves the secretion of hypotonic fluid onto the skin. Sweat glands are found in most mammal groups, though their importance and distribution varies with species. For example they are present all over the skin in primates, being especially well developed in man, whereas in dogs and cats they are confined to the pads of the feet. The physiology of sweating has been studied most in man and domesticated mammals. Sweat glands appear to be absent in marine mammals, and are completely absent from birds. Despite the covering of feathers and the absence of sweat glands, loss of water through the skin is still an important component of fluid balance in birds.

There are two main types of sweat gland, and these differ in their size, distribution on the skin and mechanism of secretion. Eccrine sweat glands are distributed generally all over the body in primates, though in other groups they are confined to the palms of the hand and the soles of the feet. They are small and they open directly onto the skin surface. The sweat they produce is derived from blood plasma, and hence it contains some electrolytes (which is why sweat tastes salty). Their primary function is thermoregulation through evaporation, but the sweat also helps prevent colonisation of the skin by pathogenic organisms such as bacteria and fungi. Apocrine sweat glands have a very limited distribution in man, being confined to the armpit, genital area, outer ear canal and eyelids. They are larger than the eccrine glands, and open into the pilary canal of the hair follicle. The fluid produced by apocrine glands is more concentrated than eccrine sweat, and contains pheromones[41].

The secretion of sweat is under the control of the sympathetic nervous system. It is controlled primarily by brain temperature, but also modulated by skin temperature. This is sometimes expressed as 'mean body temperature', which is a weighted sum of internal and average skin temperatures.

10.10.2 Panting and gular fluttering

When mammals and birds breathe, the incoming air is generally cooler and less humid than the expired air. This incoming air is warmed and humidified as it passes along the respiratory passages. The water that humidifies the inspired air evaporates from the moist membranes of the respiratory passages; evaporation of water during breathing is thus inevitable, and all mammals and birds lose heat this way. Some water condenses onto the surfaces of the respiratory turbinates during expiry. While this conserves both heat and water, there is a small net loss of both during normal breathing.

The loss of heat can, however, be increased by panting, which is a controlled increase in the frequency of breathing, coupled with a reduction in the tidal flow. The rate of panting can greatly exceed the rate of breathing under normal resting conditions; this results in an increase in the evaporation of water but runs the risk of respiratory alkalosis (an abnormally high pH of the extracellular fluids caused by removal of too much CO_2). The likelihood of alkalosis is minimised by a reduction in tidal flow, so that much of the increased air-flow does not involve an increase in gas exchange. The section of the respiratory tract involved in panting can be enriched in arterial blood vessels, which increases the loss of heat; a well-characterised example here is the rock pigeon *Columba livia*.

There is a metabolic cost to panting because the increased flow of air has to be achieved by muscular work. There is, however, some evidence that this cost is minimised by matching the frequency of panting to the resonant frequency of the respiratory passages: in the 1960s Eugene Crawford showed that domestic dogs induced to overheat panted at a frequency of 5.33 Hz, and the resonant frequency of their respiratory system was 5.28 Hz. He was also able to show that the pigeon *Columba livia* panted at the resonant frequency of their respiratory system (in this case ~ 10 Hz), suggesting that this is a general mechanism for minimising the energy cost of panting[42]. In mammals there is a general tendency for panting to be more important in larger species, and sweating to be more important in smaller species.

Birds also pant, but in addition some groups of non-passerines can supplement evaporative water loss through panting by using gular fluttering. Gular fluttering is achieved by moving the hyoid bone and allowing the pharynx to flap, and results in substantial evaporation of water from the upper respiratory tract. The rate of gular fluttering appears to be independent of external temperature and body temperature. It does, however, vary across species of different size: in the brown pelican, *Pelecanus occidentalis* (3.13 kg), it is 230–290 Hz; in double-crested cormorant, *Phalacrocorax auritus* (1.34 kg), it is 645–730 Hz and in the very much smaller mourning dove, *Zenaida macroura* (150 g), it is 680–735 Hz. However, the duration and amplitude of gular fluttering does increase with heat load. In some species, gular fluttering is synchronous with panting, in others it is decoupled. It may be that these patterns are related to the resonant frequency of the gular region, or the entire thoracic/abdominal respiratory exchange system[43].

10.10.3 Saliva spreading and other mechanisms

In mammals the spreading of saliva onto the fur is another means of using evaporation to lose heat. This is less effective than sweating, because the fur must be thoroughly wetted right through to the skin for the loss of heat to be effective in cooling the blood. First reported from rats, this has been most studied in kangaroos[44].

A recent study of the inca dove, *Columbina inca*, showed that, above an ambient temperature of 42 °C, evaporation of water from the cloaca was a significant contribution to heat balance and dissipated on average 150 mW. At this temperature the total evaporation of water was 53% through the skin, 25.4% through the respiratory tract and 21% from the cloaca. Cloacal evaporation was also shown to be important in the Eurasian quail, *Coturnix coturnix*[45]. Finally, some storks and vultures defecate onto their bare legs, which are then cooled by evaporative water loss.

10.11 The evolution of endothermy

A question which has intrigued ecologists, physiologists and palaeobiologists for over a century is how such an expensive lifestyle as endothermy could ever have evolved. Sadly, fossils preserve little direct evidence of physiology, although as we shall see, there is valuable secondary evidence. One way to start exploring this question is to compare the ecology and thermal physiology of living reptiles, birds and mammals, and see what ecological advantages that endothermy provides (Table 10.7).

This approach can be informative, but the advantages provided by endothermy now do not necessarily tell us why it evolved in the first place. Furthermore, endothermy evolved independently in mammals and birds, and we cannot assume that the selection pressures were the same in each case. Although there is a long history of speculation, the modern debate over the origin of endothermy was defined by a trio of important papers published in the late 1970s[46].

In 1978 Alfred Crompton and colleagues proposed that endothermy in mammals was achieved in two steps. The first was the acquisition of the ability to maintain a more or less constant body temperature, but lower than in modern mammals to avoid the need for evaporative cooling. This enabled the early mammals to become nocturnal

Table 10.7 Some benefits of endothermy[47].

Model	Proposed ecological or physiological benefit
Optimisation of enzyme activity	Warm and stable thermal environment for enzyme activity.
Niche expansion	Capacity for activity independent of environmental temperature, allowing nocturnal and crepuscular activity.
Aerobic capacity	Sustained aerobic activity.
Parental care	Efficient gestation and lactation, facilitating more rapid reproduction.
Parental care	Increased assimilation capacity, facilitating parental care.
Complexity	Warm and stable thermal environment allows for increased complexity, especially in the nervous system.
Brain size	Increased metabolic rate allows for an increase in brain size.
Stoichiometry	Increased metabolic rate allows burning of excess carbon from herbivorous diet low in nitrogen.
Pathogen resistance	High body temperature protects against fungal pathogens.

and take advantage of what was believed to be an underutilised niche. A body temperature of about 30 °C (typical of monotremes and placentals such as tenrecs today) would have allowed for fine control of body temperature through regulation of heat loss, and it was assumed that this was aided by the presence of fur. The second stage was the evolution of a higher metabolic rate and body temperature in taxa that returned to diurnal activity. Under this scenario a high body temperature maintained by true endothermy evolved after the therapsid to mammal transition rather than before.

Also in 1978, Brian McNab proposed that the critical factor in the evolution of mammalian endothermy was the reduction in the body size of some therapsids around the close of the Permian[48]. He argued that ancestral reptiles of body mass 30–100 kg would have been inertial homeotherms and that body temperatures would have been relatively high. The critical step was the subsequent evolution of small size: as size decreased, the physiological advantages of a warm body could have been retained by an increase in mass-specific metabolic rate. This would have required some form of insulation to retain metabolic heat, but this combination could have converted inertial homeothermy to true endothermy. The evolution of endothermy allowed the smaller proto-mammals to be active by night and thus avoid competition from larger, diurnal, reptiles. This scenario places the evolution of endothermy in the therapsid ancestors of mammals, earlier than proposed by Alfred Crompton's nocturnal niche hypothesis. An interesting feature of this hypothesis is its emphasis on endothermy having evolved initially to maintain an ancestral high body temperature, rather than to increase it.

The third proposal came from the physiologist Al Bennett and the palaeobiologist John Ruben in 1979, and was quite different. They suggested that the key factor in the evolution of endothermy was selection not for a warm body but for an enhanced aerobic capacity.

In developing their argument, Bennett and Ruben drew on earlier studies which showed that the net cost of transport is similar in terrestrial lizards and mammals. They also showed that (as we saw in Chapter 8) the mean factorial aerobic scope (the ratio of sustained to basal metabolic rate) is broadly similar in ectothermic vertebrates (fish, amphibians and reptiles), mammals and birds. The critical difference comes in stamina: mammals and birds can sustain their active metabolic rate for much longer then ectotherms, all of which tire rapidly. This difference in stamina arises because the basal or resting metabolic rate of endotherms is much higher than that of ectotherms (Figure 10.8). A similar factorial scope thus leads to a much greater absolute scope in endotherms: a mammal or a bird has much more energy available to fuel sustained aerobic performance than does a reptile.

Bennett and Ruben concluded that selection had acted to enhance aerobic capacity to support sustained locomotor activity, and that resting metabolic rate increased in consequence. This assumes a functional link (and hence a correlation) between basal and sustained metabolic rates, though the precise mechanism was not discussed (Box 10.3). While they noted that both maximal oxygen consumption and absolute aerobic scope are greater at warmer temperatures, Bennett and Ruben argued explicitly that a warmer body and endothermy were secondary consequences of selection for enhanced aerobic scope.

The Bennett and Ruben hypothesis, usually referred to as the *aerobic scope hypothesis*, is quite distinct from those built around the selective energetic advantages for an increased body temperature per se. It also distinguishes clearly between the different roles of aerobic scope and body temperature in the evolution of endothermy.

Mammals and birds are notable for the extent of parental involvement in the raising of offspring, and in 2000 two new hypotheses were proposed, each centred on the selective advantage of enhanced parental care. Colleen Farmer suggested that a regulated high body temperature would have benefited incubation and an enhanced aerobic scope would have improved capability to provide food for the growing young. This is essentially a combination of two of the earlier hypotheses, but with the emphasis on parental care as the primary focus of selection. In the same year a quite different suggestion was made by Paweł Koteja, who proposed that selection was for an enhanced ability of adults to process food for provisioning young. He argued that this drove an enlargement of the

> **Box 10.3 Linking basal and sustained metabolic rate**
>
> The aerobic scope hypothesis proposes that selection for an increase in sustained aerobic performance would bring about a concomitant increase in basal (resting) metabolism (BMR). This prompted many researchers to test the hypothesis by looking for correlations between BMR and measures of sustained or maximum metabolic rate, and correlations have been found both within and between species in both mammals and birds. While these correlations have not been detected universally, the balance of current evidence points clearly to a correlation between basal and maximal aerobic metabolism: more active species tend to have higher factorial aerobic scopes, and also a higher resting metabolism[49].
>
> A difficulty with linking these correlations to the evolution of endothermy is that in living endotherms BMR is largely a function of internal organs (brain, heart, viscera) whereas active metabolism is largely a function of skeletal muscle activity. This might suggest there is no reason to expect a direct relationship between BMR and any measure of sustained aerobic metabolism (despite this being an implicit assumption in the original aerobic scope hypothesis).
>
> There are, however, a number of factors that might provide a functional link between BMR and active metabolism.
>
> The first is that any increase in the number of mitochondria will necessarily increase the cost of maintenance and hence BMR. The evolution of endothermy will also have involved parallel changes in a range of physiological systems within the body. These will likely have included an increased permeability of membranes to facilitate metabolite transport during exercise, with a corresponding increment in Na^+/K^+-ATPase costs at rest to maintain osmotic balance, an augmentation of the cardiovascular system, and possibly also an expansion of nervous system to coordinate movement. All of these have the consequence of higher resting costs.
>
> A final point is that there must be a limit to the extent to which the activity of an individual mitochondrion can be uprated. This limit might be set by internal constraints within the mitochondrion, or limitations to the supply of oxygen or nutrients. Whatever the nature of these constraints, the result would be a limit to the maximum factorial aerobic activity that could be sustained, and thus a more or less constant ratio between BMR and sustained metabolic rate (daily energy expenditure) across species. As we saw in Chapter 8, this is what we observe.

viscera, which resulted in an increase in metabolic rate. As we saw earlier, viscera are expensive to maintain and have an inherently higher BMR than other tissues. This increase in BMR would have initiated a positive feedback, resulting in higher BMR as well as improved provisioning of young[50]. These are intriguing ideas, but they have yet to displace the aerobic scope hypothesis as the most widely accepted theory for the evolution of endothermy.

Endothermy has two features, whose selective advantages may well differ: a relatively high body temperature and a stable internal environment. A stable body temperature is often regarded as allowing for the evolution of a closely integrated and fine-tuned physiology, and also the development of more complex neural processes, though it is difficult to know how these suggestions could be quantified or tested. Almost all terrestrial ectotherms exhibit significant variability in body temperature on a wide range of timescales and so can tell us little about the advantages of thermal stability, but insights may be gained from comparative studies of aquatic organisms living in stable or variable thermal environments[51].

The selective advantages of a high body temperature are easier to quantify: a warmer body allows for higher mitochondrial activity and a greater muscle power output, as well as more rapid digestion of food and faster growth. Perhaps here lies an important key to the selection pressures that drove the evolution of endothermy: the greater aerobic capacity and improved mechanical performance of warm bodies and muscles, allowing faster growth. This would suggest a subtle but important modification to the aerobic scope hypothesis: a warmer body was not a consequence of selection for an enhanced aerobic scope, rather it was the means by which that greater scope was achieved.

This modified aerobic scope hypothesis provides a clear rationale for the selective advantage of even a small increase in body temperature. Body temperature can, however, only increase if there is insulation such as fur, feathers or subcutaneous fat. Furthermore a steady incremental increase in body temperature and aerobic scope brings with it a requirement for enhanced ability to obtain oxygen from the air and deliver it to the tissues. As Tom Kemp has long emphasised, the evolution of an endothermic metabolism must have involved the integrated change in a whole suite of characters. Discussions of the evolution of endothermy often contrast 'thermoregulation first' with 'aerobic scope first', even though the original authors were often careful to point out that an endotherm physiology represents a complex and coordinated suite of physiological and anatomical features, at least some of which must have evolved in parallel[52].

A conceptual model of the modified aerobic scope hypothesis is shown in Figure 10.14. The key features are that it shows how there are physiological and ecological benefits from even a small increase in body temperature, and the presence of a positive feedback loop that will drive body temperature upwards until it reaches a point where metabolic costs outweigh the ecological advantages. The key feature is that insulation must have been present from the start for the positive feedback to operate.

10.11.1 The fossil record of endothermy

The fossil record is frustratingly equivocal when it comes to the evolution of endothermy. One thing it does tell us very clearly, however, is that the evolution of endothermy in birds and mammals were separate events.

For the origin of endothermy in mammals, we need to look at the early synapsids ('mammal-like reptiles' in the older literature). It seems clear that the pelycosaurs of the Early Permian were ectothermic, particularly those such as *Dimetrodon* and *Edaphosaurus* whose dorsal sails appear to have been primarily for thermoregulation. Their size would have damped diurnal variations in body temperature and pelycosaurs could be viewed as the first evolutionary experiment with maintaining a stable warm body. Intriguingly, one group (Edaphosaurids) also appear to be the first terrestrial vertebrates to exhibit widespread herbivory. The pelycosaurs had a somewhat sprawling posture, but many later synapsids reveal a distinct change to an upright posture, that is with the legs held vertically beneath the body rather than splayed out to the side. This has been taken to indicate a high metabolic rate but the correlation between posture and metabolism is very loose and posture alone is very weak evidence for metabolic status. The widespread presence in synapsids of a secondary palate, which allows for continuous ventilation of the lungs (that is, eating and breathing at the same time), has also been taken as indicating partial or complete endothermy. However the secondary palate has a range of functions in mammals, and so the evidence here is again equivocal. The strongest anatomical evidence for elevated metabolic rates in the synapsids comes from the presence of respiratory turbinates in therocephalians

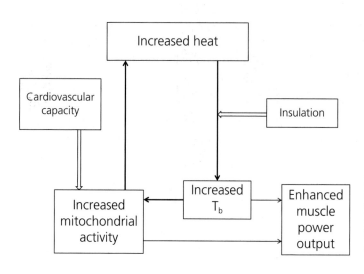

Figure 10.14 A modified aerobic scope hypothesis for the evolution of endothermy. Selection for increased mitochondrial activity (metabolic power) in the presence of insulation to retain heat causes an increase in body temperature leading to an increased aerobic scope and muscle power output[53].

from the Late Permian. Turbinates are essential in living mammals for heat exchange and regulation of evaporative water loss during respiration and their presence implies a higher metabolic rate.

The earliest true mammals such as *Morganucodon* were very small, and their anatomy suggests they were active predators of small invertebrates, probably arboreal and almost certainly nocturnal. Given the tough challenge of maintaining a constant body temperature at such a small size, it seems reasonable to conclude that these early mammals must have had fur. Evidence for fur in the fossil record is almost non-existent, because, being a protein, fur generally does not fossilise. Some recent reports, however, suggest the possibility that fur extends back to at least the Jurassic[54]. We do not know precisely when mammalian endothermy evolved, but if it pre-dated the therapsid–mammal transition, as is indicated by the fossil record of respiratory turbinates and the presence of a secondary palate, this places the evolutionary event at least as far back as the Permian[55].

The timing of the evolution of endothermy in birds is even more difficult to pin down than in mammals. Although once controversial, it is now firmly established that birds evolved from coelurosaurian dinosaurs[56]. Biophysical modelling has suggested that even during the warmest parts of the Mesozoic, small active dinosaurs such as these would have been subject to diurnal and seasonal changes in body temperature at most latitudes and would therefore have needed a eurythermal physiology[57]. They would have required some form of insulation to retain metabolic heat, and while direct evidence for feathers in fossil birds had long been known from *Archaeopteryx*, until recently there was little other fossil evidence to indicate when endothermy might have evolved in the theropod/bird lineage.

Everything changed, however, with the discovery of a range of exquisitely preserved small dinosaurs and early birds in China. These fossils have revealed a range of integumental structures in a wide variety of small birds, pre-avian and non-avian dinosaurs. Nothing makes the link between small dinosaurs and birds so vivid as to see one of these delicate fossils with a complete array of feathers preserved in the rock. Feathers or feather-like structures are now known from a rapidly increasing range of taxa spread over all the main lineages most closely related to birds. Modern techniques can even give us tantalising hints as to what colour these feathers might have been. While the precise nature and function of these structures is debated in individual taxa, the general indication is that the potential for an insulating layer arose early in the lineage that gave rise to birds. It also possible that feathers or a feather-like covering was present in a range of other theropod taxa, raising the intriguing possibility that endothermy was widespread among saurischian dinosaurs[58].

Current evidence thus points to endothermy having evolved in the theropod/bird lineage in the Mesozoic (Jurassic or Cretaceous); sadly we can be no more specific than that with current knowledge. This timing does, however, raise the question as to whether other dinosaur lineages retained endothermy from an endothermic common ancestor, or even evolved it independently.

10.11.2 Endothermy elsewhere

Dinosaurs were long regarded, at least amongst the general public and the creators of many museum displays, as cold, slow and lumbering. It was the discovery of *Deinonychus* that changed all that and sparked what is sometimes called the dinosaur renaissance or dinosaur revolution. This captured the public imagination and led to speculation that all dinosaurs might have been endothermic (Box 10.4).

Dinosaurs were the dominant terrestrial vertebrates throughout the Mesozoic. They were hugely variable in size, shape and lifestyle, and they were found from tropics to poles. With this diversity of ecologies came a diversity of physiologies. Some dinosaurs appear to have remained resolutely ectothermic (for example ankylosaurs) much like modern reptiles. The giant sauropods would have been inertial homeotherms, and warm (as confirmed by isotope palaeothermometry, see Chapter 3). Modelling of heat flow suggests that these giants could not have been endothermic for they would have overheated fatally. They represent a unique physiology and ecology, now sadly lost.

Although the metabolic status of dinosaurs has attracted much scientific attention no consensus has

Box 10.4 Hot-blooded dinosaurs

Dinosaurs are iconic animals, and those best known to the general public are probably *Diplodocus* and *Tyrannosaurus*, skeletons or reconstructions of which can be seen in many museums. In the history of dinosaur physiology, however, the icon is, without doubt, *Deinonychus*.

In August 1964, John Ostrom and his assistant Grant Meyer were exploring south-central Montana and found a small dinosaur like no other previously known, *Deinonychus* ('terrible claw'). Over the next three seasons they unearthed three individuals. Working back in the Peabody Museum at Yale, Ostrom reconstructed *Deinonychus* as a small (~80 kg), lightly built bipedal dinosaur with a posture not unlike a modern-day ostrich: the body trunk held horizontal with the neck curving upwards and the rigid tail extending out behind for counterbalance. The skeleton of the arms indicated that *Deinonychus* was able to hold and manipulate objects, and that the slashing claw that gave the animal its scientific name could be held off the ground during locomotion. Everything pointed to a mobile, active and intelligent predator, quite at odds with the popular view of dinosaurs as sluggish, scaled-up crocodiles (the scientific view at the time was more nuanced).

Assisting on the 1964 expedition was a Yale student, Robert T. (Bob) Bakker, destined to become the *enfant terrible* of vertebrate palaeontology. His first professional publication, an article in *Discovery* (a Peabody Museum journal), was entitled '*The superiority of dinosaurs*'. This was not so much a formal scientific paper as a call to arms for a new look at dinosaur physiology. Bakker is a talented artist, and his reconstruction of *Deinonychus* as an alert, dashing predator said everything about how he felt dinosaurs should be viewed. Based on this reconstruction, Ostrom and Bakker argued that *Deinonychus* and its relatives must have had an endothermic metabolism akin to modern birds and mammals.

Bakker elaborated his ideas in a popular book, liberally illustrated with his charismatic drawings. The idea of 'hot-blooded dinosaurs' rapidly captured the public imagination and became highly controversial. Arguments for and against were aired at the Annual Meeting of the American Association for the Advancement of Science held in Washington in 1978, and published as '*A cold look at the warm-blooded dinosaurs*'. The argument was never resolved, for it could not be. Almost 40 years on, we now recognise that dinosaurs exhibited a range of lifestyles: large and small, herbivorous and carnivorous, fast and slow and with a diversity of physiologies to match. This shift in our view of dinosaurs was ushered in by the careful and meticulous Ostrom, the maverick and evangelical Bakker, and *Deinonychus*[59].

Figure 10.15 Bob Bakker's reconstruction of *Deinonychus* at full speed, which appeared in John Ostrom's monograph describing the species. Modern evidence suggests *Deinonychus* may have been feathered.

emerged. Once again the problem is deciding what features of a fossil give an indication of the animal's physiology when alive. As with mammals, the clearest indications come from the presence of insulation and respiratory turbinates. Nasal anatomy suggests that neither the non-avian dinosaurs *Nanotyrannus* and *Ornithomimus* nor *Deinonychus* and the maniraptoran *Dromaeosaurus* were endothermic. In contrast, the less equivocal evidence from bone histology has been interpreted as indicating that the high growth rate and metabolic rates that characterise living birds may have evolved earlier than the common ancestor or birds and pterosaurs, possibly in the Early Triassic[60]. There is clearly much to resolve in the extent and timing of the evolution of endothermy in the theropod/bird lineage.

A particularly interesting question is whether pterosaurs were endothermic. It has been recognised for a long time that at least some pterosaurs possessed integumental structures resembling fur, and it is now known that they possessed the components of a bird-like respiratory system, including air sacs. As flying vertebrates, their metabolic status has also been argued intensely. At present the evidence is equivocal, but it remains an intriguing possibility that pterosaurs represent a third independent evolution of endothermy in a lineage now extinct[61].

10.12 Summary

1. Endothermy is the maintenance of a high and relatively constant internal body temperature, where the principal source of heat is a high metabolic rate at rest. The main sources of this heat are the visceral organs (especially the liver, spleen and gut), which tend to be larger and with greater metabolic capacity than in ectotherms. An important contribution also comes from heat produced by muscular activity during routine daily activity. Among living animals, only mammals and birds are true endotherms.
2. A range of reptiles, fish, insects and even plants can generate heat in particular tissues for limited periods. These are not endotherms, and are better referred to as heterotherms.
3. Body temperatures are generally higher in birds than in mammals, and in both groups mean body temperature varies with lineage, environmental temperature and diet.
4. Within the thermoneutral zone (TNZ) endotherms regulate their body temperature by controlling the loss of sensible heat. Below the TNZ, endotherms generate extra heat by uprating the metabolic rate of viscera, shivering, increased activity and, in some mammals, switching on a specialised heat-generating tissue (brown adipose tissue, BAT). Above the TNZ, endotherms lose heat by evaporation of water (panting, sweating in mammals, gular fluttering in some birds).
5. Endotherms vary their insulation seasonally and depending on climate. The physiological aspects of endothermy have to be viewed not in isolation, but in the context of behavioural trade-offs relating to other aspects of their energetics and ecology.
6. Endothermy evolved independently in mammals and birds, but the precise timing of its evolution is not clear in either lineage.

Notes

1. This quotation comes from the introduction to the first edition of Arthur Ramsay's book *Physiological approach to the lower animals* (Ramsay 1952). The context for this statement is that at that time biology was taught almost exclusively from the perspective of the mammal; Ramsay's aim was to introduce general principles of animal physiology as a background to exploring the biology of invertebrates.
2. Note that we have slipped into a more informal use of language here. Recall from Chapter 2 that heat is what we call energy when it is flowing. Strictly body temperature is a consequence of the internal energy content, reflecting the balance between those processes adding or removing energy. For our purposes the more familiar language will suffice, though we should recognise that we are being less than rigorous.
3. Terminology in science can often be confusing, especially when similar or identical terms are used for different things. An endothermic animal is one that generates its own heat metabolically and dissipates that heat to the environment, whereas an endothermic chemical reaction is one that absorbs heat from the environment (in contrast to an exothermic reaction, where heat is released, sometimes explosively).

4. For original sources see Cowles (1940, 1962) and Bergmann (1847). The threshold variability of 2 K to define the boundary between homeothermy and poikilothermy was suggested by Bligh & Johnson (1973), and the same threshold was suggested by Pough & Gans (1982) to define the boundary between endothermy and heterothermy (in the sense of endotherms that allow extremities to cool).
5. Brian McNab (2002) regards heterothermy as a *'nearly useless term'* for *'a heterogeneous grab bag of species'*. I agree completely. But abandoning it means we need new terms both for the 'facultative endothermy' (an equally unhelpful term) of insects and possibly for those mammals and birds that undergo localised cooling of extremities or which go into torpor. In this book I restrict use of the term insects, and possibly also for those mammals and birds that undergo localised cooling of extremities or which enter torpor. In this chapter I have restricted use of the term to those ectotherms that use a variety of mechanisms for local warming of tissues.
6. Wegner et al. (2015).
7. Plotted with data taken from Clarke & Rothery (2008).
8. A key problem in examining the role of size in ecology is that the frequency distribution of body size in mammals and birds is highly skewed: most mammals and birds are small and only a few are large. This skewness violates a key assumption underpinning the use of regression analysis, and so the body mass values need to be transformed before analysis; the usual transformation used by ecologists in these circumstances is logarithmic. The median body mass of 9991 bird taxa (over 90% of known species) is 16.4 g, and the geometric mean size 51.5 g (data downloaded from https://ag.purdue.edu/fnr/Documents/WeightBookUpdate.pdf, updating original analysis by Blackburn & Gaston 1994). For 3841 mammals (roughly 70% of known species) the median body mass is 89 g and the geometric mean 197 g (data from PanTHERIA database). The modal (most common) size is smaller: ~11 g for birds and ~33 g for mammals.
9. Modified from Clarke & Rothery (2008). In birds females and males often have different body temperatures, especially in species with marked sexual dimorphism in size. For 174 species the mean difference (female–male) was 0.12 K (SD 0.67). Where there was more than one data point for the T_b of a particular species, the higher value was used. The data for these plots were collated from published studies in the literature. Large data sets such as these are, however, not without their problems. In the case of body temperature, the data were often collected incidentally to the main focus of the study, and the measurement techniques may not have been ideal. While the data can be scrutinised carefully to remove erroneous data, more subtle problems may remain buried in the data to bias the overall picture.
10. The differences across these traditional groupings are statistically significant (one-way ANOVA with group as fixed factor). Mammals: $F_{593,2} = 31.3$, $p < 0.001$. Birds: $F_{482,1} = 170.8$, $p < 0.001$.
11. Plotted with data taken from Clarke & Rothery (2008) after updating the taxonomy.
12. Ordinary least-squares regression of the data suggests that in both mammals and birds, body temperature decreases with increasing body mass (both $p < 0.05$), but when allowance is made for taxonomic structure (as a proxy for phylogenetic effects), the scaling becomes non-significant in mammals ($p = 0.40$) but is almost significant in birds ($p = 0.052$). For a detailed analysis see Clarke & Rothery (2008).
13. For example in 1578, Johannis Hasler, a physician of Berne, tabulated the *natural degree of temperature of each man, as determined by his age, the time of year, the elevation of the pole and other influences*. By *elevation of the pole* Hasler meant latitude, and it was widely believed at the time that people living in the tropics had higher body temperatures than those living at higher latitudes. See Middleton (1966).
14. Modified from Clarke et al. (2010). Mean environmental temperature was calculated by averaging annual temperature over the range of each species; data downloaded from the 2008 version of the PanTHERIA database.
15. The seminal work here is by Brian McNab. The primary literature is McNab (1992, 2008), and there are valuable general discussions in two textbooks (McNab 2002, 2012). The statistical analysis including body temperature is reported in Clarke et al. (2010) and Clarke & O'Connor (2014).
16. Modified from Clarke & O'Connor (2014).
17. Body temperature and diet in reptiles is discussed by Pough (1973), Cooper & Vitt (2002) and Espinoza et al. (2004), and in fish by Horn (1989). The influence of body temperature on digestion in ectotherms is discussed by Lang (1979), Gaines & Lubchenco (1982), Wang et al. (2002), Floeter et al. (2005) and Petersen et al. (2011). Temperature and herbivory in insects is discussed by Wilf & Labandeira (1999), Wilf et al. (2001) and Currano et al. (2008, 2010).
18. Note that this is strictly the specific heat capacity at constant pressure, which increases sharply towards zero and more slowly towards 100 °C, with a broad minimum over the temperature range covering the body temperatures of most endotherms. Specific heat capacity at constant volume shows a quite different pattern as it decreases sharply with temperature.

19. This diagram is modified from Morowitz (1978). The turnover rate of a representative 'average' protein is taken from the thermal denaturation parameters for 20 predominantly mammalian proteins (Table 14–16 in Morowitz 1978). The body temperature data are the same as those plotted in Figure 10.2.
20. The classic demonstration of the difference in resting metabolic rate between ectotherms and endotherms is that of Hemmingsen (1950, 1960), although knowledge of this difference goes back at least to the eighteenth century Italian physiologist Lazzaro Spallanzani (1729–1799). A recent analysis by Craig White and colleagues has shown that vertebrate resting metabolism falls into two clear groups: a high level in mammals and birds and low levels in reptiles, amphibians and fish (White et al. 2006).
21. Data are for 634 species of mammal and 149 species of reptile (one data point per species), the plotted data are uncorrected for body temperature. Data from Clarke & Pörtner (2010) and Clarke et al. (2010).
22. The data in Table 10.5 were compiled from Rolfe & Brown (1997), Aschoff (1971), Munro (1969) and Pitts & Bullard (1968).
23. Data from Else & Hulbert (1981). Sample numbers for tissue mass comparison were 5 (*Mus*) and 18 (*Ctenophorus*), for mitochondrial volume density number of determinations ranged from 72 to 108. Mitochondrial surface area calculated from mean data for 18–102 determinations.
24. Plotted with data from Table 1 in Brand et al. (1991).
25. The involvement of UCPs in thermogenesis in birds is discussed by Raimbault et al. (2001), Talbot et al. (2004), Mozo et al. (2005); Teulier et al. (2010) and Rey et al. (2010) explore the role of UCPs in regulating reactive oxygen species in birds. Krauss et al. (2002) showed that UCP2 contributes significantly to basal proton leak in mouse thymocytes, and Hughes et al. (2009) discuss evolutionary aspects of the UCP family. UCPs in plants are discussed by Borecký et al. (2001) and Vercesi et al. (2006).
26. The comparison of mammals and reptiles was by Hulbert & Else (1989), and the analysis across mammals by Hulbert & Else (2005). The membrane pacemaker hypothesis was proposed by Hulbert & Else (1999, 2000). Intraspecific analyses were undertaken by Brzęk et al. (2007) and Haggerty et al. (2008).
27. Newman et al. (2013) discuss the role of muscle thermogenesis in the evolution of birds, and Rowland et al. (2015) review current knowledge.
28. The classic reference on the efficiency of muscular activity is Whipp & Wasserman (1969). Other useful discussions are Margaria (1976), Vogel (1988) and Pennycuick (1992). Margaria is authoritative, Vogel a delightful combination of penetrating insight with stylish writing, and Pennycuick succinct and to the point.
29. The studies of variation in BMR across species are Daan et al. (1990), Raichlen et al. (2010) and Wiersma et al. (2007, 2012). A recent study (Londoño et al. 2015) confirmed the lower BMR for tropical species (Panama and Peru), but did not measure organ masses. The within-species studies are Liu & Li (2006), Gallagher et al. (2006), Vézina et al. (2006), Huang et al. (2009), Konarzewski & Książek (2013) and Zheng et al. (2014).
30. See Wilson et al. (2011), van de Ven et al. (2013).
31. The thermoneutral zone (TNZ) can also be defined more formally as the range of ambient temperatures over which body temperature can be maintained solely by adjustments to sensible heat loss, without changes in metabolic heat production or in evaporative heat loss. The stipulation that there are no changes in evaporative heat loss within the TNZ is because such losses involve latent (as against sensible) heat.
32. The mouse data come from Zhu et al. (2008). The ptarmigan data are from West (1972). Both diagrams reproduced with the permission of Elsevier.
33. Plotted from data in Riek & Geiser (2013) and reproduced with the permission of Cambridge University Press. In this study phylogenetic analyses suggested that the relationship between TNZ breadth and body size was different for marsupials and placentals, though this involved forcing the relationships through the value assumed for the root of the phylogenetic tree. Analysis with a General Linear Model (which assumes that the relationship is determined entirely by physiology) suggested that there was no difference, and the line shown is fitted to all mammals. Analysis with a GLM suggested that the variation of T_{UC} and T_{LC} with mean environmental temperature were significantly different ($F = 4.02$, $p = 0.004$).
34. If you need convincing that a mammal or bird is regulating its temperature when within its TNZ, consider what would happen if it were to die. Obligate thermogenesis would cease, and the corpse would cool down (unless it were in a habitat where ambient temperature exceeds body temperature). This is because the organism is no longer producing heat, or adjusting its physiology to control heat loss. This is another example of where the terminology (jargon) of thermal ecology contains a trap for the unwary.
35. The term shivering has also been used to describe the muscular activity used by some insects to warm the thorax before flight. Although the process is similar in that muscular contractions are used for heat generation without external mechanical work, the nervous

control of this is very different in insects compared with vertebrate endotherms. The classic early paper on shivering in birds is West (1965); for mammals see Kleinebeckel & Klussmann (1990).
36. See Kessel (1976). More recent studies of daily energy expenditure in small birds wintering in northern forests are Cooper (2000), Sgueo et al. (2012) and Petit & Vézina (2014). Marchand (1987) gives a nice broad summary of winter ecology. For an excellent introduction to how birds and mammals cope with winter in northern forests see Heinrich (2003). In this beautifully written book, with the author's own delightful illustrations, Bernd Heinrich mixes careful science with engaging first-hand descriptions of the harsh winters of Maine and Vermont in the very best tradition of scientists who have kept a sense of wonder at the natural world.
37. A review by Humphries & Careau (2011) found evidence for substitution in 35 of 51 studies, and in 28 of 32 species examined. Individual studies of substitution include Bruinzeel & Piersma (1998) on red knot, Webster & Weathers (1990) on verdins, and Cooper & Sonsthagen (2007) on black-capped chickadees.
38. For discussions of the trade-offs involved in foraging see Houston et al. (1993), MacLeod et al. (2005) and Bonter et al. (2013).
39. A useful review of brown adipose tissue is Cannon & Nedergaard (2004). Hayward & Lisson (1992) showed clearly that BAT is absent from monotremes and marsupials.
40. In the cumbersome jargon of thermal physiology this is sometimes referred to as *non-shivering thermogenesis (thermoregulatory)*, to distinguish it from *non-shivering thermogenesis (obligatory)*. Thermoregulatory non-shivering thermogenesis is usually discussed solely in terms of brown fat (BAT), but this ignores the fact that many birds increase their resting metabolic rate in winter and birds do not have BAT.
41. Eccrine sweat glands are also known as atrichial (without hair) and apocrine sweat glands as apitrichial (associated with hair). Important early references are Nadel et al. (1971a, b) and Smiles et al. (1976).
42. See Crawford (1962) and Crawford & Kampe (1971).
43. Panting is discussed by Robertshaw (2006), and the water relations of desert birds by Schleucher et al. (1991) and Gerson et al. (2014). Gular fluttering appears to be known only from a few orders of birds: Galliformes (pheasants, partridges and allies), Anseriformes (geese, ducks and swans), Pelecaniformes (pelicans, gannets and boobies), Strigiformes (owls), Columbiformes (pigeons and doves) and Caprimulgiformes (nightjars and frogmouths). It is unknown in Passeriformes (songbirds). The key references here are all from the early pioneers of physiological ecology: Cowles & Dawson (1951), Lasiewski & Dawson (1964), Lasiewski & Bartholomew (1966), Bartholomew et al. (1968).
44. The initial study of saliva spreading was in rats (Hainsworth 1967); saliva spreading in kangaroos was first reported by Needham et al. (1974) and a recent study is Dawson et al. (2000).
45. See Hoffman et al. (2007).
46. The critical papers here are Crompton et al. (1978), McNab (1978) and Bennett & Ruben (1979). For earlier speculation see Martin (1903), Colbert et al. (1946) and Cowles (1946, 1958).
47. The key references here are Heinrich (1977), Crompton et al. (1978), Bennett & Ruben (1979), Avery (1979), Farmer (2000), Koteja (2000), Klaassen & Nolet (2008), Robert & Casadevall (2009) and Bergman & Casadevall (2010).
48. While some species did evolve smaller size at this time, the decrease in the mean body size of therapsids came from the extinction of larger taxa (Huttenlocker & Botha-Brink 2013, Huttenlocker 2014). Sookias et al. (2012) show that miniaturisation became more evident towards the end of the Triassic.
49. Correlations within species have been reported by Hayes & Garland (1995), Chappell & Backman (1995), Boily (2002), Sadowska et al. (2005) and Swanson et al. (2012), though significant correlations were not observed in all species. Correlations between species were reported by Dutenhoffer & Swanson (1996), Lovegrove (2000, 2004), Rezende et al. (2002) and White & Seymour (2004), though Hayes & Garland (1995) and Gomes et al. (2004) found only mixed support for a relationship, and Ricklefs et al. (1996) were unable to confirm the relationship for birds. Taigen (1983) discussed possible mechanisms for a functional correlation between BMR and active metabolism.
50. See Farmer (2000, 2003) and Koteja (2000, 2004).
51. Hans-Otto Pörtner has done this, and argues that there are important physiological differences between mitochondria isolated from organisms in stable (stenothermal) and variable (eurythermal) aquatic environments: see Pörtner (2002, 2004).
52. This modified aerobic scope hypothesis is presented by Clarke & Pörtner (2010). For thorough discussions of the integrated nature of the evolution of endothermy in mammals see Kemp (1982, 2006).
53. This diagram modified from Clarke & Pörtner (2010).
54. Evidence for fur in proto-mammals is discussed by Rougier et al. (2003) and Ji et al. (2006).
55. The incidence of respiratory turbinates in the fossil record of mammals and birds is discussed by

Hillenius (1992, 1994), Ruben (1995), Ruben et al. (2003), Hillenius & Ruben (2004) and Klaas et al. (2011). Their presence in early mammals is discussed by Lillegraven et al. (1991).

56. The evolutionary link between birds and dinosaurs was promulgated strongly by Huxley (1868, 1870). This view then fell out of favour, but was revived by John Ostrom (Ostrom 1973, 1975, 1976) and is now the mainstream view. The English scientist Thomas Henry Huxley (1825–1895) was most famous for his robust defence of Darwin.

57. See Seebacher (2003). For earlier discussions of thermal relationships of dinosaurs see Spotila et al. (1973), Spotila (1980), Dunham et al. (1989) and O'Connor & Dodson (1999).

58. For a thorough introduction to the new discoveries from China see Norell & Xu (2005), Chiappe (2007) and Chiappe & Qingjin (2016). The elucidation of the colour of *Sinosauropteryx* feathers is described by Zhang et al. (2010).

59. Ostrom later realised that a partial specimen of *Deinonychus* had been collected previously by Barnum Brown, but was never formally described and languished in the American Museum of Natural History. The original description of *Deinonychus* is Ostrom (1969), and the arguments for dinosaur endothermy are presented in Ostrom (1970, 1980), and the case was also put by Bakker (1968, 1972, 1980, 1986). Wilford (1985) provides a nice overview of the gradual change in our view of dinosaurs, and the proceedings of the AAAS symposium edited by Thomas & Olson (1980) give a full flavour of the arguments. Parsons (2001) provides a valuable historical perspective on the debate, and a modern view of dinosaur physiology is given by Fastkovsky & Weishampel (2009).

60. The bone histology evidence is discussed by Werning et al. (2012).

61. The evidence for fur in pterosaurs is discussed by Padian & Rayner (1993) and Lü (2002), and the presence of air-sacs by Butler et al. (2009).

CHAPTER 11

Torpor and hibernation

> The stark differences in overwintering biology within this one group of related animals shows that hibernation is less a strategy of avoiding the cold than of . . . avoiding famine.
>
> Bernd Heinrich[1]

In the previous chapter we explored the nature and ecological consequences of endothermy. In this chapter we will examine the circumstances under which mammals and birds may relax their commitment to a high body temperature and enter torpor.

In an endotherm, the energy to maintain a high body temperature comes principally from the heat dissipated during metabolism, but some also comes from the environment. Energy is lost through the skin and from evaporation of water during respiration. The body temperature is the result of a dynamic balance between these different flows of energy, and a number of factors can affect this balance. For example, the air is usually colder by night than by day, making the maintenance of a high body temperature (what physiologists often term 'defending a set-point') more difficult then. Indeed, many mammals and birds show a small diurnal (circadian) variation in body temperature, and a typical example is shown by the laughing-thrush *Leucodioptron canorum* (formerly *Garrulax canorus*) (Figure 11.1).

Passerine birds are visual predators. They cannot forage in darkness, so they pass the night in a roost where they can sleep. During this time they must metabolise reserves to maintain a high body temperature, and roosting birds use a wide variety of techniques for reducing these energetic costs, including the selection of a suitable warm or well-insulated roosting site, and in many small birds, huddling. Once *Leucodioptron* has gone to roost, it allows its body temperature to fall rapidly to a lower set-point, which is then maintained. In contrast, body mass decreases steadily through the night as fat reserves are used. In this study the mean circadian variation in temperature was thus ~2.6 K. In mammals and birds, allowing core temperature to drop in this way is usually referred to as *heterothermy* (Box 11.1).

Some birds, however, allow their bodies to cool to a much greater extent than *Leucodioptron*, and become torpid. Well known to aviculturists who kept hummingbirds, torpor was probably first recognised in captive birds in the 1830s by the father of American ornithology, Alexander Wilson. It was first identified in the field in South American hummingbirds, notably the Andean hillstar *Oreotrochilus estella*, and then in swifts, mousebirds and caprimulgids (nightjars and allies). These early studies showed that fully torpid hummingbirds were completely inactive and unresponsive, and that it took some time (typically 20–30 minutes) before they could resume normal activity. This makes the choice of roosting site absolutely critical, as a torpid hummingbird must be safe from predation[2].

Although torpor is widespread in hummingbirds, it is not universal. For example in western North America there are two sympatric species of hummingbird: the sedentary Anna's hummingbird, *Calypte anna*, and the highly migratory rufous hummingbird, *Selasphorus rufus*. These two species differ in the way they manage energy. Both species gain mass (principally fat stores) during the day, but to differing extents. *Calypte* regulates its energy intake behaviourally, feeding more and gaining

Principles of Thermal Ecology. Andrew Clarke, Oxford University Press (2017).
© Andrew Clarke 2017. DOI 10.1093/oso/9780199551668.001.0001

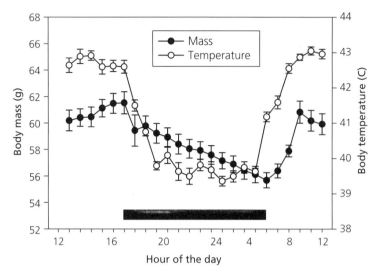

Figure 11.1 Diurnal variation in the body mass (black symbols) and body temperature (white symbols) of the laughing-thrush *Leucodioptron canorum* during winter. Data are plotted as mean and standard error (n = 9), and the black bar shows the period of darkness at the location in Zhejiang Province, eastern China, where the study was undertaken. The data shown are for winter; in summer the daytime warm period is longer, but the peak body mass is lower[35].

extra body mass on colder days. It then uses this fat to maintain a relatively high core temperature overnight, and only very occasionally enters torpor. In contrast *Selasphorus* uses torpor to reduce energy loss and depletion of fat stores overnight.

In northern forests species that live alongside one another have very different ways of coping with the challenge of winter. American red squirrels (*Tamiasciurus hudsonicus*) cache food in late summer and live off these stores in winter, whereas the Arctic ground squirrel *Spermophilus parryii* hibernates with the lowest body temperature ever recorded for a torpid mammal (−2.9 °C). In other areas of the Arctic, the habitat may be shared by obligate hibernators, facultative hibernators and food-caching non-hibernators.

In Africa there are many examples of species living in the same habitat but differing in their use of torpor. In western Madagascar the mouse-lemur *Microcebus myoxinus* uses daily torpor whereas the dwarf lemur *Cheirogaleus medius* in the same forest hibernates continuously for up to seven months, and in the arid grasslands of Namaqualand the elephant shrew *Elephantulus rupestris* makes extensive use of torpor, whereas the sympatric *E. edwardii* enters torpor only occasionally[3].

This exemplifies that even among closely related species living in the same place, the management of energy and the use of torpor can differ markedly. These studies also capture a general pattern: many birds and mammals (quite possibly all of them) show a circadian variation in body temperature. In some species, but not all, core body temperature is allowed to drop very low and the animal enters a torpid state. This poses the obvious questions of what torpor is, how can we measure it and why do not all species use it?

Box 11.1 Heterothermy revisited

Heterothermy is one of the most confusing terms in the thermal ecology literature. It has been used to describe those plants, insects and fish that warm specific tissues for limited periods of time, but it has mostly been used to describe mammals and birds that allow extremities to cool in order to retain heat (*regional heterothermy*), or which go into torpor and hibernation (*temporal heterothermy*). The latter use, often shortened to simply 'heterothermy', is so well established in the thermal literature that it will be difficult to dislodge, whatever reservations one might have over the use of the same term by different people to mean different things.

In the context of torpor and hibernation, heterothermy is a decrease in body temperature below that of a fully active individual (*euthermy*, also called *normothermy*). Note that euthermy is a different concept from *eurythermy*, which is the ability of an organism (typically an ectotherm) to tolerate a wide range of body temperatures.

11.1 Daily torpor

As always in science we need to start with a clear definition of what we are talking about. A useful definition of torpor comes from Fritz Geiser, who defined it as *a controlled reduction in body temperature and metabolic rate for less than 24 hours, accompanied by inactivity and absence of locomotion*[4]. The time stipulation differentiates daily torpor from hibernation, a distinction to which we will return in section 11.2.

This definition encompasses three elements of torpor, namely temperature, metabolic rate and activity. Clearly there are strong physiological contrasts between a slightly cool, active laughing-thrush, and a cold torpid hummingbird at roost. This poses the question of where on the spectrum between euthermy and deep torpor the key differences lie, and also whether we should define torpor in terms of temperature, metabolism or activity. Activity is easy to observe subjectively but often difficult to quantify in the field, and the measurement of metabolism generally requires bringing the animal into the laboratory (although this is changing rapidly with the development of devices to log activity and body temperature in free-ranging animals). It is therefore not surprising that quantitative studies of torpor have traditionally been framed in terms of body temperature. A number of approaches have been taken, including threshold temperatures and minimum temperatures (Box 11.2).

The onset of daily torpor is frequently defined by a threshold temperature, but there has been little consistency in the value of this threshold, literature values of which range from 26 °C to 36 °C. Field and experimental data indicate that different species enter torpor at different body temperatures, but there has been little experimental work linking the onset of physiological torpor to particular body temperatures. The small number of studies where body temperature and metabolic rate have been measured simultaneously during the entry into torpor in mammals suggest that if the threshold temperature is defined by the onset of a decline in metabolism, then in small species (< 70 g) the core temperature at this point increases with body mass (Figure 11.2)[5].

Defining the torpid state in terms of metabolic rate is physiologically rigorous, but is generally not suitable for use in the field. The advent of small temperature sensors that can monitor body temperature continuously in free-ranging mammals and birds has focussed attention on temperature thresholds as a practical measure of the onset of torpor. One proposal has been to determine the pattern of body temperature during euthermy, and define the onset of torpor as the point when temperature falls below this (for

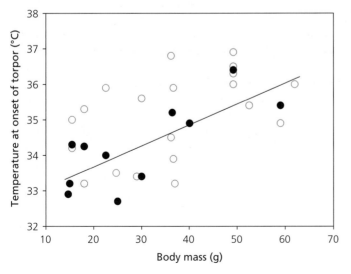

Figure 11.2 Relationship between body temperature at the onset of torpor and body mass in mammals of body mass < 70 g. The line is fitted to median data for each species (black symbols) and is significant ($p < 0.05$); the individual data points are also shown (white symbols)[36].

Box 11.2 Quantifying torpor

A variety of approaches have been taken to quantifying torpor, most of them based on body temperature (T_b)[34].

Minimum T_b. The lowest body temperature achieved during a bout of torpor.

Range (maximum T_b–minimum T_b).

Thermoregulatory scope (TS). The arithmetic difference between the mean T_b during daytime activity (eurythermy) and the minimum T_b achieved during torpor. This is also frequently referred to as the *range*.

Threshold T_b. An arbitrary T_b, above which the bird or mammal is assumed to be fully endothermic (*euthermic*), and below which it is regarded as having entered torpor. Its usefulness comes in allowing quantification of the time spent in torpor (Figure 11.3). The difficulty is that there is no agreed value for the threshold T_b.

Thermal sensitivity. For endotherms exposed to a range of air temperatures, this is the percentage change in core T_b per unit change in environmental temperature. This is of limited use for comparing different species (or even differently sized individuals of the same species) as it will be affected by thermal conductance.

Integration. This usually takes the form of a calculation of the period of time spent below the threshold T_b in degree-minutes or degree-hours.

Heterothermy index (HI). This is calculated as:

$$HI = \sqrt{\frac{\sum (T_{bopt} - T_{bi})^2}{(n-1)}}$$

This measure is analogous to a standard deviation. It was designed to allow comparison between different individuals of a species or between different studies. Here T_{bi} is the body temperature observed, and the optimal body temperature, $T_{b\,opt}$, is defined as 'the T_b that maximises fitness in a hypothetically ideal environment'. A difficulty here is that this has not been defined for any endotherm. The pragmatic alternative is to use the mean T_b that characterises normal daytime activity.

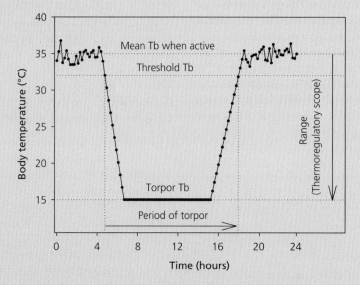

Figure 11.3 Measures of heterothermy for a single bout of night-time torpor in a hypothetical mammal with a mean body temperature during euthermy of 35 °C, an arbitrary threshold temperature of 32 °C and a body temperature during torpor of 15 °C. These numbers are arbitrary and chosen simply to illustrate the concepts.

example once it falls below the 95% or 99% confidence interval, assuming the small variation in core temperature during euthermy can be described well by a normal distribution). The end of the bout of torpor is defined as the point at which the threshold is crossed again during arousal (see Box 11.2). In a species with deep torpor, this can lead to a bimodal frequency distribution of body temperature values. A nice example of such a bimodal distribution comes from a study of free-ranging western pygmy-possums in southern Australia (Figure 11.4). The data come from implanted transmitters, and while they are for animals entering hibernation rather than short-term torpor, they illustrate the principle very nicely. For animals or birds entering short-term shallow torpor, the frequency distributions of euthermy and torpor core temperatures may overlap.

Threshold temperature values allow us to define the period of torpor, but not its intensity. Simple measures of intensity include the lowest core temperature reached during torpor, or the difference between this and the mean temperature during euthermy (Box 11.2). One approach that combines duration and intensity has been to integrate the time-course of core temperature below the threshold value, giving a measure with units of degrees-time. The heterothermy index, HI (Box 11.2), is another approach. A problem with both of these metrics is that they attempt to capture variation in two independent variables (time and temperature) with a single number. This is impossible to do unambiguously: a given value of integrated temperature or HI can result from a short bout of deep torpor, a long bout of shallow torpor or anything in between. The HI may be useful where the threshold temperature for the onset of torpor and the temperature during torpor do not vary (for example when comparing individuals of the same species) but for broader comparisons its value is limited.

11.2 Hibernation

Hibernation is essentially a long period of deep torpor[6]. Both torpor and hibernation represent a controlled reduction in body temperature and metabolic rate in response to shortage of food or water. Hibernation, however, lasts for much longer, and in smaller species is often interrupted by short periods of arousal when body temperature returns to normal. A typical example is shown by a mammal of the grasslands of North America, the thirteen-lined ground squirrel, *Spermophilus* (*Ictidomys*) *tridecemlineatus*. The ground squirrels hibernate over winter, when their core temperature falls to 5 °C. These bouts of deep torpor are, however, interrupted regularly by short periods of arousal, termed *inter-bout arousals* (IBAs) (Figure 11.5). The function of these regular arousals is not fully understood, but they must be physiologically necessary because the warming during arousal uses considerable energy. During these periodic arousals, mammals may eat, drink and urinate to eliminate wastes.

Fritz Geiser has also provided a widely used definition of hibernation, which is where body temperature drops below 10 °C and the metabolic rate falls

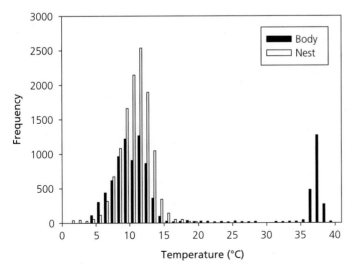

Figure 11.4 Frequency distribution of body temperature (°C) in free-ranging *Cercartetus concinnus* (western pygmy-possum) in southern Australia (black bars, n = 3). Also shown are data for the temperature in three nests (white bars)[37].

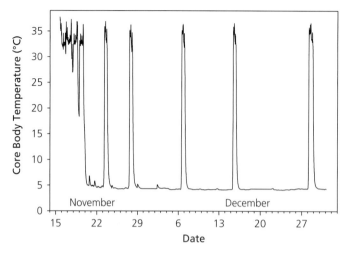

Figure 11.5 Hibernation in the thirteen-lined ground squirrel, *Spermophilus (Ictidomys) tridecemlineatus*, showing the initial period of euthermy (mean body temperature ~ 33 °C), initial short bouts of shallow torpor and then the main period of hibernation (core temperature 5 °C), interrupted by short period of arousal when body temperature returns to normal[38].

by > 90% for longer than a day[7]. This distinguishes hibernation from daily torpor on the basis of core temperature, metabolic rate and duration, but it is very much a definition for small mammals as it excludes bears. It also leaves us with no term for periods of shallow torpor lasting longer than a day.

The traditional picture of hibernation, at least in the eyes of the public, is the seasonal torpor of small mammals or bears over winter in northern latitudes. In North America there are 24 species of true hibernators known, but only six of these (all of them rodents) are found above 60 °N[8]. Most hibernators are actually found in the mid-latitudes, where it is as much a response to seasonal variations in food availability as to the seasonal cold. Hibernation is, however, also seen in mammals that inhabit warm arid environments, where the availability of food or water is often seasonally unpredictable and highly variable spatially. In these environments, hibernation is a more labile event and overlaps in some characteristics with daily torpor.

Comparative analyses indicate physiological differences between the extremes of deep hibernation and shallow torpor: mammals and birds undertaking daily torpor tend to exhibit higher core body temperatures and metabolic rates during torpor than do hibernating mammals. Metabolic rate and body temperature are not, of course, independent of each other, but the combination does differentiate daily torpor and hibernation in many cases (117 of 121 species in a recent analysis)[9].

In the arid environments of Africa and Australia the use of torpor or hibernation is widespread and varied in its timing and intensity. Here the increasing use of telemetry and data-loggers to monitor core temperature in free-ranging mammals has blurred the traditional dichotomy between torpor and hibernation. Periods of torpor may be short or long, minimum body temperature may be high or low and many species exhibit marked phenotypic variability in all features. While the extremes of long seasonal hibernation in deep torpor and short bouts of shallow torpor are clearly very different, we now recognise that there are many intermediate variations. This suggests that torpor and hibernation are best viewed as extremes of a continuum of heterothermic response to environmental challenges[10]. This is not a universal view, and it is also possible that some related physiological responses such as shallow hypothermia during rest periods may be sufficiently distinct to fall outside this continuum. As so often in ecology, as we learn more of natural history we uncover a wider range of adaptations and the old certainties become blurred.

11.3 Ecology of torpor and hibernation

The main ecological driver for the onset of torpor in bird and mammals is a limitation in the availability of food, exacerbated when combined with a drop in environmental temperature. It is often difficult to assess food availability in the wild, so the relationship between torpor and food availability has mostly been explored experimentally. Classic studies of captive pocket mice showed that the time spent in torpor is dependent strongly on the availability of food, and that the duration of this torpor had a major impact on daily energy expenditure (Figure 11.6).

These early studies established that torpor, and more extended hibernation, is a means of reducing energy use by the animal during a period of food limitation (either because the food is not there or because an environmental factor such as snow cover prevents access). Where a period of food limitation is predictable, such as in winter at high latitudes, then organisms can prepare for this ahead of time using environmental cues such as photoperiod or temperature. At high latitudes some mammals are obligate hibernators: the golden-mantled ground squirrel *Spermophilus* (*Citellus*) *lateralis* will enter hibernation even when held in constant light and at temperatures above 0 °C. This is governed by an endogenous rhythm which dictates that hibernation occurs when winter would be beginning in the wild[11].

In contrast where the environment is unpredictable, organisms need to be flexible and able to enter torpor at short notice. A nice example of this is the elephant shrew *Elephantulus myurus* studied in southern Africa by Nomakwezi Mzilikazi. Free-ranging elephant shrews were followed for a year and although they were physiologically capable of entering torpor at any point, this was most frequent in the winter. Short bouts of daily torpor were characterised by core temperature dropping to ~ 15 °C, but there were occasional bouts which lasted over 24 hours and in these core temperature dropped below 10 °C. Such flexibility is important in mammals living in regions where rainfall and food availability (in this case insects) vary unpredictably between years, driven by large-scale climatic variability. Similar flexibility is shown in a marsupial, the pygmy-possum *Cercartetus nanus* monitored by James Turner and colleagues over summer and winter in a warm-temperate area of Australia. Short periods of torpor were used in all seasons, although in summer only by males, whereas in winter both sexes underwent more extended periods of hibernation (5 to 20 days). The pygmy-possums typically entered torpor at night as the air cooled, rewarming during the afternoon when air temperature had increased. During these periods of short-term hibernation individuals mixed bouts of torpor of up to ~6 days with periods of nocturnal foraging, indicating how torpor is an

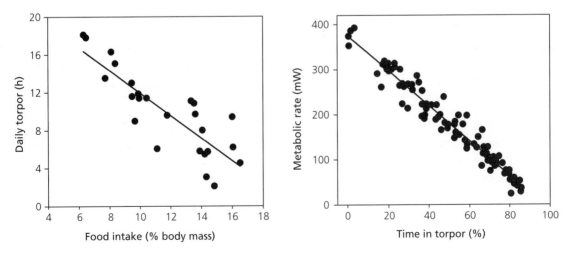

Figure 11.6 Daily torpor in pocket mice. A (left panel): relationship between food intake and duration of daily torpor in captive *Chaetodipus* (previously *Perognathus*) *californicus*. B (right panel): relationship between time spent in torpor during the hibernation season and daily energy expenditure in captive *Perognathus longimembris*[39].

important behavioural and physiological mechanism for energy management in small mammals living in a variable and unpredictable environment.

These three studies from contrasting regions and environments illustrate nicely how evolution has adjusted torpor and hibernation to the ecological circumstances. The use of flexible torpor is a widespread approach, for in the Afrotropical region in general, torpor or hibernation has now been reported in 40 species from six orders of mammals, and torpor in 16 species from seven orders of birds. Among marsupials, daily torpor or hibernation is known from ~15% of all species[12].

Many mammals lay down large fat stores in preparation for hibernation. For short bouts of torpor, fat stores are generally sufficient to provide the necessary energy, even if core temperature does not drop very low. For a long bout of seasonal torpor, however, fat stores alone may be insufficient for small animals (and protein is also utilised). This is because whereas maximum (pre-hibernation) fat stores form a constant proportion of body mass, mass-specific metabolic rate is much greater in smaller species (Table 11.1). This different scaling of fat storage and metabolic rate is a key factor in the ecology of hibernation at high latitudes: all other things being equal a large mammal that does not enter torpor can last longer on its fat reserves than can a small mammal. Torpor lowers body temperature and metabolic rate, thereby extending the period for which reserves can last. The lower the core temperature (and hence metabolic rate) the longer the fat reserves last, but the greater the cost of arousal where warming is fuelled metabolically[13].

Energy for extended hibernation can also be provided by the storage of food which is consumed during periodic arousals. These stores need to be resilient to perishing, and food hoarding hibernators tend to be granivores (seed eaters) as seeds are adapted to pass long periods in a dormant and dry state. They are also energy-rich and because such stores are always open to pilfering careful food caching is important to survival. Overall, the dichotomy between cached food and fat as a fuel for hibernation is linked to diet rather than size. Although granivory is common in passerine birds, none of these are known to hibernate or enter daily torpor. Many small birds in northern forests do, however, cache food.

In northern forests and tundra, some species continue to forage beneath winter snow cover and enter neither torpor nor hibernation, though they may huddle to conserve energy during the night. The insulation provided by the overlying snow means that the thermal environment beneath is benign, and temperatures may be around 0 °C. Under

Table 11.1 Fat stores, food stores and hibernation[49].

Species	Body mass (g)	BMR (W)	Maximum reserve size		
			Mass (g)	Relative to body mass (g/g)	Euthermic endurance (d)
Energy stored as body fat					
Myotis lucifugus (bat)	10	0.07	3.8	0.39	25
Zapus princeps (jumping mouse)	36	0.31	14.6	0.40	23
Spermophilus parryii (ground squirrel)	985	2.9	473	0.48	75
Marmota monax (woodchuck)	4900	6.7	2840	0.40	195
Ursus arctos (brown bear)	237 400	260	135 600	0.44	241
Energy stored as food cache					
Perognathus parvus (pocket mouse)	24	0.22	4400	1342	1703
Cricetulus triton (hamster)	73	0.49	35 000	3511	5955
Tamias striatus (chipmunk)	100	0.81	12 200	893	1276
Cricetus cricetus (hamster)	362	1.26	90 000	1092	3616

these circumstances small mammals such as mice and voles, and even small carnivores such as the least shrew *Cryptotis parva*, can remain active year-round; torpor and hibernation are not the only solutions to the challenge of a northern winter. These winter active small mammals are sufficiently common to provide a food resource for predators, predominantly owls in winter at these latitudes[14].

Discussion of the ecology of daily torpor often concentrates on its energetic advantages. There are, however, also risks to torpor, an obvious one being the possibility of predation while inactive. In torpid mammals and birds this means selecting a safe location: for the hummingbird *Mellisuga minima* this is the very end of a branch so delicate it cannot support the weight of any potential predator, for many mammals it is a secluded and difficult-to-locate nest. The only hibernating bird, the poorwill, *Phalaenoptilus nuttallii*, was first discovered in a rock face crevice but subsequent work has shown that, remarkably, it usually hibernates on the ground relying on camouflage to avoid predation. In mammals, however, it appears that hibernation is a relatively safe way to pass the winter: monthly survival in hibernators is, on average, higher than in non-hibernating species[15].

Body temperature affects muscle performance and this means that in a mammal emerging from torpor locomotor ability is reduced until normal body temperature is achieved. A study of three marsupials from Australia (kaluta, *Dasykaluta rosamondae*, dunnart, *Sminthopsis crassicaudata* and planigale, *Planigale gilesi*) showed that these were all able to move around at body temperatures down to < 20 °C, but that locomotor ability was strongly dependent on muscle temperature and full capacity was achieved only at normal (euthermic) body temperature (Figure 11.7). This means that mammals emerging from torpor or hibernation are exposed to an increased risk of predation during arousal and a safe location for rewarming is essential.

Flight is energetically demanding and mourning doves, *Zenaida macroura*, have been shown to have impaired flight ability following a drop in core temperature of ~5 K. An additional problem for birds may come from the burden of extra body mass when fattening up. For small passerine birds, the evasion of an ambush predator such as a hawk is dependent on the speed with which they can take off, and also the angle at which they can climb. Studies of take-off capability in a number of passerines (blackcap, *Sylvia atricapilla*, sedge warbler, *Acrocephalus schoenobaenus* and Eurasian robin, *Erithacus rubecula*) have shown that this is significantly impaired by pre-migratory fat loads. In the willow tit, *Poecile montanus*, however, fat accumulated prior to overnight roost appears not to affect take-off ability. In contrast, the natural diurnal variation in body mass in zebra finches, *Taeniopygia guttata*, did have an impact on take-off performance[16].

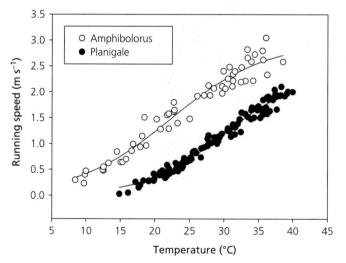

Figure 11.7 Running speed in an Australian mammal (*Planigale gilesi*) and a lizard (*Amphibolurus muricatus*) in relation to body temperature. In both species running speed increases with temperature; the solid lines are sigmoidal curves fitted by least squares. The absolute difference in running speed is related to the difference in size[40].

Whether or not a mammal or bird enters torpor is thus a more complex issue than simply energy saving. There are important trade-offs in terms of predation risk, both from impaired performance during arousal the extra mass of overnight fat stores, and the increased feeding time required to lay down fat stores. Furthermore, there are also consequences for overwinter mortality from factors such as unexpectedly low temperatures. The ecology of torpor involves a complex balance of energetic advantages and ecological risks[17].

11.3.1 Temperature and energy during torpor

The onset of torpor is marked by a reduction in body temperature. This can be achieved either by an increase in heat loss or by a decrease in heat production. Many mammals at high latitudes increase their insulation in winter, thereby reducing their heat loss. Conductance does not drop to zero, and so heat loss does contribute to the drop in temperature, but the main driver is a decrease in the heat supplied by metabolism. This can be seen clearly in the relationship between metabolic rate and core temperature during hibernation in the ground squirrel *Spermophilus* (*Ictidomys*) *tridecemlineatus* (Figure 11.8). As the ground squirrel re-enters torpor from a short bout of arousal, metabolic rate drops rapidly but body temperature decreases more slowly. What is happening here is that on entry to torpor metabolic rate is down-rated sharply (presumably following a change in set-point, though we have no idea how this is achieved), thereby decreasing the energy available to maintain core temperature. Temperature drops as thermal energy is lost to the environment faster than it is replaced by the reduced rate of heat dissipation in metabolism. Over time the rate of cooling slows as the difference between internal and external temperature gets ever smaller. The drop in core temperature can be modelled as a simple passive heat exchange with the environment, with the temperature differential as the main driver of heat loss, although in some cases (for example the Alaskan marmot, *Marmota broweri*) the pattern of cooling during entry into torpor is more complex. During re-entry to torpor, metabolic rate and core temperature are thus uncoupled and so looking for any relationship between the two is meaningless[18].

Once torpor is established again, both temperature and metabolic rate remain stable, and it is now meaningful to ask what sets the metabolic rate. Is it simply dictated by core temperature, as in an ectotherm, or is small degree of heat production continuing?

Analysis of the metabolic rate during torpor and hibernation shows that metabolic rate during torpor is higher than during hibernation (Figure 11.9). As we saw earlier, this difference is associated with a difference in core temperature: for these data the mean temperature during torpor was 18.8 °C (SE 0.58, n = 108, comprising 68 mammals and 40 birds) and for hibernation it was 6.8 °C (SE 0.75, n = 55, 54

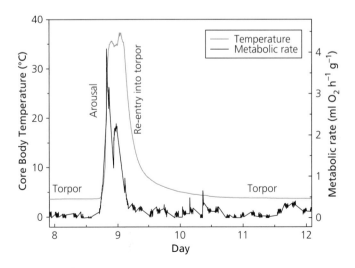

Figure 11.8 Body temperature (upper trace) and metabolic rate (lower trace) throughout a single arousal during hibernation in the thirteen-lined ground squirrel, *Spermophilus* (*Ictidomys*) *tridecemlineatus*[41].

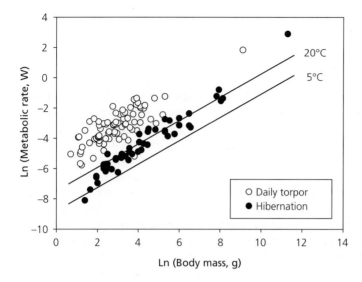

Figure 11.9 Metabolic rate of mammals and birds during daily torpor (white symbols) and hibernation (black symbols). The lines show the regression lines for the mean resting metabolic rate of reptiles with a body temperature of 5 °C and 20 °C for comparison[42].

mammals and 1 bird). The lines showing the regression relationships for resting metabolism in reptiles with body temperatures of 5 °C and 20 °C indicate that, even in hibernation, metabolism is greater than that of an ectotherm with the same core temperature. For torpor the discrepancy is even greater. This indicates that metabolic rate and core temperature remain carefully regulated during hibernation and torpor: mammals in deep hibernation are still endotherms. They have simply adjusted the set-point at which they regulate their body temperature and obligate thermogenesis continues, albeit at a lower rate.

We saw in Chapter 10 that the resting metabolic rate of mammals varies with body temperature, because it costs more to maintain a warmer body than it does a cooler one. This relationship remains evident during hibernation and torpor (Figure 11.10): there is a significant difference between daily torpor and hibernation in the elevation of the scaling relationship between metabolic rate and body mass but not in its slope or temperature sensitivity. The overall Q_{10} is 1.92 and this reflects the universal relationship between the temperature of cells and the cost of their staying alive. The Q_{10} for metabolism

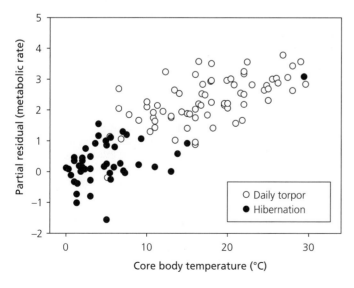

Figure 11.10 Metabolic rate and body temperature in mammals and birds during daily torpor (white symbols) and hibernation (black symbols); data plotted as partial residuals from a general linear model to remove the effect of body mass[43].

during torpor or hibernation is lower than for BMR in euthermic endotherms (which is 2.9), and this is presumably because of the greater contribution of obligate thermogenesis to resting metabolism in euthermic endotherms[19].

How much energy is saved by entering torpor or hibernation? A basic estimate can be made by comparing the scaling relationships in Figure 11.8 with relationships established for metabolic rate during euthermy in mammals and birds. This indicates that the metabolic rate during torpor is about 25% of BMR, and in hibernation < 5% BMR (Table 11.2).

Once a mammal or bird has entered torpor, maintenance costs are lower because its temperature is lower. This is not a direct effect of temperature on resting metabolism, rather it reflects the reduced metabolic costs of all those processes necessary to keep the animal alive at the lower temperature. These costs can be estimated crudely by calculating the resting metabolism for a reptile of body mass of 50 g with a body temperature equivalent to that of a similarly sized mammal when euthermic, torpid or hibernating. The difference between these estimates and the measured total metabolic rate then represents (very roughly) the costs of thermogenesis plus the costs of maintaining the extra thermogenic capacity. These thermogenic costs arise because the lower set-point still requires an input of heat from the dissipation of energy by metabolism. Putting all of this together, we arrive at the simple conceptual picture shown in Figure 11.11.

This crude model emphasises that there are differences in physiological state between torpor and hibernation, caused principally by the higher thermogenesis associated with the higher T_b in daily torpor compared with hibernation. Here, the difference between torpor and hibernation is one of degree, caused principally by the contrast in obligate thermogenesis required to defend the different core temperature values.

Table 11.2 Estimated metabolic savings in torpor and hibernation. All data estimated for a mammal of body mass 50 g. BMR: basal metabolic rate (360 mW), FMR: field metabolic rate (1180 mW)[50].

	MR (mW)	MR (%BMR)	MR (%FMR)
Torpor	89.2	25.3	7.60
Hibernation	10.7	3.0	0.9

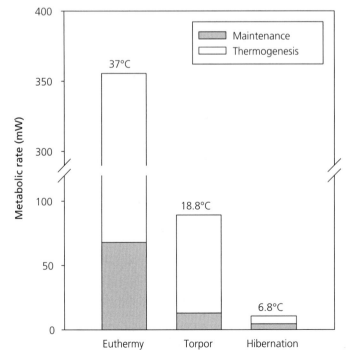

Figure 11.11 A simple conceptual model of the decrease in metabolic rate during torpor and hibernation compared with BMR during euthermy for a mammal of body mass 50 g. Maintenance is estimated from the resting metabolic rate of a reptile of the same body mass and temperature, and thermogenesis from the difference between this and the observed metabolic rate of the mammal. This will likely have underestimated maintenance costs but does not affect the physiological point that thermogenesis continues during torpor.

These broad comparisons indicate substantial energy savings when comparing the metabolic rate during torpor or hibernation with those during euthermy. These savings are, however, offset by the high energy costs of arousal: the high thermal capacity of tissue (because of its water content) means that it takes considerable energy to return to euthermy. For a hibernating chipmunk or ground squirrel in cold northern latitudes, this energy has to be supplied entirely by metabolism, as can be seen from the close match between metabolic rate and T_b during arousal (Figure 11.7). This may have led to selection for a high thermogenic capacity in these species[20].

While periodic arousals in the depths of the northern winter must be fuelled entirely by an increase in metabolism, in many parts of the world the environment can provide significant energy. The dasyurid marsupial *Pseudantechinus macdonnellensis* inhabits the arid environments of Australia and frequently enters torpor by night, using crevices for protection. In the morning, and while still torpid, it moves to more exposed locations to bask in the sun as it returns to active temperatures. Basking during arousal has been shown to be important in the elephant shrew *Elephantulus myurus*, and the mouse lemurs *Microcebus myoxinus* and *M. murinus*, and many birds can be seen to sit high in trees to absorb heat from the sun after emerging from roost. A low core temperature during torpor saves considerable energy, but is energetically expensive for arousal, whereas a high temperature during torpor is more expensive but allows a rapid return to full activity. These costs and benefits interplay in a complex way with the thermal characteristics of the environment, and also with patterns of food availability[21].

To quantify the true energy savings from torpor or hibernation we must therefore allow for the cost of arousal and the duration of these arousals. This was done by Lawrence Wang in an early study of Richardson's ground-squirrel *Spermophilus* (*Urocitellus*) *richardsonii*. Using radio-telemetry to monitor body temperature remotely, periods of torpor were found to start from mid-July in adults and from mid-September in juveniles. The maximum duration of torpor bouts was in January (19.1 days) and the duration of euthermy during arousals ranged from 5 to 25 hours. Combining metabolic data measured in captive animals in the laboratory with time-budgets determined for animals in the field, it was estimated that of the total energy used during the torpor season, entry into torpor accounted for 12.8%, deep torpor 16.6%, arousal from torpor 19% and inter-bout euthermy 51.6%. Overall, hibernation was estimated to save 88% of the energy that would otherwise be spent were the ground squirrels not to enter hibernation. The precise figures will differ in other species, and likely also in the same species in other places or in different seasons. The important general point they establish is that even allowing for the need to arouse regularly, hibernation saves a considerable amount of energy[22].

The temperature dependence of metabolic rate during torpor and hibernation has important consequences. The first is that the utilisation of reserves, and hence the decrease in body mass during hibernation, will be greater at warmer temperatures, as has been noted in hibernating European ground squirrel, *Spermophilus citellus*[23]. The second is that if the duration of an individual bout of torpor is determined by some aspect of hibernation physiology, for example, the accumulation of waste products that must be cleared by a period of arousal, then the duration of torpor should depend on body temperature during hibernation. This has been found to be the case (Figure 11.12). The temperature sensitivity of torpor bout length was found to be similar in all three species, and equivalent to a Q_{10} of 2.4. Metabolic rate during hibernation has a slightly lower Q_{10}, of 1.92; sufficiently close to suggest that the length of an individual bout of torpor may be dictated by some aspect of metabolism.

11.4 The physiology of torpor

If the core temperature of a non-hibernating mammal is lowered by even a few degrees below its euthermic set-point, it will die. In most mammals and birds the tissues suffer severe damage if cooled; for example when core temperature drops to ~20 °C, most non-hibernators develop severe arrhythmias and ventricular fibrillation leading to rapid cardiac arrest. The hearts of hibernating mammals, however, continue to work even at temperatures approaching 0 °C. This adaptation to low temperatures must involve much, if not all, of a mammal's

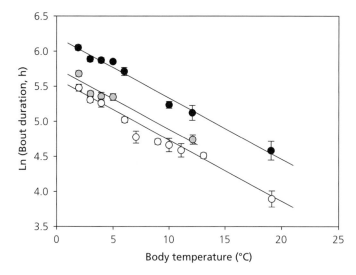

Figure 11.12 Relationship between torpor bout length (h) and body temperature (°C) during hibernation in three species of ground squirrel (black symbols: *Spermophilus columbianus*, grey symbols: *S. tridecemlineatus*, white symbols: *S. lateralis*). Data plotted as mean and SE. Regression lines fitted with a general linear model[44].

physiology: during hibernation blood must circulate, nerves and brain must function and kidneys must filter wastes from the blood. Most attention has, however, been directed at the heart.

The ability of a mammalian heart to continue working at very low temperatures is linked to its capacity to recycle Ca^{2+} into the sarcoplasmic reticulum (SR) after contraction. Critical to this recycling is the activity of SR-Ca^{2+}-ATPase (SERCA), the same enzyme that is involved in fish heater organs and muscle thermogenesis. SERCA is a membrane-bound enzyme and like many such enzymes it is sensitive to the fatty acid composition of the membrane in which it sits.

The activity of SERCA has been studied by Silvain Giroud and colleagues in the Syrian or golden hamster, *Mesocricetus auratus*. This species has a limited distribution in the arid areas of SE Turkey and adjacent Syria, and is now probably commoner as a pet and laboratory animal than in its native habitat. It is a facultative hibernator that can enter hibernation at any season, though it may require several weeks of cold temperatures to induce a period of deep torpor. Environmental cues for hibernation include photoperiod and food availability. Laboratory studies have shown that the core temperature is broadly similar in summer and winter, and also during inter-bout arousals, at 34–35 °C. During deep torpor body temperature is 5.4–7.4 °C, only 1.7 K on average above environmental temperature.

This difference is small but indicates careful control, and also that thermogenesis continues during hibernation.

The activity of SERCA was significantly higher in cardiac SR isolated from hibernating hamsters than in active animals (0.23 versus 0.18 µmol ATP hydrolysed per mg protein per minute, with all measurements carried out at 35 °C to eliminate the effects of assay temperature on activity). Furthermore SERCA activity varied with body temperature during deep torpor, and also with the fatty acid composition of the SR membrane phospholipids (Figure 11.13).

We do not know the direction of causality for the relationships in these plots. It could be that the level of SERCA activity, adjusted by membrane phospholipid fatty acid composition, depends on the set-point of the individual (itself set by other factors through the hypothalamus). Or it could be that body temperature during torpor is set by SERCA activity, itself being set by SR composition. The former seems more likely, but in truth we do not know.

11.4.1 Mitochondria during torpor

All physiological processes continuing during torpor and hibernation require ATP, and hence mitochondria must remain active no matter what their temperature. Mitochondria would also seem a likely target for organism-level control of metabolic rate required during arousal or entry into torpor.

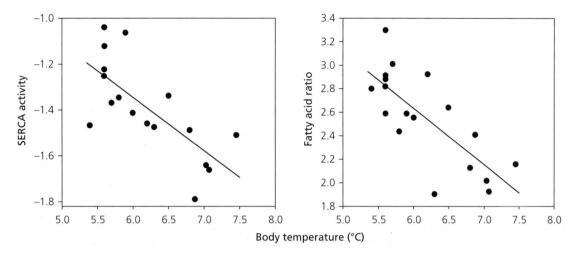

Figure 11.13 Activity of sarcoplasmic reticulum Ca^{2+}-ATPase (SERCA) from heart of the hamster, *Mesocricetus auratus*. A (left panel): activity of SERCA (μmol ATP hydrolysed per mg protein per minute) as a function of body temperature during hibernation. B (right panel): the ratio of linoleic acid (18:2ω6) to docosahexaenoic acid (22:6ω3) concentrations in sarcoplasmic reticulum phospholipids, as a function of body temperature during hibernation[45].

The maximum ATP synthesis capacity of a mitochondrion is assessed by its activity in the presence of saturating concentrations of electron supply (NADH or succinate), ADP and oxygen (as electron acceptor); this is termed state 3 respiration, as the rate of ATP synthesis is estimated by oxygen consumption. A study of the thirteen-lined ground squirrel, *Spermophilus tridecemlineatus*, by Jim Staples and colleagues showed that state 3 respiration in mitochondria isolated from hibernating animals was moderately suppressed (30%) in cardiac and skeletal muscle of torpid animals, but not suppressed in brain cortex, when compared with animals sampled when fully aroused during interbout euthermy (IBE). The greatest suppression was found in liver (70%) (Figure 11.14)[24].

The degree of suppression varied with time through a bout of torpor. Suppression was greatest during deep torpor, but mitochondrial activity increased rapidly during arousals. This pattern

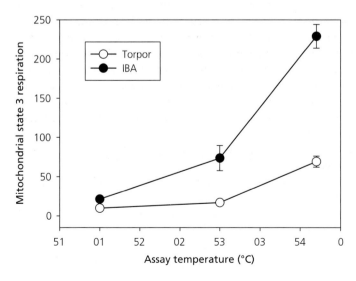

Figure 11.14 Maximum ATP synthesis activity (state 3 respiration, nmol O$_2$ mg^{-1} protein min^{-1}) in mitochondria isolated from torpid and aroused *Spermophilus tridecemlineatus*. The extent of metabolic suppression of mitochondria from torpid ground squirrels varies strongly with assay temperature, being greatest at 37 °C. Data are plotted as mean and SE.

tracked the increase in whole-organism metabolic rate and core temperature (Figure 11.7) almost exactly. During the return to deep torpor, however, mitochondrial activity was suppressed rapidly, again following the pattern of whole-organism metabolic rate but preceding the drop in body temperature. This indicates that entry into torpor is initiated by an organism-level control reducing metabolic rate at the cellular level.

The obvious question is what mechanism provides this control? Suppression of mitochondrial activity *in vitro* is greater with succinate as electron donor than pyruvate or fatty acid derivatives, suggesting that the point of control is at, or downstream of, complex II in the electron transport system of the mitochondrion. The speed with which suppression happens argues against control at the genome level by changes in transcription or translation. Not only that, but in *Spermophilus lateralis* initiation of protein synthesis ceases below 18 °C, and hence is presumably not occurring during deep torpor. A likely mechanism is some form of post-translational modification of pre-existing proteins, such as phosphorylation or acetylation[25].

There are changes in metabolite concentrations between torpor and arousals, but these could be a consequence of the mechanism of suppression as much as its cause. At present the mechanism by which metabolism is regulated during torpor and arousal is unknown. While the adjustment of the set-point temperature presumably involves the hypothalamus, we do not know how this is achieved at the molecular level. What we can say, however, is that the available evidence points to the deployment of physiological mechanisms that are common across endotherms, and possibly all vertebrates, to regulate metabolism rather than processes peculiar to hibernators.

11.5 Torpor, hibernation and size

We have seen (Figure 11.3) that core temperature at the onset of torpor is correlated with size in small mammals. What happens if we look at patterns in the incidence of torpor and hibernation across all mammals? (We will concentrate on mammals in this section because almost no birds hibernate.)

Like birds, mammals show diurnal (circadian) and seasonal rhythms in body temperature. First noted in the mid-nineteenth century, these rhythms have now been documented in over 100 bird and mammal species and are almost certainly a universal feature of endotherm physiology. The magnitude of variation is usually captured by calculating the arithmetic difference between the minimum core temperature during the daily or seasonal cycle and the maximum; this is the range or thermoregulatory scope, TS (Box 11.2). Recently Justin Boyles and colleagues have compiled data on TS for 560 mammals (~10% of all mammal species). These data show that most mammals have a narrow TS: 244 species (44%) had a TS < 2 K and 340 species (61%) had a TS < 4 K (Figure 11.15). A variability in T_b of 2 K has been proposed as a threshold between homeothermy (relatively constant body temperature) and poikilothermy (variable body temperature). Adopting this definition would classify 39% of mammals as poikilotherms, although all of them are of course are endotherms. This is a nice example of how difficult it can be in trying to define a threshold when the real world is a continuum. The frequency histogram would suggest that if a threshold is desirable to distinguish mammals with a relatively constant core temperature from those where it is more variable, then a better value would be a TS of 4 K. There are indications of smaller peaks at TS ~20 K (typically species undergoing short bouts of shallow torpor) and ~30 K (typically hibernators), but the dominant pattern is one of a continuum of TS values in endotherms undergoing torpor or hibernation[26].

There are intriguing physiological signals in these data, but as always with large data sets compiled from the literature, we have to be very wary of subtle biases. Nevertheless it does appear that TS tends to be low in families with many large species (Bovidae, Canidae, Felidae) and higher in families composed predominantly of smaller species (Sciuridae, Vespertilionidae). This suggests an influence of size.

In the early 1980s, Jürgen Aschoff examined 20 species of mammal, and showed that the difference between the core temperature by day and night was greater in smaller species; in a separate study he showed a similar pattern in birds. Subsequent studies were unable to corroborate this result, possibly

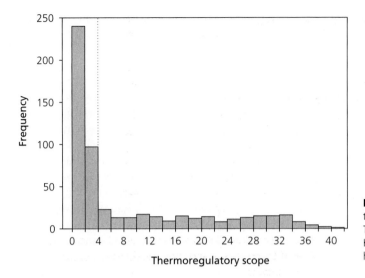

Figure 11.15 Frequency histogram of thermoregulatory scope, TS, values in mammals. The dotted line shows a threshold variability of 4 K as a division between euthermy (TS < 4 K) and heterothermy (TS > 4 K)[46].

because of a restricted range of body mass in the species examined. Then in 2004 Jacopo Mortola and Clement Lanthier collated data for 57 species and showed that while maximum (daytime) core temperature was largely independent of size, minimum (night-time) temperature was strongly dependent on body mass: as a result, the difference between day and night temperatures was greater in smaller mammals. In the much larger sample of 560 species studied by Justin Boyles and colleagues, diurnal variation in core temperature was again greater in smaller species (Figure 11.16)[27].

This result is not surprising because, all other things being equal, the physiological challenge of maintaining a set-point core temperature is greater in a small mammal where there is a large surface area from which heat can be lost relative to the volume of body where heat is generated (and as we saw in Chapter 9 there is little a small mammal can do to increase its insulation).

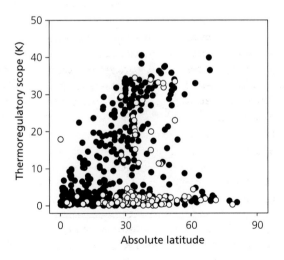
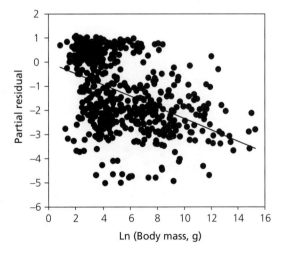

Figure 11.16 Variation in thermoregulatory scope (the difference between maximum and minimum body temperature for a given species) as a function of latitude and body mass in 560 species of mammal. A (left panel): variation in absolute scope with latitude, with data for species that hoard food (white symbols, n = 108) and those that do not (black symbols, n = 452) distinguished. B (right panel): variation in thermoregulatory scope as a function of body mass. The data are partial residuals from a general linear model[47].

There is also a relationship between TS and latitude (Figure 11.16). Although mean TS increases significantly with latitude, the pattern is more one of an increase in the upper bound (and hence the variance) of the distribution. At tropical latitudes TS is relatively low, typically < 10 K, whereas in polar latitudes TS may be as high as 40 K. This pattern reflects the need to control temperature during torpor and hibernation. Temperature can only be regulated if there is balance between energy input (heat dissipated in metabolism) and energy loss (to the environment) and energy can only be passed to the environment if core temperature exceeds ambient temperature. The upper bound to the relationship between TS and latitude is thus set by the latitudinal cline in mean environmental temperature. The extreme example of this is probably the Arctic ground squirrel, *Spermophilus parryii*. This species hibernates for eight or nine months during winter, using hibernacula in ground where temperatures may reach −15 °C and its core temperature during hibernation may fall to −2.9 °C. This is the lowest core temperature yet reported for a hibernating mammal, and one that probably requires its blood to supercool[28].

Species that cache food exhibit, on average, lower TS values than those that do not. While some food caching species do have high TS values, the majority have relatively low TS values, suggesting that a suitable food cache allows a mammal to remain active and maintain a relatively high core temperature (and hence a low TS) throughout the winter.

Minimum body temperature varies with body mass in both hibernators and species that use short periods of torpor. For any given body mass, however, hibernators have a much lower minimum core temperature, with the difference being ~ 15 K on average (Figure 11.17). The slope of the scaling relationship between minimum temperature and body mass is, however, the same in hibernators and species using torpor. The duration of hibernation increases markedly with latitude: at latitude 10° the mean length is ~2 days whereas in the far north at 70 °N it is ~35 days. In contrast, the mean duration of torpor is ~10 hours, independent of latitude. A subtle bias in these analyses is that species were classified *a priori* into hibernators and those using torpor on the basis of core temperature and torpor bout duration.

11.5.1 Do bears hibernate?

For many people the classic picture of hibernation is a bear spending the winter asleep in its den, and denning is known from all four species of *Ursus*. Denning in bears, however, meets almost none of the criteria in the definition of hibernation given

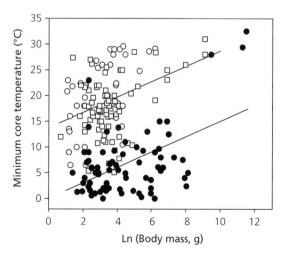 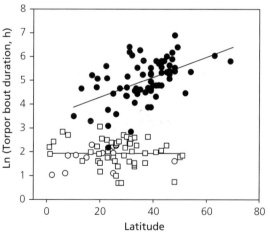

Figure 11.17 Daily torpor and hibernation in mammals. A (left panel): relationship between minimum core body temperature during torpor (white symbols, mammals as circles, birds as squares) and hibernation (black symbols). Regression lines fitted with a general linear model, indicated a mean difference in elevation of ~15 K. B (right panel): relationship between torpor bout duration and latitude for torpor and hibernation (symbols as for left panel)[48].

above: body temperatures remain high (31–36 °C compared with 37–38 °C during normal activity) and metabolic rate decreases to only ~ 70% of the normal rate. Furthermore, bears do not undergo the periodic arousals so characteristic of deep hibernation in small mammals. Instead they survive off fat stores, and although protein metabolism continues urea toxicity is prevented by the use of glycerol from fat to generate glucose, and recycling of urea into proteins, for which gut microbes appear to play an important role[29].

Does this mean that bears are not hibernators, or that our definition of hibernation is unduly restrictive? The general relationships between body size and torpor characteristics such as minimum core temperature or bout length suggest that for many features of torpor or hibernation, bears are simply at one end of a wide natural spectrum. The differences between hibernation in ground squirrels and bears are related principally to their size.

11.5.2 How should we define torpor and hibernation?

While there are obvious physiological differences between a hibernating ground squirrel and a denning bear, the comparative studies discussed above have blurred the distinction between torpor and hibernation, leaving us instead with a continuum. Any distinction we try to draw is thus arbitrary. Which is not so say that we should not draw one, just that we should recognise its arbitrary nature. Or perhaps we should not try to define particular aspects in a restrictive manner, but simply recognise torpor in a broad sense as a general response to energy restriction, but one whose characteristics vary with species, habitat and circumstances. I favour the latter.

11.5.3 Hibernation in ectotherms

The harsh northern winter is not only a problem for endotherms. Many invertebrates have life cycles which involve overwintering in highly resistant stages of the life cycle, such as the egg, larva or pupa, but anyone who has kept a pet tortoise knows that some reptiles enter a hibernation-like state in winter. This is widespread in terrestrial vertebrate ectotherms and involves a lowered metabolic rate and a reduction in activity, typically cued by changes in photoperiod and a drop in temperature. A key difference from mammalian hibernation is that core temperature, and presumably metabolism, is not actively regulated during hibernation but varies in response to environmental temperature. This difference has led to the suggestion by Wilbur Mayhew that we need a different term for winter dormancy in reptiles, and he suggested *brumation*. This term has not been adopted widely among ecologists, though it is still used by some herpetologists[30].

At present we do not know the extent to which the physiological mechanisms that lead to a lower metabolism in dormant reptiles and to torpor in mammals and birds are similar. Given the broad continuum of responses exhibited by endotherms, it is probably best to retain existing terms until we have a clearer idea of the similarities and differences between ectotherm and endotherm torpor.

Because reptiles and amphibians cannot maintain their body temperature with heat from metabolism, the selection of a suitable site for overwintering is critical to their survival (and mortality over winter can be significant in reptiles). Typical sites for overwintering include beneath decaying logs, in deep protected rock crevices and in the burrows of other animals. The key factor is the need to avoid freezing in really low temperatures (although as we saw in Chapter 6, a few vertebrate ectotherms can tolerate freezing of the extracellular water). Some reptiles also congregate to overwinter, sometimes with different species together and occasionally with many thousands of individuals. Overwintering ectotherms will often emerge to bask if the weather warms, and they also drink. They return to full activity only once the environment has warmed sufficiently to allow full activity.

11.6 A phylogenetic perspective

Ecologists have long been interested in whether there are any patterns in the occurrence of torpor and hibernation across different lineages of mammals and birds, and what these might tell us about their evolution.

One clear pattern is that while hibernation is widespread in mammals, it is almost unknown in birds. Indeed the only bird known to hibernate is the poorwill, *Phalaenoptilus nuttallii*, which was first found hibernating in the mountains of the Colorado Desert in California by Edmund Jaeger. Despite attempts to induce hibernation in other caprimulgids, this remains the only documented example of hibernation in birds[31].

Hibernating species are found in all three of the deepest branches of mammals: monotremes, marsupials and placentals. In each of these lineages, species that hibernate are interspersed with species that do not. This pattern suggests that the potential to evolve hibernation is an ancestral (that is, plesiomorphic) trait in mammals. If we take hibernation to be an extreme form of torpor (rather than something physiologically distinct from it), then to understand their evolution we should look at the distribution of torpor more generally across mammals. When the incidence of torpor is mapped onto a modern phylogeny of mammals, torpor is seen to be present in all the more phylogenetically ancient lineages. This would suggest that the ability to tolerate a wide range of body temperatures is a basal condition in mammals, inherited from their synapsid ancestors. Both the strongly seasonal and tightly regulated hibernation found in high-latitude mammals, and the strict endothermy of more recent lineages where torpor appears to be absent, then emerge as more recent, derived characteristics[32].

The incidence of torpor in birds also appears to be concentrated in more phylogenetically ancient lineages. We now know that the evolution of endothermy in mammals and birds were independent events, and yet heterothermy appears to be a primitive feature of both birds and mammals. The explanation may well be that in both groups endothermy evolved in relatively small forms, where tight control of body temperature is more difficult because of the large surface area; tolerance of a varying body temperature (that is, heterothermy) would then have been an integral physiological characteristic. We can thus picture torpor as a capacity that was retained, though undoubtedly also continually refined by evolution, in lineages where this remained advantageous. It then developed into an extreme form (hibernation) in lineages occupying particularly challenging environments, or lost in the evolution of continuous strict endothermy in other lineages[33]. It remains an intriguing mystery why hibernation is almost completely absent in birds, despite deep torpor being a feature of some lineages.

This appears at present to be the most parsimonious explanation for the pattern of occurrence of torpor and hibernation, but will undoubtedly be refined in details as the advent of instrumentation to monitor the temperature, metabolism and activity of free-ranging mammals and birds widens our knowledge of the incidence and nature of torpor in endotherms.

11.7 Summary

1. A diurnal (circadian) rhythm in body temperature is a widespread, and possibly universal, feature of endotherms. Many species also exhibit seasonal variations in temperature. The difference between mean core temperature during normal activity (euthermy) and the daily or seasonal minimum is the thermoregulatory scope (TS) or range. When TS exceeds ~ 4 K, it is termed heterothermy.

2. Some mammals and birds down-regulate their metabolic rate significantly by night, allowing their body temperature to drop sufficiently that they become inactive and enter torpor. Both the minimum temperature achieved and the duration of torpor are highly variable. Daily torpor is principally a response to reduced energy intake, and a drop in ambient temperature.

3. In some mammals torpor can last many months, with a low minimum body temperature and complete inactivity. This is hibernation and is essentially an extreme form of torpor. Small mammals hibernating at high latitudes show regular arousals during which they urinate and may feed. Bears hibernate with relatively high body temperature, and do not undergo arousal. Only one bird, the poorwill, is known to hibernate.

4. Rewarming during arousal may be fuelled exclusively by metabolism (for example in small mammals in the Arctic) or with significant energy input from basking (for example in subtropical arid areas).

5. The capacity for torpor appears to be an ancestral (plesiomorphic) character in both mammals and birds, possibly related to the origin of endothermy in small species subject to marked diurnal

and/or seasonal variation in body temperature. Both deep hibernation and strict endothermy are probably derived characteristics.

Notes

1. This quotation comes from Heinrich's description of his experience of winter in the woods of eastern North America, *Winter world* (Heinrich 2003); reproduced with permission of the author and Springer-Verlag.
2. The early studies of torpor in hummingbirds are Wilson & Bonaparte (1831), Huxley et al. (1939), Pearson (1950, 1953) and Carpenter (1974). Bartholomew et al. (1957) report torpor in other birds. Hamel (2012) discusses the selection of roost sites by hummingbirds entering night-time torpor. An important review of the early studies is Lyman (1963).
3. The study of sympatric Californian hummingbirds is Beuchat et al. (1979). Marchand (1987) and Staples (2014) describe the different overwintering ecology of mammals in northern woods, and Barnes (1989) describes hibernation in the Arctic ground squirrel. The African examples are Dausmann et al. (2004) and Dausmann (2005) for lemurs, and McKechnie & Mzilikazi (2011) for elephant shrews and a valuable general review of torpor in Afrotropical mammals and birds.
4. Geiser (2004).
5. See Barclay et al. (2001) and Willis (2007).
6. The term hibernation derives from the Latin. The noun for winter or the winter cold was *hiems*, while the verb *hiberno* meant to spend the winter (usually in warmer climes), and the *hibernaculum* was the winter quarters for the Roman army. Aristotle discussed hibernation (he believed many small birds hibernated over winter) but the first use of the term 'hibernation' to describe winter dormancy in animals is generally credited to the English polymath Erasmus Darwin (1731–1802, grandfather of Charles Darwin).
7. This definition comes from Geiser (2011).
8. This number reflects the number of well established cases; the actual number of hibernators may be higher.
9. See Ruf & Geiser (2015).
10. For recent studies using continuous monitoring of body temperature see Lee et al. (2009, 2016), Williams et al. (2011), Kisser & Goodwin (2012) and Sheriff et al. (2012).
11. The work on endogenous cycles in hibernating ground squirrels is described by Pengelley & Fisher (1961, 1963) and Pengelley et al. (1976). Variability in torpor is also shown by Merola-Zwartjes & Ligon (2000), Vuarin et al. (2013) and Kobbe et al. (2014).
12. The elephant shrew study is Mzilikazi & Lovegrove (2004), the pygmy-possum study is J. M. Turner et al. (2012) and valuable reviews of torpor and hibernation are McKechnie & Mzilikazi (2011), Ruf & Geiser (2015) and Riek & Geiser (2014).
13. An important early reference here is Lindstedt & Boyce (1985).
14. Marchand (1987) lists 13 species of small mammal in North America that, although non-colonial, often congregate during winter to conserve heat.
15. See Turbill et al. (2011). The poorwill study is Woods & Brigham (2004).
16. These studies are Metcalfe & Ure (1995), Kullberg et al. (1996), Kullberg (1998), Lind et al. (1999), Kullberg et al. (2000) and Carr & Lima (2013), and a useful review is Lind et al. (2010).
17. Trade-offs are discussed by Brodin (2001) and the relationship between hibernation and extinction by Geiser & Turbill (2009), and the Alaskan marmot study is Lee et al. (2009).
18. That heart rate and metabolism decrease before there is any significant drop in body temperature was probably first reported by Lyman (1963). Tucker (1966) modelled cooling as passive heat flow.
19. Analysis with a General Linear Model, using data from Table 1 in Ruf & Geiser (2015); slopes for hibernators and species using torpor were just significantly different (F = 4.17, p = 0.04) whereas there was a highly significant difference in elevation after fitting a common slope (F = 53.9, p < 0.001). The temperature sensivitiy for BMR in birds and mammals combined is from Clarke et al. (2010).
20. This topic is reviewed by Staples & Brown (2008) and Staples (2014). See also Heldmaier et al. (2004) and Careau (2013).
21. The first report of endogenous heating during arousal appears to be Schmid (1996). Recent studies are Schmid (2000), Schmid et al. (2000), Mzilikazi et al. (2002), Geiser & Pavey (2007), Nowack et al. (2010, 2013) and Thompson et al. (2015). See also French (1976). Trade-offs are discussed by Chruszcz et al. (2002), Humphries et al. (2003) and Schleucher (2004).
22. The ground-squirrel study is Wang (1979). See also Wang & Wolowyk (1988) and Karpovich et al. (2009). Lee et al. (2016) have shown that hibernating communally reduces energy costs in marmots. Schleucher (2002, 2004) explores the energy savings from torpor in birds.
23. Németh et al. (2009).
24. See Staples (2014). Kutsche et al. (2013) found suppression of activity in mitochondria from liver but not kidney, skeletal muscle or heart in the hamster *Phodopus sungorus*, and Grimpo et al. (2014) showed that suppression may not be evident in shallow torpor.
25. See van Breukelen & Martin (2001).
26. These terms are defined in Table 10.1 in the previous chapter. The suggestion of a 2 K threshold for poikil-

othermy was made by Bligh & Johnson (1973), and Pough & Gans (1982) suggested the same threshold for the boundary between endothermy and heterothermy; both were based on consideration of relatively large species. Schleucher (2004) suggested a threshold of 5 K for defining torpor in birds.

27. The early studies are Aschoff (1981a, 1982), Refinetti & Menaker (1992) and Refinetti (1999). The recent more comprehensive analyses are Mortola & Lanthier (2004) and Boyles et al. (2013).
28. The Arctic ground squirrel study is Barnes (1989).
29. Hellgren (1998) gives a nice summary of the physiology of hibernation in bears. See also Hochachka & Somero (1984), Tøien et al. (2011) and Stenvinkel et al. (2013).
30. Hibernation in *Anolis carolinensis* was reported by Dessauer (1953) and the term brumation was introduced by Mayhew (1965) based on his classic early study of winter dormancy in the lizard *Phrynosoma*. *Bruma* is the Latin word for the winter solstice, though it was also used to refer to winter or wintry conditions in general. Aleksuik (1976) explored the physiology underpinning dormancy in the garter snake *Thamnophis sirtalis*, and Ultsch (1989) gives a valuable review of winter dormancy in vertebrate ectotherms, in which he argues against the introduction of a new term for dormancy or hibernation in reptiles.
31. The original reports of a hibernating poorwill are Jaeger (1948, 1949). See also Marshall (1955), Bartholomew et al. (1957), Howell & Bartholomew (1959), Ligon (1970) and Woods & Brigham (2004).
32. The most recent bird phylogenies are Hackett et al. (2008), McCormack et al. (2013) and Prum et al. (2015), which make significant adjustments to the revolutionary phylogeny proposed by Sibley et al. (1988) and Sibley & Ahlquist (1991). The most recent mammal phylogenies are Beck et al. (2006), Bininda-Emonds et al. (2007) and Morgan et al. (2013).
33. The older view of the phylogenetic distribution and evolution of hibernation and torpor in endotherms is summarised nicely by Geiser (1998). More recent discussions are McKechnie & Mzilikazi (2011), who discuss the impact of changes in our view of avian phylogeny, and Lovegrove (2012) who presents a detailed evolutionary hypothesis for torpor and endothermy in mammals.
34. See Boyles et al. (2011b) and Brigham et al. (2011).
35. Plotted with data from Zhao et al. (2015).
36. Plotted with data from Willis (2007).
37. Redrawn from J. M. Turner et al. (2012); reproduced with permission of John Wiley and Sons
38. Plotted with data kindly provided by Jim Staples; reproduced with permission of the Company of Biologists.
39. Plotted with data from Tucker (1966) (left panel) and French (1976), as modified by McNab (2012) (right panel); reproduced with permission of John Wiley and Sons.
40. Modified from Rojas et al. (2012).
41. Plotted with data kindly provided by Jim Staples; reproduced with permission of the Company of Biologists.
42. Plotted with data from Ruf & Geiser (2015) with addition of reptile resting metabolic rate data from Clarke (2013).
43. Partial residuals from a general linear model using data from Table 1 in Ruf & Geiser (2015), with metabolic rate converted to SI units.
44. Diagram modified from Malan (2010); original data from Twente et al. (1977). Analysis with a GLM indicated no significant difference in the temperature sensitivity of the three species (p = 0.91), but significant differences in the elevation (p < 0.001). The line for metabolism assumes a Q_{10} of 1.92.
45. Plotted with data from Giroud et al. (2013); reproduced with permission of the Company of Biologists.
46. Plotted with data from Boyles et al. (2013); reproduced with permission of John Wiley and Sons.
47. Plotted with data from Boyles et al. (2013); reproduced with permission of John Wiley and Sons..
48. Plotted with data taken from Table 1 in Ruf & Geiser (2015).
49. Table modified from Humphries et al. (2003).
50. Data for BMR and FMR taken from scaling relationships in Clarke et al. (2010). Data for metabolic rates during torpor and hibernation taken from scaling relationships derived from data in Ruf & Geiser (2015).

CHAPTER 12

The Metabolic Theory of Ecology

> Often and often it happens that our physical knowledge is inadequate to explain the mechanical working of the organism; the phenomena are superlatively complex, the procedure is involved and entangled, and the investigation has occupied but a few short lives of men.
>
> D'Arcy Wentworth Thompson[1]

We have seen in previous chapters how temperature affects any process involving a movement of energy, and thereby all of physiology. Given that physiologists have studied the influence of temperature on many cellular processes, and done so for a long time, it is perhaps surprising that until recently no general theory of thermal physiology had emerged. Despite a good understanding of how temperature affects many individual processes, we had no synthetic theory that linked all of these results into a single coherent picture of how organisms deal with temperature.

And then in the late 1990s, two ecologists from the University of New Mexico, Jim Brown and Brian Enquist, together with a physicist from the nearby Los Alamos Laboratory, Geoffrey West, decided to take a close look at metabolism. Their initial work was concerned with how metabolic rate varies with body size, and they developed a model (or models, as there is more than one) based upon the architecture of distribution systems in plants and animals. Soon afterwards, in collaboration with Jamie Gillooly, they added a temperature term to the model to derive the fundamental equation of the *Metabolic Theory of Ecology* (MTE). From this basic theory, the originators, together with a series of students and collaborators, have examined the implications of the MTE for a wide range of ecological processes and patterns[2].

This theory has generated enormous controversy. Because it makes explicit recognition of the role of temperature in governing the flow of energy and materials through organisms, it has the potential to form a general theory of thermal ecology of wide applicability. It therefore deserves close scrutiny from any physiologist or ecologist interested in temperature. Before examining the theory in detail, however, it is helpful to establish the historical context.

12.1 The influence of size: scaling

Ask someone to describe a mouse or an elephant and it is almost certain that they will mention its size. Size, and especially size relative to ourselves, is a key feature of any object. We also know intuitively that size influences shape: an elephant does not look like a huge mouse, nor does a shrew resemble a small buffalo. These contrasts in shape are less marked in water, where a large fish can look pretty similar to a small fish. This difference between land and water has much to do with the need for terrestrial animals to support their own weight against gravity. These observations introduce a topic that has been central to the debate about the relationship between metabolic rate and size: *geometric similarity* or *size invariance*.

Two objects of the same general form show geometric similarity if the ratio of two corresponding linear measurements (say length or breadth) are the same. This concept goes back to the Ancient Greek geometers, and quite likely as far back as the

Babylonians. Euclid and Archimedes knew how to calculate the surface areas and volumes of complex shapes, but it must have been common knowledge to any engineer at that time that they could not build a large ship or bridge safely by simply scaling up a small one. In other words, the design of ships or bridges is not scale-invariant: as size increases, shape has to change. Neither is the natural world scale-invariant. This insight is traditionally ascribed to Galileo, who in a famous image contrasted the fate of a horse that falls from a small height with that of a cat which survives a much larger fall[3].

The Babylonians, Greeks and Galileo have left us with the foundations that have underpinned all subsequent discussions of scaling: in geometrically similar organisms surface area scales with length squared whereas volume (or mass) varies with length cubed, and organisms are not scale-invariant: as animals or plants get bigger or smaller, they change shape.

12.2 The scaling of metabolic rate

Interest in how metabolic rate varies with body size can be traced back to a seminal study of dogs by the German physiologist Max Rubner. Rubner selected seven adult dogs of different sizes and measured their metabolic rates. Because he was interested in comparing the fundamental metabolic rate of a unit amount of tissue in each dog, Rubner expressed his data on a mass-specific basis (that is, metabolic rate divided by mass). When plotted as absolute metabolic rate, the results showed that revealed metabolic rate increased with size although the rate of increase slowed in larger dogs (Figure 12.1). Rubner also calculated the metabolic rate as a function of surface area, and showed that this was more or less constant for all the dogs[4]. Rubner had earlier developed his *Law of Surface Area*, which was that the metabolic rate of birds and mammals maintaining a steady body temperature is proportional to body area. Although this idea can be traced back half a century earlier[5], Rubner's work was the critical demonstration of its general applicability to mammals.

These early studies related the observed metabolic rates to a clear physical theory. It had long been known that mammals and birds exhibited a high rate of metabolism, related to the need to maintain their high internal body temperature. Since heat was generated by the whole body, but lost principally through the surface, a scaling exponent (Box 12.1) of 0.67 was to be expected, whether comparing within species (as Rubner had done for domestic dogs) or across species (as he went on to do). The Surface Law would predict a scaling exponent of 0.67; Rubner's data for dogs have a scaling exponent of 0.62, and the

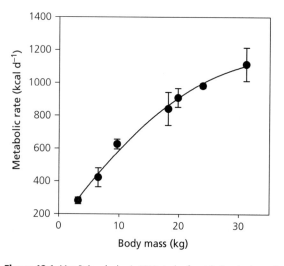
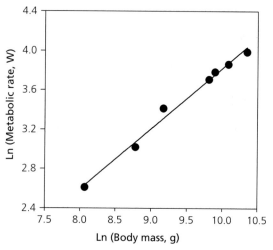

Figure 12.1 Max Rubner's classic 1883 study of metabolism in domestic dogs. A (left panel): the original data, plotted as mean and standard deviation (the number of measurements for each dog varied between 2 and 13) for seven breeds of domesticated dog. The line shows a fitted power law curve. B (right panel): the same data replotted using the mean estimate of metabolic rate for each dog, converted to SI units, both variables transformed to natural logs. The regression is highly significant ($F = 450$, $p < 0.001$), and the slope of the line is 0.62[6].

Box 12.1 Determining the scaling exponent

The development of the analytical tools for exploring the effects of size is usually traced back to a classic early study of growth by Julian Huxley[7]. Huxley showed that the change in many features of an organism with its size could be captured by a simple power law:

$$Y = aM^b$$

where Y is the variable of interest, M is some measure of body size, a is a constant and b is the scaling exponent. This relationship can be linearised by logarithmic transformation of both variables:

$$\ln(Y) = \ln(a) + b\ln(M)$$

A plot of $\ln(Y)$ against $\ln(M)$ thus yields a straight line of slope b and intercept $\ln(a)$, and these two parameters can be estimated by linear regression. This simple technique has been the mainstay of scaling analyses in biology ever since. So pervasive and routine has this technique become, that its limitations and assumptions are often forgotten.

Statistical and dimensional issues

Of particular importance is the choice of regression model to be used, as this depends critically on the distribution of error in the two variables. Most studies use ordinary least-squares regression and where body mass is measured directly and with little error, this is fine[8].

A more subtle problem arises because a logarithm is a transcendental function. A transcendental function is one that cannot be expressed in terms of a finite sequence of the algebraic operations of addition, subtraction, multiplication, division and root extraction (and hence it 'transcends' algebra). The problem comes to the fore in dimensional analysis, as transcendental functions make sense only when their argument is dimensionless. Thus $\log(M, g)$ or $\ln(M, g)$ are meaningless operations as they result in dimensional errors (as does logarithmic transformation of metabolic rate). Huxley recognised this, and his allometric equation rescaled M by dividing it by an arbitrary reference mass M^*, to produce the dimensionless ratio M/M^*. This avoids the problem, for $\ln(M/M^*)$ is then a valid operation; the difficulty for ecologists is that there is no obvious candidate for M^*[9].

Where we are using log-transformed variables in a simple plot to determine the numerical value of the scaling exponent for comparative studies, the dimensional problem is not too severe. Where the result is to be incorporated into a physical model, or a scaling relationship is to be mixed with other relationships in an ecological model, dimensions and units are critical and the problem is very real.

95% confidence intervals include the theoretical value of 0.67 (Figure 12.1)[10].

The universal nature of the Surface Law, and its basis in heat flow, was not accepted by all physiologists. In the early twentieth century August Krogh had noted that a similar exponent also applied to ectotherms such as reptiles and insects, both of whose metabolic rate (and hence heat dissipation) varied with environmental temperature. He reviewed the extensive work already published at the time but could discern no general pattern. He recognised that the scaling exponent varied in different animal groups, and concluded that there was nothing fundamental about this exponent. For Krogh, the scaling exponent was simply a number to be determined empirically for each animal group of interest[11].

A practical difficulty in testing the Surface Law was the precise measurement of surface area for an object as complex in shape as an animal. A number of ingenious devices were designed to attempt this, driven not so much by an academic interest in scaling as the practical need to understand how much energy was needed to sustain agriculturally important animals such as a cows, horses or sheep[12]. Because of the difficulties of measuring surface area precisely and accurately, attention soon shifted to the relationship between metabolic rate and the more easily measured body mass.

As further measurements were made, it became clear that resting metabolic rate often scaled with body mass more steeply than predicted by the Surface Law. Two important studies are shown in Figure 12.2, and the early work is summarised in Table 12.1. These studies indicated a scaling exponent in the range 0.71–0.79, suggesting that heat loss from the body surface was not the major factor dictating how resting metabolism varied with size in mammals and birds. The eminent physiologist Max Kleiber suggested that the exponent be rounded to 0.75, purely for ease of computation.

This value of 0.75 has become enshrined in the literature as *Kleiber's Law* or the *three-quarter power rule*; it is somewhat ironic that this most deeply entrenched of physiological rules should have its origin simply in the practical difficulties of calculation[13].

These early studies were concerned principally with domesticated or agriculturally important animals, and were undertaken predominantly in the United States. A number of factors make it difficult to compare this work directly with modern studies.

Domesticated mammals have been subject to considerable artificial selection, often for improved growth rate or milk production, and this selection may well have increased resting metabolic rate above that found in their wild ancestors. In many cases data from birds and mammals were mixed, and frequently included more than one data point per species (a procedure now regarded as statistically inadvisable). The nature of the data also varied between studies. Sometimes the data represent the average rate observed for a single individual,

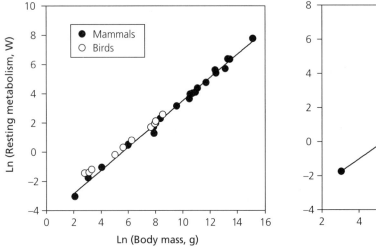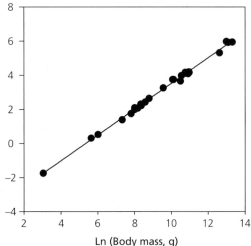

Figure 12.2 Two early studies of the scaling of resting metabolism in mammals and birds (original data converted to SI units). A (left panel): Francis Benedict, with data for birds and mammals distinguished. The slope of the line (fitted to the mammal data only) is 0.79. B (right panel): Max Kleiber, showing data for mammals only. The slope of the line is 0.76[14].

Table 12.1 Early studies of the scaling of mammalian resting metabolism. Many of these studies mixed data from mammals and birds; for these the data for mammals have been extracted and the relationship calculated from these. They also often included data for more than one individual of a species, and so both the number of data points and the number of species are shown. *b* is the scaling coefficient, and SE is its associated standard error; nd indicates that the data are not available. Studies are numbered in date order[15].

b	SE	n (data)	n (species)	Study
0.79	0.013	22	16	1
0.73	0.006	58	18	2
0.74	0.009	10	5	3
0.76	0.009	26	Nd	4
0.68	0.015	79	Nd	5
0.71	0.009	321	321	6
0.87	0.034	122 (< 3.2 kg)	nd	7
0.63	0.010	563 (> 3.2 kg)	nd	7

sometimes it is a representative value estimated from a detailed within-species study and occasionally it was estimated from a previously established regression relationship. Despite these problems the early studies showed clearly that empirical data for mammals and birds were generally not compatible with the Surface Law.

Important developments followed with the work of two Danish scientists, Axel Hemmingsen and Erik Zeuthen, who broadened the analysis to include ectotherms and unicells[16]. This extended what had become known as the 'mouse to elephant curve' to the 'bacteria to whale curve'. Together with an increasing number of studies in the 1960s and 1970s (Table 12.2), this work was influential in establishing that the general patterns established for mammals and birds could be extended to ectotherms, aquatic invertebrates and even unicells. In doing so, it was necessary to confront the problem that whereas mammals and birds maintained body temperatures over a fairly narrow range, the metabolic rates of all other organisms were measured at a wide range of temperatures, depending on the habitat in which the organism normally lived. The usual analytical approach to this problem was to rescale ('correct') the data to an arbitrary intermediate temperature, typically 20 or 25 °C, on the basis of an assumed temperature dependence of metabolism. In most of these studies, scaling exponents were typically > 0.7 and often > 0.8. Scaling was also shown to vary with the type of organism, and sometimes with size.

At the same time as the taxonomic range of organisms studied was being widened, an ever increasing body of data for mammals allowed for new analyses with much larger data sets. Heinz Bartels suggested that the scaling exponent differed for large and small mammals: the exponent for the entire data set was 0.68 (Table 12.1), but was 0.76 for large species (body mass > 260 g) and only 0.43 for species (rodents and 'insectivores') below this threshold. Brian McNab compiled all the mammalian data available in the late 1980s, editing these carefully to ensure the highest quality. The resultant database comprised 321 species, with one data point for each species. These data were not well fitted by the relationship established by Kleiber, and the overall scaling exponent was 0.71. Again, the exponent for smaller species (< 300 g) was lower (0.30) than for larger species (0.75). This study was important in showing that although an overall relationship for all mammals could be calculated there was marked variation between different groups of mammals, both at high taxonomic levels (monotremes, marsupials and placentals) and at lower levels (order), and also between different feeding types. Fred Heusner undertook a similar analysis, though with a larger data set (but here containing more than one data point per species). This study also suggested that the scaling relationships were different for larger and smaller mammals, though here the threshold was taken as 3.2 kg. Above this threshold the scaling exponent was 0.87, whereas smaller species had an exponent of only 0.63.

Table 12.2 A selection of early studies of the scaling of resting metabolism of organisms other than mammals. In all cases the analyses included data for several individuals for some species[17].

Taxa	b	SE	n	Study
Non-passerine birds	0.73	0.034	14	1
Passerine birds	0.73	0.022	17	1
Non-passerine birds	0.74	0.013	81	2
Passerine birds	0.63	0.034	49	2
Reptiles	0.80	0.008	73	3
Snakes	0.77	0.03	35	3
Snakes	0.86	0.03	49	4
Marine fishes	0.79	0.014	133	5
Crustaceans	0.73	0.043	249	6
Unicells	0.75	0.015	80	7

12.3 The West, Brown and Enquist model

In the 1980s, the field of scaling (or allometry: see Box 12.2) was summarised comprehensively in four books, published very close together, which provide a valuable summary of the knowledge of the field at that date[18]. While there was a general consensus at this time that the scaling of metabolic rate was best summarised by an exponent of ~0.75, by the late 1980s it was becoming clear that scaling in mammals and birds was often slightly lower overall, and that there was significant variation with size. There was, however, no theoretical underpinning for these observations; they were merely what John Lawton has called 'widely observable tendencies'[19].

At this point, Geoffrey West, Jim Brown and Brian Enquist (hereafter WBE for convenience) took up the challenge of providing a firm theoretical foundation for the scaling of metabolic rate. They started from the assumption that the key process was not heat loss through the body surface but delivery of oxygen and nutrients to the tissues, and consequently the scaling of metabolism with body size was dictated by the nature of the distribution network.

The model was based on three principles: firstly that the distribution system is space-filling with a fractal-like branching pattern, secondly the final branch in the network (the capillaries in a mammalian cardiovascular system) is size-invariant and thus the same size in a shrew and an elephant and thirdly that the energy required to circulate fluid is minimised. The last of these constraints can also be viewed as minimising the time for, and resistance against, the delivery of oxygen and nutrients to tissues, thereby maximising the efficiency of delivery. The WBE model is thus based on an idealised version of a vertebrate circulatory system that is similar to, but not identical with, that of a mammal[20].

WBE tested their model by using it to predict the scaling parameters for a range of variables (16 cardiovascular, 16 respiratory) associated with the mammalian cardiovascular system. Although these variables were not all fully independent, the predictions were largely upheld. One of these variables was oxygen consumption (presumably resting or basal oxygen consumption, though this was not specified), and the WBE model led to a simple equation that captured the dependence of metabolic rate on body mass:

$$B = B_0 M^b$$

where B is the metabolic rate, M the body mass and b the scaling exponent, the value of which was predicted by the model to be 0.75. B_0 is a normalisation constant which captures, among other factors, differences in cell size and metabolism between organisms; it thus effectively sets the level of metabolism. It varies, for example, between endotherms and ectotherms, and between basal (resting) and routine metabolism. It also varies between different animal groups, and within those groups, with ecology. Unlike the scaling exponent, the value of B_0 cannot be predicted by the WBE model and has to be determined empirically by fitting the model to experimental data.

Given the discussion it generated, it is worth noting how WBE described their model. They emphasised that *the present model should be viewed as an idealized zeroth-order approximation: it accounts for many of the features of distribution networks and can be used as a point of departure for more detailed analyses and models*. In other words, they recognised that the model was not complete, but was an attempt to capture the key features of the circulatory system that dictated the scaling of metabolism. While based on the structure of the vertebrate circulatory system, the WBE model was designed to be general. The mammalian circulatory system was described as *a minor variant* of the general model, but the critical statement was that WBE regarded their model as providing *a theoretical, mechanistic basis for understanding the central role of body size in all aspects of biology*[20].

The WBE model provided a description of an important central tendency in physiology, and moreover for the first time offered a coherent physical description of why this tendency should be what it is. They noted explicitly that the predicted scaling exponent held only for large animals, and that the scaling would be steeper (that is, the scaling exponent would be greater) in smaller animals. They also

---

Box 12.2 Allometry or scaling?

The study of the influence of size on morphology or physiology, nowadays usually referred to as scaling, is referred to extensively in the older literature as 'allometry'. Allometry was a term coined by Julian Huxley and Georges Teissier in 1936, to replace Huxley's original term of 'heterogony'[21]. Originally applied to studies of relative growth in anatomy, it was later extended to studies of metabolism in relation to size (and often termed *metabolic allometry*). Nowadays such studies are more usually referred to as scaling, and this term has largely replaced allometry in the literature.

The scaling exponent has been variously represented as α, k and b; in recent years, b has become favoured. Where the scaling exponent is precisely 1, the relationship between the two variables is described as *isometric*. In an isometric relationship the variable of interest increases in direct proportion to body size or mass.

commented that the scaling of metabolism would be expected to be different in organisms that were effectively two dimensional such as flatworms. These caveats were often missed in the debate that followed.

12.3.1 Comparing the model with data

To test the WBE model rigorously we need observations of the scaling of nutrient supply to cells in a range of animals of different sizes. While it is possible to measure blood flow directly, these measurements are rather few (especially for invertebrates). It is far easier to estimate the rate at which the nutrients supplied by the network are being used by the cells, by measuring the rate of oxygen consumption. This is not a direct test of a model built around the supply of nutrients to cells but it is acceptable because in most cells there is a dynamic balance between nutrient supply and demand: cells typically do not swell because they are gaining more nutrients than they use, nor do they shrink because demand outstrips supply.

An important point here is that respiration (oxygen consumption) measures the rate at which energy is being dissipated to the environment as heat. This represents only part of the energy supplied to the cell by the cardiovascular system, the balance being energy stored in fat or glycogen deposits, newly synthesised proteins and so on. Cells also lose energy in the waste products and or exported materials they pass to the cardiovascular system. The energy balance of a cell (or tissue, or organism) is thus complex, and a measure of oxygen consumption captures only part of this[23].

The predictions of the WBE model have been compared almost exclusively with empirical data for the scaling of resting or standard metabolic rate. This is convenient for there are a great many such measurements in the literature, with mammals being particularly well studied. A problem here is that almost no mammal spends any significant time in the wild operating at resting level. We should really be testing the model by looking at the scaling of routine metabolism (that is field metabolic rate, or daily energy expenditure: see Chapter 9). Testing a model of the cardiovascular system with data for resting metabolism is akin to testing the strength of a bridge at midnight when there is little traffic, rather than under the normal daily traffic flow for which it was designed. The scaling of maximum metabolic rate, such as can be achieved by running to exhaustion on a treadmill, is not a meaningful test of the WBE model as this represents a short-term excursion when normal constraints may be circumvented. In the wild such short excursions are seen when running down prey or eluding a predator, where the extra costs involved are set against survival[24].

The scaling of field metabolic rate, FMR, in mammals is steeper than that of BMR (Figure 12.3.)

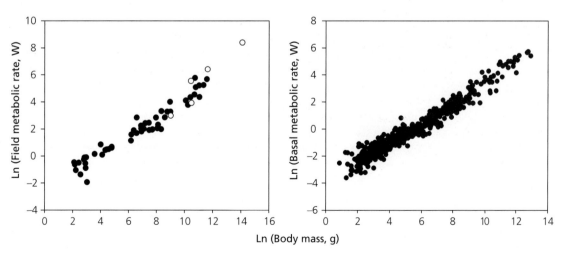

Figure 12.3 Scaling of metabolic rate (W) with body mass (g) in mammals; note logarithmic transformation (natural logs) of both variables. One data point per species. A (left panel): field metabolic rate, with terrestrial species shown with black symbols and marine species shown with white symbols. B (right panel): resting (basal) metabolic rate[25].

Box 12.3 Scaling and metabolic level

In all vertebrates so far studied, the scaling of metabolic rate steepens (that is, the scaling exponent increases) in the sequence BMR<FMR<Maximum MR (Table 12.3).

The simplest explanation for this pattern is an increasing dominance of metabolic rate by muscular work. BMR excludes muscular activity for locomotion (by definition: see Chapter 8), FMR includes normal day-to-day locomotor activity and at maximum MR muscle activity dominates oxygen requirement. The steepening of the scaling of metabolism then arises because muscle mass scales linearly with body mass (Figure 12.4), and the greater the proportion of metabolic rate represented by muscular activity, the greater the scaling exponent.

In mammals the total mitochondrial volume of muscle also scales isometrically (Figure 12.4). Since the density of inner mitochondrial membrane is invariant with body mass in mammals, the maximum ATP regeneration rate is directly proportional to mitochondrial volume. The ability of a muscle to regenerate ATP thus scales directly with its size.

Table 12.3 Scaling of terrestrial vertebrate metabolic rates. BMR: basal or resting metabolic rate, FMR: field metabolic rate, Max MR: maximum aerobic metabolic rate, n: number of species, b: scaling exponent, SE: standard error of the scaling exponent[26].

Class	n	b	SE
Mammals (> 150 g)			
BMR	280	0.75	0.110
FMR	38	0.81	0.046
Max MR	33	0.88	0.031
Birds			
BMR	83	0.64	0.018
FMR	95	0.68	0.018
Reptiles (lizards only)			
BMR	84	0.83	0.021
FMR	48	0.92	0.021

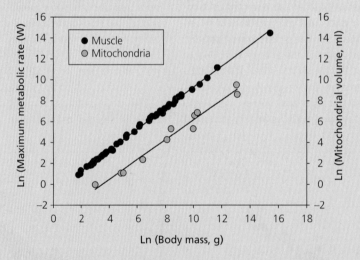

Figure 12.4 Scaling of muscle mass (black symbols) and total muscle mitochondrial volume (grey symbols) in mammals. The fitted relationships have slopes of 1.00 (muscle) and 0.95 (mitochondrial volume), and neither are significantly different from $b = 1$ (isometry)[27].

and over most of the size range of mammals it is linear with a scaling exponent of ~0.80. The scaling of maximum metabolic rate is steeper again (Box 12.3).

12.3.2 Curvature in scaling

In most studies of scaling, a simple linear regression is fitted to all the data (Box 12.1). In mammals, however, careful examination of the residuals indicates clearly that the scaling of both BMR and FMR is slightly curvilinear: smaller species tend to have a higher BMR or FMR than would be expected on the basis of the relationship for larger species. This curvature in the scaling of FMR and BMR in mammals can be captured by fitting a quadratic relationship, and can be seen most easily if the plots are confined to the smaller species (Figure 12.5). In both BMR and FMR it can be seen that smaller species tend to have higher metabolic rates than predicted by extrapolation of the relationship for larger species. Here the larger species relationship is established for species with body mass > 150 g. While this is a relatively small size, it is actually close to the median size for a mammal. In other words, half of all mammals conform to the predictions of the WBE model and half do not.

Does this invalidate the WBE model, or does it simply indicate that at small body size other factors come into play? The higher metabolic rates of smaller species suggest the possibility that an extra factor might be heat flow, with smaller species having to increase their metabolic rate in order to maintain body temperature. Ironically, this would bring surface area back into the picture. We will return to this later in the chapter, in section 12.7, after we examine the role of temperature[28].

12.3.3 The debate

Publication of the WBE model stimulated an intense and often heated discussion among ecologists. Some criticism was misplaced, in that the WBE model was criticised for not doing something it was never intended to do (such as describing the scaling of an individual species or an ecological process at the fine scale). Other critics were prompted to explore the fundamental structure of the model, asking two key questions: are the assumptions underpinning the model valid, and do the predictions follow logically from these assumptions (in other words, is the model internally consistent)? The final step is to see whether the predictions of the model are met in reality.

A decade after the WBE model was first presented the underlying assumptions were explored in detail by Van Savage and colleagues. They identified eight key assumptions underpinning the model, and examined each one in detail. They refined the predictions, pointing out that a scaling exponent of 0.75 was only achieved at infinite body mass; at realistic body mass values for a mammal, the prediction was actually 0.81. They also confirmed that the model predicted a steeper scaling for smaller species, which as can be seen from Figure 12.5 is precisely the opposite of that observed[29].

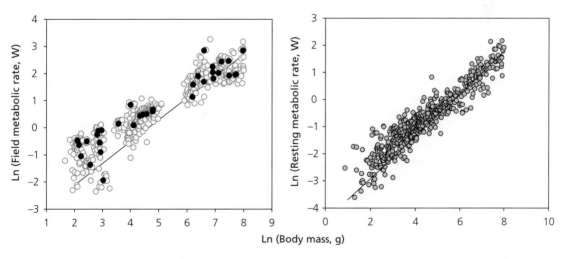

Figure 12.5 Scaling of metabolic rate (W) with body mass (g) in small mammals; note logarithmic transformation (natural logs) of both variables. The lines show least-squares linear regressions fitted to species with body mass > 150 g, extrapolated to small body masses. Note the tendency in both plots for small species to lie above the extrapolated line. A (left panel): field metabolic rate; median data for each species shown in black, all data (2–36 data points per species) shown in grey. B (right panel): resting (basal) metabolic rate.

The mathematical challenges of hydrodynamics are formidable, and at present there is no consensus as to whether the WBE model is internally consistent or not[30]. Analysis of the assumptions of the model has shown that these are generally met in the mammalian vascular system (which is the system usually studied), although the assumption of volume-filling may not be met, and the assumption that the size of capillaries is invariant across mammals has long been known to be incorrect[31]. It would appear that while some of the assumptions are consistent with real (that is, mammalian) vascular networks, others are not supported by empirical data. What is not clear at present is whether these deviations from the assumptions invalidate the WBE model, or simply require it to be modified slightly.

A number of physiologists were stimulated to tackle the problem afresh, and a series of alternative models was suggested. A summary of models developed to explain the scaling of metabolic rate is given in Table 12.4.

These models range from alternative formulations for a fractal-like vascular systems, a return to heat loss as the driving factor, evolutionary optimisations of factors other than the cost of vascular transport or quantum behaviour in the coupling between electron transport and ATP-synthase.

These new models have themselves generated considerable debate, and they pose an interesting conundrum. The traditional concept of model selection in science is derived mostly from an idealisation of what happens occasionally in physics: competing models are examined in terms of their predictions, and empirical data are used to distinguish between the models[32]. The problem with the models listed in Table 12.4 is that they all predict the same thing, namely a scaling exponent of ~ 0.75. This is because we know the answer from the start, as the scaling of resting metabolism with body mass has long been known to be ~0.75. This leaves us with the challenge of deciding between competing models with identical predictions. At present all of these models remain under consideration, though it is the WBE model that is established most firmly in the ecological literature.

Table 12.4 Summary of models that have been proposed to explain the scaling of metabolic rate with body mass[33]

Mechanism	Study
Models proposed prior to WBE	
Heat loss (Surface Law)	1
Elastic similarity	2
Fractal structure of exchange surfaces	3
Models proposed after WBE	
Body size optimisation	4
Matching supply to demand in optimal transportation networks	5
Optimisation of oxygen delivery to tissues	6
Optimisation between constraints of uptake/loss and power generation ('metabolic boundaries')	7
Efficiency in a fractal-like hierarchical supply network	8
Heat loss under thermoneutral conditions (endotherms only)	9
Avoidance of hyperthermia	10
Quantum metabolism	11

12.4 The influence of temperature

A few years after the publication of the WBE model, Jamie Gillooly and colleagues added a term to capture the dependence of metabolic rate on temperature[34]. This temperature term is a simple Boltzmann factor:

$$e^{-\frac{E}{k_B T}}$$

where E is an apparent activation energy, k_B is Boltzmann's constant and T is the absolute (thermodynamic) temperature. The Boltzmann factor is well known to physicists as it derives from the Maxwell–Boltzmann distribution of molecular energies (see Chapter 2), and it appears in the Arrhenius equation which describes the effect of temperature on reaction rate (which we examined in Chapter 7).

The use of a simple Boltzmann factor to capture the temperature dependence of metabolism was based on the argument that temperature governs metabolism through its effects on the rates of biochemical reactions, and that reaction kinetics vary with temperature according to a Boltzmann factor.

Here the apparent activation energy E was taken to represent an average value for the rate-limiting enzyme-catalysed biochemical reactions of metabolism (though these rate-limiting reactions were not identified). Based on literature values, E was assumed to vary between 0.2 and 1.2 eV, with an average of ~ 0.6 eV. This is actually a very wide range of temperature sensitivities, equivalent to a range of Q_{10} values from 1.31 to 5.06 at a median temperature of 20 °C. In later publications the range of E values predicted was narrowed to 0.6–0.7 (equivalent to Q_{10} values of 2.25–2.57 at 20 °C), with a mean of 0.65 eV[35].

The concept of activation energy was introduced by physical chemists to explain the marked sensitivity of some reactions to temperature (see Chapter 7). The concept proved very powerful, but the only way that the activation energy for a particular reaction could be estimated was from measurements of the temperature sensitivity of that reaction. The activation energy of a reaction is now formally *defined* as its temperature sensitivity[36], and so using a measured activation energy to predict temperature sensitivity is perilously close to circularity. In essence it is arguing that the temperature sensitivity of some reactions associated with respiration is equivalent to an E of ~0.65 eV, so we predict that respiration in other organisms will exhibit similar temperature sensitivity. Given the highly conserved nature of intermediary metabolism across organisms, this is a fairly safe prediction.

The use of a Boltzmann factor to capture the influence of temperature on organisms has been criticised on two grounds. The first is that, as we saw in Chapter 7, the Arrhenius equation is only useful as a descriptor of reaction kinetics for simple reactions in the gas phase. For the more complex reactions of physiology, which typically involve one or more transition states, more complex equations are needed. The second criticism is that a simple mechanistic link between temperature and metabolism would imply that organisms have little or no control over their metabolism: if the temperature increases, then so must metabolism. A simple mechanistic link runs counter to all that we know of how intermediary metabolism is controlled, and how temperature influences resting (basal) metabolic rate (see Chapter 8).

Temperature sensitivity can be estimated from experimental data by fitting a statistical model that allows the scaling of metabolism with body mass and temperature to be estimated simultaneously, and any interactions to be assessed. Temperature sensitivities for different classes of vertebrates estimated from a multiple regression model vary from a Q_{10} of 2.81 in mammals to 2.01 in fish (Table 12.5). Although this suggests that mammals have a higher temperature sensitivity than other vertebrates, the differences in temperature sensitivity across classes are not statistically significant[37]. Conversion of the fitted exponential factors (Q_{10}) to apparent activation energies indicates that some fall outside the predicted range of 0.6–0.7 eV. Does this invalidate the proposed Boltzmann factor as a general descriptor of temperature sensitivity, or does it simply reflect the evolutionary variability that characterises much of ecology?

While the temperature sensitivity of BMR appears to be fairly similar across many taxa, BMR comprises only a fraction of routine metabolism (roughly 20–25%: Chapter 8). Day-to-day energy flux is dominated by processes for which we have no data on temperature sensitivity. Indeed it is perfectly possible that behavioural thermoregulation and trade-offs mean that for many organisms average daily energy expenditure is largely independent of temperature[38].

12.5 The central equation of the Metabolic Theory of Ecology

Combining the Boltzmann temperature correction with the scaling relationship derived from the WBE model produces what has become known as the central equation of the MTE:

$$B = B_0 M^b e^{-\frac{E}{k_B T}}$$

This equation is now firmly established in the ecological literature and has been used extensively to explore the influence of body size, body temperature and metabolic rate on a wide range of physiological and ecological processes.

Table 12.5 Temperature sensitivity of resting or standard metabolism. Q_{10} calculated over the range of body temperature in the data set. The apparent activation energy was calculated for the stated reference temperature. n = number of species (one data point per species, except for those marked * where all data were used)[39].

Taxon	n	Temperature range (°C)	Q_{10}	Reference temperature (°C)	Apparent activation energy	
					eV	kJ mol^{-1}
Vertebrates						
Mammals	469	30–41	2.8	35	0.84	81.4
Birds	83	36–42	2.1	40	0.62	60.1
Reptiles	147	16–40	2.3	30	0.65	62.6
Amphibians	150	10–30	2.0	20	0.53	51.2
Fish	69	−1.5–37	2.0	20	0.52	49.9
Invertebrates						
Insects*	346	nd	2.3	20	0.61	59.2
Ants*	7	20–28	2.4	25	0.65	63.1
Bivalves*	nd	−0.3–30	2.1	20	0.52	51.3
Epipelagic copepods*	35	−1.4–30	2.0	20	0.51	49.2

12.5.1 Earlier statistical models

This was not the first equation linking a temperature term with a scaling relationship for metabolic rate and body mass. This had been done almost 15 years previously by Richard Robinson and colleagues[40]. Their equation was:

$$B = aM^b e^{cT}$$

Here B is the metabolic rate (as captured by the rate of oxygen consumption), M is the body mass (g), T is the temperature (°C), and a, b and c are constants whose values are derived by fitting to data. This differs from the MTE equation in two aspects: the scaling exponent, b, is fitted empirically rather than being defined by theory and the exponential temperature factor has a formulation akin to the Berthelot rather than Arrhenius model (see Chapter 7). Several subsequent analyses of metabolic rate, notably for marine invertebrates, also used a combination of a power law to capture the scaling with mass and an exponential factor for the variation with temperature, and were widely used by biological oceanographers before the central equation of the MTE was published[41].

These equations were all conceived as statistical models, and were typically fitted as multiple regressions using least-squares. They are simple models that capture important relationships; they differ significantly from the central equation of the MTE in that all parameters are free, and the statistical analysis allows for the significance of interactions to be assessed[42].

12.6 The status of the MTE as a theory

The Metabolic Theory of Ecology has become an important strand in modern ecology. It is widely accepted as a valid mechanistic explanation for both the scaling of metabolism and its temperature sensitivity. Its very name, the Metabolic Theory of Ecology, suggests something fundamental and many ecologists regard the MTE central equation as the product of solid physical theory, and one that can be used to explore aspects of ecology on very broad scales[43]. For others, the MTE offers nothing that is new, or of value. So where, almost two decades on, does the MTE stand now?

The predicted value for the scaling exponent follows from an idealised physical model of the vertebrate cardiovascular system and while there are several competing theories, the WBE model is generally accepted as the best mechanistic explanation we have for the scaling of metabolic rate in vertebrates. The model does not, however, provide an estimate of the normalisation constant, B_0, which has to be fitted empirically.

An unresolved aspect of the MTE is that most animals are not vertebrates and many have a circulatory system that is very different from the idealised vertebrate system that forms the basis of the WBE model. For example sponges have no circulatory system at all, and planarian flatworms have a simple gastrovascular cavity allowing both nutrients and oxygen to reach the cells by diffusion alone. Many other invertebrates have open circulatory systems in which there is no distinction between haemolymph and interstitial fluids. The haemolymph is pumped by a heart through a series of networked body cavities, bathing the organs, and then drains back into the main body cavity. This is typical of many arthropods and molluscs (though cephalopod molluscs and many crustaceans have a more complex, almost closed, system). Of particular interest are insects, where the nutrients are supplied through an open circulatory system but oxygen is supplied through the tracheae.

Despite these very different architectures, many invertebrates often have metabolic rates that scale with $b \sim 0.75$, although some groups have higher or lower exponents. For example pelagic cnidarians (jellyfish) have a simple gastrovascular cavity, but can reach very large size, and their metabolic rate has a scaling exponent of ~ 0.78[44]. In insects as a whole, the scaling exponent for resting metabolic rate is 0.82. The scaling, however, varies across insect orders and some have $b \sim 0.75$; for insects as a whole $b \sim 0.75$ after allowing for phylogenetic effects[45]. Do these results for invertebrates mean that the circulatory architecture is immaterial and the scaling of metabolism is actually determined by something else entirely, or that even invertebrates have sufficient of their circulatory system in a sufficiently fractal-like architecture that the WBE model still applies? This is an unresolved issue that goes to the heart of the generality, or lack of it, of the MTE. It has been argued that the theory can be extended to isolated cells, mitochondria, the respiratory complexes of the inner mitochondrial membrane, even down to individual citrate synthase enzyme molecules. For this to work, it must be assumed that *bacteria and mitochondria have self-similar metabolic pathways with fractal-like networks that could be real or virtual and the terminal units of which are respiratory complexes*. Quite what these real or virtual fractal-like networks are is not clear[46].

In the original presentation of their model, WBE identified bryozoans and planarian flatworms as examples of effectively two-dimensional organisms where the scaling of metabolic rate would likely differ from vertebrates. There appear to be no studies of metabolic scaling in flatworms, but in bryozoan colonies, scaling varies from 0.84 to 1.13[47]. Interestingly, unicells were at first believed to match the model, but the scaling of metabolism in these is now known to be different, with $b \sim 1$ in many cases. In active bacteria, metabolism scales with $b \sim 1.96$, in protists $b \sim 1.0$ for both inactive and growing cells, and for unicellular metazoans $b \sim 0.76$ when inactive and ~ 0.8 when active[48].

At present we lack a clear picture of which organisms do and do not match the prediction $b \sim 0.75$, but this is an important area for resolution as it may well prove to be the ground on which we can distinguish between the competing models and establish the generality of the WBE model, and hence the MTE.

12.7 MTE and mammals revisited

The WBE model was developed for a cardiovascular system essentially similar to that of a mammal. While the model predicted curvilinear scaling, this was precisely opposite to that observed[49]. The curvature is small, but it is real, and a quadratic statistical model captures this curvilinearity well. There is no indication of significant curvature in other vertebrate classes, and while slight curvature been detected in terrestrial invertebrates this is evident at only the very smallest sizes (where data points are few) and is greatly reduced once taxonomic structure is allowed for[50].

The result of this curvature is that the resting metabolic rate of small mammals (< ~ 150 g) is greater than expected on the basis of linear scaling of larger mammals (Figure 12.4). This means that these small mammals generate relatively more heat, but since they maintain a similar body temperature to larger mammals, there must be an increased loss of heat at smaller sizes. In mammals thermal conductance per unit surface area does not vary with size (see Chapter 4), and so perhaps ironically this is a return to surface area as an important factor in the scaling of resting metabolic rate in mammals, but only at the smaller sizes[51].

It is a key principle of the WBE model that the distribution network sets the metabolic rate through the delivery of nutrients to the cells. While under normal conditions there is a dynamic balance between the supply of nutrients to cells and the demand for their use, this balance poses an interesting question: is the metabolic rate of a cell determined by the rate at which nutrients are supplied to that cell, or does the metabolic rate of the cell determine the rate at which nutrients must be supplied?

The answer depends on the perspective from which the question is posed. The WBE model predicts the underlying pattern that emerges when natural selection minimises transport costs. This is a reasonable assumption and the predicted scaling is an important central tendency. The WBE model does not preclude variation about this central tendency associated with ecology, or with short-term variations in energy expenditure (and hence demand) by cells. The control of metabolism at the cellular, tissue and organismal levels (Chapter 8) is subtle and complex. It affects the level of metabolism (captured by the normalisation constant, B_0), but not the scaling.

The way metabolism is regulated within the cell does mean, however, that the influence of temperature cannot be captured mechanistically by a simple temperature correction factor, be it Boltzmann or an exponential (Berthelot) formulation. A Boltzmann factor does capture the emergent relationship between metabolic rate and temperature, and as such it is a useful statistical description of the way organisms behave on average.

12.8 Does the Metabolic Theory of Ecology constitute a general theory of thermal biology?

The WBE model provides a physical explanation for the scaling of metabolic rate with body mass in animals with a fractal-like cardiovascular system, and its predictions are largely upheld in vertebrates. What is not resolved is why many (but by no means all) invertebrates with very different circulatory systems should exhibit similar scaling patterns. Bacteria and unicellular eukaryotes appear to exhibit quite different scaling. This is perhaps to be expected, for neither have circulatory systems, but this does prescribe limits to the generality of the WBE model.

The Boltzmann factor is a useful statistical description of the emergent relationship between metabolic rate and temperature, but cannot be regarded as mechanistic. In statistical terms it is no better, or worse, than an exponential (Q_{10}) factor.

The two parts of the central equation of the Metabolic Theory of Ecology thus have rather different status. Overall the MTE offers a valuable way to capture two important central tendencies in ecology: the way metabolic rate varies with size and the way it varies with temperature. It is useful in exploring how the flow of energy through an organism might influence its overall ecology. The WBE model thus provides a description of the scaling of metabolic rate in vertebrates that is well grounded in physics, but its unclear relevance to invertebrates and the statistical rather than mechanistic nature of the Boltzmann factor mean that the MTE as a whole cannot be regarded as a physical model. Extrapolation to new arenas therefore needs to be done with care. But if nothing else, the WBE model has provided a valuable stimulus for ecologists to return to one of the important roots of their subject, the flow of energy.

12.9 Summary

1. The model of West, Brown and Enquist (WBE) is built on the assumption that the metabolic rate of cells is determined by the architecture of the vascular network that supplies them with oxygen and nutrients.

2. For a fractal-like network, and assuming that evolution has minimised cardiovascular costs, the WBE model predicts a scaling exponent, b, of 0.75 at infinite size, and $b \sim 0.8$ at realistic larger sizes. The original model also predicts that the scaling should steepen (that is, b should increase) at the smallest sizes. Despite these explicit caveats, the WBE model is usually taken to predict $b = 0.75$. Subsequent analyses have shown that not all of the necessary simplifying assumptions underpinning the WBE model are met in real vascular networks, but that this does not influence the predictions significantly.

3. Scaling exponents ~ 0.75 for standard or resting metabolic rate are observed widely, but far from universally, including in some invertebrates with cardiovascular systems very different from that assumed in the WBE model.

4. Testing the WBE model with data for standard or resting metabolic rate is convenient (there are many such data) but a more relevant test evolutionarily is the scaling of routine metabolic rate or daily energy expenditure. Data for field metabolic rate in vertebrates typically exhibit $b \sim 0.8$, which matches the WBE prediction very well.

5. Addition of a simple Boltzmann factor to capture the effects of body temperature on metabolic rate yields the central equation of the Metabolic Theory of Ecology (MTE). The MTE posits an apparent activation energy of 0.6–0.7 eV (58–68 kJ mol^{-1}), equivalent to a Q_{10} of 2.3–2.6 at 20 °C.

6. The MTE has become an important strand in ecology, and the WBE model is the most widely accepted physical explanation for the scaling of metabolic rate with body mass. Capturing the effect of temperature through a Boltzmann factor is a useful statistical description but too simple to qualify as a complete physical theory of thermal ecology.

Notes

1. This quotation comes from Chapter 1 of Thompson's classic book *On growth and form* (1961 abridged version, edited by John Tyler Bonner). The original was published in 1917.

2. Here we will concentrate on the model developed for animals (West et al. 1997). West et al. (1999a, b) describe closely similar models for plants and unicells.

3. It is also worth reading the short essay *On being the right size* by J.B.S. Haldane. Always known simply by his initials, Haldane (1892–1964) was an English mathematician and biologist, a committed socialist and one of the most gifted popularisers of science. Høyrup (2002) and Fara (2009) provide an introduction to Babylonian geometry.

4. See Rubner (1883).

5. The critical paper here is that by a professor of mathematics, Pierre Sarrus, and a doctor, Jean-François Rameaux, both from Strasbourg (then, prior to the Franco-Prussian war, part of France). Their paper was read before the Académie Royale de Médecine (now the Académie Nationale de Médecine) in Paris on 23 July 1839, having been tabled earlier in the year on 13 March. Based on the observed constancy of the extraction coefficient of oxygen in the lungs and simple biometrical considerations, they concluded that the heat generated by an organism must be proportional to the body surface area through which it is dissipated. The usual citation for this work (Sarrus and Rameaux 1839) is actually a short verbal summary of the paper, which was itself never published, though the detailed ideas were eventually presented over two decades later (Rameaux 1857, 1858). These papers are the origin of the *Surface Law* relating heat production by animals to their surface area. A valuable summary of the early stages in the conceptual development of the relationship between body mass and metabolic rate is given by Heusner (1985), and Heusner (1991) discusses the many experiments on domesticated dogs that followed Rubner's original study.

6. Plotted from original data in Rubner (1883).

7. See Huxley (1932). Julian Huxley (1887–1975) was a grandson of Thomas Henry Huxley, famous for his robust defence of Darwin, brother of the novelist Aldous Huxley and half-brother of the physiologist Andrew Huxley (Nobel laureate in Physiology & Medicine in 1963 for his work on nerve conduction).

8. See E. White et al. (2012) for a useful discussion of this problem.

9. Although this problem was known to Huxley, and discussed in his book (Huxley 1932), recognition of the problem of dimensions in transcendental functions is absent from almost all subsequent scaling analyses in the ecological literature. For an exception see Kooijman (1993, 2010).

10. The exponent of 0.67 arises because this is the ratio of surface area to volume in geometrically similar bodies, though this only applies if the dogs do not change shape as well as size.
11. Krogh summarised his work on respiration in a classic book, *The respiratory exchange of animals and man* (Krogh 1916), from which these conclusions are drawn.
12. Brody (1945) gives a thorough account of the struggle to measure the surface area of domesticated animals, and test the Surface Law.
13. Remember that this was before the development of the electronic calculators that make statistics so easy today. In Kleiber's day the use of fractional exponents required the tedious use of logarithmic tables. An exponent of 0.75 could, however, be manipulated relatively simply with a slide rule, by taking the square root of the square root of the cube of the body mass (Kleiber 1961, p. 217). In making this suggestion, Kleiber was well aware that his proposed value of 0.75 had no theoretical foundation; it was simply a useful approximation to the empirical result (Kleiber 1932). Heusner (1985) has commented that the widespread acceptance of a scaling exponent of 0.75 as a canonical value is an example of what the mathematician Georg Cantor has called the 'law of conservation of ignorance': *a false conclusion once arrived at and widely accepted is not easily dislodged and the less it is understood the more tenaciously it is held* (quoted by Kline 1980, p. 88).
14. Plotted from original data in Benedict (1938) and Kleiber (1961).
15. The studies are 1: Benedict (1938), 2: Brody (1945), 3: Kleiber (1932), 4: Kleiber (1961), 5: Bartels (1982), 6: McNab (1988), 7: Heusner (1991).
16. Both Hemmingsen and Zeuthen were students of August Krogh. The key publications are Zeuthen (1947, 1953) and Hemmingsen (1950, 1960).
17. The studies are 1: Aschoff & Pohl (1970), 2: Zar (1969), 3: Bennett & Dawson (1976), 4: Galvão et al. (1965), 5: Winberg (1960), 6: Ivleva (1980), 7: Hemmingsen (1960).
18. These books are Peters (1983), McMahon & Bonner (1983), Calder (1984) and Schmidt-Nielsen (1984). All are worth consulting. Peters and Calder provide valuable and comprehensive summaries of the scaling of metabolism as well as many other anatomical and physiological attributes, whereas Schmidt-Nielsen gives a clear and succinct account of the development of the field and the underlying physiology.
19. Lawton (1999). In this typically provocative article the British ecologist John Lawton has much of value to say about the differences between physics and biology in terms of 'laws' and the role of scale in biology.
20. This model has, confusingly, also been assigned different acronyms by other authors. These include the *Arrhenius fractal supply*, AFS, model (Downs et al. 2008), and *fractal network theory*, FNT (Agutter & Tuszynski 2011). It is difficult to see how inventing new names and acronyms for the same thing is helpful to the development of ideas.
21. Huxley & Teissier (1936), which was also published in French in the same year. Gayon (2000) provides a useful history of the concept of allometry.
22. A zeroth-order approximation is basically an educated guess, or order-of-magnitude estimate. The predicted value of 0.75 for the scaling exponent is actually more precise than a zeroth-order approximation. The quotations in italics are taken directly from the original presentation of the model (West et al. 1997).
23. The WBE model is built around minimising the cost of the supply of oxygen and nutrients to cells. The most rigorous test of its validity would therefore involve comparing the observed scaling of this supply rate with that predicted by the model, and in the initial presentation of the model (West et al. 1997) this is precisely what was done. The model predicts a scaling relationship for 'metabolism', and it is around metabolic rate that the subsequent full Metabolic Theory of Ecology was largely presented. In discussions of the WBE model, metabolic rate is typically defined very generally, for example as *the rate at which energy and materials are taken up from the environment and used for maintenance, growth, and reproduction* (Gillooly et al. 2005). For most circumstances, the fact that oxygen consumption captures only part of metabolism as so defined is relatively unproblematic. It is absolutely critical, however, when the theory is applied to growth (see Chapter 13).
24. WBE did acknowledge this point (West & Brown 2005), though subsequent discussions of the WBE model remain centred on BMR. In a mammal fleeing a predator the short-term demand for ATP may exceed the rate at which oxygen and nutrients can be supplied to the muscle cells. ATP is then generated anaerobically from storage compounds such as creatine phosphate, with the lactate that accumulates being cleared subsequently.
25. Field metabolic rate data from Hudson et al. (2013), and resting metabolic rate data from Clarke et al. (2010).
26. The BMR data for these analyses come from Clarke et al. (2010) for mammals, White et al. (2006) for birds and reptiles. The FMR data come from Hudson et al. (2013) and Nagy (2005). The maximum MR data for mammals come from Weibel & Hoppeler (2005).
27. Replotted from data in Weibel & Hoppeler (2005) and Weibel et al. (2004).

28. Interestingly, in later publications the prediction concerning curvature at smaller sizes changes. In a refinement of the WBE model (West & Brown 2004) it was argued that as size decreases the balance between pulsatile and Poiseuille flow in the cardiovascular system changes, and in consequence the scaling exponent for metabolic rate *should depend weakly* on body mass, and that this leads to the prediction that the scaling exponent *should decrease below 3/4* in smaller mammals, as observed (West & Brown 2005). The explanation is that smaller mammals dissipate more energy in their cardiovascular networks, but this changes the original prediction (that scaling exponent should increase at smaller sizes: West et al. 1997), to its exact opposite, but one that now matches empirical data.

29. Savage et al. (2008).

30. Dodds et al. (2001), Agutter & Wheatley (2004), Etienne et al. (2006), van der Meer (2006), Martínez del Rio (2008) and Apol et al. (2008).

31. See Price et al. (2012) for a full discussion of this. Huo & Kassab (2012) showed that the length ratio of blood vessels deviates from volume-filling in a number of mammals, and Dawson (2001, 2003, 2010), using data from Gehr et al. (1981), has shown that the assumption capillary size is invariant across mammals is not upheld.

32. In one of his classic Lectures on Physics, Richard Feynman describes the testing of theory as follows: *If it disagrees with experiment, it's wrong. In that simple statement is the key to science. It doesn't make any difference how beautiful your guess is, it doesn't matter how smart you are who made the guess, or what his name is … If it disagrees with experiment, it's wrong. That's all there is to it.* (Feynman et al. 1966). It isn't always that simple, of course, because theoreticians dislike abandoning their theories and the outcome is often a modification of the theory rather than its complete abandonment. Ecology is also characterised by variance: unlike molecules, atoms or fundamental particles, not all individuals or species behave the same. In consequence, estimates of ecological variables or parameters typically come with wider confidence intervals than is typical of physics or chemistry. Deciding whether or not an observation rejects a theory is not always as straightforward as portrayed by many philosophers of science (see Ziman 1978 for a valuable discussion).

33. The models are 1: Sarrus & Rameaux (1839), Rameaux (1857, 1858), 2: McMahon (1973), 3: Sernetz et al. (1985), 4: Kozłowski & Weiner (1997), 5: Banavar et al. (1999, 2002), 6: Santillán (2003), 7: Glazier (2008, 2010), 8: Banavar et al. (2010), 9: Roberts et al. (2010), 10: Speakman & Król (2010), 11: Demetrius (2003, 2006), Demetrius & Tuszynski (2010) and Agutter & Tuszynski (2011).

34. Gillooly et al. (2001).

35. In the original paper, two small data sets were used in support of the suggested apparent activation energy. These were the activation energies of citrate synthase in four species of polar crustacean (Vetter 1995), and respiratory electron chain activity in six species of unicellular algae (Raven & Geider 1988). The narrower range of predicted activation energies is presented by Brown et al. (2004).

36. This is the formal definition of activation energy by the International Union of Pure and Applied Chemistry: Laidler (1981, 1984).

37. General Linear Model; $F_{5,556} = 1.63, p = 0.203$. The lack of significance arises partly because the estimates of temperature sensitivity for birds and mammals are based on a fairly narrow range of body temperatures.

38. The subject of how organisms use behavioural mechanisms to influence their thermal biology is treated in depth by Angilletta (2009).

39. Data are from White et al. (2006), Clarke et al. (2010) for vertebrates, and Addo-Bediako et al. (2002), Ikeda et al. (2001) and Peck & Conway (2000) for invertebrates.

40. Robinson et al. (1983).

41. Ivleva (1980), Ikeda (1985), Ikeda et al. (2001).

42. This is a subtle but important point. In many studies utilising the MTE, the effect of body mass is removed by correcting the data assuming $b = 0.75$ and the data then analysed for temperature effects, or the effect of temperature is allowed for assuming $E = 0.65$ eV and body mass effects explored. This is equivalent to using residuals, which is not always a safe procedure if there are interactions (see Freckleton 2009). The use of a general linear model with all parameters free to be fitted gets around this problem.

43. Perhaps the most uncritical advocacy is Whitfield (2006), which includes the remarkable claim that the Metabolic Theory of Ecology is *triggering a revolution as potentially important to biology as Newton's insights were to physics.*

44. Ehnes et al. (2011) explore curvature of scaling in terrestrial invertebrates. The study of cnidaria by Thuesen & Childress (1994) combined data from a range of Hydromedusae and Scyphomedusae; metabolism was expressed on a mass-specific basis (wet mass, which in cnidaria is mostly water), and the scaling exponent was thus −0.22.

45. See Chown et al. (2007) and Riveros & Enquist (2011).

46. West et al. (2002). Quotations in italics are taken directly from the paper.

47. See Hughes & Hughes (1986) for *Electra pilosa*, and Peck & Barnes (2004) for studies of three Antarctic bryozoans. Bryozoans are clonal organisms, though the individual modules are connected to allow transport of nutrients.

48. See Okie (2012).
49. Curvature is mentioned in passing in the original paper (West et al. 1997) and discussed in more detail by Savage et al. (2008). As noted above, later publications reversed the prediction, and the revised prediction matches observations well. Neither the original nor revised prediction was presented quantitatively.
50. See Ehnes et al. (2011).
51. It also means that the value of a linear scaling exponent fitted to mammalian data will depend on the size range of species in the data set: the more small mammals are included the less steep the estimated scaling exponent. This effect can explain the steeper scaling estimated in the early historical studies, as these were concerned primarily with larger species (Clarke et al. 2010).

CHAPTER 13

Temperature, growth and size

Size is probably the most commonly measured and manipulated character in biology. Despite its importance, we know remarkably little about how the developmental processes underlying growth relate to natural variation in body size.

Jeff Arendt[1]

Two key features of an organism's life history are its size and how fast it grows. Both are affected by temperature, but in different ways. We will start by examining growth, and the first problem we need to resolve is how best to quantify an organism's growth rate.

13.1 How should we measure growth rate?

Many organisms grow slowly at first, with growth rate then increasing during the juvenile stages before slowing again as they become adult. This produces a sigmoidal (S-shaped) growth curve in which absolute growth rate peaks at the point of inflection, and relative growth rate (growth rate divided by body mass) declines throughout life (Figure 13.1). For organisms whose development involves a drastic change in shape (for example, holometabolous insects and many marine invertebrates), then the growth curve can assume a more complex shape[2]. In some species the final size is asymptotic or the adult has a defined size, and growth is said to be determinate; examples are mammals, birds and insects. In others, growth continues as an adult, albeit slowly, and examples of this indeterminate growth include reptiles, fish and many invertebrates.

The simplest measure of growth rate would be to divide the adult size or mass by the time taken to achieve this. This provides a growth rate averaged over the life history but obscures the changes in growth rate during ontogeny. While clearly a very crude measure, for many extinct organisms this may be the only viable approach (though it does assume we can determine how old the fossil animal was when it died, independently of its size).

As ecologists determined the pattern of growth in more and more species, they searched for a general model that might describe these. A simple and much used mathematical function that captures sigmoidal growth is that derived by August Pütter:

$$\frac{dm}{dt} = am^\alpha - bm^\beta$$

Here m is the mass at time t (so dm/dt is the instantaneous growth rate at time t), α and β are exponents that characterise the shape of the growth curve and a and b are fitted coefficients. For growth to cease at some maximum size, α must be $< \beta$. The values of a, b, α and β determine the maximum size and the point in the growth trajectory at which growth rate is maximal. Where $\alpha = 1$ and $\beta = 2$, the equation corresponds to the logistic growth model (as shown in Figure 13.1)[3]. The Pütter equation has proved enormously useful in that it allows the growth trajectory of a wide variety of animals to be captured with relatively few parameters, and the growth rates of different species to be compared simply.

In the Pütter equation, growth rate is a linear function of mass; other growth equations have growth rate as a logarithmic function of mass (the Gompertz function) or a power function of mass

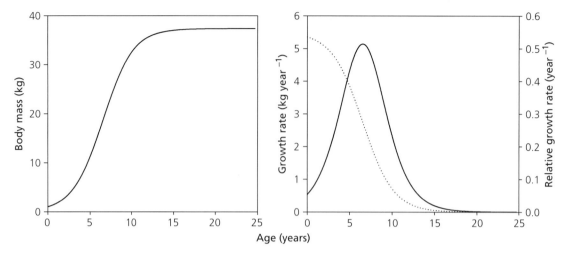

Figure 13.1 Hypothetical growth curve for an organism exhibiting simple logistic growth. A (left panel): the growth trajectory, body mass as a function of age. B (right panel): instantaneous absolute growth rate (solid curve) and relative growth rate (dotted line) for the same growth trajectory[4].

(the Richards function). These equations are essentially descriptive (what physicists call phenomenological), although some were based on physiological arguments. In particular Ludwig von Bertalanffy argued, based on Pütter's original work, that growth resulted from a balance between catabolic processes utilising reserves and anabolic processes synthesising new tissues. Assuming that anabolism (synthesis) was limited by the surface area over which the organism gains nutrients whereas catabolism was a function of body mass, von Bertalanffy suggested that $\alpha = 2/3$ (this being the ratio of surface area to body mass) and, on the basis that anabolic processes occur throughout the body, $\beta = 1$. This equation can be solved analytically and in mass terms it is:

$$M_t = M_\infty \left(1 - e^{-K(t-t_0)}\right)^3$$

Here M_t is the mass at time t, M_∞ is the asymptotic mass and t_0 is the hypothetical time at which mass is zero. K, the von Bertalanffy growth coefficient, defines the rate at which the growth curve approaches M_∞. In a more complex version of his model, von Bertalanffy recognised a series of metabolic and growth rate 'types' defined by different scaling of metabolic rate with mass. The basic model has also been modified to describe species where growth is seasonal. An alternative physiological interpretation came from Michael Reiss who proposed that the first term (am^α) represented the assimilation and the second (bm^β) the metabolic costs of existence[5].

The von Bertalanffy growth equation has long been popular with biologists interested in growth. This is despite the assumed exponents being shown to be unrepresentative of the real world: for example an early comparative study of fish growth suggested that the best estimate of α was 0.59 (rather than 2/3) and $\beta = 0.83$ (rather than 1)[6]. Recently ecologists have revisited the problem, attempting to build new growth models from physiological principles. Before we look at these we need to establish the key energetic features of growth, and how these are affected by temperature.

13.2 The energetics of growth

Growth requires both raw materials and energy. The raw materials include amino acids for proteins, simple sugars for polysaccharides, purines and pyrimidines for nucleic acids and fatty acids for lipids. The energy required to synthesise macromolecules from monomers is supplied by ATP or GTP. The nutrients that provide both the raw materials for synthesis and substrates for the regeneration of ATP by intermediary metabolism are carried to where they are required by the circulatory system. Once the nutrients reach the cell they are used either as raw material or to regenerate ATP (Figure 13.2).

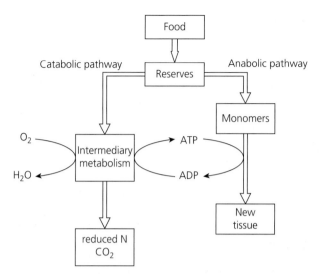

Figure 13.2 A simplified schematic of energy flow in growth. The chemical potential energy that flows through the anabolic pathway is retained in the new tissue whereas that flowing down the catabolic pathway is largely dissipated as heat. Note that while macromolecular synthesis uses other nucleotide phosphates (notably GTP in protein synthesis), these are regenerated from ATP.

We saw in Chapter 8 that only part of the chemical potential energy in nutrients entering glycolysis and the TCA cycle is captured in ATP, with the remainder being dissipated as heat. When ATP or GTP is used to link monomers into a macromolecule, again only part of the chemical potential energy is retained in the new bond, and the bulk is dissipated as heat (Box 13.1). Most of the energy that passes down the catabolic pathway is thus dissipated as heat and only a small fraction is incorporated into the newly synthesised tissue. This is not inefficiency; it drives the change in entropy that is required for the synthesis of organised macromolecules to be energetically favoured. The synthesis of macromolecules creates a local area of low entropy which is possible only if at the same time the entropy of the environment increases sufficiently for the total entropy (organism plus environment) to increase. This is achieved by the dissipation of energy as heat.

This brief description tells us something important: there are two energetic streams involved in growth, namely the chemical potential energy retained in the new tissue and the metabolic energy used to construct that new tissue. The former is traditionally estimated from the enthalpy of combustion (typically using bomb calorimetry) and the latter from oxygen consumption. A correct

---

Box 13.1 Cost of growth at the molecular level

The synthesis of proteins, polysaccharides, nucleic acids and phospholipids all have an important feature in common: the bonds linking the subunit monomers are all formed by condensation (dehydration) reactions, involving the removal of the elements of water as the bond is formed. The energy for bond formation is provided by ATP or GTP, and in all cases there is a marked decrease in entropy and an increase in Gibbs energy compared with the free monomers.

The formation of a peptide bond between two amino acids requires hydrolysis of the equivalent of four ATP molecules. Of the roughly 200 kJ mol^{-1} released under cellular conditions, about 8–16 kJ mol^{-1} is incorporated as chemical potential energy in the bond, with the remainder being dissipated as heat. Similarly, the formation of a glycosidic bond between two glucose monomers in the synthesis of glycogen requires the hydrolysis of 2 ATPs, of which ~ 17 kJ mol^{-1} is incorporated in the bond, and in the formation of triacylglycerol storage lipids or phospholipid structural lipids, only some of the free energy liberated by ATP hydrolysis is stored as chemical potential energy in the formation of the covalent bonds linking the fatty acids or phosphorylated base to the glycerol backbones[7].

treatment of the energetics of growth requires knowledge of both pathways. This has long been recognised by microbial ecologists, who routinely consider the role of entropy and heat dissipation in bacterial growth, but by relatively few ecologists concerned with the growth of plants, invertebrates or vertebrates[8].

Revisiting the balanced energy budget we explored in Chapter 4, we can express the total energy cost of synthesising new somatic tissue, E_s, as:

$$E_s = P_s + R_s$$

where P_s is the chemical potential energy sequestered in the new tissue and R_s is the metabolic cost of synthesising that tissue. R_s is termed variously the *metabolic* (or sometimes *calorimetric*) *cost of growth* or the *metabolic overhead* of synthesising new tissue. A better term might be the *thermodynamic cost of growth* because of its basis in entropy, although practical measures of R_s inevitably include metabolic costs additional to those associated with bond formation. If we assume that R_s is a constant proportion of R_s, (and we express both in the same units) then we can calculate a dimensionless fractional cost of growth, c:

$$c = \frac{R_s}{P_s}$$

and hence

$$E_s = P_s + cP_s$$

Because R_s is part of the total metabolic rate measured during growth, it is not easy to determine. While we can estimate a minimum cost of synthesising a protein from the ATP requirement for peptide bond synthesis, this ignores a large number of associated costs, such as RNA processing, transport within the cell, post-translational modification of the newly synthesised proteins, the recycling of proteins that fold incorrectly and so on. We can estimate an overall cost of growth, however, from the relationship between new tissue production and respiration. If we assume that the cost of growth, R_s is additional to the costs of tissue maintenance and also the general activity required in existence, then (after controlling for the effects of mass on both growth rate and metabolic rate) we would expect a positive linear relationship between the rate of growth and metabolic rate. The slope of this relationship reflects the thermodynamic cost of growth (a higher cost leading to a steeper slope). Data for a range of aquatic organisms suggests a value for c of ~0.32 (Figure 13.3)[9].

It is this metabolic cost of growth that underpins the widely observed correlation between respiration and production at the population level. Two

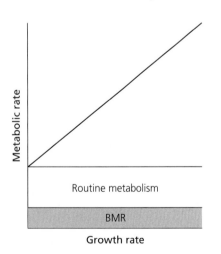

Figure 13.3 The cost of growth. A (left panel): a simple conceptual model showing the relationship between metabolic rate and growth rate for an individual organism. BMR: basal metabolic rate. The slope of the relationship between metabolic rate and growth rate reflects the thermodynamic cost of growth; a higher cost results in a steeper line. B (right panel): across-species relationship between rate of growth (production) and metabolic rate. The line is a least-squares regression with a slope of 0.32[10].

comprehensive studies have shown linear relationships between production and respiration in a range of animal populations over seven orders of magnitude of annual production. The relationships for endotherms and ectotherms exhibited similar slopes but were offset (with endotherms having a higher respiration rate for a given production rate). While measurements at a population level will be affected by a range of additional factors, such as the respiratory costs of non-growing individuals, the general result is precisely that to be expected from the thermodynamics of growth[11].

It is clear immediately that a simple estimate of the chemical potential energy content of new tissue underestimates the actual energy needed for growth by leaving out the metabolic cost of assembling that tissue. Equally a measure of energy dissipated during growth (for example by estimating metabolic rate by oxygen consumption) misses entirely the energy retained in the new tissue. The relationship between these two measures is shown schematically in Figure 13.4.

The best estimate of c is ~0.33 (Table 13.1). This exceeds the cost of peptide synthesis by a factor of about 3, indicating the importance of the associated costs. There is also an indication from work on isolated fish hepatocytes that the cost varies with the rate of protein synthesis, which suggests that while the basic cost of assembling a protein does not change (that is, the ATP cost for synthesis of peptide

Table 13.1 Estimates of the dimensionless cost of growth, c^{12}.

Study	C	Comments
1	0.12	Estimated for the marine copepod *Calanus hyperboreus* growing rapidly with growth fuelled from lipid reserves; calculated from the observed partial growth efficiency of 89%.
2	0.25–0.43	Derived from studies of a variety of domesticated and cultured vertebrates, taking extreme values of the partial growth efficiency to be 70% and 80%.
3	0.32	Estimated from the slope of the relationship between metabolic rate and growth rate in the amphibian *Bufo bufo*.
4	0.33	A consensus value derived from studies of fish and aquatic invertebrates.
5	0.32	Estimated from growth of the garter snake *Thamnophis sirtalis*. Original measure 1.67 kJ g^{-1} (wet mass) converted assuming a water content of 75% and a tissue energy content of 20.7 kJ g^{-1}.
6	0.51	Estimated from the slope of the relationship between metabolic rate and growth rate in the timber rattlesnake, *Crotalus horridus*.

bonds is invariant), the associated costs can. At present we do not know which of these associated costs changes, or why[13].

13.3 Temperature and growth

Growth at the cellular level involves a complex and carefully regulated sequence of chemical, mechanical and diffusional events. In eukaryotes this involves transcription of messenger RNA (mRNA) in the nucleus, modification of the mRNA (removal of introns), transfer of the mRNA to the ribosome in the cytoplasm, elongation of the new peptide, post-translational modification and often transport of the new protein to where it is needed. As we saw previously in discussing metabolism, the temperature dependence of such a complex system cannot be predicted theoretically; all we can do is examine the effect of temperature empirically and hope to describe this with relatively simple statistics.

One step in the chain where the temperature sensitivity has been studied in some detail is protein synthesis and an early study is shown in Figure 13.5. Here protein synthesis in isolated hepatocytes from

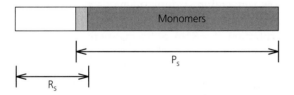

Figure 13.4 Estimating the total energy costs of growth. The chemical potential energy in the monomers is shown in dark grey, the additional chemical potential energy derived from ATP/GTP and incorporated into peptide bonds is shown in light grey and the energy dissipated in all other associated processes required for protein synthesis is shown in white. P_s and R_s are defined in the text, and the relative proportions of the three components are based on the synthesis of protein from amino acid monomers. The small overlap (double counting) arises because the standard conversion factors relating oxygen consumption to energy assume that all the chemical potential energy released from the cleavage of the phosphate bonds of ATP or GTP is dissipated as heat, whereas a small fraction is retained in the newly formed peptide bond.

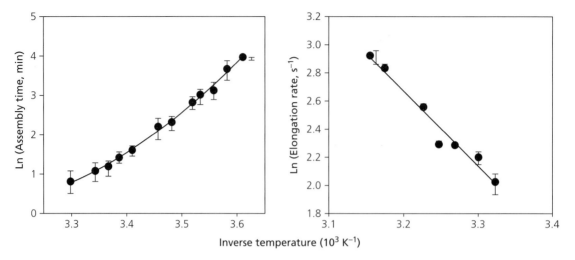

Figure 13.5 Effects of temperature on protein synthesis. A (left panel): assembly time for an average polypeptide in isolated hepatocytes from the toadfish, *Opsanus tau*. The fitted line is a quadratic (least-squares fit). The curvature indicates that the thermal sensitivity of the overall process is not described well by an Arrhenius model. B (right panel): elongation rate (amino acid residues added per second) of β-galactosidase in *Escherichia coli*. The fitted line is a least-squares linear regression. All data plotted as mean and SE[14].

the toadfish, *Opsanus tau*, proceeds faster (that is assembly time is shorter) at higher temperatures. The relationship does not, however, exhibit Arrhenius behaviour (it is curved, not straight). This study used radio-labelled amino acids and determined the rate of all protein synthesis. More recently it has become possible to examine the temperature sensitivity of synthesis of individual proteins, and the example shown is β-galactosidase in *Escherichia coli*. Here an Arrhenius model appears to describe the data well (Figure 13.5).

The overall rate of protein synthesis is influenced by a suite of factors including the rate of movement of tRNA to the ribosome, and internal thermal movements of the ribosomal protein. An extra complication is that the rate of translation is also affected by the genetic code itself: some codons promote fast translation, others slow it down. The latter are important, for the small delay they impart allows for error-checking: an incorrect tRNA codon match can bind only loosely to the mRNA and will likely be released by thermal motion before its amino acid is incorporated into the growing polypeptide. Even at the level of a single ribosome, the overall thermal dependency is dictated by a suite of factors, making the prediction of the rate dependency from first principles impossible. There are also important differences between bacteria and eukaryotes in this respect (Box 13.2).

The thermal sensitivity of polypeptide elongation means that ectothermic organisms living at low temperatures can synthesise proteins only slowly. In theory it might be possible to offset this rate limitation, at least partially, by one or more of thermal adaptation of the protein synthetic machinery (evolution of ribosomes that function faster at low temperatures), duplication of genes coding for particularly important proteins, increasing the concentration of mRNA or having more ribosomes in the cell.

At present we know little about the extent to which evolution has been able to modify the thermal characteristics of the ribosome itself. There is evidence for an increased copy number of the gene coding for antifreeze glycopeptides in Antarctic fish, and also for duplication of genes concerned with mitochondrial biogenesis and function (intermediary metabolism). A recent comparison of the Antarctic fish *Dissostichus mawsoni* with temperate and tropical teleosts indicated extensive gene duplication: 118 protein-coding genes, including many involved in protein synthesis and protein folding, were duplicated by between 3- and 300-fold.

Box 13.2 Protein synthesis in bacteria and eukaryotes

There are important differences in protein synthesis between prokaryotes (bacteria and archaeans) and eukaryotes. In eukaryotes mRNA is transcribed in the nucleus and extensively modified before being transported into the cytoplasm (introns are sliced out and a poly-adenine tail added). Once the mRNA is at the ribosome, there are at least ten initiation factors involved in transcription and overall protein synthesis is regulated by > 100 different control factors.

Bacteria have no nucleus, and so the spatial separation of transcription and translation is less. The mRNA does not have introns, and so translation can start immediately (and indeed it may do so before transcription is completed). Furthermore an individual mRNA may include the coding sequence for several genes on a given metabolic pathway, and it may be translated by many ribosomes simultaneously (forming a polyribosome or polysome). Taken together these factors mean that bacteria can synthesise a protein molecule roughly an order of magnitude faster than occurs in a eukaryote cell. Representative figures would be 10–20 peptide bonds per second in a bacterium at 37 °C, but only 2–10 in a eukaryote at the same temperature.

The median size of a eukaryote protein is ~360 amino acids (or ~400 when allowance is made for protein frequency)[15]. For a polypeptide synthesis rate of 2 amino acid residues per second, the assembly time for an average eukaryote protein is ~3 minutes.

Intriguingly, the more derived lineages in the notothenioid radiation in the Southern Ocean have larger genomes, with the more basal groups similar to the ancestral perciform fishes[16].

There is also a well-established correlation between growth rate and RNA concentration: all other things being equal, faster-growing individuals have a higher RNA content. This is usually expressed either as a higher RNA:DNA ratio or a higher RNA:protein ratio. The increased RNA could reflect any or all of an increased concentration of mRNA (through increased transcription and/or longer mRNA lifetime), increased tRNA concentration or more ribosomes. Comparative studies of ectotherms living at different temperatures indicate a strong relationship between RNA concentration (captured as RNA: protein ratio) and temperature, with RNA concentration increasing roughly three-fold from 25 °C to 0 °C, although there is much variation in the data (Figure 13.6).

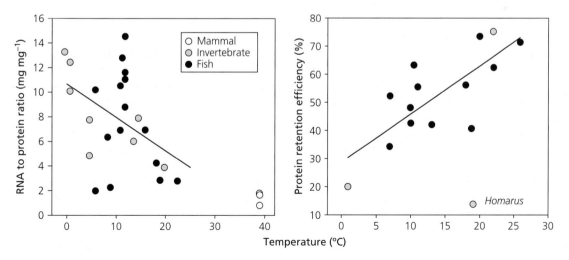

Figure 13.6 Temperature and protein synthesis. A (left panel): RNA to protein ratio in a range of fish and marine invertebrates, with data for *Rattus* as comparison. Least-squares regression line (slope = −0.27, $p < 0.001$) fitted to all data excluding *Rattus*. B (right panel): protein retention efficiency (protein retained as a fraction of total protein synthesis) as a function of temperature for fish and marine invertebrates. Least-squares regression (slope = 1.94, $p < 0.001$) fitted to all data excluding *Homarus*[17].

The across-species relationship is the exact reverse of that observed in within-species studies of growth rate, but there is a common logic: increasing RNA concentration allows for a faster rate of protein synthesis. Within species, faster growth rates are supported by a higher RNA concentration in the cell, whereas the relationship across species shown in Figure 13.6 arises through an increase in RNA to offset the low rate of polypeptide elongation at low temperature[18].

Once a polypeptide has been synthesised, it must fold into its active configuration if it is to become functional. This folding is typically entropy-driven (see Chapter 5), involving the burying of hydrophobic residues within the core of the protein and exposing charged residues that can hydrogen bond to water on the outside. Folding can start before the polypeptide is fully synthesised and is still attached to the ribosome. The folding is the result of random thermal movements, although quantum fluctuations (tunnelling) may be important, and folding is aided by specific proteins which increase the chance of correct folding. These molecular chaperones are many and varied; they include the so-called trigger factor, which is bound to the ribosome, chaperonins which encase a newly synthesised protein while it folds and heat-shock proteins such as Hsp70. Should a protein misfold it is rapidly identified and recycled. There are a variety of pathways for protein degradation in the cell, but a particularly important one is ubiquitination, whereby a protein that has misfolded, become unfolded or is otherwise no longer required, is tagged with the protein ubiquitin and broken down in the proteasome. This is important both in terms of cellular economy, but also because misfolded proteins can be dangerous (indeed a number of human diseases are caused by the accumulation of misfolded proteins, including Alzheimer's disease, Parkinson's disease, motor neurone diseases and muscular dystrophies)[19]. Protein recycling also serves to regulate the overall cellular protein concentration, manage amino acid availability and remove toxins. The homeostatic balance between protein synthesis and recycling is usually referred to as *proteostasis*.

The identification and recycling of misfolded or unfolded proteins represents an energetic cost to both routine existence (cellular maintenance) and the synthesis of new tissue: a fraction of all newly synthesised proteins are non-functional and immediately degraded. Both recognition and recycling require energy in the form of ATP, and the energy invested in synthesising the mis-folded proteins does not contribute to the energy sequestered in the new tissue. The extent of this recycling is usually quantified as the fraction of newly synthesised proteins that are retained in new tissue, and expressed as a protein retention efficiency (PRE). Data from aquatic ectotherms indicates a marked temperature dependence to PRE which drops from ~70% in fish living at 25 °C to ~30% in polar fish at 0 °C (Figure 13.6b): organisms living at low temperatures retain only a small fraction of newly synthesised protein compared with those with higher body temperatures.

This low protein retention efficiency represents a largely unrecognised metabolic cost to living at lower temperatures. As far as we know, the synthesis of a polypeptide by the ribosome costs no more at low temperatures (that is, the number of ATP and GTP molecules required to synthesise a peptide bond is independent of temperature), but for some reason fewer of the newly synthesised proteins are retained. At present we do not know why, but a number of studies point to protein folding being less effective at low temperatures. These include a higher constitutive expression of chaperone proteins in fish and marine invertebrates from Antarctica, and a higher incidence of ubiquinated proteins[20].

As with the processes underpinning maintenance costs discussed previously (Chapter 8), the pattern in Figure 13.6 is not simply a direct consequence of temperature. The primary structure (that is, the amino acid sequence) of proteins expressed in organisms living at low temperatures will be different from those in warm-temperature organisms, to ensure that their thermal characteristics match their environment: their structure is modified to ensure function. The observed PRE/temperature relationship is thus the result of a three-way trade-off between structure, function and thermal sensitivity. It would appear that a consequence of this is a reduced tendency to fold correctly at low temperatures. That is, proteins adapted to function at low temperature are more likely to misfold when synthesised than proteins adapted to function at a higher temperature, when both are synthesised at the temperature to which they are adapted to function. One possible factor here is the thermal sensitivity of the hydrophobic interaction itself, which is weaker at lower temperatures (Chapter 5). The relationship in Figure 13.6 suggests that this is a significant feature of energetics and worth further exploration at the molecular level.

So far we have explored the effect of temperature on the mechanisms underpinning growth in a cellular context. We now need to consider the whole organism growing in the wild, where growth rate is affected by factors additional to temperature, such as food availability and interactions with other organisms.

13.4 Growth and temperature in the natural environment

It is a commonplace observation that organisms grow faster when they are warmer. A typical example of this is shown in Figure 13.7. Here growth of *Manduca sexta* caterpillars in the laboratory increases with temperature up to ~35 °C, above which growth decreases, presumably because of sub-lethal effects of temperature on some aspect of the growth mechanism.

While the relationship is what we might expect from the temperature sensitivity of protein synthesis, we do not know what other factors may be at work. It is perfectly possible, for example, that temperature also limits the ability of the larva to gather food, or to digest and absorb the food it has eaten. Jonathan Jeschke and colleagues have suggested that the processing of food in the gut may act as a bottleneck limiting energy intake in some organisms[21].

The thermal sensitivity of growth exhibited by *Manduca* caterpillars follows from the effect of temperature on a suite of interacting physiological processes. For organisms in the wild, growth is affected by additional factors, both environmental and ecological. While it is difficult to see how these could produce a growth rate faster than the maximum rate observed under laboratory conditions, they may well reduce them. A critical ecological factor is the availability of food: without energy intake an organism cannot grow (although storage of lipid or glycogen means that food intake and growth can be separated in time). For example a polar organism whose annual growth is slow may have its growth rate constrained by the low temperature. Equally, however, it may be growing fast when it has resources, but then not growing (or even shrinking) in periods when food availability is low[22].

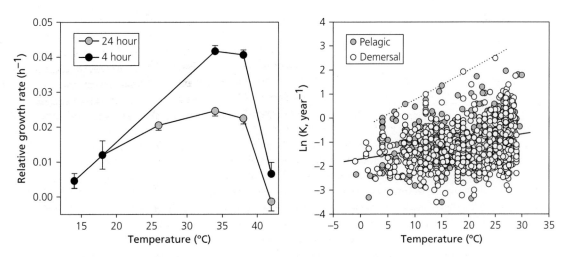

Figure 13.7 Growth rate and temperature. A (left panel): relative growth rate, expressed as mass-specific growth rate in fifth-instar caterpillars of the moth *Manduca sexta*, measured over a 4-hour or 24-hour period, as a function of experimental temperature. Data show mean ± standard error. B (right panel): mean growth rate (*K*, determined from von Bertalanffy growth models) in the wild for 1219 species of pelagic and demersal marine teleost fish, as a function of mean environmental temperature, with fitted least-squares regression line. The dotted line is a least-squares fit to the maximum growth rates observed in bins of 5 K (−10 °C to −5 °C, and so on). Note logarithmic transformation of growth rate data[23].

Organisms living at different latitudes typically have to deal with two environmental variables that are likely to affect their growth rate: variation in mean temperature and differences in the period over which resources are available (seasonality). Particularly important in disentangling the effects of these have been studies of species whose distribution extends over a range of temperature and seasonality.

13.4.1 Within-species studies

The bivalve genus *Leukoma* (previously *Protothaca*) is widely distributed along the Pacific coast of America. As might be expected, annual growth rate declines northwards from California to the Gulf of Alaska. Fine-scale analysis of increments in the shell reveals the growth history of individuals, and shows that the slower annual growth at higher latitudes arises because the animals are only able to feed for a shorter period of the year. Combining data for the first year of life (before the onset of reproductive maturity, which complicates the growth trajectory) in two closely related species (*Leukoma staminea* and *L. grata*) indicates a smooth cline in the duration of the summer feeding period from ~150 days in the Gulf of Alaska to ~330 days in Panama. While this demonstrates a clear role for seasonality in constraining the growth of *Leukoma* at higher latitudes, what we do not know is whether there is also any evolutionary adjustment in the physiological capacity for growth. Mean seawater temperature decreases towards high latitudes and this could decrease growth independently of the increased seasonality of food[24].

Evidence for adjustments to growth rate has come from the work of David Conover and colleagues on several species of nearshore fish along the Atlantic seaboard of America. American shad, *Alosa sapidissima*, striped bass *Morone saxatilis* and the killifish or mummichog *Fundulus heteroclitus* have ranges that span ~29° N to ~46° N, across which the summer growing season decreases by a factor of about 2.5 Despite this variation, in all three species juvenile body size at the end of the first growing season is the same at all latitudes. Common-garden experiments showed that intrinsic growth rate was highest in the northern populations and lowest in the southern, and that these differences were genetically based. This striking pattern has been termed *countergradient variation* (Box 13.3).

Box 13.3 Countergradient variation

Countergradient variation is a term coined by Richard Levins to describe situations where genetic variation within species counteracts the effects of environmental variation such that the phenotype is (relatively) invariant across the environment. It followed from studies of *Drosophila* along an altitudinal gradient, and has been much used by David Conover and colleagues in their extensive studies of growth rate in fish.

The concept is based in classical quantitative genetics, in which the phenotypic variance, V_P, can be partitioned additively between components of variance resulting from genetic, V_G, and environmental, V_E, effects:

$$V_P = V_G + V_E + V_{GxE} + 2CoV(G, E)$$

where V_{GxE} is the non-additive interaction between genotypic and environmental effects and $CoV(G, E)$ is the covariance between genetic and environmental sources of variance.

In the experimental determination of heritable differences among different strains it is important to eliminate the confounding effects of covariation, typically by the random allocation of genotypes across treatments. In the wild, however, genotypes are not distributed randomly and here countergradient variation is defined formally as where $CoV(G, E)$ is negative; that is, genetic and environmental influences oppose one another[25].

An obvious question is how does this differ from simple evolutionary adaptation to temperature, as for example in the example of *Fundulus heteroclitus* LDH discussed in Chapter 7? Here evolutionary (genetic) change results in physiological performance that is more independent of temperature across the range than would otherwise be the case. This is a case of two terms, originating in different fields of enquiry, describing the same thing.

These evolutionary adjustments in growth rate are best viewed as a response to the variation in seasonality with latitude, but it necessarily involves evolutionary adaptation to temperature: northern populations in colder water grow as fast as southern populations in warmer water. Common-garden experiments reveal a shift in reaction norm for growth (that is the optimal temperature for growth varies across latitude)[26].

The ecological driver for achieving a large size at the end of the first growing season is winter mortality, which is greater in smaller fish. There are, however, fitness costs to faster growth at higher latitudes, such as reduced physiological performance, greater mortality and an enhanced susceptibility to predation. The important message here is that we must always view evolutionary adaptation to temperature in its wider ecological context (as we also saw in the work on *Colias* butterflies discussed in Chapter 7)[27].

Countergradient variation has largely been explored within species. Comparisons across species are also important, but here studies have to deal with extra layers of variation introduced by differences between species in phylogeny and ecology.

13.4.2 Across-species studies

Analysis of the growth in the wild of marine teleost fish reveals a highly significant but not particular strong relationship between temperature and growth rate (Figure 13.7b). As well as a shallow slope, the data also exhibit considerable variability: at any given environmental temperature fish can be found with a wide range of growth rates, associated with differences in habitat and lifestyle. A least-squares fit to the entire data set has a slope equivalent to a Q_{10} of 1.42, with no significant difference between pelagic and demersal species. A difficulty with the von Bertalanffy growth coefficient, K, is that it varies with size and so the estimated slope will be biased by any systematic variation in final adult size with temperature. Analysis with a general linear model to allow for the effects of size, and any interactions, reduces the slope to one equivalent to a Q_{10} of 1.98. The slope of the upper bound of the distribution is equivalent to a Q_{10} of 3.38. A more recent study of 136 species which incorporated body mass and phylogenetic structure into the analysis suggested a temperature sensitivity of teleost fish growth equivalent to a Q_{10} of 1.52[28].

A comprehensive review of zooplankton growth rates was undertaken by Andrew Hirst and colleagues. They analysed data from seven phyla; excluding data where the analysis involved fewer than four species, or covered a temperature range < 10 K, the Q_{10} values ranged from 0.38 to 3.86, with a median value of 1.86. Pooling the data into logarithmic size bins (with all species pooled) yielded Q_{10} values ranging from 2.12 to 5.62, with a median value of 2.31. The range of values (roughly an order of magnitude) reflects, at least in part, the practical difficulty of determining growth rates in marine zooplankton, but it would appear that the temperature sensitivity of zooplankton growth is broadly similar to that observed in fish[29].

A recent wide-ranging analysis by Ross Corkrey and colleagues compared growth rates across all three domains of life, Archaea, Bacteria and Eukarya. In all cases the relationship between relative growth rate and temperature exhibited a large amount of scatter, though there was also a clear upper bound to each of the distributions (Figure 13.8). The maximum temperature for growth is much lower in eukaryotes (~55 °C) than in either bacteria (~100 °C) or archaeans (122 °C); we will explore these differences further in Chapter 14.

The data for bacteria show a steep increase in maximum observed growth rate from –10 °C to 40 °C. This upper bound presumably reflects the maximum bacterial growth rate that can be achieved at any given temperature, and an exponential fit to these data suggests a temperature sensitivity to this maximum growth rate equivalent to a Q_{10} of 2.29. Similar analyses for eukaryotes and archaeans yielded Q_{10} values of 1.81 and 3.39 respectively. It is tempting to relate these differences to the slower protein synthesis rates in eukaryotes compared with bacteria, but this may be too simplistic. The value for eukaryotes is close to the value of 1.88 determined by Richard Eppley in his classic study of growth rates of phytoplankton[30].

An intriguing feature of the bacterial data is the sharp drop in maximum growth rate above 42 °C, and the suggestion of a second peak in growth rates at ~70 °C. This pattern suggests that 42 °C may

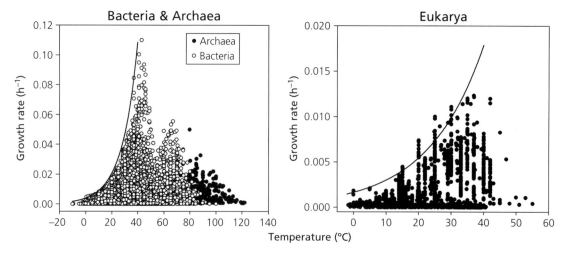

Figure 13.8 Growth rate and temperature. A (left panel): Bacteria and Archaea. B (right panel): Eukarya. All data are relative growth rates, and the lines show exponential fits to the maximum growth rates observed in bins of 5 K (−10 °C to −5 °C, and so on)[31].

mark a genuine threshold between mesophile and thermophile bacteria, and the temperature range from 42 °C to 60 °C has been termed the *mesophile-thermophile gap*.

13.5 Modelling growth

The growth of an organism is such an important feature of its interaction with its environment and other organisms that capturing the main features in a simple manner is critical to ecosystem modelling. As with any model, the aim is to capture the key elements of growth in as few parameters as possible without leaving out anything important. The enormous variation in observed growth rates of all organisms might suggest that there are a large number of factors at work, and that capturing the main drivers in a simple model would be difficult. Two models (strictly they are groups of models) that have attracted much recent attention are the *dynamic energy budget* and the *ontogenetic growth model*. These two models differ greatly in their aims, structure and formulation.

13.5.1 The dynamic energy budget

The dynamic energy budget (DEB) tracks the energy content of an organism through its life cycle from embryo to adult. As we saw in Chapter 4 it can describe growth well, but its disadvantage for many ecologists is that the data needed to estimate the core variables and parameters are not always available. DEB models are thermodynamically rigorous and comprehensive, but there applicability can be limited by our lack of knowledge of the system being modelled[32].

While a powerful tool for modelling the energetics of an individual species and capturing how this varies across its range, or might alter as climate changes, DEB models are not always the most useful way to capture general features of the growth of whole assemblages, such as is needed for ecosystem models. For this we need more general models.

13.5.2 The ontogenetic growth model

The ontogenetic growth model, OGM, was developed by Geoffrey West, James Brown and Brian Enquist, and is a development of the Metabolic Theory of Ecology, MTE, which we explored in Chapter 12. The OGM starts with the assumption that the total metabolic rate of an organism is the sum of energy devoted to growth and the energy required to maintain existing biomass:

$$B = E_m \frac{dm}{dt} + B_m m$$

where B is the total metabolic rate (W), m the body mass at time t, E_m the energy required to synthesise new tissue (J g^{-1}) and B_m the energy required to maintain a unit of biomass (W g^{-1}). Allowing for the scaling of metabolic rate as formulated in the MTE (where b_0 is a coefficient that varies with taxon and metabolic level and has to be derived empirically):

$$B = b_0 m^\alpha$$

and rearranging produces an equation identical in form to the Pütter equation:

$$\frac{dm}{dt} = am^\alpha - bm$$

where $a = \dfrac{b_o}{E_m}, b = \dfrac{B_m}{E_m}$ and $\alpha = 0.75$ as predicted by the MTE. While the OGM is a simple Pütter equation, the authors argue that it is derived more rigorously than any previous growth model with coefficients that are related directly to physiological variables[33].

Precise and unambiguous definition is an essential component of rigour. In the OGM B is defined as *the incoming rate of energy flow* and quantified as *the average resting metabolic rate of the whole organism at time t*, and B_m is defined as the *power needed to sustain the organism in all of its activities*. This introduces an unfortunate ambiguity, because it would seem to imply that B is equivalent to basal or resting metabolic rate, BMR, and B_m equivalent to daily energy expenditure, DEE. In a typical mammal DEE exceeds BMR by a factor of about 3–4, and yet the OGM equation requires $B > B_m$. We saw in Chapter 8 that resting metabolic rate is a measure of energy dissipation. While this dissipated energy includes the thermodynamic cost of growth (R_s in Figure 13.4), it does not include the chemical potential energy retained in the new tissue (P_s in Figure 13.4); indeed it cannot without contravening the conservation of energy. In the OGM, however, the cost of producing new tissue E_m is estimated from the *energy content of mammalian tissue*, which is a measure of P_s rather than R_s (Figure 13.4). Neither is it clear how the OGM allows for metabolic expenditure required for daily activity (since B is defined as *resting* metabolic rate). The lack of clarity over definitions and appropriate estimation makes it very difficult to evaluate the model further[34].

The original OGM took no account of temperature, but a temperature term was introduced subsequently; as with the MTE itself, this was a simple Boltzmann factor[35]. The implications of incorporating temperature were explored for the growth of juvenile stages well below adult size, where it could be assumed that maintenance costs are negligible. The model described embryonic development of a range of organisms well, and the fit was improved further by including a term of stoichiometry (specifically C:P ratio, which captures, in part the concentration of RNA in the cell). The Boltzmann factor was applied to both terms in the OGM equation, which meant that adult mass was independent of temperature. In nature, however, there is a striking relationship between body size and temperature, which we explore below. This indicates that the relationship between temperature and growth is more subtle and complex than was captured in the OGM by a simple Boltzmann factor.

A subsequent revised version of the OGM redefined some terms[36]. In this version B is described as the *rate of energy assimilation*, which is quite different from its definition in the original OGM as the rate of energy dissipation. More importantly E_m is now defined as *the total metabolic work the organism expends to create biomass from preformed organic molecules*, and explicitly does not include the energy content of the new tissue. This is a clear definition of the thermodynamic cost of growth (R_s in Figure 13.4) and marks a fundamental shift in the conceptual basis for E_m from the original OGM. Their estimates of E_m are derived by comparing metabolic rate with growth rate in embryos and juveniles of vertebrates, when maintenance costs are assumed to be small. Estimates of E_m for juveniles ranged from 4.0–7.5 kJ g^{-1}, based on tissue growth as wet mass. Conversion to dry mass yields values that exceed previous estimates of the thermodynamic cost of growth (Table 13.1) by up to four-fold.

The model was revised further by Chen Hou and colleagues[37], and this version (the extended OGM) distinguishes explicitly between the use of assimilated food to provide energy for maintenance, energy for activity, energy for synthesis and the raw materials for new tissue. This model thus has a structure identical to the balanced energy budget developed in the early part of the twentieth century (see Chapter 4), though it also has the valuable development of including scaling. The extended OGM model captures the growth of endothermic vertebrates very well, but the authors acknowledge that it has yet to be tested in ectothermic vertebrates or invertebrates.

An important aspect of the OGM is that it suggests that beneath the variety of growth trajectories observed in nature lies a 'universal growth model'. This arises when a dimensionless mass ratio, r, is plotted as a function of a dimensionless time variable, τ, where

$$r = \left(\frac{m}{M}\right)^{1/4}$$

and

$$\tau = \frac{at}{4M^{1/4}} - \ln\left[1 - \left(\frac{m_0}{M}\right)^{1/4}\right]$$

Here m_0 is the mass at birth and M the maximum body mass (other variables are as defined above). The universal growth curve is then:

$$r = e^{-\tau}$$

This captures the growth rate of a variety of vertebrates very well, including both ectotherms and endotherms and also species with determinate or indeterminate growth (Figure 13.9). Although one species of invertebrate (the planktonic mysid shrimp *Mysis mixta*) was included, subsequent more extensive analyses suggested that the universal growth model was not a good description of growth in organisms where the life history includes a fundamental change in morphology between larval and adult stages, such as insects and aquatic invertebrates[38].

The obvious question is why, if the logical foundation and parameterisation of the OGM is unclear, should the universal growth model derived from it describe vertebrate growth data so well? The answer would appear to be that the form of the universal growth model is dictated principally by the scaling parameters in the OGM. Variations in the growth and metabolic variables have much less impact, as suggested by the tightness of the data about the model. If so, this points to the importance of the structure of the cardiovascular system in determining not just the scaling of metabolism in vertebrates, but also of growth. This is perhaps not so surprising because the same cardiovascular system supplies the tissue with oxygen for metabolism and raw material for growth. It may also explain why the universal growth model is not so good at capturing the growth of invertebrates, with their very different circulatory systems.

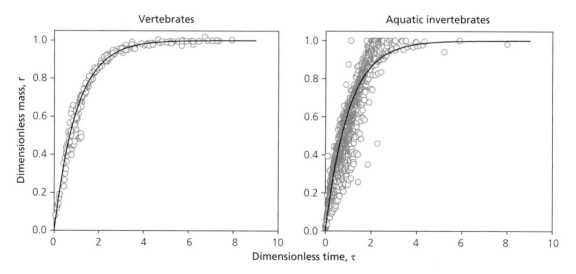

Figure 13.9 The universal growth model, UGM. A (left panel): UGM fitted to data for vertebrate growth. B (right panel): UGM fitted to data for aquatic invertebrates. In both cases the raw growth data are shown as circles and the fitted UGM as a solid line[39].

13.6 Temperature and size

It has long been recognised that environmental temperature and organism size are related, though the relationship is complex. Two important areas of discussion have been Bergmann's rule in endotherms and the temperature-size rule in ectotherms.

13.6.1 Bergmann's rule

Bergmann's rule is one of the most quoted of all ecological rules yet is frequently misunderstood, probably because so few ecologists have read what Carl Bergmann actually said. This is understandable given that the original extends to over 110 pages of mid-nineteenth century German, and has yet to be translated in full. Recently Volker Salewski and Cortney Watt have provided a valuable service to the ecological community by publishing a thorough digest of the paper, correcting many long-held misconceptions[40].

Bergmann was concerned exclusively with thermoregulation in mammals and birds. He assumed that heat is lost through the body surface but generated throughout the body volume, and he speculated on the importance of muscle in this. Noting that as body size increases, surface area increases more slowly than volume, Bergmann concluded that a larger mammal or bird needs less heat, relative to its size, to increase body temperature by a given amount above ambient temperature. This is the biophysical statement which Bergmann called a 'law' (*Gesetz*, which also translates as 'principle' or 'rule').

A consequence of this biophysical rule is that if factors such as insulation are ignored, larger endotherms should be warmer than smaller ones. Bergmann noted that this was not true (we saw in Chapter 10 that Bergmann's conclusion remains valid over a century on) and also that larger and smaller species were often found together. He also suggested that, when all else is equal, such as insulating fur, habitat or behaviour, smaller animals should live in warmer areas. He tested his idea by examining >300 species of bird from 86 genera. His approach was to examine each genus to see whether smaller species live further south than larger species. Although avian taxonomy has undergone major revisions since Bergmann's time, and hence some of his congeneric comparisons are not as closely related as was once thought, this must be one of the earliest examples of evolutionary relatedness being explicitly included in a comparative physiology study (and Bergmann's paper was published twelve years before the *Origin of species*). He also tested his idea intraspecifically using data from domesticated mammals.

The key features of Bergmann's study are that he was concerned explicitly and exclusively with endotherms, that his law or rule concerns a biophysical mechanism and that the suggestion there might be a cline in the size of mammals or birds was a hypothesis that followed as a consequence of the biophysical mechanism. A correct statement of this hypothesis is:

Within species and among closely related species of endotherms, a larger size is often achieved in colder climates[41].

While this is more accurately described as Bergmann's hypothesis, it is likely that after a century and a half of use it will continue to be referred to as Bergmann's rule. Bergmann's rule follows from a biophysical mechanism that is confined to endotherms. Extending the concept to ectotherms is thus inappropriate, for the biophysical mechanism he was considering does not apply to them.

Bergmann's biophysical arguments suggest the existence in endotherms of a cline in body size with latitude, driven by the influence of environmental temperature on thermoregulation. The problem we face in evaluating Bergmann's hypothesis is that should we find such a cline, there are other factors that could equally be responsible; these include resource availability, population density, character displacement and predation. Bergmann himself examined patterns of size variation with latitude both within species (domesticated mammals) and across species (birds).

A nice example of a strong cline in size within species is that of North American ungulates (Figure 13.10). This pattern has been interpreted as disproving Bergmann's rule on two grounds: size decreases at latitudes above ~60° N, and where there is a positive relationship between size and latitude (10° N to 50° N) the slope is too shallow. The latter point is trivial, because Bergmann was at pains to point

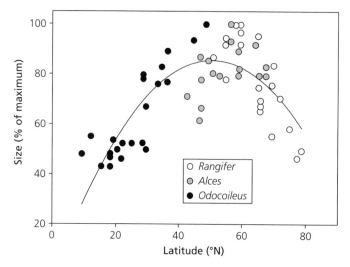

Figure 13.10 A test of Bergmann's hypothesis: variation in body size of three North American herbivores in relation to latitude. Data are plotted separately for moose (*Alces americana*), mule deer (*Odocoileus hemionus*) and reindeer/caribou (*Rangifer tarandus*). The line is a quadratic fitted to all data[42].

out that for the hypothesised cline to exist, all other things must be equal. In most cases endotherms in cooler climates will adjust the rate of heat loss by increasing insulation, and so all other things are not equal. The decrease in size in northernmost latitudes is undoubtedly related to food availability; while larger individuals will retain their heat and avoid starvation better than smaller ones, their greater size means that they still require more food.

Further south, white-tailed deer, *Odocoileus virginianus*, exhibit a cline in size as predicted by Bergmann, but at the same time there is latitudinal variation in net primary productivity and population density. It is likely that the cline in size we observe is the result of several factors operating together, of which only one is thermoregulation. This does not, as has been suggested, invalidate Bergmann's biophysical mechanism; it merely indicates that ecology is complex[43].

The most recent broad-scale (macroecological) studies of body size in mammals and birds have started to tease apart this complexity by using statistical models to explore the effects of more than one factor on body size. In mammals this has indicated important roles for temperature (for which latitude is a rough but useful proxy) and hence thermoregulation, and starvation resistance. Furthermore, the extent to which a cline is evident varies geographically, being stronger in colder environments; at lower latitudes factors such as primary productivity exert a greater influence. In birds as a whole, temperature is the strongest environmental correlate of body size, but with resource availability and local species richness also being important[44]. The resultant pattern of variation in body size with latitude is shown in Figure 13.11. There is relatively little variation in median body mass across temperate and tropical latitudes, but marked clines towards high latitudes. The cline is steeper at southerly latitudes, and this is undoubtedly related in part to the lack of passerines (which are typically small).

We can conclude that there is good evidence in support of Bergmann's hypothesis from both within- and between-species studies, but that as we might expect, the body size of an endotherm is influenced by a suite of ecological factors of which thermoregulation is only one. The relative importance of these factors varies from place to place, but current evidence suggests that thermoregulatory and food availability factors are particularly important at high latitudes, whereas at lower latitudes the key factors are food availability and ecological interactions with other species in the local assemblage.

13.6.2 Dehnel's phenomenon

Mammals living in seasonal environments but which neither enter torpor nor hibernate are faced with a severe reduction in food availability over winter. Some small mammals such as arvicoline rodents (*Microtus* and *Myodes* species) respond by

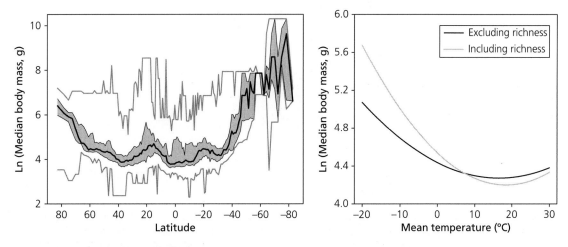

Figure 13.11 Avian body size and latitude. A (left panel): relationship between median body size (black line) and latitude (by convention southerly latitudes are negative); also shown are the interquartile range (grey shading) and the extreme values (grey lines). B (right panel): modelled relationship between median body mass and mean environmental temperature for birds. Predictions come from generalised least-squares models excluding (black line) or including (grey line) a species richness term[45].

reducing their body mass in autumn (fall). This has also been recorded in *Antechinus* from Australia, where variations in food availability are related to rainfall rather than temperature. In some shrews this reduction involves not just fat stores or tissue mass but also the skeleton and brain. This was first reported in three species of *Sorex* by August Dehnel, and to date the phenomenon appears to be restricted to shrews where it has become known as Dehnel's phenomenon[46].

13.6.3 The temperature-size rule

Some vertebrate ectotherms also exhibit clines in size, with larger species at higher latitudes. This has engendered a debate over the extent to which reptiles and amphibians do, or do not, confirm to 'Bergmann's rule'. However, as we saw above, Bergmann was concerned with a mechanism that applies only to endotherms and hence the rule named after him is relevant to mammals and birds alone; we need another term to describe size clines in ectotherms, if any exist. Many terrestrial invertebrates also exhibit clines in size, although here the largest species tend to be in the tropics. The lack of a coherent pattern relating size to latitude in ectotherms suggests there is no general mechanism at work, and this size of ectotherms is dictated by different factors in different species[47].

In contrast, a clear pattern has emerged when looking at variations in adult size within species. Here a large body of experimental work indicates a widespread tendency for organisms raised in cool conditions to reach a larger adult size than conspecifics raised at higher temperatures. In the 1990s David Atkinson synthesised the available data and concluded that over 80% of species for which there were data tended to be larger when reared at cooler temperatures, a pattern he termed the temperature-size rule (TSR) and which was later expressed succinctly as '*Hotter is smaller*'[48].

The basic features of the TSR are shown in Figure 13.12a. Here growth curves are shown for a hypothetical species raised at two different temperatures. Individuals raised at the warmer temperature grow more rapidly but achieve a smaller adult size than those grown in a cooler environment. Also shown are data from an early experimental study of two species of bryozoans (Figure 13.12). Bryozoans are ideal organisms for such work because they are clonal and the individual zooids have a skeletal framework that preserves both the size of the zooid and the life history of the colony. In this study the size of the zooids was smaller when the colonies

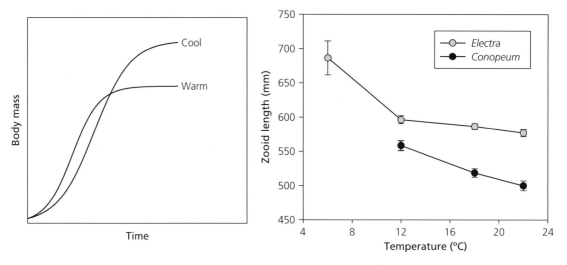

Figure 13.12 The temperature-size rule (TSR). A (left panel): schematic diagram showing difference in growth trajectories of an organism following the TSR raised in a warm and cool environment. B (right panel): size of zooids in colonies of two species of bryozoans (*Electra pilosa* and *Conopeum reticulum*) grown at different temperatures. Date are mean zooid length with standard error (n = 50 zooids in each case)[49].

were grown in warmer water. Later experimental work enabled the effects of temperature and food availability to be distinguished, showing that temperature affects zooid size but food availability dictates the rate of colony growth. Experiments with bryozoans provide important insights because genetic effects can be controlled for by using replicate treatments of identical clones[50].

The temperature-size rule is exhibited by a wide range of organisms, including many unicells such as diatoms, amoebae, ciliates, flagellates and dinoflagellates. In these taxa decreases in linear size of 2.5% K^{-1} (95% confidence intervals 1.7–3.3) or ~4% K^{-1} in volume have been reported[51]. When we observe a pattern to be widespread across many different organisms, and moreover one that is manifest in marine, freshwater and terrestrial organisms, it suggests something fundamental at work. The question is, what?

There are two contrasting interpretations of this widespread pattern. One is that it is an inevitable consequence of temperature, driven by a physical process which evolution has been unable to modify. The second is that it is an adaptive response, in that a smaller body size (or a correlate of small body size, such as rapid development) brings fitness benefits in a warmer environment.

A possibility that has attracted considerable attention is whether the temperature sensitivity for development and growth might differ. This has been explored theoretically, starting with the simplifying assumption that the rates of growth and differentiation are both controlled by a single flux-limiting enzyme, and that the temperature sensitivity of these enzymes (and hence the two processes) were different. While the first of these assumptions is clearly a gross simplification, this basic model has proved valuable for exploration of the evolution of the reaction norm for growth. The model has mainly been tested in crustaceans and insects, where development can be quantified independently of growth by the progression through instars or larval stages. In the brine shrimp *Artemia franciscana* the temperature sensitivity of growth rate is stronger in earlier stages than later, whereas development exhibited a constant temperate sensitivity. The consequence is that *Artemia* exhibits the classic temperature-size response only in the later stages of its growth trajectory, a feature which has also been found in other crustaceans. Recent studies of *Manduca sexta* have shown that diet quality determines whether or not growth at different temperatures matches the temperature-size picture. Given that growth and development are both highly complex processes under careful genetic control, it would be surprising indeed if the temperature-size rule were simply an emergent property following from a rigid response of single enzyme systems to temperature[52].

An intriguing alternative possibility is that shifts in body size follow from a response to growth temperature at the cellular level. A smaller cell has a larger surface area to volume ratio, which has consequences for the rate at which the cell can exchange gases, take up nutrients or excrete wastes. Warmer cells have higher maintenance requirements (Chapter 8) and it may be that a smaller size is needed to allow nutrient and oxygen supply to meet metabolic demand at these higher temperatures. There are well-documented evolutionary relationships between genome size, cell size, metabolic rate and body size, and some of these also underpin the phenotypic response to growth temperature (although clearly not genome size in the case of a phenotypic change during development)[53].

Body size is integral to life history because it influences, among other things, fecundity, the likelihood of predation and metabolic energy requirements. The temperature-size rule poses a puzzle to life-history modellers, because temperature is exerting opposing effects on growth rate and adult size, suggesting there may be a trade-off between these two. Life-history modelling suggests that body size is likely to respond to factors that alter growth rate and juvenile survival, but as yet we have no clear picture of how these factors might operate to produce the temperature-size rule[54].

Although examples of the temperature-size rule are seen across all habitats and the effect is well established qualitatively, its magnitude varies considerably. Attention is now shifting to an exploration of the many factors which influence the strength of the response, and in particular the striking differences in the magnitude and direction of the effect on land and in the sea. A recent review suggested that in aquatic environments the temperature-size effect averaged a decrease of ~5% K^{-1}, whereas in terrestrial environments is ~0.5% K^{-1}, a difference of an order of magnitude[55]. This suggests that there are additional factors at play, and a much-discussed possibility is oxygen. This brings us to a closely related topic, but one so distinctive it has been given its own name, *polar gigantism*.

13.6.4 Polar gigantism

In 1830 James Eights was a naturalist aboard the brigs *Annawan* and *Seraph*, as they explored the South Shetland Islands off the Antarctic Peninsula. He made numerous important geological observations, including the first recorded discovery of fossil wood, but he is best remembered for his careful description of a series of remarkable Antarctic marine invertebrates. These included *Ceratoserolis trilobitoides*, an isopod looking for all the world like a living trilobite, the large isopod *Glyptonotus antarcticus* and a ten-legged pycnogonid (the first ever reported), *Decolopoda australis*, which is even more notable for its enormous size. James Eights had discovered polar gigantism[56].

Gigantism is not a feature of all polar marine organisms. Rather it is confined to a few invertebrate groups, notably pycnogonids and crustaceans, where a small number of species are strikingly large. In contrast, polar molluscs tend to be small, unornamented and thin-shelled, possibly as a result of the greater metabolic cost of producing a calcareous shell at low temperatures (see Chapter 5). These are broad generalisations to which, as always in ecology, there are exceptions: most Antarctic pycnogonids are actually rather small, and the Southern Ocean has a few large bivalves.

The best quantitative data on polar gigantism come from a careful analysis of the distribution of adult body size in entire assemblages of amphipods by Gauthier Chapelle and Lloyd Peck. They showed that while the size of the smallest species was pretty much the same everywhere, there was a striking relationship between the size of the largest species in the assemblage and the oxygen content of the water (Figure 13.13). The interesting aspect of this study is that while there is a strong (and slightly curvilinear) relationship between maximum size and water temperature, data for Lake Baikal lie away from the line. In contrast, the relationship with oxygen content is tighter, and linear. The results that are significant in terms of mechanism are those from freshwater, where the low salinity increases oxygen solubility, and particularly Lake Titicaca (which has a lower oxygen content because of the reduced partial pressure in the atmosphere at altitude)[57].

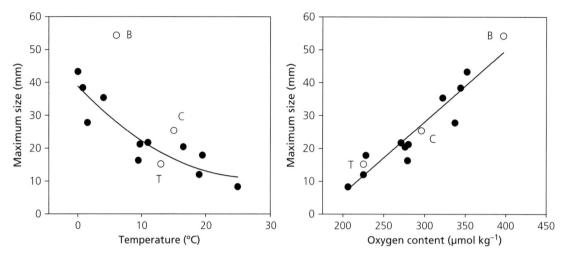

Figure 13.13 Temperature, oxygen content and size in assemblages of amphipod crustaceans. A (left panel): maximum size as a function of mean annual temperature in marine (black symbols) and freshwater (open symbols) habitats. The line is a quadratic fitted to the marine data alone. B (right panel): maximum size as a function of oxygen content. The line is a least-squares regression fitted to all data. (B: Lake Baikal, C: Caspian Sea, T: Lake Titicaca)[58].

While there is general agreement that there is a strong coupling between maximum size and oxygen in many aquatic taxa, the mechanism is less understood. Oxygen solubility is greater in cold waters, but at equilibrium the partial pressure will be identical with that in the atmosphere, which is more or less the same at sea level everywhere. The rate at which oxygen diffuses into tissue is dictated by the permeability of the membrane surface across which it is diffusing, the partial pressure gradient across the membrane, but also the solubility coefficient. The latter factor is strongly temperature-dependent, but temperature also influences oxygen demand (metabolic rates are lower at colder temperatures). While the precise mechanism by which oxygen influences size has yet to be fully teased apart, it is clear that, at least in aquatic environments, oxygen plays a key role in influencing the maximum size that an organism can achieve[59].

Oxygen currently makes up ~21% of the atmosphere, but in the Carboniferous and Permian periods it reached much higher levels (> 30%). At this time there were dragonflies much larger than any known today, with wing-lengths exceeding 30 cm; many terrestrial invertebrates were also larger. The correlation between insect size and oxygen through geological time is convincing, though there are indications that latitude was also an important factor, suggesting a possible role for temperature. That this correlation has a mechanistic basis is indicated by experiments rearing insects in high and low oxygen conditions[60].

13.6.5 Abyssal gigantism

About 50 years after James Eights' discoveries, large pycnogonids were also reported from the deep sea by Henry Moseley. These were collected by HMS *Challenger* during one of the greatest scientific voyages in history. Further work established that here too there were species that were unusually large, and this became known as *abyssal gigantism*. As both the deep sea and polar regions have seawater that is cold and oxygen-rich, it is likely that the same mechanism underpins gigantism in both regions, although in the deep sea the situation is complicated by the effects of hydrostatic pressure[61].

13.7 Summary

1. Growth involves two flows of energy. The first is chemical potential energy in the monomers used to construct the proteins, lipids, polysaccharides and nucleic acids forming the new tissue. The

second is the metabolic energy (ATP or GTP) used to construct the new tissue; this is the metabolic (thermodynamic) cost of growth.
2. The metabolic cost of growth can be expressed as a dimensionless fraction of the energy retained in the new tissue, and its value is ~0.33. That is, the synthesis of 100 kJ of new tissue from monomers requires the dissipation of ~33 kJ energy as heat. It is this dissipation which ensures that the synthesis of low-entropy macromolecules is energetically favoured.
3. Sigmoidal growth is described well by a simple Pütter equation. The effect of temperature on growth cannot be predicted from theory, and can only be described statistically. Typical temperature sensitivities for growth in the wild lie in the range Q_{10} 1.5–3. Within species there may be evolutionary adjustments to growth rate to offset the effects of temperature, though these involve trade-offs with other physiological factors affecting fitness.
4. Outside the tropics, many mammals and birds exhibit a cline in size, with larger species at higher latitudes (Bergmann's rule). Carl Bergmann predicted such a cline from biophysical arguments based on endotherm thermoregulatory costs. Bergmann's rule thus applies only to mammals and birds.
5. Many ectotherms grow more slowly but attain a larger adult size when grown at lower temperatures (the temperature-size rule). We still lack a full mechanistic or evolutionary understanding of this.
6. The large size of some aquatic invertebrates at lower temperatures (notably in the polar regions and the deep sea) is associated with a higher oxygen content of the water. The precise physiological mechanism is not fully understood.

Notes

1. This comes from the introduction to a review of body size in relation to cell size and cell number (Arendt 2007). A decade on there is still much we do not know.
2. Holometabolous insects are those that undergo complete metamorphosis, where the life cycle involves egg, larva, pupa and adult (imago). Examples are butterflies and moths (Lepidoptera), flies (Diptera) and bees, wasps and ants (Hymenoptera). Incomplete metamorphosis (hemimetabolism) involves growing through a series of nymphal stages, each separated by a moult, with the nymph becoming more adult-like at each moult. An example here would be dragonflies and damselflies (Odonata), where the nymphs are aquatic but the adults essentially terrestrial. Many marine invertebrates have planktonic larval stages that are quite unlike the benthic (bottom-living) adult stages (Barrington 1967).
3. A difficulty with the Pütter equation is that it cannot be solved analytically for all values of a, b, α and β (Hozumi 1985) (but see Ohnishi et al. 2014).
4. For this example a simple logistic growth model was parameterised with data for the extinct dinosaur *Psittacosaurus lujiatunensis* from Erickson et al. (2009): growth rate 0.55, maximum adult size 37.38 kg, assumed to be reached at age 20.
5. The key publications are Pütter (1920), von Bertalanffy (1938, 1951, 1957) and Reiss (1989). Useful reviews of growth rate models are Kaufmann (1981) and Wootton (1990).
6. See Ursin (1967, 1979). These mismatches have prompted regular calls for the von Bertalanffy model to be abandoned (see for example Roff 1980 and Day & Taylor 1997).
7. A casual glance at the literature will reveal a range of values for the number of ATP molecules hydrolysed in peptide bond formation. This is because the number depends on precisely what stages of protein synthesis are being considered. The value of four comes from 2 ATP equivalents for the activation of the amino acid (that is, its attachment to the relevant transfer-RNA) and then one each for the two GTP molecules used during ribosomal-based attachment of an amino acid to the growing polypeptide chain: one for tRNA binding and one for tRNA translocation. We refer to these as *ATP equivalents* because ATP is used to regenerate GTP from GDP, and also because the activation of the amino acid actually involves the removal of two phosphate groups from one ATP molecule (that is, the ATP is hydrolysed to AMP rather than ADP).
8. The thermodynamics of microbial growth are discussed by Forrest (1970) and Forrest & Walker (1971), and recent perspectives provided by von Stockar et al. (2006), Roden & Jin (2011), Battley (2013) and Desmond-Le Quéméner & Bouchez (2014).
9. The physiological basis of the cost of growth was discussed in detail by Parry (1983) and Wieser (1985, 1994). Vahl (1984) showed the cost of growth appeared in the increase in metabolic rate that accompanies feeding (so-called specific dynamic action, SDA), Jobling (1981, 1983, 1985) reviewed the energetics of growth in fish and Secor (2009) provides a thorough review of the processes underpinning SDA. The discussion

here has been framed in terms of the cost of protein synthesis, but this is fine because well over half the organic constituents of a typical cell are proteins and the costs of protein synthesis dominate growth costs under most conditions (Rogers & Fraser 2007).
10. Data taken from Wieser (1994) and converted to SI units before plotting. The data are for the blue mussel *Mytilus edulis*, the amphibian *Bufo bufo* and six species of teleost fish. The regression is significant (b = 0.32, SE = 0.02, $p < 0.001$).
11. McNeill & Lawton (1970), Humphreys (1979).
12. Partial growth efficiency is calculated from the energy actually assimilated (that is, consumption minus faeces) rather than consumption. Studies are 1: Conover (1978), 2: Parry (1983), 3: Wieser (1994), 4: Jørgensen (1988), 5: Peterson et al. (1999), 6: Beaupre & Zaidan (2012).
14. *Opsanus* data from Mathews & Haschmeyer (1978), *E. coli* data from Farewell & Neidhardt (1998).
15. Brocchieri & Karlin (2005) discuss the frequency distribution of polypeptide length in proteins from eukaryotes, bacteria and archaeans.13.See Pannevis & Houlihan (1992).
16. Hsaio et al. (1990) report high copy number in Antarctic fish antifreeze genes, Chen et al. (2008) report extensive gene duplication in an Antarctic fish, Detrich & Amemiya (2010) report on the genome sizes in Antarctic notothenioids and Coppe et al. (2013) describe duplication of mitochondrial genes in icefish. The notothenioid radiation in the Southern Ocean is described by Clarke & Johnston (1996) and Eastman & Clarke (1998).
17. Data courtesy Keiron Fraser. All data standardised for mass before analysis. Original sources in Rogers & Fraser (2007).
18. Clarke et al. (1989) demonstrated a strong relationship between RNA:DNA ratio and growth in the cuttlefish *Sepia officinalis*, and Koch (1970) provides a nice example for bacteria.
19. Chaperone function is reviewed by Hartl et al. (2011) and Kim et al. (2013). See also Behnke et al. (2016) for the role of Hsp70 and Cuellar et al. (2014) discuss adaptation of chaperones to temperature in Antarctic fish. The importance of the role of ubiquitin in proteolytic pathways was acknowledged by the award of the 2004 Nobel Prize in Chemistry to Aaron Ciechanover, Avram Hershko and Irwin Rose for their studies of this system.
20. Buckley et al. (2004), Todgham et al. (2007), Clark et al. (2008).
21. Jeschke et al. (2002).
22. Clarke (1991). In plants there is often a delay between peak photosynthesis and peak growth, with the latter sometimes being faster by night.
23. The *Manduca sexta* data are redrawn from Kingsolver & Woods (1997). The fish data are for marine teleosts, downloaded from Fishbase (www.fishbase.org) on 23 October 2013. Data for freshwater species were removed, and a mean growth rate and temperature calculated for each of 1219 species, so each is represented by only a single data point. The upper bound relationship was fitted to transformed data (natural logs) following Blackburn et al. (1992).
24. The bivalve growth rate study is Harrington (1987); the genus was previously *Protothaca*.
25. The original definition is Levins (1969). A useful short review is Conover & Schultz (1995).
26. See Conover & Schultz (1995), Schultz et al. (1996), Conover et al. (1997), Brown et al. (1998). Interestingly, Brown et al. (2012) report the absence of countergradient variation in the oceanic silverside *Leuresthes tenuis*, presumably related to high gene flow limiting local adaptation.
27. See Conover (1990), Hurst & Conover (1998), Munch & Conover (2003).
28. Sibly et al. (2015).
29. Hirst et al. (2003). The taxa analysed were Medusozoa (Cnidaria), Ctenophora, Chaetognatha, Crustacea, Gastropoda (Mollusca), Polychaeta and Thaliacea (Chordata).
30. The original study is Eppley (1972). See also Bissinger et al. (2008) who used a larger database and improved statistical analysis to confirm a value of Q_{10} = 1.88 for the maximum growth rate of phytoplankton.
31. Plotted with data from Corkrey et al. (2016). The upper bound relationship was fitted following Blackburn et al. (1992).
32. For an example of a DEB model of the growth of an insect with a complex life cycle, see Maino & Kearney (2015).
33. The original OGM model is presented in West et al. (2001). The model was formulated in terms of energy per cell summed over all cells; the presentation here is based on West et al. (2004).
34. The quotations in italics are taken verbatim from West et al. (2001). Makarieva et al. (2004) provide a rigorous thermodynamic critique of the OGM. A more subtle difficulty is that the model assumes the cost of maintenance per cell to be independent of body mass, whereas BMR (the usual measure of maintenance costs) scales with body mass with an exponent ~ 0.75.
35. The Boltzmann temperature factor was introduced to the OGM by Gillooly et al. (2002).
36. The revised model is Moses et al. (2008), and again quotations in italics are taken verbatim from the paper. A range of E_m values of 4–7.5 kJ g^{-1} wet mass converts to 16–30 kJ g^{-1} on a dry mass basis. Assuming a tissue en-

ergy content of 20.7 kJ g^{-1} yields a dimensionless cost of growth of 0.77–1.45. These estimates are much higher than Wieser's consensus value of 0.33, and values determined for amphibians and snakes (Table 13.1).

37. The third version of the model is Hou et al. (2008).
38. Hirst & Forster (2013); Maino & Kearney (2015).
39. Plotted with vertebrate growth data from supplementary information in West et al. (2001) and with aquatic invertebrate growth data kindly supplied by Andrew Hirst (see Hirst & Forster 2013).
40. This is Salewski & Watt (2017), and it is essential reading for any ecologist interested in size and temperature. An earlier important review of the ideas is Blackburn et al. (1999). See also Watt et al. (2009). The original is Bergmann (1847).
41. This definition is based on Salewski & Watt (2017). Because of the debate over whether Bergmann's rule applies only within species, or across closely related species, Blackburn et al. (1999) suggested a more rigorous definition of 'closely related' as species 'within a monophyletic higher taxon'.
42. Plotted with data from Geist (1987); reproduced with permission of NRC Research Press.
43. The white-tailed deer study is Wolverton et al. (2009); see also Huston & Wolverton (2011).
44. The macroecological studies are Blackburn & Hawkins (2004) and Rodríguez et al. (2006) for mammals and Olson et al. (2009) for birds. In the latter study the pattern was evident at the species, genus and family level. Other important studies are McNab (1971), Ashton et al. (2000), Ashton (2002), Meiri & Dayan (2003) and Freckleton et al. (2003); McNab (2012) provides a valuable overview.
45. Plotted with data from Olson et al. (2009). Data kindly supplied by David Orme and rescaled before plotting. Reproduced with permission of John Wiley and Sons.
46. Seasonal reduction in the body mass of small mammals from high northern latitudes has been reported by Iverson & Brian (1974) and Merritt & Zegers (1991), and for *Antechinus* in Australia by Green (2001). Seasonal skeletal reduction in *Sorex* was reported by Dehnel (1949), Pucek (1963) and Mezhzherin (1964); see also Churchfield et al. (2012) and Taylor et al. (2013).
47. Discussions of 'Bergmann's rule' in ectotherms include Ashton (2001, 2002), Ashton & Feldman (2003), Angilletta et al. (2004a, b) and Adams & Church (2008). Makarieva et al. (2005a, b) review size clines in terrestrial invertebrates.
48. The key early papers are Ray (1960) and Atkinson (1994); the aphorism comes from Kingsolver & Huey (2008). Hessen et al. (2013) refer to the TSR as '*a family of empirically based ecological rules*'.
49. Plotted with data taken from Table 1 of Menon (1972).
50. See Hunter & Hughes (1994), O'Dea & Okamura (1999), Atkinson et al. (2006).
51. Montagnes & Franklin (2001), Atkinson et al. (2003).
52. The theoretical proposal was van der Have & de Jong (1996), and tests of this include Diamond & Kingsolver (2010) Forster et al. (2011) and Forster & Hirst (2012).
53. See Hessen et al. (2013).
54. Sibly & Atkinson (1994), Atkinson & Sibly (1997).
55. Forster et al. (2012), Horne et al. (2015); see also Daufresne et al. (2009).
56. James Eights (1798–1882) was an American physician, geologist and naturalist. He described his Antarctic discoveries in a series of papers in the *Journal of the Boston Natural History Society*, but appears not to have undertaken any further work as a naturalist. He died in relative obscurity, but is recognised topographically through the Eights Coast region of Antarctica. His life and work are described by Clarke (1916) and Calman (1937).
57. Chapelle & Peck (1999), Peck & Chapelle (2003). To allow for incomplete sampling of the amphipod assemblages, they took the largest individual sampled for each species, and used the median size of the fifth and sixth largest species as their measure of the largest species in the assemblage. They designated this measure 95/5. The results are unaffected if other measures of maximum size are used.
58. Plotted with data kindly supplied by Lloyd Peck.
59. For the debate concerning mechanism see Spicer & Gaston (1999), Verberk et al. (2011), Verberk & Bilton (2011) and Hoefnagel & Verberk (2015).
60. The key historical references are Graham et al. (1995) and Dudley (1998). Recent analyses include Kaiser et al. (2007), Harrison et al. (2010) and Clapham & Karr (2012). Experimental studies are Peck & Maddrell (2005) and Klok et al. (2009). Differences between the influence of oxygen on invertebrate size in the aquatic and terrestrial realms are discussed by Verberk & Bilton (2011) and Horne et al. (2015).
61. Moseley (1880).

CHAPTER 14

Global temperature and life

> Climate plays an important part in determining the average number of a species, and periodical seasons of extreme cold or drought, I believe to be the most effective of all checks.
>
> **Charles Darwin**[1]

Life evolved in the sea, and for much of its early evolution appears to have been confined to large bodies of water, or perhaps their margins[2]. It has long been recognised that water and life are intimately linked and we explored aspects of this relationship in Chapter 5. It is often stated that life can be found on Earth wherever there is liquid water and while this is not quite true, as we shall see below, life has certainly successfully established itself almost everywhere there is water.

The emergence of plants then animals onto the land was one of the most important steps in evolution. The newly emerged plants and animals took their internal marine-like environment with them, but were now exposed to an entirely new set of thermal challenges, for the temperature characteristics of terrestrial and aquatic environments are very different. On land temperatures can reach higher and drop lower than in the ocean, and they also change much faster.

This introduces a simple evolutionary question: does temperature set limits to life as we know it on Earth? Ecologists generally approach such questions by first looking for patterns, so we can start by comparing the distribution of animals and plants on Earth with the distribution of water and temperature. If we find liquid water with no living things, then its temperature may tell us something about the thermal limits to life. Alternatively if we cannot find thermal environments without life, there is the possibility that life on Earth has not extended to cover the widest range of temperatures that could be achieved evolutionarily.

There are thus two patterns we must need to determine: firstly the range and distribution of temperature on Earth and secondly the fraction of this range over which life is to be found. We will explore these two in turn, starting with the distribution of temperature. For this we need to consider the marine and terrestrial realms separately because of their very different thermal characteristics.

14.1 The global distribution of temperature on land

It is common experience that temperature varies throughout the day, and that it generally gets cooler as we climb a mountain. Clearly if we wish to compare the temperature in different places, or at different times, we need to standardise how these temperatures are measured.

14.1.1 Measuring temperatures on land

For meteorologists, the standard is a temperature measurement taken at 1.5m above the ground in a louvred thermometer screen (see Chapter 3). The screen ensures the thermometer is shielded from direct solar radiation and is ventilated, and the measurement is referred to as the *surface air temperature*. This reflects the thermal environment experienced

---

Principles of Thermal Ecology. Andrew Clarke, Oxford University Press (2017).
© Andrew Clarke 2017. DOI 10.1093/oso/9780199551668.001.0001

by us as humans (which is why it was so designed) but also by larger plants and many vertebrates such as reptiles, mammals and birds. Most organisms, however, are small or very small. For these the relevant temperature is not the meteorological air temperature, but the temperature where they actually live, which is typically close to, on, or beneath the ground[3].

The World Meteorological Organisation (WMO) operates over 11 000 meteorological stations providing measurements of surface air temperature all over the world. This sounds a lot, and indeed it is, but they provide on average only one measurement for every 13 000 km², and the instruments are not evenly spread over the Earth. In particular there are few instruments in the hottest and coldest places, mainly because these tend to be remote and difficult of access. Much better coverage is provided by satellites, but these do not measure air temperature. Instead they measure the radiometric surface temperature (or skin temperature), which is a very different thing. Under a clear sky at night the ground surface can be significantly colder than the overlying air, and bare soil under full solar illumination and when wind speed is low can be distinctly hotter than the overlying air. Despite these difficulties, remotely sensed temperature data from satellites have been enormously important in refining our understanding of how terrestrial temperature varies spatially across the globe, and with time of day and season[4].

14.1.2 Temperature extremes on land

The lowest meteorological surface air temperature yet measured is −89.2 °C (184 K), recorded at Vostok Station, Antarctica on 21 July 1983. The mean monthly temperatures for Vostok in autumn and winter range from −65 °C in April to −68 °C in August, indicating that this was a truly unusual event. A compilation of six-hourly temperatures for Vostok Station over the period 1958–2006 shows a modal temperature of −71 °C, close to the monthly mean temperatures in winter (Figure 14.1).

Vostok is on the Antarctic plateau and at an elevation of 3420m. The atmospheric conditions leading to this record low included an absence of solar radiation (July is in the Antarctic winter), clear skies leading to a loss of heat by radiation and unusually low mid-tropospheric temperatures, coupled with little vertical mixing and calm air for a long period. It is possible that at nearby locations similar circumstances could lead to surface air temperatures as low as −100 °C[5]. Temperatures close to this have

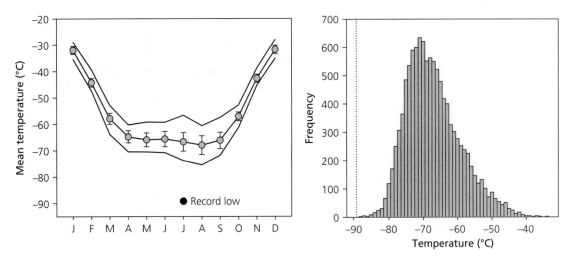

Figure 14.1 The coldest known place on Earth: Vostok Station, Antarctica. A (left panel): mean monthly surface air temperatures at Vostok Station. Data are presented as monthly mean ± standard deviation, for the period 1958–2012, with lines showing the mean monthly high and low temperatures and the record low shown as a black symbol. B (right panel): frequency histogram of six-hourly surface air temperature at Vostok Station, 1958–2006, with the record low shown as a dotted line[6].

been recorded by satellites, with the current record being −93.2 °C recorded by the NASA MODIS satellite on 19 August 2013 (though at the time of writing this is a provisional figure, and may be subject to revision). This, however, is the temperature of the ice surface, and radiative cooling can lead to surface temperatures lower than the meteorological air temperature. Direct comparisons from the Antarctic plateau have shown that the ice surface temperature recorded by satellites is typically 2–4 K cooler than the overlying surface air temperature. The Vostok measurement remains the lowest meteorological surface air temperature ever recorded on Earth.

Other sites on the Antarctic plateau have recorded very low temperatures (Table 14.1). Data from Plateau Station, which was occupied for about three years until January 1969, provided recorded the world's lowest air temperature for a continuous month, which was −73.2 °C in July 1968.

Vostok, Plateau and Amundsen-Scott Stations are all deep in the interior of Antarctica, on the high ice plateau. Although most of continental Antarctica is ice-covered, there are a number of mountain ranges that are exposed, typically as a series of nunataks[7]. These nunataks support life, including lichens, mosses, microbial algae and fungi together with bacteria and a range of microscopic invertebrates such as nematodes, rotifers and tardigrades (Plate 12). It is likely that the surface air temperatures around these nunataks are similar to those widespread within continental Antarctica, but as yet we do not know. While data from occupied stations and automatic weather stations deployed on the Antarctic plateau indicate a fairly uniform pattern of temperatures across the Antarctic continent, meteorologists put their instruments where the local topography does not disrupt measurement. In consequence we lack temperature data from many of the areas of continental Antarctica where organisms actually live (Box 14.1). While we might assume the surface air temperatures around nunataks are similar to those recorded on the plateau, there is always the possibility that the extra height where nunataks emerge through the ice plateau produces a different thermal regime.

The organisms found in these isolated places are exclusively low growing or live within the soil. The darker colour of the soil will absorb more incident radiation, warming the habitat so the temperatures of these habitats will be different from the surface air temperature (indeed they would have to be, or liquid water, and hence life, could not exist there). It would be extremely valuable to have records of habitat temperature from these isolated places, but as yet we lack these data for the more isolated locations deep in Antarctica.

In the northern hemisphere, the native terrestrial flora and fauna extends to the northerly edge of the land, and here the surface air temperature has real ecological relevance. The lowest surface air temperature recorded for a northern hemisphere location is −67.7 °C measured at Oymyakon in Siberia (Table 14.1). A very similar temperature has been

Table 14.1 Record low temperatures. All data are meteorological surface air temperatures[8].

Place	Lat/Long	Temperature (°C)	Comments
Southern Hemisphere			
Vostok Station, Antarctica	77° 32′ S, 106° 40′ E	−89.2	21 July 1983
Amundsen-Scott Station, Antarctica	South Pole	−88.3	24 August 1960
Plateau Station, Antarctica	79° 30′ S, 40° 00′ E	−86.2	20 July 1968
Northern Hemisphere			
Oymyakon, Siberia	67° 33′ N, 133° 23′ E	−67.7	6 Feb 1933
Verkhoyansk, Siberia	63° 28′ N, 142° 23′ E	−67.6	5 and 7 Feb 1892 (previous records of −69.8 °C and −68.8 °C in 1892 are now regarded as unreliable)
Prospect Creek, Alaska	66° 49′ N, 150° 39′ W	−62.2	23 Jan 1971
Snag airport, Yukon	62° 23′ N, 140° 22′ W	−63.4	3 Feb 1947

Box 14.1 Life in Antarctica

Less than 0.2% of the surface of Antarctica is exposed land (free of permanent snow and ice). A recent study examined how much of this exposed land had any recorded plants (mosses, lichens or flowering plants). The entire continent was divided into a series of boxes, each 1° of latitude by 1° of longitude. Because the Earth is a sphere, these boxes are trapezoid in shape rather than rectangular, and their area gets smaller towards the pole. Most of these boxes contain nothing but ice, but some contain exposed rock or soil and hence have the potential to support plant life. Boxes were classified as containing ice-free ground when the total area of exposed rock within that box exceeded 10^3 m^2 (a typical football field is five times this area).

Plants are known from only 30% of boxes containing exposed land, although these comprise roughly 50% of the known ice-free area. The remaining area of exposed land is unexplored, rather than necessarily being free of biota. In other words, about half of exposed land in Antarctica has no record of any plant, because no ecologist has ever been there (Figure 14.2).

This map shows clearly how terrestrial habitats are concentrated at the edges of the continent. There are, however, significant areas of exposed land in the interior of the continent, particularly along the Transantarctic Mountains. Some isolated nunataks have been explored for their biota, and although some individual samples were apparently devoid of life, all locations investigated to date have microbial and invertebrate life[9].

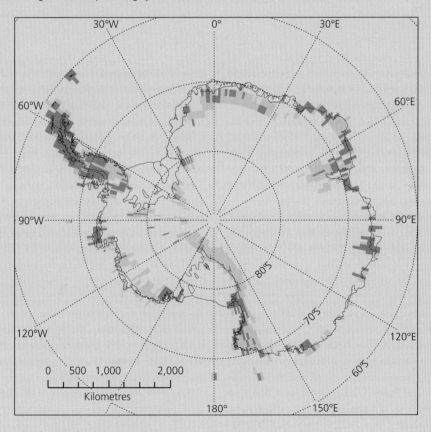

Figure 14.2 Map of Antarctica showing lat/long boxes where there is exposed land. Dark grey boxes are those where there is at least one plant record, light grey boxes are where there is exposed land but no record of any plant.

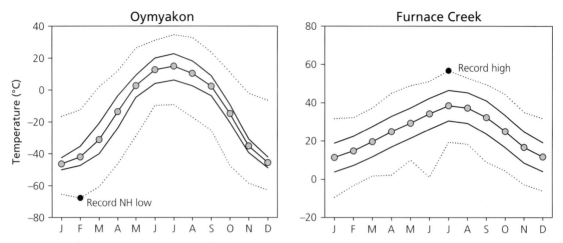

Figure 14.3 Seasonal variation in meteorological surface air temperatures at the location of the global record high and low surface air temperatures for the northern hemisphere. A (left panel): mean monthly air temperature (grey symbols), with mean monthly high and low temperatures (solid lanes) and monthly extreme temperatures (dotted lines), for Oymyakon, Siberia; the record low surface air temperature for the northern hemisphere is shown (black symbol). B (right panel): mean monthly air temperature for Furnace Creek, California; presentation as for left panel, with the record high surface air temperature shown[10].

recorded at Verkhoyansk, and both sites are in the Sakha (Yakutia) Republic of Russia. The daily mean temperatures for January at these sites are –45.5 °C (Verkhoyansk) and –46.4 °C (Oymyakon). The interior of Alaska also gets very cold in winter, and the record low here is only just above the record low for North America as a whole (Table 14.1). As with Antarctica, these record low temperatures are well below the normal seasonal pattern (Figure 14.3).

Although record temperatures attract considerable interest, and are relevant in that they capture the occasional extreme events to which organisms may be subject, it is the normal seasonal pattern that is more relevant to ecology. Seasonal data for Siberia indicate that in a typical winter, the northern hemisphere flora and fauna will be subject to temperature minima around –30 °C, although the seasonal pattern does vary from place to place.

The highest surface air temperature ever recorded is 56.7 °C, measured at Furnace Creek Ranch (formerly Greenland Ranch) in Death Valley, California on 10 July 1913 (Table 14.2). A previous record of 58 °C from El Azizia, Libya, is now regarded as unreliable. The two highest records are also regarded as unreliable by some meteorologists, but are currently accepted by the WMO. It is possible that the recent observations from Kuwait and Iraq will eventually be accepted as the highest reliable recorded surface air temperatures, and it is also possible that as the Earth continues to warm these will be exceeded[11].

Many desert organisms are small and live on, in, or close to, the ground surface. Here temperatures can be very different, and the early desert ecologists recorded beetles and lizards scurrying across sand with surface temperatures as high as 80 °C. Direct measurements of desert surface temperature with thermometers or thermocouples have recorded values up to 93.9 °C (Death Valley, 15 July 1972), with similar values reported from Sudan (83.9 °C at Wadi Halfa) and Congo (83.9 °C at Loango). Accurate measurement of the surface temperature of bare soil is, however, very difficult to achieve. A particular problem is error from the sensor being exposed to solar radiation. For these reasons, these very highest surface temperature measurements for desert soils need to be treated with caution[12].

Fewer errors are associated with radiometric determinations from satellites, which use the infrared radiation emitted from the soil surface to estimate *land surface temperature* (sometimes referred to as *skin temperature*). Satellites do, however, integrate temperatures over a large area (typically 5 km² for current

Table 14.2 Record high temperatures. All data are meteorological surface air temperatures.

Place	Lat/Long	Temperature (°C)	Comments
Northern Hemisphere			
Furnace Creek, California	36° 27' N, 116° 51' W	56.7	10 July 1913. 54.0 °C recorded here on 30 June 2013
Kebili, Tunisia	33° 42' N 8° 58' E	55.0	7 July 1931
Mitribah, Kuwait	29° 82' N, 47° 35' E	54.0	21 July 2016. Yet to be confirmed by WMO
Basra, Iraq	30° 34' N, 47° 47' E	53.9	22 July 2016. Yet to be confirmed by WMO
Southern Hemisphere			
Oodnadatta, Australia	27° 32' S, 135° 26' E	50.7	2 Jan 1960
Dunbrody, Eastern Cape, South Africa	33° 28' S, 25° 32' E	50.0	3 Nov 1918

measurements), and so may miss localised hot-spots. The hottest land surface temperatures on Earth determined remotely from satellites come from the Lut Desert in Iran, where in 2005 a record of 70.7 °C was measured. Similar high desert surface temperatures have been measured for the Turpan Basin in western China (66.8 °C in 2008) and in Queensland, Australia (69.3 °C in 2003). A global analysis indicated that annual maximum surface temperatures were typically 50–65 °C for desert and sparse shrublands, 35–50 °C for grasslands and savannahs and 25–35 °C for forests. The ice- and snow-covered lands of polar regions and mountain ranges typically had annual maximum temperatures in the range –30–+ 5 °C[13].

14.1.3 Geothermal areas

There are many places on Earth where geothermally heated water reaches the land surface, including Iceland, New Zealand, Kamchatka and even in Antarctica. Organisms living near these geothermal areas can experience temperatures higher than anywhere else in the terrestrial environment and, as we shall see, have provided powerful insights into the thermal limits to life.

14.1.4 Mean annual temperature and seasonality on land

The uneven distribution of stations measuring surface air temperatures on land means that synoptic coverage is best achieved with satellite data, bearing in mind that satellites are measuring land surface temperature rather than meteorological surface air temperature.

Mean temperature decreases polewards, but the distribution of high and low temperatures across the land surface is far from even or regular. The highest annual mean temperatures are found in the Sahara Desert in north Africa, eastwards through Saudi Arabia to south-central Asia, the Namib and Kalahari deserts in southern Africa and Australia, with smaller areas of the deserts of America (Chihuahua, Sonora and Mojave in North America and Atacama in South America) (Plate 13). In contrast much of the northern hemisphere has a moderate or low annual mean temperature. Since most (~68%) of the world's land is in the northern hemisphere, this leads to a distinct asymmetry in the pattern of mean temperature on land.

This asymmetry can be seen clearly in a latitudinal (north–south) transect of mean temperatures through the Americas (Figure 14.4). As would be expected, mean temperature is highest in the tropics and declines both to the north and south. The rate of decline with latitude differs, however, being steeper in the northern hemisphere. There is also a marked difference in the seasonal range of temperature, this being far greater in the northern hemisphere. Both of these reflect the powerful influence of the ocean on climate. The ocean ameliorates climate extremes, and it is the large land mass of Eurasia that results in low winter temperatures and high seasonality.

A map of the seasonality of land surface temperature (Plate 14) shows that a strong seasonality

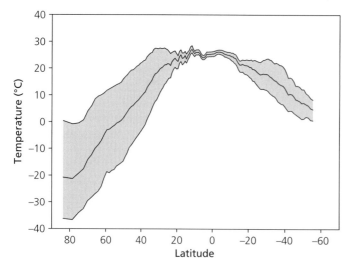

Figure 14.4 Latitudinal clines in surface air temperature in the Americas. The black line symbols show the mean annual land surface temperature for the period 1961–1990 in bands of 1° of latitude (by convention southern latitudes are negative), and the grey area shows the mean annual maximum and minimum temperatures for each latitudinal band. Note that as these data are limited to the American continent, data from Antarctica are not included[14].

in temperature is predominantly a feature of the northern hemisphere. The most thermally seasonal terrestrial environments lie in a broad band across the mid northern latitudes, encompassing most of the boreal forest (taiga) and the tundra to its north. The degree of seasonality is driven predominantly by a difference in minimum temperature. For most latitudes summer temperatures are similar, but latitude for latitude the winter minimum temperature is much lower in the northern hemisphere. In part this is the result of the greater land mass in the northern hemisphere, reducing the ameliorating effect of oceanic water and allowing the temperature to fall further in winter, leading to a lower mean annual temperature.

14.2 The global distribution of oceanic temperatures

The ocean is a challenging place to work. As with the land, our view of oceanic temperature has been revolutionised by satellite remote sensing. These satellites only measure the very top of the ocean surface (down to 20 micrometres for infrared and a few millimetres for microwave radiometers); this is usually referred to as *sea surface temperature* (SST). While satellites have been of enormous significance in providing detailed global pictures of temperatures at the surface, where the oceans and atmosphere interact closely, they cannot see into the interior of the oceans, which is where the life is. For this, modern oceanographic research relies on the platinum resistance thermometer, usually deployed in conjunction with a conductivity meter to measure salinity and a pressure sensor to register depth (the whole package being abbreviated to CTD). These are either deployed on wires from ships, or increasingly in floats or autonomous underwater vehicles such as gliders.

The overall range of temperature of bulk seawater in the oceans is much narrower than for terrestrial habitats and the high thermal capacity of water means that rates of change, both across space and in time, are much lower than in air. Oceanic waters vary in temperature with depth, latitude and season. The enormous size of the oceans, coupled with their great depth, make it difficult to calculate an overall temperature for seawater. Below the seasonally warmed and cooled surface layers, seawater remains permanently between –2 and +4 °C, except for the deep Mediterranean and Red Sea (which are warm) and the immediate environs of hydrothermal vents (which can be very hot). In the bulk of the ocean the water cannot cool further simply because there is nowhere for the thermal energy to go: the cool deep water is capped by the thermocline above, and geothermal energy enters the ocean from the seabed below. If we take a mean depth for the extent of seasonal warming (the thermocline) to be ~1000 m, then the seasonally variable surface

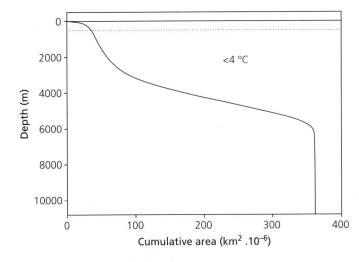

Figure 14.5 Depth and area in the ocean (the hypsographic curve). The solid line shows the ocean surface and the dotted line the mean depth of the thermocline. All oceanic water below this depth has a temperature of −1 to +3 °C.

waters comprise ~11.8% of ocean volume and thermally stable deep ocean water the remaining 88.2% (Figure 14.5). It is clear that the mean temperature of oceanic water overall is < 4 °C. The largest habitat on the face of the Earth, deep ocean water, is decidedly cool. There are a few places where geothermally heated waters enter the ocean, in the hydrothermal vent fields associated with the spreading ridge systems of the deep ocean where new oceanic crust is being formed. The emerging mineral-rich water can be very hot (> 300 °C) but it cools rapidly as it mixes with cold ocean-bottom water. This causes the minerals to precipitate out and build large chimney-like structures. These vents are localised in area, but support very rich chemosynthetic communities.

Surface oceanic waters can reach ~30 °C in tropical regions and −1.98 °C (the freezing point of seawater of salinity 36 at atmospheric pressure) in high latitudes (Plate 15). Unlike the complex patterns of temperature on land, mean oceanic temperatures vary in a broad simple pattern. The warmest waters are around the equator, and mean temperature decreases steadily towards either pole. A critical detail within this broad pattern is the presence of upwellings on the western coasts of Africa and America, where cold waters rise to the surface under the influence of Eckman transport induced by a combination of surface winds and the Coriolis effect. This draws cold and nutrient-rich waters from depth and these upwellings are often highly productive, supporting important fisheries. They also typically have deserts immediately inland. The most important of these upwellings are in the Benguela and Canary currents off the west coast of Africa, and the Humboldt and Californian currents off the west coast of the Americas.

Whereas the pattern of mean oceanic surface temperature is fairly simple, the pattern of seasonal variation is complex (Plate 16). Seasonality in temperature is mostly a feature of the northern hemisphere and of coastal regions (though the North Pacific is a striking exception to this generalisation). The strongest seasonality of sea surface temperature in the southern hemisphere is in the coastal region of the Argentine Basin.

The patterns of mean temperature and seasonality are thus decoupled in the sea, with the greatest seasonality evident in intermediate (temperate) latitudes (Figure 14.6). In the Atlantic Ocean seasonality is much stronger in the northern hemisphere than the southern, whereas in the Pacific Ocean the contrast is not so marked. We can thus divide the surface oceans of the world into three broad categories based on the temperature: the tropics (high mean temperature, low seasonality), the temperature regions (intermediate mean temperature, very high seasonality) and the polar regions (low seasonality, low mean temperature). The bulk of the ocean, below the seasonal thermocline, is similar to the surface polar oceans in being thermally stable and cold. The polar oceans also exhibit a small but ecologically important seasonality of surface salinity caused by the seasonal freezing and melting of sea-ice[15].

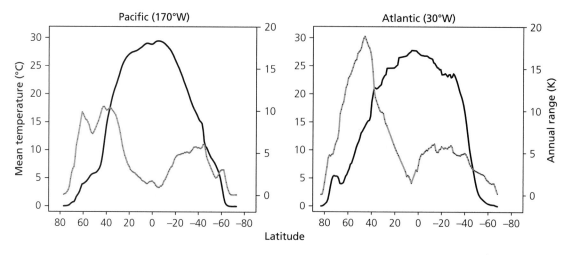

Figure 14.6 Mean (black line) and mean annual range (grey line) of sea-surface temperature (SST) for the period 1983–2003 in two ocean basins. A (left panel): along 170° W in the Pacific Ocean. B (right panel) along 30° W in the Atlantic Ocean. Note how the largest seasonal variations are in the temperate latitudes[16].

14.3 Are there thermal limits to life?

Having established the distribution of temperature on land and sea, we can return to our question of whether there are thermal limits to life as we know it on Earth. First, however, we must define what we mean by 'life'. This is not simply because it is good scientific practice to define any terms we use, but also because ecologists interested in limits to adaptation or survival often mean different things when referring to 'life'. If we are to discuss life as a general concept, we need to know exactly what we are talking about.

14.3.1 Defining 'life'

Life is an elusive concept; like time, it is something we feel we understand intuitively and yet can find difficult to explain to others. Philosophers have been discussing the nature of life since at least the time of the ancient Greeks, and we can be fairly sure all earlier civilisations did so too. Despite all this debate we still lack a universally agreed definition of life, a clear indication that the problem is exceptionally difficult. What philosophers have agreed upon, however, is that we can produce a thorough and useful *description* of life as we know it (in other words, life on Earth). This is too complex a discussion to concern us here, but there are a few key concepts that it is important to touch upon[17].

Any general description of a living organism needs to capture the essential elements of everything that is alive, from microbes to whales (Box 14.2). A general description such as this also counters naive arguments that flames are alive because they give off heat and utilise oxygen, or that crystals are alive because under defined circumstances they can grow. It also implies constraints on the physical size of a living entity, which cannot be too small (to avoid fatal stochastic imbalances in the internal environment) or too large (such that internal integration becomes impossible).

Almost all discussions of the nature of life refer back to Erwin Schrödinger's classic short book *What is life?*[18]. In this book, Schrödinger emphasised the role of free (Gibbs) energy and entropy in maintaining the living state, and explored the nature of the chemistry underpinning the coding of genetic information. Although prescient in his analysis of the thermodynamic aspects of life, Schrödinger confounded two important features of heredity, namely the code itself and the translation and execution of that code. At that time, however, the precise chemical nature of the genetic material was unknown and genes were largely suspected to be proteins. It was only later that they were shown to be nucleic acid,

> **Box 14.2 A description of life**
>
> While we cannot define life, we can describe life as we know it here on Earth[19]. A simple description of a living organism might be that:
>
> 1. It is isolated from the environment by a barrier or partition.
> 2. It has an internal composition that is non-equilibrium thermodynamically, and which is maintained by a flow of materials and the utilisation of free energy.
> 3. It reacts to changes in the external world.
> 4. It reproduces itself, based on a set of internal instructions.
>
> This is sometimes referred to as a *containment, metabolism and program* (CMP) description of life.
>
> An important consideration is that while containing the organism, the barrier must allow the flow of selected materials inwards from the environment, and the release of waste products. This flow and modification of materials means that where life is widespread and common it may modify the composition of its environment.
>
> Many other definitions, both shorter and much longer, can be found in the literature. The trick is to have a definition which describes not just the one form of life we know (here on Earth), but which would allow us to recognise life elsewhere, should we ever encounter it.

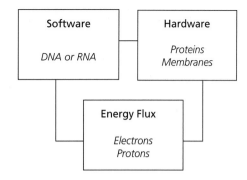

Figure 14.7 A tripartite description of life on Earth, based on Schrödinger and von Neumann. All three components are necessary, but not sufficient: for a living entity to perpetuate itself, it must also replicate.

and the nature both of the code and its manner of replication recognised with the elucidation of the structure of DNA[20].

John von Neumann recognised and corrected Schrödinger's error, emphasising that the genetic material must both be copied, so that its integrity is preserved and translated (that is, interpreted)[21]. Von Neumann's analysis thus recognised the critical distinction between what we would now call software (the coded instructions embodied in nucleic acids) and hardware (the protein machinery produced by translating and interpreting that software).

Taking these insights together, we can identify a tripartite description of life (Figure 14.7). A living entity on Earth comprises three essential features: the genetic information encoded in nucleic acids (software), the structural elements of the cell in the protein and lipid architecture, and the protein functional elements (hardware), and energy flux. The interior of the cell is maintained in a thermodynamically non-equilibrium state and as Schrödinger emphasised, for this locally low entropy environment to be maintained there must be a continuous flow of energy. Without such flow, the internal composition of the cell cannot be maintained, and the organism dies; Gibbs energy is then dissipated and the internal composition drifts close to equilibrium as the organism decays.

An organism is thus alive only if it exhibits all three of these features. A virus has only software: it contains the genetic instructions for the construction of new viruses, but lacks the machinery to do this and exhibits no energy flow. Viruses are obligate parasites of cells, which they utilise for both the machinery of construction and the energy flux to drive that construction. On the basis on our working description of life, viruses are not alive[22].

A resting spore, such as those of bacteria, fungi, eukaryotic algae or the brine-shrimp *Artemia salina* contains both the software (DNA) and the hardware (proteins, membranes), but exhibits no energy flux. Like tardigrade tuns, dehydrated nematodes and many plant seeds, *Artemia* spores have the potential (or capacity) to be alive when circumstances change, but they are not alive in themselves. This context-dependence of the living state sets an interesting philosophical challenge for those wrestling with the definition of life.

A living entity thus has software, hardware and energy flux (metabolism). While necessary, however, these are not sufficient. No living entity can perpetuate itself forever (all living things age) and the fourth essential component of life is reproduction. This reflects a second key distinction identified by von Neumann, namely between metabolism and replication. All living entities on Earth reproduce themselves; for a bacterium or archaean this means cell division to produce daughter cells, for a sexually reproducing eukaryote it means completion of the cycle from zygote to zygote.

14.3.2 An aside: the NASA definition of life

It is impossible for the instructions embodied in nucleic acids to be copied with complete accuracy, differences will always be introduced each time the DNA is replicated as the cell divides. Where these errors (mutations) occur in germ cells they provide the raw material for evolution, and so, in a resource-limited world where not all individuals can survive, this process will lead inexorably to evolutionary change. This has suggested to some that life might be defined in terms of evolution. Indeed this is the approach embraced by NASA, who define life as *'a self-sustaining chemical system capable of Darwinian evolution'*[23].

A subtle but important feature here is that the basis of the description of life in Box 14.2 is the individual entity, and an individual organism does not evolve. Evolution is a change in the frequency of genes in a defined group of individuals: it is a feature of populations, not of individuals. Darwinian evolution is thus a population consequence of the general features of life, rather than a defining characteristic of an individual living entity. While attractive in its simplicity, the NASA definition confounds a consequence of life with its definition[24].

14.3.3 A second aside: detecting life elsewhere

Because living organisms take in materials to build or maintain their bodies, and discard waste products, they have the capacity to modify the chemical nature of their environment. Indeed, where life is sufficiently abundant it can modify the entire atmosphere of a planet, producing an atmospheric composition that is away from thermodynamic equilibrium. An example from Earth is the co-existence in the atmosphere of oxygen with small quantities of biogenically produced methane. Without abundant life the atmosphere of Earth would contain only traces of oxygen and nitrogen, but large quantities of carbon dioxide and water vapour. This led Jim Lovelock to propose that life could be detected on distant planets simply by examining the atmospheric composition spectroscopically: any planet exhibiting a large chemical free energy gradient between surface matter and the atmosphere in contact with it, and a significant chemical disequilibrium in the atmosphere itself, can be assumed to support abundant life. Using these criteria, we can conclude that neither Venus nor Mars currently harbour abundant life, without even visiting[25].

14.4 Temperature thresholds for life

The discussion above may seem somewhat arcane, but it is central to any consideration of the relationship between temperature and life. In particular it tells us that we need to draw an important distinction between two different threshold temperatures: a threshold for completion of the life cycle, T_L, and a threshold for metabolism, T_M (Figure 14.8). This distinction is important because many organisms live in thermal environments where they are metabolically active but unable to complete their life cycle. This might be because of insufficient energy to produce gametes or it might be caused by a temperature-related failure of a key physiological process involved in reproduction but which is not in itself lethal.

It is important to recognise that organisms can exist outside the temperature range over which they metabolise. Resting stages such as cysts or spores, or dehydrated microinvertebrates such as tardigrade tuns, can survive in a state of suspended animation. By our definition of life above, these resting stages are not alive in themselves, but they have the capacity to resume living once circumstances change. A limit to such survival is always reached at high temperature, where it often coincides with the thermal limit to metabolism. The lower limit to survival, however, may be very much colder than the limit for metabolism. Indeed the recovery of living tardigrades from tuns exposed to liquid helium indicate the lower survival limits for resting stages may extend almost to absolute zero[26].

GLOBAL TEMPERATURE AND LIFE 319

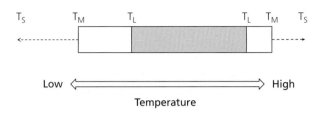

Figure 14.8 Temperature thresholds for life on Earth. T_L: thermal limits for completion of the life cycle; T_M: thermal limits for metabolism. The shaded portion shows the temperature range over which the life cycle can be completed, and defines the thermal limits for the continued existence of a species over generations. The clear portions show the temperature ranges over which the organism is metabolising but cannot reproduce. The temperature ranges defined by the dotted lines are those where the organism is in suspended animation[27].

The important point that emerges from this analysis is that it is the upper and lower T_L thresholds that mark the limits to life on Earth. Existence outside these boundaries may be possible for periods of time but does not allow for completion of the life cycle. While studies of extremophile microbes have tended to concentrate on establishing T_L by determining the thermal limits to cell division, studies of plants and animals have tended to concentrate on limits to survival.

Failure to recognise this distinction has produced considerable confusion in the literature concerned with limits to life. Temperature constraints on survival, metabolism and reproduction are confounded, and sometimes it is not even clear which threshold is being considered. As a result we still lack a clear idea of the temperature constraints on the completion of the life cycle in some important groups of organism.

Living organisms on Earth range in size and complexity from bacteria and small mycoplasma to whales and redwoods. These organisms show a wide range of internal structure and, in the case of multicellular organisms, tissue architecture. We should therefore not expect them all to respond to temperature in the same way, and indeed they do not. We can, however, divide them into a small number of functional groups based on features likely to influence their response to temperature (Table 14.3).

14.5 Low temperature limits

A general lower limit for life in free-living unicells would appear to be set by the temperature at which freeze-concentration of the external environment dehydrates the cell interior and drives vitrification (Chapter 6). While vitrification defines a lower limit for metabolism, it is possible that one or more other factor may limit completion of the life cycle at temperatures above that at which the cell vitrifies.

A thorough survey of studies of microbial growth at low temperatures suggests that none grow below −20 °C (Figure 14.9). There are a few reports in the

Table 14.3 The three major domains of life on Earth, with Eukarya subdivided into categories with differing features of potential importance to their thermal ecology

Domain	Key features
Archaea	No nucleus, free ribosomes, ether-linked membrane lipids
Bacteria	No nucleus, free ribosomes, acyl-linked membrane lipids
Eukarya	Nucleus, mitochondria, ribosomes on internal membranes, acyl-linked membrane lipids
Unicellular eukaryotes	Cell in direct contact with external environment
Multicellular eukaryotes	Cells exposed to internal body environment
Lichens	Symbiosis between fungal hyphae and algae; often highly resistant to dehydration
Fungi	Heterotrophic, non-vascular, absorbing nutrients from the environment through hyphae
Mosses and liverworts	Non-vascular (no xylem), absorbing water and nutrients through leaf surfaces
Higher plants	Vascular tissues for distributing resources through plant
Invertebrates	No backbone; cardiovascular system for distributing nutrients
Ectothermic vertebrates (fish, amphibians, reptiles)	Cardiovascular system for distributing nutrients
Endothermic vertebrates (mammals, birds)	Maintain internal body temperature above ~30 °C

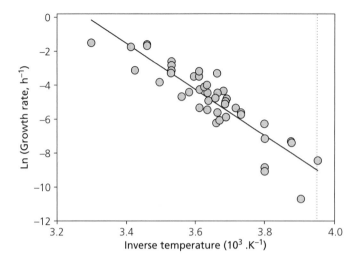

Figure 14.9 The lower temperature limit for growth in unicellular organisms; growth rate (natural log-transformed) plotted as a function of inverse temperature. The dotted line shows the lowest temperature at which cell division has been reported (−20 °C)[28].

literature of microbial activity at temperatures below −20 °C but in these studies growth or metabolism are usually inferred from proxies (such as the release of apparent metabolic products) and not direct measures of cell number as a function of time. It is possible that in these cases geochemical processes or concentrated solution chemistry is mimicking metabolism. Indeed, despite a century of refrigeration technology, there are no reports of spoilage organisms growing below −20 °C. It is clear that the lower limit for completion of the life cycle in free-living unicells (bacteria, archaeans, unicellular eukaryotes) is around −20 °C[29].

The lower thermal limit for life in marine organisms is set by the lowest temperature at which seawater can remain liquid. The equilibrium freezing point of seawater at atmospheric pressure is −1.86 to −1.92 °C, depending on salinity, and this increases with depth (see Chapter 6). Seawater bathing the continental shelf seabed around parts of Antarctica is close to the freezing point year-round and so the rich communities of marine invertebrates and fish that live there must complete their life cycles at this temperature (T_L ~−2 °C).

When sea-ice forms, salt is excluded from the growing ice. As a result sea-ice contains many channels filled with brine, and these can reach very low temperatures (~ −20 °C) without freezing. Diatoms and many other unicellular eukaryotes, as well as some invertebrates, live in these channels. It is possible that these assemblages contain taxa that can complete their life cycle at lower temperatures than in the surrounding seawater: for example the Arctic sea-ice diatom *Nitzschia frigida* can grow down to −8 °C, though the doubling time is very long (60 days)[30]. The lower thermal limit for cell division in these habitats is not, however, well defined.

It has long been known that fungal infections can affect crops under snow and mushrooms that emerge through snow in northern forests are well known. The lower thermal limit for completion of the life cycle in free-living fungi and yeasts remains unknown but appears to be about 0 °C, and while some moulds and rusts can grow at low water activities on refrigerated foods, none are known to grow below −20 °C[31].

Lichens are among the most tolerant of organisms, being found in habitats where higher plants are absent. They are able to dehydrate extensively, and in this state can tolerate very low temperatures. Among Antarctic lichens photosynthesis has been recorded down to −16.5 °C in *Xanthoria candelaria*, −17 °C in *Umbilicaria aprina* and −18 °C in *Neuropogon acromelanus*. The lowest recorded temperature for photosynthetic carbon fixation is −24 °C. In *Umbilicaria aprina* dark respiration and growth ceased at higher temperatures than photosynthesis, suggesting that T_L is ~−10 °C. Lichens are a symbiotic relationship between an alga (the photobiont) and a fungus (the mycobiont). Interestingly, the symbiotic organism (the lichen) appears to withstand more extreme conditions than either the phytobiont or mycobiont alone[32].

In most cases, terrestrial organisms able to survive extreme cold can only complete their life cycle once temperatures have risen again in summer. So while some very impressive examples of low temperature survival are known, it is much more difficult to assign a value to the low temperature threshold for completion of the life cycle. The lowest limits documented so far are for invertebrates in meltwater on glaciers, such as the enchytraeid annelid 'ice worms' of the genus *Mesenchytraeus* and chironomid midges of the genus *Diamesa* where T_L is ~ 0 °C[34]. This is slightly higher than the T_L for marine invertebrates, but it may be that we simply lack documentary evidence of lower T_L values for terrestrial plants and invertebrates. The marked diurnal and seasonal variations in environmental temperature will, however, make such data difficult to obtain.

Current knowledge of the lower thermal limits to life on Earth is summarised in Table 14.4.

14.6 The special case of endotherms

Two lineages of vertebrates, mammals and birds, have independently evolved endothermy, the capacity to maintain a high and constant body temperature. Typically endotherm body temperatures are in the range 30–45 °C (Chapter 10), with birds tending to be warmer than mammals.

While endotherms are found from the hottest deserts to the polar regions, their cell temperatures are confined to a narrow range. To take mammals as an example, the range of body temperatures is 30–41 °C, but they live in areas with annual mean temperatures ranging from −11 °C to 26°C. The extreme endotherm example is, however, probably a bird: the emperor penguin, *Aptenodytes forsteri*, which raises its chick on sea-ice in the depths of the Antarctic winter, when temperatures are −20 °C or below[35].

This poses the question of what is the correct value of T_L for an endotherm: is it the internal body temperature at which the cellular physiology operates, or is it the environmental temperature? For humans with cultural adaptations (clothing, housing), the range of environmental temperatures that define the T_L range is the widest of any species on Earth.

14.7 High temperature limits to life

The earliest record of life in hot springs appears to be that of Pliny the Elder, who noted in his *Natural History* (probably written around 77–79 AD) that green plants could be found growing in the hot springs at Padua. These springs are still active, and undeveloped sections contain cyanobacterial mats (which are presumably the 'green plants' recorded by Pliny).

Table 14.4 Low temperature limits for life on Earth. T_L: temperature threshold for completion of the life cycle; T_S: temperature threshold for survival; nd: no data; *: unclear whether the entire life cycle is completed at this temperature[33].

Taxon	T_L (°C)	T_S (°C)	Comments
Archaea			
Species/strain unknown	−16.5	nd	Methanogenesis in permafrost cores
Bacteria			
Species/strain unknown	~−20	<−196	Microbial respiration in permafrost cores
Eukarya			
Unicellular algae	~−8	nd	Diatom *Nitzschia frigida* in brine channels
Yeasts	~−18	<−196	*Rhodotorula glutinis* on stored peas
Lichens	−10*	<−80	Dark respiration in *Umbilicaria aprina* in Antarctica
Mosses	nd	−30	No recent data
Angiosperms	~0*	~−70	No recent data
Terrestrial invertebrates	~0	<−196	Nematodes, tardigrades and rotifers in the dehydrated state can survive extremely low temperatures
Freshwater invertebrates	~0	nd	Annelids and insects in glacial meltwater
Marine invertebrates	~−2	~−2	Antarctic benthos; small metazoans (crustaceans) in brine channels may have lower T_L values
Ectothermic vertebrates	~−2	~−2	Antarctic fish (many species)
Endothermic vertebrates	nd	nd	Cell temperatures in range ~30–45 °C

The basis of our understanding of microbial life at high temperatures stems from the pioneering work of Thomas Brock and colleagues in Yellowstone National Park in Wyoming. Brock first reported microbes growing at high temperatures in the 1960s. Shortly afterwards, together with a colleague Hudson Freeze, he isolated a bacterium growing at 70 °C which they described as *Thermus aquaticus*. This organism has proved to be of enormous significance as the source of the enzyme (TAQ polymerase) which underpins the genomic revolution. Based on his experiences of life in hot springs in Yellowstone and elsewhere, Brock predicted that life would be found wherever water was liquid, a prediction that was vindicated spectacularly by the discovery of microbial life at very high temperatures and pressures associated with hydrothermal vents in the 1970s[36].

The very hot water, often at temperatures > 300 °C, that emerges from hydrothermal vents mixes with the local seawater and this leads to very strong thermal gradients. The very hottest waters appear to be abiotic, but areas where the water has cooled are characterised by extensive microbial mats. These have yielded a wide range of hyperthermophiles (organisms with an optimal growth temperature at or above 80 °C), both archaeans and bacteria. Hyperthermophiles require liquid water, so growth above 100 °C is possible only where high pressure keeps the water liquid. They are found in a wide variety of terrestrial and marine habitats, all associated with geothermal sources of heat. The low solubility of oxygen at these high temperatures and the frequent presence of large amounts of reducing gases mean that most habitats for hyperthermophiles are anaerobic. Most hyperthermophiles utilise inorganic redox reactions as sources of energy, and CO_2 as the sole carbon source.

The current record for high temperature growth is *Methanopyrus kandleri*, originally isolated from a vent in the Gulf of California and found to grow between 84 and 110 °C. However a strain of *Methanopyrus kandleri* isolated subsequently from the Kairei vent field on the Central Indian Ridge was found to grow at 122 °C under 40 MPa pressure, just surpassing the previous record for 121 °C for *Geogemma barossii*. Microbes growing at the very highest growth temperatures all appear to be archaeans, but there are some bacteria which are able to grow to ~100 °C, with the current record being *Geothermobacterium ferrireducens*, which was isolated from Obsidian Pool in Yellowstone. Two other taxa, *Aquifex aeolicus* and *Thermotoga maritima*, can grow at 90 °C or above, and there are a range of Fe(III)-reducing thermophilic bacteria with T_L values in the range 65–75 °C. This difference between archaea and bacteria in sensitivity to high temperatures is evident in the distribution of the two groups within active vents, where there can be a transition from a mixed assemblage of archaea and bacteria near the cooler exterior of the chimney, to primarily archaea in the hotter interior[37].

The steep and highly variable thermal gradients surrounding hydrothermal vents make it extremely difficult to quantify the precise thermal environment of the microbes growing there. Early experimental work suggested that some of these microbes could grow in culture at 250 atm (26.85 MPa) pressure and temperatures of 250 °C. These studies raised the upper thermal limit for hyperthermophiles by a staggering 140 K, and initiated an intense debate centred on the possibility of artefacts or contamination and the instability of many biological molecules at such high temperatures. To date these results have not been replicated, and the currently accepted upper limit for microbial growth is 122 °C[38].

14.7.1 Eukaryotes at high temperature

Eukaryotes are unable to live at the very highest temperatures that characterise geothermal waters; these are the domain of bacteria and archaeans alone. The highest T_L for a unicellular eukaryote appears to be 55–56 °C, which is the upper limit for the rhodophyte *Cyanidium caldarium*, although its optimal (maximum) growth was at 45 °C. Over a century and a quarter ago, however, William Dallinger reported a remarkable experiment in which he raised the temperature of a culture of 'monads' (unicellular flagellates, including *Tetramitus rostratus*, *Monas dallingeri* and *Dallingera drysdali*), in a series of small steps, inspecting the cultures for morphology, activity, fission and sexual fusion after allowing the cultures to acclimate following each small increment in temperature. The experiment ran from 1880 to 1886 and Dallinger reported that the flagellates were still active and reproducing at 70 °C; sadly the experiment ended at this point when the apparatus was accidentally destroyed[39]. This experiment needs repeating with replication, modern means of temperature control and documentation of growth from cell counts, as it may well establish a new thermal maximum for growth in unicellular eukaryotes.

Slightly higher temperatures appear to be tolerated by filamentous fungi, and a survey of a range of high temperature habitats revealed species able to grow at 55–60 °C. Higher plants can be found growing close to hot springs wherever they occur, including in Antarctica. In the perennial grass *Dichanthelium lanuginosum* (intriguingly named 'hot springs panic grass') in Yellowstone, the thermal tolerance is mediated through a mutualistic endophytic fungus *Curvularia protuberata* and a mycovirus. With both the fungus and mycovirus present, plants can grow in soils up to 65 °C; with either missing the plants are unable to grow above 38 °C[40].

Hot springs also provide the hottest habitats inhabited by invertebrates and vertebrates. Temperature in these springs may reach over 50 °C, and the fauna includes crustaceans, chironomid larvae, nematodes and molluscs, as well as fish. It is difficult to establish T_L values for these for although many secondary and anecdotal sources quote a range of temperatures for hot springs, there are few primary sources with data for both temperature and fauna. Two nematodes, *Rhabditis terrestris* and *Udonchus tenuicaudatus*, appear to be ubiquitous in thermal springs and have been recorded as living up to 42.8 °C in Granada, Spain; in addition, *Aphelenchoides parientus* lives in hot springs up to 51 °C, and *Dorylaimus thermae* is found in Yellowstone in waters up to 53 °C. Hot springs contain a range of other aquatic invertebrates, including crustaceans such as the isopod *Thermosphaeroma subequalum*, insect larvae (especially chironomids) and molluscs such as the springsnail *Tryonia julimensis*; all of these will presumably have similar T_L values. The highest temperature for completion of the life cycle in an invertebrate may be for nematodes of the genus *Aphelenchoides* and *Panagrolaimus* which tolerate temperatures of 60 °C in compost heaps[41].

Hot springs also have fish, and the classic high temperature fish are the desert pupfish of the genus *Cyprinodon*. These live in shallow geothermal streams, where the temperatures are high but vary both spatially and throughout the day and with season. *Cyprinodon pachycephalus* from the hot springs of San Diego de Alcalá, Chihuahua, México, lives in waters of 39.2 to 43.8 °C, and *Cyprinodon julimes* recently described from the hot springs of Julimes, Chihuahua, México, lives at temperatures between 38 and 46 °C. In contrast to terrestrial vertebrates which can use shade to avoid the heat of the sun, and which cool off by night, desert pupfish spend their entire life at these high temperatures. While the water temperatures do vary a little diurnally and the fish often select the cooler water, these two species of *Cyprinodon* are believed to be the fish with the highest T_L on Earth. They are also limited to a few small springs, and are consequently highly endangered[42].

The hottest aquatic environments of all are hydrothermal vents, and these have a spectacularly rich and abundant fauna that includes a range of crustaceans, molluscs and worms. As with the microbial flora within the vent chimneys, the very steep thermal gradients make it difficult to assess precisely what temperatures any given animal is experiencing. Behavioural observations and associated temperature measurements suggest that many motile vent animals select warm but not hot locations, and that they are very sensitive to changes in temperature.

The most studied vent animal in relation to temperature is the Pompeii worm, *Alvinella pompeiana*. This polychaete lives in a tube through which vent fluids pass, and from which it emerges to forage. Recordings with a temperature probe indicated that at the base of the worm (nearest the vent) the temperature averaged 61 °C, though there were occasional spikes up to 81 °C, whereas at the mouth of the tube temperatures averaged 22 °C. These data indicate that *Alvinella* is subject to a quite remarkable thermal gradient along its body (roughly 60 K). However it is difficult to assess its T_L, because the worm leaves its tube to forage in much cooler water (~2 °C). A recent study has shown that long-term survival, as assessed by a two-hour ramped thermal exposure, is above 42 °C but below 50 °C. Similarly, another vent polychaete *Paralvinella sulfincola*, can be found in waters up to 88 °C, but has an upper incipient lethal temperature (at which 50% of the population cannot survive indefinitely) of only 45 °C. Although the precise T_L values for *Alvinella* or *Paralvinella* are unknown, current data suggest that they may hold the record T_L for an aquatic animal, and are also probably some of the most eurythermal metazoans on the planet[43].

Away from geothermal areas, the hottest environments are deserts. Deserts are, however, often only hot during the day; by night and under a clear sky temperatures can drop below freezing. This combination of high daytime temperatures and low night-time temperatures poses severe physiological problems for organisms living there. Some motile forms are active by day and can tolerate brief periods of very high temperatures. For example the Saharan silver ant *Cataglyphis bombycina* forages for very short periods in air temperatures up to 55 °C. Similarly, *Ocymyrmex barbiger*, an ant from the Namib Desert, forages at temperatures up to 67 °C. Being small these ants have a very low thermal mass and in consequence they heat up and cool down quickly. These and other small arthropods active in the desert heat minimise the period of time for which they are exposed to the highest temperatures, and they climb frequently up stems of vegetation where the air is cooler[44].

Desert plants exposed to direct solar heating can reach temperatures well above that of the surrounding air, but cannot move about to alleviate the direct effects of heat. The record appears to be held by the cactus *Opuntia*, several species of which can reach internal temperatures up to 65 °C[45]. One must assume that some plants living in the hottest deserts have thermal tolerances at least comparable with those living around geothermal springs, but data on the maximum temperatures at which plants can complete their life cycle are very difficult to find.

Current knowledge of the upper thermal limits to life is summarised in Table 14.5. Although there is a diverse literature on thermal limits to survival, data on the thermal limits to the completion of the life cycle are far more difficult to obtain, and for some groups the data in Table 14.5 rely principally on data collated over 40 years ago.

Table 14.5 High temperature limits for life on Earth. T_L: temperature threshold for completion of the life cycle; T_S: temperature threshold for survival; nd: no data; *: unclear whether the entire life cycle is completed at this temperature[46].

Taxon	T_L (°C)	T_S (°C)	Comments
Archaea			
Methanopyrus kandleri	122	<130	Strain isolated from Kairei hydrothermal field, Central Indian Ridge
Bacteria			
Geothermobacterium ferrireducens	100	nd	Isolated from Obsidian Pool, Yellowstone Park
Eukarya			
Unicellular algae	60	nd	*Cyanidium caldarium*, isolated from acid hot springs in Yellowstone
Yeasts	60–62	nd	Filamentous fungi isolated from a range of geothermal sites in Yellowstone
Lichens	~45*	nd	Thermal tolerance depends on state of hydration; no recent data
Macroalgae	~45	nd	No recent data
Mosses	~50*	nd	No recent data
Angiosperms	65	nd	*Dichanthelium lanuginosum* in Yellowstone national park
Terrestrial invertebrates	~60	~70	Nematodes in compost heaps
Freshwater invertebrates	~46	nd	Crustaceans and molluscs living alongside pupfish in hot springs
Marine invertebrates	>42*	~90	Polychaete *Alvinella pompejana* from hydrothermal vents
Ectothermic vertebrates	~46	nd	Desert pupfish, *Cyprinodon* species
Endothermic vertebrates	nd	nd	Cell temperatures in range ~30–45 °C

14.8 The thermal limits to life on Earth

The discussion above shows clearly that the upper and lower thermal limits to life vary markedly across the different domains of life on Earth (Figure 14.10). Bacteria and Archaea can exist over a much broader range of temperatures than can

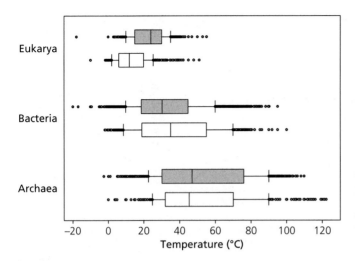

Figure 14.10 Temperature range for life. Box-plots of temperature ranges for Archaea, Bacteria and Eukarya, with autotrophic species shown in white and heterotrophic or mixotrophic species in grey. The box shows the median and interquartile range, the lines cover the central 95% of data and the symbols show individual outliers[47].

Eukarya, with present data suggesting Archaea are the most tolerant of high temperatures and bacteria of low temperatures. The narrower range of tolerance of Eukarya suggests that there is something about their more complex cellular architecture that limits their thermal tolerance, though at present we do not know what this is.

We can now return to the question posed at the start of this chapter. The low temperature limit to life on Earth appears to be set by vitrification of the cell interior, although it is possible that some unicells can metabolise and even grow within undercooled water droplets in clouds down to −40 °C. The low temperature limit for life on Earth is thus associated closely with the existence of liquid water. In contrast, there are aquatic environments on Earth which appear to be too hot to sustain life, most notably the hottest waters emanating from hydrothermal vents. The high temperature limit to life appears to be by biophysical constraints on the structure, stability and function of biomolecules, rather than the presence of liquid water itself.

14.9 Summary

1. The extreme meteorological surface air temperatures recorded to date are −89.2 °C in Antarctica, and 56.7 °C in Death Valley, California. Ground (skin) temperatures can be higher or lower than these air temperatures.
2. The asymmetric distribution of land and ocean on Earth produces a hemispheric difference in the pattern of mean temperature and seasonality. The hottest land areas are generally in the equatorial regions or the southern hemisphere, whereas the most seasonal environments are the northern boreal forests and tundra.
3. The bulk of oceanic water is cold (< 4 °C) and thermally stable. Surface temperature shows a symmetrical distribution of mean temperature about the tropics, but the most seasonal surface waters are in the northern hemisphere and mostly close to shore.
4. The upper and lower thermal limit to life vary markedly across the domains of life on Earth. While data on limits to survival attract considerable attention, the thermal limits to completion of the life cycle (which define the limits to life) are far more difficult to determine and hence are much less well known.
5. Currently identified upper thermal limits for growth are 122 °C for archaeans, 100 °C for bacteria and ~60 °C for unicellular eukaryotes. No unicells appear to grow below −20 °C, a limit that is probably set by dehydration-linked vitrification of the cell interior.
6. The range of temperatures over which multicellular eukaryotes can complete their life cycle is much narrower than for unicells. The upper limits on land would appear to be exhibited by nem-

atodes in hot springs and compost heaps (~ 60 °C), and in the sea by polychaetes associated with hydrothermal vents (> 40 °C, but poorly defined). The upper limit for an ectothermic vertebrate would appear to be 38–46 °C for desert pupfish.

7. The lower thermal limits for survival in multicellular organisms extend in the natural world to at least −70 °C, the winter minimum temperature of the Arctic tundra and taiga. However in all cases known to date, completion of the life cycle requires summer warmth and the lowest temperature for completion of the life cycle appears to be ~0 °C for invertebrates in glacial meltwater and ~−2 °C for marine invertebrates and fish living on the continental shelves around Antarctica.

8. In lichens and the grass *Dichanthelium lanuginosum*, extreme temperature tolerance is conferred by symbiosis or mutualism. The mechanism by which the symbiotic organism gains enhanced temperature tolerance in comparison with the isolated individual components remains obscure.

Notes

1. This quotation comes from p. 68 of the first edition of *The origin of species* (Darwin 1859).
2. Research on the origin of life is currently undergoing something of a renaissance with the recognition that certain types of hydrothermal vents (the cooler, alkaline vents typically found away from the main spreading axis of mid-ocean ridges) could provide the right physical, chemical and thermal conditions for the origin of biological chemistry. The key papers here are Russell et al. (1993, 2010, 2013), and Nick Lane provides a thorough but highly readable introduction to the current state of the field (Lane 2015).
3. Obviously, as we have seen in previous chapters, organisms exposed to direct sunlight by day, or a clear sky by night, can gain or lose heat by radiation. Meteorological surface air temperature is nevertheless a useful first-order measure of the thermal environment experienced by relatively large organisms such as vertebrates or larger plants.
4. Satellites can determine the large-scale variation of temperature through the atmosphere by using a range of wavelengths that penetrate the atmosphere to differing extents. However current satellite sensor technology is unable to provide a precise, remotely sensed, measure of air temperature at 1.5 m above the ground. Current meteorological satellites record radiometric skin temperature at a resolution of 1 km^2 over the entire Earth land surface up to twice each day under cloud-free conditions.
5. The atmospheric circumstances leading to the record low temperature at Vostok are discussed by Turner et al. (2009a).
6. Plotted with data kindly supplied by John Turner (British Antarctic Survey).
7. Nunatak is an Inuit word for a mountain exposed through the surface of an ice sheet. Nunataks are effectively glacial islands.
8. The Vostok temperature was reported by Budretsky (1984). Cerveny et al. (2007) give this temperature as −89.4 °C in their Table 2, quoting Krause and Flood (1997). The error was probably the result of a conversion from Celsius to Fahrenheit, rounding to −129 °F, then back-conversion to Celsius. The data for Oymyakon and Verkhoyansk are from a careful analysis by Nina Stepanova (1958). These temperatures are both reported as −67.8 °C by Cerveny et al. (2007), and also by the WMO. A sign at Oymyakon records a temperature of −71.2 °C in 1924, but this is completely unofficial. Somewhat ironically, Oymyakon translates as 'non-freezing water'; the settlement was presumably so named because there is a geothermal spring nearby.
9. Burton-Johnson et al. (2016) estimated the ice-free area of Antarctica to be 0.18% (less than half previous estimates). The plant study is Peat et al. (2007). Records of biota from isolated nunataks in Antarctica include Sohlenius et al. (1995), Broady & Weinstein (1998) and Sohlenius & Boström (2005).
10. Plotted with data from the World Meteorological Association.
11. El Fadli et al. (2013) discuss the validity of the data for Libya.
12. Desert temperatures up to 80 °C were reported by Ward & Seeley (1966). The very high desert surface temperatures are reported and discussed by Kubecka (2001). The difficulties associated with such direct measurements are discussed by Fuchs & Tanner (1968).
13. Mildrexler et al. (2011). The values > 0 °C for snow and ice areas come from pixels containing some areas of exposed ground (ice and snow temperature obviously cannot exceed 0 °C).
14. Annual mean, maximum and minimum surface air temperature for North and South America over the period 1961–1990 at 10-min resolution interpolated from station means and resampled to an equal-area (see Clarke & Gaston 2006 for details). The data source was the Climate Research Unit, University of

15. East Anglia, and data were plotted after pooling into bins of 1° of latitude.
15. Longhurst (1998) provides a finely detailed classification of the surface waters of the ocean, based on ecological factors (light, chlorophyll and nutrients) in addition to the temperature considered here.
16. The SST data were from the AVHRR Pathfinder v. 5.0, obtained from the Physical Oceanography Distributed Active Archive Centre (PO.DAAC) at the NASA Jet Propulsion Laboratory, Pasadena, California.
17. Philosophical discussions of the definition of life are often difficult and obscure. Useful discussions for ecologists are Benner (2010), Bedau (2010) and Bains (2014).
18. Schrödinger (1944). This book is based on a short series of lectures delivered in Dublin in 1943. Erwin Schrödinger (1887–1961) was an Austrian physicist who worked in Berlin until he felt the need to leave in 1933, largely because of the strengthening anti-Semitism of the rising Nazi Party. He moved to Oxford, where his somewhat unconventional living arrangements proved a problem, and after visits to Princeton and Graz he eventually moved to Dublin in 1940. Here he established an Institute for Advanced Studies and remained for 17 years before moving finally to Vienna, where he died in 1961. He was awarded the 1933 Nobel Prize in Physics for his work on quantum mechanics, and the formulation of what is now known as the Schrödinger equation.
19. This description of life is taken from Clarke (2014).
20. The demonstration that genes were nucleic acid was by Avery et al. (1944), and the structure of DNA was reported by Watson & Crick (1953).
21. See von Neumann (1951, 1966). John von Neumann (1903–1957) was a Hungarian- American mathematician and physicist who did important work on quantum mechanics and was a seminal figure in the development of cybernetics, computing and artificial intelligence.
22. The recent discovery of very large viruses (*Megavirus, Mimivirus, Mamavirus*) with extensive genomes has invigorated the debate over the evolutionary origin of viruses in general. The genomes of these giant viruses include genes for 750–1100 proteins, including a range of aminoacyl tRNA synthases. This points clearly to derivation from an ancestral cellular genome, and the evolution of these viruses appears to have involved a simplification of the original genome by deletions, a process paralleling that of many parasites, but here carried to an extreme degree. For recent work on giant viruses see Fischer et al. (2010) and Arslan et al. (2011).
23. See Luisi (1998) and Jakosky (1998).
24. The NASA definition of life, taken to the absurd limit, would also mean that a sterile hybrid such as a mule is not alive, or that while a male and female rabbit together are alive but either one alone is not. The segregation of somatic and reproductive cells differs between animals and plants, with important consequences for the way mutations generate variety, but this does not affect the general argument here.
25. The key papers developing this idea are Lovelock (1965), Hitchcock & Lovelock (1967) and Lovelock & Giffin (1969). These papers laid the groundwork for what has become known as Earth System Science, and also led Lovelock to propose the Gaia hypothesis (Margulis & Lovelock, 1974). Lovelock made important early contributions to cryobiology and cryopreservation, in particular our understanding of how low temperature and freezing affects cells (see Chapter 4). He also invented the electron-capture detector which revolutionised our ability to detect contaminants such as pesticides at very low levels in the natural environment.
26. Morowitz (1968).
27. Modified from Clarke et al. (2013).
28. Plotted with data from Price & Sowers (2004).
29. Metabolism and growth at low temperature is discussed by Clarke et al. (2013) and the growth of spoilage organisms by Geiges (1996).
30. Thomas (2012).
31. Geiges (1996), Hoshino et al. (2003).
32. Kappen (1993), Schroeter et al. (1994), de Vera et al. (2008).
33. Table modified from Clarke (2014); data sources Precht et al. (1973), Schroeter et al. (1994), Rivkina et al. (2000, 2007).
34. Kohshima (1984), Farrell et al. (2004), Hagvar (2010).
35. Endotherms in polar regions, or in winter, may have cold extremities (ears, nose, feet), but their core temperatures remain high.
36. The work in Yellowstone National Park is described by Brock & Brock (1966), Brock (1967) and Brock & Freeze (1969), and summarised in Brock (1978).
37. See Huber et al. (1998), Kurr et al. (1991), Kashefi et al. (2002), Kashefi & Lovley (2003), Schrenk et al. (2003), Sokolova et al. (2006) and Takai et al. (2008).
38. The original reports were Baross et al. (1982), Baross & Deming (1983) and Deming (1986). The subsequent critiques were Trent et al. (1984), White (1984), Bernhardt et al. (1984) and Lang (1986).
39. This experiment, apparently suggested to William Dallinger by Charles Darwin, was reported in Dallinger (1887).

40. Tansey & Brock (1972), Redman et al. (2002), Márquez et al. (2007).
41. Hoeppli (1926), Hoeppli & Chu (1932), Ocaña (1991), Wharton (2002), Steel et al. (2013).
42. Minckley & Minckley (1986), Miller et al. (2005), Montejano & Absálon (2009).
43. Cary et al. (1998), Desbruyères et al. (1998), Dilly et al. (2012), Lutz (2012), Ravaux et al. (2013).
44. Marsh (1985), Heurtault & Vannier (1990) and Wehner et al. (1992).
45. Smith et al. (1984).
46. Table modified from Clarke (2014); data sources Doemel & Brock (1970, 1971), Tansey & Brock (1972), Precht et al. (1973), Cary et al. (1998), Redman et al. (2002), Miller et al. (2005), Márquez et al. (2007), Kashefi et al. (2002), Takai et al. (2008), Montejano & Absálon (2009), Ravaux et al. (2013), Steel et al. (2013).
47. Plotted with data from Corkrey et al. (2016), with additional data from Schroeter et al. (1994), Rivkina et al. (2000) and Carpenter et al. (2000).

CHAPTER 15

Temperature and diversity

> Those who are capable of surveying nature with a comprehensive glance, and abstract their attention from local phenomena, cannot fail to observe that organic development and abundance of vitality gradually increase from the poles to the equator, in proportion to the increase of animating heat.
>
> **Alexander von Humboldt**[1]

We have long known that plants and animals are not distributed evenly over the planet. Novel plants and animals had been reaching Europe from far-off places for centuries, but it was the travels of the great explorer-naturalists such as Alexander von Humboldt, Alfred Russell Wallace, Joseph Banks and Charles Darwin that revealed the true richness of the tropics. Knowledge of the ocean lagged behind that on land and it was not until the pioneering oceanographic voyages of the late nineteenth and early twentieth centuries that we became aware of large-scale biogeographic patterns in the sea. We now recognise that, as a broad generalisation, diversity on land and in the ocean attains its highest values in some (but not all) tropical areas and is lowest at the poles, with temperature regions often intermediate[2].

Concern with man's impact on the living world has meant that 'biodiversity' has now firmly entered public consciousness, and the activities of conservation organisations and pressure groups have given it a powerful emotional and political dimension. As is so often the case, this has been accompanied by subtle changes in language. We therefore need to start by defining clearly what we mean by 'diversity'[3].

15.1 What is biological diversity?

Even a cursory glance at the literature will reveal many definitions of biological diversity. A widely used definition is that enshrined within Article 2 of the Convention on Biological Diversity[4], which defines biological diversity as:

> ... the variability among living organisms from all sources including, inter alia, terrestrial, marine and other aquatic ecosystems and the ecological complexes of which they are part; this includes diversity within species, between species and of ecosystems.'

While useful in a political context, such all-encompassing definitions tend not to help the development of science. This definition explicitly links genomic, species and ecosystem diversity, but at the same time it makes 'biodiversity' rather broad and vague. An alternative, far narrower, definition is that suggested by Anne Magurran, who defined biological diversity as:

> ... the variety and abundance of species in a defined unit of study.

This definition is clear and succinct; it also captures a major concern of the early studies of biological diversity which was the problem of devising a statistic which simultaneously captured both the number of species and the distribution of individuals within those species. The problem here is that it is impossible to capture the values of two independent variables unambiguously in a single number. Species number and abundance are, however, not completely independent, and the nature of the relationship between them is of importance to understanding the evolution of diversity. However,

Principles of Thermal Ecology. Andrew Clarke, Oxford University Press (2017).
© Andrew Clarke 2017. DOI 10.1093/oso/9780199551668.001.0001

attempts to incorporate both species number and species abundance into a single diversity index have resulted in no single agreed approach[5].

Many of the more sophisticated measures of diversity require a considerable amount of information, for example on functional traits or the abundance of all species. Such data can be time-consuming and expensive to acquire but provide powerful insights. However, for many requirements a simple richness measure (that is, the number of species) is sufficient. Taxon richness does, of course, run the risk of sampling bias, but when used with care it has the value of being quick, simple and informative, particularly when assessing diversity over large spatial scales where relative abundance data are rarely available.

15.2 Some practical issues

The fundamental unit of diversity is usually taken to be the species. This generally works well for the organisms we are most familiar with, although problems may arise with cryptic species, or in defining species in rapidly differentiating lineages. It runs into real difficulties, however, when we start considering unicellular organisms, where morphological differences can be small and hard to detect, and with microbes (Bacteria and Archaea) where the traditional concept of species may not apply.

In some cases species-level data may be unavailable or unreliable; here genus or family richness can provide a working indication of underlying species richness. A good example here is in palaeobiology where it may be possible to assign a fossil only to a genus or family[6].

Before we can ask meaningful questions about large-scale patterns of diversity there are two important issues that we must confront. The first is sampling, and the second is the influence of area.

15.2.1 Sampling

A moment's reflection tells us that the bigger our sample, the more species we would expect to find. For small samples the number of species we detect must also be small, for we cannot have more species than individuals. The relationship between sample size and observed richness extends to large sample sizes, and the reason is that most species are rare. Common species are collected rapidly, but scarce ones are collected only occasionally. We might expect the relationship to reach a plateau once we have sampled all the species, but this only happens if the area being sampled is so discrete or remote that no transient colonisers arise, or no itinerant species wander through. More typically the curve continues to climb for as long as sampling proceeds.

The ubiquity of rare species means that few species inventories, if any, are complete. If we are to compare samples from different areas in a meaningful way we have to allow for this incompleteness. Two main techniques have been used for this. The first is rarefaction, in which the observed data are resampled randomly many times by a computer and a relationship derived which captures the average number of species that would be expected for a given sample size. We can then compare different samples by calculating, for example, the number of species that would be expected in a sample of 50, 100 or 500 individuals. This technique has been much used in marine diversity work, and an example for Antarctic marine molluscs is shown in Figure 15.1.

Here the rarefaction curves can be used to estimate the number of species to be expected in 500 samples. The difference in the shape of the rarefaction curves is caused by a combination of differences in the abundance structure (the distribution of species across individuals) and the distribution of species across samples. This is a major drawback, for it can lead to a rarefaction curve which predicts a very different species richness from the actual (observed) accumulation of species with individuals in the real sample (the sample accumulation curve)[7].

Despite these difficulties, correcting for sampling effort is critical. It is the only way to distinguish *there are not many species here* from *there may be many species here but we have found very few of them*, and is essential in comparing results from different studies or different areas. It is particularly important in palaeobiology, where the observed diversity of a particular taxon depends on how much rock has been sampled for fossils.

15.2.2 Area

In the mid-nineteenth century the English botanist Hewett Watson noted that the number of species of plant recorded increased with the size of area being surveyed. Michael Rosenzweig has described this

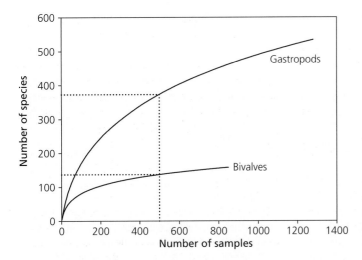

Figure 15.1 Rarefaction plots for bivalve and gastropod mollusc samples from the Antarctic continental shelf. The dotted lines show the number of species expected in 500 samples of bivalves (137 species) and gastropods (373 species)[8].

as *'the world's oldest known empirical example of an ecological pattern'*[9].

Subsequent work has established this to be a fundamental ecological pattern and the overall relationship is termed the species–area curve. Theoretical analyses show that where population size is proportional to area, then a log-series abundance distribution will lead to an exponential species–area relationship, whereas a log-normal abundance distribution will approximate to a power law relationship. These days, a simple power law relationship is typically assumed:

$$S = c.A^z$$

where S is the number of species in area A. The scaling exponent z typically lies in the range 0.1–0.2 for species–area curves determined within a biogeographic province.

There are a number of proposed ecological explanations for the species–area relationship, but as yet no consensus. At present we are left with the empirical observation that the bigger the area we survey, the more species we find, and a power law as a useful statistical fit to these data[10].

15.2.3 Spatial scale

The spatial scale over which we measure diversity is critical, not just because the larger the area studied the more species are detected, but also through its connection with the underlying processes. At small scales the composition of a local assemblage is dictated principally by ecological factors such as predation, competition or dispersal, which influence which members of the regional species pool are present. At large scales, the composition of the regional species pool is determined predominantly by evolutionary processes (speciation and extinction); indeed this is one definition of a biogeographic province. The relationship between richness and spatial scale is usually approached using the scheme proposed by Robert Whittaker (Box 15.1).

15.3 Global diversity: land and ocean

The sea covers two-thirds of the Earth's surface, making seawater the single largest habitat there is. Life originated in the sea and today diversity at higher taxonomic levels (class, phylum) in the sea significantly exceeds that on land (Table 15.1).

At lower taxonomic levels the contrast between land and sea is very much reversed: the vast majority of described species are terrestrial. It has been variously speculated that the deep sea may harbour a vast number of undescribed species, that undetected cryptic species may be masking a large fraction of marine diversity and that coral reefs alone may harbour half a million or more species. The

Table 15.1 Global diversity of metazoan animal phyla and classes on land and in the ocean[11].

	Ocean	Land
Phylum	36	12
Class	90	33

> **Box 15.1 Diversity and spatial scale**
>
> Robert H. Whittaker proposed partitioning diversity into three components, which he termed alpha, beta and gamma diversities.
>
> *Alpha diversity*
>
> This is the diversity in a small area, and is sometimes called point diversity. The size of the 'small area' is frequently undefined, and anyway will vary with the organism. The underlying idea is that this is the diversity of a single habitat or assemblage. It is often taken to be the diversity of a sample (though if the sample is taken blind, such as marine samples often are, there is no guarantee that only a single habitat or assemblage has been sampled).
>
> *Gamma diversity*
>
> This is the diversity of a large area, encompassing many habitats. It is often taken to represent the regional species pool from which the species comprising a local assemblage will be drawn. The nature or extent of the regional species pool is rarely defined objectively, but the assumption is that it represents an evolutionary unit that has been relatively isolated from other such units for a long period of time (this is one definition of a biogeographic province).
>
> Robert J. Whittaker has proposed replacing these terms with 'local' (*alpha*), 'landscape' (*gamma*) diversity, and these terms are being used increasingly in the diversity literature.
>
> *Beta diversity*
>
> This is a quite different type of diversity, being the turnover of species as one moves across the landscape. It is thus a measure of the way alpha diversity changes spatially within a region or province. It has a large number of mathematical formulations[12].

suggestion that there may be a vast number of undescribed species in the deep sea was prompted by the seminal study of Fred Grassle and Nancy Maciolek who analysed 233 box cores taken along a 176 km transect along the foot of the continental slope (~2000 m depth) off New Jersey. They recovered 798 species, of which half were new to science, and a combination of rarefaction and extrapolation led to the suggestion that the deep sea might contain as many as 10 million species (almost all of which would be undescribed). There followed an intense, and as yet unresolved, debate over the extent to which the deep sea contains a largely undescribed fauna[13].

While it is clear that we really have no idea how many species there may be in the sea, the conventional wisdom is that we are not missing a group or groups that approach even remotely the diversity exhibited by terrestrial arthropods. The striking contrast between the land and the sea at the species level is unlikely to be altered by future work.

15.3.1 Global diversity: how many species are there?

This is actually two questions: how many species have we described and how many might there be in total?

It is not at all straightforward to estimate the number of species described to date but several attempts have converged on a figure of around 1.8 million, although a significant fraction of these are probably synonyms. Only about 200 000 described species are marine whereas approximately 1 million are insects (an almost exclusively terrestrial group). This striking difference is driven in large part by the intense species richness of some insect groups (notably Hymenoptera, Diptera, Coleoptera, Lepidoptera and Homoptera), but it may also reflect our ignorance of the ocean[14].

Most recent estimates of the total number of species on Earth have either used some form of statistical extrapolation, or canvassed the opinions of experts in particular groups. These estimates suggest that there might be anywhere between 10 and 30 million species of insect alone, depending on the assumptions underpinning the extrapolation, and that fungi (*sensu lato*) might total 1.5 million species. The most recent estimate suggested a total of 8.7 ± 1.3 million species of eukaryote, of which 2.2 ± 0.18 million are marine, but in truth this is little more than an educated guess[15].

Before we can start to explore the role of temperature in biological diversity we need to establish the broad features of diversity on the planet. As so often in ecology, we look first to the patterns and then to the explanation.

15.4 Global patterns of diversity

In exploring global patterns of diversity, we need to decide what scale is relevant. Ideally we should be looking at global patterns in the richness of biogeographic provinces, as we can then use statistical models to allow for confounding effects of area or other factors. A nice example of this approach is a study of New World bats by Michael Willig and Christopher Bloch, which revealed a strong cline in richness with latitude, but no effect of area[16].

More typically, species inventories are simply pooled over conveniently large areas, and the data often then presented as the number of species per unit area. The species–area relationship makes rigorous comparison between such studies difficult but robust patterns do emerge, and these patterns differ between the terrestrial and marine realms.

15.4.1 The terrestrial realm

As was established by the early explorer-naturalists, flowering plants exhibit their greatest diversity in the tropics, with richness dropping away towards either pole. At the family level, the pattern is bold and striking though even at this very broad scale the decrease in family diversity towards high latitudes differs between the two hemispheres and within each latitudinal band there are regions with very high and very low family diversity, particularly in the tropical regions (Figure 15.2).

This strong, clear pattern might suggest a simple explanation[17]. If, however, we examine the pattern at a finer taxonomic level, species rather than families, and at smaller geographic scales, then much complicating detail emerges (Plate 17). Of the 867 ecoregions defined for the terrestrial environment, 51 contain > 5000 species of vascular plants. All but five of these are tropical or subtropical and dominated by moist broadleaved forest. The five centres of highest plant diversity are Costa Rica, tropical East Andes, northern Borneo and New Guinea, all close to the equator, and Atlantic Brazil which is 20°–30° S. Other areas of very high richness identified are all temperate forest ecosystems, and include the montane fynbos of South Africa. Clearly, the areas of highest vascular plant diversity are mostly tropical, but some temperate areas are also highly diverse. Furthermore, some tropical or subtropical areas are very low in diversity, such as the deserts of Africa, South America and Australia. This trivial observation is important when we come to consider the mechanisms that drive the global pattern[18].

Similar patterns are shown by many terrestrial animals: birds, mammals, reptiles, amphibians and insects are all most diverse in the tropics. A few

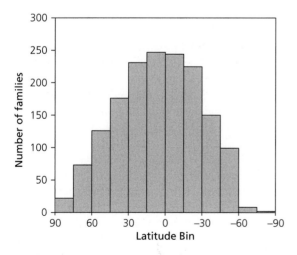
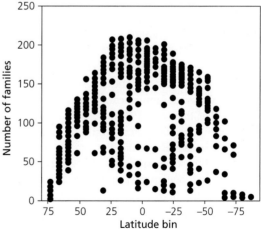

Figure 15.2 Angiosperm (flowering plant) diversity and latitude. A (left panel): the number angiosperm families pooled by latitude bin (0–15° N, 15–30° N and so on). By convention southern latitudes are shown as negative. B (right panel): angiosperm family diversity by latitude, but here plotted as diversity within boxes of area 6.1×10^5 km^2; the shape of the Earth means that the number of boxes increases towards lower (tropical) latitudes[19].

groups, however, buck the trend; for example bumblebees (solitary bees of the genus *Bombus*) and other groups of Hymenoptera reach their greatest diversity in the cool temperate latitudes of the northern hemisphere, whereas spiders have their highest diversity in the temperate latitudes of the southern hemisphere. While there is general agreement that most vertebrates achieve their highest diversities in the tropics, we must be wary of a megafaunal bias, as the patterns in the majority of terrestrial invertebrate groups have yet to be determined[20].

15.4.2 The special case of microbial diversity

Microbes are different. Archaea and Bacteria are morphologically fairly uniform and it is only since we have been able to study them at the molecular level that the full range of their variety, physiology and ecology has become apparent. In addition their tendency to exchange genetic material means that it is not clear how we should define a microbial species. Extrapolation using the relationship between the number of species and the number of individuals, coupled with data on genetic diversity, has suggested a global total of 1 trillion (10^{12}) microbial 'species'[21].

It was suggested in the 1930s that microbial biogeography might be fundamentally different from that of larger organisms, when the Dutch microbiologist Louis Bass Becking proposed that microbial dispersal was global and the microbial assemblage at a particular location was determined only by the local environmental conditions. He summarised this in the memorable aphorism *Alles is overal: maar het milieu selecteert* (*Everything is everywhere: but the environment selects*, italics as in the Dutch original). Later Bland Finlay and Tom Fenchel argued that organisms smaller than about 1 μm (bacteria, archaeans and unicellular eukaryotes) would be ubiquitous, with the consequences that the species–area relationship for these organisms would be flat and there would be no biogeographic structure such as a latitudinal or altitudinal gradient in richness[22].

Recent studies have thrown doubt on this idea, for there is increasing evidence for strong biogeographic structure in microbial assemblages, produced by dispersal limitation at large spatial scales, coupled with ecological factors operating at more local scales. A key problem here is that disentangling the effect of dispersal limitation and local selection is difficult because a greater distance between sites typically also means a larger environmental dissimilarity[23].

15.4.3 The marine realm

The presence of a latitudinal diversity cline in the sea was probably first suggested by the Danish marine ecologist Gunnar Thorson in the 1950s. Working in the North Atlantic, Thorson described an increase in the species richness of the epifauna on shallow-water hard substrata from polar through temperate to tropical regions, but could detect no comparable differences in the soft-bottom fauna. Subsequent work on bivalves revealed a distinct cline in diversity from tropics to polar regions, with a clear centre of diversity in the Indo-West Pacific. Together with an earlier study of marine gastropods and a slightly later one for foraminiferans, these studies formed the entire basis for the assumption of a latitudinal diversity cline in the shallow seas until the 1990s[24].

It had long been recognised that reef-building (hermatypic) scleractinian corals are confined to warm, clear tropical waters, and also that both gastropod and bivalve molluscs exhibit their highest diversity on these reefs, as do fish. The diversity of coral reefs has often been compared with that of tropical rain forests but in truth we have no idea how many species there are on reefs, though the better-known groups such as molluscs, crustaceans and fish all point to reefs being the richest habitats in the sea. Two estimates of the total species richness of the world's reefs were made in the 1990s, both involving extrapolation from small-scale richness data and assuming a species–area relationship determined for rain forests. The two estimates were very different: 950 000 and 2.6 million species[25].

From the 1990s, sufficient studies have covered a broad latitudinal range (at least one hemisphere) to establish the existence of a latitudinal cline in shallow-water diversity in a variety of taxa, including gastropods, bivalves, bryozoans and decapod crustaceans (Figure 15.3). A notable feature of several of these studies is that diversity does not decline in a simple fashion from tropics to poles; rather it is often uniformly high in the tropics, and starts to decline only around 20–30° N. In some cases, the area of highest diversity, although technically tropical, is not always centred on the equator but displaced slightly to higher latitudes[26].

As on land, there are some marine groups that do not exhibit a strong latitudinal cline in diversity. There is no systematic variation in the species richness of macroalgae between 60° N and 60° S, but richness is reduced at very high latitudes because of the effects of ice scour. Sea anemones (actinarians) show a more complex pattern, with the highest richness found at 30–40° N and at 30–40° S; tropical regions had lower diversities, and the two polar regions lower still. The species richness of sponges (here including hexactinellids which are often assigned their own phylum) when analysed by the 232 recognised marine ecoregions shows a fairly uniform pattern with areas of highest richness clearly associated with areas that have been studied most thoroughly (North-west Europe, Caribbean, southern India, Japan and Australia)[27].

The marine benthos is difficult to study, for most of the ocean is deep. Recent developments in technology, notably remotely operated vehicles and improved imaging, have revolutionised our view of deep-sea diversity. In particular they have demonstrated the high diversity of previously unexplored hard substratum assemblages on seamounts and hydrothermal vents. While vents are associated predominantly with mid-ocean ridges and spreading centres, seamounts are distributed globally in the ocean. These important habitats are only beginning to be explored, and it remains to be seen whether the fauna they support will change the large-scale patterns established from the more easily accessible habitats close to shore[28].

Although hard substrata often support rich epifaunal assemblages, the majority of the seabed is soft sediment. Thorson's suggestion that variation in richness with latitude is less distinct in this habitat has largely been supported by later studies. Comparative studies of nearshore sediments from Svalbard, Norway, the European continental shelf and estuaries have revealed no consistent pattern, and any relationship between richness and latitude in these habitats is clearly very weak[29].

As we move away from shallow waters across the continental shelf and into the deep sea, species richness and abundance generally peaks at intermediate depths. The deep sea is, however, often described as being very diverse, a view which can be traced back to a seminal study of an area of the North Atlantic by Howard Sanders. The deep sea tends to have fewer individuals per species than comparable shallow-water areas, whereas species density (number of species per unit area of sediment) is higher in the deep sea at local scales but similar to shallow waters at large scales[30].

Recent work has provided evidence for a strong latitudinal gradient in species richness in gastropods, bivalves and isopods in the deep North Atlantic but the gradients were far less steep in the

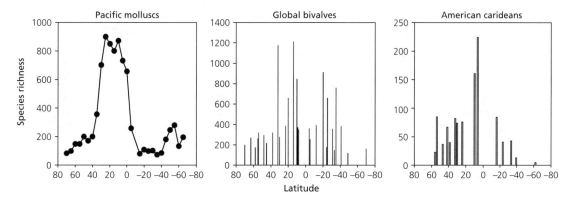

Figure 15.3 Latitudinal clines in diversity (species richness) of three marine invertebrate groups: molluscs along the Pacific coast of America pooled in bins of 10° latitude, global bivalves pooled by province, caridean shrimps along the Atlantic coast of America pooled by province. Data plotted at mid-point latitude of bin or province, and by convention southern latitudes are negative.

southern hemisphere (Figure 15.4). In contrast, a study of North Atlantic deep-sea nematodes revealed an inverse gradient, with richness increasing northwards from 13° N to 56° N. Benthic foraminifera differ again, in exhibiting their highest richness in tropical regions, declining towards both poles[31].

Despite the enormous area of the oceans, and their great depth, planktonic communities are far less diverse than benthic assemblages, and the holozooplankton (animals that are planktonic for their entire life cycle) may contain as few as 5000 species overall. In general the greatest diversity is attained in warm temperate and tropical latitudes, as illustrated by epipelagic calanoid copepods (Figure 15.4). A pattern of higher diversity in the tropical regions is also seen in other groups including chaetognaths, pterodods and euphausiids, although more detailed analyses indicate clearly that zooplankton biogeography is related most strongly to the dominant water masses that have long been known to oceanographers. Alan Longhurst has used a variety of physical and biological criteria to divide the ocean into four primary biomes: westerlies, trades, coastal and polar; these in turn are subdivided into 51 provinces. There is increasing evidence that zooplankton biogeography and diversity is related primarily to these biomes rather than to any single abiotic factor such as temperature[32].

15.4.4 Summarising the patterns

A century of ecological and biogeographic work has largely confirmed the broad picture established by the early naturalists: many plants and animal groups exhibit their highest species richness in, or close to, the tropics. However it has also established that the picture is far from universal: patterns differ on land and in the sea, they differ between taxa and they are often not symmetrical about the equator (that is, the northern and southern hemispheres often show different patterns). With this background we can now explore the question of what might be the mechanism(s) that dictate global patterns of diversity.

15.5 Mechanism

The early naturalists identified the factors they believed to be driving speciation and extinction, and thus diversity. In the famous final paragraph of the *Origin of species*, Darwin draws a vivid picture of how competitive interactions lead to a struggle for existence and thereby selection for improved phenotype. Here Darwin was focussing on ecological interactions on a local scale, but elsewhere in the *Origin* he explicitly recognised the role for occasional climatic extremes in limiting

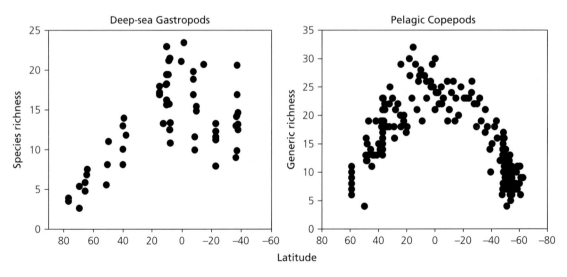

Figure 15.4 Latitudinal variation in species richness of marine taxa. A (left panel): diversity of gastropods in the deep sea, estimated by rarefaction and shown as the number of species expected in a sample of 50 individuals. B (right panel): latitudinal variation in richness (genera) of epipelagic calanoid copepods. By convention southern latitudes are negative[33].

the overall number of species through extinction. At about the same time Wallace had identified changes in sea level driving vicariant speciation as an important factor in the high diversity he had observed in the Indo-West Pacific, and also pointed to recent glacial history as a key factor at high latitudes. Joseph Hooker had speculated in a letter to Darwin that fragmentation of a great southern continent might explain the disjunct distribution of plants such as the southern beech *Nothofagus* and John Gulick had demonstrated the importance of vicariance in his seminal study of land snails in Hawai'i[34].

In recent years strong arguments have been advanced for the primacy of area, habitat heterogeneity, neutral processes of dispersal and extinction, and energy in determining species richness. The problem is that it is very difficult to devise tests that distinguish convincingly between these competing explanations. We must also recognise that patterns of diversity may not be the result of a single driver, but may represent the outcome of several factors interacting simultaneously[35]. In this chapter we will examine just two: temperature, because it is the subject of this book, and energy, because its influence is often confounded with that of temperature.

15.5.1 Temperature, climate and environmental toughness

A long-standing assumption that permeates our view of the relationship between organisms and environment, sometimes explicit but more often implicit, is that some places are tough to live, and hence few organisms can survive there. We can illustrate this idea nicely with a classic study of the diversity in geothermal springs by Thomas Brock and colleagues (Figure 15.5).

In the geothermally heated pools of Yellowstone National Park, the number of diving beetle species declines almost linearly as water temperature increases from 30 °C to 45 °C. Similarly the diversity of microbial taxa that can exist in these waters declines with increasing water temperature above ~30 °C and the very highest temperatures support only a small number of extremophile taxa. Recently Christine Sharp and colleagues detected a similar pattern in microbial diversity (assessed as operational taxonomic units, OTUs, from 16S rRNA sequences) in geothermal areas from Canada and New Zealand. At first glance this pattern might seem entirely sensible: as the environment gets tougher fewer species can live there. Indeed we often make a subjective assessment of the harshness of an environment on the basis

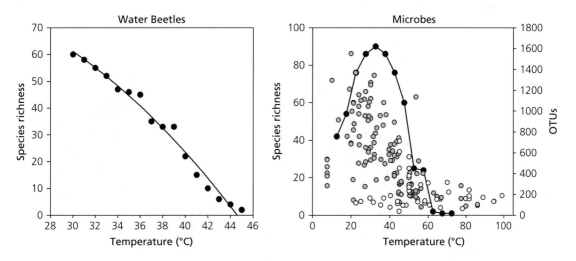

Figure 15.5 Diversity and temperature in geothermal pools. A (left panel): species richness of water beetles (Coleoptera) in relation to water temperature in pools at Yellowstone National Park. The line is a quadratic, fitted by least-squares regression. B (right panel): diversity of microbes in a range of geothermally heated pools: cyanobacteria from Yellowstone, plotted as the median temperature of the pool (black symbols with line), and microbial operational taxonomic units (OTUs) determined from 16S rRNA sequencing, in geothermal pools from Canada (grey symbols) and New Zealand (white symbols)[36].

of how many (or how few) types of organism live there. This reasoning is, of course, inherently circular: we explain the low diversity of organisms on the basis of the perceived harshness of the environment, but we also assess the harshness of an environment from the diversity of organisms that live there.

The question that this begs is what precisely constitutes toughness in an environment? Early judgements were clearly anthropocentric: an environment was benign if it seemed pleasant to us as a poorly insulated mammal. This is the intuitive feeling underpinning the quotation from Alexander von Humboldt at the head of this chapter: warmer habitats are easier places to make a living than colder ones, and hence more species live there. There is, however, as yet no compelling evidence that a polar ectotherm is any more (or less) physiologically stressed or disadvantaged than a tropical one, although as we shall see below temperature does influence what organisms can do.

15.6 Temperature and diversity

While the idea that the tropics are a more amenable place for life in general has largely slipped away, variation in temperature (or more broadly, climate) remains a favoured explanation for patterns of diversity. The basic idea is simple: where the climate is warmer, more individuals and hence more species can live. This is often called the *'ambient energy hypothesis'* but this is misleading, for while environmental temperature exerts considerable influence over ectotherm activity, temperature is not energy, nor can it be used as such by organisms.

We will explore this point in more detail shortly, but it is nevertheless a widespread observation that for many ectotherms the number of species is higher in warmer areas. For example at small scales in the UK the number of species of butterfly is correlated strongly with mean summer temperature, as is that of breeding birds, and these relationships remain when patterns are examined at larger spatial scales (continental)[37]. Two examples from North America are shown in Figure 15.6.

Although these relationships typically contain considerable scatter, the central tendency is clear and strong. They are, however, simply correlations and many discussions of the relationship between diversity and climate are somewhat vague when it comes to mechanism, appealing to loose concepts such as 'favourableness of climate'. An explicit physical mechanism has, however, been proposed by Drew Allen and colleagues, who argued that diversity is linked directly and mechanistically to temperature through *'the generally faster biological rates observed at higher temperatures'*. Combining the energy equivalence rule, which posits that the total energy flux of a population per unit area is invariant with respect to body size, with the relationship between metabolic rate, body size and temperature derived within

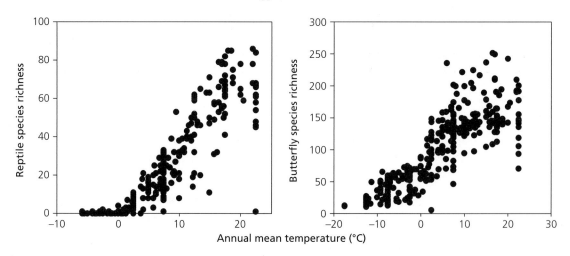

Figure 15.6 Relationship between species richness and environmental temperature for North American reptiles (left panel) and butterflies (right panel)[38].

the Metabolic Theory of Ecology (Chapter 12) they predicted a 50-fold increase in diversity from poles to tropics, though with the important caveats that this was a very general, first-order prediction that would not necessarily apply to narrowly defined taxonomic groups, or organisms that *'are strongly regulated in body size and abundance by temperature'*, such as reptiles. They tested the model with data for trees, amphibians, continental shelf marine gastropods and fish parasites, and concluded that there was good agreement with the predicted relationship between richness and temperature[39].

The model was subsequently recast, abandoning the dependence on energy equivalence for which there is only mixed empirical evidence. The second model links two factors: a direct effect of temperature on evolutionary rate and an effect of population size on speciation rate. A key assumption in the model is that the rate of speciation is linked directly to temperature through the generation of mutations by oxidative damage during metabolism (Box 15.2). The model assumes that speciation simply follows from the accumulation of genetic divergence and it predicts a linear relationship between the natural logarithm of species richness and inverse temperature (that is, an Arrhenius relationship), accepting that this pattern could be blurred by other ecological processes at some times and in some places[40]. This model has become known as the *Metabolic Theory of Biodiversity*.

While there have been many suggestions that diversity is related in some way to climate or temperature, the Metabolic Theory of Biodiversity is the first attempt to link diversity and temperature directly with an explicit physical mechanism. How well is it supported by evidence?

15.6.1 Testing the Metabolic Theory of Biodiversity

A theory linking diversity to temperature through metabolism-driven mutation rate has two clear predictions:

1. Diversity should covary with body temperature. In ectotherms body temperature is dictated primarily by ambient temperature, and so strong

---

Box 15.2 Temperature, mutation and speciation

Early comparative studies of nucleotide substitution rates identified two factors important in determining the pace of molecular evolution in animals: the number of germ-line replication events and DNA damage from the by-products of oxidative metabolism (especially reactive oxygen species, ROS). The rate of germ-line replication events is dictated by generation time, and it has been proposed that damage from ROS is linked to metabolic rate (the *metabolic rate hypothesis* developed by Andrew Martin and Stephen Palumbi). These two effects are not independent, of course, because organisms with shorter generation times tend to have higher metabolic rates. The effect of metabolic rate was suggested from comparative studies of DNA substitution in sharks and mammals, although differences in generation time were not controlled for[41].

The primary mechanism for the generation of ROS is believed to be the leakage of electrons from the mitochondrial electron transport chain, which then combine with oxygen to form a superoxide radical that can damage cell membranes and nucleic acids. The cell has mechanisms for limiting this damage (notably the enzymes superoxide dismutase and catalase), and ROS generally do not enter the nucleus, which is where most of the genome resides.

For changes to become fixed, they must be transmitted to the next generation. General somatic mutations arising from ROS are thus not relevant[42]; it is only mutations in the mitochondrial genome (which in most organisms is transmitted matrilinearly) and in the germ-line nuclear DNA that count. There are many links in the chain between generation of ROS in mitochondria and damage to germ-line DNA, and strong evidence for the metabolic rate hypothesis has remained elusive[43].

While it is tempting to link the rate of molecular evolution to speciation, this step in the chain is far from understood at a molecular level. Molecular divergence is typically assessed in genes that would appear to have little to do with speciation. Indeed the ecological processes that drive speciation may be unconnected with the physiological processes producing variation in DNA, and there is little evidence that the rate of speciation is limited by the rate at which mutations arise. Neither the mechanistic link between metabolic rate and DNA evolution, nor the link to speciation (and hence diversity) is clear.

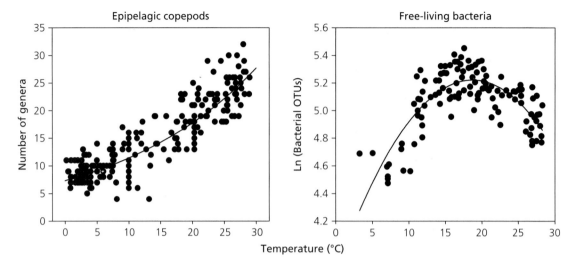

Figure 15.7 Oceanic diversity and temperature. A (left panel): relationship between generic richness per sample and sea-surface temperature at the time of collection in epipelagic marine copepods. The line is an exponential relationship, fitted with least-squares regression. B (right panel): relationship between diversity of free-living bacteria (operational taxonomic units, OTUs, determined from 16S rRNA sequences, natural log transformed) and surface temperature. The fitted line is a quadratic, suggesting that diversity peaks around 18 °C[44].

gradients in environmental temperature should promote strong gradients in ectotherm diversity.
2. Endotherms, having a limited range of body temperatures, should exhibit a poor or non-existent relationship between diversity and environmental temperature.

A clear example of a tight spatial correlation between diversity and temperature is provided by epipelagic marine copepods, where there is a strong linear relationship between temperature and generic richness in samples taken from the North and South Atlantic from 59° N to 63° S. In contrast data for oceanic microbes show an increase in richness with temperature up to ~20°C, but a decline at warmer temperatures (Figure 15.7).

Benthic assemblages shows a far less coherent relationship between diversity and temperature. The most detailed data for shallow-water benthic diversity are for continental shelf gastropods and bivalves of North America. Here the patterns of richness with latitude are remarkably similar along the Atlantic and Pacific continental shelves but the contrasting oceanographic regimes in these two basins results in very different relationships between richness and temperature (Figure 15.8). Along the Pacific continental shelf, the relationship is monotonic and roughly exponential. The relationship for the Atlantic continental shelf is quite different: diversity is constant at ~100 species from 0 °C to 20°C, but then increases sharply to ~1000 species at 28 °C. While species richness is higher in the tropical waters of both oceans, there is no common relationship between richness and temperature. If, however, an Arrhenius relationship is fitted to the combined data for the Atlantic and Pacific shelves, then the overall slope is −0.56; the predicted value from the Metabolic Theory of Biodiversity is −0.65, which falls outside the 95% confidence intervals (−0.52 to −0.61). This analysis begs two questions: is the fitted slope close enough to be taken as support for the theory (or if not, does it then constitute refutation of the theory), and given the two very different patterns in the Atlantic and Pacific basins, is it valid to fit an Arrhenius model to the combined data in the first place[45]?

A powerful test of the nature of any functional relationship between diversity and temperature comes from the deep sea. This is because, with the exception of a few areas such as the Red Sea, the temperature of the deep sea is fairly uniform over the globe, being dictated by the production of cold bottom water at high latitudes. Any theory that

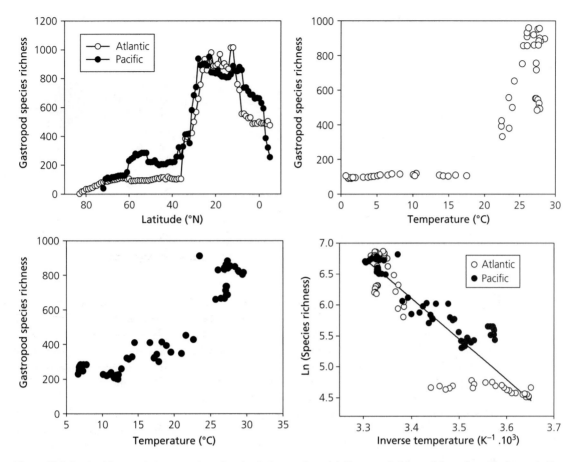

Figure 15.8 Species richness and temperature in northern hemisphere continental shelf gastropods. A (upper left panel): species data pooled in bins of 1° of latitude, assuming species are found in all bins between range maximum and minimum. B (upper right panel): relationship between diversity and temperature for shallow-water gastropods from the Atlantic continental shelves of North America. C (lower left panel): similar, for Pacific continental shelves. D (lower right panel): Arrhenius plot of species richness as a function of inverse temperature; line fitted by ordinary least-squares regression to all data[46].

links diversity directly to temperature would thus necessarily predict a uniform distribution of diversity in the deep sea. This, however, is not what we see: deep-sea macrofaunal diversity shows strong regional differentiation, with a marked cline in the northern hemisphere but little evidence for any cline in the southern hemisphere (Figure 15.4).

Overall we are forced to conclude that there is no strong evidence for a direct relationship between temperature and diversity in the sea: strong patterns in diversity exist where there is no variation in temperature, and in some areas strong gradients in temperature are accompanied by more or less uniform diversity.

On land, species richness tends to be highest in the tropics, and there is a strong correlation between richness and temperature for many terrestrial organisms[47]. In the context of the Metabolic Theory of Biodiversity the critical variable for an endotherm is not the ambient environmental temperature but the internal body temperature. The pattern of diversity in mammals and birds thus runs counter to that predicted: despite only a small variation in body temperature, endotherms show a strong variation in richness with ambient (environmental) temperature (Figure 15.9). Overall on both land and sea, there is only limited support for the theory[48].

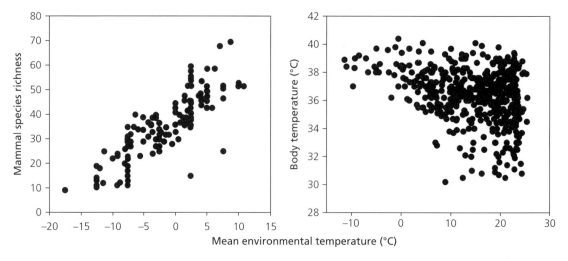

Figure 15.9 Mammal species richness and temperature. A (left panel): relationship between species richness and mean annual temperature for Canadian mammals. B (right panel): body temperature as a function of mean environmental temperature in mammals[49].

The explanation must lie elsewhere. A quite different hypothesis, but one that is often confounded with a dependence on temperature, is that diversity is related to energy.

15.7 Diversity and energy

Discussion of the role of energy flow in diversity has its roots in the development of the trophic structure of ecosystems by Raymond Lindemann and Evelyn Hutchinson. In 1983 David Wright formalised an explicit link between energy flow and diversity by modifying classic island biogeography theory, replacing area with available energy. Specifically Wright argued that the diversity of one trophic level was determined by the amount of energy available from the level below. Wright developed his ideas for a very specific set of circumstances, namely islands, but the theory has subsequently been extended to both marine and terrestrial ecosystems in general[50].

The first step in the transfer of energy through any ecosystem is primary production, the capture of light energy by plants in photosynthesis and the storage of this as chemical potential energy in new tissue[51]. If photon flux were all that was involved in determining the abundance or diversity of higher plants then we would expect a more even distribution of that diversity across the globe than is actually observed. Despite a strong latitudinal variation in the seasonality and intensity of light input, when averaged over the year the difference between the energy available to plants at the tropics and poles is only about fourfold[52], whereas the variation in plant diversity is very much greater (Plate 17).

The reason that the diversity of terrestrial plants is not coupled tightly to patterns of received light energy is that for a plant to make use of the light it receives, it also needs water. Water not only provides the electrons and protons needed by photosynthesis, but it is the movement of water through the plant that allows it to absorb nutrients and to move materials.

The influence of water on plant physiology is captured by evapotranspiration, which is a measure of the transfer of water to the atmosphere from plants and soil; it can be thought of as a measure of the simultaneous availability of water and solar energy (Box 15.3). Based on earlier work by Jack Major, Michael Rosenzweig showed that AET was an excellent predictor of above-ground plant productivity for 24 habitats ranging from barren desert to tropical rainforest (Figure 15.10)[53].

Wright used AET to estimate the energy available for plants, and showed that this was also a good predictor of plant species richness. It is important to recognise that this mechanistic link combines two distinct processes: more energy results in more individuals (productivity is higher), and more individuals lead to more species. The mechanism linking the number of individuals to diversity (the second link in the chain) is generally believed to be that species with larger populations have a lower likelihood of extinction. If abundance structure (the distribution of individuals across species) is broadly independent of the number of individuals, the larger the population the more species can exist above the threshold for stochastic extinction. This is generally known as the *more individuals mechanism*[54].

The chemical potential energy stored in plant tissue is what fuels the rest of the food-web. The basic mechanism that allows for a greater diversity of herbivores is directly analogous to that for the plants themselves: more plants to eat allows for a higher population of herbivores, and a larger number of herbivores allows for more species. The next step in the transfer of energy through the food web is from herbivores to carnivores, and here the same processes are active: more energy from the food allows for more carnivores, which translates into a greater species richness of carnivores.

The fundamental energy flows are shown in Figure 15.11. Only the very first step, the capture of light energy in photosynthesis that drives primary production, is regulated directly by AET. It is this step that underpins the relationship between primary production and temperature (or more broadly climate) and also, through the more individuals mechanism, the relationship between

---

Box 15.3 Estimating energy input to the biosphere

Plants convert the energy of visible light to chemical potential energy in new tissue, and this chemical energy fuels the rest of the food-web.

Net primary production

This is a measure of the amount of new plant tissue produced per unit time; it is essentially the difference between the energy gained by the plant in photosynthesis minus the energy lost in respiration. It is usually measured experimentally by the uptake of radio-labelled CO_2 or the rate of accumulation of biomass, although many traditional techniques underestimate important components of below-ground production. The direct measurement of plant productivity is time-consuming and typically small-scale; ecologists interested in broad-scale patterns of primary productivity have therefore tended to use indirect estimates from meteorological or remotely sensed data.

Evapotranspiration

Plants may be bathed in light, but their ability to use this is limited by water. The energy required to evaporate water is provided by solar energy and, to a lesser extent, the surrounding air. The rate of evapotranspiration thus depends on the intensity of solar radiation, air temperature, humidity and wind speed, and is conventionally estimated indirectly from meteorological temperature. *Potential evapotranspiration* (PET) is a measure of the maximum amount of water that *could potentially* be moved from the soil to the atmosphere by evaporation and transpiration, assuming sufficient water is present in the system. In most areas of the world water availability is insufficient to meet PET.

Actual evapotranspiration (AET) is defined as precipitation minus water lost as runoff and percolation into soils. It was described memorably by Michael Rosenzweig as 'the inverse of rain'[55]. PET and AET are measured in units of water flux (volume per unit area per unit time, which is equivalent to water depth per unit time). They measure the capacity for productivity, not productivity itself.

Normalised Difference Vegetation Index (NDVI)

Living plants absorb visible light strongly, but reflect infrared radiation. They therefore appear relatively dark in the visible and relatively bright in the near-infrared. Satellites can measure the reflectance within visible (0.55–0.70 μm, R_{VIS}) and near-infrared (0.73–1 μm, R_{NIR}) wavelength bands. Reflectance is the ratio of the reflected to the incoming radiation in each spectral band, and hence values range between 0.0 and 1.0. The NDVI is then:

$$NDVI = \frac{(R_{NIR} - R_{VIS})}{(R_{NIR} + R_{VIS})}$$

NDVI itself thus varies between −1.0 and +1.0, a value close to 1 indicating a pixel dominated by actively photosynthesizing plants (low negative values indicate water, and values around zero typically bare ground). Being a ratio, NDVI has neither dimensions nor units. NDVI correlates with photosynthetic activity and physical characteristics such as canopy density (i.e. leaf area or percent cover) or total biomass. Where water is a limiting factor, NDVI will correlate strongly with plant-available soil moisture (and hence AET).

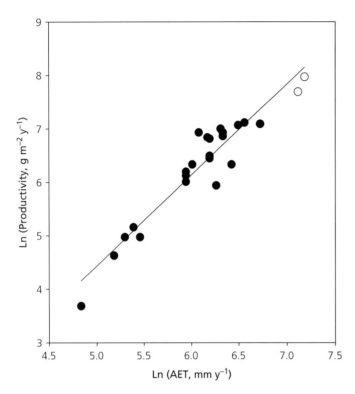

Figure 15.10 Relationship between actual evapotranspiration (AET) and above-ground primary production (expressed as dry matter) for 22 sites in North America (black symbols) ranging from desert through prairie to deciduous forest, and 2 tropical rain forest sites in west Africa (white symbols). The line is an ordinary least-squares regression fitted to all data[56].

plant richness and climate. All other steps in the food-web are flows of chemical potential energy. The flow of energy through the food-web combined with the more individuals mechanism working at each level means that higher trophic levels may still exhibit an emergent correlation with AET, but these will likely become less strong the more removed the organisms are from primary production.

Wright tested his theory by examining the relationship between the number of breeding non-marine bird species and plant productivity for 28 islands. The relationship is strong and makes a convincing case for the flux of energy being a key factor in determining species richness. Wright noted in passing that an area effect was also important, and recently David Storch and colleagues have developed a more detailed theory relating species richness to both area and energy input[57].

Following Wright's work a number of large-scale analyses have confirmed the strong relationship between water availability and plant richness. These have also shown that diversity at higher levels in

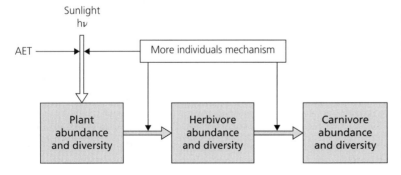

Figure 15.11 The flow of energy through a food web. Light energy ($h\nu$) is shown by the clear arrow, chemical potential energy by the grey boxes and arrows. Note that this simplified flow ignores several important pathways (detritivores, parasites, saprophytes and the microbial web). Line arrows show the influence of AET and the action of the more individuals mechanism linking higher abundance to higher species richness.

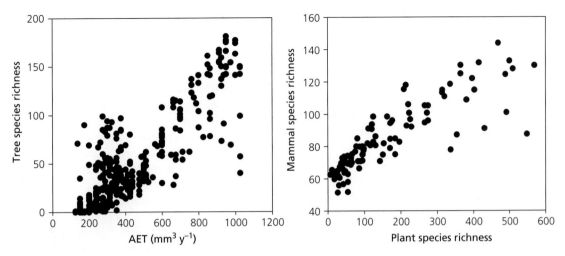

Figure 15.12 Diversity and energy. A (left panel): tree species richness and AET in North America. B (right panel): mammal species richness as a function of plant species richness in South Africa[58].

the food-web is often also related to AET, reflecting how the influence of plant diversity permeates the rest of the food-web[59]. Examples of this, one from North America and one from Africa are shown in Figure 15.12.

Attempts to quantify the relationship between plant richness and AET have perforce been statistical. This is because we do not have a quantitative mechanistic theory that can predict plant species richness from rainfall and temperature. A detailed statistical model was developed by Eileen O'Brien linking what she called 'water-energy dynamics' to woody plant richness. The initial model was developed in the context of South Africa, but has subsequently been generalised to the full range of water dynamics found on Earth. The model describes woody plant richness as a linear function of rainfall and a parabolic function of solar energy (expressed as the minimum mean monthly PET value for the location), and has been successful at predicting woody plant diversity on several continents[60].

In the ocean, water is freely available but macronutrients (N, P, together with Si for diatoms and some other taxa) are not, and the availability of light is influenced strongly by depth (as well as the density of other light-absorbing phytoplankton, and also surface ice and snow in polar regions). Oceanographers have developed a suite of mechanistic models linking the rate of primary production by phytoplankton to light, depth, temperature and nutrient availability. In these models the availability of light is influenced by turbulent mixing (which carries the phytoplankton cells into deeper, less-well-illuminated waters), nutrient availability and temperature (which affects the rate of photosynthesis). Linking productivity to diversity in phytoplankton, however, is extremely challenging, not least because what we call 'phytoplankton' is actually a phylogenetically diverse assemblage of organisms covering a huge range of sizes. There is as yet no theory that can help here, but experimental studies are starting to document changes in diversity associated with warming[61].

15.8 Confounding energy with temperature

Despite the clarity with which Wright posed and tested his hypothesis, the subsequent literature on energy and diversity is beset with loose terminology and confusion. The key problem has been a confusion of energy with temperature, for example where 'biological energy' is expressed as mean annual temperature, or where sea-surface temperature is used as a proxy for productivity or available chemical potential energy in the ocean.

The essential points in testing species–energy theory are:

1. The energy for primary productivity in almost all ecosystems is visible light. For land plants the extent to which this can be utilised is captured by AET, and for oceanic phytoplankton the important factor is depth. In both environments it is light alone that provides the energy captured in photosynthesis.
2. The energy available for all heterotrophic organisms (herbivores, carnivores, parasites) in the food-web, regardless of trophic level, is the chemical potential energy in the plant or animal tissues consumed.
3. Temperature affects the rate at which energy is transferred (and perhaps also the efficiency of transfer). It is not energy itself, nor is it a meaningful proxy for energy.
4. Plots of species richness as a function of temperature are not testing species–energy theory in any meaningful way.

If you are unconvinced by this, think about what happens to a plant in the dark. Even if supplied with all the water it needs and kept warm, it will die. The energy for plant survival and growth is light. Or consider the fate of a cow kept warm but unfed; it too will die because the energy for heterotrophs comes in the tissues they eat. In neither case does temperature do anything other than dictating when they die.

A subtle codicil here is the increasing tendency in the ecological literature to plot species richness as a function of $1/k_B T$, where T is absolute temperature and k_B is Boltzmann's constant. This is typically labelled 'temperature', or sometimes 'Boltzmann temperature'. Neither is correct. Boltzmann's constant is a conversion factor between energy and temperature, so the units of $1/k_B T$ are eV^{-1} or J^{-1}, depending on the formulation of Boltzmann's constant used. As we saw in Chapter 2, the variable $1/k_B T$ is sometimes termed *thermodynamic beta* and it captures the mean energy of a molecule of an ideal gas in equilibrium at temperature T. It has absolutely nothing to do with the chemical potential energy available to herbivores or carnivores in a food web[62].

15.9 Can energy explain everything?

While many studies have revealed a strong relationship between energy and diversity, there is always considerable scatter in the plots (see for example Figure 15.10). Clearly other factors are at play.

Ecologists have established two further mechanisms that will act to increase diversity. The first is that many species (indeed, as far as we know, all species) are subject to an array of predators, pathogens and parasites, many of them highly specific and some of which are themselves parasitised. The addition of one insect or mammal species, for example, to an assemblage thus has the potential to increase local diversity by much more than unity. This process is often expressed as *'diversity begets diversity'*.

The second mechanism is that many species create habitat or niches for other species (*niches beget niches*). The more dramatic of these, such as termites on land or corals in the sea, are sometimes called 'ecosystem engineers'. This process can be very important; for example the creation of reefs provides a wealth of habitat and niches for a vast array of other species and is a key factor in reefs being so rich and diverse[63].

A third factor that has attracted relatively little attention is the role of physiology. Here temperature does play a role influencing diversity, because body temperature affects what an organism can, or cannot, do. As we saw in chapters 9 and 10, an active lifestyle requiring high metabolic power or maximum muscle power output is possible only with a warm body. The lifestyles available through the interaction of body temperature with physiology might be termed 'metabolic niches' (Box 15.4).

It is clear that the diversity of an assemblage, local or regional, depends on a number of factors. Energy appears to set the underlying relationship, and a suite of other factors can act to increase diversity. There are, however, factors that act to lower diversity, or at least prevent local diversity from realizing the level it otherwise might. The most important of these is time.

15.10 A final comment: the role of time

As we have seen earlier, for a long time the favoured explanation for global patterns of diversity has been climate (on land) or simply temperature (in

Box 15.4 The metabolic niche

The relationship between physiology and temperature has subtle implications for diversity. Many physiological processes are able to proceed faster at warmer body temperatures, and for ectotherms this allows for lifestyles that are not possible in colder climes. For example, in teleost fish highly active predators are found only in warmer waters, and herbivory is also largely a feature of warmer-water fish. These different lifestyles are associated with differences in resting metabolism (SMR), reflecting the varying maintenance costs of an anatomy and physiology supporting different ways of making a living. This can be seen well in the resting metabolism of teleost fish (Figure 15.13). At low water temperatures the range of SMR values is low, whereas in tropical waters fish can be found with a broad range of SMR. Data for terrestrial ectotherms indicate that relative aerobic scope is largely invariant with water temperature, which indicates that the absolute scope for activity (that is, the metabolic power available to fuel activity, growth and reproduction) is greater at warmer temperatures. In other words, there are more energetic ways of making a living at warm temperatures than at cold temperatures, and hence a potential opportunity for higher diversity.

Recently Christina Anderson and Walter Jetz have shown a related pattern for birds and mammals, whereby the range of metabolic rates (in this case field metabolic rates) is wider at lower latitudes. Resting metabolic rate tends to be lower in mammals from tropical habitats, and this may be linked to lower thermoregulatory costs. The mechanism thus differs between ectotherms and endotherms, though both are related directly to the correlation between environmental temperature and resting metabolic rate. Nevertheless there are strong parallels, in that environmental temperature is affecting diversity through its influence on the range of lifestyles that can be supported. In warmer environments there are more metabolic niches available, and this affords a mechanism by which diversity may be linked positively to environmental temperature[64].

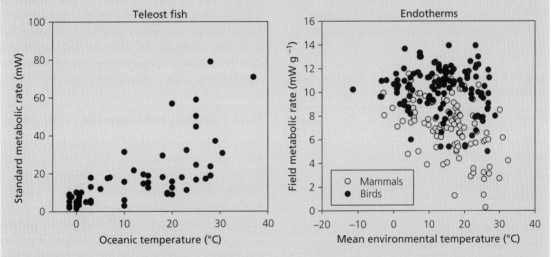

Figure 15.13 The metabolic niche. A (left panel): standard metabolic rate of teleost fish as a function of water temperature. B (right panel): field metabolic rate of mammals and birds as a function of environmental temperature[65].

the sea). While correlations between diversity and temperature are often strong, the only convincing mechanistic link that has been established is that between terrestrial plant diversity and a combination of temperature with water availability (AET). This association permeates the food-web, getting weaker with distance from primary production.

We cannot, however, simply assume that these patterns have a biological explanation. A necessary first step is to eliminate the possibility that these patterns may be generated by purely random or neutral processes. Statistical null models suggest that placing species of randomly sized ranges at random on a globe would generate a latitudinal cline with no other mechanisms involved. While these models cannot reproduce the detail of the patterns we see in nature, especially the marked differences between northern and southern hemispheres, they provide a valuable tool in highlighting the deviations from a random model which require a non-neutral explanation[66].

Non-neutral explanations fall into four broad categories, although these are not necessarily mutually exclusive (Table 15.2). The processes we have considered above are largely equilibrium explanations. In other words if we observe the system for a long period of time individual species may come and go, but the richness will remain broadly the same. There are, however, also a set of important non-equilibrium explanations.

Recent climate history suggests we live in a non-equilibrium world, for we are still recovering from the last glacial maximum (LGM). Many higher latitude habitats have only recently become available for colonisation, with the time since the LGM being shorter at higher latitudes. In contrast the tropical regions have probably been in existence for hundreds of millions of years, though undoubtedly they have varied widely in extent over this period.

Existing plant assemblages have sufficiently well-defined environmental correlates that they can be classified into clearly defined biomes, but this does not tell us whether any of these biomes is saturated with species. The key question for low diversity areas such as the recently re-colonised northern polar regions, is: if we wait, will more species colonise, or do the Arctic tundra or boreal forest have as many species now as they can accommodate[67]?

Patterns of marine diversity also suggest a strong role for history and a non-equilibrium world. The marine biota of the Arctic was eliminated during the last glacial event; the current fauna is thus relatively young and recolonisation of the Arctic basin and northern North Atlantic is continuing, principally from the North Pacific. In contrast, the shallow-water fauna of the Antarctic continental shelf appears to have survived the last glacial maximum, either in refugia or by moving into deeper water, and diversity appears to be higher than in the younger Arctic fauna. The asymmetry in patterns of diversity between the northern and southern hemisphere is a clear indication of an important role for history in shaping large-scale patterns of diversity[68].

Table 15.2 A classification of explanations for variations in diversity across the globe.

Mechanism	Comments
Null models	
Geometry	Mid-domain effect.
Neutral biological models	
Balance between speciation, extinction and dispersal	Applies to assemblages in balance with regional species pool.
Equilibrium explanations	
Carrying capacity	AET or productivity allows higher diversity in tropical areas.
Diversification	Greater speciation relative to extinction (can also be non-equilibrium if steady state has not been reached).
Non-equilibrium explanations	
Time	Greater time for speciation (tropics are older than temperate or polar regions).
Glacial history	Higher northern latitudes still recovering from glaciation and recolonisation is still incomplete.

15.10.1 Diversity and temperature in deep time

From the viewpoint of what palaeontologists call deep time, we live in an unusual world. Latitudinal clines in climate as extreme as those we see today are not typical of the history of life as a whole, and the present cline developed during the Cenozoic (essentially the last 65 million years since the K/Pg mass extinction event). During this period the trajectory of both terrestrial and marine diversity has been almost exclusively upward and the present diversity of life on Earth is quite possibly the highest it has ever been. Neither correction for sampling intensity (that is, volume of rock) nor taxonomic revision has significantly altered the shape of this curve[69].

The upward trend in Cenozoic diversity reflects a steady recovery from the K/Pg mass extinction event. Michael Foote has shown that the post-extinction diversification rate was high immediately after the extinction, as would be expected, but then stabilised. Marine diversification during the Cenozoic has involved principally diversification within some provinces, although there may also have been a simultaneous increase in the number of marine provinces as the latitudinal temperature gradient steepened and oceanographic patterns changed. In marine molluscs the intensification of the Cenozoic diversity cline was driven almost entirely by younger clades diversifying in the tropics, whereas in some mid-latitude provinces richness has been fairly constant. This is not to say that the high latitudes have not been the site of evolutionary novelty, and many taxa have indeed radiated there including amphipods, isopods, buccinid gastropods and some teleost fish. Nevertheless, the development of strong diversity clines indicates that diversification at high latitudes has been less marked than tropical radiations[70].

The Cenozoic increase in marine diversity has coincided with a steady decrease in bulk oceanic temperature. During the Cenozoic the broad-scale relationship between standing global diversity and global oceanic temperature has thus been inverse: global marine diversity has increased as mean ocean temperature has decreased. A complication here is that while bulk oceanic temperature is dictated largely by deep water, most of the fossils come from shallow-water habitats. An analysis for the entire Phanerozoic (the past 600 million years) using tropical surface-water temperatures by Peter Mayhew and colleagues has shown that global marine benthic diversity (at least that preserved as fossils) tends to be greater at times of higher surface temperature[71]. The correlation between marine diversity and oceanic temperature is thus equivocal when examined over very long temporal scales.

It is clear that no single explanation will suffice to explain global patterns of biological diversity. We can identify a clear role for climate in influencing plant diversity, and there are good ecological reasons to expect this relationship to propagate through the food-web. However the marked asymmetry between the northern and southern hemispheres, as well as powerful differences between continental land masses, points to an important role for evolutionary history. It is likely that progress will come from attempts to integrate these various factors, but what we can say is that simple explanations based on a single factor such as temperature are insufficient[72].

15.11 Summary

1. The diversity (species richness) of plants and animals is typically highest in the tropics. This pattern is not universal and often differs markedly between the northern and southern hemispheres.
2. The strongest environmental correlate of species richness is climate. Plant production is fuelled by solar energy, but is governed jointly by temperature and the availability of water (as captured by actual evapotranspiration, AET). Greater production is then linked to higher diversity because larger population size protects against stochastic extinction (the *more individuals mechanism*). A greater biomass and diversity of plants allows for a greater diversity of herbivores and so on through the food-web, though the correlation with climate (AET) gets progressively weaker at higher trophic levels. This is the basis of the species–energy theory of diversity.
3. The Metabolic Theory of Biodiversity posits a mechanistic explanation for higher diversity in warmer places mediated through an enhanced generation of mutations as a by-product of the faster metabolic rate associated with a higher body temperature. Evidence for this is equivocal, and this mechanism cannot explain the strong association between endotherm species richness and climate.
4. Temperature and energy and often confused in relation to diversity. Temperature is not energy, nor is it a meaningful proxy for energy. The source of energy for plants is sunlight, and for all other levels of the food-web it is the chemical potential of the plant or animal tissues being consumed.
5. The striking differences between the northern and southern hemispheres point to an important role for history, particularly recent glacial history, in influencing current patterns of diversity.
6. We still lack a comprehensive theory of biological diversity, but evidence points to a complex series of factors being important, with the dominant ones being energy and time (history).

Notes

1. This quotation comes from an essay of 1807 entitled *'Ideas for a physiognomy of plants'*.
2. The classic narratives of the explorer-naturalists are von Humboldt (1808), Wallace (1876) and Darwin (1839). Particularly important to our understanding of the oceans were the voyages of HMS *Challenger* (1872–1876) and the *Discovery* Investigations (1923–1951). Alexander von Humboldt (1769–1859) was a Prussian explorer and naturalist who undertook pioneering collecting work in South America and subsequently Russia; his life and work is described by Wulf (2015). Alfred Russell Wallace (1823–1913) was a British explorer and naturalist who travelled extensively in the Far East and arrived independently at the idea of evolution; his travels and scientific contribution are described by Raby (2001). Charles Darwin (1809–1882) surely needs no introduction; an excellent biography is Desmond & Moore (1991).
3. The neologism 'biodiversity' originates with the National Forum on BioDiversity, held in Washington, D.C. on 21–24 September 1986, under the auspices of National Academy of Sciences and the Smithsonian Institution. The proceedings from this meeting (Wilson & Peter 1988) were highly influential in establishing the term in the scientific and public consciousness.
4. This far-reaching treaty was signed by 156 nations on 5 June 1992 at the United Nations Conference on Environment and Development (UNCED) in Rio de Janeiro.
5. Magurran (2004) provides an excellent introduction to the measurement of biological diversity.
6. Williams & Gaston (1994), Williams et al. (1994), Roy et al., (1996) and Lee (1997) discuss the use of higher taxonomic levels in the study of diversity.
7. The problems inherent in rarefaction curves are discussed by Gray (2002), Gray et al. (2004) and Ugland et al. 2003.
8. Redrawn from Clarke et al. (2007); reproduced with permission of the Royal Society of London.
9. Watson (1835). The quotation comes from Rosenzweig (1995).
10. May (1975) and Connor & McCoy (1979) provide a rigorous background to the nature of the species–area relationship. Important early references are Arrhenius (1921), Williams (1964), Preston (1948, 1960, 1962). See also Tjørve (2003). Harte et al. (2005) present a null model for the species–area relationship.
11. The original definitions come from Robert H. Whittaker (Whittaker, 1960, 1972). A valuable modern perspective is provided by Robert J. Whittaker and colleagues (Whittaker et al. 2001), and an excellent introduction is Magurran (2004). The various diversity indices are discussed by Jost (2006, 2007), and the nature of the regional species pool by Carstensen et al. (2013). Measures of turnover are reviewed by Gaston et al. (2007).
12. Data are from Nicol (1971) as summarised by May (1994) for the land, and Clarke & Johnston (2003) for the ocean. The precise number of phyla and classes depends on the taxonomy used but the comparison remains robust to variations in higher level taxonomy; here it is Barnes (1998).
13. Briggs (1994) provides an early perspective on the contrast between terrestrial and marine diversity. The studies suggesting a largely undescribed deep-sea fauna are Grassle (2001) and Grassle & Maciolek (1992). Our understanding of marine diversity has recently received a huge boost from the Census of Marine Life (Williams et al. 2010, Snelgrove 2010).
14. Alessandro Minelli has provided an insightful review of the nature of the data on which such estimates are based, and provided his own estimate of 1.8 million described species (Minelli 1993). Alroy (2002) presents a rigorous analysis of the degree of synonymy that might permeate estimates of described species, and Weiser et al. (2007) and Boyle et al. (2013) have estimated the extent for errors in the names of plants: in an inventory of 12 980 names, 42% proved to be erroneous, obsolete or misspelled.
15. Erwin (1982) estimated beetle diversity (but see Bartlett et al. 1999, Ødegaard et al. 2000, Dolphin & Quicke 2001 and Novotny et al. 2002 for alternative reviews of insect diversity, and Walter & Behan-Pelletier 1999 for high diversity in forest canopy mites), Hawksworth (2001) reviewed fungal diversity (but see Fröhlich & Hyde 1999 and Arnold et al. 2000 for alternative views of fungal diversity), Finlay et al. (1998) and Finlay & Fenchel (1999) reviewed global protist diversity and Mora et al. (2011) reviewed total eukaryote diversity.
16. The importance of scale is emphasised by Willis & Whittaker (2002). The bat study is Willig & Bloch (2006).
17. It was John Lawton who commented that 'without bold, regular patterns ecologists do not have anything very interesting to explain' (Lawton 1996). The challenge for ecologists is that bold simple patterns rarely have simple explanations.
18. The global maps of vascular plant distribution were produced by Barthlott et al. (1996, 2007) and Kier et al. (2005). They are based on careful analysis of 1800 geographic units (countries, protected areas, natural vegetation units), pooled into the 867 ecoregions of the classification produced by Olson et al. (2001).

19. Plotted with data from Woodward (1987) and Gaston et al. (1995); reproduced with permission of Cambridge University Press.
20. Platnick (1991), Eggleton (1994), Gaston & Williams (1996).
21. The nature of microbial species is discussed by Cohan (2002), Fenchel (2005) and Konstantinidis et al. (2006), and an example of the use of genetic data in this context is Chan et al. (2012). The extrapolation study is Locey & Lennon (2016). A valuable overview of the issue of microbial diversity patterns is given by Green & Bohannan (2007).
22. Baas Becking (1934). See also De Wit & Bouvier (2006) and O'Malley (2008) for modern perspectives. The microbial ubiquity hypothesis was given impetus by Bland Finlay and Tom Fenchel (Fenchel et al. 1997, Finlay & Clarke, 1999, Finlay et al. 1999, Finlay & Fenchel 2004 and Fenchel & Finlay 2004).
23. Examples of biogeographic structure in microbial assemblages are Pommier et al. (2007), Fuhrman et al. (2008), Stomp et al. (2011), Ghighlione et al. (2012), Pointing et al. (2015) and Cox et al. (2016). Mittelbach & Schemske (2015) provide a valuable general overview.
24. Thorson (1957), Fischer (1960), Stehli et al. (1967, 1972). Given the paucity of hard evidence, Clarke (1992) questioned whether it was safe to assume the existence of a global latitudinal cline in diversity in all marine taxa. Gunnar Thorson (1906–1971) was one of the great marine ecologists. He overwintered in Greenland in the early 1930s as part of the Three-year Greenland Expedition led by Lauge Koch, and made seminal contributions to our understanding of the larval biology of marine invertebrates, establishing the conceptual framework we use to this day.
25. The coral reef/rainforest comparison can be traced back to a seminal paper by Joe Connell (Connell 1978), although the aim of this paper was to discuss mechanisms not actual diversities. The reef diversity extrapolations are by Reaka-Kudla (1996) and Small et al. (1998). Bouchet et al. (2002) provide the most thorough assessment of molluscan diversity on tropical reefs.
26. Molluscs: Roy et al. (1994, 1998), Crame (2000a, b) and Valdovinos et al. (2003); bryozoans: Clarke & Lidgard (2000); crustaceans: Boschi (2000). See also Macpherson (2002) for a comparative study of latitudinal gradients in richness of molluscs, crustaceans and fish along the eastern and western Atlantic Ocean.
27. Macroalgae: Santelices et al. (2009); sponges: Van Soest et al. (2012); actinarians: Fautin et al. (2013). The most recent marine ecoregions classification is Spalding et al. (2007).
28. Van Dover (2000) provides a valuable review of hydrothermal vents and their associated fauna, Pitcher et al. (2007) review the biology of seamounts and Yesson et al. (2011) document current knowledge of the global distribution of seamounts.
29. The comparative studies of soft sediment assemblage richness are Kendall & Aschan (1993), Kendall (1996), Ellingsen & Gray (2002) and Renaud et al. (2009). The estuarine studies are Engle & Summers (1999) and Attrill et al. (2001).
30. The original study is Sanders (1968). Gray (1994, 2001) and Gray et al. (1997) summarise the differences between the deep and shallow soft sediments.
31. The deep-sea gastropod data are from Rex et al. (2000, 2005) and Stuart et al. (2003), the nematode data are from Lambshead et al. (2000) and the foraminifera study is Culver & Buzas (2000).
32. The copepod study is Woodd-Walker et al. (2002) and the other zooplankton studies are from Pierrot-Bults (1997). The classification of the oceans into biomes and provinces is Longhurst (1998).
33. Plotted with data kindly provided by Michael Rex (deep-sea gastropods) and Rachel Woodd-Walker (oceanic copepods). Copepod richness was determined at the level of genus because of taxonomic difficulties at the species level.
34. Gulick (1872).
35. Important contributions to the debate are Rosenzweig (1995) who argues strongly for area, Huston (1979, 1994) who places emphasis on habitat heterogeneity, Wright (1983) who examines the role of energy and Hubbell (2001) who derives a purely neutral theory of community assembly and diversity.
36. Plotted with data from Table 3.2 in Brock (1978) and Supplementary Information in Sharp et al. (2014).
37. Turner et al. (1987, 1988) analysed patterns within the UK. Hawkins & Porter (2003) extended the analysis to the whole of western Europe.
38. See Algar et al. (2007). Plotted with data kindly provided by Adam Algar.
39. This model is presented in Allen et al. (2002), and the quotations in italics are taken verbatim from the paper. The energy equivalence hypothesis was suggested by Damuth (1987), though empirical support for the idea has proved elusive (Marquet et al. 1995, Russo et al. 2003, Ackerman et al. 2004).
40. The second model is Allen et al. (2007). See also Allen & Gillooly (2006). The model has been further developed by Stegen et al. (2009).
41. Kohne (1970) outlines the role of generation time. The role of metabolic rate in DNA evolution was proposed by Martin et al. (1992), Martin & Palumbi (1993), Martin (1995, 1999).
42. This is only true of animals; in plants reproductive tissue differentiates from meristem tissue and mutations here can thus be promulgated to the next generation.

43. Lanfear et al. (2007), Goldie et al. (2011). See also Held (2001) for effects of temperature, and Lanfear et al. (2010a, b), Thomas et al. (2010) and Bromham (2011) for life-history effects.
44. Epipelagic copepod data kindly provided by Rachel Woodd-Walker. Free-living bacteria data from Milici et al. (2016).
45. These are the same data used by Allen et al. (2002) to test their original model; they concluded that the slope of an Arrhenius model matched their prediction.
46. Plotted with data kindly supplied by Kaustov Roy.
47. David Currie has shown this cline very clearly; see Currie (1991). Hillebrand (2004) provides a detailed and thorough review of these patterns.
48. For tests of the Metabolic Theory of Biodiversity see Algar et al. (2007), Hawkins et al. (2007a, b) and Latimer (2007).
49. Plotted with data from Kerr & Packer (1997) and Clarke et al. (2010).
50. Lindemann (1942). The theory of island biogeography was developed by MacArthur & Wilson (1963, 1967). Although Brown (1981) had argued persuasively for the role of energy in regulating diversity, the formal analysis is Wright (1983).
51. Note that we are ignoring any input from chemosynthetic energy here. This is generally an insignificant contribution to energy flow in the terrestrial realm, which is where much ecological theory has been developed, but it can be very important in some deep-sea benthic environments.
52. Öpik & Rolfe (2005).
53. Major (1963), Rosenzweig (1968). Michael Rosenzweig's paper is worth reading for the clarity with which it lays bare the difficulties facing ecologists trying to derive even simple relationships from studies done in different ways by different people at different locations.
54. The assumptions implicit in species–energy theory have been explored by Evans & Gaston (2005), Evans et al. (2005) and Hurlbert & Stegen (2014).
55. The calculation of PET from meteorological temperature was suggested by Holdridge (1959). Rosenzweig (1968) discussed the derivation and usefulness of AET.
56. Data taken from Table 1 in Rosenzweig (1968), plotted after rescaling to natural logarithms.
57. The reason for excluding marine species is that while these breed on land they feed at sea. The revision of the theory is by Storch & Šizling (2007).
58. Tree species richness data replotted from Currie & Paquin (1987), plotted with data kindly provided by David Currie. Mammal species richness replotted from Andrews & O'Brien (2000); reproduced with the permission of John Wiley and Sons.
59. See, for example, Currie (1991), Stephenson (1998), Waide et al. (1999), Francis & Currie (2003), Hawkins et al. (2003), Currie et al. (2004), Davies et al. (2004), Kreft & Jetz (2007), Gillman et al. (2015) for plants, Kerr & Currie (1999), Andrews & O'Brien (2000) and Li et al. (2013) for animals in North America, South Africa and China respectively, and Davies et al. (2007) for global birds.
60. O'Brien (1993, 1998) Field et al. (2005).
61. See, for example, Yvon-Durocher et al. (2015).
62. In fundamental terms, thermodynamic beta captures the relationship between a statistical description of a system in terms of its entropy and the thermodynamics in terms of its energy: it expresses the response of the entropy of a system to an increase in its energy (note that the units of Boltzmann's constant are those of entropy).
63. See Mayer & Pimm (1997) and Janz et al. (2006) for examples of the first mechanism, and Jones et al. (1994, 1997) and Alper (1998) for discussion of ecosystem engineers.
64. The arguments here were first presented in Clarke & Gaston (2006). The fish data are discussed by Clarke & Johnston (1999) and Clarke (2003). The endotherm data are from Anderson & Jetz (2005). See also Lovegrove (2003).
65. Fish data taken from Clarke & Johnston (1999), rescaled to SI units for plotting. Mammal data replotted from Anderson & Jetz (2005); reproduced with the permission of John Wiley and Sons.
66. The first null model was Colwell & Hurtt (1994). The most influential of these null models has undoubtedly been the mid-domain model of Colwell & Lees (2000). See also Colwell et al. (2004, 2005), Connolly et al. (2003) and Connolly (2005). For critiques of this approach see Zapata et al. (2003, 2005) and Davies et al. (2005). A recent null model is Gross & Snyder-Beattie (2016).
67. The question of whether the Arctic currently supports as many species as it can was posed memorably by Evelyn Hutchinson in a short paper that has rightly become a classic of ecology (Hutchinson 1959). Qian & Ricklefs (1999, 2000) and Xiang et al. (2004) discuss the role of history in determining patterns of plant diversity. For a marine perspective see Crame & Clarke (1997) and Clarke (2009).
68. The history of the Arctic marine fauna and its on-going colonisation are described by Dunton (1992) and Vermeij (1991). The history and current diversity of the Antarctic marine fauna is described by Clarke &

Crame (1989, 1992, 1997, 2003, 2010), Crame & Clarke (1997), Clarke & Johnston (2003), Marko (2004) and Thatje et al. (2008).

69. The shape of the deep-time diversity curve is discussed by Signor (1985), Foote & Sepkoski (1999), Crampton et al. (2003), Jablonski et al. (2003) and Bush et al. (2004). Alroy et al. (2001) suggested that the long-established pattern of marine diversity through time may be an artefact of biased sampling. This was not supported by further analysis (Crampton et al. 2003; Jablonski et al. 2003; Bush et al. 2004).

70. Tropical and polar diversifications are discussed by Crame (2001), Briggs (2003, 2004), Clarke & Johnston (2003), Goldberg et al. (2005), Foote (2005), Crampton et al. (2006), Krug et al. (2009) and Clarke & Crame (2010). Clarke (2007) explores the role of history in diversity.

71. The Cenozoic studies of provincial diversity are Valentine (1968), Bambach (1977) and Valentine et al. (1978). The Phanerozoic study is Mayhew et al. (2012), which reverses the pattern reported in the original analysis (Mayhew et al. 2008).

72. The differences between the northern and southern hemispheres are often overlooked; see Chown et al. (2004). Pontarp & Wiens (2016) present an attempt to integrate a number of factors that are more traditionally viewed singly into a single model of global diversity.

CHAPTER 16

Global climate change and its ecological consequences

> The aqueous vapour constitutes a local dam, by which the temperature at the Earth's surface is deepened; the dam, however, finally overflows, and we give to space all that we receive from the sun.
>
> **John Tyndall**[1]

When the renowned biological oceanographer Victor Smetacek was growing up in the Himalayan hill country around Nainital, he kept detailed notes of the birds and other wildlife he saw about him while roaming the local countryside. He left India to pursue his education in Germany, but returned to Nainital regularly where he continued to observe the local natural history. He noticed that the local bird fauna was being joined by an increasing number of species from lower elevations, and that a few of the higher elevation species were no longer to be seen locally. Having considered habitat change as a possible cause, he concluded that the likely driver was a slow change in the local climate. In the early 1970s he published his observations as a short note, which ended with the prescient remark *it was my aim to draw attention of others to what is perhaps a new problem*[2].

This was indeed a new problem, in that it was then largely unrecognised, but the Earth's climate has always changed. The rate has varied throughout geological history, but change has been a continual feature of climate. This poses two questions: why does climate change, and what governs the rate of change? There is also the important issue of how these changes affect organisms. In this chapter we will examine both the causes and consequences of climate change, starting with the physical aspects of climate and then looking at the effects on organisms. To set the context, we first need to say something about the atmosphere.

16.1 The Earth's atmosphere

The thermal structure of the atmosphere is a consequence of both dynamic and radiative transfer processes, and the atmosphere can be divided into four layers based on differences in their dynamics and photochemistry. These layers are shown in Figure 16.1, together with the temperature reversals that define their boundaries.

Climate is essentially a feature of the lower atmosphere, the troposphere[3]. The troposphere contains the bulk of the atmosphere (over 75% of the mass), as shown by the rapid decrease in pressure with altitude. It also holds almost all the water vapour, including most clouds, and is where the weather takes place. The average composition of the troposphere is shown in Table 16.1. The troposphere is heated primarily from the ground, and so the temperature decreases with altitude, with the rate of change (the *lapse rate*) being 5–7 K km^{-1}. The top of the troposphere is marked by the tropopause, which lies at a height of between ~8 km in the polar regions and ~15 km at the equator.

Above the tropopause lies the stratosphere, which contains most of the atmospheric ozone (O_3). Because this ozone absorbs UV radiation from the Sun, the stratosphere is heated from the top, producing a temperature gradient which is the opposite of that in the troposphere. Although the temperatures almost reach those of the lower troposphere, the thin atmosphere means that the total energy content is low.

Principles of Thermal Ecology. Andrew Clarke, Oxford University Press (2017).
© Andrew Clarke 2017. DOI 10.1093/oso/9780199551668.001.0001

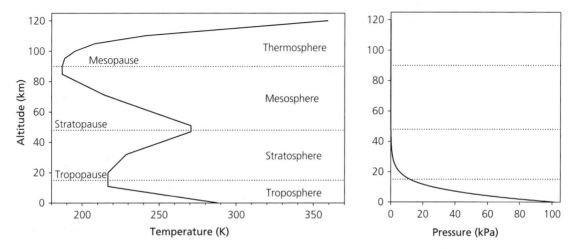

Figure 16.1 The vertical structure of the atmosphere, showing the variation in temperature (left) and pressure (right) with altitude, and the approximate position of the temperature reversals (dotted lines) that mark the division of the atmosphere into four layers[4].

Table 16.1 Composition of the troposphere, expressed as the mixing ratio, which is the molar ratio by volume for each component (ppmv: parts per million by volume, ppbv: parts per billion by volume, CFC: chlorofluorocarbon). Radiatively active components are shown (R), and some minor components have been omitted[5].

Component	Tropospheric mixing ratio	
Nitrogen (N_2)	78.08%	
Oxygen (O_2)	20.95%	
Water vapour (H_2O)	< 3.00% (highly variable)	R
Argon (Ar)	0.93%	
Carbon dioxide (CO_2)	408 ppmv (2016; increasing)	R
Ozone (O_3)	10 ppmv (variable)	R
Methane (CH_4)	1.84 ppmv (2016; increasing)	R
Nitrous oxide (N_2O)	337 ppbv (2016; increasing)	R
Carbon monoxide (CO)	120 ppbv	R
CFC11 and CFC12	< 0.8 ppbv	R

The temperature is highest at the stratopause, approximately 50 km in altitude, then falls linearly again in the mesosphere as ozone heating diminishes. Above the mesopause is the thermosphere, where solar radiation ionises the molecules to form a plasma, and aurorae are located. Apart from the important protection from incoming solar UV radiation afforded by stratospheric ozone, these upper layers of the atmosphere have little immediate influence on living organisms.

16.2 Why is the Earth's temperature what it is?

The temperature of the Earth allows for the presence of liquid water, and hence life to exist. To see why the Earth's temperature is what it is, we need to start with some basic physics.

The Earth receives essentially all of its energy from the Sun[6]. The incoming solar irradiance at the top of the Earth's atmosphere is ~1361 W m^{-2}, integrated over all wavelengths[7]. Since only half the Earth is illuminated at any one time, and the strength of incident radiation also depends strongly on latitude, the annual irradiance received at the top of the atmosphere averages ~341 W m^{-2}. Some of this incident radiation is reflected back into space by the atmosphere (~22%, principally from clouds and aerosols) or the ground (~9%, principally by ice), and the average energy absorbed by the atmosphere and surface (continents and oceans) is thus ~235 W m^{-2}. For the Earth to remain in long-term energy balance, it must also radiate ~235 W m^{-2} back into space; were this not so, the Earth would long ago have frozen solid or boiled dry[8].

The reflected sunlight is dominated by visible wavelengths, whereas the radiation emitted by the Earth is predominantly in the infrared (Figure 16.2). The radiation emitted by an object depends on its temperature, as captured by the Stefan–Boltzmann

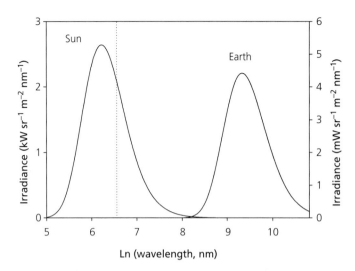

Figure 16.2 A comparison of the radiation spectra emitted by a black body at the temperature of the surface of the Sun (5780 K, left axis) and at the effective temperature of the Earth (255 K, right axis). Note the logarithmic wavelength axis and the large difference in irradiance scales: the two spectra differ by a factor of ~10^6 in their peak intensity. The dotted line marks the conventional boundary between the visible and near infrared wavelengths (700 nm); peak irradiance in the Sun's output is in the visible spectrum, whereas all of the radiation emitted by the Earth is in the infrared.

Law (see Chapter 3). For Earth the outgoing infrared radiation of 235 W m^{-2} corresponds to a blackbody at 255 K (–18 °C). This is the temperature of the Earth that would be seen by a spacecraft looking at Earth and measuring its temperature from its infrared spectrum[9].

Using a number of simplifying assumptions, we can estimate the effective temperature of planet of a given albedo, with or without rotation, and without an atmosphere. For a rotating planet in Earth's orbit, with an albedo of 0.3, the effective temperature is ~255 K, identical to the black-body temperature estimated from radiation balance[10]. The mean temperature of the Earth at ground level is, of course, much warmer than this, at ~15 °C (~288 K)[11]. This indicates that the presence of an atmosphere containing gases that absorb infrared radiation maintains the temperature of the Earth's surface some 33 K warmer than it would otherwise be. This is the *greenhouse effect*, and it is what makes the Earth habitable, for without it the surface of the planet would be permanently frozen.

16.3 The greenhouse effect

The greenhouse effect is a simple consequence of an atmosphere containing gases that are transparent to visible light but which absorb infrared radiation. Absorption at infrared wavelengths occurs in any asymmetrical diatomic molecule or any molecule with three or more atoms. The gases in the Earth's atmosphere absorbing infrared radiation are H_2O, CO_2, N_2O, CH_4 and O_3, together with a number of gases added by mankind's industrial activities such as CFCs (chlorofluorocarbons); collectively these are referred to as *radiatively active gases* or *'greenhouse' gases*. The gases that comprise the bulk of the Earth's atmosphere, namely N_2, O_2 and Ar, are completely transparent to infrared radiation. The effect of absorption by the radiatively active gases can be seen in the spectrum of light received at the surface of the Earth (Figure 16.3). In contrast to the smooth spectrum emitted by the Sun and received at the top of the atmosphere, at ground level there are very strong absorption bands in the near infrared part of the solar spectrum caused largely by CO_2 and water vapour.

The spectrum of incoming solar radiation is dominated by visible light, to which the atmosphere is transparent but which it scatters. This Rayleigh scattering is wavelength-dependent, and is the reason that the sky is blue and the setting Sun red[12]. Much of the energy of sunlight (~47% of the total energy) passes through the atmosphere to be absorbed by seawater, plants or bare ground. This absorbed energy warms the surface of the Earth and the upper layers of the ocean.

The land and ocean also emit radiation. This emitted radiation, however, is all in the infrared, and covers wavelength bands that are absorbed strongly by radiatively active gases in the lower atmosphere (Figure 16.4). Absorption of the outgoing radiation is virtually complete at wavelengths

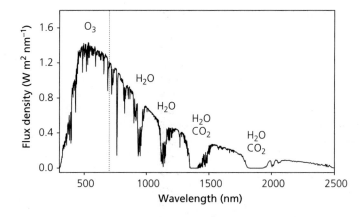

Figure 16.3 Spectrum of sunlight received at sea level on the surface of the Earth. The dotted line shows the conventional boundary between near infrared and visible wavelengths (700 nm). Note the small absorption bands in the visible caused by ozone (O_3) and the deep gaps in the near infrared, resulting from absorption by CO_2 and water vapour[13].

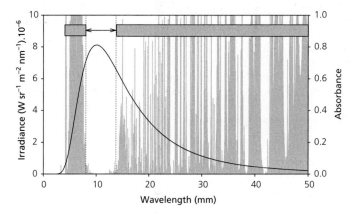

Figure 16.4 The radiation spectra (black line) for a black body at the mean temperature of the Earth's surface (288 K) showing the powerful absorption of the emitted radiation by atmospheric CO_2 and water vapour (vertical grey bars) at longer wavelengths (13 μm and above) and by N_2O, CH_4, CO_2 and water vapour at shorter wavelengths (below 7 μm). Absorption is scaled from 0 (transparent atmosphere) to 1 (complete absorption). The bars show the range over which absorption is effectively complete. This absorption means that of the long-wave infrared radiation emitted from the Earth's surface, only wavelengths between ~7 and ~13 μm (shown by the dotted lines and arrow) pass directly through the atmosphere to space; the rest is absorbed[14].

below ~8 μm and above ~13 μm. At shorter wavelengths absorption is by N_2O, CO_2, CH_4 and water vapour, whereas at longer wavelengths it is predominantly by CO_2 and water vapour. Between these is an important window (often called the '10 μm window', though the range is actually about 8–13 μm). This window allows between 15 and 30% of the emitted energy to pass straight through the atmosphere, depending on atmospheric conditions; the remainder is absorbed and contributes to the greenhouse effect.

The best way to understand the mechanism underpinning the greenhouse effect is to view the atmosphere as a series of layers. As the long-wave infrared radiation emitted by the ground or the ocean moves upwards, some is absorbed in each layer by molecules of CO_2, H_2O or other radiatively active gases. These molecules either re-emit this radiation or transfer the energy to other molecules, thereby warming the atmosphere as a whole (Box 16.1).

Each layer of the atmosphere also emits radiation, the intensity and spectral characteristics of which depend on its temperature. Radiation is emitted in all directions (isotropically), so some goes back towards the Earth's surface and some goes upwards to higher levels. At ever higher levels, the atmosphere becomes thinner, but importantly it also holds less water. The dryness increases the absorption by CO_2 relative to that by H_2O (which dominates at lower levels), and the thinness increases the fraction of radiation that can escape into space.

The temperature of the atmosphere varies with height, and the height at which the atmospheric black-body radiation is finally lost to space is that where the temperature equals the effective temperature of the Earth (255 K, −18 °C). If the concentration

> **Box 16.1 The greenhouse effect at the molecular level**
>
> When a photon hits a molecule of a radiatively active gas, it will either be absorbed or scattered. Absorption is probabilistic, with the probability of absorption being higher the closer the photon energy is to a vibrational energy level (resonance) in the molecule. Photons with energy that resonates with vibrational energy levels in molecules of H_2O or CO_2 correspond to the infrared region of the spectrum. Typically there are so many energy level resonances that absorption takes place over a range of wavelengths (absorption bands) rather than a narrow line, and there are also a number of physical processes that broaden the wavelength range absorbed. Each of the radiatively active gases in the atmosphere has a range of energies over which it can absorb energy. In consequence much of the long-wave radiation emitted by the surface of the Earth is absorbed in the troposphere, and only a fraction passes uninterrupted into space.
>
> When a photon is absorbed, the molecule increases in energy (it becomes 'excited') as the relevant vibrational energy level is filled. An isolated molecule will decay back to a lower energy level by emitting a photon whose energy is the difference between the two energy levels. In the troposphere, however, the frequency of collision with other gas molecules is so high (billions of collisions per second) that almost always the energy is distributed to other molecules before a photon can be re-emitted. This rapid transfer of energy leads to an equilibrium distribution of energy across all independent degrees of freedom (this is the equipartition principle that we met in Chapter 2). Some of these degrees of freedom are translational, and the consequent faster average speed of the molecules means that the temperature of the atmosphere has increased.
>
> If this were the only process involved, the emission of infrared from the Earth's surface would cause the atmosphere to continually increase in temperature. This does not happen because the atmosphere, like all bodies above absolute zero, also loses energy by radiation, the intensity and spectral distribution of which is a function of the atmospheric temperature (Chapter 3). The important point is that while only a small fraction of the molecules in the atmosphere absorb infrared radiation, the rapid redistribution of the absorbed energy means that it is the atmosphere as a whole that is warmed and radiates.

of radiatively active gases in the atmosphere increases, then absorption of the radiation will be extended to higher levels and the height from which most of the radiation finally leaves Earth increases. This increase in height comes about because the equilibration of energy in the atmosphere is between radiation, kinetic energy (temperature and molecular vibration) and potential energy (height).

The long-term average state of the atmosphere is thus dictated by a balance between absorption of the infrared radiation being emitted from the Earth's surface, and radiation from the top of the atmosphere into space. The troposphere temperature at which this balance is achieved is set by the concentration of greenhouse gases, which acts analogously to a thermostat. While water vapour is important in absorbing infrared radiation, and passing this energy to the rest of the lower atmosphere, it is predominantly the concentration of carbon dioxide that sets the tropospheric temperature through its influence on radiative losses from the top of the atmosphere where water vapour is absent[15].

16.3.1 The greenhouse effect and real greenhouses

The oceans and the surface of the Earth are warmed by absorbing solar radiation (principally but not exclusively at visible wavelengths). Their temperature is coupled tightly with that of the lower atmosphere by powerful flows of sensible and latent heat. The Earth's surface also emits energy and because the lower atmosphere contains radiatively active gases, it slows the rate at which the Earth loses this energy and the troposphere is warmer than would otherwise be the case.

This has long been known as the *greenhouse effect* (see Box 16.2), but the mechanism is quite different from the way a horticultural greenhouse works. A greenhouse allows the interior to warm by absorbing the visible radiation from the Sun. Although the glass does prevent long-wave infrared radiation from escaping, the strongest effect by far is to trap the warm air inside by preventing convection. This is why greenhouses have vents in the roof: if the

> **Box 16.2 The discovery of the greenhouse effect**
>
> The origins of our understanding of the role of the atmosphere in regulating the Earth's climate lie predominantly with four key individuals.
>
> The French mathematician Jean-Baptiste Joseph Fourier was probably the first to recognise (in 1824) the differing absorption by the atmosphere of incoming solar radiation and outgoing long-wave radiation (though he did not use those terms). However it was another French physicist, Claude Pouillet, who in 1838 attributed this effect to carbon dioxide and water vapour.
>
> The Irish physicist John Tyndall conducted important experiments on the absorption of radiant heat by various gases. In 1860 he demonstrated that O_2, N_2 and H_2 were transparent to thermal radiation, but H_2O, CO_2 and O_3 absorbed it strongly. He concluded that water vapour was the most important gas controlling the Earth's surface temperature, noting that *its tendency is to preserve to the Earth a portion of heat which would otherwise be radiated into space*.
>
> In 1920 the Swedish physical chemist Svante Arrhenius demonstrated by detailed calculation that variation in the concentration of CO_2 in the atmosphere could affect the overall energy budget of the planet, and that the strength of this effect varied with latitude. He also showed that without this effect the temperature of the Earth's surface would be very much colder, and estimated the change in temperature for a decrease or an increase in the concentration of CO_2 in the atmosphere. In doing so he explicitly considered that transport of energy by the atmosphere and oceans, and cloudiness, were, on average, constant.
>
> Despite the work of Arrhenius, for much of the first half of the twentieth century scientific attention concentrated on the control of climate by astronomical or geophysical processes. It was only in the 1940s and 1950s that our understanding of the radiative properties of the atmosphere advanced significantly for it to be recognised that an increase in CO_2 as a result of man's activities was indeed changing the Earth's climate and attention was directed once again at what was then widely referred to as the '*greenhouse effect*'[16].
>
> The origin of the term '*greenhouse effect*' is unclear. In 1908 Arrhenius published a popular science book, in which he described the development by Fourier, Pouillet and Tyndall of what he called the 'hot-house' theory of the atmosphere. He described the way the atmosphere was transparent to the visible light but absorbs the '*dark heat rays*' (the infrared) and compares this with the glass of greenhouses. Although Fourier had discussed much earlier the mechanisms by which enclosed spaces (including hot houses) can warm, perhaps here lies the origin of modern use of the term '*greenhouse effect*'.

temperature inside gets too great, the vents can be opened to allow warm air to escape.

The 'greenhouse effect' is thus another example of a scientific term that is useful as a rough analogy, but misleading if taken too literally. A better analogy would be that of a dam or weir, which holds back water while not affecting the flow. Indeed this precise analogy was used by John Tyndall roughly 150 years ago to illustrate how radiatively active gases retain energy while not affecting the flux of energy into space. In this analogy, the height of the weir is the concentration of radiative gases: increase the height (concentration) and more water (energy) is retained.

Because the 'greenhouse effect' is a consequence of an atmosphere containing radiatively active gases, a better term would perhaps be the 'atmosphere effect', but the term 'greenhouse effect' is so firmly established it is probably here to stay.

16.4 The temperature at the Earth's surface

The energy balance of the Earth as a whole is simple: at equilibrium the energy received from the Sun is balanced by the energy emitted. The detail of what happens to this energy in the atmosphere and ocean is, however, complex (Figure 16.5).

While we have a good understanding of the basic flows of energy, the confidence we can attach to the estimates of the various fluxes varies greatly. The most thorough attempts to achieve an overall picture have been made by Kevin Trenberth and colleagues, who estimated that of the incoming solar radiation (341.3 W m^{-2}), only about half (161 W m^{-2}, 47%) actually reaches the Earth's surface. This is (almost) balanced by a flux of infrared radiation from the top of the atmosphere of 283.5 W m^{-2}, where the temperature is ~ 255 K. The temperature

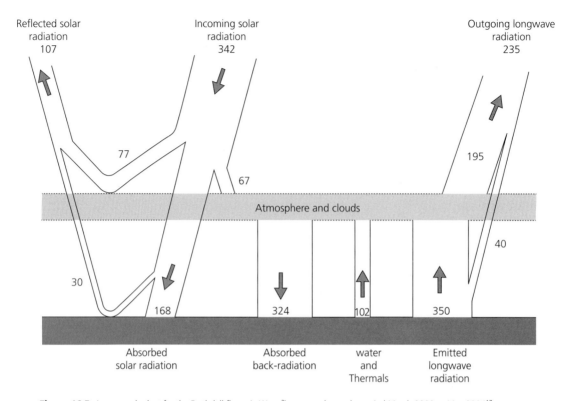

Figure 16.5 An energy budget for the Earth (all fluxes in W m^{-2}), averaged over the period March 2000 to May 2004[17].

of the lower troposphere, which is that experienced by organisms, is higher and dominated by complex patterns of energy exchange and distribution. These include powerful convective and advective processes moving large quantities of energy (such as weather systems and Hadley cells in the atmosphere, and the thermohaline circulation in the ocean). Latent heat is also important because water in the lower atmosphere is changing constantly between its solid, liquid and gaseous phases. A consequence of all this is that the temperature of the lower atmosphere and the Earth's surface are tightly coupled.

The many flows of energy shown in Figure 16.5 indicate just how difficult it is to model the global movement of energy. Adding oceanic processes (not shown in Figure 16.5) makes these global climate models (GCMs) fiendishly complex, requiring considerable computing power to run them. These GCMs are very good at capturing the broad features of the climate, and are getting better all the time. Any model, however, involves a simplification of the system under study; the skill in building one comes in knowing what to leave out, while ensuring that the essential detail is included but without making the model too complex, unwieldy or expensive to run[18].

A significant challenge for such climate models is testing them. One approach is to start the model with a known past climate, then run the model forward to the present day and compare the predictions with reality (a process modellers call *hindcasting*). If the models predict the past correctly, we can have confidence that they will predict the future accurately, and hindcasting has proved to be an extremely valuable technique for developing and improving GCMs.

A much greater challenge is to explore the Earth's climate when it was very different from recent times, and important insights into how the climate system works have come from looking at ice ages.

16.5 Ice ages

Geologists and physical geographers have long known that large parts of northern Europe, Asia and North America have only recently emerged from beneath retreating ice-sheets. The signs are everywhere in the shape of hills and valleys, remnant moraines, eskers and erratic boulders, and ice-scoured rocks. Putting all this evidence together into a coherent framework, however, proved extremely difficult, not least because each advance of an ice-sheet obliterates much of the evidence of previous advances and the intervening warmer interglacial periods.

All of this changed with the analysis of ice cores from Antarctica and Greenland. These have provided a window into the Earth's climate over the past 800 000 years and revealed a sequence of recent glacial cycles with a periodicity of ~100 000 years. Sediment cores from the deep ocean have extended our knowledge even further back, and we now know in some detail the broad features of the Earth's climate over the past 70 million years. Perhaps the most significant feature of the ice ages revealed by these cores is their frequency: the period of ~100 000 years between the most recent ice ages points clearly to the influence of orbital cycles.

There are three dominant cycles in the Earth's orbit and rotation, and the influence of these on the Earth's climate was first demonstrated by Milutin Milanković (Box 16.3). Orbital cycles cause variations in the duration, intensity and distribution of the seasons. Each of the various cycles has a different frequency; sometimes the cycles reinforce one another and at others times they partially cancel out. Milanković showed that of particular importance in the most recent glaciations has been the energy received at high northern latitudes. The northern latitudes are critical because most land is in the

---

Box 16.3 Milanković cycles

Orbital variations are an inevitable consequence of gravitational forces in a system containing more than one planet. Variations in the Earth's orbit are caused principally by the Moon and Venus (because they are close), Jupiter and Saturn (because they are large). The three dominant effects are variations in eccentricity, obliquity and precession.

Eccentricity
The elliptical orbit of the Earth was discovered by Johannes Kepler; eccentricity measures how elliptical the orbit is (that is, the extent to which the orbit departs from circular). Two centuries later the French astronomer Urbain Le Verrier discovered that the Earth's orbit fluctuates between a more circular (eccentricity ~0) and a more elliptical (eccentricity 0.0679) shape, with a major period of 413 000 years and minor periods of 95 000 and 125 000 years (usually combined as ~100 000 years). The current value of eccentricity is 0.017, and is decreasing. Eccentricity affects the variation in solar radiation received at closest approach to the Sun (perihelion) and furthest distance (aphelion). Currently this difference is about 6.8%; at maximum eccentricity the difference is 23%.

Obliquity
Obliquity (axial tilt) varies between 22.1° and 24.5°, with a mean period of ~41 000 years; the current value is 23.27°. Obliquity governs the latitudinal distribution of solar energy received on Earth. If the obliquity were zero, the Earth would have no seasons.

Precession
The precession of the equinoxes is the movement of the Earth's axis of rotation relative to the fixed stars, and was known to the ancient astronomers. This is a gyroscopic motion, under the influence of the Moon. It is quasi-periodic with dominant cycles at ~19 000 and ~23 000 years. Precession affects the position of the equinoxes and solstices, and thereby the timing of the seasons.

The importance of eccentricity to Earth's climate was recognised in the eighteenth century. In 1875 the Scottish scientist James Croll combined the cycles of eccentricity and precession to produce a pattern resembling the history of the ice ages (which was just becoming known). Half a century later the Serbian engineer Milutin Milanković included the obliquity cycle and showed that the dominant effect on climate came from precession and obliquity. His ideas were not accepted at the time, but were vindicated by the discovery of cyclical variations in ocean temperature recorded in the tests of foraminifera in deep ocean sediments. Orbital variations with periods in the range 19 000 to 400 000 years are now known as Milanković cycles[19].

northern hemisphere, and temperature changes in the southern hemisphere are buffered by the high thermal mass of the greater expanse of ocean there.

Milanković orbital cyclicity has been a constant feature of Earth's history, but ice ages have not (Table 16.2). This is because glaciation requires a specific set of circumstances. Of particular importance is an arrangement of continents that restricts the transport of warm water from the tropics to the poles. At present, for example, we have a continental land mass sitting over one pole (Antarctica) and a polar ocean with limited exchange of water with lower latitudes at the other (the Arctic). In earlier glaciations the arrangement of continents was different, but the key feature was a constraint on the distribution of energy by the oceans.

Once glaciation starts, a suite of feedbacks allow ice-sheets to build. This is typically a slow process, with intermittent warm periods (interstadials) along the way. In contrast the switch from glacial to interglacial is often fast. Indeed fossil insect remains have shown that at the most recent deglaciation in some northern locations conditions switched from a cold continental climate to a warm oceanic climate within the space of a human lifetime[20]. This rapid movement of climatic zones represents a significant shift in the distribution of energy on the surface of the Earth, but not necessarily a change in the total amount.

Thus while the pacemaker for the waxing and waning of the ice ages has been the small changes in solar energy received on Earth, feedbacks in the Earth's climate system amplify these and shift the climate dramatically from one state to another. The long-term record of glacial cycles revealed by ice and marine sediment cores has shown that variations in the size of the Antarctic continental ice-sheet at orbital frequencies extend back to at least the Oligocene–Miocene boundary (~ 24 Ma BP). Until about 900 000 years ago the dominant period was ~40 000 years, since then it has been ~100 000 years; we do not know why[21].

The most detailed picture comes from the recent glacial cycle from which we have only recently emerged. Evidence in ice cores from Greenland shows that neither glacial nor interglacial periods were uniform or stable. The most recent glacial in the northern hemisphere was characterised by frequent brief warming events, called Dansgaard–Oeschger (D–O) events after the two scientists who discovered them, and also periods of rapid ice advance that released large numbers of icebergs (Heinrich events). D–O events typically show a rapid warming phase lasting a few decades, followed by a slower cooling lasting hundreds of years. Over the past 75 000 years there have been 20 D–O warming events and 6 Heinrich cooling events. The most recent of these ushered in the last cold episode to affect the northern hemisphere, the so-called Younger Dryas, which represents the final great shudder of the climate before the arrival of the milder Holocene (Figure 16.6)[22].

The dynamic nature of glacial periods reveals the importance of the oceans in the climate system, and in particular shifts in the strength, direction and magnitude of the great currents that carry energy around the globe. Towards the end of the last glacial, oceanic circulation in the North Atlantic underwent rapid shifts, which drove substantial movements of climatic zones across the globe. Just how important ocean currents are to local climate can be seen by comparing the mild climate of northwest Europe, which receives energy from the North Atlantic Drift, and the much harsher climate of Labrador, which does not[23].

16.5.1 The ecological lessons of the Ice Ages

I live by the North Sea, not far from the site of the oldest known human occupation of northern Europe at Happisburgh. Here excavation of sediments exposed by rapid coastal erosion has revealed a

Table 16.2 Ice ages in Earth's history. Ma: millions of years; BP: before present.

Ice age	Dates (Ma BP)
Proterozoic	
Huronian	2400–2100
Cryogenian	850–630
Phanerozoic	
Andean-Saharan (Late Ordovician/Silurian)	460–420
Karoo (Carboniferous/Early Permian)	360–260
Quaternary (Pleistocene)	2.58 to date

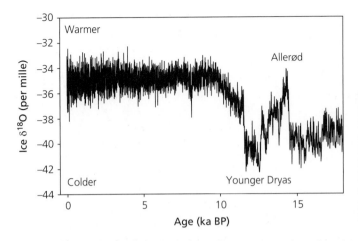

Figure 16.6 The temperature recorded in the oxygen isotope composition of ice from the GRIP (Greenland Ice Project) core. The isotope data are plotted in the conventional per mille notation relative to Vienna-SMOW (see Chapter 3). ka: thousands of years; BP: before present. Note the relatively stable climate signal back through the Holocene to the Younger Dryas cold episode (roughly 11.7 to 12.9 ka BP), the earlier short period of warming (Allerød oscillation) and then the last glacial itself (14.5 ka and older at this location)[24].

large number of flint artefacts and also human footprints. These date from ~800 000 years ago, or, put another way, some eight glacial cycles back in history; they are to date the oldest human artefacts known from northern Europe and the oldest human footprints outside Africa. At that time much of the shallow North Sea was land covered in boreal forest and the fauna included mammoths, woolly rhinoceros and sabre-tooths. Since then the ice-sheets have advanced and retreated many times, and the repeated shifts of climatic zones will have forced the flora and fauna to do the same[25].

An important consequence of the dynamic nature of the climate revealed by the ice ages is that on ecological and evolutionary timescales the world is rarely stable. Climate changes all the time, and with it so must the plants and animals (including humans). This has a profound message for ecology. It tells us that for ecological processes which take time, the world is never constant. Evolutionary and ecological theories that are predicated on stability and equilibrium are likely to be incomplete at best, and at worst wrong.

16.6 Recent climate change

The current interglacial has already lasted an unusually long time and all other things being equal, Milanković variability should be moving us towards a new glaciation. Indeed between the 1940s and 1970s the Earth's climate did experience a short cooling phase, prompting widespread media speculation that a new ice age was imminent.

We saw in section 16.3 how the long-term average state of the atmosphere is set by the concentration of radiatively active gases. While the models that describe the radiation balance of the atmosphere are complex, the results are simple and clear: increase the concentration of radiatively active gases in the atmosphere and more energy is retained, warming the troposphere. This relationship is important, because man has been altering the composition of the atmosphere by adding radiatively active gases for over two centuries[26]. We are now experiencing the consequences: the Earth is not cooling, but warming.

The concentration of CO_2 in the lower atmosphere has been monitored since 1958 at Mauna Loa observatory in Hawai'i. These measurements were started by Charles Keeling of Scripps Institution of Oceanography, and continue to this day. Hawai'i is ideal for measurements such as these because of its isolation in the Pacific, far removed from any cities or other human influences. The steady rise in atmospheric CO_2 revealed by these measurements is the clearest indication we have of the way that mankind has altered the atmosphere, and is continuing to do so. This rise in atmospheric CO_2 is often referred to as the Keeling curve (Figure 16.7)[27].

The monthly data show a clear seasonal pattern. In spring CO_2 concentration decreases as plants in the northern hemisphere photosynthesise and grow. CO_2 concentration reaches a minimum in September and levels then rise again as respiration and decay (which release CO_2) outpace photosynthesis. This seasonal variation is about 5 ppmv. The

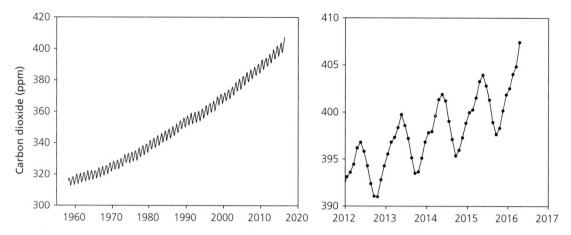

Figure 16.7 The Keeling curve: atmospheric carbon dioxide as measured at Mauna Loa observatory. A (left panel): the complete data series from the start of observations in March 1958 to April 2016. Data are plotted as monthly mean values. B (right panel): detail for period Jan 2012 to April 2016, showing seasonal variation driven principally by changes in the balance between photosynthesis and respiration in the biosphere[28].

dominance of northern hemisphere processes in the seasonal pattern arises partly because most of the Earth's land (~68%) is in the northern hemisphere, but also because a significant fraction of land in the southern hemisphere is arid with relatively low plant productivity[29].

Although the Keeling curve from Hawai'i has become the most widely reproduced record of atmospheric CO_2, there is also a long-term record from air sealed into flasks at the South Pole. This work was also started by Charles Keeling in 1958 and it shows similar seasonal and long-term trends. There are now over 100 stations monitoring atmospheric CO_2, in a network run alongside measurements by the US National Oceanic and Atmospheric Administration (NOAA).

Analysis of ice cores from Antarctica have also shown that over the past 800 000 years the CO_2 content of the lower atmosphere has undergone cycles correlated with the waxing and waning of the continental ice-sheets. Over this period, CO_2 cycled between ~280 ppmv during interglacial periods and ~180 ppmv at the height of the glacial[30]. The current CO_2 concentration exceeds the highest interglacial concentration we can measure by a considerable degree. Methane shows a similar pattern cycling between ~400 and 800 ppbv during the last eight glacial cycles, but is now at >1800 ppbv and is climbing rapidly. From these directly measured air samples, we know that the concentration of CO_2 and CH_4 in the lower atmosphere is unprecedented in the last 800 000 years and (quite possibly for the last 20 million years).

Because the equilibrium temperature of the troposphere is set by the concentration of radiatively active gases, we would expect the recent increase in the concentration of these gases to have resulted in an increased surface temperature. This has indeed happened (Figure 16.8) and is referred to by a variety of terms including global warming, global climate change and anthropogenic climate change. The mechanism is also referred to more generally as the enhanced greenhouse effect (the adjective 'enhanced' is important, though often left off, for as we saw above, the greenhouse effect was operative before man started to modify the atmosphere, and indeed it is what makes the Earth habitable at all).

The variable plotted here is the Land–Ocean Temperature Index (LOTI), which is a composite measure combining surface air temperatures (SATs) from meteorological stations on land with sea-surface temperatures (SSTs) from oceanic areas without sea-ice. Because air cools and warms much faster than water, SATs and SSTs may be quite different; their anomalies are, however, very similar (if water temperature is 1 K above normal, the air immediately above it will also be 1 K above normal). This allows climate scientists to calculate a value for the temperature anomaly averaged over the entire planet, which is what the LOTI is.

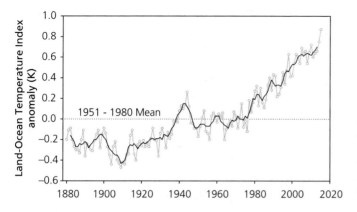

Figure 16.8 Global mean temperature relative to the mean temperature 1951–1980. Data for individual years are shown in grey, and the five-year running mean in black. With the exception of 1989, the ten warmest years in the 134-year record have all occurred since 2000, and 2015 was the warmest year in the series to 2015[31].

The LOTI is a valuable indicator of how the increased retention of energy is warming the lower atmosphere, but it is not very useful in other contexts. This is because it is a two-dimensional field, whereas the measure of importance, the change in energy in the Earth system, involves integrating over height (in the atmosphere) and depth (in the oceans)[32].

16.6.1 Feedbacks

A warmer atmosphere contains more water, because evaporation from the oceans increases and the capacity of the atmosphere to hold water is also enhanced. This extra water absorbs infrared radiation, and thereby provides a positive feedback. This feedback means that the lower atmosphere is warmer than it would be were it perfectly dry. Other important feedbacks in the climate system come from clouds, ice and snow, and changes in vegetation on land. There are also negative feedbacks, the most important of which is infrared radiation: a hotter planet radiates more energy into space[33].

16.6.2 Other factors influencing climate

While the concentration of radiatively active gases affects how energy is retained in the troposphere and oceans, the intensity of the incoming solar radiation itself varies and this exerts an influence on climate. There has been a steady increase in the energy emitted by the Sun over geological time but this has been far too slow to have driven recent climate change. Of more importance are short-term cycles in solar activity. The most well known of these is the sunspot cycle which is has a mean period of ~11 years, though cycle length has varied from 9 to 13.5 years, with a long period of inactivity (the Maunder Minimum) between about 1645 and about 1715[34].

A combination of observational data and modelling enables us to determine the relative importance of the different factors that have contributed to recent climate change. Some of these factors acted to increase troposphere temperature, others to decrease it. Overall, however, the dominant factor by far has been the concentration of radiatively active gases. The change in energy balance caused by anthropogenic factors, relative to the pre-industrial era (1750) has been estimated by the IPCC as 0.57 W m^{-2} in 1950, 1.25 W m^{-2} in 1980 and 2.29 W m^{-2} in 2011, showing the upward trend as man continues to modify the atmosphere.

16.6.3 Where does the extra energy go?

In discussions of global climate change, most attention has been directed at the increased temperature of the lower atmosphere, this being the temperature we experience as humans. The vast majority of the extra energy trapped by radiatively active gases, however, actually ends up in the ocean (Table 16.3).

16.6.4 Changes in ocean heat content

Energy passes into the oceans at the surface. While the seasonal variation in ocean energy content (and hence temperature) is confined to the uppermost

Table 16.3 Relative contribution of ocean, atmosphere and land sinks for the extra energy retained on Earth over the period 1961–2003[35].

Energy sink	Energy content change (W m^{-2})	%
Oceans	14.2	89.3
Atmosphere	0.5	3.1
Continents	0.76	4.8
Glaciers and ice-caps	0.22	1.4
Greenland ice-sheet	0.02	0.1
Arctic sea-ice	0.15	0.9
Antarctic ice-sheet	0.06	0.4
Total	15.91	

layers, energy is transferred deep into the ocean by diffusion and the global thermohaline circulation which, together with more local upwellings and downwellings, move large quantities of energy around the oceans and between depths. Without these vast movements of water, oxygen could not get from the surface to the abyss, nor could nutrients regenerated at depth reach the surface to fuel phytoplankton blooms.

Recent data indicate a steady warming of the ocean down to 2000 m (Figure 16.9), and there are now clear indications of warming at abyssal depths (< 4000 m), at least in the Southern Ocean. These are very difficult estimates to make, because the deep sea is vast and observations are few, but they are important for constraining the overall energy budget of the planet[36].

As would be expected, because the net energy flow is from the atmosphere to the ocean, the strongest warming is found near the ocean surface, which has warmed by over 0.1 K per decade since the early 1970s and has likely been warming since the 1950s. At 700 m depth the warming has been ~ 0.015 K per decade. The linear trend in energy increase over the period 1955–2013 is equivalent to 0.27 W m^{-2} over the ocean surface down to 700 m, and 0.39 W m^{-2} when integrated down to 2000 m[38].

16.6.5 The other CO_2 problem: ocean acidification

As atmospheric CO_2 increases, more enters the oceans. This reduces the concentration of CO_2 in the lower atmosphere, thereby slowing the rate at which energy accumulates there and hence the rate of climate change. In the oceans, however, the consequences are less than desirable: the CO_2 reacts with water to form carbonic acid, which then dissociates rapidly to form a bicarbonate ion (HCO_3^-) and a proton (H^+). Proton concentration is quantified by pH (Chapter 5), and since pre-industrial times the pH of the surface ocean has decreased by 0.1 units; this may seem small but pH is an inverse logarithmic scale and this change equates to an increase in the proton concentration of ~26%.

These changes affect the entire carbonate system of the ocean, resulting in an increase in the concentration of HCO_3^- and a decrease in that of CO_3^{2-}. It is therefore not surprising that the main impact

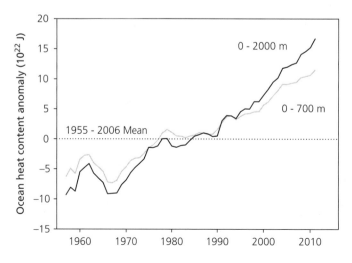

Figure 16.9 Ocean heat content. Data are five-year running means for ocean heat integrated over 0–700 m (grey trace) and 0–2000 m (black trace) plotted as the difference (anomaly) from the 1955–2006 mean (dotted line). Note how over time more and more energy is reaching deep oceanic water[37].

of increasing acidity in the ocean falls upon organisms that build their skeletons from calcium carbonate (calcifiers). These organisms face two problems. The first is that the basic reaction for calcification involves the release of a proton from bicarbonate:

$$HCO_3^- \rightarrow CO_3^{2-} + H^+$$

The carbonate ion then combines with calcium and skeletal carbonate is precipitated:

$$CO_3^{2-} + Ca^{2+} \rightarrow CaCO_3 \text{ (solid)}$$

The source of the bicarbonate differs between organisms; it may be environmental or derived from metabolic CO_2 through the action of the enzyme carbonic anhydrase. Organisms are able to manipulate the environment at the site of calcification and the key physiological challenge is not the availability of Ca^{2+} or HCO_3^- but the proton which must be removed to maintain internal acid–base balance. This is more difficult (that is, it costs more energy) if the proton is passed to the external environment, the ocean, when this is itself more acid. Indeed there is some evidence that the metabolic cost of acid–base regulation increases for many organisms, not just calcifiers, in a more acid ocean[39].

The second problem is that a more acid ocean is more likely to dissolve any existing carbonate skeleton. In some cases the carbonate is protected by an external organic layer (for example in molluscs) in others it is not and so is exposed directly to the increased acidity (for example coral reefs). Calcium carbonate exists in a number of crystalline forms of which the two most important biologically are calcite and aragonite. These differ in their solubility characteristics and the tendency to dissolution depends critically on the saturation state of the surrounding seawater (Chapter 5).

The spectre of the loss of marine organisms from ocean acidification has prompted a surge of research. This has indicated a range of physiological effects, including lower growth rates and reduced survival, in diverse marine organisms when exposed to seawater of increased acidity, but also examples of no effect (for example in some brachiopods). The exigencies of research funding mean that these tend to be short-term acute studies. However, where experiments have been allowed to run across generations, the observed impacts are typically much reduced. This suggests that where changes are slow, organisms have the capacity to acclimatise[40].

These long-term studies indicate that important insights may be provided by the fossil record of previous ocean acidification events. For the four mass extinction events with the clearest indication of associated ocean acidification, all show evidence of widespread loss of calcifying organisms. In all cases, however, the rate of change of ocean acidity was much slower than that being experienced today; the current ocean acidification event is already the most rapid in the past 65 million years, and possibly longer[41].

16.6.6 Changes in atmospheric temperature

The temperature of the lower atmosphere has been rising over the last half-century (Figure 16.8), but the rate of increase has varied widely across the planet. In some places the atmosphere has been warming rapidly, in others the increase has been slow. The three areas that have warmed most rapidly are all in the polar regions: north-western North America and the Siberian plateau in the north, and the western Antarctic Peninsula in the south. In much of continental North America, low latitude Eurasia, and Africa the warming has been much less intense. These differences indicate that the way climate change is expressed depends critically on local factors. This can be seen clearly if we compare an area that has recently undergone rapid warming (the western Antarctic Peninsula) with one where the warming has been more moderate (central England).

16.6.7 Regional warming of the Antarctic Peninsula

The climate of the western Antarctic Peninsula has warmed rapidly over the period for which we have instrumental records, the longest such record being for Faraday (now Vernadsky) Station in the Argentine Islands (Figure 16.10). Here there has been a significant increase in mean surface air temperature (SAT) over the second half of the twentieth century. The trend line, equivalent to an increase in mean annual temperature of ~2.9 K per decade is plotted only as far as 1998, because recent analyses have suggested that the warming trend has ceased and 1998 is the best statistical estimate of the point

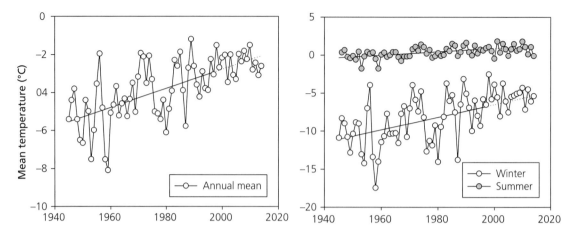

Figure 16.10 Increase in surface air temperature at Faraday (now Vernadsky) Station on the western Antarctic Peninsula, with trend lines calculated for 1945–1997 (solid lines) extrapolated to 2015 (dotted lines). A (left panel): annual mean temperature. B (right panel): mean temperature in winter (June to August: white symbols) and summer (December to February: grey symbols)[42].

at which this happened. Since then the SAT of the western Antarctic Peninsula has cooled slightly[43].

The twentieth century trend in mean annual SAT along the western Antarctic Peninsula is striking, but it mixes a strong warming over winter with a less intense warming in summer (Figure 16.10). At Faraday/Vernadsky, mean winter SAT increased from 1950 to 1998 at a rate of 1.09 K per decade, whereas over the same period mean summer SAT increased by only 0.23 K per decade.

The causes of this regional warming are complex and incompletely understood. Unravelling the mechanism(s) is made difficult by the limited duration of the SAT record, the large interannual variability of the system and its sensitivity to local interactions between the atmosphere, ocean and sea-ice. What we do know is that changes in local winds have shortened the duration of winter sea-ice along the west coast of the Antarctic Peninsula. In this region the Drake Passage constricts the flow of the powerful Antarctic Circumpolar Current, forcing the (relatively) warm Circumpolar Deep Water against the continental slope and onto the continental shelf. The reduction in winter sea-ice cover then allows an increased flux of energy from ocean to atmosphere, driving the rapid increase in winter temperature. In addition many glaciers and ice-shelves have retreated as a consequence of increased basal melting[44].

The SAT record for Antarctica is frustratingly short but a valuable historical context is provided by the isotope signal in ice cores, which reveals periods of warming of similar scope and duration to the present one at several times during the past 300 years, driven by large scale connections (*teleconnections*) to oceanographic conditions in the central Pacific Ocean. This intense variability makes it very difficult to tease out any long-term change related to anthropogenic effects (Box 16.4.), and the dramatic recent warming of the western Antarctic Peninsula has yet to exceed the extent of past natural variability[45].

Box 16.4 Separating cycles and trends

If, as is very often the case with climate, there are naturally occurring cycles, then to detect any underlying trend against the background variability, the data need to extend for at least two cycles (this is the *Nyquist frequency criterion*), and preferably longer. Typical short-term cycles affecting global climate include the El Niño Southern Oscillation (ENSO), the North Atlantic Oscillation (NAO) and the solar sunspot cycle.

For analysing the intensity of frequency of extreme events, statistical techniques such as analysis by ARIMA (autoregressive integrated moving average) are valuable, but these require very long runs of data to ensure there are sufficient such events present in the data for the analysis to provide useful results.

The climate of the east coast of the Antarctic Peninsula is very different, as is the oceanographic context, for it is isolated from the warming influence of the Antarctic Circumpolar Current. This area is difficult to reach, and meteorological records are sparse. What records there are, coupled with satellite measurements, indicate that there has been an increase in summer temperatures. This warming may have been driven by a shift in the circumpolar winds, which has led to more frequent föhn winds carrying warm air to the east of the Antarctic Peninsula. A historical context is provided by data from an ice core drilled on James Ross Island, which suggest that warming is evident from the 1920s and is unusual in the context of the Holocene climate history of this region. A period of cooling from 2500 to 600 years ago stimulated the development of substantial ice-shelves along the east coast of the Antarctic Peninsula. In recent years there has been a steady warming, with increased summer melting. In contrast to the western Antarctic Peninsula, this warming can be attributed to global anthropogenic effects. A consequence of this warming has been the dramatic collapse of ice-shelves as atmospheric temperatures exceed the stability threshold, and in the past 21 years we have seen the disappearance of ice-shelves that are at least 2000 years old[46].

The Antarctic Peninsula thus presents us with two differing pictures. The eastern side has warmed in summer, and the loss of ice-shelves suggests that this is unprecedented in the recent Holocene. The western side has warmed dramatically, mostly in winter, but this warming cannot be distinguished from the natural variability evident in the historical record. There are, however, factors that suggest the current warming may be different from previous such episodes, namely the continuing increase in radiatively active gases in the atmosphere. The ozone hole (Box 16.5) may also be playing a role in preventing the increase in radiatively active gases in the atmosphere from increasing surface air temperatures over the main continent of Antarctica. At present SATs in continental Antarctica show no discernible warming, and some areas are even cooling slightly[47].

16.6.8 Central England temperatures

Despite its frequent appearance in discussion of climate change, the Antarctic presents us with a complex picture where the effect of man on global climate is not always easy to see. We can, however, discern the anthropogenic signal in climate very clearly where the data series are longer and the extent of natural variability less marked. The longest SAT data series in the world is that for central England, and here the signal of climate change can be seen distinctly (Figure 16.12). The temperature of central England has broadly been rising since the late nineteenth century, although there was a period of cooling from about 1950 to 1970, after which warming recommenced. Despite the marked variability and a suggestion of a 20–30 year periodicity, the recent warming is statistically highly significant.

A recent analysis of global climate over the past 2000 years has established clear signs of warming in all areas of the globe except Antarctica. They also found that warming was roughly twice as rapid in the northern compared with the southern hemisphere, and that fastest warming was in areas of the Arctic, where there is strong feedback from ice and snow albedo. As before, the longer-term historical record is important to provide context, and over the past 2000 years, until about 100 years ago, the Earth was broadly cooling. There were warmer periods, such as the Medieval Warm Period, but different regions warmed at different times, and the peak temperatures then have now been exceeded by the current warming[48].

16.6.9 Sceptics, deniers and the lunatic fringe

It is impossible to explore the subject of climate change and not come across a variety of sources claiming that the climate is not changing, or if it is the cause is not man, or it was much warmer in the past anyway (sometimes with less CO_2 than today, sometimes with more). These sources vary in their motivation, language and scientific credibility. Typically, however, they misunderstand (or misrepresent) how the climate system actually works, blindly mix local with regional or global processes, present short runs of carefully selected data that distort trends and in the worst cases misrepresent

Box 16.5 The ozone hole

Ozone (O_3) is formed and destroyed by high energy ultraviolet photons in the stratosphere. The concentration of ozone is a balance between production and loss processes, which together comprise the Chapman cycle. This cycle converts the energy of solar ultraviolet photons to heat with no net loss of ozone and it has been a feature of the atmosphere since oxygen first appeared there[49].

Long-term monitoring of total ozone over the British research station at Halley in Antarctica revealed a decline in spring-time ozone from the 1970s (Figure 16.11). This loss resulted from loss of ozone catalysed by Cl^- and Br^-, principally from chlorinated fluorocarbons (CFCs) produced by man for aerosols and the refrigeration industry. Although such losses had been predicted theoretically, it was expected that ozone destruction would be largely a tropical phenomenon. The ozone hole came as a surprise because it was found in the wrong place at the wrong time[50].

The ozone hole is a seasonal phenomenon. In winter an atmospheric vortex develops under the influence of the Earth's rotation (the Coriolis effect), and the isolated air cools sufficiently for stratospheric clouds to form. Chlorine and bromine are held on the surface of the ice crystals in these clouds and when the Sun returns in spring these halogens catalyse the destruction of ozone. As spring progresses the polar vortex disperses, the ozone-depleted air over Antarctica mixes with undepleted air from lower latitudes and ozone levels over Antarctica are replenished.

The ozone hole was reported in 1985, and sparked widespread international concern. The Montreal Protocol was agreed in September 1987 and came into force in 1989. With cooperation of the industries involved, CFCs have been replaced with chemicals far less damaging, and there are signs that the ozone layer is beginning to recover. The Montreal Protocol has been described as the most successful international agreement in history[51].

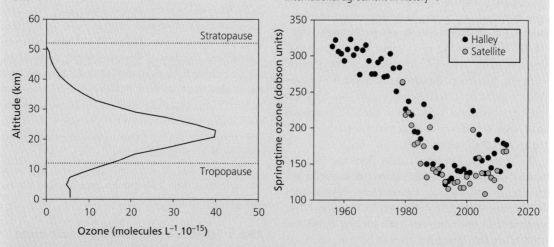

Figure 16.11 Stratospheric ozone. A (left panel): concentration of ozone in the atmosphere from a balloon ascent. The position of the tropopause and stratopause (dotted lines) are approximate. Note how the ozone is concentrated in a layer between about 10 and 30 km above the ground. B (right panel): long-term measurements of stratospheric ozone from Halley, Antarctica, show the rapid loss of spring-time ozone from the 1970s to the 1990s (black symbols). Also shown are satellite measurements for the same period (grey symbols). One Dobson unit is equivalent to a layer of pure ozone 10 μm thick at standard temperature and pressure[52].

statistical fits. They can be worth exploring to understand how and why climate change is so misunderstood in some quarters, but the uninformed or even cavalier approach to analysis can be dispiriting and the tenor of discussion can make much political debate appear rational and polite. Science is a human activity and it must also be acknowledged that not all of those presenting the detailed science have themselves used the most balanced or temperate language. The argument has become polemical and politicised, often to the detriment of scientific rigour. The only rational approach is to concentrate

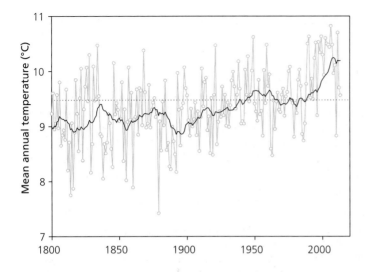

Figure 16.12 Mean annual temperature in central England from 1800 to the present day (2015). Data for individual years are shown in grey, and the 20-year running mean in black. The dotted line shows the mean temperature for the period 1961–1990[53].

on the hard science (and some of the science of climate change is genuinely difficult), including an honest recognition of the gaps and uncertainties in our knowledge.

The unfortunate reality is that whether one 'believes' in climate change or not, humankind will have to deal with its consequences for generations to come.

16.7 Organism response to climate change

Organisms respond to both weather and climate, and this sets ecologists the same challenge in distinguishing the long-term response from short-term variation as it does climatologists looking to define secular climatic change (Box 16.6).

Faced with a change in climate, organisms can respond in two ways: they can move, shifting their distribution to track the climate to which they are adapted, or they can stay put. In the latter case the organism must be able to cope with the new circumstances, through tolerance or phenotypic plasticity, or they may adapt through a shift in population genetic structure, or they may simply fail to cope and may become locally or globally extinct. These options are not mutually exclusive: a species may shift its distribution and also evolve a new population genetic structure. They do, however, offer a useful framework within which to explore possible responses to recent climate change. Recently a tripartite view of organism response to climate change has become popular: organisms may change their distribution, phenology or size[54].

As with climate change itself, looking back in time gives us a valuable context. Seeing how organisms have responded to climate change in the past can provide insights into what might be happening now, or might happen in the future. Two informative examples come from the movement of climatic zones

---

Box 16.6 Weather and climate

What we call weather is the day-to-day manifestation of the state of the atmosphere at a specific location. This includes windiness, cloudiness, humidity and precipitation as well as temperature. Climate is the long-term average condition, but also includes the extent to which we can expect this to vary seasonally. The difference between weather and climate is captured in the aphorism traditionally ascribed to the American humourist Mark Twain[55]:

'Climate is what you expect, weather is what you get.'

The obvious question is how long do we need to observe the weather to define climate? The meteorological standard for climate is a minimum of 30 years, although this is only a rule of thumb based on the extent to which weather varies from day to day and year to year. A corollary of this is that to detect an ecological response to climate, as against weather, we need long runs of data.

16.7.1 Pleistocene molluscs of the eastern Pacific

The Pleistocene molluscan fauna of the eastern Pacific continental shelf is well known, with rich fossil collections from over 400 locations. This fossil record shows that the movement of climatic zones during the Pleistocene was characterised by extensive latitudinal shifts in geographic distribution, with little extinction. Of the bivalves and gastropods currently found on the continental shelf (~ 700 species), 77% are known as Pleistocene fossils, and only ~20% of the known Pleistocene fauna has become extinct[56].

The vagaries of the fossil record mean that when a given taxon disappears from a particular rock sequence it can be difficult to distinguish between extinction and a simple shift in distribution. It is only when the record is exceptionally good, as in the Californian Pleistocene, that changes in distribution can be documented. The value of this excellent record is that we can see how species move individually, not as coherent assemblages. Good as the record is, however, it does not have sufficient resolution to capture range shifts associated with the more rapid changes in climate, and these may explain why some species can appear to be 'out of place' (thermally anomalous). Slow shifts in the distribution of molluscs along the Pacific coast continue to this day[57].

The marine fauna from the same period preserved as fossils on the other side of the Pacific in Japan also show little extinction. In contrast, there was a major pulse of marine extinction in the western Atlantic around 1–2 million years ago, coincident with the intensification of glaciation in the northern hemisphere. There was widespread extinction of shallow-water marine bivalves, with tropical species being hit particularly hard, and there was a similarly extensive extinction event in Caribbean corals[58].

The Pliocene to Pleistocene marine fossil record thus provides us with mixed messages. At some times and in some places, changes in the local marine environment were accompanied by widespread extinction of shallow-water benthic organisms; in others there was little or no extinction and species simply shifted in distribution. An important factor here is clearly the existence of somewhere to move to: the Pacific coast of North America is essentially a long linear habitat, whereas the Caribbean is a scattered archipelago of islands separated by unsuitable deep-water habitat, and so populations are more isolated. Sea level will have been different then and while changes in sea level appear to have been unimportant along linear coasts, in archipelagos they can be important agents of extinction and speciation[59].

16.7.2 Vegetation changes following the Last Glacial Maximum

The retreat of the vast continental ice-sheets of northern Europe, Asia and North America since the Last Glacial Maximum has been associated with large-scale changes in the distribution of plants. These changes can be followed from fossil pollen preserved in peat bogs and lake sediments, as well as the occasional macrofossil (leaves or wood), and more recently by reconstruction of genetic history and distribution from molecular data (phylogeography). These various lines of evidence suggest that during the Late Pleistocene the dominant plant assemblage of the unglaciated northern hemisphere was steppe-tundra, a treeless sweep of grasses and low-growing herbs that has no modern day analogue[60].

It might be expected that as the ice-sheets retreated there would have been a general movement northwards of less cold-tolerant plants from their southern refugia. It turns out that recolonisation was complex and individualistic, with different species moving at different rates and in different directions. In North America, some species moved predominantly east to west, whereas others did the reverse. Some species appear to have moved without being limited by climate, whereas others (or sometimes the same species in a different location) seem to have been restricted by climate. These complex patterns characterised by the individual movement of species resulted in assemblages being more akin to random assortments of species and not the coherent movement of integrated communities. It also results in plant distributions being apparently out of equilibrium with the climate, in a manner analogous to the thermally anomalous marine molluscs of the Pacific coast[61].

One intriguing aspect of post-glacial recolonisation is that the rate at which some plants appear to have shifted their distribution greatly exceeds dispersal rates estimated from living plants. This has been called *Reid's paradox*, after the British plant ecologist Clement Reid who commented that for oak trees to have achieved their present postglacial distribution in Britain would have required over a million years without external aid, whereas in fact this was achieved in only thousands of years. Part of the resolution of Reid's paradox comes from the identification of very rare long-range dispersal events, and the recognition of previously unknown refugia much closer to the ice. The latter comes from sparse macrofossil data that can be radiocarbon dated, and increasingly from phylogeography (particularly the geographical distribution of rare alleles in populations)[62].

16.7.3 What does the historical record tell us?

The fossil record generally has a low temporal resolution: it can be very good at tracking slow, large-scale changes in distribution but it is not so good when the climate changes rapidly. Nevertheless, the record of change following the Last Glacial Maximum carries some ecological messages that are important for understanding what might be happening now.

Firstly, when circumstances allow, species will shift distribution in response to the movement of climatic zones. Secondly these shifts in distribution are individual and idiosyncratic. Thirdly where topographic constraints limit the ability to disperse and colonise new habitat, local or global extinction may result.

While the association between climate change and extinction in the fossil record is clear, the precise mechanism by which extinction occurs is not.

16.7.4 How does climate change cause extinction?

The history of life has been punctuated by periods when relatively rapid climate change has been accompanied by widespread extinction. The rate of extinction has varied through time, but statistical analysis has revealed a number of periods of enhanced extinction. These '*mass extinctions*' have attracted a great deal of attention but there is still no general consensus as to whether they are simply extremes of a continuum of extinction rates, definable only in statistical terms, or whether they are different in kind from background extinction. Although the mass extinction at the end of the Cretaceous is coincident with the impact of a large bolide, climate change remains a favoured explanation for many periods of enhanced extinction[63].

In the marine system, periods of enhanced extinction are often correlated with times of climate cooling and in many cases tropical species are particularly badly hit. A well-documented example of this comes from Steven Stanley's analysis of the Pliocene extinctions in western Atlantic molluscs mentioned above. Here, of 361 Early Pliocene species, 65% had become extinct by the Early Pleistocene. Of the 57 species that remain extant, all have ranges that extend into non-tropical areas, whereas all the Pliocene species that were exclusively tropical became extinct. This suggests a role for temperature change, and it has been argued that this has been the proximal cause of many marine extinctions. This would seem logical, and yet even at times of rapid climate change the rate of change of mean temperature is lower, by several orders of magnitude, than that experienced daily, seasonally or even annually by marine organisms living today. While a change in temperature may be the ultimate cause, extinction comes about not by all individuals dying from direct thermal challenge such as heat stress or cold, but by other factors[64].

While a few species have become extinct in the time that ecologists have been working, in general we have to turn to the fossil record to see what these factors might be. Such studies are necessarily correlative and also likely to be biased by selective preservation of larger, more numerous, widely distributed taxa. They are also predominantly marine. Nevertheless there are consistent indications that wide geographical range, high abundance and a dispersing larval stage confer resistance to extinction. These traits often co-vary, of course (for example a species with a wide range tends to have a large population), but the patterns are strong and convincing. Interestingly these patterns are characteristic of the long periods of background extinction; they disappear during mass extinctions, suggesting that at these extreme times other factors come into play.

The immediate cause of extinction must lie within the realm of population dynamics. A generally accepted model for extinction is that when a population is small, stochastic environmental or genetic processes result in the death of all individuals. These processes, however, simply provide the *coup de grâce* to populations already at low abundance; they are quite distinct from those processes which reduced population size in the first place[65].

16.7.5 Climate change and speciation

When the distribution of a plant or animal is driven south by a deepening glaciation, its range may fragment (we are taking a northern hemisphere perspective here, because this is where the effects of glacial advance and retreat are most evident in the terrestrial biota). Fragmentation could be caused by areas of suitable habitat being partitioned by mountain ranges or other inhospitable terrain (as in continental North America, or Asia) or by refugia being confined to a series of isolated peninsulas (as in Mediterranean Europe). Under such circumstances the populations isolated in different refugia may diverge genetically, and this in turn may lead to differences in morphology or behaviour. In some cases the divergence may be sufficient to prevent interbreeding when the populations come into contact once again following deglaciation (that is, speciation has occurred).

Nice examples of this are found in birds (but are also known from amphibians, insects and plants). For example in North America some species have distinct forms in the east and west which reflect isolation into separate disjunct refugia during glacial maxima. Thus the yellow-rumped warbler is now generally treated as two species, Audubon's warbler, *Setophaga auduboni*, in the west and myrtle warbler, *Setophaga coronata*, widely distributed across North America. Similar patterns are found in species of flicker (*Colaptes*), orioles (*Icterus*), grosbeaks (*Pheucticus*) and buntings (*Passerina, Junco*). In Europe there are eastern and western forms of Bonelli's warbler (*Phylloscopus*), olivaceous warbler (*Iduna*) and subalpine and orphean warblers (*Sylvia*) which are now generally treated as separate species based on differences in call, song and genetics; again these divergences reflect isolation into geographically distinct refugia during glacial maxima[66].

Range fragmentation has also been important in the sea. As we saw earlier, changes in sea level driven by vast quantities of water being locked into ice-sheets at glacial maxima and released during interglacials have been particularly important in driving speciation in shallow water marine organisms living in archipelagos. Around the Antarctic continent, periodic extensions of the continental ice-sheet over the continental shelf will have fragmented the distribution of many shallow-water organisms, and also forced some species down-slope into deeper water. These cycles of fragmentation and recombination have driven extinction and speciation in the shallow-water marine realm, just as it has on land[67].

16.8 Responses to recent climate change

Although climate change has only been evident in the surface air temperature (SAT) record for about a century, there are clear indications of a response in the distribution of many animals and plants. Many species are responding to climate change in ways that make general sense, by moving polewards and, as Victor Smetacek noted, upwards on mountains (Table 16.4).

In the terrestrial realm groups that have been particularly well studied include insects (especially butterflies), birds and plants. These all have the advantage of a long history of investigation by professional ecologists and amateur natural historians alike, which provides a valuable reference against which to measure change. Camille Parmesan and colleagues have collated a large number of studies and the observational evidence for widespread changes in distribution is now clear and unequivocal: in environments ranging from poles to tropics, the Earth's flora and fauna is responding to present-day climate change by shifting in distribution[68].

Changes in distribution are also evident in the ocean, where in places the movement of climate zones is even faster than on land. An early indication that things were changing in the sea came from a study of the Californian rocky intertidal by Jim Barry and colleagues. They re-examined a shore transect at Hopkins Marine Station that had been first surveyed in 1931–33, and found that eight of nine southern (warmer water) species had increased in abundance while five of eight northern (cooler water) species had decreased. They interpreted this as indicating a shift northwards of distributions, in exactly the same place and exactly the same way as shown by Jim Valentine and colleagues to have happened during Pleistocene climate change. Important as this early work was, a study at a single point is limited in its ability to demonstrate shifts in distribution.

There is also a long historical legacy of work in the rocky intertidal around UK, and recent survey work has revealed widespread changes in distribution. Although these changes are in the direction to be expected, interpreting changes in intertidal assemblages is complicated by their ecology. The adult animals are exposed to both air and water temperatures, but reproduction involves dispersal exclusively by sea and hence nearshore oceanography exerts a powerful influence on distribution. This may be why not all rocky shore assemblages are responding to climate change in the expected manner: along the east coast of Australia the subtropical rocky shore fauna shows no changes over 60 years despite oceanic temperatures having warmed. This is likely because local oceanographic conditions are more important here than temperature.

Changes are also evident in the open ocean. A broad-scale, long-term study of zooplankton by Grégory Beaugrand and colleagues showed that North Atlantic copepods have shown a strong biogeographic shift of 10° in latitude between 1960 and 1999. Importantly these shifts are related both to long-term warming and the variability associated with the North Atlantic Oscillation. Numerous other studies have attested to widespread and significant shifts in the distribution of northern hemisphere plankton, fish and benthos[69]. Recent distribution changes are summarised in Table 16.4.

These studies are important in telling us what is happening now; a significant challenge is to predict what the future may hold.

Table 16.4 Some examples of changes in distribution associated with recent climate change[70].

Biota	Response
On land	
Tree line	Moving polewards or rising in altitude (Europe, Alaska, New Zealand)
Shrub vegetation	Expanding range into tundra (Alaska, Siberia)
Alpine plants	Altitude rise of 1–4 m per decade (Europe)
Bird distribution	Average of 19 km northwards
Invertebrates	274 species from 11 groups of UK terrestrial invertebrates, 10 groups had moved significantly north (34–104 km) and 9 had increased altitude (13–62 m).
Reptiles and amphibians	3 UK species had moved significantly southwards (mean 83 km)
Butterflies	35 species of non-migratory UK butterflies, 22 have ranges that have shifted north by 35–240 km during twentieth century and 1 has shifted south
In the sea	
Arctic biota	Range shifts and changes in abundance in various marine taxa, with northward extension of benthos and zooplankton and fishes into the Arctic basin
Fish	Range of 15 of 36 North Sea species has moved significantly northwards between 1977 and 2001
Rocky shore invertebrates	9 UK species have extended range northwards, with 3 retreating and 3 showing no change
Rocky shore invertebrates	Over 60-year period at one California site, abundance of 8 of 9 southern species increased and 5 of 8 northern species decreased
Zooplankton	14 species of North Atlantic zooplankton have moved 1100 km northwards between 1958 and 1999

16.8.1 Predicting species' future distributions

In the late 1950s, the palaeoecologist Russell Coope was working with a late Quaternary assemblage of fossil beetles from Chelford in UK. Many of the fossils were of species still living and so he was able to reconstruct the likely climate at the time the beetles were alive by using meteorological data from stations within the modern range of the species. In 1982 Coope joined with Timothy Atkinson and Keith Briffa of the Climate Research Unit of the University of East Anglia to refine this approach. Principal components analysis established that the climate within the range of every species in the fossil assemblage could be captured effectively with just two measures: the mean temperature of the warmest month and the difference between the mean temperature of the warmest and coldest months (a measure of seasonality). The climate indicated by the fossil assemblage was that defined by the overlap of all species in the assemblage (Figure 16.13).

This was termed the *mutual climatic range* technique and comparative tests between different living insect assemblages indicated that it worked well. The method proved important in reconstructing past climates, and was used to demonstrate the dramatic change in local climate during the switch from glacial to interglacial conditions, many years before this was confirmed by ice-core data. This seminal work was also the foundation of what is today called *climate envelope modelling*, which has become a cornerstone of predicting how species distributions may shift in future in response to recent climate change[71].

Like palaeoclimate reconstruction, climate envelope modelling relies on establishing a statistical relationship between distribution and climate. However instead of estimating climate from organism distribution, it is predicting changes in distribution from the predicted movement of climatic zones. The first stage is to construct the climate envelope model. Typically a minimal climate envelope is fitted to the selected climate variables for the present range of the species. The models vary in the climate variables included, and the statistical technique used to fit the envelope. These include generalised additive models, locally weighted regressions and classification trees; the latter can capture non-additive effects and complex interactions, but equally can produce overly complex models.

These models set challenges for calibration and validation. A given species has only one distribution, but a model cannot be tested with the data that were used to construct it. A valid test requires truly independent data. A commonly used approach is to split the distribution data randomly into calibration and validation data sets before fitting the model. An alternative approach is to compare fits with a suitable null model[72].

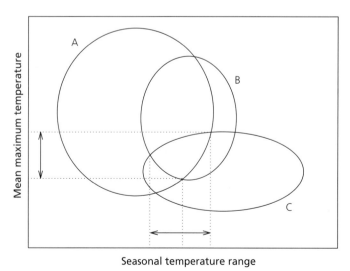

Figure 16.13 Diagrammatic illustration of the mutual climatic range method for palaeoclimatic reconstruction. Here the modern climate zones of three living species (A, B, C) that are also found as fossils are defined by the mean maximum temperature and the seasonal temperature range. The reconstructed palaeoclimate for the fossil location is defined by the area of mutual overlap, with the limits shown as dotted lines. The arrows thus mark the range of the most probable climate conditions for the assemblage.

A key assumption of climate envelope modelling is that it is climate that actually sets the distribution. In other words the variation in performance across the biogeographic range should reflect the physiological climate niche. The concept of the niche has a long history in ecology but the most influential modern approach stems from Evelyn Hutchinson, who defined the *fundamental niche* as the total range of conditions under which the individual or population can live and reproduce itself. Furthermore he distinguished this fundamental niche from the *realised niche*, which is the actual set of conditions under which the organism is found in nature because of constraints from competition, predation and so on. Hutchinson envisaged the niche as being defined by a range of environmental conditions, and hence it could be represented as a hypervolume in multidimensional space. While this can be handled mathematically, it difficult to visualise or represent in two dimensions[73].

If we simplify the niche to a single environmental variable, for example temperature, then the fundamental and realised niches can be illustrated conceptually with the thermal performance curve discussed previously (Chapter 7). In this conceptual model (Figure 16.14) the fundamental niche is defined by the thermal performance curve, with the boundaries shown by dashed lines set (arbitrarily) at 1% of maximum performance. In the white areas it is assumed that reduced performance renders the organism uncompetitive with other species that operate closer to their performance maximum in these environments. The realised thermal niche is then the central portion (grey) where performance is sufficiently close to maximum for the species to be competitive and hence persist. The boundary here is (again arbitrarily) set at 50% maximum performance.

This simple conceptual model would imply that reproductive success, and perhaps also abundance, would be greater in the centre of the distribution, and lower towards the edge. Documenting such patterns in nature takes considerable effort, but data do exist for some species. Some organisms exhibit the pattern expected from the simple niche model, but many do not. Plants for example, often remain abundant right up to the very edge of their range (think of the sharpness of the tree line). The patterns observed will also be influenced by factors such as local adaptation, the variation in relevant environmental factors across space and the degree of coupling between environmental factors and abundance.

The asymmetric shape that characterises many thermal performance curves implies that the variation in performance is steeper at high temperatures than at low. In turn, this might suggest that the factors setting range boundaries may differ from place to place. There is still much debate over precisely what factors set range boundaries in organisms. Many correlational studies suggest

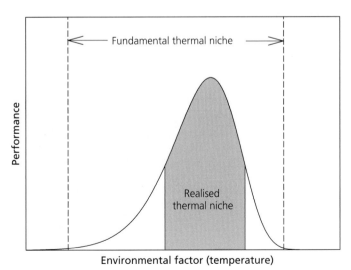

Figure 16.14 The thermal performance curve revisited. In this conceptual model the fundamental niche is defined by the thermal performance curve, with the boundaries shown by dashed lines set (arbitrarily) at 1% of maximum performance. In the white areas it is assumed that reduced performance renders the organism uncompetitive with other species that operate closer to their performance maximum in these environments. The realised thermal niche is then the central portion (grey) where performance is sufficiently close to maximum for the species to be competitive and hence persist. The boundary here is (again arbitrarily) set at 50% maximum performance.

a key role for abiotic factors in setting range limits, of which temperature is a common one. On the other hand, there are is also abundant evidence that species interactions are important. In plants a number of studies have shown that the northern (poleward) boundaries are set by abiotic factors, whereas southern limits are typically set by species interactions, as transplant experiments show that fitness remains high even outside the current range. This recalls speculation by Darwin that one might expect more northerly range limits to be set by physical constraints and more southerly limits by competitive ability. As yet we have no consensus as to what factors set range boundaries, but it is the range boundaries that mark the moving edge which responds to climate. Chris Thomas studied the recent change in the range boundaries of 329 species of terrestrial organism (vertebrates and invertebrates) in UK and found that 84% had moved polewards as the climate has warmed, suggesting that climate is a major factor in determining range boundaries[74].

The bulk of a given organism's distribution is rarely uniform. Areas of the population where performance and reproductive success are high (source populations) may sustain others where performance is sub-optimal (sink populations). There are also indications that the genetic population structure and behaviour (for example tendency to dispersal) can be different between the centre and margin of a population. In species whose distributions are discontinuous or fragmented because of a patchy distribution of suitable habitat, some sub-populations may suffer local extinction from stochastic events, and can only re-establish by immigration. The population as a whole is then made up of a network of smaller populations, the overall behaviour of which can be described by metapopulation dynamics, an approach which has been particularly successful with fragmented populations of butterflies and some plants[75]. The potential response of such a population to a change in climate is shown schematically in Figure 16.15.

A change in the environment means that the climate niche moves (in the northern hemisphere under climate warming this would be a shift northwards). For a species whose range boundaries are set by climatic factors, this dictates the need to colonise new, more northerly habitat, by dispersal and probably also extinction of populations further south. Difficulties arise, however, where dispersal to new habitat involves crossing inhospitable terrain. Clearly the impact of such a barrier will depend on factors such as its size, the ability of the organism to disperse (and also its dispersal mode) as well as the rate and variability of climate change itself (for

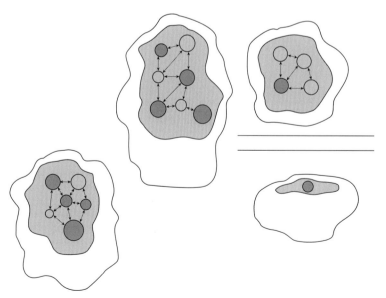

Figure 16.15 A simple conceptual model of distribution shifts driven by global warming. A (left): the species has a realised niche (light grey) which is smaller than the fundamental niche (white), and which exists as a series of small populations forming a metapopulation. Here suitable patches that are occupied are shown in dark grey and suitable habitat which is not occupied as light grey. B (middle): the species has responded to a shift in the environment by a smooth change in distribution. C (right): the range shift is constrained by a physical barrier such as a large river, a mountain range or extensive man-made habitat, and movement to the new area requires a jump dispersal.

example, a variable climate may restrict the ability to establish into a new habitat). There are also more subtle population dynamic factors that might come into play: a recent study of seven species of migrant warblers in UK showed that although the species have shifted north in recent years, productivity of the more northerly populations remains low[76].

It is beyond the scope of this book to explore these various aspects of range shift in detail. The important principle that emerges from recent work is that while the nature of the thermal performance curve is important in terms of the fitness of the population within a particular environment, the response to climate change is also influenced by a wealth of other, ecological, factors. A detailed, physiologically based model of the type we explored in Chapter 4 can be extremely valuable in exploring how organisms use their thermal environment, but these tend to be time-consuming to construct and are overly detailed for the more general task of predicting the response to climate change. They are important, however, in validating the correlational approaches used in constructing climate envelope models. While climate envelope models are crude, with many limitations, they are our best approach if we wish to predict in very general terms how a particular species may respond to climate change. But even the best of such models are essentially only correlations, and we cannot be certain that other factors affecting that correlation will be the same in the new range. Furthermore the process of getting to the new place will be affected by an array of processes that the models do not capture. But at the moment, in a world where answers are needed quickly, they remain an important tool[77].

A warming climate does not only influence where species can live, it also affects when they do things. This is the subject of phenology.

16.8.2 Phenology

Phenology is the study of the timing of recurring seasonal biological events[78]. Humans have a long history of noting the annual appearance of flowers or leaves in spring, the arrival and departure of migrant birds or the first frost-damaged leaves in autumn. This must have been partly out of curiosity about the world, but it was also of immense practical importance: ancient Greek farmers used the timing of leaf fall to decide when to plant their winter crops.

The appearance of cherry and other fruit blossom in spring is woven deeply into the culture of China and Japan, and in Asia records of the time of flowering extend back to the sixth century. The earliest phenological records known so far from the UK date from 1684, but the most important early work was done by Robert Marsham, who recorded up to 27 'indicators of spring' every year from 1736 until his death in 1798. These indicators included flowering dates of plants, leafing dates of trees, the arrival or first song of migrant birds and signs of the breeding activity of birds, frogs and toads. These findings were reported to the Royal Society in 1789, the same year as the publication of Gilbert White's *'Natural History of Selborne'*. Recording was continued by successive members of the family until 1958, and constitutes the longest phenological record in UK.

There are also extensive phenological records from Germany and Norway. The record is far less extensive in the USA, although here flowering dates for lilac and honeysuckle extend back to the early twentieth century. There is also a huge archive of bird migration data which started in 1881, emanating from what is now the North American Bird Phenology Program[79]. The recognition of climate change and its possible effects on the timing of ecological events has led to a proliferation of studies worldwide and phenology is now a leading research topic, for which observations by amateur natural historians, gardeners and citizen scientists are central.

Every animal and plant develops through a sequence of life-history stages. In seasonal environments the transition between these stages must be timed to match suitable environmental conditions: a seed must germinate, a butterfly emerge or a bird hatch at the appropriate time, or it will die. For example any seedling that emerges early runs the risk of frost damage whereas one that emerges late may have insufficient time to mature and set seed before winter arrives. In a stable climate the timing of germination and emergence is thus under intense stabilising selection, with the spread of emergence times (the phenotypic variance) reflecting the extent to which weather varies from year to year. Similarly in the sea, production of eggs and larvae needs to match the availability of food for the first feeding stages.

There is now widespread evidence of a shift in the timing of events such as germination, emergence, leafing, flowering and leaf fall in response to the changing climate. For example the alpine mustard *Boechera stricta* (Drummond's rockcress) from the US Rocky Mountains emerges soon after snowmelt, and a 48-year field study has shown that the timing of flowering is correlated with the date of snowmelt (Figure 16.16). Since snowmelt is now occurring earlier, and air temperatures are warmer, flowering is also earlier.

One of the powerful lessons of post-glacial plant recolonisation, however, is that species react individualistically. While studies of individual species are important at teasing out process, demonstration of a general response to climate change comes from analysis of whole floras (or at least a significant proportion of a flora).

An example of this comes from the UK, where analysis of a 58-year record of flowering in 11 plant species showed that in all cases the date of flowering was related significantly to central England temperature. This work was then extended to derive an assemblage-level index of first flowering dates for 405 plant species in UK, with data extending back 250 years. This revealed a strong correlation with the February to April mean temperature in central England (Figure 16.17), with flowering on average 5 days earlier for every 1 K increase in mean temperature. The recent increase in central England temperature (Figure 16.12) means that the index for the most recent 25 years is 2.2 to 12.7 days earlier than any other consecutive 25-year period since 1760. Plant phenological records are far less extensive in North America, but a spring index compiled from data for lilac and honeysuckle flowering does extend back to 1900 and reveals recent advances of 4–8 days, as well as previous delays of similar extent[81].

The tight coupling of plant phenology to temperature means that plants will respond not only to the long-term change in mean SAT, but also to shorter-term cycles and variability. Of particular importance in the northern hemisphere is the North Atlantic Oscillation (NAO), which is a measure of the seasonal mean sea-level air pressure difference between the Iceland low and the Azores high, variability in which exerts a widespread influence over European flora and fauna. The effect of such regional climatic factors needs to be teased out before we can demonstrate any underlying response to long-term climate change[82].

Among animals, earlier emergence in recent years has been recorded for Lepidoptera (butterflies and moths) and Odonata (dragonflies and damselflies), and earlier breeding in amphibians and birds. For example, UK Odonata have tended to emerge earlier with the start of the main flight period advancing by an average of 1.5 days per decade between 1960

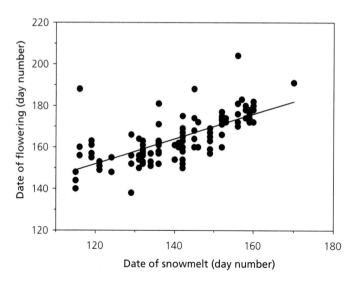

Figure 16.16 Date of flowering of the mustard *Boechera stricta* in relation to the date of snowmelt in the Rocky Mountains, USA. The relationship is significant ($p < 0.01$) and indicates that earlier snowmelt is associated with earlier flowering[80].

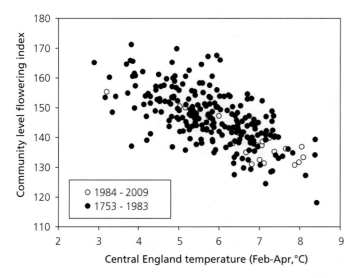

Figure 16.17 Relationship between the flowering index compiled from 405 species of plant and the mean February to April temperature in central England. Data for the most recent years (1984–2009) are identified by open symbols. The relationship is significant ($p < 0.001$), indicating a decrease in the index (earlier flowering) of 5 days for a 1 K increase in mean temperature[83].

and 2004, or 3.0 days per decade when phylogeny is included in the statistical model. Also in the UK, the first emergence of most butterflies has advanced over the past two decades, as has the date of peak abundance, and in multivoltine species (those with more than one generation in a season) also the overall flight period. Similarly, in the central valley of California, the mean date of first appearance of 23 butterflies has advanced by an average of 24 days over 30 years. Migrating butterflies are also responding, but here the pattern is driven more by the NAO. Analysis of a 113-year record for 29 species of migrant Lepidoptera in the UK showed that when the NAO causes higher temperatures in France, more migrant butterflies and moths arrive in the UK[84].

Not all species react similarly, and the phenological response of a species to higher temperature is influenced strongly by its ecology. For example, analysis of a data set covering >150 years for 566 European butterflies and moths showed that the strength of the change in phenology was related strongly to the seasonality and palatability of the larval food plant[85].

The varying phenological responses to warming by different groups of plants and insects raises the possibility that climate change may disrupt plant–herbivore or plant–pollinator relationships. From 1884 to 1916 Charles Robertson recorded all the interactions he observed between 429 species of plant and 1420 species of their pollinator in western Illinois. These data have been used to produce a highly resolved network of plant–pollinator interactions, and modelled simulation of differing phenological responses by plants to a warming climate indicated that the ancestral flight periods of the pollinators would result in a significantly reduced match and marked decrease in diet breadth and pollinator success[86].

A clear example of a phenological shift in the breeding of birds comes from the pied flycatcher, *Ficedula hypoleuca*. This is one of a complex of closely related black-and-white flycatchers characteristic of broadleaved woodlands across Europe, western Asia and North Africa. At East Dartmoor National Nature Reserve in the UK the mean date of the first egg advanced by 11 days between 1967 and 2016 (Figure 16.18). Across Europe, the timing of pied flycatcher reproduction has also advanced, with breeding earlier in areas where warming has been greatest. Between 1971 and 1995, 51 of 65 species of UK bird advanced the start of the breeding season, by an average of 9 days in those species showing a phenological response. This is a general result for European birds and there is now compelling evidence for earlier breeding in a wide range of birds[87].

The pied flycatcher is a migrant, arriving in spring from wintering grounds in sub-Saharan Africa. In order to breed earlier, it arrives earlier and many migrant birds have started arriving earlier during the last few decades. The development of

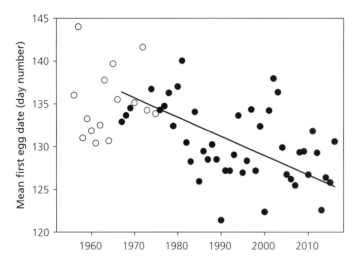

Figure 16.18 Mean date of first egg in a population of pied flycatcher, *Ficedula hypoleuca*, over a 50-year period, at East Dartmoor National Nature Reserve, UK. The open symbols indicate years when the first date was estimated from fewer than 10 pairs. The line fitted to data from years with >10 pairs is highly significant ($p < 0.001$) and indicates that breeding has advanced by ~10 days in the past half-century[88].

bird observatories with extensive ringing (banding) programmes and the advent of birding as a popular recreational activity, all in the first half of the twentieth century, resulted in a rapid increase in our knowledge of bird migration. While year-to-year variation in the timing of migration had long been known, a general advancement in the arrival time of some migrant birds was evident in the 1980s. An early analysis looked at the arrival dates of summer migrants into Leicestershire in England and established differences between species: some were arriving earlier but some were delaying migration. A benefit of the long history of bird recording is that data series are often long, >50 years in many cases, and this enables us to see how migration dates have shifted previously in history. In the UK, fifteen species arrived earlier in the 1940s; some were later in the cooler springs of the 1960s and 1970s but are now advancing again. A more recent analysis of arrival times of 20 species of trans-Saharan migrant in Oxfordshire between 1971 and 2000 showed that, on average, arrival time had advanced by ~8 days[89].

While first arrival dates (FADs) are a convenient measure, and eagerly recorded by amateur naturalists, they may not be representative of overall population behaviour. In many birds the first individuals to arrive are often males returning to claim territories, and the bulk of migration occurs significantly later. A study in Massachusetts showed that in 32 species of migrant North American passerines (songbirds) between 1970 and 2002 FADs had advanced by 0.2 days per decade, whereas bulk passage had advanced more strongly, by 0.78 days per decade[90].

Although a general pattern of early arrival has been established for many migrant songbirds in both Europe and North America over the past few decades, the magnitude of the change is related to migration distance. In Europe long-distance migrants (those arriving from wintering grounds in sub-Saharan Africa) tend to have advanced their arrival more than some short-distance migrants (which typically winter in southern Europe or North Africa), although the extent of the difference varies from study to study, presumably reflecting spatial differences across Europe. Differences are also evident in the arrival times of migrants into North America, though here short-distance migrants (those wintering within North America) have tended to advance more than long-distance migrants wintering in South America[91].

This difference between long- and short-distance migrants is likely related to conditions on the wintering grounds, which affects when the birds decide to move north, but also en route, which influences how long they spend at intermediate stopover locations. Both of these are influenced by regional-scale climatic variability, and several studies have revealed strong correlations between the timing of migration and the NAO. Where conditions on the wintering grounds encourage a later departure but those on the breeding grounds require an earlier start, the possibility exists for a disastrous phenological mismatch. In the

Nearctic (North America) the extent of this mismatch has been linked to population decline, whereas in the Palaearctic (Eurasia) such declines appear to be associated more with migration distance[92].

In the northern hemisphere, bird migration is predominantly north–south, reflecting the glacial and climatic history of the region. In Australia, while shorebirds are typically long-distance migrants from breeding grounds in northern Asia, passerines move mostly within the continent. In both south-east and south-west Australia some migrants are arriving significantly earlier, but some later, and there are again differences depending on migration distance. Furthermore, in this predominantly arid continent shifts in the phenology of passerines are as much a response to shifting patterns of precipitation as they are temperature[93].

The change in the timing of migration and breeding in birds raises the possibility, as discussed for insect pollinators above, of a phenological mismatch between predator and prey. This is particularly so when a bird relies on a food resource that is seasonal and available for only a limited period of time. An example here would be passerines that are reliant on lepidopteran larvae (caterpillars) to feed their chicks. In pied flycatchers the potential for mismatch with caterpillar abundance is greater in oak than in other woodland habitats, and is critical in that the rate of nestling growth is tied to the fraction of caterpillars in the diet. Mismatch with caterpillar availability has also been shown in species of tit (Paridae) breeding in broad-leaved woodland in Europe[94].

The need for a predator to couple its life history to prey availability is obvious, and the timing of phenological events must have been adjusted continuously in response to past climatic variability, such as that driven by the NAO. Analysis of long-term data sets has suggested that in at least some species the timing of migration has responded to this variability by shifting earlier or later in concert with climate. Birds use a variety of environmental clues to adjust the timing of reproduction, although there must also be sufficient food available for the female to make eggs. Photoperiod and temperature are important cues for stimulating the onset of reproduction, but fine-scale cues are also important. For example in great tits, *Parus major*, caterpillars on oak trees are an important food resource for breeding and birds adjust the timing of egg-laying according to the leafing of oaks in the immediate vicinity of their territories, thereby maximising the chances of a rich and reliable food supply for their chicks[95].

16.8.3 Decrease in size

While ecologists have directed considerable attention at how organisms have responded to recent climate change by changes in distribution and phenology, there is increasing evidence for a third response: a change in size.

Nice examples of this are the high Arctic butterflies *Boloria chariclea* (Arctic fritillary) and *Colias hecla* (northern clouded yellow) which have been monitored at Zackenberg in Greenland since 1996. Over this period mean summer temperatures have risen by ~2 K, snowmelt has advanced by ~15 days and mean wing length in both sexes of both species has decreased significantly. The suggestion is that warmer temperatures elevate metabolic costs (specifically maintenance metabolism) and that this increase can be offset by a smaller body size, which reduces total maintenance costs. This will in turn affect dispersal capability and fecundity, as both are functions of body size.

An explanation based on the temperature sensitivity of metabolic costs might apply generally to ectotherms. Reductions in size associated with recent climate change have, however, also been reported widely in mammals and birds. The widely observed relationship between temperature and body size (Chapter 13) would predict a decrease in body size with a warming climate, but at present we do not understand the mechanism[96].

16.9 Final comments

Observation shows clearly that many organisms have responded to recent climate change by changes in one or more of their distribution, phenology and size. It is also clear that not all organisms, even those living alongside one another, react in the same way: the nature of the response of any given species is mediated strongly through its ecology. Phenotypic flexibility is important in allowing species to adjust continually to their environment, but the outcomes are influenced by trade-offs. Species will differ in their ability to adjust and hence we can expect continual change in the composition of ecological assemblages caused by shifts in the competitive balance. All this is normal, and will have been a feature of assemblages everywhere throughout history.

This might suggest that we should not be concerned with the changes we see happening today, as changes in distribution, phenology, size and competitive ability will have been a continual feature of the natural world. Sadly this is not so, and the reason is that the pace of recent climate change is faster than at any time since at least the past 800 000 years, and it may be outpacing the ability of phenotypic flexibility alone to cope. It is distinctly possible that if climate continues to change at its current rate (and given the time-lags inherent in the climate system there is every indication that it will do so for decades to come), then genetic adaptation will be necessary.

Although evolutionary response to climate change can be explored with models, and these are valuable in defining possibilities, there are few studies where the genetic component of an observed response to climate change has been examined. Quantitative genetic studies have shown that the shift in the flowering time of *Boechera stricta* (Figure 16.16) has been achieved partly by phenotypic flexibility but that genetic adaptation must also have been involved. We have a plethora of examples of changes in distribution, phenology and size associated with recent climate change. What we lack in most cases is the mechanistic understanding to see precisely how temperature has its effect[97]. In a rapidly warming world this understanding is needed urgently.

16.10 Summary

1. The greenhouse effect is a simple consequence of an atmosphere containing gases that are transparent to visible light but which absorb infrared radiation. These include H_2O, CO_2, N_2O, O_3 and CH_4, collectively known as radiatively active gases (greenhouse gases).

2. The temperature of the lower troposphere (the temperature we experience as humans) is set by the radiation balance at the top of the atmosphere, and is currently determined predominantly by the CO_2 concentration.

3. Man has been adding radiatively active gases to the atmosphere since the Industrial Revolution, and the increase in concentration of these gases has led to an increase in the energy in the lower atmosphere, and thus a rise in its temperature. The bulk of the extra energy (~90%) has entered the ocean, which has also warmed significantly over the past century.

4. The rate and extent of warming varies across the planet, depending on local circumstances. In some places regional warming is not yet distinguishable from natural variability, in others it is clear and distinct.

5. Palaeoecological studies have shown that changes in distribution have been a frequent response to climate change, though this requires somewhere for the organisms to move to. Where no suitable habitat exists, extinction may result. Species move individualistically, resulting in the formation of new assemblages.

6. A wide range of organisms have been recorded as shifting their distribution in response to recent climate change. These include plants, insects and birds on land, and benthic and planktonic invertebrates, fish, seabirds and marine mammals in the ocean. In the northern hemisphere, these movements are predominantly south to north. There are also altitudinal shifts in mountain regions, and in the arid areas of the southern hemisphere movements are as much a response to shifting patterns of precipitation as to temperature.

7. Many organisms have also shifted the timing of life-cycle events (phenology), with migration, breeding in animals, and germination, emergence, leafing and flowering in plants all occurring earlier in some (but not all) species. There are also changes in size, with some species becoming smaller as the climate warms.

8. While fluctuations have been a continual feature of Earth's climate, to which organisms have responded, present climate change appears to be unprecedented in its rate. Phenotypic plasticity

may not be sufficient, and where genetic change (evolution) cannot keep pace, extinction may result.

Notes

1. This metaphor comes from a lecture delivered by John Tyndall to the Royal Institution in London on 23 January 1863, entitled 'On radiation through the Earth's atmosphere', and is quoted by Fleming (1998).
2. Smetacek (1974).
3. The name derives from the Greek τροπος (tropos), meaning turning, because in this region of the atmosphere convective processes dominate radiative factors.
4. The US Standard Atmosphere 1976. This assumes a dry atmosphere of defined composition and standardised conditions at sea level; it represents an average state and composition of the atmosphere rather than representing any specific place or time.
5. Data from Hartmann (1994), with carbon dioxide data for May 2016 from Mauna Loa observatory, Hawai'i and methane and nitrous oxide data for February 2016 from NOAA.
6. There is a small contribution from geothermal energy derived from radioactive decay (principally of ^{40}K, ^{232}Th, ^{235}U and ^{238}U) in the mantle.
7. The strength of solar energy input is captured by the solar constant, S, which is the solar irradiance, integrated across all wavelengths, per unit area incident on a plane perpendicular to the rays, at a distance of one astronomical unit (AU) from the Sun (roughly the mean distance from the Sun to the Earth: 1.496×10^8 km). Recent satellite estimates indicate a value for S of 1360.8 ± 1.3 W m^{-2} at solar minimum and approximately 1.6 W m^{-2} (0.12%) greater at solar maximum (Kopp & Lean 2011). There are also small differences in the spectral distribution of energy between solar maxima and minima (Haigh et al. 2010).
8. These numbers come from the careful assessments of the Earth's radiation budget by Kiehl & Trenberth (1997) and Trenberth et al. (2009).
9. The total spectrum of light detected by a spacecraft looking at Earth is actually complex, with a peak in the visible spectrum from reflected sunlight, and another in the infrared from the long-wave radiation emitted from the upper atmosphere. In the past century we have also added microwave and radiowave frequencies to the spectrum.
10. These estimates are from Smith (2008).
11. The European Centre for Medium-Range Weather Forecasts reanalysis project estimated the global mean surface temperature for 2012 to be 14.53 °C (287.7 K).
12. Rayleigh scattering is the elastic scattering of light by particles much smaller than the wavelength of the light (in the atmosphere these are the gas molecules). The scattering is wavelength-dependent with scattering inversely proportional to the fourth power of the wavelength. Blue light (450 nm) is scattered roughly four times as much as red light (600 nm), giving rise to the diffuse radiation that makes the sky appear blue. At sunset, the much greater path-length for sunlight means that blue wavelengths are effectively completely scattered, leaving the sun's disc appearing orange or red. Where other particles are present, such as volcanic dust after a large eruption, then the increased scattering from these particles can make the sky appear red rather than blue. The 1783 eruption of Laki in Iceland caused a notable change in the climate of England, with many deaths, and the 'blood-coloured Sun' was recorded by Gilbert White (Letter 65 to the Honorable Daines Barrington, White 1789). The eruption of Tambura in 1815, probably the largest volcanic eruption in recorded history, may well explain the vivid red skies painted by the English artist J. M. W. Turner at that time. Robinson (2005) and Zerefos et al. (2007) discuss the representation of weather and climate in European art.
13. The spectrum plotted is the American Society for Testing and Materials (ASTM) AM1.5 Reference Spectrum.
14. The absorption line spectra for CO_2 and water are from the HITRAN model (Smithsonian Center for Astrophysics), calculated for a mean latitude in summer and an atmosphere of 288 K.
15. For recent useful discussions of this important aspect of the greenhouse effect see Lacis et al. (2010) and Schmidt et al. (2010).
16. The key historical papers are Fourier (1824, 1827), Tyndall (1861a, b) and Arrhenius (1896). The key later research was by the US military, who were interested in the optical properties of the high atmosphere in relation to the ability to detect infrared radiation from aircraft and missiles. This work drove major advances in our theoretical understanding of absorption of radiation, and also prompted laboratory measurements of much greater sensitivity than were possible in Tyndall's day. Together these revolutionised our understanding of how the atmosphere absorbs radiation.
17. Diagram modified from Trenberth et al. (2009)
18. These were initially referred to as global circulation models, but are now usually known as global climate models.

19. Johannes Kepler (1571–1630) formulated the laws of planetary motion, showing that the orbits of the planets were elliptical. Urbain Jean Joseph Le Verrier (1811–1877) was most famous for his prediction of the existence and position of the then unknown planet Neptune. The discovery of the precession of the equinoxes is traditionally ascribed to the Greek astronomer Hipparchos of Niceae (ca 190–ca 120 BC), though it is likely that it was known to the ancient Babylonian astronomers, of whose work Hipparchos was aware. James Croll (1821–1890) was an early pioneer at attempting to determine the effect of orbital variation on climate, but the problem was solved by Milutin Milanković (often spelt Milankovitch, 1879–1958). The critical discovery that confirmed Milanković's theory was that of Hays et al. (1976), and the importance of Milanković cyclicity to the ice ages was explored by Imbrie et al. (1992, 1993). Imbrie & Imbrie (1979) provide an early popular account of the main ideas, and Burroughs (2005) gives an excellent modern summary.
20. The indications from fossil beetles of rapid shifts in climatic zones at the end of the last glacial maximum were reported by Coope & Brophy (1972), many decades before these were confirmed by the analyses of ice cores from Greenland. See also Coope et al. (1998) and Elias (1994). The ice core evidence for rapid change is reported by Steffensen et al. (2008).
21. See Elkibbi & Rial (2001). Naish et al. (2001) report orbital cyclicity in the size of the Antarctic ice-sheet at the Oligocene–Miocene boundary, and Naish et al. (2009) in the Pliocene. Both studies are based on marine sediment cores.
22. In the northern hemisphere, two recent cooler periods are named after a characteristic plant of the Arctic tundra, *Dryas octopetala* (Mountain Avens), which moves south as the temperature drops and the ice advances. I took the vivid metaphor (because I could not improve on it) and the details of D–O and Heinrich events from Burroughs (2005).
23. It is now clear that D–O events are strongly coupled to changes in the global thermohaline circulation of oceanic water, and particularly the Atlantic meridional overturning circulation. The movement of energy by ocean currents also links these events with warming events in the Antarctic climate record (EPICA Community Members: Barbante et al. 2006).
24. Johnsen et al. (1997). Data downloaded from the NOAA National Climatic Data Center on 4 June 2016.
25. The finds at Happisburgh are reported by Parfitt et al. (2010) and Ashton et al. (2014). See also Parfitt et al. (2005) for slightly younger artefacts from interglacial sediments at a nearby location. These were not modern man, *Homo sapiens*, but an earlier hominid, probably *Homo heidelbergensis*.
26. The important change was the Industrial Revolution, which started around 1760, and which was fuelled by the burning of coal. Since then man has released increasing quantities of radiatively active gases from the burning of fossil fuels, cement production and changes in land use.
27. Charles Keeling died in 2005, and supervision of the monitoring project is now in the hands of his son, Ralph Keeling, himself a Professor of Oceanography at Scripps Institution. The classic early papers are Keeling et al. (1976a, b).
28. Data courtesy of Dr. Pieter Tans, NOAA/ESRL and Dr. Ralph Keeling, Scripps Institution of Oceanography.
29. Roughly two-thirds of annual global primary productivity is by plants on the land, and one-third in the oceans. This is why the land signal dominates the Keeling curve.
30. The Vostok ice core provided data back to 420 000 years BP (Petit et al. 1999). This record was extended to 740 000 years, covering the eight most recent glacial cycles, with data from the ice core drilled at Dome C, Antarctica by the European Project for Ice Coring in Antarctica (EPICA Community Members: Augustin et al. 2004). See also Jouzel et al. (2007).
31. Data are the annual mean anomalies (relative to the 1959–1980 mean) for the Land–Ocean Temperature Index computed by the NASA Goddard Institute for Space Studies. Data courtesy NASA GISS. Very similar trends are shown by the data series compiled independently by NOAA, the UK Meteorological Office and the Climatic Research Unit of the University of East Anglia.
32. There is also the subtle issue that the lower atmosphere is not in thermodynamic equilibrium, and hence in formal terms a temperature has no meaning (see Chapter 3). In practice, when averaging over large temporal and spatial scales the system may be regarded as being in quasi-equilibrium and so the LOTI becomes a useful comparative measure.
33. The additional warming caused by water vapour is sometimes described as 'amplification' (because the warming is greater than would be the case in a dry atmosphere). This feedback is better regarded as simply the extra capacity to retain energy because of the additional absorption of infrared radiation by the water vapour.
34. Models of solar evolution suggest an increase of about 3–4% in total solar luminosity over the Phanerozoic (the last 600 Ma) (Gough 1981). The data for solar cycle length come from the last 23 cycles. A much-publicised study suggested a correlation between

solar cycle length and climate (Friis-Christensen & Lassen 1991). Subsequent work, correcting some errors in the original data, showed that the recent climate warming cannot be explained by changes in solar cycle length (Damon & Laut 2004).

35. Data from Bindoff et al. (2007). More recent estimates put the oceanic contribution as > 90%.
36. Palmer et al. (2011) discuss the importance of the deep ocean to the Earth's energy budget, and Purkey & Johnson (2010) estimate the warming of the deep ocean from the sparse deep-ocean CTD data. In contrast to these direct measurements, Llovel et al. (2014) were unable to detect any signal of deep ocean warming (below 2000 m) from a modelling study of surface heat and overall sea level rise.
37. Data from NOAA, downloaded from European Environment Agency, 9 June 2016.
38. Lyman et al. (2010), Levitus et al. (2012), Balmaseda et al. (2013). Durack et al. (2014) suggest that these estimates may be too low, because of poor data coverage in the southern hemisphere.
39. Wittmann & Pörtner (2013) review the impacts of acidification on a range of marine organisms. Other useful reviews are Doney et al. (2009), Wicks & Roberts (2012). The identification of the problem of ocean acidification is usually ascribed to two publications: Feely et al. (2004) and Sabine et al. (2004).
40. Cross et al. (2015, 2016) report studies of brachiopods, Suckling et al. (2015) and Rodríguez-Romero et al. (2016) present long-term studies of the effects of acidification on marine invertebrates and Cyronak et al. (2016) discuss the importance of the ocean carbonate cycle on calcification.
41. The geological record of acidification and its impacts on the marine biota are discussed by Kiessling & Simpson (2011) and Hönisch et al. (2012).
42. Faraday Station is at 65° 15′ S, 64° 16′ W on the west coast of the Antarctic Peninsula. From 1945 to 1977 it was known as Argentine Islands, and in 1996 operation was transferred to Ukraine, who continue observations to this day. Data courtesy British Antarctic Survey.
43. The recent cooling is reported by Turner et al. (2016).
44. The late twentieth century warming of the western Antarctic Peninsula is described by Vaughan et al. (2003) and Turner et al. (2005), and the factors driving this discussed by Marshall et al. (2006), Marshall (2007) and J. Turner et al. (2012). Meredith & King (2005) report the warming of the surface waters along the western Antarctic Peninsula. The consequences for glaciers, predominantly along the west coast of the Antarctic Peninsula, are described by Cook et al. (2005, 2016). Recent work on the link between oceanography and ice-shelf dynamics is reported by Jenkins et al. (2010), Jacobs et al. (2011), Stanton et al. (2013) and Paolo et al. (2015).
45. See Thomas et al. (2013) and Steig et al. (2013). The collapse of major ice-shelves along the east coast of the Antarctic Peninsula is described by Vaughan & Doake (1996).
46. The ice core data are reported by Mulvaney et al. (2012) and Abram et al. (2013). The attribution studies are Marshall et al. (2006) and Gillett et al. (2008). See also Elvidge et al. (2016).
47. Thompson & Solomon (2002), Turner et al. (2009).
48. The central England temperature series was constructed initially by Manley (1953), then updated and improved by Parker et al. (1992), Parker & Horton (2005). The historical context is provided by the PAGES (Past Global Changes) analysis of global temperatures over the past 2000 years (Ahmed et al. 2013).
49. Named after the British mathematician and geophysicist Sydney Chapman (1888–1970) who worked out the processes leading to ozone production and loss (Chapman 1930).
50. The loss of ozone over Antarctica was reported by Joe Farman, Brian Gardiner and Jonathan Shanklin of the British Antarctic Survey (Farman et al. 1985), and springtime levels continued to fall (Jones & Shanklin 1995). The loss of Antarctic ozone had been independently noted by US researchers, but the British publication came first and so in the way of science, the UK team were awarded priority. The hypothesis that stratospheric ozone was at risk from destruction by anthropogenic halogens had been developed much earlier by Mario Molina, and Sherwood Rowland (Molina & Rowland 1970), while Paul Crutzen was drawing attention to the importance of nitrous oxide to ozone destruction. At the same time Jim Lovelock had shown that man-made CFCs were entering the atmosphere, where they had a very long residence time. Crutzen, Molina and Rowland shared the 1995 Nobel Prize in Chemistry for their work on ozone.
51. The *Montreal Protocol on Substances that Deplete the Ozone Layer* (a protocol to the *Vienna Convention for the Protection of the Ozone Layer*) was agreed on 16 September 1987, and entered into force on 1 January 1989. Since then, it has undergone eight revisions. As a result of this international agreement, the ozone hole in Antarctica is starting to recover and current model projections indicate that the ozone layer will return to 1980 levels between 2050 and 2070. The widespread adoption and implementation of the Montreal Protocol led to its being hailed as an example of exceptional international co-operation, with

Kofi Annan (then Secretary General of the United Nations) describing it as 'perhaps the single most successful international agreement to date'.

52. Ozone profile data for 1 July 2011, Table Mountain Test facility, Boulder Colorado, obtained from NOAA Earth System Research facility. Halley data courtesy British Antarctic Survey, Cambridge.

53. The data plotted are the Hadley Centre Central England Temperature (HadCET) dataset. The original compilation is Manley (1953), updated by Parker et al. (1992) and with uncertainties discussed by Parker & Horton (2005). Data downloaded from http://www.metoffice.gov.uk/hadobs/hadcet/ 27 July 2016.

54. Clarke (1996) presents the leave/stay dichotomy, Bowden et al. (2015) discuss the three-way response.

55. In fact there is no evidence that Mark Twain (real name Samuel Langhorne Clemens) ever actually said or wrote this, but he did include a similar remark ('climate lasts all the time and weather only a few days') in an essay of 1887, reviewing a book collating answers by students to classroom questions. It does, however, appear verbatim in Robert Heinlein's 1973 novel *Time enough for love*.

56. The Californian marine fossil record has been studied in detail by Jim Valentine and colleagues over many years. See Valentine (1961, 1989), Valentine & Jablonski (1991) and Roy et al. (1995, 1996b, 2001).

57. See Barry et al. (1995) and Roy et al. (1996a).

58. The western Atlantic Pliocene extinction is described by Stanley (1986) and that in the Caribbean by Jackson & Johnson (2000). The differing pattern of extinction in the two ocean basins argues against the astronomical cause suggested by Benítez et al. (2002).

59. See Palumbi (1996, 1997) for a nice example of the role of sea level in speciation in the eastern Pacific.

60. Today we regard steppe and tundra as quite different plant assemblages. The term steppe-tundra was coined by Russian palaeoecologists working in Beringia (the area of land that linked present-day Alaska and eastern Siberia) to describe the complex vegetation that existed in Late Pleistocene times, and in which large herbivores such as mammoths were a key factor in maintaining the vegetation structure. See Hibberd (1982) and Zimov et al. (1995).

61. For reviews of post-glacial plant recolonisation see Gates (1993) and Prentice et al. (1991). Normand et al. (2011) discuss the role of climate and dispersal in European plants, and the nature of climate disequilibrium in plant distributions. The history of modern plant assemblages is one of the strong lines of evidence in favour of the individualistic view of plant ecology argued by the American botanist Henry Gleason (1882–1975). This contrasted with the view prevailing in the early part of the twentieth century proposed by another American botanist, Frederic Clements (1874–1945), which was that plant succession led to a coherent, well-defined climax community. See Clements (1916) and Gleason (1917, 1926, 1939). A useful review of the development of these ideas is McIntosh (1967).

62. Clement Reid discussed this mismatch between observed and theoretical postglacial dispersal rates in his classic book, *The origin of the British flora* (Reid 1899). A valuable review of the ideas is Clark et al. (1998). The possible role of cryptic northern refugia in post-glacial recolonisation is discussed by Willis et al. (2000), Stewart & Lister (2001), Stewart et al. (2010), Tzedakis et al. (2013) and Roberts & Hamann (2015).

63. There are five or nine periods of enhanced extinction in the fossil record, depending on the threshold criteria applied. Useful discussions of the nature of mass extinctions are Hubbard & Gilinsky (1992), Raup (1991), Jablonski (1986a, b) and Payne & Finnegan (2007). The bolide impact theory was presented by Alvarez et al. (1980) and although controversial at the time is now widely accepted as the ultimate cause of the Cretaceous extinction. Donovan (1989) lists climate change as a potential causative agent for at least five of the nine largest extinction events known.

64. Clarke (1993, 1996) discusses the mismatch between rates of climate change associated with extinction in the fossil record, and those experienced by marine organisms living today.

65. Lawton (1993, 1995) provides useful introductions to the relations between range, abundance and extinction. See also Gaston (2003) and Thomas et al. (2004).

66. Coyne & Orr (2004) provide a general view of speciation; Finlayson (2011) discusses Palaearctic birds in terms of the climatic and geological history of the entire Cenozoic.

67. The role of range fragmentation and recombination in shallow-water marine organisms driven by glacial cycles was discussed by Clarke & Crame (1989, 1992, 2003). More recently attention was drawn to cycles of range shift on land by Dynesius & Jansson (2000), who coined the term '*orbitally forced species' range dynamics*' (ORD), though the term has not caught on.

68. Parmesan et al. (1999), Walther et al. (2002), Parmesan & Yohe (2003), Root et al. (2003), Deutsch et al. (2008).

69. Burrows et al. (2011) analyse the shift in oceanic climate. Barry et al. (1995) analysed the Californian rocky intertidal, and Beaugrand et al. (2002) North Atlantic copepods.

70. From Hickling et al. (2006), Perry et al. (2005), Hawkins et al. (2009), Mieszkowska et al. (2006, 2007),

Walther et al. (2002), Wassmann et al. (2010), Grebmeier (2012), Richardson (2008), Barry et al. (1995), Perry et al. (2005), Beaugrand et al. (2002), Beaugrand (2003), Beaugrand & Reid (2003).

71. See Coope (1959), Atkinson et al. (1987) for descriptions of the mutual climatic range technique, and Elias (1994, 1997) for useful discussions of its application.

72. A null model is one where some elements are held constant but the variable of interest is allowed to vary randomly. Any pattern that emerges can then be compared with that in the observed data, and they may be regarded as a test that 'nothing has happened' (Gotelli & Graves 1996). Beale et al. (2008) used a null model approach which suggested that matching of the distribution to climate was no better than random chance in 68 of 100 species of European bird.

73. See Hutchinson (1957). Pianka (1978) gives an excellent summary of the history of the niche concept.

74. Darwin speculated on the different factors that might set range boundaries in Chapter 3 of the *Origin of species* (Darwin 1859; p. 69 of the first edition), and Ettinger et al. (2011) provide a nice example for plants. Gaston (2003) provides an excellent summary of the factors affecting the biogeographic range of organisms. See also Sexton et al. (2009), Thomas (2010) and Louthan et al. (2015).

75. Hanski (1996) gives a valuable historical introduction to the main concepts of metapopulation ecology, and Hanski & Gilpin (1997) provide a comprehensive overview of the subject.

76. Early & Sax (2011) show how constraints on the climate path may affect range change. Eglington et al. (2015) and Morrison et al. (2016) discuss the importance of local adaptation.

77. Kearney et al. (2009) discuss how behavioural thermoregulation may modify species response to climate change. Buckley et al. (2010) provide a valuable comparison of mechanistic models with species distribution models. Useful critiques of simplistic climate envelope modelling are Pearson & Dawson (2003), Hampe (2004) and Heikkinen et al. (2006).

78. The term is probably a contraction of 'phenomenology', and is derived from Greek φαινω (phaino), meaning to show or appear.

79. Robert Marsham (1708–1798) was an English naturalist who lived at Stratton Strawless, near Norwich, in Norfolk. His observations were published by the Royal Society (Marsham 1789) and soon afterwards he began a correspondence with Gilbert White. The Marsham phenology data series has been analysed by Sparks & Carey (1995). The use of autumn leaf fall by Greek farmers was noted by Pliny the Elder (Bostock & Riley 1855–1857), and the early records of flowering in Asia are discussed by Hameed & Gong (1994) for the Yangtse Valley in China, and Aono & Kazui (2008) for Japan. Zelt et al. (2012) describe the revived North American Bird Phenology Program.

80. Replotted with data from Anderson et al. (2012); reproduced with permission of the Royal Society of London. A similar relation to snowmelt is shown by the glacier lily *Erythronium grandiflorum* (Lambert et al. 2010)

81. The initial analysis is provided by Sparks et al. (2000), Amano et al. (2010) describe the 250-year assemblage level analysis and Schwartz et al. (2013) describe the North American spring index.

82. For example Walther et al. (2002) examined 50 years of data for 13 plant species at 137 northern hemisphere locations, and detected an influence of the NAO in 71% of populations.

83. Replotted with data from Amano et al. (2010); reproduced with permission of the Royal Society of London.

84. The analysis of UK Odonata is Hassall (2007). Phenological changes in UK butterflies are described by Roy & Sparks (2000), Roy & Thomas (2003) and Sparks et al. (2005), and in Californian butterflies by Forister & Shapiro (2003). See also Menzel et al. (2006).

85. Altermatt (2010).

86. Robertson (1929) describes the original fieldwork, Memmott et al. (2007) the simulation model.

87. Crick (2004) documented earlier breeding in UK birds, Both et al. (2004) documented variation in the timing of breeding in pied flycatchers, and general summaries are provided by Walther et al. (2002), Parmesan & Yohe (2003) and Charmantier & Gienapp (2014).

88. Data kindly provided by Malcolm Burgess (PiedFly.net).

89. Mason (1995), Cotton (2003).

90. Miller-Rushing et al. (2008).

91. Jenni & Kéry (2003), Lehikoinen et al. (2004), Tryjanowski et al. (2005), Jonzén et al. (2006), Kullberg et al. (2015).

92. The influence of the NAO is discussed by Hüppop & Hüppop (2003), Vähätalo et al. (2004) and Gordo et al. (2005). The links between migration phenology and population decline are explored by Møller et al. (2008), Jones & Cresswell (2010) and Gilroy et al. (2016).

93. Migration in Australian birds is discussed by Beaumont et al. (2006), Chambers (2008), Smith & Smith (2012) and Chambers et al. (2014).

94. Phenological mismatch in pied flycatchers is discussed by Burger et al. (2012), and in great tits by Vis-

ser et al. (1998, 2006) and Thomas et al. (2001). See also Hinks et al. (2015).

95. Mason (1995) showed how arrival times of migrant songbirds into UK has shifted in response to decadal-scale variability in climate. The fine-scale adaptation of great tit phenology to variability in their environment is discussed by Charmantier et al. (2008) and Reed et al. (2013a, b).

96. Bowden et al. (2015) report the study of *Boloria* and *Colias*, while Sheridan & Bickford (2011), Gardner et al. (2011) and Forster et al. (2012) discuss changes in body size associated with climate change.

97. Anderson et al. (2012) explored the genetic underpinning of the change in flowering phenology of *Boechera stricta*, and Charmantier & Gienapp (2013) discuss the role of evolutionary change in birds.

CHAPTER 17

Ten principles of thermal ecology

Now that we have reached the end of this book, it is worth summarising the key principles that, in my view, underpin thermal ecology and which I hope have emerged in the previous chapters.

1. Temperature affects everything. All processes that involve a movement of energy are affected by temperature and because of this its influence on physiology and ecology is all-pervasive.

An important consequence of this, often under-appreciated, is that it is impossible to construct a fully controlled experiment to investigate the role of temperature. That is, we cannot devise an experimental set-up where the only thing that differs between the treatment and the control is temperature. This is because when temperature differs, so does everything else: water at different temperatures also differs in density, viscosity, ionisation, gas solubility and so on, while air at different temperatures varies in water content, gas partial pressures and density.

It also means that a correlation of a particular feature with temperature, even when that correlation is strong, does not necessarily indicate cause and effect. The observed relationship may be down to something else entirely but which happens to vary itself with temperature.

2. Thermodynamics matters. Thermodynamics dictates how, and in which direction, energy will flow. It is therefore fundamental to both physiology and ecology. It does not, however, say how fast energy will flow; that is the role of kinetics.

Thermodynamics tells us that heat and temperature are different things, that temperature is not energy, that entropy underpins all movement of energy and hence its consequences need to be appreciated and that energy budgets (as well as the equations that describe them) must balance exactly. If they don't, something has been left out.

3. Hotter is better, but at a cost. Of critical importance to organisms is the temperature-dependence of the maximum rate at which mitochondria can regenerate ATP. The maximum performance of contractile proteins, protein synthesis by ribosomes and many other cellular processes is also greater at warmer temperatures. All other things being equal, organisms perform better (and their fitness is greater) when they are warmer. However organisms operate at maximum capacity only rarely; the actual rate at which physiological processes proceed is down to other factors. Thermal physiology can only be understood fully in its ecological context.

Warmer cell temperatures, however, bring with them a physiological cost: it takes more energy for an organism to keep a warm cell alive than it does a cool cell. At the cellular level these maintenance costs are predominantly those associated with protein turnover, homeostasis of the intracellular ionic environment and the integrity of membranes. There are also routine costs associated with whole-organism maintenance; these include the function of the circulatory, neural and hormonal systems and in many organisms also muscular function to maintain posture. The relationship between the costs of maintenance (usually estimated by resting metabolic rate) and temperature is indirect: warmer cells require more maintenance, which costs more; these costs may offset the performance advantages of a warmer cell.

An important ecological and evolutionary consequence is that where the environment allows for a warm body, a greater range of lifestyles is possible; there are more thermal niches available in the tropics than at the poles.

Principles of Thermal Ecology. Andrew Clarke, Oxford University Press (2017).
© Andrew Clarke 2017. DOI 10.1093/oso/9780199551668.001.0001

4. Not all physiological processes exhibit compensation for temperature. This follows directly from the previous two principles. While there are good examples of evolutionary adaptation in enzymes to offset the slower kinetic energy at lower temperatures, many physiological processes are still strongly temperature-dependent because of thermodynamic constraints about which evolution can do little or nothing. In these cases evolution has to make the best of a bad job. Nevertheless, despite the strong temperature dependence of many physiological processes, organisms have adapted to a wide range of thermal and metabolic niches.

5. Ecological context is important, so simple patterns can have complex explanations. Laboratory experiments can demonstrate fundamental relationships between physiological processes and temperature, but they do not necessarily tell us what happens in the real world. Where organisms actually live, trade-offs abound, organisms interact with one another and the emergent patterns we see may be determined by several factors acting together, either synergistically or antagonistically. To understand how temperature affects physiology we need experiments in the laboratory, to understand how this translates to ecology we need to look at Nature.

6. Rates of temperature change are critical. The rate at which temperature changes, be it daily, seasonal or in long-term climate change, is critical to how organisms can respond. It is therefore essential for experimental studies to use rates of temperature change that are relevant to the physiological or ecological processes of interest. The acute experiments dictated by short-term funding do not necessarily tell us what happens to organisms exposed to a long, slow, secular change. In relating laboratory studies to the real world we must also bear in mind that occasional extreme temperatures may have more physiological or ecological impact than a shift in the mean.

7. 'Correction' for temperature is rarely sensible. You probably do not know the underlying relationship, and 'correction' using the wrong temperature coefficient introduces error rather than removing an effect. The same applies to 'correction' for mass or phylogeny. It is better to use multivariate statistical techniques and allow the coefficients that capture the effects of temperature, mass or any other factor of interest to vary freely and be estimated (with their error quantified) from the data.

8. Metabolic rate is a cost, not a benefit. Oxygen consumption is a measure of the rate at which an organism is using resources (food or reserves) to regenerate ATP that has been used to power a physiological process, resources which then need to be replaced. The fitness benefit comes from the process (gathering food, avoiding a predator, growing, producing eggs), and the oxygen used by intermediary metabolism quantifies the cost of that process. The causal relationship is thus that the process dictates the rate of oxygen consumption, not the other way around. In other words, metabolic rate does not dictate what an animal can do; what an animal is doing dictates its metabolic rate.

Thermodynamics tells us that the energy dissipated as heat in metabolism is not wasted; it ensures the change in entropy which dictates that the metabolic processes are energetically downhill. Endotherms can trap some of this heat as it passes to the environment, and use it to keep themselves warm, but energy dissipated as heat in metabolism cannot be used for anything else (or at least not without contravening the First Law of Thermodynamics).

9. Size matters. An organism's size exerts a huge influence over its physiology and ecology, but the relationship is rarely directly proportional. Of particular importance to thermal ecology is the change in surface area to volume ratio as an organism changes in size, and that the scaling of metabolic rate with body mass typically has an exponent in the range 0.7–0.9, depending on which level of metabolism (resting, routine, maximum) is involved.

10. Individuals are different. Animals adapted to live at different body temperatures will themselves be different. In particular they will have different proteins so that, as far as evolutionarily possibly, function is matched to the temperature(s) at which they need to operate. We cannot therefore extrapolate meaningfully from the thermal sensitivity or physiological performance of a tropical organism to polar temperatures (or vice versa). When we explore relationships across space, we must recognise that these are primarily statistical in nature: data for species living at different temperatures differ in more than temperature. Individuals, or species, cannot be treated mathematically as though they were identical; their individuality (which includes their evolutionary history) has to be allowed for.

17.1. A few final thoughts

The principles outlined above provide a basis for doing thermal ecology. But there are also a few subtle biases in the way we do thermal ecology that need to be recognised, and avoided, if our results are to have wide applicability.

Firstly, we need to be careful to avoid anthropocentric judgements. Just because we as humans find polar regions cold and deserts hot does not mean that organisms adapted to live there are any more physiologically stressed than those living in the humid tropics or mild temperate regions. A tough environment for us is not necessarily a tough environment for a microbe, animal or plant adapted to live there (it may be, of course, but we need to decide that objectively and that is not always easy).

Most ecologists train, and many of them work, in the northern hemisphere. The northern and southern hemispheres differ greatly in the distribution of land and sea, patterns of annual temperature and glacial history. We should not necessarily expect the thermal ecology of organisms in the two hemispheres to be the same. The thermodynamic challenges may be identical, but the evolutionary responses may not be. Studies undertaken in North America or Europe do not necessarily apply without modification to Africa or Australia.

Exploring how temperature affects physiology in the laboratory is enormously satisfying (when it works!), but there is no substitute for seeing what animals do in the real world. Thermal ecology provides wonderful opportunities for both.

References

Aarset, A. V. (1982). Freezing tolerance in intertidal invertebrates: a review. *Comparative Biochemistry and Physiology, Part A, 73*(4), 571–580.

Abram, N. J., Mulvaney, R., Wolff, E. W., et al. (2013). Acceleration of snow melt in an Antarctic Peninsula ice core during the twentieth century. *Nature Geoscience, 6*(5), 404–411.

Ackerman, J. L., Bellwood, D. R., & Brown, J. H. (2004). The contribution of small individuals to density-body size relationships: examination of energetic equivalence in reef fishes. *Oecologia, 139*(4), 568–571.

Adams, D. C., & Church, J. O. (2008). Amphibians do not follow Bergmann's rule. *Evolution, 62*(2), 413–420.

Addo-Bediako, A., Chown, S. L., & Gaston, K. J. (2002). Metabolic cold adaptation in insects: a large-scale perspective. *Functional Ecology, 16*(3), 332–338.

Agutter, P. S., & Tuszynski, J. A. (2011). Analytic theories of allometric scaling. *Journal of Experimental Biology, 214*(7), 1055–1062.

Agutter, P. S., & Wheatley, D. N. (2004). Metabolic scaling: consensus or controversy? *Theoretical Biology and Medical Modelling, 1*, Article 13(11 pages).

Ahmed, M., Anchukaitis, K. J., Asrat, A., et al. (2013). Continental-scale temperature variability during the past two millennia. *Nature Geoscience, 6*(5), 339–346.

Albe, K. R., Butler, M. H., & Wright, B. E. (1990). Cellular concentrations of enzymes and their substrates. *Journal of Theoretical Biology, 143*(2), 163–195.

Aleksiuk, M. (1976). Reptilian hibernation: evidence of adaptive strategies in *Thamnophis sirtalis parietalis*. *Copeia, 1976*(1), 170–178.

Alexander, M. (1966). *The earliest English poems*. Harmondsworth: Penguin Books.

Alexander, R. L. (1996). Evidence of brain-warming in the mobulid rays, *Mobula tarapacana* and *Manta birostris* (Chondrichthyes: Elasmobranchii: Batoidea: Myliobatiformes). *Zoological Journal of the Linnean Society, 118*(2), 151–164.

Alexander, R. M. (1968). *Animal mechanics*. London: Sidgwick & Jackson.

Algar, A. C., Kerr, J. T., & Currie, D. J. (2007). A test of Metabolic Theory as the mechanism underlying broad-scale species-richness gradients. *Global Ecology and Biogeography, 16*(2), 170–178.

Ali, F., & Wharton, D. A. (2014). Intracellular freezing in the infective juveniles of *Steinernema feltiae*: an entomopathogenic nematode. *PLoS One, 9*(4), e94179.

Aliev, M. K., Dos Santos, P., Hoerter, J. A., Soboll, S., Tikhonov, A. N., & Saks, V. A. (2002). Water content and its intracellular distribution in intact and saline perfused rat hearts revisited. *Cardiovascular Research, 53*(1), 48–58.

Al-Khalili, J. (2010). *Pathfinders: the golden age of Arabic science*. London: Allen Lane.

Allen, A. P., Brown, J. H., & Gillooly, J. F. (2002). Global biodiversity, biochemical kinetics, and the energy-equivalence rule. *Science, 297*(5586), 1545–1548.

Allen, A. P., & Gillooly, J. F. (2006). Assessing latitudinal gradients in speciation rates and biodiversity at the global scale. *Ecology Letters, 9*(8), 947–954.

Allen, A. P., Gillooly, J. F., & Brown, J. H. (2007). Recasting the species-energy hypothesis: the different roles of kinetic and potential energy in regulating diversity. In D. Storch, P. A. Marquet, & J. H. Brown (Eds.), *Scaling biodiversity* (pp.283–299). Cambridge: Cambridge University Press.

Alper, J. (1998). Ecology: ecosystem 'engineers' shape habitats for other species. *Science, 280*(5367), 1195–1196.

Alroy, J. (2002). How many named species are valid? *Proceedings of the National Academy of Sciences of the United States of America, 99*(6), 3706–3711.

Alroy, J., Marshall, C. R., Bambach, R. K., et al. (2001). Effects of sampling standardisation on estimates of Phanerozoic marine diversification. *Proceedings of the National Academy of Sciences of the United States of America, 98*(11), 6261–6266.

Altermatt, F. (2010). Climatic warming increases voltinism in European butterflies and moths. *Proceedings of the Royal Society of London, Series B, 277*(1685), 1281–1287.

Alvarez, L. W., Alvarez, W., Asaro, F., & Michel, H. V. (1980). Extraterrestrial cause for the Cretaceous-Tertiary extinction. *Science, 208*(4480), 1095–1108.

Amano, T., Smithers, R. J., Sparks, T. H., & Sutherland, W. J. (2010). A 250-year index of first flowering dates and its response to temperature changes. *Proceedings of the Royal Society of London, Series B, 277*(1693), 2451–2457.

Amthor, J. S. (2000). The McCree-de Wit-Penning de Vries-Thornley respiration paradigms: 30 years later. *Annals of Botany, 86*(1), 1–20.

Anand, P., Elderfield, H., & Conte, M. H. (2003). Calibration of Mg/Ca thermometry in planktonic foraminifera from a sediment trap time series. *Paleoceanography, 18*(2), article 1050.

Anderson, J. T., Inouye, D. W., McKinney, A. M., Colautti, R. I., & Mitchell-Olds, T. (2012). Phenotypic plasticity and adaptive evolution contribute to advancing flowering phenology in response to climate change. *Proceedings of the Royal Society of London, Series B, 279*(1743), 3843–3852.

Anderson, K. J., & Jetz, W. (2005). The broad-scale ecology of energy expenditure of endotherms. *Ecology Letters, 8*(3), 310–318.

Andrews, P., & O'Brien, E. M. (2000). Climate, vegetation, and predictable gradients in mammal species richness in southern Africa. *Journal of Zoology, 251*(2), 205–231.

Angilletta, M. J. (2001a). Thermal and physiological constraints on energy assimilation in a widespread lizard (*Sceloporus undulatus*). *Ecology, 82*(1), 3044–3056.

Angilletta, M. J. (2001b). Variation in metabolic rate between populations of a geographically widespread lizard. *Physiological and Biochemical Zoology, 74*(1), 11–21.

Angilletta, M. J. (2006). Estimating and comparing thermal performance curves. *Journal of Thermal Biology, 31*(541–545).

Angilletta, M. J. (2009). *Thermal adaptation: a theoretical and empirical synthesis*. Oxford: Oxford University Press.

Angilletta, M. J., Niewiarowski, P. H., Dunham, A. E., Leaché, A. D., & Porter, W. P. (2004a). Bergmann's clines in ectotherms: illustrating a life-history perspective with sceloporine lizards. *American Naturalist, 164*(6), E168–E183.

Angilletta, M. J., Niewiarowski, P. H., & Navas, C. A. (2002). The evolution of thermal physiology in ectotherms. *Journal of Thermal Biology, 27*(3), 249–268.

Angilletta, M. J., Steury, T. D., & Sears, M. W. (2004b). Temperature, growth rate, and body size in ectotherms: fitting pieces of a life-history puzzle. *Integrative and Comparative Biology, 44*(6), 498–459.

Anholt, B. R., Hotz, H., Guex, G.-D., & Semelitsch, R. D. (2003). Overwinter survival of *Rana lessonae* and its hemiclonal associate *Rana esculenta*. *Ecology, 84*(2), 391–397.

Aono, Y., & Kazui, K. (2008). Phenological data series of cherry tree flowering in Kyoto, Japan, and its application to reconstruction of springtime temperatures since the 9th century. *International Journal of Climatology, 28*(7), 905–914.

Apol, M. E. F., Etienne, R. S., & Olff, H. (2008). Revisiting the evolutionary origin of allometric metabolic scaling in biology. *Functional Ecology, 22*(6), 1070–1080.

Arendt, J. (2007). Ecological correlates of body size in relation to cell size and cell number: patterns in flies, fish, fruits and foliage. *Biological Reviews, 82*(2), 241–256.

Arnold, A. E., Maynard, Z., Gilbert, G. S., Coley, P. D., & Kursar, T. A. (2000). Are tropical fungal endophytes hyperdiverse? *Ecology Letters, 3*(4), 267–274.

Arrhenius, O. (1921). Species and area. *Journal of Ecology, 9*(1), 95–99.

Arrhenius, S. (1889a). Über die Dissociationswärme und den Einfluß der Temperatur auf den Dissociationsgrad der Elektrolyte. *Zeitschrift für Physikalische Chemie, 4*, 96–116.

Arrhenius, S. (1889b). Über die Reaktionsgeschwindigkeit bei der Inversion von Rohrzucker durch Säuren. *Zeitschrift für Physikalische Chemie, 4*, 226–248.

Arrhenius, S. (1896). On the influence of carbonic acid in the air upon the temperature of the ground. *The London, Edinburgh, and Dublin Philosophical Magazine and Journal of Science, Series 5, 4*(251), 237–276.

Arslan, D., Legendre, M., Seltzer, V., Abergel, C., & Claverie, J.-M. (2011). Distant Mimivirus relative with a larger genome highlights the fundamental features of Megaviridae. *Proceedings of the National Academy of Sciences of the United States of America, 108*(42), 17486–17491.

Aschoff, J. (1971). Temperature regulation. In J. Aschoff, B. Günther, & K. Kramer (Eds.), *Energiehaushalt und Temperaturregulation (Physiologie des Menschen, Bd 2)* (pp. 43–116). München: Urban & Schwarzenberg.

Aschoff, J. (1981a). Der Tagesgang der Körpertemperatur von Vögeln als Funktion des Körpergewichtes (The 24 hour rhythm of body temperature in birds as a function of body weight). *Journal für Ornithologie, 122*(2), 129–152.

Aschoff, J. (1981b). Thermal conductance in mammals and birds: its dependence on body size and circadian phase. *Comparative Biochemistry and Physiology, Part A, 69*(4), 611–619.

Aschoff, J. (1982). The circadian rhythm of body temperature as a function of body size. In C. R. Taylor, K. Johansen, & L. Bolis (Eds.), *A companion to animal physiology* (pp.173–188). Cambridge: Cambridge University Press.

Aschoff, J., & Pohl, H. (1970). Rhythmic variations in energy metabolism. *Federation Proceedings, 29*(4), 1541–1552.

Ashton, K. G. (2001). Are ecological and evolutionary rules being dismissed prematurely? *Diversity and Distributions, 7*(6), 289–295.

Ashton, K. G. (2002). Patterns of within-species body size variation of birds: strong evidence for Bergmann's rule. *Global Ecology and Biogeography, 11*(6), 505–523.

Ashton, K. G., & Feldman, C. R. (2003). Bergmann's rule in nonavian reptiles: turtles follow it, lizards and snakes reverse it. *Evolution, 57*(5), 1151–1163.

Ashton, K. G., Tracy, M. C., & de Queiroz, A. (2000). Is Bergmann's rule valid for mammals? *American Naturalist, 156*(4), 390–415.

Ashton, N., Lewis, S. G., De Groote, I., et al. (2014). Hominin footprints from Early Pleistocene deposits at Happisburgh, UK. *PLoS One*, *9*(2), e88329.

Atchley, W. R., Gaskins, C. T., & Anderson, D. (1976). Statistical properties of ratios. I. Empirical results. *Systematic Zoology*, *25*, 137–148.

Atici, Ö., & Nalbantoğlu, B. (2003). Antifreeze proteins in higher plants. *Phytochemistry*, *64*(7), 1187–1196.

Atkins, P. (2007). *Four laws that drive the universe*. Oxford: Oxford University Press.

Atkins, P., & de Paula, J. (2011). *Physical chemistry for the life sciences* (Second ed.). Oxford: Oxford University Press.

Atkinson, D. (1994). Temperature and organism size—a biological law for ectotherms? *Advances in Ecological Research*, *25*, 1–58.

Atkinson, D., Ciotti, B. J., & Montagnes, D. J. S. (2003). Protists decrease in size linearly with temperature: ca. 2.5% $°C^{-1}$. *Proceedings of the Royal Society of London, Series B*, *270*(1533), 2605–2611.

Atkinson, D., Morley, S. A., & Hughes, R. N. (2006). From cells to colonies: at what levels of body organisation does the 'temperature-size rule' apply? *Evolution & Development*, *8*(2), 202–214.

Atkinson, D., & Sibly, R. M. (1997). Why are organisms usually bigger in colder environments? Making sense of a life history puzzle. *Trends in Ecology and Evolution*, *12*(6), 235–239.

Atkinson, T. C., Briffa, K. R., & Coope, G. R. (1987). Seasonal temperatures in Britain during the past 22,000 years, reconstructed using beetle remains. *Nature*, *325*(6105), 587–592.

Attrill, M. J., Stafford, R., & Rowden, A. A. (2001). Latitudinal diversity patterns in estuarine tidal flats: indications of a global cline. *Ecography*, *24*(3), 318–324.

Augustin, L., Barbante, C., Barnes, P. R. F., et al. (2004). Eight glacial cycles from an Antarctic ice core. *Nature*, *429*(6992), 623–628.

Avery, O. T., MacLeod, C. M., & McCarty, M. (1944). Studies on the chemical nature of the substance inducing transformation of Pneumococcal types: induction of transformation by a deoxyribonucleic acid fraction isolated from *Pneumococcus* Type III. *Journal of Experimental Medicine*, *79*, 137–159.

Avery, R. A. (1979). *Lizards: a study in thermoregulation (Studies in Biology no 109)*. London: Edward Arnold.

Baardsnes, J., & Davies, P. L. (2001). Sialic acid synthase: the origin of fish type III antifreeze protein? *Trends in Biochemical Sciences*, *26*(8), 468–469.

Baas Becking, L. G. M. (1934). *Geobiologie of inleiding tot de milieukunde*. The Hague: W.P. Van Stockum & Zoon.

Bagriantsev, S. N., & Gracheva, E. O. (2015). Molecular mechanisms of temperature adaptation. *Journal of Physiology*, *593*(16), 3483–3491.

Bains, W. (2014). What do we think life is? A simple illustration and its consequences. *International Journal of Astrobiology*, *13*(2), 101–111.

Baker, M. A. (1982). Brain cooling in endotherms in heat and exercise. *Annual Review of Physiology*, *44*, 85–96.

Bakken, G. S., & Angilletta, M. J. (2014). How to avoid errors when quantifying thermal environments. *Functional Ecology*, *28*(1), 96–107.

Bakken, G. S., & Gates, D. M. (1975). Heat-transfer analysis of animals: some implications for field ecology. In D. M. Gates & R. B. Schmerl (Eds.), *Perspectives of biophysical ecology* (pp. 255–290). New York: Springer-Verlag.

Bakker, R. (1986). *The dinosaur heresies: a revolutionary view of dinosaurs*. Harlow, Essex: Longman.

Bakker, R. T. (1968). The superiority of dinosaurs. *Discovery (Peabody Museum, Yale)*, *3*(2), 11–22.

Bakker, R. T. (1972). Anatomical and ecological evidence of endothermy in dinosaurs. *Nature*, *238*(5359), 81–85.

Bakker, R. T. (1980). Dinosaur heresy—dinosaur renaissance: why we need endothermic archosaurs for a comprehensive theory of bioenergetic evolution. In R. D. K. Thomas & E. C. Olson (Eds.), *A cold look at the warm-blooded dinosaurs* (Vol. 28, pp. 351–462). Washington, D.C.: Westview Press, for the American Association for the Advancement of Science.

Baldwin, R. L. (1986). Temperature dependence of the hydrophobic interaction in protein folding. *Proceedings of the National Academy of Sciences of the United States of America*, *83*(21), 8069–8072.

Baldwin, R. L., & Smith, N. E. (1974). Molecular control of energy metabolism. In J. D. Sink (Ed.), *The control of metabolism* (pp. 17–25). Pennsylvania: Pennsylvania State University Press.

Baldwin, R. L., Smith, N. E., Taylor, J., & Sharp, M. (1980). Manipulating metabolic parameters to improve growth rate and milk secretion. *Journal of Animal Science*, *51*(6), 1416–1428.

Bale, J. S. (1980). Seasonal variation in cold hardiness of the adult beech leaf mining weevil *Rhynchaenus fagi* L. in Great Britain. *Cryo-Letters*, *1*(11), 372–383.

Bale, J. S. (1993). Classes of insect cold hardiness. *Functional Ecology*, *7*, 751–753.

Bale, J. S. (1996). Insect cold hardiness: a matter of life and death. *European Journal of Entomology*, *93*(3), 369–382.

Ball, P. (1999). *H_2O: a biography of water*. London: Weidenfield & Nicolson.

Balmaseda, M. A., Trenberth, K. E., & Källén, E. (2013). Distinctive climate signals in reanalysis of global ocean heat content. *Geophysical Research Letters*, *40*(9), 1754–1759.

Banavar, J. R., Damuth, J., Maritan, A., & Rinaldo, A. (2002). Supply-demand balance and metabolic scaling. *Proceedings of the National Academy of Sciences of the United States of America*, *99*(16), 10506–10509.

Banavar, J. R., Maritan, A., & Rinaldo, A. (1999). Size and form in efficient transportation networks. *Nature*, *399*(6732), 130–132.

Banavar, J. R., Moses, M. E., Brown, J. H., et al. (2010). A general basis for quarter-power scaling in animals. *Proceedings of the National Academy of Sciences of the United States of America*, *107*(36), 15816–15820.

Bar Dolev, M., Braslavsky, I., & Davies, P. L. (2016). Ice-binding proteins and their function. *Annual Review of Biochemistry*, *85*(1), 515–542.

Barbante, C., Barnola, J.-M., Becagli, S., et al. (2006). One-to-one coupling of glacial climate variability in Greenland and Antarctica. *Nature*, *444*(7116), 195–198.

Barclay, R. M. R., Lausen, C. L., & Hollis, L. (2001). What's hot and what's not: defining torpor in free-ranging birds and mammals. *Canadian Journal of Zoology*, *79*(10), 1885–1890.

Barker, S., Cacho, I., Benway, H., & Tachikawa, K. (2005). Planktonic foraminiferal Mg/Ca as a proxy for past oceanic temperatures: a methodological overview and data compilation for the Last Glacial Maximum. *Quaternary Science Reviews*, *24*, 821–834.

Barnes, B. M. (1989). Freeze avoidance in a mammal: body temperatures below 0 °C in an Arctic hibernator. *Science*, *244*(4912), 1593–1595.

Barnes, R. S. K. (Ed) (1998). *The diversity of living organisms*. Oxford: Blackwell Science.

Baross, J. A., & Deming, J. W. (1983). Growth of 'black smoker' bacteria at temperatures of at least 250 °C. *Nature*, *303*(5916), 423–426.

Baross, J. A., Lilley, M. D., & Gordon, L. I. (1982). Is the CH_4, H_2 and CO venting from submarine hydrothermal systems produced by thermophilic bacteria? *Nature*, *298*(5872), 366–368.

Barrington, E. J. W. (1967). *Invertebrate structure and function*. London: Thoms Nelson & Sons.

Barry, J. F., McCarron, D. J., Norrgard, E. B., Steinacker, M. H., & DeMille, D. (2014). Magneto-optical trapping of a diatomic molecule. *Nature*, *512*(7514), 286–289.

Barry, J. P., Baxter, C. H., Sagarin, R. D., & Gilman, S. E. (1995). Climate-related, long-term faunal changes in a California rocky intertidal community. *Science*, *267*(5198), 672–675.

Bartels, H. (1982). Metabolic rate of mammals equals the 0.75 power of their body weight. *Experimental Biology and Medicine. Monographs on Interdisciplinary Topics*, *7*, 1–11.

Barthlott, W., Hostert, A., Kier, G., et al. (2007). Geographic patterns of vascular plant diversity at continental to global scales. *Erdkunde*, *61*(4), 305–315.

Barthlott, W., Lauer, W., & Placke, A. (1996). Global distribution of species diversity in vascular plants: towards a world map of phytodiversity. *Erdkunde*, *50*(4), 317–327.

Bartholomew, G. A., & Heinrich, B. (1973). A field study of flight temperatures in moths in relation to body weight and wing loading. *Journal of Experimental Biology*, *58*(1), 123–135.

Bartholomew, G. A., Howell, T. R., & Cade, T. J. (1957). Torpidity in the white-throated swift, Anna hummingbird, and poor-will. *Condor*, *59*(3), 145–155.

Bartholomew, G. A., & Hudson, J. W. (1962). Hibernation, estivation, temperature regulation, evaporative water loss, and heart rate of the pigmy possum, *Cercartetus nanus*. *Physiological Zoology*, *35*(1), 94–107.

Bartholomew, G. A., Lasiewski, R. C., & Crawford, E. C. (1968). Patterns of panting and gular fluttering in cormorants, pelicans, owls, and doves. *Condor*, *70*(1), 31–34.

Bartlett, R., Pickering, J., Gauld, I., & Windsor, D. (1999). Estimating global biodiversity: tropical beetles and wasps send different signals. *Ecological Entomology*, *24*(1), 118–121.

Battaglia, M., Olvera-Carillo, Y., Garciarrubio, A., Campos, F., & Covarrubias, A. (2008). The enigmatic LEA proteins and other hydrophilins. *Plant Physiology*, *148*(1), 6–24.

Battley, E. H. (2013). A theoretical study of the thermodynamics of microbial growth using *Saccharomyces cerevisiae* and a different free energy equation. *Quarterly Review of Biology*, *88*(2), 69–96.

Beale, C., Lennon, J. J., & Gimona, A. (2008). Opening the climate envelope reveals no macroscale associations with climate in European birds. *Proceedings of the National Academy of Sciences of the United States of America*, *105*(39), 14908–14912.

Beamish, F. W. H., & Mookherjii, P. S. (1964). Respiration of fishes with species emphasis on standard oxygen consumption: I. Influence of weight and temperature on respiration of goldfish, *Carassius auratus* L. *Canadian Journal of Zoology*, *42*(2), 161–175.

Beaugrand, G. (2003). Long-term changes in copepod abundance and diversity in the north-east Atlantic in relation to fluctuations in the hydroclimatic environment. *Fisheries Oceanography*, *12*(4–5), 270–283.

Beaugrand, G., & Reid, P. C. (2003). Long-term changes in phytoplankton, zooplankton and salmon related to climate. *Global Change Biology*, *9*(6), 801–817.

Beaugrand, G., Reid, P. C., Ibañez, F., Lindley, J. A., & Edwards, M. (2002). Reorganisation of North Atlantic marine copepod biodiversity and climate. *Science*, *296*(5573), 1692–1694.

Beaumont, L. J., McAllan, I. A. W., & Hughes, L. (2006). A matter of timing: changes in the first data of arrival and last date of departure of Australian migratory birds. *Global Change Biology*, *12*(7), 1339–1354.

Beaupre, S. J. (2005). Ratio representations of specific dynamic action (mass-specific SDA and SDA coefficient) do not standardise for body mass and meal size. *Physiological and Biochemical Zoology*, 78(1), 126–131.

Beaupre, S. J., & Zaidan, F. (2012). Digestive performance in the timber rattlesnake (*Crotalus horridus*) with reference to temperature dependence and bioenergetic cost of growth. *Journal of Herpetology*, 46(4), 637–642.

Beck, R. M. D., Bininda-Emonds, O. R. P., Cardillo, M., Liu, F.-G. R., & Purvis, A. (2006). A higher-level MRP supertree of placental mammals. *BMC Evolutionary Biology*, 6, article 93.

Bedau, M. A. (2010). An Aristotelian account of minimal chemical life. *Astrobiology*, 10(10), 1011–1020.

Behnke, J., Mann, M. J., Scruggs, F.-L., Feige, M. J., & Hendershot, L. M. (2016). Members of the hsp70 family recognize distinct types of sequences to execute ER quality control. *Molecular Cell*, 63(5), 739–752.

Beilin, B. J., Knight, G. J., Munro-Faure, A. D., & Anderson, J. F. (1966). The sodium, potassium, and water contents of red blood cells of healthy human adults. *Journal of Clinical Investigation*, 45(11), 1817–1825.

Bell, T. M., Strand, A. E., & Sotka, E. A. (2014). The adaptive cline at LDH (Lactate Dehydrogenase) in killifish *Fundulus heteroclitus* remains stationary after 40 years of warming estuaries. *Journal of Heredity*, 105(4), 566–571.

Benedict, F. G. (1915). Factors affecting basal metabolism. *Journal of Biological Chemistry*, 20, 263–299.

Benedict, F. G. (1938). *Vital energetics: a study in comparative basal metabolism* (Publication 503). Washington, D.C.: Carnegie Institute of Washington.

Benedict, F. G., Fox, E. L., & Coropatchinsky, V. (1932). The incubating python: a temperature study. *Proceedings of the National Academy of Sciences of the United States of America*, 18(2), 209–212.

Benítez, N., Maíz-Appellániz, J., & Canelles, M. (2002). Evidence for nearby supernova explosions. *Physical Review Letters*, 88, article 081101.

Ben-Naim, A. (2012). *Entropy and the second law*. New Jersey: World Scientific.

Benner, S. A. (2010). Defining life. *Astrobiology*, 10(10), 1021–1030.

Bennett, A. F. (1987). Evolution of the control of body temperature: is warmer better? In P. Dejours, L. Bolis, C. R. Taylor, & E. R. Weibel (Eds.), *Comparative physiology: life on water and on land* (pp. 421–431). Padova: Liviana Press.

Bennett, A. F., & Dawson, W. R. (1976). Metabolism. In C. Gans & W. R. Dawson (Eds.), *Biology of the Reptilia* Vol. 5,(pp.127–223). New York: Academic Press.

Bennett, A. F., & Ruben, J. A. (1979). Endothermy and activity in vertebrates. *Science*, 206, 649–654.

Bennett, N. C., Jarvis, J. U. M., & Davies, K. C. (1988). Daily and seasonal temperatures in the burrows of African rodent moles. *South African Journal of Zoology*, 23(3), 189–195.

Bennett, V. A., Lee, R. E., Nauman, J. S., & Kukal, O. (2003). Selection of overwintering microhabitats used by the Arctic woollybear caterpillar, *Gynaephora groenlandica*. *Cryo-Letters*, 24(3), 191–200.

Bergman, A., & Casadevall, A. (2010). Mammalian endothermy optimally restricts fungi and metabolic costs. *mBio*, 1(5), Article e00212–00210.

Bergmann, C. (1847). Ueber die Verhältnisse der Wärmeökonomie der Thiere zu ihrer Grösse. *Göttingen Studien, Part 1*, 595–708.

Berman, D. I., Leirikh, A. L., & Mikhailova, E. I. (1984). Winter hibernation of the Siberian salamander *Hynobius keyserlingi* (In Russian, with English summary). *Zhurnal Evolyutsionnoi Biokhimii i Fiziologii*, 20(3), 323–327.

Bernal, D., Dickson, K. A., Shadwick, R. E., & Graham, J. B. (2001). Review: Analysis of the evolutionary convergence for high performance swimming in lamnid sharks and tunas. *Comparative Biochemistry and Physiology, Part A*, 129(2–3), 695–726.

Bernhardt, G., Lüdemann, H.-D., Jaenicke, R., König, H., & Stetter, K. O. (1984). Biomolecules are unstable under black smoker conditions. *Naturwissenschaften*, 71(11), 583–586.

Bernoulli, D. (1968). *Hydrodynamica (Hydrodynamics, 1738, translated by Thomas Carmody and Helmut Kobus, with preface by Hunter Rouse)* (T. Carmody & H. Kobus). New York: Dover Publications.

Beuchat, C. A., Chaplin, S. B., & Morton, M. L. (1979). Ambient temperature and the daily energetics of two species of hummingbirds, *Calypte anna* and *Selasphorus rufus*. *Physiological Zoology*, 52(3), 280–295.

Biewener, A. A. (2003). *Animal locomotion*. Oxford: Oxford University Press.

Bilham, E. G. (1937). A screen for sheathed thermometers. *Quarterly Journal of the Meteorological Society*, 63(271), 309–322.

Bindoff, N. L., Willebrand, J., Artale, V., et al. (2007). Observations: oceanic climate change and sea level. In S. Solomon, D. Qin, M. Manning, Z. Chen, M. Marquis, K. B. Averyt, M. M. B. Tignor, & H. L. Miller (Eds.), *Climate change 2007: the physical science basis. Contribution of Working Group I to the Fourth Assessment Report of the Intergovernmental Panel on Climate Change* (pp. 1009). Cambridge: Cambridge University Press.

Bininda-Emonds, O. R. P., Cardillo, M., Jones, K. E., et al. (2007). The delayed rise of present-day mammals. *Nature*, 507, 507–512.

Bishop, C. M. (1999). The maximum oxygen consumption and aerobic scope of birds and mammals: getting to the heart of the matter. *Proceedings of the Zoological Society of London, Series B*, 266(1453), 2275–2281.

Bissinger, J. E., Montagnes, D. J. S., Sharples, J., & Atkinson, D. (2008). Predicting marine phytoplankton maximum growth rates from temperature: improving on the Eppley curve using quantile regression. *Limnology & Oceanography, 53*(2), 487–493.

Blackburn, T. M., & Gaston, K. J. (1994). Animal body size distributions: patterns, mechanisms and implications. *Trends in Ecology & Evolution, 9*(12), 471–474.

Blackburn, T. M., Gaston, K. J., & Loder, N. (1999). Geographic gradients in body size: a clarification of Bergmann's rule. *Diversity and Distributions, 5*(4), 165–174.

Blackburn, T. M., & Hawkins, B. A. (2004). Bergmann's rule and the mammal fauna of northern North America. *Ecography, 27*(6), 715–724.

Blackburn, T. M., Lawton, J. H., & Perry, J. N. (1992). A method for estimating the slope of upper-bounds of plots of body size and abundance in natural animal assemblages. *Oikos, 65*(1), 107–112.

Blaxter, K. L. (1967). *The energy metabolism of ruminants* (Second ed.). London: Hutchinson.

Blaxter, K. L. (1989). *Energy metabolism in animals and man*. Cambridge: Cambridge University Press.

Bligh, J., & Johnson, K. G. (1973). Glossary of terms for thermal physiology. *Journal of Applied Physiology, 35*(6), 941–961.

Block, B. A. (2011). Endothermy in tunas, billfishes, and sharks. In A. P. Farrell (Ed.), *Encyclopedia of fish physiology: from genome to environment* (Vol. 3: Energetics, interactions with the environment, lifestyles, and applications), pp.1914–1920). San Diego: Academic Press.

Block, B. A., & Finnerty, J. R. (1994). Endothermy in fishes: a phylogenetic analysis of constraints, predispositions, and selection pressures. *Environmental Biology of Fishes, 40*(3), 283–302.

Block, W., Webb, N. R., Coulson, S., Hodkinson, I. D., & Worland, M. R. (1994). Thermal adatation in the Arctic collembolan *Onychiurus arcticus* (Tullberg). *Journal of Insect Physiology, 40*(8), 715–722.

Boily, P. (2002). Individual variation in metabolic traits of wild nine-banded armadillos (*Dasypus novemcinctus*), and the aerobic capacity model for the evolution of endothermy. *Journal of Experimental Biology, 205*, 3207–3214.

Bonter, D. N., Zuckerberg, B., Sedgwick, C. W., & Hochachka, W. M. (2013). Daily foraging patterns in free-living birds: exploring the predation-starvation trade-off. *Proceedings of the Royal Society of London, Series B, 280*, 20123087.

Borecký, J., Maia, I. G., & Arruda, P. (2001). Mitochondrial uncoupling proteins in mammals and plants. *Bioscience Reports, 21*(2), 201–212.

Boschi, E. E. (2000). Species of decapod crustaceans and their distribution in the American marine zoogeographic provinces. *Revista de Investigación y Desarrollo Pesquero, 13*(1), 1–136.

Bostock, J., & Riley, H. T. (Eds). (1855–1857). *The natural history of Pliny, translated, with copious notes and illustrations, by the late John Bostock and H. T. Riley*. London: H G Bohn.

Both, C., Artemyev, A. V., Blaauw, B., et al. (2004). Large-scale geographical variation confirms that climate change causes birds to lay earlier. *Proceedings of the Royal Society of London, Series B, 271*(1549), 1657–1662.

Bouchet, P., Lozouet, P., Maestrati, P., & Heros, V. (2002). Assessing the magnitude of species richness in tropical marine environments: exceptionally high numbers of molluscs at a New Caledonia site. *Biological Journal of the Linnean Society, 75*(4), 421–436.

Bowden, J. J., Eskildsen, A., Hansen, R. R., Olsen, K., Kurle, C. M., & Høye, T. T. (2015). High-Arctic butterflies become smaller with rising temperatures. *Biology Letters, 11*(10), article 20150574.

Boyd, R. B., & DeVries, A. L. (1983). The seasonal distribution of anionic binding sites in the basement membrane of the kidney glomerulus of the winter flounder *Pseudopleuronectes americanus*. *Cell and Tissue Research, 234*(2), 271–277.

Boyle, B., Hopkins, N., Lu, Z., et al. (2013). The taxonomic name resolution service: an online tool for automated standardization of plant names. *BMC Bioinformatics, 14*, article 16.

Boyle, R. (1661). *The sceptical chymist: or chymico-physical doubts & paradoxes, touching the spagyrist's principles commonly call'd hypostatical, as they are wont to be propos'd and defended by the generality of alchymists*. London: J. Cadwell.

Boyles, J. G., & Bakken, G. S. (2007). Seasonal changes and wind dependence of thermal conductance in dorsal fur from two small mammal species (*Peromyscus leucopus* and *Microtus pennsylvanicus*). *Journal of Thermal Biology, 32*(7–8), 383–387.

Boyles, J. G., Smit, B., & McKechnie, A. E. (2011a). Does use of the torpor cut-off method to analyze variation in body temperature cause more problems than it solves? *Journal of Thermal Biology, 36*(7), 373–375.

Boyles, J. G., Smit, B., & McKechnie, A. E. (2011b). A new comparative metric for estimating heterothermy in endotherms. *Physiological and Biochemical Zoology, 84*(1), 115–123.

Boyles, J. G., Thompson, A. B., McKechnie, A. E., Malan, E., Humphries, M. M., & Careau, V. (2013). A global heterothermic continuum in mammals. *Global Ecology and Biogeography, 22*(9), 1029–1039.

Bradley, S. R., & Deavers, D. R. (1980). A re-examination of the relationship between thermal conductance and body weight in mammals. *Comparative Biochemistry and Physiology, Part A, 65*(4), 465–476.

Brafield, A. E. (1985). Laboratory studies of energy budgets. In P. Tytler & P. Calow (Eds)., *Fish energetics: new perspectives* (pp.257–281). London: Croom Helm.

Brafield, A. E., & Llewellyn, M. J. (1982). *Animal energetics*. Glasgow & London: Blackie.

Brakefield, P. M. (1984a). Ecological studies on the polymorphic ladybird *Adalia bipunctata* in the Netherlands. I. Population biology and geographical variation of melanism. *Journal of Animal Ecology, 53*(3), 761–774.

Brakefield, P. M. (1984b). Ecological studies on the polymorphic ladybird *Adalia bipunctata* in the Netherlands. II. Population dynamics, differential timing of reproduction and thermal melanism. *Journal of Animal Ecology, 53*(3), 775–790.

Brand, M. D., Couture, P., Else, P. L., Withers, K. W., & Hulbert, A. J. (1991). Evolution of energy metabolism: proton permeability of the inner membrane of liver mitochondria is greater in a mammal than in a reptile. *Biochemical Journal, 275*(1), 81–86.

Brashears, J. A., & DeNardo, D. F. (2013). Revisiting python thermogenesis: brooding Burmese pythons (*Python bivittatus*) cue on body, not clutch, temperature. *Journal of Herpetology, 47*(3), 440–444.

Bravo, L. A., & Griffith, M. (2005). Characterization of antifreeze activity in Antarctic plants. *Journal of Experimental Botany, 56*(414), 1189–1196.

Brett, J. R. (1964). The respiratory metabolism and swimming performance of young sockeye salmon. *Journal of the Fisheries Research Board of Canada, 21*(5), 1183–1226.

Brett, J. R. (1965). The relation of size to rate of oxygen consumption and sustained swimming speed of sockeye salmon (*Oncorhynchus nerka*). *Journal of the Fisheries Research Board of Canada, 22*(6), 1491–1501.

Brett, J. R. (1972). The metabolic demand for oxygen in fish, particularly salmonids, and a comparison with other vertebrates. *Respiration Physiology, 14*(1–2), 151–170.

Brett, J. R. (1986). Production energetics of a population of sockeye salmon, *Oncorhynchus nerka*. *Canadian Journal of Zoology, 64*, 555–564.

Brett, J. R., & Groves, T. D. D. (1979). Physiological energetics. In W. S. Hoar, D. J. Randall, & J. R. Brett (Eds.), *Fish physiology* (Vol. 8: Bioenergetics and growth, pp.279–352). New York: Academic Press.

Briggs, J. C. (1994). Species diversity: land and sea compared. *Systematic Biology, 43*(1), 130–135.

Briggs, J. C. (2003). Marine centres of origin as evolutionary engines. *Journal of Biogeography, 30*(1), 1–18.

Briggs, J. C. (2004). Older species: a rejuvenation on coral reefs? *Journal of Biogeography, 31*(4), 5252–5530.

Brigham, M. M., Willis, C. K. R., Geiser, F., & Mzilikazi, N. (2011). Baby in the bathwater: should we abandon the use of body temperature thresholds to quantify expression of torpor? *Journal of Thermal Biology, 36*(7), 376–379.

Broady, P. A., & Weinstein, R. N. (1998). Algae, lichens and fungi in La Gorce Mountains, Antarctica. *Antarctic Science, 10*(4), 376–385.

Brocchieri, L., & Karlin, S. (2005). Protein length in eukaryotic and prokaryotic proteomes. *Nucleic Acids Research, 33*(10), 3390–3400.

Brock, T. D. (1967). Micro-organisms adapted to high temperatures. *Nature, 214*(5091), 882–885.

Brock, T. D. (1978). *Thermophilic microorganisms and life at high temperatures*. New York: Springer-Verlag.

Brock, T. D., & Brock, M. L. (1966). Temperature optima for algal development in Yellowstone and Iceland hot springs. *Nature, 209*, 733–734.

Brock, T. D., & Freeze, H. (1969). *Thermus aquaticus* gen.n. and sp. n., a nonsporulating extreme thermophile. *Journal of Bacteriology, 98*(1), 289–297.

Brockington, S., & Clarke, A. (2001). The relative influence of temperature and food on the metabolism of a marine invertebrate. *Journal of Experimental Marine Biology and Ecology, 258*(1), 87–99.

Brodin, A. (2001). Mass-dependent predation and metabolic expenditure in wintering birds: is there a trade-off between different forms of predation? *Animal Behaviour, 62*(5), 993–999.

Brody, S. (1945). *Bioenergetics and growth, with special reference to the efficiency complex in domestic animals* (First ed.). New York: Rienhold Publishing Corporation. (Reprinted 1974 by Hafner Press, New York.)

Bromham, L. (2011). The genome as a life-history character: why rate of molecular evolution varies between mammal species. *Philosophical Transactions of the Royal Society of London, Series B, 366*(1577), 2503–2513.

Brooks, D. R., & Wiley, E. O. (1986). *Evolution as entropy*. Chicago: University of Chicago Press.

Brown, E. E., Baumann, H., & Conover, D. O. (2012). Absence of countergradient and cogradient variation in an oceanic silverside, the California grunion *Leuresthes tenuis*. *Marine Ecology Progress Series, 461*, 175–186.

Brown, J. H. (1981). Two decades of homage to Santa Rosalia: toward a general theory of diversity. *American Zoologist, 21*(4), 877–888.

Brown, J. H. (1995). *Macroecology*. Chicago: University of Chicago Press.

Brown, J. H., Gillooly, J. F., Allen, A. P., Savage, V. M., & West, G. B. (2004). Toward a metabolic theory of ecology. *Ecology, 85*(7), 1771–1789.

Brown, J. J., Ehtisham, A., & Conover, D. O. (1998). Variation in larval growth rate among striped bass from different latitudes. *Transactions of the American Fisheries Society, 127*(4), 598–610.

Bruinzeel, L. W., & Piersma, T. (1998). Cost reduction in the cold: heat generated by terrestrial locomotion partly substitutes for thermoregulation costs in Knot *Calidris canutus*. *Ibis, 140*, 323–328.

Brzęk, P., Bielawska, K., Książek, A., & Konarzewski, M. (2007). Anatomic and molecular correlates of divergent selection for basal metabolic rate in laboratory mice. *Physiological and Biochemical Zoology, 80*(5), 491–499.

Buckley, B. A., Place, S. P., & Hofmann, G. E. (2004). Regulation of heat shock genes in isolated hepatocytes from an Antarctic fish, *Trematomus bernacchii*. *Journal of Experimental Biology*, 207(21), 3649–3656.

Buckley, L. B. (2008). Linking traits to energetics and population dynamics to predict lizard ranges in changing environments. *American Naturalist*, 171(1), E1–E9.

Buckley, L. B., Urban, M. C., Angilletta, M. J., Crozier, L. G., Rissler, L. J., & Sears, M. W. (2010). Can mechanism inform species' distribution models? *Ecology Letters*, 13(8), 1041–1054.

Budretsky, A. B. (1984). New absolute minimum of air temperature (in Russian). *Information Bulletin of Soviet Antarctic Expedition (Leningrad, Hydrometeoizdat)*, 105.

Buffenstein, R., & Yahav, S. (1991). Is the naked mole-rat *Heterocephalus glaber* an endothermic yet poikilothermic mammal? *Journal of Thermal Biology*, 16(4), 227–232.

Burger, C., Belskii, E., Laaksonen, T., et al. (2012). Climate change, breeding date and nestling diet: how temperature differentially affects seasonal changes in pied flycatcher diet depending on habitat variation. *Journal of Animal Ecology*, 81(4), 926–936.

Burroughs, W. J. (2005). *Climate change in prehistory: the end of the reign of chaos*. Cambridge: Cambridge University Press.

Burrows, M. T., Schoeman, D. S., Buckley, L. B., et al. (2011). The pace of shifting climate in marine and terrestrial ecosystems. *Science*, 334(6056), 652–655.

Burton, R. F. (2002). Temperature and acid–base balance in ectothermic vertebrates: the imidazole alphastat hypotheses and beyond. *Journal of Experimental Biology*, 205(23), 3587–3600.

Burton-Johnson, A., Black, M., Fretwell, P. T., & Kaluza-Gilbert, J. (2016). An automated methodology for differentiating rock from snow, clouds and sea in Antarctica from Landsat 8 imagery: a new rock outcrop map and area estimation for the entire Antarctic continent. *The Cryosphere*, 10, 1665–1677.

Bush, A. M., Markey, M. J., & Marshall, C. R. (2004). Removing bias from diversity curves: the effects of spatially organized biodiversity on sampling-standardization. *Paleobiology*, 30(4), 666–686.

Butler, R. J., Barrett, P. M., & Gower, D. J. (2009). Postcranial skeletal pneumaticity and air-sacs in the earliest pterosaurs. *Biology Letters*, 5(4), 557–560.

Buxton, P. A. (1923). *Animal life in deserts: a study of the fauna in relation to the environment*. London: Edward Arnold.

Buxton, P. A. (1924). Heat, moisture, and animal life in deserts. *Proceedings of the Royal Society of London, Series B*, 96(673), 123–131.

Calder, W. A. (1984). *Size, function and life history*. Cambridge, Massachusetts: Harvard University Press.

Callendar, H. L. (1887). On the practical measurement of temperature: experiments made at the Cavendish Laboratory, Cambridge. *Philosophical Transactions of the Royal Society of London, Series A*, 178, 161–230.

Calman, W. T. (1937). Presidential address: James Eights, a pioneer Antarctic naturalist. *Proceedings of the Linnean Society of London*, 149(4), 171–184.

Campbell, A. L., Naik, R. R., Sowards, L., & Stone, M. O. (2002). Biolgical infrared imaging and sensing. *Micron*, 33(2), 211–225.

Cannon, B., & Nedergaard, J. (2004). Brown adipose tissue: function and physiological significance. *Physiological Reviews*, 84, 227–359.

Careau, V. (2013). Basal metabolic rate, maximum thermogenic capacity and aerobic scope in rodents: interaction between environmental temperature and torpor use. *Biology Letters*, 9(2), article 20121104.

Carey, F. G., & Lawson, K. D. (1973). Temperature regulation in free-swimming bluefin tuna. *Comparative Biochemistry and Physiology, Part A*, 44(2), 375–392.

Carey, F. G., & Teal, J. M. (1966). Heat conservation in tuna fish muscle. *Proceedings of the National Academy of Sciences of the United States of America*, 56(5), 1464–1469.

Carey, F. G., & Teal, J. M. (1969). Mako and porbeagle: warm-bodied sharks. *Comparative Biochemistry and Physiology*, 28(1), 199–204.

Carpenter, E. J., Lin, S., & Capone, D. G. (2000). Bacterial activity in South Pole snow. *Applied and Environmental Microbiology*, 66(10), 4514–4517.

Carpenter, F. L. (1974). Torpor in an Andean hummingbird: its ecological significance. *Science*, 183(4124), 545–547.

Carr, J. M., & Lima, S. L. (2013). Nocturnal hypothermia impairs flight ability in birds: a cost of being cool. *Proceedings of the Royal Society of London, Series B*, 280(1772), article 20131846.

Carroll, J. J., Slupsky, J. D., & Mather, A. E. (1991). The solubility of carbon dioxide in water at low pressure. *Journal of Physical and Chemical Reference Data*, 20(6), 1201–1209.

Carstensen, D. W., Lessard, J.-P., Holt, B. G., Borregaard, M. K., & Rahbek, C. (2013). Introducing the biogeographic species pool. *Ecography*, 36(1), 1–9.

Carter, J. G. (1980). Environmental and biological controls of bivalve shell mineralogy and microstructure. In D. C. Rhoads & R. A. Lutz (Eds.), *Skeletal growth of aquatic organisms: biological records of environmental change* (pp. 69–113). New York: Plenum Press.

Carter, P. A., & Watt, W. B. (1988). Adaptations at specific loci. V. Metabolically adjacent enzyme loci may have very distinct experiences of selective pressures. *Genetics*, 119(4), 913–924.

Cary, S. C., Shank, T., & Stein, J. (1998). Worms bask in extreme temperatures. *Nature*, 391(6667), 545–546.

Catling, D., & Kasting, J. F. (2007). Planetary atmospheres and life. In W. T. Sullivan & J. A. Baross (Eds.), *Planets and life: the emerging science of astrobiology* (pp. 91–116). Cambridge: Cambridge University Press.

Caulton, M. S. (1978). The importance of habitat temperatures for growth in the tropical cichlid *Tilapia rendalli* Boulenger. *Journal of Fish Biology*, 13(1), 99–112.

Cavicchioli, R., Siddiqui, K. S., Andrews, D., & Sowers, K. S. (2002). Low-temperature extremophiles and their application. *Current Opinion in Biotechnology*, 13(3), 253–261.

Cavicchioli, R., Thomas, T., & Curmi, P. M. G. (2000). Cold stress response in Archaea. *Extremophiles*, 4(6), 321–331.

Cerveny, R. S., Lawrimore, J., Edwards, R., & Landsea, C. (2007). Extreme weather records: compilation, adjudication, and publication. *Bulletin of the American Meteorological Society*, 88(6), 853–860.

Chambers, L. E. (2008). Trends in timing of migration of south-western Australian birds and their relationship to climate. *Emu*, 108(1), 1–14.

Chambers, L. E., Beaumont, L. J., & Hudson, I. L. (2014). Continental scale analysis of bird migration timing: influences of climate and life history traits-a generalized mixture model clustering and discriminant approach. *International Journal of Biometeorology*, 58(6), 1147–1162.

Chambers, W. H. (1952). Max Rubner: (June 2, 1854–April 22, 1932). *Journal of Nutrition*, 48(1), 1–12.

Chan, J. Z.-M., Halachev, M. R., Lomain, N. J., Constantinidou, C., & Pallen, M. J. (2012). Defining bacterial species in the genomic era: insights from the genus *Acinetobacter*. *BMC Microbiology*, 12, article 302.

Chang, H. (2004). *Inventing temperature: measurement and scientific progress*. Oxford: Oxford University Press.

Chapelle, G., & Peck, L. S. (1999). Polar gigantism dictated by oxygen availability. *Nature*, 399(6732), 114–115.

Chaplin, M. (2006). Do we underestimate the importance of water in cell biology? *Nature Reviews Molecular Cell Biology*, 7(11), 861–866.

Chaplin, M. F. (2000). A proposal for the structuring of water. *Biophysical Chemistry*, 83(3), 211–221.

Chapman, S. (1930). A theory of upper atmosphere ozone. *Memoirs of the Royal Meteorological Society*, 3(26), 103–125.

Chappell, M. A., & Backman, G. C. (1995). Aerobic performance in Belding's ground squirrels (*Spermophilus beldingi*): variance, ontogeny, and the aerobic capacity model of endothermy. *Physiological Zoology*, 68(3), 421–442.

Charmantier, A., & Gienapp, P. (2014). Climate change and timing of avian breeding and migration: evolutionary versus plastic changes. *Evolutionary Applications*, 7(1), 15–28.

Charmantier, A., McCleery, R. H., Cole, L. R., Perrins, C., Kruuk, L. E. B., & Sheldon, B. C. (2008). Adaptive phenotypic plasticity in response to climate change in a wild bird population. *Science*, 320(5877), 800–803.

Chen, L. B., DeVries, A. L., & Cheng, C.-H. C. (1997a). Evolution of antifreeze glycoprotein gene from a trypsinogen gene in Antarctic notothenioid fish. *Proceedings of the National Academy of Sciences of the United States of America*, 94(8), 3811–3816.

Chen, L. B., DeVries, A. L., & Cheng, C.-H. C. (1997b). Convergent evolution of antifreeze glycoproteins in Antarctic notothenioid fish and Arctic cod. *Proceedings of the National Academy of Sciences of the United States of America*, 94(8), 3817–3822.

Chen, Z., Cheng, C.-H. C., Zhang, J., et al. (2008). Transcriptomic and genomic evolution under constant cold in Antarctic notothenioid fish. *Proceedings of the National Academy of Sciences of the United States of America*, 105(35), 12944–12949.

Cheng, C.-H. C. (1998). Origin and mechanism of evolution of antifreeze glycoproteins in polar fishes. In G. di Prisco, E. Pisano, & A. Clarke (Eds.), *Fishes of Antarctica: a biological overview* (pp. 311–328). Berlin: Springer-Verlag.

Cheng, C.-H. C., & Chen, L. B. (1999). Evolution of an antifreeze glycoprotein. *Nature*, 401(6752), 443–444.

Cheng, C.-H. C., Chen, L. B., Near, T. J., & Jun, Y. M. (2003). Functional antifreeze glycoprotein genes in temperate water New Zealand notothenioid fish infer anAntarctic evolutionary origin. *Molecular Biology and Evolution*, 20(11), 1897–1908.

Cheng, K. C., & Fuji, T. (1998). Isaac Newton and heat transfer. *Heat Transfer Engineering*, 19(4), 9–21.

Chiappe, L. M. (2007). *Glorified dinosaurs: the origin and early evolution of birds*. Hoboken, New Jersey: John Wiley & Sons.

Chiappe, L. M., & Qingjin, M. (2016). *Birds of stone: Chinese avian fossils from the age of dinosaurs*. Baltimore: Johns Hopkins University Press.

Chown, S. L., Marais, E., Terblanche, J. S., Klok, C. J., Lighton, J. R. B., & Blackburn, T. M. (2007). Scaling of insect metabolic rate is inconsistent with the nutrient supply network model. *Functional Ecology*, 21(2), 282–290.

Chown, S. L., Scholtz, C. H., Klok, C. J., Joubert, F. J., & Coles, K. S. (1995). Ecophysiology, range contraction and survival of a geographically restricted African dung beetle (Coleoptera: Scarabaeidae). *Functional Ecology*, 9(1), 30–39.

Chown, S. L., & Sinclair, B. J. (2010). The macrophysiology of insect cold-hardiness. In D. L. Denlinger & R. E. Lee (Eds.), *Low temperature biology of insects* (pp. 191–222). Cambridge: Cambridge University Press.

Chown, S. L., Sinclair, B. J., Leinas, H. P., & Gaston, K. J. (2004). Hemispheric asymmetries in biodiversity—a serious matter for ecology. *PLoS Biology*, 2(11), 1701–1707.

Christian, K. A., Baudinette, R. V., & Pamula, Y. (1997). Energetic costs of activity by lizards in the field. *Functional Ecology*, 11, 392–397.

Christian, K. A., & Bedford, G. S. (1995). Seasonal changes in thermoregulation by the frillneck lizard, *Chlamydosaurus kingii*, in tropical Australia. *Ecology*, 76(1), 124–132.

Christian, K. A.,&Weavers, B. W.(1996). Thermoregulation of monitor lizards in Australia: an evaluation of methods of thermal ecology. *Ecological Monographs, 66*(2), 139–157.

Chruszcz, B. J.,&Barclay, R. M.R.(2002).Thermoregulatory ecology of a solitary bat, Myotis evotis, roosting in rock crevices. Functional Ecology, 16(1), 18–26.

Churchfield, S., Rychlik, L., & Taylor, J. R. E. (2012). Food resources and foraging habits of the common shrew, *Sorex araneus*: does winter food shortage explain Dehnel's phenomenon? *Oikos, 121*(10), 1593–1602.

Clapeyron, B. P. É. (1834). Mémoire sur la puissance motrice de la chaleur. *Journal de l'École Royale Polytechnique, 23*(14), 153–190.

Clapham, M. E., & Karr, J. A. (2012). Environmental and biotic controls on the evolutionary history of insect body size. *Proceedings of the National Academy of Sciences of the United States of America, 109*(27), 10927–10930.

Clark, J. S., Fastie, C., Hurtt, G., et al. (1998). Reid's paradox of plant migration. *Bioscience, 48*(1), 13–24.

Clark, M. S., Fraser, K. P. P., & Peck, L. S. (2008). Antarctic marine molluscs do have an hsp70 heat shock response. *Cell Stress & Chaperones, 13*(1), 39–49.

Clarke, A. (1980). A reappraisal of the concept of metabolic cold adaptation in polar marine invertebrates. *Biological Journal of the Linnean Society, 14*(1), 77–92.

Clarke, A. (1991). What is cold adaptation and how should we measure it? *American Zoologist, 31*(1), 81–92.

Clarke, A. (1992). Is there a latitudinal diversity cline in the sea? *Trends in Ecology & Evolution, 7*(5), 286–287.

Clarke, A. (1993). Temperature and extinction in the sea: a physiologist's view. *Paleobiology, 19*(4), 499–518.

Clarke, A. (1996). The influence of climate change on the distribution and evolution of organisms. In I. A. Johnston & A. F. Bennett (Eds)., *Animals and temperature: phenotypic and evolutionary adaptation* (Vol. 59, pp. 375–407). Cambridge: Cambridge University Press.

Clarke, A. (2003). Costs and consequences of evolutionary temperature adaptation. *Trends in Ecology & Evolution, 18*(11), 573–581.

Clarke, A. (2007). Climate and diversity: the role of history. In D. Storch, P. A. Marquet, & J. H. Brown (Eds)., *Scaling biodiversity* (pp.225–245). Cambridge: Cambridge University Press.

Clarke, A. (2008). Ecological stoichiometry in six species of Antarctic marine benthos. *Marine Ecology Progress Series, 369*, 25–37.

Clarke, A. (2009). Temperature and marine macroecology. In J. D. Witman & K. Roy (Eds)., *Marine macroecology* (pp.250–278). Chicago: University of Chicago Press.

Clarke, A. (2013). Dinosaur energetics: setting the bounds on feasible physiologies and ecologies. *American Naturalist, 182*(3), 283–297.

Clarke, A. (2014). The thermal limits to life on Earth. *International Journal of Astrobiology, 13*(Special Issue 02), 141–154.

Clarke, A., & Crame, J. A. (1989). The origin of the Southern Ocean marine fauna. In J. A. Crame (Ed)., *Origins and evolution of the Antarctic biota* (pp. 253–268). London: The Geological Society. Geological Society Special Publications, 47.

Clarke, A., & Crame, J. A. (1992). The Southern Ocean benthic fauna and climate change: a historical perspective. *Philosophical Transactions of the Royal Society of London, Series B, 338*(1285), 299–309.

Clarke, A., & Crame, J. A. (1997). Diversity, latitude and time: patterns in the shallow sea. In R. F. G. Ormond, J. D. Gage, & M. V. Angel (Eds)., *Marine biodiversity: causes and consequences* (pp.122–147). Cambridge: Cambridge University Press.

Clarke, A., & Crame, J. A. (2003). The importance of historical processes in global patterns of diversity. In T. M. Blackburn & K. J. Gaston (Eds)., *Macroecology: concepts and consequences* (pp.130–151). Oxford: Blackwell.

Clarke, A., & Crame, J. A. (2010). Evolutionary dynamics at high latitudes: speciation and extinction in polar marine faunas. *Philosophical Transactions of the Royal Society of London, Series B, 365*(1558), 3655–3666.

Clarke, A., & Gaston, K. J. (2006). Temperature, energy and diversity. *Proceedings of the Royal Society of London, Series B, 273*(1599), 2257–2266.

Clarke, A., Griffiths, H. J., Linse, K., Barnes, D. K. A., & Crame, J. A. (2007). How well do we know the Antarctic marine fauna? A preliminary study of macroecological and biogeographical patterns in Southern Ocean gastropod and bivalve molluscs. *Diversity and Distributions, 13*(5), 620–632.

Clarke, A., Holmes, L. J., & Gore, D. J. (1992). Proximate and elemental composition of gelatinous zooplankton from the Southern Ocean. *Journal of Experimental Marine Biology and Ecology, 155*(1), 55–68.

Clarke, A., & Johnston, I. A. (1996). Evolution and adaptive radiation of Antarctic fishes. *Trends in Ecology & Evolution, 11*(5), 212–218.

Clarke, A., & Johnston, N. M. (1999). Scaling of metabolic rate with body mass and temperature in teleost fish. *Journal of Animal Ecology, 68*(5), 893–905.

Clarke, A., & Johnston, N. M. (2003). Antarctic marine benthic diversity. *Oceanography and Marine Biology: an Annual Review, 41*, 47–114.

Clarke, A., & Lidgard, S. (2000). Spatial patterns of diversity in the sea: bryozoan species richness in the North Atlantic. *Journal of Animal Ecology, 69*(5), 799–814.

Clarke, A., Morris, G. J., Fonseca, F., Murray, B. J., Acton, E., & Price, H. C. (2013). A low temperature limit for life on Earth. *PLoS One, 8*(6), e66207.

Clarke, A., & O'Connor, M. I. (2014). Diet and body temperature in mammals and birds. *Global Ecology and Biogeography, 23*(9), 1000–1008.

Clarke, A., & Pörtner, H. O. (2010). Temperature, metabolic power and the evolution of endothermy. *Biological Reviews, 85*(4), 703–727.

Clarke, A., & Prothero-Thomas, E. (1997). The influence of feeding on oxygen consumption and nitrogen excretion in the Antarctic nemertean *Parborlasia corrugatus*. *Physiological Zoology, 70*(6), 639–649.

Clarke, A., Prothero-Thomas, E., & Whitehouse, M. J. (1994). Nitrogen excretion in the Antarctic limpet *Nacella concinna* (Strebel, 1908). *Journal of Molluscan Studies, 60*(2), 141–147.

Clarke, A., Rodhouse, P. G., Holmes, L. J., & Pascoe, P. L. (1989). Growth rate and nucleic acid ratio in cultured cuttlefish *Sepia officinalis* (Mollusca: Cephalopoda). *Journal of Experimental Marine Biology and Ecology, 133*(3), 229–240.

Clarke, A., & Rothery, P. (2008). Scaling of body temperature in birds and mammals. *Functional Ecology, 22*(1), 58–67.

Clarke, A., Rothery, P., & Isaac, N. J. B. (2010). Scaling of basal metabolic rate with body mass and temperature in mammals. *Journal of Animal Ecology, 79*(3), 610–619.

Clarke, J. M. (1916). The reincarnation of James Eights, Antarctic explorer. *The Scientific Monthly, 2*(2), 189–202.

Clausius, R. (1879). *The mechanical theory of heat, with its application to the steam engine and to the physical properties of bodies (translation of Die Mechanische Wärmethe, 1865, edited by Thomas Archer Hirst, with an introduction by John Tyndall)* (J. Tyndall, Trans.). Paternoster Row, London: John van Voorst.

Clegg, J. S., & Drost-Hansen, W. (1991). On the biochemistry and cell physiology of water. In P. W. Hochachka & T. P. Mommsen (Eds)., *Biochemistry and molecular biology of fishes, Volume 1: Phylogenetic and biochemical perspectives* (Vol. 1, pp. 1–23). Amsterdam: Elsevier.

Clements, F. E. (1916). Plant succession: an analysis of the development of vegetation (pp. 512). Washington, D.C.: Carnegie Institution.

Clench, H. K. (1966). Behavioral thermoregulation in butterflies. *Ecology, 47*(6), 1021–1034.

Clusella-Trullas, S., Terblanche, J. S., Blackburn, T. M., & Chown, S. L. (2008). Testing the thermal melanism hypothesis: a macrophysiological approach. *Functional Ecology, 22*(2), 232–238.

Clusella-Trullas, S., van Wyk, J. H., & Spotila, J. R. (2007). Thermal melanism in ectotherms. *Journal of Thermal Biology, 32*, 235–245.

Cohan, F. M. (2002). What are bacterial species? *Annual Review of Microbiology, 56*, 457–487.

Colbert, E. H., Cowles, R. B., & Bogert, C. M. (1946). Temperature tolerances in the American alligator and their bearing on the habits, evolution, and extinction of the dinosaurs. *Bulletin of the American Museum of Natural History, 86*, 329–373.

Colwell, R. K., & Hurtt, G. C. (1994). Nonbiological gradients in species richness and a spurious Rapoport effect. *American Naturalist, 144*, 570–595.

Colwell, R. K., & Lees, D. C. (2000). The mid-domain effect: geometric constraints on the geography of species richness. *Trends in Ecology & Evolution, 15*(2), 70–76.

Colwell, R. K., Rahbek, C., & Gotelli, N. J. (2004). The mid-domain effect and species richness patterns: what have we learned so far? *American Naturalist, 163*(3), E1–E23.

Colwell, R. K., Rahbek, C., & Gotelli, N. J. (2005). The mid-domain effect: there's a baby in the bathwater. *American Naturalist, 166*(5), E149–E154.

Connell, J. H. (1978). Diversity in tropical rain forests and coral reefs: high diversity of trees and corals is maintained only in a nonequilibrium state. *Science, 199*(4335), 1302–1310.

Connolly, S. R. (2005). Process-based models of species distributions and the mid-domain effect. *American Naturalist, 166*(1), 1–11.

Connolly, S. R., Bellwood, D. R., & Hughes, T. P. (2003). Indo-Pacific biodiversity of coral reefs: deviations from a mid-domain model. *Ecology, 84*(8), 2178–2190.

Connor, E. F., & McCoy, E. D. (1979). The statistics and biology of the species–area relationship. *American Naturalist, 113*(6), 791–833.

Conover, D. O. (1990). The relation between capacity for growth and length of growing season: evidence for and implications of countergradient variation. *Transactions of the American Fisheries Society, 119*(3), 416–430.

Conover, D. O., Brown, J. J., & Ehtisham, A. (1997). Countergradient variation in growth of young striped bass (*Morone saxatilis*) from different latitudes. *Canadian Journal of Fisheries and Aquatic Sciences, 54*(10), 2401–2409.

Conover, D. O., & Schultz, E. T. (1995). Phenotypic similarity and the evolutionary significance of countergradient variation. *Trends in Ecology & Evolution, 10*(6), 248–252.

Conover, R. J. (1978). Transformation of organic matter. In O. Kinne (Ed)., *Marine ecology (Volume 4: Dynamics)* (pp.221–499). Chichester: John Wiley & Sons.

Cook, A. J., Fox, A. J., Vaughan, D. G., & Ferrigno, J. G. (2005). Retreating glacier fronts on the Antarctic Peninsula over the past half-century. *Science, 308*(5721), 541–544.

Cook, A. J., Holland, P. R., Meredith, M. P., Murray, T., Luckman, A., & Vaughan, D. G. (2016). Ocean forcing of glacier retreat in the western Antarctic Peninsula. *Science, 353*(6296), 283–286.

Cook, D., Fowler, S., Fiehn, O., & Thomashow, M. F. (2004). A prominent role for the CBF cold response pathway in configuring the low-temperature metabolome of *Arabidopsis*.

Proceedings of the National Academy of Sciences of the United States of America, 101(42), 15243–15248.

Coope, G. R. (1959). A Late Pleistocene insect fauna from Chelford, Cheshire. *Proceedings of the Royal Society of London, Series B, 151*(942), 70–86.

Coope, G. R., & Brophy, J. A. (1972). Late glacial environmental changes indicated by a coleopteran succession from North Wales. *Boreas (Oslo), 1*(2), 97–142.

Coope, G. R., Gibbard, P. L., Hall, A. R., Preece, R. C., Robinson, J. E., & Sutcliffe, A. J. (1997). Climatic and environmental reconstructions based on fossil assemblages from Middle Devensian (Weichselian) deposits of the River Thames at South Kensington, Central London, UK. *Quaternary Science Reviews, 16*, 1163–1195.

Coope, G. R., Lemdahl, G., Lowe, J. J., & Walkling, A. (1998). Temperature gradients in northern Europe during the last glacial-Holocene transition (14–19 C-14 kyr BP) interpreted from coleopteran assemblages. *Journal of Quaternary Science, 13*, 419–433.

Cooper, S. (2000). Seasonal energetics of Mountain Chickadees and Juniper Titmice. *The Condor, 102*(3), 635–644.

Cooper, S., & Sonsthagen, S. (2007). Heat production from foraging activity contributes to thermoregulation in Black-capped Chickadees. *The Condor, 109*(2), 446–451.

Cooper, W. E., & Vitt, L. J. (2002). Distribution, extent, and evolution of plant consumption by lizards. *Journal of Zoology, 257*(4), 487–517.

Coopersmith, J. (2010). *Energy: the subtle concept*. Oxford: Oxford University Press.

Coppe, A., Agostini, C., Marino, I. A. M., et al. (2013). Genome evolution in the cold: Antarctic icefish muscle transcriptome reveals selective duplications increasing mitochondrial function. *Genome Biology and Evolution, 5*(1), 45–60.

Corkrey, R., McMeekin, T. A., Bowman, J. P., Ratkowsky, D. A., Olley, J., & Ross, T. (2016). The biokinetic spectrum for temperature. *PLoS One, 11*(4), e0153343.

Cornish-Bowden, A. (2002). Enthalpy–entropy compensation: a phantom phenomenon. *Journal of Bioscience, 27*(2), 121–126.

Cossins, A. R., Bowler, K., & Prosser, C. L. (1981). Homeoviscous adaptation and its effects upon membrane-bound enzymes. *Journal of Thermal Biology, 6*(4), 183–187.

Cossins, A. R., & Prosser, C. L. (1978). Evolutionary adaptation of membranes to temperature. *Proceedings of the National Academy of Sciences of the United States of America, 75*(4), 2040–2043.

Costanzo, J. P., & Lee, R. E. (2013). Avoidance and tolerance of freezing in ectothermic vertebrates. *Journal of Experimental Biology, 216*(11), 1961–1967.

Costanzo, J. P., Reynolds, A. M., do Amaral, M. C. F., Rosendale, A. J., & Lee, R. E. (2015). Cryoprotectants and extreme freeze tolerance in a subarctic population of the wood frog. *PLoS One, 10*(2), e0117234.

Cott, H. B. (1940). *Adaptive coloration in animals*. London: Methuen.

Cotton, P. A. (2003). Avian migration phenology and global climate change. *Proceedings of the National Academy of Sciences of the United States of America, 100*(21), 12219–12222.

Coulson, R. A., Hernandez, T., & Herbert, J. D. (1977). Metabolic rate, enzyme kinetics *in vivo*. *Comparative Biochemistry and Physiology, Part A, 56*(3), 251–262.

Cowles, R. B. (1940). Additional implications of reptilian sensitivity to high temperatures. *American Naturalist, 74*, 542–561.

Cowles, R. B. (1946). Fur and feathers: a result of high temperatures? *Science, 103*, 74–75.

Cowles, R. B. (1958). Possible origin of dermal temperature regulation. *Evolution, 12*, 347–357.

Cowles, R. B. (1962). Semantics in biothermal studies. *Science, 135*(3504), 670.

Cowles, R. B., & Dawson, W. R. (1951). A cooling mechanism for the Texas Nighthawk. *Condor, 53*(1), 19–22.

Cox, F., Newsham, K. K., Bol, R., Dungait, J. A. J., & Robinson, C. H. (2016). Not poles apart: Antarctic soil fungal communities show similarities to those of the distant Arctic. *Ecology Letters, 19*(5), 528–536.

Coyne, J. A., & Orr, H. A. (2004). *Speciation*. Sunderland, MA: Sinauer.

Crame, J. A. (2000a). Evolution of taxonomic diversity gradients in the marine realm: evidence from the composition of Recent bivalve faunas. *Paleobiology, 26*(2), 188–214.

Crame, J. A. (2000b). The nature and origin of taxonomic diversity gradients in marine bivalves. In E. M. Harper, J. D. Taylor, & J. A. Crame (Eds.)., *The evolutionary biology of the Bivalvia* (Vol. 177, pp.347–360). London: The Geological Society.

Crame, J. A. (2001). Taxonomic diversity gradients through geological time. *Diversity and Distributions, 7*, 175–189.

Crame, J. A., & Clarke, A. (1997). The historical component of taxonomic diversity gradients. In R. F. G. Ormond, J. D. Gage, & M. V. Angel (Eds.)., *Marine biodiversity: causes and consequences* (pp. 258–273). Cambridge: Cambridge University Press.

Crampton, J. S., Beu, A. G., Cooper, R. A., Jones, C. M., Marshall, B., & Maxwell, P. A. (2003). Estimating the rock volume bias in paleobiodiversity studies. *Science, 301*(5631), 358–360.

Crampton, J. S., Foote, M., Beu, A. G., et al. (2006). The ark was full! Constant to declining Cenozoic shallow marine biodiversity on an isolated midlatitude continent. *Paleobiology, 32*(4), 509–532.

Crawford, D. L., & Powers, D. A. (1989). Molecular basis of evolutionary adaptation at the lactate dehydrogenase-B locus in the fish *Fundulus heteroclitus*. *Proceedings of the National Academy of Sciences of the United States of America, 86*(23), 9365–9369.

Crawford, E. C. (1962). Mechanical aspects of panting in dogs. *Journal of Applied Physiology, 17*(2), 249–251.

Crawford, E. C., & Kampe, G. (1971). Resonant panting in pigeons. *Comparative Biochemistry and Physiology, Part A, 40*(2), 549–552.

Crawshaw, L. I., & Hammel, H. T. (1971). Behavioral thermoregulation in two species of Antarctic fish. *Life Science, 10,* 1009–1020.

Crick, H. Q. P. (2004). The impact of climate change on birds. *Ibis, 146*(Supplement 1), 48–56.

Crisp, D. J., Davenport, J., & Gabbott, P. A. (1977). Freezing tolerance in *Balanus balanoides*. *Comparative Biochemistry and Physiology, Part A, 57*(3), 359–361.

Crompton, A. W., Taylor, C. R., & Jagger, J. A. (1978). Evolution of endothermy in mammals. *Nature, 272,* 333–336.

Cross, E. L., Peck, L. S., & Harper, E. M. (2015). Ocean acidification does not impact shell growth or repair of the Antarctic brachiopod *Liothyrella uva* (Broderip, 1833). *Journal of Experimental Marine Biology and Ecology, 462*(1), 29–35.

Cross, E. L., Peck, L. S., Lamare, M. D., & Harper, E. M. (2016). No ocean acidification effects on shell growth and repair in the New Zealand brachiopod *Calloria inconspicua* (Sowerby, 1846). *ICES Journal of Marine Science, 73*(3), 920–926.

Cruz, A. L. B., Hebly, M., Duong, G.-H., et al. (2012). Similar temperature dependencies of glycolytic enzymes: an evolutionary adaptation to temperature dynamics? *BMC Systems Biology, 6,* article 151.

Cuellar, J., Yébenes, H., Parker, S. K., et al. (2014). Assisted protein folding at low temperature: evolutionary adaptation of the Antarctic fish chaperonin CCT and its client proteins. *Biology Open, 3*(4), 261–270.

Culver, S. J., & Buzas, M. A. (2000). Global latitudinal species diversity gradient in deep-sea benthic foraminifera. *Deep-Sea Research, Part I: Oceanographic Research Papers, 47*(2), 259–275.

Currano, E. D., Labandeira, C. C., & Wilf, P. (2010). Fossil insect folivory tracks paleotemperature for six million years. *Ecological Monographs, 80*(4), 547–567.

Currano, E. D., Wilf, P., Wing, S. L., Labandeira, C. C., Lovelock, E. C., & Royer, D. L. (2008). Sharply increased insect herbivory during the Paleocene-Eocene thermal maximum. *Proceedings of the National Academy of Sciences of the United States of America, 105*(4), 1960–1964.

Currie, D. J. (1991). Energy and large-scale patterns of animal and plant species richness. *American Naturalist, 137,* 27–49.

Currie, D. J., Mittelbach, G. G., Cornell, H. V., et al. (2004). Predictions and tests of climate-based hypotheses of broad-scale variation in taxonomic richness. *Ecology Letters, 7*(12), 1121–1134.

Currie, D. J., & Paquin, V. (1987). Large-scale biogeographical patterns of species richness of trees. *Nature, 329*(6137), 326–327.

Cyronak, T., Schulz, K. G., & Jokiel, P. L. (2016). The Omega myth: what really drives lower calcification rates in an acidifying ocean. *ICES Journal of Marine Science, 73*(3), 558–562.

Cziko, P. A., DeVries, A. L., Evans, C. W., & Cheng, C.-H. C. (2014). Antifreeze protein-induced superheating of ice inside Antarctic notothenioid fishes inhibits melting during summer warming. *Proceedings of the National Academy of Sciences of the United States of America, 111*(40), 14583–14588.

Daan, S., Masman, D., & Groenewold, A. (1990). Avian basal metabolic rates: their association with body composition and energy expenditure in nature. *American Journal of Physiology, 259*(2), R333–R340.

Dallinger, W. H. (1887). The president's address. *Journal of the Royal Microscopical Society, 7*(2), 185–199.

D'Amico, S., Collins, T., Marx, J.-C., Feller, G., & Gerday, C. (2006). Psychrophilic microorganisms: challenges for life. *EMBO Reports, 7*(4), 385–389.

Damon, P. E., & Laut, P. (2004). Pattern of strange errors plagues solar activity and terrestrial climate data. *Eos, Transactions, American Geophysical Union, 85*(39), 370–374.

Damuth, J. (1987). Interspecific allometry of population density in mammals and other animals: the independence of body mass and population energy use. *Biological Journal of the Linnean Society, 31*(3), 193–246.

Daniel, R. M., & Danson, M. J. (2010). A new understanding of how temperature affects the catalytic activity of enzymes. *Trends in Biochemical Sciences, 35*(10), 584–591.

Daniel, R. M., Danson, M. J., Eisenthal, R., Lee, C. K., & Peterson, M. E. (2007). New parameters controlling the effect of temperature on enzyme activity. *Biochemical Society Transactions, 35,* 1543–1546.

Darwin, C. (1839). *Journal of researches into the natural history and geology of the countries visited during the voyage round the world of H.M.S. 'Beagle' under the command of Captain Fitzroy, R.N.* (Vol. 3). London: John Murray.

Darwin, C. (1859). *On the origin of species by means of natural selection, or the preservation of favoured races in the struggle for life.* London: John Murray.

Daufresne, M., Lengfellner, K., & Sommer, U. (2009). Global warming benefits the small in aquatic ecosystems. *Proceedings of the National Academy of Sciences of the United States of America, 106*(31), 12788–12793.

Dausmann, K. H. (2005). Hibernation in the tropics: lessons from a primate. *Journal of Comparative Physiology, B, 157*(3), 147–155.

Dausmann, K. H., Glos, J., Ganzhorn, J. U., & Heldmaier, G. (2004). Hibernation in a tropical primate. *Nature, 429*(6994), 825–826.

Davenport, J. (1992). *Animal life at low temperature*. London: Chapman & Hall.

Davies, R. G., Orme, C. D. L., Storch, D., et al. (2007). Topography, energy and the global distribution of bird species richness. *Proceedings of the Royal Society of London, Series B, 274*(1614), 1189–1197.

Davies, T. J., Barraclough, T. G., Savolainen, V., & Chase, M. W. (2004). Environmental causes for plant biodiversity gradients. *Philosophical Transactions of the Royal Society of London, Series B, 359*(1450), 1645–1656.

Davies, T. J., Grenyer, R., & Gittleman, J. L. (2005). Phylogeny can make the mid-domain effect an inappropriate null model. *Biology Letters, 1*(2), 143–146.

Davis, D. J., & Lee, R. E. (2001). Intracellular freezing, viability, and composition of fat body cells from freeze-intolerant larvae of *Sarcophaga crassipalpis*. *Archives of Insect Biochemistry and Physiology, 48*(4), 199–205.

Davy, J. (1835). On the temperature of some fishes in the genus *Thynnus*. *Edinburgh New Philosophical Journal, 19*(1), 325–330.

Dawson, T. H. (2001). Similitude in the cardiovascular system of mammals. *Journal of Experimental Biology, 204*(3), 395–407.

Dawson, T. H. (2003). Scaling laws for capillary vessels of mammals at rest and in exercise. *Proceedings of the Royal Society of London, Series B, 270*(1516), 755–763.

Dawson, T. H. (2010). Scaling laws for plasma concentrations and tolerable doses of anticancer drugs. *Cancer Research, 70*(12), 4801–4808.

Dawson, T. J., Blaney, C. E., Munn, A. J., Krockenberger, A., & Maloney, S. K. (2000). Thermoregulation by kangaroos from mesic and arid habitats: influence of temperature on routes of heat loss in eastern grey kangaroos (*Macropus giganteus*) and red kangaroos (*Macropus rufus*). *Physiological and Biochemical Zoology, 73*(3), 374–381.

Dawson, T. J., Webster, K. N., & Maloney, S. K. (2014). The fur of mammals in exposed environments; do crypsis and thermal needs necessarily conflict? The polar bear and marsupial koala compared. *Journal of Comparative Physiology, B, 184*(2), 273–284.

Dawson, W. R., & Bartholomew, G. A. (1956). Relation of oxygen consumption to body weight, temperature, and temperature acclimation in lizards *Uta stansburiana* and *Sceloporus occidentalis*. *Physiological Zoology, 29*(1), 40–51.

Day, T., & Taylor, P. D. (1997). Von Bertalanffy's growth equation should not be used to model age and size at maturity. *American Naturalist, 149*(2), 381–393.

Dayton, P. K., Robilliard, G. A., & DeVries, A. L. (1969). Anchor ice formation in McMurdo Sound, Antarctica, and its biological effects. *Science, 163*, 273–274.

de Cock Buning, T. (1983). Thermal sensitivity as a specialization for prey capture and feeding in snakes. *American Zoologist, 23*(2), 363–375.

de Jong, P. W., Gussekloo, S. W. S., & Brakefield, P. M. (1996). Differences in thermal balance, body temperature and activity between non-melanic and melanic two-spot ladybird beetles (*Adalia bipunctata*) under controlled conditions. *Journal of Experimental Biology, 199*(12), 2655–2666.

de Lamarck, J.-B. (1778). *Flore Française (in 3 volumes)*. Paris: Imprimerie royale.

de Lavoisier, A.-L., & Laplace, P.-S. (1784). Mémoire sur la chaleur. *Mémoires de l'Académie des Sciences, 1780*, 355–408.

de Lavoisier, A.-L., & Séguin, A. (1793). Premier mémoire sur la respiration des animaux. *Mémoires de l'Académie Royale des Sciences, 1789*.

de Lavoisier, A.-L., & Séguin, A. (1814). Second mémoire sur la respiration. *Annales du Chimie, 91*, 318–334.

de Réaumur, R.-A. F. (1736). *Mémoires pour servir à l'histoire des insectes* (Vol. 2). Paris: L'imprimerie Royale.

de Vera, J.-P., Rettberg, P., & Ott, S. (2008). Life at the limits: capacities of isolated and cultured lichen symbionts to resist extreme environmental stresses. *Origins of Life and Evolution of Biospheres, 38*(5), 457–468.

De Wit, R., & Bouvier, T. (2006). '*Everything is everywhere, but, the environment selects*'; what did Baas Becking and Beijerinck really say? *Environmental Microbiology, 8*(4), 755–758.

Debenedetti, P. G. (1996). *Metastable liquids: concepts and principles*. Princeton, New Jersey: Princeton University Press.

Dehnel, A. (1949). Studies on the genus *Sorex* (In Polish with English summary). *Annales Universititatis Mariae Curie-Skłodowska, Sectio C, Biologia, 4*(2), 17–97.

Dell, A. I., Pawar, S. S., & Savage, V. M. (2011). Systematic variation in the temperature dependence of physiological and ecological traits. *Proceedings of the National Academy of Sciences of the United States of America, 108*(26), 10591–10596.

DeLong, K. L., Quinn, T. M., & Taylor, F. W. (2007). Reconstructing twentieth-century sea surface temperature variability in the southwest Pacific: a replication study using multiple coral Sr/Ca records from New Caledonia. *Paleoceanography, 22*(4), PA4212.

Demetrius, L. (2003). Quantum statistics and allometric scaling of organisms. *Physica A—Statistical Mechanics and its Applications, 322*(1–4), 477–490.

Demetrius, L. (2006). The origin of allometric scaling laws in biology. *Journal of Theoretical Biology, 243*(4), 455–467.

Demetrius, L., & Tuszynski, J. A. (2010). Quantum metabolism explains the allometric scaling of metabolic rates. *Journal of the Royal Society Interface, 7*(44), 507–514.

Deming, J. W. (1986). Thermophilic bacteria associated with black smokers along the East Pacific Rise. *Actes de Colloques, 3*, 325–332.

Deng, C., Cheng, C.-H. C., Ye, H., He, X., & Chen, L. (2010). Evolution of an antifreeze protein by neofunctionalization under escape from adaptive conflict. *Proceedings of the National Academy of Sciences of the United States of America, 107*(50), 21593–21598.

Denny, M. W. (1993). *Air and water: the biology and physics of life's media*. Princeton, New Jersey: Princeton University Press.

Desbruyères, D., Chevaldonné, P., Alayse, A.-M., et al. (1998). Biology and ecology of the 'Pompeii worm' (*Alvinella pompejana* Desbruyères and Laubier), a normal dweller of an extreme deep-sea environment: a synthesis of current knowledge and recent developments. *Deep-Sea Research, Part II: Topical Studies in Oceanography, 45*(1–3), 383–422.

Desmond, A. J., & Moore, J. (1991). *Darwin*. London: Michael Joseph.

Desmond-Le Quéméner, E., & Bouchez, T. (2014). A thermodynamic theory of microbial growth. *ISME Journal, 8*(8), 1747–1751.

Dessauer, H. C. (1953). Hibernation of the lizard *Anolis carolinensis*. *Experimental Biology and Medicine, 82*(2), 351–353.

Detrich, H. W., & Amemiya, C. T. (2010). Antarctic notothenioid fishes: genomic resources and strategies for analyzing an adaptive radiation. *Integrative and Comparative Biology, 50*(6), 1009–1017.

Deutsch, C. A., Tewksbury, J. J., Huey, R. B., et al. (2008). Impacts of climate warming on terrestrial ectotherms across latitude. *Proceedings of the National Academy of Sciences of the United States of America, 105*(18), 6668–6672.

DeVries, A. L., Komatsu, S. K., & Feeney, R. E. (1970). Chemical and physical properties of freezing point-depressing glycoproteins from Antarctic fishes. *Journal of Biological Chemistry, 245*(11), 2901–2908.

DeVries, A. L., Vandenheede, J., & Feeney, R. E. (1971). Primary structure of freezing point-depressing glycoproteins. *Journal of Biological Chemistry, 246*(2), 305–308.

DeVries, A. L., & Wohlschlag, D. E. (1969). Freezing resistance in some Antarctic fishes. *Science, 163*(3871), 1973–1075.

Diamond, S. E., & Kingsolver, J. G. (2010). Environmental dependence of thermal reaction norms: host plant quality can reverse the temperature-size rule. *American Naturalist, 175*(1), 1–10.

Dias, C. L., Ala-Nissila, T., Wong-ekkabut, J., Vattulainen, I., Grant, M., & Karttunen, M. (2010). The hydrophobic effect and its role in cold denaturation. *Cryobiology, 60*(1 (S1)), 91–99.

Dickson, K. A., & Graham, J. B. (2004). Evolution and consequences of endothermy in fishes. *Physiological and Biochemical Zoology, 77*(6), 998–1018.

Dilly, G. F., Young, C. R., Lane, W. S., Pangilinan, J., & Girguis, P. R. (2012). Exploring the limit of metazoan thermal tolerance via comparative proteomics: thermally induced changes in protein abundance by two hydrothermal vent polychaetes. *Proceedings of the Royal Society of London, Series B, 279*(1741), 3347–3356.

Dimmick, R. L., Straat, P. A., Wolochow, H., Levin, G. V., Chatigny, M. A., & Schrot, J. R. (1975). Evidence for metabolic activity of airborne bacteria. *Journal of Aerosol Science, 6*(6), 387–393.

Dodds, P. S., Rothman, D. H., & Weitz, J. S. (2001). Re-examination of the '3/4-law' of metabolism. *Journal of Theoretical Biology, 209*(1), 9–27.

Doemel, W. N., & Brock, T. D. (1970). Upper temperature limit of *Cyanidium caldarium*. *Archiv für Mikrobiologie, 72*(4), 326–332.

Doemel, W. N., & Brock, T. D. (1971). Physiological ecology of *Cyanidium caldarium*. *Journal of General Microbiology, 67*(1), 17–32.

Dolphin, K., & Quicke, D. L. J. (2001). Estimating the global species richness of an incompletely described taxon: an example using parasitoid wasps (Hymenoptera: Braconidae). *Biological Journal of the Linnean Society, 73*(3), 279–286.

Doney, S. C., Fabry, V. J., Feeley, R. A., & Kleypas, J. A. (2009). Ocean acidification: the other CO_2 problem. *Annual Review of Marine Science, 1*, 169–192.

Donovan, S. K. (Ed).. (1989). *Mass extinctions: processes and evidence*. London: Belhaven Press.

Dotterweich, H. (1928). Beiträge zur Nervenphysiologie der Insekten. *Zoologische Jahrbücher. Abteilung für allgemeine Zoologie und Physiologie der Tiere, 44*, 399–450.

Doucet, C. J., Byass, L., Elias, L., Worrall, D., Smallwood, M., & Bowles, D. J. (2000). Distribution and characterization of recrystallization inhibitor activity in plant and lichen species from UK and maritime Antarctic. *Cryobiology, 40*(3), 218–227.

Dougherty, R. C. (1998). Temperature and pressure dependence of hydrogen bond strength: A perturbation molecular orbital approach. *Journal of Chemical Physics, 109*(17), 7372–7378.

Downs, C. J., Hayes, J. P., & Tracy, C. R. (2008). Scaling metabolic rate with body mass and inverse body temperature: a test of the Arrhenius fractal supply model. *Functional Ecology, 22*(2), 239–244.

Drost-Hansen, W. (1982). The occurrence and extent of vicinal water. In F. Franks & S. F. Mathias (Eds)., *Biophysics of water* (pp.163–169). Chichester: John Wiley & Sons.

DuBois, E. F. (1927). *Basal metabolism in health and disease* (2nd ed.). Philadelphia: Lea & Febiger.

Dudley, R. (1998). Atmospheric oxygen, giant Paleozoic insects and the evolution of aerial locomotor performance. *Journal of Experimental Biology, 201*(8), 1043–1050.

Duman, J. G. (1977). The role of macromolecular antifreeze in the darkling beetle, *Meracantha contracta*. *Journal of Comparative Physiology, 115*(2), 279–286.

Duman, J. G. (1993). Purification and characterization of a thermal hysteresis protein from a plant, the bittersweet nightshade *Solanum dulcamara*. *Biochimica et Biophysica Acta*, *1206*(1), 129–135.

Duman, J. G. (2001). Antifreeze and ice nucleator proteins in terrestrial arthropods. *Annual Review of Physiology*, *63*, 327–357.

Duman, J. G. (2014). An early classic study of freeze avoidance in fish. *Journal of Experimental Biology*, *217*(6), 820–823.

Duman, J. G. (2015). Animal ice-binding (antifreeze) proteins and glycolipids: an overview with emphasis on physiological function. *Journal of Experimental Biology*, *218*(12), 1846–1855.

Duman, J. G., & DeVries, A. L. (1974). Freezing resistance in winter flounder *Pseudopleuronectes americanus*. *Nature*, *247*(5438), 237–238.

Duman, J. G., & DeVries, A. L. (1976). Isolation, characterization, and physical properties of protein antifreezes from the winter flounder, *Pseudopleuronectes americanus*. *Comparative Biochemistry and Physiology, Part B*, *54*(3), 375–380.

Duman, J. G., & Olsen, T. M. (1993). Thermal hysteresis activity in bacteria, fungi, and phyogenetically diverse plants. *Cryobiology*, *30*(3), 322–328.

Duman, J. G., Walters, K. R., Sformo, T., et al. (2010). Antifreeze and ice-nucleator proteins. In D. L. Denlinger & R. E. Lee (Eds.), *Low temperature biology of insects* (pp.59–90). Cambridge: Cambridge University Press.

Dunham, A. E., Grant, B. W., & Overall, K. L. (1989). Interfaces between biophysical and physiological ecology and the population ecology of terrestrial vertebrate ectotherms. *Physiological Zoology*, *62*(2), 335–355.

Dunitz, J. D. (1995). Win some, lose some: enthalpy–entropy compensation in weak intermolecular interactions. *Chemistry & Biology*, *2*, 709–712.

Dunton, K. (1992). Arctic biogeography: the paradox of the marine benthic fauna and flora. *Trends in Ecology & Evolution*, *7*(6), 183–189.

Durack, P. J., Glecker, P. J., Landerer, F. W., & Taylor, K. E. (2014). Quantifying underestimates of long-term upper-ocean warming. *Nature Climate Change*, *4*(11), 999–1005.

Dure, L., & Chlan, C. (1981). Developmental biochemistry of cottonseed embryogenesis and germination. XII. Purification and properties of principal storage proteins. *Plant Physiology*, *68*(1), 180–186.

Dutenhoffer, M. S., & Swanson, D. L. (1996). Relationship of basal to summit metabolic rate in passerine birds and the aerobic capacity model for the evolution of endothermy. *Physiological Zoology*, *69*(5), 1232–1254.

Dynesius, M., & Jansson, R. (2000). Evolutionary consequences of changes in species' geographical distributions driven by Milankovitch climate oscillations. *Proceedings of the National Academy of Sciences of the United States of America*, *97*(16), 9115–9120.

Dzialowski, E. M. (2005). Use of operative temperature and standard operative temperature models in thermal biology. *Journal of Thermal Biology*, *30*(4), 317–334.

Eagle, R. A., Schauble, E. A., Tripati, A. K., Tütken, T., Hulbert, R. C., & Eiler, J. M. (2010). Body temperatures of modern and extinct vertebrates from ^{13}C–^{18}O bond abundances in bioapatite. *Proceedings of the National Academy of Sciences of the United States of America*, *107*(23), 10377–10382.

Eagle, R. A., Tütken, T., Martin, T. S., et al. (2011). Dinosaur body temperatures determined from isotopic (^{13}C–^{18}O) ordering in fossil biominerals. *Science*, *333*(6041), 443–445.

Early, R., & Sax, D. F. (2011). Analysis of climate paths reveals potential limitations on species range shifts. *Ecology Letters*, *14*(11), 1125–1133.

Eastman, J. T. (1993). *Antarctic fish biology: evolution in a unique environment*. San Diego, California: Academic Press.

Eastman, J. T. (2005). The nature of the diversity of Antarctic fishes. *Polar Biology*, *28*(2), 93–107.

Eastman, J. T., Boyd, R. B., & DeVries, A. L. (1987). Renal corpuscle development in boreal fishes with and without antifreezes. *Fish Physiology and Biochemistry*, *4*(2), 89–100.

Eastman, J. T., & Clarke, A. (1998). A comparison of adaptive radiations of Antarctic fish with those of nonantarctic fish. In G. di Prisco, E. Pisano, & A. Clarke (Eds.), *Fishes of Antarctica: a biological overview* (pp. 3–26). Berlin: Springer-Verlag.

Eastman, J. T., & Eakin, R. R. (2000). An updated species list for notothenioid fish (Perciformes; Notothenioidei), with comments on Antarctic species. *Archive of Fishery and Marine Research*, *48*(1), 11–20.

Eddington, A. S. (1928). *The nature of the physical world*. Cambridge: Cambridge University Press.

Edmund, J. M., & Gieskes, J. M. T. M. (1970). On the calculation of the degree of saturation of sea water with respect to calcium carbonate under *in situ* conditions. *Geochimica et Cosmochimica Acta*, *34*(12), 1261–1291.

Ege, R., & Krogh, A. (1914). On the relation between the temperature and the respiratory exchange in fishes. *Internationale Revue der gesamten Hydrobiologie und Hydrographie*, *7*(1), 48–55.

Egginton, S., Cordiner, S., & Skilbeck, C. (2000). Thermal compensation of peripheral oxygen transport in skeletal muscle of seasonally acclimatized trout. *American Journal of Physiology—Regulatory, Integrative and Comparative Physiology*, *279*(2), R375–R388.

Egginton, S., & Sidell, B. D. (1989). Thermal acclimation induces adaptive changes in subcellular structure of fish skeletal muscle. *American Journal of Physiology—Regulatory, Integrative and Comparative Physiology, 256*(1), R1–R9.

Eggleton, P. (1994). Termites live in a pear-shaped world: a response to Platnick. *Journal of Natural History, 28*(5), 1209–1212.

Eglington, S. M., Julliard, R., Gargallo, G., et al. (2015). Latitudinal gradients in the productivity of European migrant warblers have not shifted northwards during a period of climate change. *Global Ecology and Biogeography, 24*(4), 427–436.

Ehnes, R. B., Rall, B. C., & Brose, U. (2011). Phylogenetic grouping, curvature and metabolic scaling in terrestrial invertebrates. *Ecology Letters, 14*(10), 993–1000.

Eiler, J. M. (2007). 'Clumped-isotope' geochemistry—The study of naturally-occurring, multiply-substituted isotopologues. *Earth and Planetary Science Letters, 262*, 309–327.

Eiler, J. M. (2011). Paleoclimate reconstruction using carbonate clumped isotope thermometry. *Quaternary Science Review, 30*(25–26), 3575–3588.

El Fadli, K. I., Cerveny, R. S., Burt, C. C., et al. (2013). World Meteorological Association assessment of the purported world record 58 °C temperature extreme at Al Azizia, Libya (13 September 1922). *Bulletin of the American Meteorological Society, 94*(2), 199–204.

Elias, M., Wieczorek, G., Rosenne, S., & Tawfik, D. S. (2014). The universality of enzymatic rate-temperature dependency. *Trends in Biochemical Sciences, 39*(1), 1–7.

Elias, S. A. (1994). *Quaternary insects and their environments*. Washington, D.C.: The Smithsonian Institute.

Elias, S. A. (1997). The mutual climatic range method of palaeoclimate reconstruction based on insect fossils: new applications and interhemispheric comparisons. *Quaternary Science Reviews, 16*(10), 1217–1225.

Elias, S. A. (2013). Beetle records. In S. A. Elias (Ed)., *Encyclopedia of Quaternary Science* (Second ed., pp. 161–172). Amsterdam: Elsevier.

Elkibbi, M., & Rial, J. A. (2001). An outsider's review of the astronomical theory of the climate: is the eccentricity-driven insolation the main driver of the ice ages? *Earth-Science Reviews, 56*(1–4), 161–177.

Ellers, J., & Boggs, C. L. (2004). Functional ecological implications of intraspecific differences in wing melanization in *Colias* butterflies. *Biological Journal of the Linnean Society, 82*(1), 79–87.

Ellingsen, K. E., & Gray, J. S. (2002). Spatial patterns of benthic diversity: is there a latitudinal gradient along the Norwegian continental shelf? *Journal of Animal Ecology, 71*(3), 373–389.

Elliott, J. M. (1981). Some aspects of thermal stress on freshwater teleosts. In A. D. Pickering (Ed)., *Stress and fish* (pp. 209–245). London: Academic Press.

Ellis, R. J. (2001). Macromolecular crowding: obvious but unappreciated. *Trends in Biochemical Sciences, 26*(10), 597–604.

Else, P. L., & Hulbert, A. J. (1981). Comparison of the 'mammal machine' and the 'reptile machine': energy production. *American Journal of Physiology, 240*, R3–R9.

Elvidge, A. D., Renfrew, I. A., King, J. C., et al. (2016). Foehn jets over the Larsen C Ice Shelf, Antarctica. *Quarterly Journal of the Royal Meteorological Society, 141*(688), 698–713.

Engle, V. D., & Summers, J. K. (1999). Latitudinal gradients in benthic community composition in Western Atlantic estuaries. *Journal of Biogeography, 26*(5), 1007–1023.

Eppley, R. W. (1972). Temperature and phytoplankton growth in the sea. *Fishery Bulletin, 70*(4), 1063–1085.

Epstein, S., Buchsbaum, R., Lowenstam, H. A., & Urey, H. C. (1951). Carbonate-water isotopic temperature scale. *Bulletin of the Geological Society of America, 62*, 417–426.

Epstein, S., Buchsbaum, R., Lowenstam, H. A., & Urey, H. C. (1953). Revised carbonate-temperature isotopic temperature scale. *Bulletin of the Geological Society of America, 64*, 1315–1326.

Erickson, G. M., Makovicky, P. J., Inouye, B. D., Zhou, C.-F., & Gao, K.-Q. (2009). A life table for *Psittacosaurus lujiatunensis*: initial insights into ornithischian dinosaur population biology. *The Anatomical Record, 292*(10), 1514–1521.

Erwin, T. L. (1982). Tropical forests: their richness in Coleoptera and other arthropod species. *The Coleopterists Bulletin, 36*(1), 74–75.

Espinoza, R. E., Wiens, J. J., & Tracy, C. R. (2004). Recurrent evolution of herbivory in small, cold-climate lizards: Breaking the ecophysiological rules of reptilian herbivory. *Proceedings of the National Academy of Sciences of the United States of America, 101*(48), 16819–16824.

Etienne, R. S., Apol, M. E. F., & Olff, H. (2006). Demystifying the West, Brown & Enquist model of the allometry of metabolism. *Functional Ecology, 20*(2), 394–399.

Ettinger, A. K., Ford, K. R., & HilleRisLambers, J. (2011). Climate determines upper, but not lower, altitudinal range limits of Pacific Northwest conifers. *Ecology, 92*(6), 1323–1331.

Evans, C. W., Gubala, V., Nooney, R., Williams, D. E., Brimble, M. A., & DeVries, A. L. (2011). How do Antarctic notothenioid fishes cope with internal ice? A novel function for antifreeze glycoproteins. *Antarctic Science, 23*(1), 57–64.

Evans, K. L., & Gaston, K. J. (2005). Can the evolutionary-rates hypothesis explain species-energy relationships? *Functional Ecology, 19*(6), 899–915.

Evans, K. L., Greenwood, J. J. D., & Gaston, K. J. (2005). Dissecting the species-energy relationship. *Proceedings of the Royal Society of London, Series B, 272*(1577), 2155–2163.

Evans, M. G., & Polanyi, M. (1935). Some applications of the transition state method to the calculation of reaction velocities, especially in solution. *Transactions of the Faraday Society, 31*, 875–894.

Evans, W. G. (1964). Infra-red receptors in *Melanophila acuminata* DeGeer. *Nature, 202*(4928), 211.

Evans, W. G. (1966a). Morphology of the infrared sense organs of *Melanophila acuminata* (Buprestidae: Coleoptera). *Annals of the Entomological Society of America, 59*(5), 873–877.

Evans, W. G. (1966b). Perception of infrared radiation from forest fires by *Melanophila acuminata* de Geer (Buprestidae, Coleoptera). *Ecology, 47*(6), 1061–1065.

Exner, O. (1964a). Concerning the isokinetic relationship. *Nature, 201*, 488–490.

Exner, O. (1964b). On the enthalpy–entropy relationship. *Collection of Czechoslovak Chemical Communications, 29*(5), 1094–1113.

Exner, O. (1970). Determination of the isokinetic temperature. *Nature, 227*(366–367).

Eyring, H. (1935). The activated complex in chemical reactions. *Journal of Chemical Physics, 3*, 107.

Fahrenheit, D. G. (1724). Experimenta & observationes de congelatione aquae in vacuo factae a D. G. Fahrenheit, R. S. S. *Philosophical Transactions of the Royal Society of London, 33*, 78–84.

Fara, P. (2009). Science: a four thousand year history. Oxford: Oxford University Press.

Farewell, A., & Neidhardt, F. C. (1998). Effect of temperature on *in vivo* protein synthetic capacity of *Escherichia coli*. *Journal of Bacteriology, 180*(17), 4704–4710.

Farman, J. C., Gardiner, B. G., & Shanklin, J. D. (1985). Large losses of total ozone in Antarctica reveal seasonal ClOx/NOx interaction. *Nature, 315*(6016), 207–210.

Farmer, C. G. (2000). Parental care: the key to understanding endothermy and other convergent features in birds and mammals. *American Naturalist, 155*(3), 326–334.

Farmer, C. G. (2003). Reproduction: the adaptive significance of endothermy. *American Naturalist, 162*(6), 826–840.

Farrell, A. H., Hohenstein, K. A., & Shain, D. H. (2004). Molecular adaptation in the ice worm, *Mesenchytraeus solifugus*: divergence of energetic-associated genes. *Journal of Molecular Evolution, 59*(5), 666–673.

Fastovsky, D. E., & Weishampel, D. B. (2009). *Dinosaurs: a concise natural history*. Cambridge: Cambridge University Press.

Fautin, D. G., Malarky, L., & Soberón, J. (2013). Latitudinal diversity of sea anemones (Cnidaria: Actiniaria). *Biological Bulletin, 224*(2), 89–98.

Feely, R. A., Sabine, C. L., Lee, K., et al. (2004). Impact of anthropogenic CO_2 on the $CaCO_3$ system in the oceans. *Science, 305*(5682), 362–366.

Feild, T. S., & Brodribb, T. (2001). Stem water transport and freeze-thaw xylem embolism in conifers and angiosperms in a Tasmanian treeline heath. *Oecologia, 127*(3), 314–320.

Feller, G. (2007). Life at low temperatures: is disorder the driving force? *Extremophiles, 11*(2), 211–216.

Feller, G. (2010). Protein stability and enzyme activity at extreme biological temperatures. *Journal of Physics— Condensed Matter, 22*(32), article 323101.

Fenchel, T. (2005). Cosmopolitan microbes and their 'cryptic' species. *Aquatic Microbial Ecology, 41*(1), 49–54.

Fenchel, T., Esteban, G. F., & Finlay, B. J. (1997). Local versus global diversity of microorganisms: cryptic diversity of ciliated protozoa. *Oikos, 80*(2), 220–225.

Fenchel, T., & Finlay, B. J. (2004). The ubiquity of small species: patterns of local and global diversity. *BioScience, 54*(8), 777–784.

Feynman, R. P. (1998). *Six easy pieces: the fundamentals of physics explained*. London: Penguin Books.

Feynman, R. P., Leighton, R. B., & Sands, M. (1966). *The Feynman lectures on physics*: Addison Wesley Publishing Company.

Field, R., O'Brien, E. M., & Whittaker, R. J. (2005). Global models for predicting woody plant richness from climate: development and evaluation. *Ecology, 86*(9), 2263–2277.

Fields, P. A. (2001). Protein function at thermal extremes: balancing stability and flexibility. *Comparative Biochemistry and Physiology, Part A, 129*(2–3), 417–431.

Fields, P. A., Dong, Y., Meng, X., & Somero, G. N. (2015). Adaptations of protein structure and function to temperature: there is more than one way to 'skin a cat'. *Journal of Experimental Biology, 218*(12), 1801–1811.

Finlay, B. J., & Clarke, K. J. (1999). Ubiquitous dispersal of microbial species. *Nature, 400*(6747), 828.

Finlay, B. J., Esteban, G. F., & Fenchel, T. (1998). Protozoan diversity: converging estimates of the global number of free-living ciliate species. *Protist, 149*(1), 29–37.

Finlay, B. J., Esteban, G. F., Olmo, J. L., & Tyler, P. A. (1999). Global distribution of free-living microbial species. *Ecography, 22*(2), 138–144.

Finlay, B. J., & Fenchel, T. (1999). Divergent perspectives on protist species richness. *Protist, 150*(3), 229–233.

Finlay, B. J., & Fenchel, T. (2004). Cosmopolitan metapopulations of free-living microbial eukaryotes. *Protist, 155*(2), 237–244.

Finlayson, C. (2011). *Avian survivors: the history and biogeography of Palearctic birds*. London: T & AD Poyser (Bloomsbury Publishing).

Finney, J. L. (1982). Towards a molecular picture of liquid water. In F. Franks & S. F. Mathias (Eds.), *Biophysics of water* (pp. 73–95). Chichester: John Wiley & Sons.

Fischer, A. G. (1960). Latitudinal variation in organic diversity. *Evolution, 14*(1), 64–81.

Fischer, M., Allen, M. J., Wilson, W. H., & Suttle, C. A. (2010). Giant virus with a remarkable complement of genes infects marine zooplankton. *Proceedings of the National Academy of Sciences of the United States of America, 197*(45), 19508–19513.

Fisher, R. A., Corbet, A. S., & Williams, C. B. (1943). The relation between the number of species and the number of individuals in a random sample of an animal population. *Journal of Animal Ecology, 12*(1), 42–58.

Fitzgerald, M., Shine, R., & Lemckert, F. (2003). A reluctant heliotherm: thermal ecology of the arboreal snake *Hoplocephalus stephensii* (Elapidae) in dense forest. *Journal of Thermal Biology, 28*(6–7), 515–524.

Fixsen, D. J. (2009). The temperature of the cosmic microwave background. *The Astrophysical Journal, 707*(2), 916–920.

Fleming, J. R. (1998). *Historical perspectives on climate change*. Oxford: Oxford University Press.

Fletcher, G. L., King, M. J. K., Ming H., & Shears, M. A. (1989). Antifreeze proteins in the urine of marine fish. *Fish Physiology and Biochemistry, 6*(2), 121–127.

Floeter, S. R., Behrens, M. D., Ferreira, C. E. L., Paddack, M. J., & Horn, M. H. (2005). Geographic gradients of marine herbivorous fishes: patterns and processes. *Marine Biology, 147*(6), 1435–1447.

Fonseca, F., Marin, M., & Morris, G. J. (2006). Stabilization of frozen *Lactobacillus bulgaricus* in glycerol suspensions: freezing kinetics and storage temperature effects. *Applied and Environmental Microbiology, 72*(10), 6474–6482.

Foote, M. (2005). Pulsed origination and extinction in the marine realm. *Paleobiology, 31*(1), 6–20.

Foote, M., & Sepkoski, J. J. (1999). Absolute measures of the completeness of the fossil record. *Nature, 398*(6726), 415–417.

Forister, M. L., & Shapiro, A. M. (2003). Climatic trends and advancing spring flight of butterflies in lowland California. *Global Change Biology, 9*(7), 1130–1135.

Forrest, W. W. (1970). Entropy changes during biosynthesis. *Nature, 225*(5238), 1165–1166.

Forrest, W. W., & Walker, D. J. (1971). The generation and utilization of energy during growth. In A. H. Rose & J. F. Wilkinson (Eds)., *Advances in Microbial Ecology. V.* (Vol. 5, pp. 213–274). London: Academic Press.

Forster, J., & Hirst, A. G. (2012). The temperature-size rule emerges from ontogenetic differences between growth and development rates. *Functional Ecology, 26*(2), 483–492.

Forster, J., Hirst, A. G., & Atkinson, D. A. (2012). Warming-induced reductions in body size are greater in aquatic than terrestrial species. *Proceedings of the National Academy of Sciences of the United States of America, 109*(47), 19310–19314.

Forster, J., Hirst, A. G., & Woodward, G. (2011). Growth and development rates have different thermal responses. *American Naturalist, 178*(5), 668–678.

Fourier, J.-B. J. (1824). Remarques générales sur les températures du globe terrestre et des espaces planétaires. *Annales de chimie et de physique, 27*(1824), 136–167.

Fourier, J.-B. J. (1827). Mémoire sur les températures du globe terrestre et des espaces planétaires. *Mémoires de l'Académie Royale des Sciences de l'Institut de France, 7*(1827), 569–604.

Fowler, R. H., & Guggenheim, E. A. (1939). *Statistical thermodynamics: a version of statistical mechanics for students of physics and chemistry*. Cambridge: Cambridge University Press.

Francis, A. P., & Currie, D. J. (2003). A globally consistent richness-climate relationship for angiosperms. *American Naturalist, 161*(4), 523–536.

Franks, F. (1982). The properties of aqueous solutions at subzero temperatures. In F. Franks (Ed)., *Water and aqueous solutions at subzero temperatures (Water: a comprehensive treatise, Volume 7)* (pp. 215–338). New York: Plenum Press.

Franks, F. (1985). *Biophysics and biochemistry at low temperatures*. Cambridge: Cambridge University Press.

Fraser, K. P. P., Clarke, A., & Peck, L. S. (2002). Feast and famine in Antarctica: seasonal physiology in the limpet *Nacella concinna*. *Marine Ecology Progress Series, 242*, 169–177.

Fraser, K. P. P., Clarke, A., & Peck, L. S. (2007). Growth in the slow lane: protein synthesis in the Antarctic limpet *Nacella concinna* (Strebel, 1908). *Journal of Experimental Biology, 210*, 2691–2699.

Freckleton, R. P. (2002). On the misuse of residuals in ecology: regression of residuals vs. multiple regression. *Journal of Animal Ecology, 71*(3), 542–545.

Freckleton, R. P. (2009). The seven deadly sins of comparative analysis. *Journal of Evolutionary Biology, 22*(7), 1367–1375.

Freckleton, R. P., Harvey, P. H., & Pagel, M. (2003). Bergmann's rule and body size in mammals. *American Naturalist, 161*(5), 821–825.

Freely, J. (2012a). *Before Galileo: the birth of modern science in Medieval Europe*. London: Oliver Duckworth.

Freely, J. (2012b). *The flame of Miletus*. New York: I B Tauris & Co.

French, A. R. (1976). Selection of high temperatures for hibernation by the pocket mouse, *Perognathus longimembris*: ecological advantages and energetic consequences. *Ecology, 57*(1), 185–191.

Friis-Christensen, E., & Lassen, K. (1991). Length of the solar cycle: an indicator of solar activity closely associated with climate. *Science, 254*(5032), 698–700.

Fröhlich, J., & Hyde, K. D. (1999). Biodiversity of palm fungi in the tropics: are global fungal diversity estimates realistic? *Biodiversity and Conservation, 8*(7), 977–1004.

Fuchs, M., & Tanner, C. B. (1968). Surface temperature measurements of bare soils. *Journal of Applied Meteorology, 7*(2), 303–305.

Fuhrman, J. A., Steele, J. A., Hewson, I., et al. (2008). A latiutudinal diversity gradient in planktonic marine bacteria. *Proceedings of the National Academy of Sciences of the United States of America*, *105*(22), 7774–7778.

Gaines, S. D., & Lubchenco, J. (1982). A unified approach to marine plant–herbivore interactions. 2. Biogeography. *Annual Review of Ecology and Systematics*, *13*, 111–138.

Gallagher, D., Albu, J., He, Q., et al. (2006). Small organs with high metabolic rate explain lower resting energy expenditure in African American than in white adults. *American Journal of Clinical Nutrition*, *83*(5), 1062–1067.

Galvão, P. E., Tarasantchi, J., & Guertzenstein, P. (1965). Heat production of tropical snakes in relation to body weight and body surface. *American Journal of Physiology*, *209*(3), 501–506.

Garay-Arroyo, A., Colmenero-Flores, J. M., Garciarrubio, A., & Covarrubias, A. A. (2000). Highly hydrophilic proteins in prokaryotes and eukaryotes are common during conditions of water deficit. *Journal of Biological Chemistry*, *275*(8), 5668–5674.

Gardner, J. L., Peters, A., Kearney, M. R., Joseph, L., & Heinsohn, R. (2011). Declining body size: a third universal response to warming. *Trends in Ecology & Evolution*, *26*(6), 285–291.

Gaston, K. J. (2003). *The structure and dynamics of geographic ranges*. Oxford: Oxford University Press.

Gaston, K. J., Chown, S. L., Calosi, P., et al. (2009). Macrophysiology: a conceptual re-unification. American Naturalist, 174(5), 595–612.

Gaston, K. J., Evans, K. L., & Lennon, J. J. (2007). The scaling of spatial turnover; pruning the thicket. In D. Storch, P. A. Marquet, & J. H. Brown (Eds.), *Scaling biodiversity* (pp. 181–222). Cambridge: Cambridge University Press.

Gaston, K. J., & Williams, P. H. (1996). Spatial patterns in taxonomic diversity. In K. J. Gaston (Ed.), *Biodiversity: a biology of numbers and differences* (pp. 202–229). Oxford: Blackwell Science.

Gaston, K. J., Williams, P. H., & Humphries, C. J. (1995). Large-scale patterns of biodiversity: spatial variation in family richness. *Proceedings of the Royal Society of London, Series B*, *260*(1358), 149–154.

Gates, D. M. (1980). *Biophysical ecology*. New York: Springer-Verlag.

Gates, D. M. (1993). *Climate change and its biological consequences*. Sunderland, Massachusetts: Sinauer Associates.

Gayon, J. (2000). History of the concept of allometry. *American Zoologist*, *40*(4), 748–758.

Ge, X., & Wang, X. (2009). Calculations of freezing point depression, boiling point elevation, vapor pressure and enthalpies of vaporization of electrolyte solutions by a modified three-characteristic parameter correlation model. *Journal of Solution Chemistry*, *38*(9), 1097–1117.

Gehr, P., Mwangi, D. K., Ammann, A., Maloiy, G. M. O., Taylor, C. R., & Weibel, E. R. (1981). Design of the mammalian respiratory system. V. Scaling morphometric pulmonary diffusing capacity to body mass: wild and domesticated mammals. *Respiration Physiology*, *44*(1), 61–86.

Geiges, O. (1996). Microbial processes in frozen food. *Advances in Space Research*, *18*(12), 109–118.

Geiser, F. (1998). Evolution of daily torpor and hibernation in birds and mammals: importance of body size. *Clinical and Experimental Pharmacology and Physiology*, *25*(9), 736–739.

Geiser, F. (2004). Metabolic rate and body temperature reduction during hibernation and daily torpor. *Annual Review of Physiology*, *66*(1), 239–274.

Geiser, F. (2011). Hibernation: endotherms *eLS* (pp.1–11): Wiley Online Library.

Geiser, F., & Pavey, C. R. (2007). Basking and torpor in a rock-dwelling desert marsupial: survival strategies in a resource-poor environment. *Journal of Comparative Physiology*, *177*(8), 885–892.

Geiser, F., & Turbill, C. (2009). Hibernation and daily torpor minimize mammalian extinctions. *Naturwissenschaften*, *96*(10), 1235–1240.

Geist, V. (1987). Bergmann's rule is invalid. *Canadian Journal of Zoology*, *65*(4), 1035–1038.

Gennis, R. B. (1989). Membrane dynamics and protein–lipid interactions. In R. B. Gennis (Ed.), *Biomembranes: molecular structure and function* (pp. 166–198). Berlin: Springer-Verlag.

Georlette, D., Blaise, V., Collins, T., et al. (2004). Some like it hot: biocatalysis at low temperatures. *FEMS Microbiology Reviews*, *28*(1), 25–42.

Gerday, C. (2013a). Catalysis and protein folding in psychrophiles. In I. Yumoto (Ed.), *Cold-adapted microorganisms* (pp. 137–157). Caister, Norfolk, UK: Caister Academic Press.

Gerday, C. (2013b). Psychrophiles and catalysis. *Biology*, *2*(2), 719–741.

Gerday, C., Aittaleb, M., Bentahir, M., et al. (2000). Cold-adapted enzymes: from fundamentals to biotechnology. *Trends in Biotechnology*, *18*(3), 103–107.

Gerson, A. R., Smith, E. K., Smit, B., McKechnie, A. E., & Wolf, B. O. (2014). The impact of humidity on evaporative cooling in small desert birds exposed to high air temperature. *Physiological and Biochemical Zoology*, *87*(6), 782–795.

Ghighlione, J.-F., Galand, P. E., Pommier, T., et al. (2012). Pole-to-pole biogeography of surface and deep marine bacterial communities. *Proceedings of the National Academy of Sciences of the United States of America*, *109*(43), 17633–17638.

Gilbert, J. A., Davies, P. L., & Laybourn-Parry, J. (2005). A hyperactive, Ca^{2+}-dependent antifreeze protein in an Antarctic bacterium. *FEMS Microbiology Letters*, *245*(1), 67–72.

Gilbert, J. A., Hill, P. J., Dodd, C. E. R., & Laybourn-Parry, J. (2004). Demonstration of antifreeze protein activity in Antarctic lake bacteria. *Microbiology, 150*(1), 171–181.

Gillespie, C. C. (1960). *The edge of objectivity: an essay in the history of scientific ideas*. Princeton, New Jersey: Princeton University Press.

Gillett, N. P., Stone, D. A., Stott, P. A., et al. (2008). Attribution of polar warming to human influence. *Nature Geoscience, 1*(11), 750–754.

Gillman, L. N., Wright, S. D., Cusens, J., McBride, P. D., Malhi, Y., & Whittaker, R. J. (2015). Latitude, productivity and species richness. *Global Ecology and Biogeography, 24*(1), 107–117.

Gillooly, J. F., Allen, A. P., West, G. B., & Brown, J. H. (2005). The rate of DNA evolution: effects of body size and temperature on the molecular clock. *Proceedings of the National Academy of Sciences of the United States of America, 102*(1), 140–145.

Gillooly, J. F., Brown, J. H., West, G. B., Savage, V. M., & Charnov, E. L. (2001). Effects of size and temperature on metabolic rate. *Science, 293*, 2248–2251.

Gillooly, J. F., Charnov, E. L., West, G. B., Savage, V. M., & Brown, J. H. (2002). Effects of size and temperature on developmental time. *Nature, 417*(6884), 70–73.

Gilroy, J. J., Gill, J. A., Butchart, S. H. M., Jones, V., & Franco, A. M. A. (2016). Migratory diversity predicts population declines in birds. *Ecology Letters, 19*(3), 306–317.

Giroud, S., Frare, C., Strijkstra, A., Boerema, A., Arnold, W., & Ruf, T. (2013). Membrane phospholipid fatty acid composition regulates cardiac SERCA activity in a hibernator, the Syrian hamster (*Mesocricetus auratus*). *PLoS One, 8*(5), e63111.

Glazier, D. S. (2008). Effects of metabolic level on the body size scaling of metabolic rate in birds and mammals. *Proceedings of the Zoological Society of London, Series B, 275*, 1405–1410.

Glazier, D. S. (2010). A unifying explanation for diverse metabolic scaling in animals and plants. *Biological Reviews, 85*(1), 111–138.

Glazier, D. S. (2013). Log-transformation is useful for examining proportional relationships in allometric scaling. *Journal of Theoretical Biology, 334*, 200–203.

Gleason, H. A. (1917). The structure and development of the plant association. *Bulletin of the Torrey Botanical Club, 44*(10), 463–481.

Gleason, H. A. (1926). The individualistic concept of the plant association. *Bulletin of the Torrey Botanical Club, 56*(1), 7–26.

Gleason, H. A. (1939). The individualistic concept of the plant association. *American Midland Naturalist, 21*(1), 92–110.

Gnaiger, E., Shick, J. M., & Widdows, J. (1989). Metabolic microcalorimetry and respirometry of aquatic animals. In C. R. Bridges & P. J. Butler (Eds.), *Techniques in comparative respiratory physiology: an experimental approach (Society for Experimental Biology Seminar Series 37)* (pp. 113–135). Cambridge: Cambridge University Press.

Goldberg, E. E., Roy, K., Lande, R., & Jablonski, D. (2005). Diversity, endemism, and age distributions in macroevolutionary sources and sinks. *American Naturalist, 165*(6), 623–633.

Goldie, X., Lanfear, R., & Bromham, L. (2011). Diversification and the rate of molecular evolution: no evidence of a link in mammals. *BMC Evolutionary Biology, 11*, article 286.

Gomes, F. R., Chaui-Berlinck, J. G., Bicudo, J. E. P. W., & Navas, C. A. (2004). Intraspecific relationships between resting and activity metabolism in anuran amphibians. *Physiological and Biochemical Zoology, 77*, 197–208.

Goodchild, A., Raftery, M., Saunders, N. F. W., Guilhaus, M., & Cavicchioli, R. (2004). Biology of the cold adapted archaeon, *Methanococcoides burtonii* determined by proteomics using liquid chromatography–tandem mass spectrometry. *Journal of Proteome Research, 3*(6), 1164–1176.

Goodsell, D. S. (2009). *The machinery of life* (Second ed.). New York: Copernicus Books.

Gordo, O., Brotons, L., Ferrer, X., & Comas, P. (2005). Do changes in climate patterns in wintering areas affect the timing of the spring arrival of trans-Saharan migrant birds? *Global Change Biology, 11*(1), 12–21.

Gotelli, N. J., & Graves, G. R. (1996). *Null models in ecology*. Washington and London: Smithsonian Institution Press.

Gough, D. O. (1981). Solar interior structure and luminosity variations. *Solar Physics, 74*(1), 21–34.

Goyal, K., Walton, L. J., & Tunnacliffe, A. (2005). LEA proteins prevent protein aggregation due to water stress. *Biochemical Journal, 388*(1), 151–157.

Gracheva, E. O., Ingolia, N. T., Kelly, Y. M., et al. (2010). Molecular basis of infrared detection by snakes. *Nature, 464*(7291), 1006–1011.

Graham, J. B., Dudley, R., Aguilar, N. M., & Gans, C. (1995). Implications of the late Paleozoic oxygen pulse for physiology and evolution. *Nature, 375*(6527), 17–120.

Graham, L. A., Hobbs, R. S., Fletcher, G. L., & Davies, P. L. (2013). Helical antifreeze proteins have independently evolved in fishes on four occasions. *PLoS One, 8*(12), e81285.

Graham, L. A., Li, J., Davidson, W. S., & Davies, P. L. (2012). Smelt was the likely beneficiary of an antifreeze gene laterally transferred between fishes. *BMC Evolutionary Biology, 12*, article 190.

Graham, L. A., Lougheed, S. C., Ewart, K. V., & Davies, P. L. (2008). Lateral transfer of a lectin-like antifreeze protein gene in fishes. *PLoS One, 3*(7), e2616.

Grant, B. W., & Porter, W. P. (1992). Modeling global macroclimatic constraints on ectotherm energy budgets. *American Zoologist, 32*(2), 154–178.

Grassle, J. F. (2001). Marine ecosystems. In S. A. Levin (Ed)., *Encyclopedia of biodiversity* (Vol. 4, pp. 13–25). San Diego: Academic Press.

Grassle, J. F., & Maciolek, N. J. (1992). Deep sea species richness: regional and local diversity estimates from quantitative bottom samples. *American Naturalist, 139,* 313–341.

Graus, R. R. (1974). Latitudinal trends in the shell characteristics of marine gastropods. *Lethaia, 7*(4), 303–314.

Graves, J. E., & Somero, G. N. (1982). Electrophoretic and functional enzymic evolution in four species of eastern Pacific barracudas from different thermal environments. *Evolution, 36,* 97–106.

Gray, J. S. (1994). Is deep-sea species diversity really so high? Species diversity of the Norwegian coastal shelf. *Marine Ecology Progress Series, 112*(1–2), 205–209.

Gray, J. S. (2001). Antarctic marine benthic biodiversity in a world-wide latitudinal context. *Polar Biology, 24*(9), 633–641.

Gray, J. S. (2002). Species richness of marine soft sediments. *Marine Ecology Progress Series, 244,* 285–297.

Gray, J. S., Poore, G. C. B., Ugland, K. I., Wilson, R. D., Olsgard, F., & Johannessen, Ø. (1997). Coastal and deep-sea benthic diversities compared. *Marine Ecology Progress Series, 159,* 97–103.

Gray, J. S., Ugland, K. I., & Lambshead, J. (2004).

Gray, J. S., Ugland, K. I., & Lambshead, J. (2004). Species accumulation and species area curves—a comment on Scheiner (2003). *Global Ecology and Biogeography, 13*(5), 473–476.

Grebmeier, J. M. (2012). Shifting patterns of life in the Pacific Arctic and Sub-Arctic seas. *Annual Review of Marine Science, 4,* 63–78.

Green, J., & Bohannan, B. J. M. (2007). Biodiversity scaling relationships: are microorganisms fundamentally different? In D. Storch, P. A. Marquet, & J. H. Brown (Eds)., *Scaling biodiversity* (pp. 129–149). Cambridge: Cambridge University Press.

Green, J. A., Aitken-Simpson, E. J., White, C. R., Bunce, A., Butler, P. J., & Frappell, P. B. (2013). An increase in minimum metabolic rate and not activity explains field metabolic rate changes in a breeding seabird. Journal of Experimental Biology, 216(9), 1726–1735.

Green, K. (2001). Autumnal body mass reduction in *Antechinus swainsonii* (Dasyuridae) in the Snowy Mountains. *Australian Mammalogy, 23*(1), 31–36.

Griffith, M., & Yaish, M. W. F. (2004). Antifreeze proteins in overwintering plants: a tale of two activities. *Trends in Plant Science, 9*(8), 399–405.

Grimpo, K., Kutschke, M., Kastl, A., et al. (2014). Metabolic depression during warm torpor in the Golden spiny mouse (*Acomys russatus*) does not affect mitochondrial respiration and hydrogen peroxide release. *Comparative Biochemistry and Physiology, Part A, 167*(1), 7–14.

Grimstone, A. V., Mullinger, A. M., & Ramsay, J. A. (1968). Further studies on the rectal complex of the mealworm *Tenebrio molitor*, L. (Coleoptera, Tenebrionidae). *Philosophical Transactions of the Royal Society of London, Series B, 253*(788), 343–382.

Gross, K., & Snyder-Beattie, A. (2016). A general, synthetic model for predicting biodiversity gradients from environmental geometry. *American Naturalist, 188*(4), E85–E97.

Gu, J., & Hilser, V. J. (2009). Sequence-based analysis of protein energy landscapes reveals nonuniform thermal adaptation within the proteome. *Molecular Biology and Evolution, 26*(10), 2217–2227.

Guerin, G. R., Wen, H., & Lowe, A. J. (2012). Leaf morphology shift linked to climate change. *Biology Letters, 8,* 882–886.

Gulick, J. T. (1872). On the variation of species as related to their geographical distribution, illustrated by the Achatinellinae. *Nature, 6,* 222–224.

Guo, S., Garnham, C. P., Whitney, J. C., Graham, L. A., & Davies, P. L. (2012). Re-evaluation of a bacterial antifreeze protein as an adhesin with ice-binding activity. *PLoS One, 7*(11), e48805.

Gurian-Sherman, D., & Lindow, S. E. (1993). Bacterial ice nucleation: sigfnificance and molecular basis. *FASEB Journal, 7*(14), 1338–1343.

Gvoždík, L. (2003). Postprandial thermophily in the Danube crested newt, *Triturus dobrogicus*. *Journal of Thermal Biology, 28*(8), 545–550.

Gwak, Y., Jung, W., Lee, Y., et al. (2014). An intracellular antifreeze protein from an Antarctic microalga that responds to various environmental stresses. *FASEB Journal, 28*(11), 4924–4935.

Haake, V., Cook, D., Riechmann, J. L. s., Pineda, O., Thomashow, M. F., & Zhang, J. Z. (2002). Transcription factor CBF4 is a regulator of drought adaptation in *Arabidopsis*. *Plant Physiology, 130*(2), 639–648.

Hackett, S. J., Kimball, R. T., Reddy, S., et al. (2008). A phylogenomic study of birds reveals their evolutionary history. *Science, 320*(5884), 1763–1768.

Hagen, J. B. (1992). *An entangled bank: the origins of ecosystem ecology*. New Brunswick, New Jersey: Rutgers University Press.

Hagvar, S. (2010). A review of Fennoscandian arthropods living on and in snow. *European Journal of Entomology, 107*(3), 281–298.

Haigh, J. D., Winning, A. R., Toumi, R., & Harder, J. W. (2010). An influence of solar spectral variations on radiative forcing of climate. *Nature, 467*(7316), 696–699.

Hainsworth, F. R. (1967). Saliva spreading, activity, and body temperature regulation in the rat. *American Journal of Physiology, 212*(6), 1288–1292.

Hameed, S., & Gong, G. (1994). Variation of spring climate in lower-middle Yangtse River Valley and its relation with solar-cycle length. *Geophysical Research Letters*, 21(24), 2693–2696.

Hamel, P. B. (2012). First observed roost site of the Vervain Hummingbird (*Mellisuga minima*). *Journal of Caribbean Ornithology*, 25(2), 95–97.

Hamilton, W. J. (1973). *Life's color code*. New York: McGraw-Hill.

Hammen, C. S. (1979). Metabolic rates of marine bivalves molluscs determined by calorimetry. *Comparative Biochemistry and Physiology, Part A*, 62(4), 955–959.

Hammen, C. S. (1980). Total energy metabolism of marine bivalve mollusks in anaerobic and aerobic states. *Comparative Biochemistry and Physiology, Part A*, 67(4), 617–621.

Hammen, C. S. (1984). Anaerobic energy metabolism of invertebrates. *Federation Proceedings*, 43(2), 220–225.

Hammond, K. A., & Diamond, J. (1997). Maximal sustained energy budgets in humans and animals. *Nature*, 386, 457–462.

Hampe, A. (2004). Bioclimate envelope models: what they detect and what they hide. *Global Ecology and Biogeography*, 13(5), 469–471.

Hanada, Y., Nishimiya, Y., Miura, A., Tsuda, S., & Kondo, H. (2014). Hyperactive antifreeze protein from an Antarctic sea ice bacterium *Colwellia* sp has a compound ice-binding site without repetitive sequences. *FEBS Journal*, 281(16), 3576–3590.

Hand, S. C., Menze, M. A., Toner, M., Boswell, L., & Moore, D. (2011). LEA proteins during water stress: not just for plants anymore. *Annual Review of Physiology*, 73, 115–124.

Hannam, J. (2009). *God's philosophers: how the Medieval world laid the foundations of modern science*. London: Icon Books.

Hansen, T. (1973). Variations in glycerol content in relation to cold-hardiness on the larvae of *Petrova resinella* L. (Lepidoptera: Tortricidae). *Eesti NSV Teaduste Akadeemia toimetised. Bioloogia*, 22, 105–111.

Hanski, I. (1996). Metapopulation ecology. In O. E. Rhodes, R. K. Chesser, & M. H. Smith (Eds)., *Population dynamics in ecological space and time* (pp. 13–43). Chicago: University of Chicago Press.

Hanski, I. A., & Gilpin, M. E. (Eds.). (1997). *Metapopulation biology: ecology, genetics and evolution*. San Diego: Academic Press.

Harcourt, A. G. V., & Esson, W. (1895). On the laws of connexion between conditions of a chemical change and its amount. III. Further researches on the reaction of hydrogen dioxide and hydrogen iodide. *Philosophical Transactions of the Royal Society of London, Series A*, 186(1895), 817–895.

Hargens, A. R., & Shabica, S. V. (1973). Protection against lethal freezing temperatures by mucus in an Antarctic limpet. *Cryobiology*, 10(4), 331–337.

Harrington, R. J. (1987). Skeletal growth histories of *Protothaca staminea* (Conrad) and *Protothaca grata* (Say) throughout their geographic ranges, northeastern Pacific. *Veliger*, 30(2), 148–158.

Harris, R. M., McQuillan, P., & Hughes, L. (2013). A test of the thermal melanism hypothesis in the wingless grasshopper *Phaulacridium vittatum*. *Journal of Insect Science*, 13, Article 51.

Harrison, J. F., Kaiser, A., & VandenBrooks, J. M. (2010). Atmospheric oxygen level and the evolution of insect body size. *Proceedings of the Royal Society of London, Series B*, 277(1690), 1837–1946.

Harte, J., Conlisk, E., Ostling, A., Green, J. L., & Smith, A. B. (2005). A theory of spatial structure in ecological communities at multiple spatial scales. *Ecological Monographs*, 75(2), 179–197.

Harte, J., Zillio, T., Conlisk, E., & Smith, A. B. (2008). Maximum entropy and the state-variable approach to macroecology. *Ecology*, 89(10), 2700–2711.

Hartl, F. U., Bracher, A., & Hayer-Hartl, M. (2011). Molecular chaperones in protein folding and proteostasis. *Nature*, 475(7356), 324–332.

Hartmann, D. L. (1994). *Global physical climatology*. San Diego: Academic Press.

Hartmann, M. (2006). Minimal length scales for the existence of local temperature. *Contemporary Physics*, 47(2), 89–102.

Hassall, C. (2007). Historical changes in the phenology of British Odonata are related to climate. *Global Change Biology*, 13(5), 933–941.

Hawes, T. C., Worland, M. R., & Bale, J. S. (2010). Freezing in the Antarctic limpet, *Nacella concinna*. *Cryobiology*, 61(1), 128–132.

Hawkins, B. A., Albuquerque, F. S., Araújo, M. B., et al. (2007a). A global evaluation of metabolic theory as an explanation for terrestrial species richness gradients. *Ecology*, 88(8), 1877–1888.

Hawkins, B. A., Diniz-Filho, J. A. F., Bini, L. M., et al. (2007b). Metabolic theory and diversity gradients: where do we go from here? *Ecology*, 88(8), 1898–1902.

Hawkins, B. A., Field, R., Cornell, H. V., et al. (2003). Energy, water, and broad-scale geographic patterns of species richness. *Ecology*, 84(12), 3105–3117.

Hawkins, B. A., & Porter, E. E. (2003). Water–energy balance and the geographic pattern of species richness of western Palearctic butterflies. *Ecological Entomology*, 28(6), 678–686.

Hawkins, S. J., Sugden, H. E., Mieszkowska, N., et al. (2009). Consequences of climate-driven biodiversity changes for ecosystem functioning of North European rocky shores. *Marine Ecology Progress Series*, 396, 245–259.

Hawkins, W. W. (1972). The calorie, the joule. *Journal of Nutrition*, 102, 1553–1554.

Hawksworth, D. L. (2001). The magnitude of fungal diversity: the 1.5 million species estimate revisited. *Mycological Research*, *105*(12), 1422–1432.

Hayes, J. P., & Garland, T. (1995). The evolution of endothermy: testing the aerobic scope model. *Evolution*, *49*, 836–847.

Haynie, D. T. (2001). *Biological thermodynamics*. Cambridge: Cambridge University Press.

Hays, J. D., Imbrie, J., & Shackleton, N. J. (1976). Variations in the Earth's orbit: pacemaker of the Ice Ages. *Science*, *194*(4270), 1121–1132.

Hayward, J. S., & Lisson, P. A. (1992). Evolution of brown fat: its absence in marsupials and monotremes. *Canadian Journal of Zoology*, *70*(1), 171–179.

Hazel, J. R. (1995). Thermal adaptation of biological membranes: is homeoviscous adaptation the explanation? *Annual Review of Physiology*, *57*, 19–42.

Heath, J. E. (1964). Reptilian thermoregulation: evaluation of field studies. *Science*, *146*(3645), 784–785.

Hegel, J. R., & Casey, T. M. (1982). Thermoregulation and control of head temperature in the sphinx moth, *Manduca sexta*. *Journal of Experimental Biology*, *101*(1), 1–15.

Heikkinen, R. K., Luoto, M., Araújo, M. B., Virkkala, R., Thuiller, W., & Sykes, M. T. (2006). Methods and uncertainties in bioclimatic envelope modelling under climate change. *Progress in Physical Geography*, *30*(6), 751–777.

Heinrich, B. (1977). Why have some animals evolved to regulate a high body temperature? *American Naturalist*, *111*(980), 623–640.

Heinrich, B. (1979). *Bumblebee economics*. Cambridge, Massachusetts: Harvard University Press.

Heinrich, B. (1990). Is 'reflectance' basking real? *Journal of Experimental Biology*, *154*(1), 31–43.

Heinrich, B. (1993). *The hot-blooded insects: strategies and mechanisms of thermoregulation*. Berlin: Springer-Verlag.

Heinrich, B. (2003). *Winter world: the ingenuity of animal survival*. New York: Harper Collins.

Heinrich, R., & Rapoport, T. A. (1974). A linear steady-state treatment of enzymatic chains: critique of crossover theorem and a general procedure to identify interaction sites with an effector. *European Journal of Biochemistry*, *42*(1), 97–105.

Heisler, N. (1986). Comparative aspects of acid–base regulation. In N. Heisler (Ed.), *Acid–base regulation in animals* (pp. 397–450). Amsterdam: Elsevier.

Held, C. (2001). No evidence for slow-down of molecular substitution rates at subzero temperatures in Antarctic serolid isopods (Crustacea, Isopoda, Serolidae). *Polar Biology*, *24*(7), 497–501.

Heldmaier, G., Ortmann, S., & Elvert, R. (2004). Natural hypometabolism during hibernation and daily torpor in mammals. *Respiratory Physiology & Neurobiology*, *141*(3), 317–329.

Hellgren, E. C. (1998). Physiology of hibernation in bears. *Ursus*, *10*, 467–477.

Helmuth, B. (2002). How do we measure the environment? Linking intertidal thermal physiology and ecology through biophysics. *Integrative and Comparative Biology*, *42*(4), 837–845.

Helmuth, B. S. T., & Hofmann, G. E. (2001). Microhabitats, thermal heterogeneity, and patterns of physiological stress in the rocky intertidal zone. *Biological Bulletin*, *201*(3), 374–384.

Hemmingsen, A. M. (1950). The relation of standard (basal) energy metabolism to total fresh weight of living organisms. *Reports of the Steno Memorial Hospital and the Nordisk Insulinlaboratorium*, *4*(1), 1–48.

Hemmingsen, A. M. (1960). Energy metabolism as related to body size and respiratory surfaces, and its evolution. *Reports of the Steno Memorial Hospital and the Nordisk Insulinlaboratorium*, *9*(1), 1–110.

Henry, M. (2005). The state of water in living systems: from the liquid to the jellyfish. *Cellular and Molecular Biology*, *51*(7), 677–702.

Hepler, L. G., & Woolley, E. M. (1973). Hydration effects and acid–base equilibria. In F. Franks (Ed.), *Water—a comprehensive treatise* (Vol. 3: Aqueous solutions of simple electrolytes, pp. 145–172). New York: Plenum Press.

Herczeg, G., Gonda, A., Saarikivi, J., & Merilä, J. (2006). Experimental support for the cost–benefit model of lizard thermoregulation. *Behavioral Ecology and Sociobiology*, *60*(3), 405–414.

Herczeg, G., Herrero, A., Saarikivi, J., Gonda, A., Jäntti, M., & Merila, J. (2008). Experimental support for the cost–benefit model of lizard thermoregulation: the effects of predation risk and food supply. *Oecologia*, *155*(1), 1–10.

Herreid, C. F., & Kessel, B. (1967). Themal conductance in birds and mammals. *Comparative Biochemistry and Physiology*, *21*, 405–414.

Hertz, P. E., Huey, R. B., & Stevenson, R. D. (1993). Evaluating temperature regulation by field-active ectotherms: the fallacy of the inappropriate question. *American Naturalist*, *142*(5), 796–818.

Hertzberg, J. E., & Schmidt, M. W. (2013). Refining *Globigerinoides ruber* Mg/Ca paleothermometry in the Atlantic Ocean. *Earth and Planetary Science Letters*, *383*, 123–133.

Hessen, D. O., Daufresne, M., & Leinaas, H. P. (2013). Temperature-size relations from the cellular-genomic perspective. *Biological Reviews*, *88*(2), 476–489.

Heurtault, J., & Vannier, G. (1990). Thermorésistance chez deux pseudoscorpions (Garypidae), l'un du désert de Namibie, l'autre de la région de Gênes (Italie). *Acta Zoologica Fennica*, *190*, 165–172.

Heusner, A. A. (1985). Body size and energy metabolism. *Annual Review of Nutrition*, *5*, 267–293.

Heusner, A. A. (1991). Body mass, maintenance and basal metabolism in dogs. *Journal of Nutrition*, 121, S8–S17.

Hew, C. L., Slaughter, D., Joshi, S. B., Fletcher, G. L., & Ananthanarayanan, V. S. (1984). Antifreeze polypeptides from the Newfoundland ocean pout, *Macrozoarces americanus*: presence of multiple and compositional diverse components. *Journal of Comparative Physiology B Biochemical Systematic and Environmental Physiology*, 155(1), 81–88.

Hibberd, D. (1982). History of the steppe-tundra concept. In D. M. Hopkins, J. V. Matthews, C. E. Schweger, & S. B. Young (Eds.), *Paleoecology of Beringia* (pp. 153–156). New York: Academic Press.

Hickling, R., Roy, D. B., Hill, J. K., Fox, R., & Thomas, C. D. (2006). The distributions of a wide range of taxonomic groups are expanding polewards. *Global Change Biology*, 12(3), 450–455.

Hill, K. A., Shepson, P. B., Galbavy, E. S., et al. (2007). Processing of atmospheric nitrogen by clouds above a forest environment. *Journal of Geophysical Research, Atmospheres*, 112(D11), D11301.

Hill, L., & Taylor, H. J. (1933). Locusts in sunlight. *Nature*, 132(3329), 276.

Hill, R. W., Wyse, G. A., & Anderson, M. (2008). *Animal physiology* (Second ed.). Sunderland, Massachusetts: Sinauer Associates.

Hillebrand, H. (2004). On the generality of the latitudinal diversity gradient. *American Naturalist*, 163(2), 192–211.

Hillenius, W. J. (1992). The evolution of nasal turbinates and mammalian endothermy. *Paleobiology*, 18(1), 17–29.

Hillenius, W. J. (1994). Turbinates in therapsids: evidence for Late Permian origins of mammalian endothermy. *Evolution*, 48(2), 207–229.

Hillenius, W. J., & Ruben, J. A. (2004). The evolution of endothermy in terrestrial vertebrates: who? when? why? *Physiological and Biochemical Zoology*, 77(6), 1019–1024.

Hillman, S. S., Hancock, T. V., & Hedrick, M. S. (2013). A comparartive meta-analysis of maximal aerobic metabolism of vertebrates: implications for respiratory and cardiovascular limits to gas exchange. *Journal of Comparative Physiology, B*, 183(2), 167–179.

Hillman, S. S., Withers, P. C., Drewes, R. C., & Hillyard, S. D. (2009). *Ecological and environmental physiology of amphibians*. Oxford: Oxford University Press.

Hinks, A. E., Cole, E. F., Daniels, K. J., Wilkin, T. A., Nakagawa, S., & Sheldon, B. C. (2015). Scale-dependent phenological synchrony between songbirds and their caterpillar food source. *American Naturalist*, 186(1), 84–97.

Hirche, H.-J. (1984). Temperature and metabolism of plankton. 1. Respiration of Antarctic zooplankton at different temperatures with a comparison of Antarctic and Nordic krill. *Comparative Biochemistry and Physiology, Part A*, 77(2), 361–368.

Hirsh, A. G. (1987). Vitrification in plants as a natural form of cryoprotection. *Cryobiology*, 24(3), 214–228.

Hirsh, A. G., Williams, R. J., & Meryman, H. T. (1985). A novel method of natural cryoprotection. Intracellular glass formation in deeply frozen *Populus*. *Plant Physiology*, 79(1), 41–56.

Hirst, A. G., & Forster, J. (2013). When growth models are not universal: evidence from marine invertebrates. *Proceedings of the Royal Society of London, Series B*, 280(1768), article 20131546.

Hirst, A. G., Roff, J. C., & Lampitt, R. S. (2003). A synthesis of growth rates in marine epipelagic invertebrate zooplankton. *Advances in Marine Biology*, 44, 1–142.

Hitchcock, D. R., & Lovelock, J. E. (1967). Life detection by atmospheric analysis. *Icarus*, 7(1–3), 149–159.

Hobbs, J. K., Jiao, W., Easter, A. D., Parker, E. J., Schipper, L. A., & Arcus, V. L. (2013). Change in heat capacity for enzyme catalysis determines temperature dependence of enzyme catalysed rates. *ACS Chemical Biology*, 8(11), 2388–2393.

Hobbs, P. V. (1974). *Ice physics*. Oxford: Oxford University Press.

Hochachka, P. W., & Somero, G. N. (2002). *Biochemical adaptation: mechanism and process in physiological evolution*. Oxford: Oxford University Press.

Hodkinson, I. D. (2003). Metabolic cold adaptation in arthropods: a smaller scale perspective. *Functional Ecology*, 17(4), 562–572.

Hoefnagel, K. N., & Verberk, W. C. E. P. (2015). Is the temperature-size rule mediated by oxygen in aquatic ectotherms? *Journal of Thermal Biology*, 54(S1), 56–65.

Hoeppli, R., & Chu, H. J. (1932). Free-living nematodes from hot springs in China and Formosa. *Hong Kong Naturalist, Supplement 1*, 15–28.

Hoeppli, R. J. C. (1926). Studies of free-living nematodes from the thermal waters of Yellowstone Park. *Transactions of the American Microscopical Society*, 45, 234–255.

Hoffman, T. C. M., Walsberg, G. E., & DeNardo, D. F. (2007). Cloacal evaporation: an important and previously undescribed mechanism for avian thermoregulation. *Journal of Experimental Biology*, 210(5), 741–749.

Holdridge, L. R. (1959). Simple method for determining potential evapotranspiration from temperature data. *Science*, 130(3375), 572.

Holeton, G. F. (1970). Oxygen uptake and circulation by a hemoglobin-less Antarctic fish (*Chaenocephalus aceratus* Lonnberg) compared with three red-blooded Antarctic fish. *Comparative Biochemistry and Physiology*, 34(2), 457–471.

Holeton, G. F. (1972). Gas exchange in fish with and without hemoglobin. *Respiration Physiology*, 14(1–2), 142–150.

Holeton, G. F. (1973). Respiration of Arctic char (*Salvelinus alpinus*) from a high Arctic lake. *Journal of the Fisheries Research Board of Canada*, 30, 717–723.

Holeton, G. F. (1974). Metabolic cold adaptation of polar fish: fact or artefact? *Physiological Zoology, 47*, 137–152.

Holmes, F. L. (1985). *Lavoisier and the chemistry of life: an exploration of scientific creativity*. Madison, Wisconsin: University of Wisconsin Press.

Hönisch, B., Ridgwell, A., Schmidt, D. N., et al. (2012). The geological record of ocean acidification. *Science, 335*(6072), 1058–1063.

Horn, M. H. (1989). Biology of marine herbivorous fishes. *Oceanography and Marine Biology: an Annual Review, 27*, 167–272.

Horne, C. R., Hirst, A. G., & Atkinson, D. (2015). Temperature-size responses match latitudinal-size clines in arthropods, revealing critical differences between aquatic and terrestrial species. *Ecology Letters, 18*(4), 327–335.

Horwath, K. L., & Duman, J. G. (1984). Yearly variations in the overwintering mechanisms of the cold-hardy beetle *Dendroides canadensis*. *Physiological Zoology, 57*(1), 40–45.

Hoshino, T., Kiriaki, M., & Nakajima, T. (2003). Novel thermal hysteresis proteins from low temperature basidiomycete, *Coprinus psychromorbidus*. *Cryo-Letters, 24*(3), 135–142.

Hou, C., Zuo, W., Moses, M. E., Woodruff, W. H., Brown, J. H., & West, G. B. (2008). Energy uptake and allocation during ontogeny. *Science, 322*(5902), 736–739.

Houston, A. I., McNamara, J. M., & Hutchinson, J. M. C. (1993). General results concerning the trade-off between gaining energy and avoiding predation. *Philosophical Transactions of the Royal Society of London, Series B, 341*(1298), 375–397.

Howard, I. K. (2002). H is for Enthalpy, thanks to Heike Kamerlingh Onnes and Alfred W. Porter. *Journal of Chemical Education, 79*(6), 697–698.

Howell, B. J., Rahn, H., Goodfellow, D., & Herreid, C. (1973). Acid–base regulation and temperature in selected invertebrates as a function of temperature. *American Zoologist, 13*(2), 557–563.

Howell, T. R., & Bartholomew, G. A. (1959). Further experiments on torpidity in the poor-will. *Condor, 61*(3), 180–185.

Høyrup, J. (2002). *Lengths, widths, surfaces: a portrait of Old Babylonian algebra and its kin*. Berlin: Springer-Verlag.

Hozumi, K. (1985). Phase diagrammatic approach to the analysis of growth curve using the u-w diagram: basic aspects. *The Botanical Magazine, Tokyo, 98*(3), 239–250.

Hsaio, K. C., Cheng, C.-H. C., Fernandes, I. E., Detrich, H. W., & DeVries, A. L. (1990). An antifreeze glycopeptide gene from the antarctic cod *Notothenia coriiceps neglecta* encodes a polyprotein of high peptide copy number. *Proceedings of the National Academy of Sciences of the United States of America, 87*(23), 9265–9269.

Huang, K.-J., Yu, A.-W., Yu, C.-C., & Liu, J.-S. (2009). Correlation between basal metabolic rate and organ masses among four passerine birds in Wenzhou, Zhejiang Province. *Sichuan Journal of Zoology, 28*(1), 44–48.

Hubbard, A. E., & Gilinsky, N. L. (1992). Mass extinctions as statistical phenomena: an examination of the evidence using χ^2 tests and bootstrapping. *Paleobiology, 18*(2), 148–160.

Hubbell, S. P. (2001). *The unified neutral theory of biodiversity and biogeography*. Princeton, New Jersey: Princeton University Press.

Huber, R., Eder, W., Hedlwein, S., et al. (1998). *Thermocrinis ruber* gen. nov., sp. nov., a pink-filament-forming hyperthermophilic bacterium isolated from Yellowstone National Park. *Applied and Environmental Microbiology, 64*(10), 3576–3583.

Hubley, M. J., Locke, B. R., & Moerland, T. S. (1996). The effects of temperature, pH, and magnesium on the diffusion coefficient of ATP in solutions of physiological ionic strength. *Biochimica et Biophysica Acta—General Subjects, 1291*(2), 115–121.

Hudson, L. N., Isaac, N. J. B., & Reuman, D. C. (2013). The relationship between body mass and field metabolic rate among individual birds and mammals. *Journal of Animal Ecology, 82*(5), 1009–1020.

Huey, R. B., & Kingsolver, J. G. (1989). Evolution of thermal sensitivity of ectotherm performance. *Trends in Ecology & Evolution, 4*(5), 131–135.

Huey, R. B., & Slatkin, M. (1976). Costs and benefits of lizard thermoregulation. *Quarterly Review of Biology, 51*(3), 363–384.

Hughes, D. A., Jastroch, M., Stoneking, M., & Klingenspor, M. (2009). Molecular evolution of UCP1 and the evolutionary history of mammalian non-shivering thermogenesis. *BMC Evolutionary Biology, 9*, Article 4 (13 pages).

Hughes, D. J., & Hughes, R. N. (1986). Metabolic implications of modularity: studies on the respiration and growth of *Electra pilosa*. *Philosophical Transactions of the Royal Society of London, 313*(1159), 23–29.

Hughes, R. N. (1986). *A functional biology of marine gastropods*. London: Croom Helm.

Hughes, S. J. M., Ruhl, H. A., Hawkins, L. E., Hauton, C., Boorman, B., & Billett, D. S. M. (2011). Deep-sea echinoderm oxygen consumption rates and an interclass comparison of metabolic rates in Asteroidea, Crinoidea, Echinoidea, Holothuroidea and Ophiuroidea. *Journal of Experimental Biology, 214*(15), 2512–2521.

Hulbert, A. J., & Else, P. L. (1989). The evolution of endothermic metabolism: mitochondrial activity and changes in cellular composition. *American Journal of Physiology, 256*(25), R1200–R1208.

Hulbert, A. J., & Else, P. L. (1999). Membranes as possible pacemakers of metabolism. *Journal of Theoretical Biology, 199*(3), 257–274.

Hulbert, A. J., & Else, P. L. (2000). Mechanisms underlying the cost of living in animals. *Annual Review of Physiology, 62*, 207–235.

Hulbert, A. J., & Else, P. L. (2005). Membranes and the setting of energy demand. *Journal of Experimental Biology, 208*(9), 1593–1599.

Humphreys, W. F. (1979). Production and respiration in animal populations. *Journal of Animal Ecology, 48*(2), 427–453.

Humphries, M. M., & Careau, V. (2011). Heat for nothing or activity for free? Evidence and implications of activity-thermoregulatory heat substitution. *Integrative and Comparative Biology, 51*(3), 419–431.

Humphries, M. M., Thomas, D. W., & Kramer, D. L. (2003). The role of energy availability in mammalian hibernation: a cost-benefit approach. *Physiological and Biochemical Zoology, 76*(2), 165–179.

Hunter, E., & Hughes, R. N. (1994). The influence of temperature, food ration and genotype on zooid size in *Celleporella hyalina* (L.). In P. J. Hayward, J. S. Ryland, & P. D. Taylor (Eds.), *Biology and palaeobiology of bryozoans (Proceedings of the 9th International Bryozoology Conference, School of Biological Sciences, University of Wales, Swansea, 1992)* (pp.83–86). Fredensborg: Olsen & Olsen.

Huo, Y., & Kassab, G. S. (2012). Intraspecific scaling laws of vascular trees. *Journal of the Royal Society Interface, 9*(66), 190–200.

Hüppop, O., & Hüppop, K. (2003). North Atlantic Oscillation and timing of spring migration in birds. *Proceedings of the Royal Society of London, Series B, 270*(1512), 233–240.

Hurlbert, A. H., & Stegen, J. C. (2014). When should species richness be energy limited, and how would we know? *Ecology Letters, 17*(4), 401–413.

Hurst, T. P., & Conover, D. O. (1998). Winter mortality of young-of-the-year Hudson River striped bass (*Morone saxatilis*): size-dependent patterns and effects on recruitment. *Canadian Journal of Fisheries and Aquatic Sciences, 55*(5), 1122–1130.

Huston, M. (1979). A general hypothesis of species diversity. *American Naturalist, 113*(1), 81–101.

Huston, M. A. (1994). *Biological diversity: the coexistence of species on changing landscapes*. Cambridge: Cambridge University Press.

Huston, M. A., & Wolverton, S. (2011). Regulation of animal size by eNPP, Bergmann's rule, and related phenomena. *Ecological Monographs, 81*(3), 349–405.

Hutchinson, G. E. (1957). Concluding remarks. *Cold Spring Harbour Symposia on Quantitative Biology, 22*, 415–427.

Hutchinson, G. E. (1959). Homage to Santa Rosalia, or why are there so many kinds of animals? *American Naturalist, 93*(870), 145–159.

Hutchinson, V. H., Dowling, H. G., & Vinegar, A. (1966). Thermoregulation in a brooding female Indian python, *Python molurus bivittatus*. *Science, 151*(3711), 694–695.

Huttenlocker, A. K. (2014). Body size reductions in non-mammalian eutheriodont therapsids (Synapsida) during the end-Permian mass extinction. *PLoS One, 9*(2), e87553.

Huttenlocker, A. K., & Botha-Brink, J. (2013). Body size and growth patterns in the therocephalian *Moschorhinus kitchingi* (Therpasida: Eutheriodontia) before and after the end-Permian extinction in South Africa. *Paleobiology, 39*(2), 253–277.

Huxley, J. S. (1932). *Problems of relative growth*. London: Methuen.

Huxley, J. S., & Teissier, G. (1936). Terminology of relative growth. *Nature, 137*, 780–781.

Huxley, J. S., Webb, C. S., & Best, A. T. (1939). Temporary poikilothermy in birds. *Nature, 143*(3625), 683–684.

Huxley, T. H. (1868). On the animals which are most nearly intermediate between birds and reptiles. *Annals and Magazine of Natural History, 4th Series, 2*, 66–75.

Huxley, T. H. (1870). Further evidence of the affinity between the dinosaurian reptiles and birds. *Quarterly Journal of the Geological Society of London, 26*(1–2), 12–31.

Ikeda, T. (1985). Metabolic rates of epipelagic marine zooplankton as a function of body mass and temperature. *Marine Biology, 85*(1), 1–11.

Ikeda, T., Kanno, Y., Ozaki, K., & Shinada, A. (2001). Metabolic rates of epipelagic marine copepods as a function of body mass and temperature. *Marine Biology, 139*(3), 587–596.

Imbrie, J., Berger, A., Boyle, E. A., et al. (1993). On the structure and origin of major glaciation cycles 2. The 100,000-year cycle. *Paleoceanography, 8*(6), 699–735.

Imbrie, J., Boyle, E. A., Clemens, S. C., et al. (1992). On the structure and origin of major glaciation cycles 1. Linear responses to Milankovitch forcing. *Paleoceanography, 7*(6), 701–738.

Imbrie, J., & Imbrie, K. P. (1979). *Ice ages: solving the mystery*. London: Macmillan.

Irving, L., & Hart, J. S. (1957). The metabolism and insulation of seals as bare-skinned mammals in cold water. *Canadian Journal of Zoology, 35*(4), 497–511.

Iverson, S. L. T., Turner Brian N. (1974). Winter weight dynamics in *Microtus pennsylvanicus*. *Ecology, 55*(5), 1030–1041.

Ivleva, I. V. (1980). The dependence of crustacean respiration on body mass and habitat temperature. *Internationale Revue der gesamten Hydrobiologie, 65*(1), 1–47.

Jablonski, D. (1986a). Background and mass extinctions: the alternation of macroevolutionary regimes. *Science, 231*(4734), 129–133.

Jablonski, D. (1986b). Larval ecology and macroevolution and marine invertebrates. *Bulletin of Marine Science, 39*(2), 565–587.

Jablonski, D., Roy, K., Valentine, J. W., Price, R. M., & Anderson, P. S. (2003). The impact of the pull of the recent on the history of marine diversity. *Science, 300*, 1133–1135.

Jackson, J. B. C., & Johnson, K. G. (2000). Life in the last few million years. In D. H. Erwin & S. L. Wing (Eds.), *Deep time: paleobiology's perspective. Supplement to Paleobiology, Volume 26(4)* (pp. 221–235). Lawrence: The Paleontological Society.

Jacobs, S. S., Jenkins, A., Giulivi, C. F., & Dutrieux, P. (2011). Stronger ocean circulation and increased melting under Pine Island Glacier ice shelf. *Nature Geoscience, 4*(8), 519–523.

Jaeger, E. C. (1948). Does the poor-will 'hibernate'? *Condor, 50*(1), 45–46.

Jaeger, E. C. (1949). Further observations on the hibernation of the poor-will. *Condor, 51*(3), 105–109.

Jaenicke, R., & Böhm, G. (1998). The stability of proteins at high temperatures. *Current Opinion in Structural Biology, 8*, 738–748.

Jakosky, B. M. (1998). *The search for life on other planets*. Cambridge: Cambridge University Press.

Janech, M. G., Krell, A., Mock, T., Kang, J.-S., & Raymond, J. A. (2006). Ice-binding proteins from sea ice diatoms (Bacillariophyceae). *Journal of Phycology, 42*(2), 410–416.

Janz, N., Nylin, S., & Wahlberg, N. (2006). Diversity begets diversity: host expansions and the diversification of plant-feeding insects. *BMC Evolutionary Biology, 6*, article 4.

Jaynes, E. (1957). Information theory and statistical mechanics. *Physical Review, 106*, 620–630.

Jenkins, A., Dutrieux, P., Jacobs, S. S., et al. (2010). Observations beneath Pine Island Glacier in West Antarctica and implications for its retreat. *Nature Geoscience, 3*(7), 468–472.

Jenni, L., & Kéry, M. (2003). Timing of autumn bird migration under climate change: advances in long-distance migrants, delays in short-distance migrants. *Proceedings of the Royal Society of London, Series B, 270*(1523), 1467–1471.

Jennings, S. (1988). The mean free path in air. *Journal of Aerosol Science, 19*(2), 159–166.

Jensen, R. A. (2001). Orthologs and paralogs—we need to get it right. *Genome Biology, 2*(8), interactions 1002.1001–interactions 1002.1003.

Jeschke, J. M., Kopp, M., & Tollrian, R. (2002). Predator functional responses: discriminating between handling and digesting prey. *Ecological Monographs, 72*(1), 95–112.

Ji, Q., Luo, Z.-X., Yuan, C.-L., & Tabrum, A. R. (2006). A swimming mammaliaform from the Middle Jurassic and ecomorphological diversification of early mammals. *Science, 311*(5764), 1123–1127.

Jobling, M. (1981). The influences of feeding on the metabolic rate of fishes: a short review. *Journal of Fish Biology, 18*, 385–400.

Jobling, M. (1983). Towards an explanation of specific dynamic action (SDA). *Journal of Fish Biology, 23*, 549–555.

Jobling, M. (1985). Growth. In P. Tytler & P. Calow (Eds.), *Fish energetics: new perspectives* (pp. 213–230). London: Croom Helm.

Johnsen, S. (2001). Hidden in plain sight: the ecology and physiology of organismal transparency. *Biological Bulletin, 201*(3), 301–318.

Johnsen, S. J., Clausen, H. B., Dansgaard, W., et al. (1997). The $\delta^{18}O$ record along the Greenland Ice Core Project deep ice core and the problem of possible Eemian climatic instability. *Journal of Geophysical Research—Oceans, 102*(C12), 26397–26410.

Johnson, G. (2005). *Miss Leavitt's stars: the untold story of the woman who discovered how to measure the universe*. New York & London: W. W. Norton & Company.

Johnston, I. A., Calvo, J., Guderley, H. E., Fernandez, D., & Palmer, L. (1998). Latitudinal variation in the abundance and oxidative capacities of muscle mitochondria in perciform fishes. *Journal of Experimental Biology, 201*(1), 1–12.

Johnston, I. A., & Goldspink, G. (1975). Thermodynamic activation parameters of fish myofibrillar ATPase enzyme and evolutionary adaptations to temperature. *Nature, 257*(5527), 620–622.

Johnston, I. A., & Walesby, N. J. (1977). Molecular mechanisms of temperature adaptation in fish myofibrillar adenosine triphosphatase. *Journal of Comparative Physiology, 119*, 195–206.

Jokumsen, A., Wells, R. M. G., Ellerton, H. D., & Weber, R. E. (1981). Hemocyanin of the giant Antarctic isopod *Glyptonotus antarcticus*: structure and effects of temperature and pH on its oxygen affinity. *Comparative Biochemistry and Physiology, Part A, 70*(1), 91–95.

Jones, A. E., & Shanklin, J. D. (1995). Continued decline of total ozone over Halley, Antarctica, since 1985. *Nature, 376*(6539), 409–411.

Jones, C. G., Lawton, J. H., & Shachak, M. (1994). Organisms as ecosystem engineers. *Oikos, 69*(3), 373–386.

Jones, C. G., Lawton, J. H., & Shachak, M. (1997). Positive and negative effects of organisms as physical ecosystem engineers. *Ecology, 78*(7), 1946–1957.

Jones, T., & Cresswell, W. (2010). The phenology mismatch hypothesis: are declines of migrant birds linked to uneven global climate change? *Journal of Animal Ecology, 79*(1), 98–108.

Jonzén, N., Lindén, A., Ergon, T., et al. (2006). Rapid advance of spring arrival dates in long-distance migratory birds. *Science, 312*(5782), 1959–1961.

Jordt, S.-E., McKemy, D. D., & Julius, D. (2003). Lessons from peppers and peppermint: the molecular logic of thermosensation. *Current Opinion in Neurobiology, 13*(4), 487–492.

Jørgensen, C. B. (1988). Metabolic costs of growth and maintenance in the toad, Bufo bufo. *Journal of Experimental Biology*, *138*(1), 319–331.

Jost, L. (2006). Entropy and diversity. *Oikos*, *113*(2), 363–375.

Jost, L. (2007). Partitioning diversity into independent alpha and beta components. *Ecology*, *88*(10), 2427–2439.

Jouzel, J., Masson-Delmotte, V., Cattani, O., et al. (2007). Orbital and millennial Antarctic climate variability over the past 800,000 years. *Science*, *317*(5839), 793–796.

Jung, W., Gwak, Y., Davies, P. L., Kim, H. J., & Jin, E. (2014). Isolation and characterization of antifreeze proteins from the Antarctic marine microalga *Pyramimonas gelidicola*. *Marine Biotechnology*, *16*(5), 502–512.

Juszczuk, I. M., & Rychter, A. M. (2003). Alternative oxidase in higher plants. *Acta Biochimica Polonica*, *50*(4), 1257–1271.

Kacser, H., & Burns, J. A. (1973). The control of flux. In D. D. Davies (Ed.), *Rate control of biological processes (Symposia of the Society for Experimental Biology 27)* (pp. 65–104). Cambridge: Cambridge University Press.

Kaiser, A., Klok, C. J., Socha, J. J., Lee, W.-K., Quinlan, M. C., & Harrison, J. F. (2007). Increase in tracheal investment with beetle size supports hypothesis of oxygen limitation on insect gigantism. *Proceedings of the National Academy of Sciences of the United States of America*, *104*(32), 13198–13203.

Kakizaki, Y., Moore, A. L., & Ito, K. (2012). Different molecular bases underlie the mitochondrial respiratory activity in the homoeothermic spadices of *Symplocarpus renifolius* and the transiently thermogenic appendices of *Arum maculatum*. *Biochemical Journal*, *445*(2), 237–246.

Kammer, A. E. (1968). Motor patterns during flight and warm-up in Lepidoptera. *Journal of Experimental Biology*, *48*(1), 277–295.

Kanwisher, J. (1959). Histology and metabolism of frozen intertidal animals. *Biological Bulletin, Marine Biological Laboratory, Woods Hole*, *116*(2), 258–264.

Kanwisher, J. W. (1955). Freezing in intertidal animals. *Biological Bulletin, Marine Biological Laboratory, Woods Hole*, *109*(1), 56–63.

Kappen, L. (1993). Lichens in the Antarctic region. In E. I. Friedmann (Ed.), *Antarctic microbiology* (pp. 433–490). New York: Wiley-Liss.

Karpovich, S. A., Tøien, Ø., Buck, C. L., & Barnes, B. M. (2009). Energetics of arousal episodes in hibernating arctic ground squirrels. *Journal of Comparative Physiology, B*, *179*(6), 691–700.

Kashefi, K., Holmes, D. E., Reysenbach, A.-L., & Lovley, D. R. (2002). Use of Fe(III) as an electron acceptor to recover previously uncultured hyperthermophiles: isolation and characterization of *Geothermobacterium ferrireducens* gen. nov., sp. nov. *Applied and Environmental Microbiology*, *68*(4), 1735–1742.

Kashefi, K., & Lovley, D. R. (2003). Extending the upper temperature limit for life. *Science*, *301*(5635), 934–934.

Kaufmann, K. W. (1981). Fitting and using growth curves. *Oecologia*, *49*(3), 293–299.

Kawahara, H., Iwanaka, Y., Higa, S., et al. (2007). A novel, intracellular antifreeze protein in an Antarctic bacterium, *Flavobacterium xanthum*. *Cryo-Letters*, *28*(1), 39–49.

Kearney, M. (2012). Metabolic theory, life history and the distribution of a terrestrial ectotherm. *Functional Ecology*, *26*(1), 167–179.

Kearney, M., Shine, R., & Porter, W. P. (2009). The potential for behavioral thermoregulation to buffer 'cold-blooded' animals against climate warming. *Proceedings of the National Academy of Sciences of the United States of America*, *106*(10), 3835–3840.

Keeler, J., & Wothers, P. (2003). *Why chemical reactions happen*. Oxford: Oxford University Press.

Keeling, C. D., Adams, J. A., Ekdahl, C. A., & Guenther, P. R. (1976a). Atmospheric carbon dioxide variations at the South Pole. *Tellus*, *28*(6), 552–564.

Keeling, C. D., Bacastow, R. B. B., Arnold, E., Ekdahl, C. A., Guenther, P. R., & Waterman, L. S. (1976b). Atmospheric carbon dioxide variations at Mauna Loa Observatory, Hawaii. *Tellus*, *28*(6), 538–551.

Kemp, T. S. (1982). *Mammal-like reptiles and the origin of mammals*. London: Academic Press.

Kemp, T. S. (2006). The origin of mammalian endothermy: a paradigm for the evolution of complex structure. *Zoological Journal of the Linnean Society*, *147*(4), 473–488.

Kendall, M. A. (1996). Are Arctic soft sediment macrobenthic communities impoverished? *Polar Biology*, *16*(6), 393–399.

Kendall, M. A., & Aschan, M. (1993). Latitudinal gradients in the structure of macrobenthic communities: a comparison of Arctic, temperate and tropical sites. *Journal of Experimental Marine Biology and Ecology*, *171*(1–2), 157–169.

Kerr, J. T., & Currie, D. J. (1999). The relative importance of evolutionary and environmental controls on broad-scale patterns of species richness in North America. *Ecoscience*, *6*(3), 329–337.

Kerr, J. T., & Packer, L. (1997). Habitat heterogeneity as a determinant of mammal species richness in high-energy regions. *Nature*, *385*(6613), 252–254.

Kessel, B. (1976). Winter activity patterns of Black-capped Chickadees in interior Alaska. *The Wilson Bulletin*, *88*(1), 36–61.

Keszei, E. (2003). Michael Polanyi's pioneering contribution to the most successful theory in chemical kinetics. *Polanyiana*, *12*(1–2), 63–74.

Kharasch, M. S. (1929). Heats of combustion of organic compounds. *Bureau of Standards Journal of Research*, *2*(2), 359–430.

Kiehl, J. T., & Trenberth, K. E. (1997). Earth's annual global mean energy budget. *Bulletin of the American Meteorological Society*, *78*(2), 197–208.

Kier, G., Mutke, J., Dinerstein, E., et al. (2005). Global patterns of plant diversity and floristic knowledge. *Journal of Biogeography*, *32*, 1107–1116.

Kiernan, M. C., Cikurel, K., & Bostock, H. (2001). Effects of temperature on the excitability properties of human motor axons. *Brain*, *124*, 816–825.

Kiessling, W., & Simpson, C. (2011). On the potential for ocean acidification to be a general cause of ancient reef crises. *Global Change Biology*, *17*(1), 56–67.

Killen, S. S., Atkinson, D., & Glazier, D. S. (2010). The intraspecific scaling of metabolic rate with body mass in fishes depends on lifestyle and temperature. *Ecology Letters*, *13*(2), 184–193.

Kim, Y. E., Hipp, M. S., Bracher, A., Hayer-Hartl, M., & Hartl, F. U. (2013). Molecular chaperone functions in protein folding and proteostasis. *Annual Review of Biochemistry*, *82*, 323–355.

Kingsolver, J. G. (1985c). Butterfly thermoregulation: organismic mechanisms and population consequences. *Journal of Research on the Lepidoptera*, *24*(1), 1–20.

Kingsolver, J. G. (1985a). Thermal ecology of *Pieris* butterflies (Lepidoptera: Pieridae): a new mechanism of behavioral thermoregulation. *Oecologia*, *66*(4), 540–545.

Kingsolver, J. G. (1985b). Thermoregulatory significance of wing melanization in *Pieris* butterflies (Lepidoptera: Pieridae): physics, posture, and pattern. *Oecologia*, *66*(4), 546–553.

Kingsolver, J. G. (1987). Evolution and coadaptation of thermoregulatory behavior and wing pigmentation pattern in pierid butterflies. *Evolution*, *41*(3), 472–490.

Kingsolver, J. G., & Huey, R. B. (2008). Size, temperature, and fitness: three rules. *Evolutionary Ecology Research*, *10*(2), 251–268.

Kingsolver, J. G., & Woods, H. A. (1997). Thermal sensitivity of growth and feeding in *Manduca sexta* caterpillars. *Physiological Zoology*, *70*, 631–638.

Kiørboe, T. (2013). Zooplankton body composition. *Limnology & Oceanography*, *58*(5), 1843–1850.

Kisser, B., & Goodwin, H. T. (2012). Hibernation and overwinter body temperatures in free-ranging thirteen-lined ground squirrels, *Ictidomys tridecemlineatus*. *American Midland Naturalist*, *167*(2), 396–409.

Klaas, M., Hampe, O., Schuldack, M., Holff, C., Kardjilov, N., & Hilger, A. (2011). New insights into the respiration and metabolic physiology of *Lystrosaurus*. *Acta Zoologica*, *92*(4), 363–371.

Klaassen, M., & Nolet, B. A. (2008). Stoichiometry of endothermy: shifting the quest from nitrogen to carbon. *Ecology Letters*, *11*(8), 785–792.

Kleiber, M. (1932). Body size and metabolism. *Hilgardia*, *6*(11), 315–353.

Kleiber, M. (1947). Body size and metabolic rate. *Physiological Reviews*, *27*, 511–541.

Kleiber, M. (1961). *The fire of life: an introduction to animal energetics*. New York: John Wiley & Sons.

Kleiber, M. (1972). Joules vs. calories in nutrition. *Journal of Nutrition*, *102*, 309–312.

Kleinebeckel, D., & Klussmann, F. W. (1990). Shivering. In E. Schöbaum & P. Lomax (Eds.), *Thermoregulation: physiology and biochemistry* (Vol. 506). New York: Pergamon Press.

Kline, M. (1980). *Mathematics: the loss of certainty*. Oxford: Oxford University Press.

Klok, C. J., Hubb, A. J., & Harrison, J. F. (2009). Single and multigenerational responses of body mass to atmospheric oxygen concentrations in *Drosophila melanogaster*: evidence for roles of plasticity and evolution. *Journal of Evolutionary Biology*, *22*(12), 2496–2504.

Knoll, A. H. (2003). Biomineralization and evolutionary history. In P. M. Dove, J. J. DeYoreo, & S. Weiner (Eds.), *Biomineralization* (Vol. 54, pp. 329–356). Washington, D.C.: Mineralogical Society of America.

Knutson, R. M. (1972). Temperature measurements of spadix of *Symplocarpus foetidus* (L.) Nutt. *American Midland Naturalist*, *88*(1), 251–254.

Knutson, R. M. (1974). Heat production and temperature regulation in Eastern Skunk Cabbage. *Science*, *186*(4165), 746–747.

Knutson, R. M. (1979). Plants in heat. *Natural History*, *88*(3), 42–47.

Kobbe, S., Nowack, J., & Dausmann, K. H. (2014). Torpor is not the only option: seasonal variations of the thermoneutral zone in a small primate. *Journal of Comparative Physiology B*, *184*(6), 789–797.

Koch, A. L. (1970). Overall controls on the biosynthesis of ribosomes in growing bacteria. *Journal of Theoretical Biology*, *28*(2), 203–231.

Kohne, D. E. (1970). Evolution of higher-organism DNA. *Quarterly Review of Biophysics*, *3*(3), 327–375.

Kohshima, S. (1984). A novel cold-tolerant insect found in a Himalayan glacier. *Nature*, *310*(5974), 225–227.

Konarzewski, M., & Książek, A. (2013). Determinants of intra-specific variation in basal metabolic rate. *Journal of Comparative Physiology B*, *183*(1), 27–41.

Konstantinidis, K. T., Ramette, A., & Tiedje, J. M. (2006). The bacterial species definition in the genomic era. *Philosophical Transactions of the Royal Society of London, Series B*, *361*(1475), 1929–1940.

Kooijman, S. A. L. M. (1993). *Dynamic energy budgets in biological systems: theory and applications in ecotoxicology*. Cambridge: Cambridge University Press.

Kooijman, S. A. L. M. (2000). *Dynamic energy and mass budgets in biological systems* (Second ed.). Cambridge: Cambridge University Press.

Kooijman, S. A. L. M. (2010). *Dynamic energy budget theory for metabolic organisation* (Third ed.). Cambridge: Cambridge University Press.

Kopp, G., & Lean, J. L. (2011). A new, lower value of total solar irradiance: evidence and climate signifiance. *Geophysical Research Letters, 38*, L01706.

Koteja, P. (2000). Energy assimilation, parental care and the evolution of endothermy. *Proceedings of the Royal Society of London, Series B, 267*(1442), 479–484.

Koteja, P. (2004). The evolution of concepts on the evolution of endothermy in birds and mammals. *Physiological and Biochemical Zoology, 77*(6), 1043–1050.

Kozłowski, J., & Weiner, J. (1997). Interspecific allometries are by-products of body size optimization. *American Naturalist, 149*(2), 352–380.

Kragh, H. (1999). *Quantum generations: a history of physics in the twentieth century*. Princeton, New Jersey: Princeton University Press.

Krause, P. F., & Flood, K. L. (1997). *Weather and climate extremes*. Alexandria, Virginia: U.S. Army Corps of Engineers Topographic Engineering Center Report TEC–0099.

Krauss, S., Zhang, C.-Y., & Lowell, B. B. (2002). A significant portion of mitochondrial proton leak in intact thymocytes depends on expression of UCP2. *Proceedings of the National Academy of Sciences of the United States of America, 99*(1), 118–122.

Kreft, H., & Jetz, W. (2007). Global patterns and determinants of vascular plant diversity. *Proceedings of the National Academy of Sciences of the United States of America, 104*(14), 5925–5930.

Kreyling, J., Schmid, S., & Aas, G. (2015). Cold tolerance of tree species is related to the climate of their native ranges. *Journal of Biogeography, 42*(1), 156–166.

Krog, J. O., Zachariassen, K. E., Larsen, B., & Smidsrød, O. (1979). Thermal buffering in Afro-alpine plants due to nucleating agent-induced water freezing. *Nature, 282*(5736), 300–301.

Krogh, A. (1914). The quantitative relation between temperature and standard metabolism in animals. *Internationale Zeitschrift für physikalisch-chemische Biologie, 1*, 491–508.

Krogh, A. (1916). *The respiratory exchange of animals and man*. London: Longmans Green.

Krogh, A. (1929). The progress of physiology. *American Journal of Physiology, 90*(2), 243–251.

Krogh, A., & Zeuthen, E. (1941). The mechanism of flight preparation in some insects. *Journal of Experimental Biology, 18*(1), 1–10.

Krug, A. Z., Jablonski, D., & Valentine, J. W. (2009). Signature of the End-Cretaceous mass extinction in the modern biota. *Science, 323*(5915), 767–771.

Krug, R. R., Hunter, W. G., & Grieger, R. A. (1976a). Enthalpy–entropy compensation. 1. Some fundamental statistical problems associated with the analysis of van't Hoff and Arrhenius data. *Journal of Physical Chemistry, 80*(21), 2335–2341.

Krug, R. R., Hunter, W. G., & Grieger, R. A. (1976b). Enthalpy–entropy compensation. 2. Separation of the chemical from the statistical effect. *Journal of Physical Chemistry, 80*(21), 2341–2351.

Krug, R. R., Hunter, W. G., & Grieger, R. A. (1976c). Statistical interpretation of enthalpy–entropy compensation. *Nature, 261*, 566–567.

Kubecka, P. (2001). A possible world record maximum natural ground surface temperature. *Weather, 56*(7), 218–221.

Kuhn, T. S. (1978). *Black-body theory and the quantum discontinuity, 1894–1912*. Chicago: Chicago University Press.

Kukal, O., & Duman, J. G. (1989). Switch in the overwintering strategy of two insect species and latitudinal differences in cold hardiness. *Canadian Journal of Zoology, 67*(4), 825–827.

Kullberg, C. (1998). Does diurnal variation in body mass affect take-off ability in wintering willow tits? *Animal Behaviour, 56*(1), 227–233.

Kullberg, C., Fransson, T., Hedlund, J., et al. (2015). Change in spring arrival of migratory birds under an era of climate change, Swedish data from the last 140 years. *Ambio, 44*(Supplement 1), S69–S77.

Kullberg, C., Fransson, T., & Jakobsson, S. (1996). Impaired predator evasion in fat blackcaps (*Sylvia atricapilla*). *Proceedings of the Royal Society of London, Series B, 263*(1377), 1671–1675.

Kullberg, C., Jakobsson, S., & Fransson, T. (2000). High migratory fuel loads impair predator evasion in Sedge Warblers. *Auk, 117*(4), 1034–1038.

Kurr, M., Huber, R., König, H., et al. (1991). *Methanopyrus kandleri*, gen. and sp. nov. represents a novel group of hyperthermophilic methanogens growing at 110 °C. *Archives of Microbiology, 156*(4), 239–247.

Kürten, L., & Schmidt, U. (1982). Thermoperception in the common vampire bat (*Desmodus rotundus*). *Journal of Comparative Physiology, 146*(2), 223–228.

Kutschke, M., Grimpo, K., Kastl, A., et al. (2013). Depression of mitochondrial respiration during daily torpor of the Djungarian hamster, *Phodopus sungorus*, is specific for liver and correlates with body temperature. *Comparative Biochemistry and Physiology, Part A, 164*(4), 584–589.

Lacis, A. A., Schmidt, G. A., Rind, D., & Ruedy, R. A. (2010). Atmospheric CO_2: principal control knob governing Earth's temperature. *Science, 330*(6002), 356–359.

Laidler, K. J. (1981). Symbolism and terminology in chemical kinetics. *Pure and Applied Chemistry, 53*(3), 753–771.

Laidler, K. J. (1984). The development of the Arrhenius equation. *Journal of Chemical Education, 61*(6), 494–498.

Laidler, K. J., & King, M. C. (1983). The development of transition-state theory. *Journal of Physical Chemistry, 87*(15), 2657–2664.

Lamare-Picquot, C.-A. (1835). Scéances académiques. *L'institut, Journal Général des Sociétés et Travaux Scientifiques de la France et de L'Étranger, 3*(95), 69–71.

Lambert, A. M., Miller-Rushing, A. J., & Inouye, D. W. (2010). Changes in snowmelt date and summer precipitation affect the flowering phenology of *Erythronium grandiflorum* (glacier lily; Liliaceae). *American Journal of Botany, 97*(9), 1431–1437.

Lambrinos, J. G., & Kleier, C. C. (2003). Thermoregulation of juvenile Andean toads (*Bufo spinulosus*) at 4300 m. *Journal of Thermal Biology, 28*(1), 15–19.

Lambshead, P. J. D., Tietjen, J., Ferrero, T. J., & Jensen, P. (2000). Latitudinal diversity gradients in the deep sea with special reference to North Atlantic nematodes. *Marine Ecology Progress Series, 194*, 159–167.

Lane, N. (2005). *Power, sex, suicide: mitochondria and the meaning of life*. Oxford: Oxford University Press.

Lane, N. (2015). *The vital question: energy, evolution, and the origins of complex life*. London: Norton.

Lanfear, R., Ho, S. Y. W., Love, D., & Bromham, L. (2010a). Mutation rate is linked to diversification in birds. *Proceedings of the National Academy of Sciences of the United States of America, 107*(47), 20423–20428.

Lanfear, R., Thomas, J. A., Welch, J. J., Brey, T., & Bromham, L. (2007). Metabolic rate does not calibrate the molecular clock. *Proceedings of the National Academy of Sciences of the United States of America, 104*(39), 15388–15393.

Lanfear, R., Welch, J. J., & Bromham, L. (2010b). Watching the clock: studying variation in rates of molecular evolution between species. *Trends in Ecology & Evolution, 25*(9), 495–503.

Lang, E. W. (1986). Physical-chemical limits for the stability of biomolecules. *Advances in Space Research, 6*(12), 251–255.

Lang, J. W. (1979). Thermophilic response of the American alligator and the American crocodile to feeding. *Copeia, 1979*(1), 48–59.

Larson, D. J., Middle, L., Vu, H., et al. (2014). Wood frog adaptations to overwintering in Alaska: new limits to freezing tolerance. *Journal of Experimental Biology, 217*(12), 2193–2200.

Lasiewski, R. C., & Bartholomew, G. A. (1966). Evaporative cooling in the Poor-will and Tawny Frogmouth. *Condor, 68*, 253–262.

Lasiewski, R. C., & Dawson, W. R. (1964). Physiological responses to temperature in the Common Nighthawk. *Condor, 66*, 477–490.

Latimer, A. M. (2007). Geography and resource limitation complicate metabolism-based predictions of species richness. *Ecology, 88*(8), 1895–1898.

Lawton, J. H. (1993). Range, population size and conservation. *Trends in Ecology & Evolution, 8*(11), 409–413.

Lawton, J. H. (1995). Population dynamic principles. In J. H. Lawton & R. M. May (Eds.), *Extinction rates* (pp. 147–163). Oxford: Oxford University Press.

Lawton, J. H. (1996). Patterns in ecology. *Oikos, 75*(2), 145–147.

Lawton, J. H. (1999). Are there general laws in ecology? *Oikos, 84*(2), 177–192.

Leanhardt, A. E., Pasquini, T. A., Saba, M., et al. (2003). Cooling Bose–Einstein condensates below 500 picokelvin. *Science, 301*(5639), 1513–1515.

Lear, C. H., Rosenthal, Y., & Slowey, N. (2002). Benthic foraminiferal Mg/Ca-paleothermometry: a revised core-top calibration. *Geochimica et Cosmochimica Acta, 66*(19), 3375–3387.

Lederman, L. M., & Hill, C. T. (2004). *Symmetry and the beautiful universe*. Amherst, New York: Prometheus Books.

Lee, C. K., Daniel, R. M., Shepherd, C., et al. (2007). Eurythermalism and the temperature dependence of enzyme activity. *FASEB Journal, 21*(8), 1934–1941.

Lee, M. S. Y. (1997). Documenting present and past biodiversity: conservation biology meets palaeontology. *Trends in Ecology & Evolution, 12*(4), 132–133.

Lee, T. N., Barnes, B. M., & Buck, C. L. (2009). Body temperature patterns during hibernation in a free-living Alaska marmot (*Marmota broweri*). *Ethology Ecology & Evolution, 21*(3–4), 403–413.

Lee, T. N., Kohl, F., Buck, C. L., & Barnes, B. M. (2016). Hibernation strategies and patterns in sympatric arctic species, the Alaska marmot and the arctic ground squirrel. *Journal of Mammalogy, 97*(1), 135–144.

Lehikoinen, E., Sparks, T. H., & Zalakevicius, M. (2004). Arrival and departure dates. *Advances in Ecological Research, 35*, 1–31.

Leigh, A., Sevanto, S., Ball, M. C., et al. (2012). Do thick leaves avoid thermal damage in critically low wind speeds? *New Phytologist, 194*(2), 477–487.

Lenton, T. M., & Watson, A. J. (2011). *Revolutions that made the Earth*. Oxford: Oxford University Press.

Levins, R. (1969). Thermal acclimation and heat resistance in *Drosophila* species. *American Naturalist, 103*(933), 483–499.

Levitus, S., Antonov, J. I., Boyer, T. P., et al. (2012). World ocean heat content and thermosteric sea level change (0–2000 m), 1955–2010. *Geophysical Research Letters*, 39, L10603.

Li, L., Wang, Z., Zerbe, S., et al. (2013). Species richness patterns and water–energy dynamics in the drylands of Northwest China. *PLoS One*, 8(6), article 66450.

Li, X.-M., Trinh, K.-Y., Hew, C. L., Buettner, B., Baenziger, J., & Davies, P. L. (1985). Structure of an antifreeze polypeptide and its precursor from the ocean pout, *Macrozoarces americanus*. *Journal of Biological Chemistry*, 260(24), 12904–12909.

Lifson, N., Gordon, G. B., & McClintock, R. (1955). Measurement of total carbon dioxide prduction by means of $D_2{}^{18}O$. *Journal of Applied Physiology*, 7, 704–710.

Lifson, N., Gordon, G. B., Visscher, M. B., & Nier, A. O. (1949). The fate of utilized molecular oxygen and the source of the respiratory carbon dioxide, studied with the aid of heavy oxygen. *Journal of Biological Chemistry*, 180, 804–811.

Lifson, N., & McClintock, R. (1966). Theory of use of the turnover rates of body water for measuring energy and material balance. *Journal of Theoretical Biology*, 12(1), 46–74.

Ligon, J. D. (1970). Still more responses of the poor-will to low temperatures. *Condor*, 72(4), 496–498.

Lind, J., Fransson, T., Jakobsen, S., & Kullberg, C. (1999). Reduced take-off ability in robins (*Erithacus rubecula*) due to migratory fuel load. *Behavioral Ecology and Sociobiology*, 46(1), 65–70.

Lind, J., Jakobsen, S., & Kullberg, C. (2010). Impaired predator evasion in the life history of birds: behavioral and physiological adaptations to reduced flight ability. In C. F. Thompson (Ed.), *Current Ornithology Volume 17* (pp. 1–30). Berlin: Springer Verlag.

Lindemann, R. L. (1942). The trophic-dynamic aspect of ecology. *Ecology*, 23, 399–413.

Lindow, S. E., Arny, D. C., & Upper, C. D. (1978a). Distribution of ice nucleation-active bacteria on plants in nature. *Applied and Environmental Microbiology*, 36(6), 831–838.

Lindow, S. E., Arny, D. C., & Upper, C. D. (1978b). *Erwinia herbicola*: a bacterial ice nucleus active in increasing frost injury to corn. *Phytopathology*, 68(3), 523–527.

Lindow, S. E., Arny, D. C., & Upper, C. D. (1982). Bacterial ice nucleation: a factor in frost injury in plants. *Plant Physiology*, 70(4), 1084–1089.

Lindstedt, S. L., & Boyce, M. S. (1985). Seasonality, fasting endurance, and body size in mammals. *American Naturalist*, 125(6), 873–878.

Lindstedt, S. L., Hokanson, J. F., Wells, D. J., Swain, S. D., Hoppeler, H., & Navarro, V. (1991). Running energetics in the pronghorn antelope. *Nature*, 353(6346), 748–750.

Ling, G. (2004). What determines the normal water content of a living cell? *Physiological Chemistry and Physics and Medical NMR*, 36(1), 1–19.

Liu, J., Nicholson, C. E., & Cooper, S. J. (2007). Direct measurement of critical nucleus size in confined volumes. *Langmuir*, 23(13), 7286–7292.

Liu, J.-S., & Li, M. (2006). Phenotypic flexibility of metabolic rate and organ masses among tree sparrows *Passer montanus* in seasonal acclimatization. *Acta Zoologica Sinica*, 52(3), 469–477.

Llovel, W., Willis, J. K., Landerer, F. W., & Fukumori, I. (2014). Deep-ocean contribution to sea level and energy budget not detectable over the past decade. *Nature Climate Change*, 4(11), 1031–1035.

Locey, K. J., & Lennon, J. T. (2016). Scaling laws predict global microbial diversity. *Proceedings of the National Academy of Sciences of the United States of America*, 113(21), 5970–5975.

Lock, G. S. H. (1990). *The growth and decay of ice*. Cambridge: Cambridge University Press.

Londoño, G. A., Chappell, M. A., Castañeda, M. d. R., Jankowski, J. E., & Robinson, S. K. (2015). Basal metabolism in tropical birds: latitude, altitude and the 'pace of life'. *Functional Ecology*, 29(3), 338–346.

Longhurst, A. (1998). *Ecological geography of the sea*. San Diego: Academic Press.

Lord, F. M. (1953). On the statistical treatment of football numbers. *American Psychologist*, 8, 750–751.

Louthan, A. M., Doak, D. F., & Angert, A. L. (2015). Where and when do species interactions set range limits? *Trends in Ecology & Evolution*, 30(12), 780–792.

Lovegrove, B. G. (2000). The zoogeography of mammalian basal metabolic rate. *American Naturalist*, 156(2), 201–219.

Lovegrove, B. G. (2003). The influence of climate on the basal metabolic rate of small mammals: a slow-fast metabolic continuum. *Journal of Comparative Physiology B*, 173(2), 87–112.

Lovegrove, B. G. (2004). Locomotor mode, maximum running speed, and basal metabolic rate in placental mammals. *Physiological and Biochemical Zoology*, 77(6), 916–928.

Lovegrove, B. G. (2012). The evolution of mammalian body temperature: the Cenozoic supraendothermic pulses. *Journal of Comparative Physiology, B*, 182(4), 579–589.

Lovelock, J. E. (1953). The haemolysis of human red bloodcells by freezing and thawing. *Biochimica et Biophysica Acta*, 10, 414–426.

Lovelock, J. E. (1965). A physical basis for life detection experiments. *Nature*, 207(997), 568–570.

Lovelock, J. E., & Giffin, C. E. (1969). Planetary atmospheres: compositional and other changes associated with the presence of life. *Advances in the Aeronautical Sciences, 25,* 179–193.

Lowe, C. H., & Heath, W. G. (1969). Behavioral and physiological responses to temperature in the desert pupfish *Cyprinodon macularis. Physiological Zoology, 42*(1), 53–59.

Lowenstam, H. A. (1954). Factors affecting the aragonite–calcite ratios in carbonate-secreting marine organisms. *Journal of Geology, 62*(3), 285–322.

Lü, J.-C. (2002). Soft tissue in an early Cretaceous pterosaur from Liaoning Province, China. *Memoir of the Fukui Prefectural Dinosaur Museum, 1*(1), 19–28.

Lucas, A. (1996). *Bioenergetics of aquatic animals* (J. J. Watson, Trans.). London: Taylor & Francis.

Luisi, P. L. (1998). About various definitions of life. *Origins of Life and Evolution of Biospheres, 28*(4–6), 613–622.

Lumry, R., & Rajender, S. (1970). Enthalpy–entropy compensation phenomena in water solutions of proteins and small molecules: a ubiquitous property of water. *Biopolymers, 9*(10), 1125–1227.

Lutz, R. A. (2012). Deep-sea hydrothermal vents. In E. M. Bell (Ed.), *Life at extremes: environments, organisms and strategies for survival* (pp. 242–270). Wallingford, Oxfordshire: CAB International.

Luxmoore, R. A. (1984). A comparison of the respiration rates of some Antarctic isopods with species from lower latitudes. *British Antarctic Survey Bulletin, 62,* 53–65.

Luyet, B. J. (1937). The vitrification of organic colloids and of protoplasm. *Biodynamica, 1*(29), 1–14.

Luyet, B. J., & Gehenio, P. M. (1940). *Life and death at low temperatures.* Normandy, Missouri: Biodynamica.

Lyman, C. P. (1963). Hibernation in mammals and birds. *American Scientist, 51*(2), 127–138.

Lyman, J. (1969). Redefinition of salinity and chlorinity. *Limnology & Oceanography, 14*(6), 928–929.

Lyman, J. M., Good, S. A., Gouretski, V. V., et al. (2010). Robust warming of the global upper ocean. *Nature, 465*(7296), 334–337.

Lyth, M. (1982). Water-content of slugs (Gastropoda: Pulmonata) maintained in standardised culture conditions. *Journal of Molluscan Studies, 48*(2), 214–218.

MacArthur, R. H., & Wilson, E. O. (1963). An equilibrium theory of insular zoogeography. *Evolution, 17,* 373–387.

MacArthur, R. H., & Wilson, E. O. (1967). *The theory of island biogeography.* Princeton: Princeton University Press.

Macleod, R., Barnett, P., Clark, J. A., & Cresswell, W. (2005). Body mass change strategies in blackbirds *Turdus merula*: the starvation–predation risk trade-off. *Journal of Animal Ecology, 74*(2), 292–302.

MacMillan, H. A., & Sinclair, B. J. (2011). Mechanisms underlying insect chill-coma. *Journal of Insect Physiology, 57*(1), 12–20.

Macpherson, E. (2002). Large-scale species-richness gradients in the Atlantic Ocean. *Proceedings of the Royal Society of London, Series B, 269*(1501), 1715–1720.

Madison, D. L., Scrofano, M. M., Ireland, R. C., & Loomis, S. H. (1991). Purification and partial characterisation of an ice nucleator protein from the intertidal gastropod, *Melampus bidentatus. Cryobiology, 28*(5), 483–490.

Magurran, A. E. (2004). *Measuring ecological diversity.* Oxford: Blackwell Science.

Maino, J. L., & Kearney, M. R. (2015). Testing mechanistic models of growth in insects. *Proceedings of the Royal Society of London, Series B, 282*(1819), article 20151973.

Maino, J. L., Kearney, M. R., Nisbet, R. M., & Kooijman, S. A. L. M. (2014). Reconciling theories for metabolic scaling. *Journal of Animal Ecology, 83*(1), 20–29.

Major, J. (1963). A climatic index to vascular plant activity. *Ecology, 44*(3), 485–498.

Makarieva, A. M., Gorshkov, V. G., & Li, B.-L. (2004). Ontogenetic growth: models and theory. *Ecological Modelling, 176*(1), 15–26.

Makarieva, A. M., Gorshkov, V. G., & Li, B.-L. (2005a). Gigantism, temperature and metabolic rate in terrestrial poikilotherms. *Proceedings of the Royal Society of London, Series B, 272*(1578), 2325–2328.

Makarieva, A. M., Gorshkov, V. G., & Li, B.-L. (2005b). Temperature-associated upper limits to body size in terrestrial poikilotherms. *Oikos, 111*(3), 425–436.

Makarieva, A. M., Gorshkov, V. G., Li, B.-L., Chown, S. L., Reich, P. B., & Gavrilov, V. M. (2008). Mean mass-specific metabolic rates are strikingly similar across life's major domains: evidence for life's metabolic optimum. *Proceedings of the National Academy of Sciences of the United States of America, 105*(44), 16994–16999.

Malan, A. (1980). Enzyme regulation, metabolic rate and acid–base state in hibernation. In R. Gilles (Ed.), *Animals and environmental fitness* (pp. 487–501). Oxford: Pergamon Press.

Malan, A. (2010). Is the torpor–arousal cycle of hibernation controlled by a non-temperature-compensated circadian clock? *Journal of Biological Rhythms, 25*(3), 166–175.

Malan, A., Wilson, T. L., & Reeves, R. B. (1976). Intracellular pH in cold-blooded vertebrates as a function of body temperature. *Respiration Physiology, 28*(1), 29–47.

Manley, G. (1953). The mean temperature of central England, 1698–1952. *Quarterly Journal of the Royal Meteorological Society, 79*(340), 242–261.

Marchand, P. J. (1987). *Life in the cold: an introduction to winter ecology.* Hanover and London: University Press of New England.

Margaria, R. (1976). *Biomechanics and energetics of muscular exercise.* Oxford: Clarendon Press.

Margulis, L., & Lovelock, J. E. (1974). Biological modulation of the Earth's atmosphere. *Icarus, 21,* 471–489.

Mark, F. C., Bock, C., & Pörtner, H. O. (2002). Oxygen-limited thermal tolerance in Antarctic fish investigated by MRI and ^{31}P-MRS. *American Journal of Physiology, 283*(5), R1254–R1262.

Marko, P. B. (2004). 'What's larvae got to do with it?' Disparate patterns of post-glacial population structure in two benthic marine gastropods with identical dispersal potential. *Molecular Ecology, 13*, 597–611.

Marlowe, I. T., Green, J. C., Neal, A. C., Brassell, S. C., Eglinton, G., & Course, P. A. (1984). Long chain (n-C_{37}–C_{39}) alkenones in the Prymnesiophyceae. Distribution of alkenones and other lipids and their taxonomic significance. *British Phycological Journal, 19*(3), 203–216.

Marquet, P. A., Navarrete, S. A., & Castilla, J. C. (1995). Body size, population density and the energy-equivalence rule. *Journal of Applied Ecology, 64*(3), 325–332.

Márquez, L. M., Redman, R. S., Rodriguez, R. J., & Roosinck, M. J. (2007). A virus in a fungus in a plant: three-way symbiosis required for thermal tolerance. *Science, 315*(5811), 513–515.

Marsh, A. C. (1985). Microclimatic factors influencing foraging patterns and success of the thermophilic desert ant, *Ocymyrmex barbiger*. *Insectes Sociaux, 32*(3), 286–296.

Marsh, A. C. (1985). Thermal responses and temperature tolerance in a diurnal desert ant, *Ocymyrmex barbiger*. *Physiological Zoology, 58*(6), 629–636.

Marshall, G. J. (2007). Half-century seasonal relationships between the Southern Annular Mode and Antarctic temperatures. *International Journal of Climatology, 27*(3), 373–383.

Marshall, G. J., Orr, A., van Lipzig, N. P. M., & King, J. C. (2006). The impact of a changing Southern Hemisphere Annular Mode on Antarctic Peninsula summer temperatures. *Journal of Climate, 19*(20), 5388–5404.

Marshall, J. F., & McCulloch, M. T. (2002). An assessment of the Sr/Ca ratio in shallow water hermatypic corals as a proxy for sea surface temperature. *Geochimica et Cosmochimica Acta, 66*(18), 3263–3280.

Marshall, J. T. (1955). Hibernation in captive goatsuckers. *Condor, 57*(3), 129–134.

Marsham, R. (1789). Indications of spring, observed by Robert Marsham, Esquire, F.R.S. of Stratton in Norfolk. Latitude 52° 45'. *Philosophical Transactions of the Royal Society of London, 79*, 154–156.

Martin, A. P. (1995). Metabolic rate and directional nucleotide substitution in animal mitochondrial DNA. *Molecular Biology and Evolution, 12*(6), 1124–1131.

Martin, A. P. (1999). Substitution rates of organelle and nuclear genes in sharks: Implicating metabolic rate (again). *Molecular Biology and Evolution, 16*(7), 996–1002.

Martin, A. P., Naylor, G. J. P., & Palumbi, S. R. (1992). Rates of mitochondrial DNA evolution in sharks are slow compared with mammals. *Nature, 357*(6374), 153–155.

Martin, A. P., & Palumbi, S. R. (1993). Body size, metabolic rate, generation time, and the molecular clock. *Proceedings of the National Academy of Sciences of the United States of America, 90*(9), 4087–4091.

Martin, C. J. (1903). Thermal adjustment and respiratory exchange in monotremes and mammals. *Philosophical Transactions of the Royal Society of London, Series B, 195*(1), 1–37.

Martínez del Rio, C. (2008). Metabolic theory or metabolic models? *Trends in Ecology & Evolution, 23*(5), 256–260.

Mason, C. F. (1995). Long-term trends in the arrival dates of spring migrants. *Bird Study, 42*(3), 182–189.

Mathews, R. W., & Haschemeyer, A. E. V. (1978). Temperature dependency of protein synthesis in toadfish liver *in vivo*. *Comparative Biochemistry and Physiology, Part B, 61*(4), 479–484.

Matsumoto, M., Saito, S., & Ohmine, I. (2002). Molecular dynamics simulation of the ice nucleation and growth process leading to water freezing. *Nature, 416*(6879), 409–413.

Matthews, L. H. (1931). *South Georgia: the British Empire's Subantarctic outpost*. Bristol: John Wright & Sons.

Maurer, B. A. (1999). *Untangling ecological complexity: the macroscopic perspective*. Chicago: University of Chicago Press.

Maxwell, J. C. (1871). *Theory of heat*. London: Longmans, Green.

May, R. M. (1975). Patterns of species abundance and diversity. In M. L. Cody & J. M. Diamond (Eds.), *Ecology and evolution of communities* (pp. 81–120). Cambridge, MA: Harvard University Press.

May, R. M. (1994). Biological diversity: differences between land and sea. *Philosophical Transactions of the Royal Society of London, Series B, 343*(1303), 105–111.

Mayer, A. L., & Pimm, S. L. (1997). Tropical rainforest: diversity begets diversity. *Current Biology, 7*(7), R430–R432.

Mayhew, P. J., Bell, M. A., Benton, T. G., & McGowan, A. J. (2012). Biodiversity tracks temperature over time. *Proceedings of the National Academy of Sciences of the United States of America, 109*(38), 15141–15145.

Mayhew, P. J., Jenkins, G. B., & Benton, T. G. (2008). A long-term association between global temperature and biodiversity, origination and extinction in the fossil record. *Proceedings of the Royal Society of London, Series B, 275*(1630), 47–53.

Mayhew, W. W. (1965). Hibernation in the horned lizard, *Phrynosoma m'calli*. *Comparative Biochemistry and Physiology, 16*(1), 103–119.

Maynard Smith, J. (2000). The concept of information in biology. *Philosophy of Science, 67*(2), 177–194.

Mazur, P. (1963). Kinetics of water loss from cells at subzero temperatures and the likelihood of intracellular freezing. *Journal of General Physiology, 47*(2), 347–369.

Mazur, P. (2004). Principles of cryobiology. In B. J. Fuller, N. Lane, & E. E. Benson (Eds.), *Life in the frozen state* (pp. 3–65). Boca Raton: CRC Press.

Mazur, P., Leibo, S. P., & Chu, E. H. (1972). A two-factor hypothesis of freezing injury. Evidence from Chinese hamster tissue-culture cells. *Experimental Cell Research*, 71(2), 345–355.

McCafferty, D. J. (2007). The value of infrared thermography for research on mammals: previous applications and future directions. *Mammal Review*, 37(3), 207–223.

McCafferty, D. J., Gilbert, C., Paterson, W., et al. (2011). Estimating metabolic heat loss in birds and mammals by combining infrared thermography with biophysical modelling. *Comparative Biochemistry and Physiology, Part A*, 158(3), 337–345.

McCafferty, D. J., Gilbert, C., Thierry, A.-M., Currie, J., Le Maho, Y., & Ancel, A. (2013). Emperor penguin body surfaces cool below air temperature. *Biology Letters*, 9, 20121192.

McCormack, J. E., Harvey, M. G., Faircloth, B. C., Crawford, N. G., Glenn, T. C., & Brumfield, R. T. (2013). A phylogeny of birds based on over 1,500 loci collected by target enrichment and high-throughput sequencing. *PLoS One*, 8(1), e54848.

McFall-Ngai, M. J., & Horwitz, J. (1990). A comparative study of the thermal stability of the vertebrate eye lens: Antarctic ice fish to the desert iguana. *Experimental Eye Research*, 50(6), 703–709.

McIntosh, R. P. (1967). The continuum concept of vegetation. *The Botanical Review*, 33(2), 130–187.

McKechnie, A. E., & Mzilikazi, N. (2011). Heterothermy in Afrotropical mammals and birds: a review. *Integrative and Comparative Biology*, 51(3), 349–363.

McKechnie, A. E., & Swanson, D. L. (2010). Sources and significance of variation in basal, summit and maximal metabolic rate in birds. *Current Zoology*, 56(6), 741–758.

McKie, D., & Heathcote, N. H. d. V. (1935). *The discovery of specific and latent heats*. London: Edward Arnold.

McLaren, I. A. (1963). Effects of temperature on the growth of zooplankton and the adaptive value of vertical migration. *Journal of the Fisheries Research Board of Canada*, 20(3), 685–727.

McMahon, T. (1973). Size and shape in biology. *Science*, 179(4079), 1201–1204.

McMahon, T. A., & Bonner, J. T. (1983). *On size and life*. New York: Scientific American Library.

McNab, B. K. (1970). Body weight and the energetics of temperature regulation. *Journal of Experimental Biology*, 53(2), 329–348.

McNab, B. K. (1971). On the ecological significance of Bergmann's rule. *Ecology*, 52(5), 845–854.

McNab, B. K. (1973). Body-weight, energetics, and the determination of body temperature. *Journal of Experimental Biology*, 58(2), 277–280.

McNab, B. K. (1978). The evolution of endothermy in the phylogeny of mammals. *American Naturalist*, 112(1), 1–21.

McNab, B. K. (1980). On estimating thermal cinductance in endotherms. *Physiological Zoology*, 53(2), 145–156.

McNab, B. K. (1988). Complications inherent in scaling the basal rate of metabolism of mammals. *Quarterly Review of Biology*, 63(1), 25–54.

McNab, B. K. (1992). A statistical analysis of mammalian rates of metabolism. *Functional Ecology*, 6(6), 672–679.

McNab, B. K. (2002). *The physiological ecology of vertebrates: a view from energetics*. Ithaca and London: Cornell University Press.

McNab, B. K. (2008). An analysis of the factors that influence the level and scaling of mammalian BMR. *Comparative Biochemistry and Physiology, Part A*, 151(1), 5–28.

McNab, B. K. (2012). *Extreme measures: the ecological energetics of birds and mammals*. Chicago: University of Chicago Press.

McNeill, S., & Lawton, J. H. (1970). Annual production and respiration in animal populations. *Nature*, 225(5231), 472–474.

Medley, P., Weld, D. M., Miyake, H., Pritchard, D. E., & Ketterle, W. (2011). Spin gradient demagnetization cooling of ultracold atoms. *Physical Review Letters*, 106, 195301.

Meinertzhagen, R. (1954). *Birds of Arabia*. London & Edinburgh: Oliver & Boyd.

Meiri, S., & Dayan, T. (2003). On the validity of Bergmann's rule. *Journal of Biogeography*, 30(3), 331–351.

Memmott, J., Craze, P. G., Waser, N., & Price, M. V. (2007). Global warming and the disruption of plant-pollinator interactions. *Ecology Letters*, 10(8), 710–717.

Mendelsohn, E. (1964). *Heat and life: the development of the theory of animal heat*. Cambridge, Massachusetts: Harvard University Press.

Menon, N. R. (1972). Heat tolerance, growth and regeneration in three North Sea bryozoans exposed to different constant temperatures. *Marine Biology*, 15(1), 1–11.

Menzel, A., Sparks, T. H., Estrella, N., et al. (2006). European phenological response to climate change matches the warming pattern. *Global Change Biology*, 12(10), 1969–1976.

Meredith, M. P., & King, J. C. (2005). Rapid climate change in the ocean west of the Antarctic Peninsula during the second half of the 20th century. *Geophysical Research Letters*, 32, L19604.

Merola-Zwartjes, M., & Ligon, J. D. (2000). Ecological energetics of the Puerto Rican Tody: heterothermy, torpor, and intra-island variation. *Ecology*, 81(4), 990–1003.

Merritt, J. F., & Zegers, D. A. (1991). Seasonal thermogenesis and body-mass dynamics of *Clethrionomys gapperi*. *Canadian Journal of Zoology*, 69(11), 2771–2777.

Meryman, H. T. (1968). Modified model for the mechanism of freezing injury in erythrocytes. *Nature*, 218, 333–336.

Metcalfe, N. B., & Ure, S. E. (1995). Diurnal variation in flight performance and hence potential predation risk in small birds. *Proceedings of the Royal Society of London, Series B, 261*(1362), 395–400.

Meyer, K. (1910). Ole Römer and the thermometer. *Nature, 82*, 296–298.

Mezhzherin, V. A. (1964). Dehnel's phenomenon and its possible explanation (In Russian with English summary). *Acta Theriologica, 7*(6), 95–114.

Middleton, W. E. K. (1966). *A history of the thermometer and its use in meteorology*. Baltimore, Maryland: Johns Hopkins Press.

Mieszkowska, N., Hawkins, S. J., Burrows, M. T., & Kendall, M. A. (2007). Long-term changes in the geographic distribution and population structures of *Osilinus lineatus* (Gastropoda: Trochidae) in Britain and Ireland. *Journal of the Marine Biological Association of the United Kingdom, 87*(2), 537–545.

Mieszkowska, N., Kendall, M. A., Hawkins, S. J., et al. (2006). Changes in the range of some common rocky shore species in Britain: a response to climate change? *Hydrobiologia, 555*, 241–251.

Mildrexler, D. J., Zhao, M., & Running, S. W. (2011). Satellite finds highest land surface temperatures on Earth. *Bulletin of the American Meteorological Society, 92*(7), 855–860.

Milici, M., Tomasch, J., Wos-Oxley, M. L., et al. (2016). Low diversity of planktonic bacteria in the tropical ocean. *Scientific Reports, 6*, article 19054.

Miller, R. R., Minckley, W. L., & Norris, S. M. (2005). *Freshwater fishes of Mexico*. Chicago: University of Chicago Press.

Miller-Rushing, A. J., Lloyd-Evans, T. L., Primack, R. B., & Satzinger, P. (2008). Bird migration times, climate change, and changing population sizes. *Global Change Biology, 14*(9), 1959–1972.

Milligan, L. P., & Summers, M. (1986). The biological basis of maintenance and its relevance to assessing responses to nutrients. *Proceedings of the Nutrition Society, 45*(2), 185–193.

Minckley, W. L., & Minckley, C. O. (1986). *Cyprinodon pachycephalus*, a new species of pupfish (Cyprinodontidae) from the Chihauhuan desert of northern México. *Copeia, 1986*(1), 184–192.

Minelli, A. (1993). *Biological systematics: the state of the art*. London: Chapman & Hall.

Mink, J. R., Blumenschine, R. J., & Adams, D. B. (1981). Ratio of central nervous system to body metabolism in vertebrates: its constancy and functional basis. *American Journal of Physiology, 241*(3), R203–R212.

Mitchell, P. D. (1961). Coupling of phosphorylation to electron and hydrogen transfer by a chemi-osmotic type of mechanism. *Nature, 191*(4784), 144–148.

Mittelbach, G. G., & Schemske, D. W. (2015). Ecological and evolutionary perspectives on community assembly. *Trends in Ecology & Evolution, 30*(5), 241–247.

Molina, M. J., & Rowland, F. S. (1970). Stratospheric sink for chlorofluoromethanes: chlorine atom catalysed destruction of ozone. *Nature, 249*(5460), 810–812.

Møller, A. P., Rubolini, D., & Lehikoinen, A. (2008). Populations of migratory bird species that did not show a phenological response to climate change are declining. *Proceedings of the National Academy of Sciences of the United States of America, 105*(42), 16195–16200.

Montagnes, D. J. S., & Franklin, D. J. (2001). Effect of temperature on diatom volume, growth rate, and carbon and nitrogen content: reconsidering some paradigms. *Limnology & Oceanography, 46*(8), 2008–2018.

Montejano, G., & Absalón, I. B. (2009). *Caracterización del hábitat acuático asociado al pez Cyprinodon (nsp.) julimes (in Spanish)*. Coyoacán, México: Laboratorio de Ficología, Facultad de Ciencias, Universidad Nacional Autónoma de México.

Mora, C., Tittensor, D. P., Adl, S., Simpson, A. G. B., & Worm, B. (2011). How many species are there on Earth and in the ocean? *PLoS Biology, 9*(8), e1001127.

Morgan, C. C., Foster, P. G., Webb, A. E., Pisani, D., McInerny, J. O., & O'Connell, M. J. (2013). Heterogeneous models place the root of the placental mammal phylogeny. *Molecular Biology and Evolution, 30*(9), 2145–2156.

Morgan, K. R. (1985). Body temperature regulation and terrestrial activity in the ectothermic beetle *Cicindela tranquebarica*. *Physiological Zoology, 58*(1), 29–37.

Morgan, K. R. (1987). Temperature regulation, energy metabolism and mate-searching in rain beetles (*Plecoma* spp.), winter-active, endothermic scarabs (Coleoptera). *Journal of Experimental Biology, 128*(1), 107–122.

Morley, S. A., Lurman, G. J., Skepper, J. N., Pörtner, H. O., & Peck, L. S. (2009). Thermal plasticity of mitochondria: a latitudinal comparison between Southern Ocean molluscs. *Comparative Biochemistry and Physiology Part A, 152*(3), 423–430.

Morowitz, H. J. (1968). *Energy flow in biology: biological organization as a problem in thermal physics*. New York: Academic Press.

Morowitz, H. J. (1978). *Foundations of bioenergetics*. New York: Academic Press.

Morowitz, H. J. (1999). A theory of biochemical organization, metabolic pathways, and evolution. *Complexity, 4*(6), 39–53.

Morris, G. J. (1987). Direct chilling injury. In B. W. W. Grout & G. J. Morris (Eds.), *The effects of low temperatures on biological systems* (pp. 120–146). London: Edward Arnold.

Morris, G. J., Coulson, G. E., & Clarke, A. (1979). The cryopreservation of *Chlamydomonas*. *Cryobiology, 16*(4), 401–410.

Morrison, C. A., Robinson, R. A., Clarke, J. A., & Gill, J. A. (2016). Causes and consequences of spatial variation in sex ratios in a declining bird species. *Journal of Animal Ecology, 85*(5), 1298–1306.

Mortola, J. P., & Lanthier, C. (2004). Scaling the amplitudes of the circadian pattern of resting oxygen consumption, body temperature and heart rate in mammals. *Comparative Biochemistry and Physiology, Part A*, *139*(1), 83–95.

Moseley, H. N. (1880). Deep-sea dredging and life in the deep sea. *Nature*, *21*(545), 543–547.

Moses, M. E., Hou, C., Woodruff, W. H., et al. (2008). Revisiting a model of ontogenetic growth: estimating model parameters from theory and data. *American Naturalist*, *171*(5), 632–645.

Moylan, T. J., & Sidell, B. D. (2000). Concentrations of myoglobin and myoglobin mRNA in heart ventricles from Antarctic fishes. *Journal of Experimental Biology*, *203*(8), 1277–1286.

Mozo, J., Emre, Y., Bouillaud, F., Ricquier, D., & Criscuolo, F. (2005). Thermoregulation: what role for UCPs in mammals and birds? *Bioscience Reports*, *25*(3/4), 227–249.

Mukaiyama, A., Koga, Y., Takano, K., & Kanaya, S. (2008). Osmolyte effect on the stability and folding of a hyperthermophilic protein. *Proteins: Structure, Function, and Bioinformatics*, *71*(1), 110–118.

Muldrew, K., Acker, J. P., Elliott, J. A. W., & McGann, L. E. (2004). The water to ice transition: implications for living cells. In B. J. Fuller, N. Lane, & E. E. Benson (Eds.), *Life in the frozen state* (pp. 67–108). Boca Raton: CRC Press.

Mulvaney, R., Abram, N. J., Hindmarsh, R. C., et al. (2012). Recent Antarctic Peninsula warming relative to Holocene climate and ice-shelf history. *Nature*, *489*(7414), 141–144.

Munch, S. B., & Conover, D. O. (2003). Rapid growth results in increased susceptibility to predation in *Menidia menidia*. *Evolution*, *57*(9), 2119–2127.

Munro, H. N. (1969). Evolution of protein metabolism in mammals. In H. N. Munro (Ed.), *Mammalian protein metabolism, Vol. 3* (Vol. 3, pp. 133–182). New York: Academic Press.

Murray, B. J., O'Sullivan, D., Atkinson, J. D., & Webb, M. E. (2012). Ice nucleation by particles immersed in supercooled cloud droplets. *Chemical Society Reviews*, *41*(19), 6519–6554.

Mzilikazi, N., & Lovegrove, B. G. (2004). Daily torpor in free-ranging elephant shrews, *Elephantulus myurus*: a year-long study. *Physiological and Biochemical Zoology*, *77*(2), 285–296.

Mzilikazi, N., Lovegrove, B. G., & Ribble, D. O. (2002). Exogenous passive heating during torpor arousal in free-ranging rock elephant shrews, *Elephantulus myurus*. *Oecologia*, *133*(3), 307–314.

Nadel, E. R., Bullard, R. W., & Stolwojk, J. A. J. (1971a). Importance of skin temperature in the regulation of sweating. *Journal of Applied Physiology*, *31*(1), 80–87.

Nadel, E. R., Mitchell, J. W., Saltin, B., & Stolwojk, J. A. J. (1971b). Peripheral modifications to the central drive for sweating. *Journal of Applied Physiology*, *31*(6), 828–833.

Nagy, K. A. (2005). Field metabolic rate and body size. *Journal of Experimental Biology*, *208*, 1621–1625.

Nagy, K. A., Odell, D. K., & Seymour, R. S. (1972). Temperature regulation by the inflorescence of *Philodendron*. *Science*, *178*(4066), 1195–1197.

Naish, T., Powell, R., Levy, R., et al. (2009). Obliquity-paced Pliocene West Antarctic ice sheet oscillations. *Nature*, *458*, 322–328.

Naish, T. R., Woolfe, K. J., Barrett, P. J., et al. (2001). Orbitally induced oscillations in the East Antarctic ice sheet at the Oligocene–Miocene boundary. *Nature*, *413*(6857), 719–723.

Naya, D. E., Spangenberg, L., Naya, H., & Bozinovic, F. (2013). Thermal conductance and basal metabolic rate are part of a coordinated system for heat transfer regulation. *Proceedings of the Royal Society of London, Series B*, *280*(1767), article 20131629.

Near, T. J., Dornburg, A., Kuhn, K. L., et al. (2012). Ancient climate change, antifreeze, and the evolutionary diversification of Antarctic fishes. *Proceedings of the National Academy of Sciences of the United States of America*, *109*(9), 3434–3439.

Near, T. J., Parker, S. K., & Detrich, H. W. (2006). A genomic fossil reveals key steps in hemoglobin loss by the Antarctic icefishes. *Molecular Biology and Evolution*, *23*(11), 2008–2016.

Needham, A. D., Dawson, T. J., & Hales, J. R. S. (1974). Forelimb blood-flow and saliva spreading in the thermoregulation of Red Kangaroo, *Megaleia rufa*. *Comparative Biochemistry and Physiology, Part A*, *49*(3A), 555–565.

Németh, I., Nyitrai, V., & Altbäcker, V. (2009). Ambient temperature and annual timing affect torpor bouts and euthermic phases of hibernating European ground squirrels (*Spermophilus citellus*). *Canadian Journal of Zoology*, *87*(3), 204–210.

Newman, E. A., & Hartline, P. H. (1981). Integration of visual and infrared information in bimodal neurons in the rattlesnake optic tectum. *Science*, *213*(4509), 789–791.

Newman, S. A., Mezentseva, N., & Badyaev, A. V. (2013). Gene loss, thermogenesis, and the origin of birds. *Annals of the New York Academy of Sciences*, *1289*, 36–47.

Nicholas, J. V., & White, D. R. (1994). *Traceable temperatures: an introduction to temperature measurement and calibration*. Chichester: John Wiley & Sons.

Nicholls, D. G., & Ferguson, S. J. (2002). *Bioenergetics 3*. London: Academic Press.

Nicol, D. (1967). Some characteristics of cold-water marine pelecypods. *Journal of Paleontology*, *41*(6), 1330–1340.

Nicol, D. (1971). Species, class and phylum diversity of animals. *Quarterly Journal of the Florida Academy of Sciences*, *34*, 191–194.

Nicol, J. A. C. (1960). *The biology of marine animals*. London: Sir Isaac Pitman & Sons.

Niven, J. E., & Laughlin, S. B. (2008). Energy limitation as a selective pressure on the evolution of sensory systems. *Journal of Experimental Biology*, *211*(11), 1792–1804.

Norell, M. A., & Xu, X. (2005). Feathered dinosaurs. *Annual Review of Earth and Planetary Sciences*, *33*, 277–299.

Normand, S., Ricklefs, R. E., Skov, F., Bladt, J., Tackenberg, O., & Svenning, J.-C. (2011). Postglacial migration supplements climate in determining plant species ranges in Europe. *Proceedings of the Royal Society of London, Series B*, *278*(1725), 3644–3653.

Novotny, V., Basset, Y., Miller, S. E., et al. (2002). Low host specificity of herbivorous insects in a tropical forest. *Nature*, *416*(6883), 841–844.

Nowack, J., Mzilikazi, N., & Dausmann, K. H. (2010). Torpor on demand: heterothermy in the non-lemur primate *Galago moholi*. *PLoS One*, *5*(5), e10797.

Nowack, J., Wippich, M., Mzilikazi, N., & Dausmann, K. H. (2013). Surviving the cold, dry period in Africa: behavioral adjustments as an alternative to heterothermy in the African lesser bushbaby (*Galago moholi*). *International Journal of Primatology*, *34*(1), 49–64.

Nybelin, O. (1947). Antarctic fishes. In O. Holtedahl (Ed.), *Scientific results of the Norwegian Antarctic Expeditions 1927–1928* (Vol. 2, Report 26, pp. 21–76). Oslo: Det Norske Videnskaps-Akademi i Oslo.

O'Brien, E. M. (1993). Climatic gradients in woody plant species richness: towards an explanation based on an analysis of southern Africa's woody flora. *Journal of Biogeography*, *20*, 181–198.

O'Brien, E. M. (1998). Water–energy dynamics, climate, and prediction of woody plant species richness: an interim general model. *Journal of Biogeography*, *25*(2), 379–398.

O'Brien, K. M., & Sidell, B. D. (2000). The interplay among cardiac ultrastructure, metabolism and the expression of oxygen-binding proteins in Antarctic fishes. *Journal of Experimental Biology*, *203*(8), 1287–1297.

Ocaña, A. (1991). A redescription of two nematode species found in hot springs. *Nematologia Mediterranea*, *19*(2), 173–175.

Och, L. M., & Shields-Zhou, G. A. (2012). The Neoproterozoic oxygenation event: environmental perturbations and biogeochemical cycling. *Earth Science Reviews*, *110*(1–4), 26–57.

O'Connor, M. P., & Dodson, P. (1999). Biophysical constraints on the thermal ecology of dinosaurs. *Paleobiology*, *25*(3), 341–368.

O'Dea, A., & Okamura, B. (1999). Influence of seasonal variation in temperature, salinity and food availability on module size and colony growth of the estuarine bryozoan *Conopeum seurati*. *Marine Biology*, *135*(4), 581–588.

Ødegaard, F., Diserud, O. H., Engen, S., & Aagaard, K. (2000). The magnitude of local host specificity for phytophagous insects and its implications for estimates of global species richness. *Conservation Biology*, *14*(4), 1182–1186.

Oertli, J. J. (1989). Relationship of wing beat frequency and temperature during take-off flight in temperate-zone beetles. *Journal of Experimental Biology*, *145*(1), 321–338.

Ohnishi, S., Yamakawa, T., & Akamine, T. (2014). On the analytical solution for the Pütter–Bertalanffy growth equation. *Journal of Theoretical Biology*, *343*, 174–177.

Okie, J. G. (2012). Microorganisms. In R. M. Sibly, J. H. Brown, & A. Kodric-Brown (Eds.), *Metabolic ecology: a scaling approach* (pp. 135–153). Oxford: Wiley-Blackwell.

Oleksiak, M. F., Churchill, G. A., & Crawford, D. L. (2002). Variation in gene expression within and among natural populations. *Nature Genetics*, *32*(2), 261–266.

Oleksiak, M. F., Roach, J. L., & Crawford, D. L. (2005). Natural variation in cardiac metabolism and gene expression in *Fundulus heteroclitus*. *Nature Genetics*, *37*(1), 67–72.

Olson, D. M., Dinerstein, E., Wikramanayake, E. D., et al. (2001). Terrestrial ecoregions of the world: a new map of life on Earth. *BioScience*, *51*(11), 933–938.

Olson, V. A., Davies, R. G., Orme, C. D. L., et al. (2009). Global biogeography and ecology of body size in birds. *Ecology Letters*, *12*(3), 249–259.

Olsson, T. S. G., Ladbury, J. E., Pitt, W. R., & Williams, M. A. (2011). Extent of enthalpy–entropy compensation in protein–ligand interactions. *Protein Science*, *20*(9), 1607–1618.

O'Malley, M. A. (2008). '*Everything is everywhere*: but *the environment selects*': ubiquitous distribution and ecological determinism in microbial biogeography. *Studies in History and Philosophy of Science Part C: Studies in History and Philosophy of Biological and Biomedical Sciences*, *39*(3), 314–325.

Oosthuizen, M. J. (1939). The body temperature of *Samia cecropia* Linn. (Lepidoptera, Saturniidae) as influenced by muscular activity. *Journal of the Entomological Society of South Africa*, *2*(1), 63–73.

Öpik, H., & Rolfe, S. A. (2005). *The physiology of flowering plants* (Fourth ed.). Cambridge: Cambridge University Press.

Ostrom, J. H. (1969). Osteology of *Deinonychus antirrhopus*, an unusual theropod from the Lower Cretaceaous of Montana. *Bulletin of the Peabody Museum of Natural History, Yale University*, *30*, 1–165.

Ostrom, J. H. (1970). Terrestrial vertebrates as indicators of Mesozoic climates. *Proceedings of the North American Paleontological Convention, Chicago 1969*, *A*(4 (Paleoclimatology)), 347–376.

Ostrom, J. H. (1973). The ancestry of birds. *Nature*, *242*(5393), 136.

Ostrom, J. H. (1975). The origin of birds. *Annual Review of Earth and Planetary Sciences*, *3*(1), 55–77.

Ostrom, J. H. (1976). *Archaeopteryx* and the origin of birds. *Biological Journal of the Linnean Society, 8*(2), 91–82.

Ostrom, J. H. (1980). The evidence for endothermy in dinosaurs. In R. D. K. Thomas & E. C. Olson (Eds.), *A cold look at the warm-blooded dinosaurs* (Vol. 28, pp. 15–54). Washington, D.C.: Westview Press, for the American Association for the Advancement of Science.

Ostrowski, S., & Williams, J. B. (2006). Heterothermy of free-living Arabian sand gazelles (*Gazella subgutturosa marica*) in a desert environment. *Journal of Experimental Biology, 209*(8), 1421–1429.

Packard, G. C. (2009). On the use of logarithmic transformations in allometric analyses. *Journal of Theoretical Biology, 257*(1), 515–518.

Packard, G. C., & Birchard, G. F. (2008). Traditional allometric analysis fails to provide a valid predictive model for mammalian metabolic rates. *Journal of Experimental Biology, 211*(22), 3581–3587.

Packard, G. C., & Boardman, T. J. (2008). Model selection and logarithmic transformation in allometric analysis. *Physiological and Biochemical Zoology, 81*(4), 496–507.

Packard, G. C., & Boardman, T. J. (2009). Bias in interspecific allometry: examples from morphological scaling in varanid lizards. *Biological Journal of the Linnean Society, 96*(2), 296–305.

Padian, K., & Rayner, J. M. V. (1993). The wings of pterosaurs. *American Journal of Science, 293 A*, 91–166.

Pais, A. (1982). *Subtle is the Lord … : the science and the life of Albert Einstein*. Oxford: Oxford University Press.

Palmer, A. R. (1992). Calcification in marine molluscs: how costly is it? *Proceedings of the National Academy of Sciences of the United States of America, 89*(4), 1379–1382.

Palmer, M. D., McNeall, D. J., & Dunstone, N. J. (2011). Importance of the deep ocean for estimating decadal changes in Earth's radiation balance. *Geophysical Research Letters, 38*, L13707.

Palumbi, S. R. (1996). What can molecular genetics contribute to marine biogeography? An urchin's tale. *Journal of Experimental Marine Biology and Ecology, 203*(1), 75–92.

Palumbi, S. R. (1997). Molecular biogeography of the Pacific. *Coral Reefs, 16*, S47–S52.

Pamatmat, M. M. (1978). Oxygen uptake and heat production in a metabolic conformer (*Littorina irrorata*) and a metabolic regulator (*Uca pugnax*). *Marine Biology, 48*(4), 317–325.

Pamatmat, M. M. (1983). Measuring aerobic and anaerobic metabolism of benthic infauna under natural conditions. *Journal of Experimental Zoology, 228*(3), 405–413.

Pannevis, M. C., & Houlihan, D. F. (1992). The energetic cost of protein synthesis in isolated hepatocytes of rainbow trout (*Oncorhynchus mykiss*). *Journal of Comparative Physiology, B, 162*(5), 393–400.

Paolo, F. S., Fricker, H. A., & Padman, L. (2015). Volume loss from Antarctic ice shelves is accelerating. *Science, 348*(6232), 327–331.

Parfitt, S. A., Ashton, N. M., Lewis, S. G., et al. (2010). Early Pleistocene human occupation at the edge of the boreal zone in northwest Europe. *Nature, 466*(7303), 229–233.

Parfitt, S. A., Barendregt, R. W., Breda, M., et al. (2005). The earliest record of human activity in northern Europe. *Nature, 438*(7070), 1008–1012.

Park, K. S., Do, H., Lee, J. H., et al. (2012). Characterization of the ice-binding protein from Arctic yeast *Leucosporidium* sp AY30. *Cryobiology, 64*(3), 286–296.

Parker, D., & Horton, B. (2005). Uncertainties in central England temperature 1878–2003 and some improvements to the maximum and minimum series. *International Journal of Climatology, 25*(9), 1173–1188.

Parker, D., Legg, T. P., & Folland, C. K. (1992). A new daily central England temperature series, 1772–1991. *International Journal of Climatology, 12*(4), 317–342.

Parmesan, C., Ryrholm, N., Stefanescu, C., et al. (1999). Poleward shifts in geographical ranges of butterfly species associated with regional warming. *Nature, 399*, 579–583.

Parmesan, C., & Yohe, G. (2003). A globally coherent fingerprint of climate change impacts across natural systems. *Nature, 421*(6918), 37–42.

Parry, G. D. (1983). The influence of the cost of growth on ectotherm metabolism. *Journal of Theoretical Biology, 101*(3), 453–477.

Parsons, K. M. (2001). *Drawing out Leviathan: dinosaurs and the science wars*. Bloomington: Indiana University Press.

Patiño, S., Grace, J., & Bänziger, H. (2000). Endothermy by flowers of *Rhizanthes lowii* (Rafflesiaceae). *Oecologia, 124*(2), 149–155.

Pauling, L. (1939). *The nature of the chemical bond and the structure of molecules and crystals: an introduction to modern structural chemistry*. Ithaca, New York: Cornell University Press.

Pauling, L. (1946). Molecular architecture and biological reactions. *Chemical and Engineering News, 24*(10), 1375–1377.

Payne, J. L., & Finnegan, S. (2007). The effect of geographic range on extinction risk during background and mass extinction. *Proceedings of the National Academy of Sciences of the United States of America, 104*(25), 10506–10511.

Pearce, R. S. (2004). Adaptation of higher plants to freezing. In B. J. Fuller, N. Lane, & E. E. Benson (Eds.), *Life in the frozen state* (pp. 171–203). Boca Raton: CRC Press.

Pearman, G. I., Weaver, H. L., & Tanner, C. B. (1971). Boundary layer heat transfer coefficients under field conditions. *Agricultural Meteorology, 10*(1), 83–92.

Pearson, O. P. (1950). The metabolism of hummingbirds. *Condor, 52*(4), 145–152.

Pearson, O. P. (1953). Use of caves by hummingbirds and other species at high altitudes in Peru. *Condor, 55*(1), 17–20.

Pearson, R. G., & Dawson, T. P. (2003). Predicting the impacts of climate change on the distribution of species: are bioclimate envelope models useful? *Global Ecology and Biogeography, 12*(5), 361–371.

Peat, H. J., Clarke, A., & Convey, P. (2007). Diversity and biogeography of the Antarctic flora. *Journal of Biogeography, 34*(1), 132–146.

Peck, L. S., & Barnes, D. K. A. (2004). Metabolic flexibility: the key to long-term evolutionary success in Bryozoa? *Proceedings of the Royal Society of London, Series B, 271*(Supplement 3), S18–S21.

Peck, L. S., & Chapelle, G. (2003). Reduced oxygen at high altitude limits maximum size. *Proceedings of the Royal Society of London, Series B (Supplement), 270*(Supplement 2), S166–S167.

Peck, L. S., & Conway, L. Z. (2000). The myth of cold adaptation: oxygen consumption in stenothermal Antarctic bivalves. In E. M. Harper, J. D. Taylor, & J. A. Crame (Eds.), *The evolutionary biology of the Bivalvia* (Vol. 177, pp. 441–450). London: The Geological Society.

Peck, L. S., & Maddrell, S. H. P. (2005). Limitation of size by hypoxia in the fruit fly *Drosophila melanogaster*. *Journal of Experimental Zoology, 303A*(11), 968–975.

Pengelley, E. T., Asmundson, S. J., Barnes, B., & Aloia, R. C. (1976). Relationship of light internsity and photoperiod to circannual rhythmicity in the hibernating ground squirrel, *Citellus lateralis*. *Comparative Biochemistry and Physiology, Part A, 53*(3), 273–277.

Pengelley, E. T., & Fisher, K. C. (1961). Rhythmical arousal from hibernation in the golden-mantled ground squirrel, *Citellus lateralis* Tescorum. *Canadian Journal of Zoology, 39*(1), 105–120.

Pengelley, E. T., & Fisher, K. C. (1963). The effect of temperature and photoperiod on the yearly behavior of captive golden-mantled ground squirrels (*Citellus lateralis* Tescorum). *Canadian Journal of Zoology, 41*(6), 1103–1120.

Pennycuick, C. J. (1992). *Newton rules biology: a physical approach to biological problems*. Oxford: Oxford University Press.

Peppe, D. J., Royer, D. L., Cariglino, B., et al. (2011). Sensitivity of leaf size and shape to climate: global patterns and paleoclimate applications. *New Phytologist, 190*, 724–739.

Pereyra, R. G., Szleifer, I., & Carignano, M. A. (2011). Temperature dependence of ice critical nucleus size. *Journal of Chemical Physics, 135*(3), article 034508.

Perry, A. L., Low, P. J., Ellis, J. R., & Reynolds, J. D. (2005). Climate change and distribution shifts in fishes. *Science, 308*(5730), 1912–1915.

Peters, R. H. (1983). *The ecological implications of body size*. Cambridge: Cambridge University Press.

Petersen, A. M., Chin, W., Feilich, K. L., et al. (2011). Leeches run cold, then hot. *Biology Letters, 7*(6), 941–943.

Peterson, C. C., Nagy, K. A., & Diamond, J. (1990). Sustained metabolic scope. *Proceedings of the National Academy of Sciences of the United States of America, 87*(6), 2324–2328.

Peterson, C. C., Walton, B. M., & Bennett, A. F. (1999). Metabolic costs of growth in free-living garter snakes and the energy budgets of ectotherms. *Functional Ecology, 13*(4), 500–507.

Peterson, M. E., Daniel, R. M., Danson, M. J., & Eisenthal, R. (2007). The dependence of enzyme activity on temperature: determination and validation of parameters. *Biochemical Journal, 402*(2), 331–337.

Petit, J. R., Jouzel, J., Raynaud, D., et al. (1999). Climate and atmospheric history of the past 420,000 years from the Vostok ice core, Antarctica. *Nature, 399*(6735), 429–436.

Petit, M., & Vézina, F. (2014). Reaction norms in natural conditions: how does metabolic performance respond to weather variations in a small endotherm facing cold environments? *PLoS One, 9*(11), e113617.

Petrenko, V. F., & Whitworth, R. W. (1999). *Physics of ice*. Oxford: Oxford University Press.

Petrusewicz, K., & Macfadyen, A. (1970). *Productivity of terrestrial animals: principles and methods* (Vol. No 13). Oxford: Blackwell Scientific Publications.

Petzel, D. H., Reisman, H. M., & DeVries, A. L. (1980). Seasonal variation of antifreeze peptide in the winter flounder, *Pseudopleuronectes americanus*. *Journal of Experimental Zoology, 211*(1), 63–69.

Pianka, E. R. (1978). *Evolutionary ecology* (Second ed.). New York: Harper & Row.

Pianka, E. R. (1985). Some intercontinental comparisons of desert lizards. *National Geographic Research and Exploration, 1*(4), 490–504.

Pianka, E. R. (1986). *Ecology and natural history of desert lizards: analyses of the ecological niche and community structure*. Princeton, New Jersey: Princeton University Press.

Pierrot-Bults, A. C. (1997). Biological diversity in oceanic macrozooplankton: more than counting species. In R. F. G. Ormond, J. D. Gage, & M. V. Angel (Eds.), *Marine biodiversity: causes and consequences* (pp. 69–93). Cambridge: Cambridge University Press.

Pitcher, T. J., Morato, T., Hart, P. J. B., Clark, M. R., Haggan, N., & Santos, R. S. (Eds.). (2007). *Seamounts: ecology, fisheries and conservation*. Oxford: John Wiley & Sons (Wiley-Blackwell).

Pitts, G. C., & Bullard, T. R. (1968). Some interspecific aspects of body composition in mammals. In: *Body composition in animals and man* (Vol. 1598, pp. 45–70). Washington, D.C., National Academy of Science.

Place, A. R., & Powers, D. A. (1979). Genetic variation and relative catalytic efficiencies: lactate dehydrogenase B allozymes of *Fundulus heteroclitus*. *Proceedings of the National Academy of Sciences of the United States of America*, 76(5), 2354–2358.

Platnick, N. I. (1991). Patterns of biodiversity: tropical *vs* temperate. *Journal of Natural History*, 25(5), 1083–1088.

Podrabsky, J. E., & Somero, G. N. (2004). Changes in gene expression associated with acclimation to constant temperatures and fluctuating daily temperatures in an annual killifish *Austrofundulus limnaeus*. *Journal of Experimental Biology*, 207, 2237–2254.

Pointing, S. B., Büdel, B., Convey, P., et al. (2015). Biogeography of photoautotrophs in the high polar biome. *Frontiers in Plant Science*, 6, article 692.

Polanyi, M., & Eyring, H. (1932). Über einfache Gasreaktionen. *Zeitschrift für Physikalische Chemie, Abteilung B*, 12, 279–311.

Pommier, T., Canbäck, B., Riemann, L., et al. (2007). Global patterns of diversity and community structure in marine bacterioplankton. *Molecular Ecology*, 16(4), 867–880.

Pontarp, M., & Wiens, J. J. (2017). The origin of species richness patterns along environmental gradients: uniting explanations based on time, diversification rate and carrying capacity. *Journal of Biogeography*, 44(4), 722–735.

Pontzer, H. (2007). Effective limb length and the scaling of locomotor cost in terrestrial animals. *Journal of Experimental Biology*, 210(10), 1752–1761.

Porter, W. P., & Gates, D. M. (1969). Thermodynamic equilibria of animals with environment. *Ecological Monographs*, 39(3), 227–244.

Porter, W. P., Mitchell, J. W., Beckman, W. A., & DeWitt, C. B. (1973). Behavioral implications of mechanistic ecology: thermal and behavioral modeling of desert ectotherms and their microenvironment. *Oecologia*, 13(1), 1–54.

Pörtner, H.-O. (2001). Climate change and temperature-dependent biogeography: oxygen limitation of thermal tolerance in animals. *Naturwissenschaften*, 88(4), 137–146.

Pörtner, H.-O. (2002). Climate variations and the physiological basis of temperature dependent biogeography: systemic to molecular hierarchy of thermal tolerance in animals. *Comparative Biochemistry and Physiology, Part A*, 132, 739–761.

Pörtner, H.-O. (2004). Climate variability and the energetic pathways of evolution: the origin of endothermy in mammals and birds. *Physiological and Biochemical Zoology*, 77(6), 959–981.

Pörtner, H.-O., Bennett, A. F., Bozinovic, F., et al. (2006). Trade-offs in thermal adaptation: the need for a molecular to ecological integration. *Physiological and Biochemical Zoology*, 79(2), 295–313.

Pörtner, H.-O., & Knust, R. (2007). Climate change affects marine fishes through the oxygen limitation of thermal tolerance. *Science*, 315(5808), 95–97.

Pough, F. H. (1973). Lizard energetics and diet. *Ecology*, 54(4), 837–844.

Pough, F. H., & Andrews, R. A. (1985). Use of anaerobic metabolism by free-ranging lizards. *Physiological Zoology*, 58(2), 205–213.

Pough, F. H., & Gans, C. (1982). The vocabulary of reptilian thermoregulation. In C. Gans & F. H. Pough (Eds.), *Biology of the Reptilia. Volume 12: Physiological ecology* (pp. 17–23). London: Academic Press.

Powers, D. A., & Place, A. R. (1978). Biochemical genetics of *Fundulus heteroclitus* (L.). I. Temporal and spatial variation in gene frequencies of Ldh-B, Mdh-A, Gpi-B and Pgm-A. *Biochemical Genetics*, 16(5–6), 593–607.

Prahl, F. G., Mix, A. C., & Sparrow, M. A. (2006). Alkenone palaeothermometry: biological lessons from marine sediment records off western South America. *Geochimica et Cosmochimica Acta*, 70(1), 101–117.

Prahl, F. G., Rontani, J.-F., Zabeti, N., Walinski, S. E., & Sparrow, M. A. (2010). Systematic pattern in—Temperature residuals for surface sediments from high latitude and other oceanographic settings. *Geochimica et Cosmochimica Acta*, 74(1), 131–143.

Prahl, F. G., & Wakeham, S. G. (1987). Calibration of unsaturation patterns in long-chain ketone compositions for paleotemperature assessment. *Nature*, 330(6146), 367–369.

Precht, H., Christophersen, J., Hensel, H., & Larcher, W. (1973). *Temperature and life*. New York, Heidelberg & Berlin: Springer-Verlag.

Prentice, I. C., Bartlein, P. J., & Webb, T. (1991). Vegetation and climate change in Eastern North America since the Last Glacial Maximum. *Ecology*, 72(6), 2038–2056.

Preston, F. W. (1948). The commonness, and rarity, of species. *Ecology*, 29(3), 254–283.

Preston, F. W. (1960). Time and space and the variation of species. *Ecology*, 41(4), 611–627.

Preston, F. W. (1962). The canonical distribution of commonness and rarity. *Ecology*, 43, 185–212.

Price, C. A., Weitz, J. S., Savage, V. M., et al. (2012). Testing the metabolic theory of ecology. *Ecology Letters*, 15(12), 1465–1474.

Price, P. B., & Sowers, T. (2004). Temperature dependence of metabolic rates for microbial growth, maintenance, and survival. *Proceedings of the National Academy of Sciences of the United States of America*, 101(13), 4631–4636.

Price, T. D. (2010). The roles of time and ecology in the continental radiation of the Old World leaf warblers (*Phylloscopus* and *Seicercus*). *Philosophical Transactions of the Royal Society of London, Series B*, 365(1547), 1749–1762.

Priede, I. G. (1985). Metabolic scope in fishes. In P. Tytler & P. Calow (Eds.), *Fish energetics: new perspectives* (pp. 33–64). London: Croom Helm.

Prum, R. O., Berv, J. S., Dornburg, A., et al. (2015). A comprehensive phylogeny of birds (Aves) using targeted next-generation DNA sequencing. *Nature, 526*(7574), 569–573.

Prusiner, S., & Poe, M. (1968). Thermodynamic considerations of mammalian thermogenesis. *Nature, 220*, 235–237.

Pucek, Z. (1963). Seasonal changes in the braincase of some representatives of the genus *Sorex* from the Palearctic. *Journal of Mammalogy, 44*(4), 523–536.

Purkey, S. G., & Johnson, G. C. (2010). Warming of global abyssal and deep Southern Ocean waters between the 1990s and 2000s: contributions to global heat and sea level rise budgets. *Journal of Climate, 23*(23), 6336–6351.

Pütter, A. (1920). Studien über physiologische Ähnlichkeit. V I. Wachstumsähnlichkeiten. *Pflüger's Archive für die gesamte Physiologie des Menschen und der Tiere, 180*(1), 298–340.

Qian, H., & Ricklefs, R. E. (1999). A comparison of the taxonomic richness of vascular plants in China and the United States. *American Naturalist, 154*(2), 160–181.

Qian, H., & Ricklefs, R. E. (2000). Large-scale processes and the Asian bias in species diversity of temperate plants. *Nature, 407*(6801), 180–182.

Qualls, C. P., & Andrews, R. M. (1999). Cold climates and the evolution of viviparity in reptiles: cold incubation temperatures produce poor-quality offspring in the lizard, *Sceloporus virgatus*. *Biological Journal of the Linnean Society, 67*(3), 353–376.

Qvist, J., Weber, R. E., DeVries, A. L., & Zapol, W. M. (1977). pH and haemoglobin oxygen affinity in blood from the Antarctic cod *Dissostichus mawsoni*. *Journal of Experimental Biology, 67*(1), 77–88.

Raby, P. (2001). *Alfred Russel Wallace: a life*. London: Chatto & Windus.

Rahn, H. (1966). Aquatic gas exchange: theory. *Respiration Physiology, 1*(1), 1–12.

Rahn, H., & Baumgardner, F. W. (1972). Temperature and acid-base regulation in fish. *Respiration Physiology, 14*(1–2), 171–182.

Raichlen, D. A., Gordon, A. D., Muchlinski, M. N., & Snodgrass, J. J. (2010). Causes and significance of variation in mammalian basal metabolism. *Journal of Comparative Physiology, B, 210*(2), 301–311.

Raimbault, S., Dridi, S., Denjean, F., et al. (2001). An uncoupling protein homologue putatively involved in facultative muscle thermogenesis in birds. *Biochemical Journal, 353*(3), 441–444.

Ralph, R., & Everson, I. (1968). The respiratory metabolism of some Antarctic fish. *Comparative Biochemistry and Physiology, 27*(1), 299–307.

Rameaux, J.-F. (1857). Des lois, suivant lesquelles les dimensions du corps, dans certaines classes d'animaux, déterminent la capacité et les mouvements fonctionnels des poumons et du coeur. *Bulletins de l'Académie Royale des Sciences, des Lettres et des Beaux Arts de Belgique, 3*, 94–104.

Rameaux, J.-F. (1858). Des lois, suivant lesquelles les dimensions du corps, dans certaines classes d'animaux, déterminent la capacité et les mouvements fonctionnels des poumons et du coeur. *Mémoires Couronnés et Mémoires des Savants Etrangers publiés par l'Académie Royale des Sciences, des Lettres et des Beaux Arts de Belgique, 29*, 1–64.

Ramires, M. L. V., Nieto de Castro, C. A., Nagasaka, Y., Nagashima, A., & Wakeham, W. A. (1995). Standard reference data for the thermal conductivity of water. *Journal of Physical and Chemical Reference Data, 24*(3), 1377–1381.

Ramsay, J. A. (1952). *Physiological approach to the lower animals*. London: Cambridge University Press.

Ramsay, J. A. (1964). The rectal complex of the mealworm *Tenebrio molitor*, L. (Coleoptera, Tenebrionidae). *Philosophical Transactions of the Royal Society of London, Series B, 248*(748), 279–314.

Rapatz, G. L., Menz, L. J., & Luyet, B. J. (1966). Freezing process in biological materials. In H. T. Meryman (Ed.), *Cryobiology* (pp. 147–148). London: Academic Press.

Raup, D. M. (1991). A kill curve for Phanerozoic marine species. *Paleobiology, 17*(1), 37–48.

Ravaux, J., Hamel, G., Zbinden, M., et al. (2013). Thermal limit for metazoan life in question: *in vivo* heat tolerance of the pompeii worm. *PLoS One, 8*(5), e64074 (64076 pages).

Raven, J. A., & Geider, R. J. (1988). Temperature and algal growth. *New Phytologist, 110*, 441–461.

Rawlins, J. E. (1980). Thermoregulation by the black swallowtail butterfly, *Papilio polyxenes* (Lepidoptera: Papilionidae). *Ecology, 61*(2), 345–357.

Ray, C. (1960). The application of Bergmann's and Allen's rules to the poikilotherms. *Journal of Morphology, 106*(1), 85–108.

Raymond, J. A., & DeVries, A. L. (1977). Adsorption inhibition as a mechanism of freezing resistance in polar fishes. *Proceedings of the National Academy of Sciences of the United States of America, 74*(6), 2589–2593.

Raymond, J. A., Fritsen, C. H., & Shen, K. (2007). An ice-binding protein from an Antarctic sea ice bacterium. *FEMS Microbiology Ecology, 61*(2), 214–221.

Raymond, J. A., Janech, M. G., & Fritsen, C. H. (2009). Novel ice-binding proteins from a psychrophilic Antarctic alga (Chlamydomonaceae, Chlorophyceae). *Journal of Phycology, 45*(1), 130–136.

Raymond, J. A., & Morgan-Kiss, R. (2013). Separate origins of ice-binding proteins in Antarctic *Chlamydomonas* species. *PLoS One, 8*(3), e59186.

Raymond, J. A., Wilson, P., & DeVries, A. L. (1989). Inhibition of growth of nonbasal planes in ice by fish antifreezes. *Proceedings of the National Academy of Sciences of the United States of America, 88*(3), 881–885.

Reaka-Kudla, M. L. (1996). The global diversity of coral reefs: a comparison with rain forests. In M. L. Reaka-Kudla (Ed.), *Biodiversity II: understanding and protecting our biological resources* (pp. 83–108). Washington, D.C.: National Academy of Sciences.

Redman, R. S., Sheehan, K. B., Stout, R. G., Rodriguez, R. J., & Henson, J. M. (2002). Thermotolerance generated by plant/fungal symbiosis. *Science, 298*(5598), 1581.

Reed, T. E., Grøtan, V., Jenouvrier, S., & Sæther, B.-E. (2013a). Population growth in a wild bird is buffered against phenological mismatch. *Science, 340*(6131), 488–491.

Reed, T. E., Jenouvrier, S., & Visser, M. E. (2013b). Phenological mismatch strongly affects individual fitness but not population demography in a woodland passerine. *Journal of Animal Ecology, 82*(1), 131–144.

Reeves, R. B. (1969). Role of body temperature in determining the acid–base state in vertebrates. *Federation Proceedings, 28*(3), 1204–1208.

Reeves, R. B. (1972). An imidazole alphastat hypothesis for vertebrate acid–base regulation: tissue carbon dioxide content and body temperature in bullfrogs. *Respiration Physiology, 14*(1–2), 219–236.

Reeves, R. B. (1977). The interaction of body temperature and acid–base balance in ectothermic vertebrates. *Annual Review of Physiology, 39*, 559–586.

Reeves, R. B. (1985). Alphastat regulation of intracellular acid–base state? In R. Gilles (Ed.), *Circulation, respiration, and metabolism: current comparative approaches (First International Congress of Comparative Physiology and Biochemistry, Liège, Belgium, August 1984)* (pp. 414–423). Berlin: Springer-Verlag.

Refinetti, R. (1999). Amplitude of the daily rhythm of body temperature in eleven mammalian species. *Journal of Thermal Biology, 24*(5–6), 477–481.

Refinetti, R., & Menaker, M. (1992). The circadian rhythm of body temperature. *Physiology & Behavior, 51*(3), 613–637.

Regnault, V. (1847). Relations des expériences entreprises par ordre de Monsieur le Ministre des Travaux Publics et sur la proposition de la commission centrale des machines à vapeur, pour déterminer les principales lois et les données numériques qui entrent dans le calcul des machines à vapeur. *Mémoires de l'Académie des Sciences de l'Institut de France, 21*(1), 3–767.

Reid, C. (1899). *The origin of the British flora*. London: Dulau.

Reiss, M. J. (1989). *The allometry of growth and reproduction*. Cambridge: Cambridge University Press.

Renaud, P. E., Webb, T. J., Bjørgesæter, A., et al. (2009). Continental-scale patterns in benthic invertebrate diversity: insights from the MacroBen database. *Marine Ecology Progress Series, 382*, 239–252.

Rex, M. A., Crame, J. A., Stuart, C. T., & Clarke, A. (2005). Large scale biogeographic patterns in marine mollusks: a confluence of history and productivity. *Ecology, 86*(9), 2288–2297.

Rex, M. A., Stuart, C. T., & Coyne, G. (2000). Latitudinal gradients of species richness in the deep-sea benthos of the North Atlantic. *Proceedings of the National Academy of Sciences of the United States of America, 97*(8), 4082–4085.

Rey, B., Roussel, D., Romestaing, C., et al. (2010). Up-regulation of avian uncoupling protein in cold-acclimated and hyperthyroid ducklings prevents reactive oxygen species production by skeletal muscle mitochondria. *BMC Physiology, 10*, Article 5 (12 pages).

Rezende, E. L., Swanson, D. L., Novoa, F. F., & Bozinovic, F. (2002). Passerines versus nonpasserines: so far, no statistical differences in the scaling of avian energetics. *Journal of Experimental Biology, 205*(1), 101–107.

Richardson, A. J. (2008). In hot water: zooplankton and climate change. *ICES Journal of Marine Science, 65*(3), 279–295.

Ricklefs, R. E., Konarzewski, M., & Daan, S. (1996). The relationship between basal metabolic rate and daily energy expenditure in birds and mammals. *American Naturalist, 147*(6), 1047–1071.

Riek, A., & Geiser, F. (2013). Allometry of thermal variables in mammals: consequences of body size and phylogeny. *Biological Reviews, 88*(3), 564–572.

Riek, A., & Geiser, F. (2014). Heterothermy in pouched mammals—a review. *Journal of Zoology, 292*(1), 74–85.

Riley, J. P., & Skirrow, G. (Eds.) (1975). *Chemical oceanography, volume 2* (Second ed.). New York and London: Academic Press.

Riveros, A. J., & Enquist, B. J. (2011). Metabolic scaling in insects supports the predictions of the WBE model. *Journal of Insect Physiology, 57*(6), 688–693.

Rivkina, E., Shcherbakova, V., Laurinavichius, K., et al. (2007). Biogeochemistry of methane and methanogenic archaea in permafrost. *FEMS Microbiology Ecology, 61*(1), 1–15.

Rivkina, E. M., Friedmann, E. I., McKay, C. P., & Gilichinsky, D. (2000). Metabolic activity of permafrost bacteria below the freezing point. *Applied and Environmental Microbiology, 66*(8), 3230–3233.

Robert, V. A., & Casadevall, A. (2009). Vertebrate endothermy restricts most fungi as potential pathogens. *Journal of Infectious Diseases, 200*(10), 1623–1626.

Roberts, D. R., & Hamann, A. (2015). Glacial refugia and modern genetic diversity of 22 western North American tree species. *Proceedings of the Royal Society of London, Series B, 282*(1804), article 20142903.

Roberts, M. F., Lightfoot, E. N., & Porter, W. P. (2010). A new model for the body size–metabolism relationship. *Physiological and Biochemical Zoology, 83*(3), 395–405.

Roberts, N. (1998). *The Holocene: an environmental history* (Second ed.). Oxford: Backwell.

Robertshaw, D. (2006). Mechanisms for the control of respiratory evaporative heat loss in panting animals. *Journal of Applied Physiology, 101*(2), 664–668.

Robertson, C. (1929). *Flowers and insects: lists of visitors of four hundred and fifty-three flowers*. Carlinville, Illinois: The Science Press Printing Company, Lancaster, Pennsylvania.

Robinson, P. J. (2005). Ice and snow in paintings of Little Ice Age winters. *Weather, 60*(2), 37–41.

Robinson, W. R., Peters, R. H., & Zimmerman, J. (1983). The effects of body size and temperature on metabolic rate of organisms. *Canadian Journal of Zoology, 61*(2), 281–288.

Roden, E. E., & Jin, Q. (2011). Thermodynamics of microbial growth coupled to metabolism of glucose, ethanol, short-chain organic acids, and hydrogen. *Applied and Environmental Microbiology, 77*(5), 1907–1909.

Rodríguez, M. Á., López-Sañudo, I. L., & Hawkins, B. A. (2006). The geographic distribution of mammal body size in Europe. *Global Ecology and Biogeography, 15*(2), 173–181.

Rodríguez-Romero, A., Jarrold, M. D., Massamba-N'Siala, G., Spicer, J. I., & Calosi, P. (2016). Multi-generational responses of a marine polychaete to a rapid change in seawater pCO_2. *Evolutionary Applications, 9*(9), 1082–1095.

Roff, D. A. (1980). A motion for the retirement of the von Bertalanffy function. *Canadian Journal of Fisheries and Aquatic Sciences, 37*(1), 127–129.

Rogers, A. D., & Fraser, K. P. P. (2007). Protein metabolism in marine animals: the underlying mechanism of growth. *Advances in Marine Biology, 52*, 267–362.

Rojas, A. D., Körtner, G., & Geiser, F. (2012). Cool running: locomotor performance at low body temperature in mammals. *Biology Letters, 8*(5), 868–870.

Rolfe, D. F. S., & Brown, G. C. (1997). Cellular energy metabolism and molecular origin of standard metabolic rate in mammals. *Physiological Reviews, 77*(3), 731–758.

Rolfe, D. F. S., Newman, J. M. B., Buckingham, J. A., Clark, M. G., & Brand, M. D. (1999). Contribution of mitochondrial proton leak to respiration rate in working skeletal muscle and liver and to SMR. *American Journal of Physiology, 276*(3), C692–C699.

Rollinson, H. (2007). *Early Earth systems: a geochemical approach*. Oxford: Blackwell Publishing.

Roosevelt, T. (1911). Revealing and concealing coloration in birds and mammals. *Bulletin of the American Museum of Natural History, 30*, 121–231 (Article 128).

Root, T. L., Price, J. T., Hall, K. R., Schneider, S. H., Rosenzweig, C., & Pounds, J. A. (2003). Fingerprints of global warming on wild animals and plants. *Nature, 421*(6918), 57–60.

Rosell-Mele, A., & Prahl, F. G. (2013). Seasonality of $U^{K'}_{37}$ temperature estimates as inferred from sediment trap data. *Quaternary Science Review, 72*, 128–136.

Rosenzweig, M. L. (1968). Net primary productivity of terrestrial communities: prediction from climatological data. *American Naturalist, 102*(923), 67–74.

Rosenzweig, M. L. (1995). *Species diversity in space and time*. Cambridge: Cambridge University Press.

Rougier, G. W., Ji, Q., & Novacek, M. J. (2003). A new symmetrodont mammal with fur impressions from the Mesozoic of China. *Acta Geologica Sinica (English Edition), 77*(1), 7–14.

Rowland, L. A., Bal, N. C., & Periasamy, M. (2015). The role of skeletal-based thermogenic mechanisms in vertebrate endothermy. *Biological Reviews, 90*(4), 1279–1297.

Roy, D. B., & Sparks, T. H. (2000). Phenology of British butterflies and climate change. *Global Change Biology, 6*(4), 407–416.

Roy, D. B., & Thomas, J. A. (2003). Seasonal variation in the niche, habitat availability and population fluctuations of a bivoltine thermophilous insect near its range margin. *Oecologia, 134*(3), 439–444.

Roy, K., Jablonski, D., & Valentine, J. W. (1994). Eastern Pacific molluscan provinces and latitudinal diversity gradient: no evidence for 'Rapoport's rule'. *Ecology, 91*, 8871–8874.

Roy, K., Jablonski, D., & Valentine, J. W. (1995). Thermally anomalous assemblages revisited: patterns in the extraprovincial latitudinal range shifts of Pleistocene marine mollusks. *Geology, 23*(12), 1071–1074.

Roy, K., Jablonski, D., & Valentine, J. W. (1996a). Higher taxa in biodiversity studies: patterns from eastern Pacific marine molluscs. *Philosophical Transactions of the Royal Society of London, Series B, 351*, 1605–1613.

Roy, K., Jablonski, D., & Valentine, J. W. (2001). Climate change, species range limits and body size in marine bivalves. *Ecology Letters, 4*(4), 366–370.

Roy, K., Jablonski, D., Valentine, J. W., & Rosenberg, G. (1998). Marine latitudinal diversity gradients: tests of causal hypotheses. *Proceedings of the National Academy of Sciences of the United States of America, 95*, 3699–3702.

Roy, K., Valentine, J. W., Jablonski, D., & Kidwell, S. M. (1996b). Scales of climatic variability and time averaging in Pleistocene biotas: implications for ecology and evolution. *Trends in Ecology & Evolution, 11*(11), 458–463.

Royer, D. L., Wilf, P., Janesko, D. A., Kowalski, E. A., & Dilcher, D. L. (2005). Correlations of climate and plant ecology to leaf size and shape: potential proxies for the fossil record. *American Journal of Botany, 92*(7), 1141–1151.

Ruben, J. (1995). The evolution of endothermy in mammals and birds: from physiology to fossils. *Annual Review of Physiology, 57*, 69–95.

Ruben, J. A., Jones, T. D., & Geist, N. R. (2003). Respiratory and reproductive paleophysiology of dinosaurs and early birds. *Physiological and Biochemical Zoology*, *76*(2), 141–164.

Rubner, M. (1883). Ueber den Einfluss der Körpergrösse auf Stoff- und Kraftwechsel. *Zeitschrift für Biologie*, *19*(4), 535–562.

Ruf, T., & Geiser, F. (2015). Daily torpor and hibernation in birds and mammals. *Biological Reviews*, *90*(3), 891–926.

Ruffner, J. A. (1963). Reinterpretation of the genesis of Newton's 'Law of Cooling'. *Archive for History of the Exact Sciences*, *2*(2), 138–152.

Rumford, B. (1798). An inquiry concerning the source of heat which is excited by friction. *Philosophical Transactions of the Royal Society of London*, *88*, 80–102.

Russell, M. J., Daniel, R. M., & Hall, A. J. (1993). On the emergence of life via catalytic iron-sulphide membranes *Terra Nova*, *5*(4), 343–347.

Russell, M. J., Hall, A. J., & Martin, W. (2010). Serpentinization as a source of energy at the origin of life. *Geobiology*, *8*(5), 355–371.

Russell, M. J., Nitschke, W., & Branscombe, E. (2013). The inevitable journey to being. *Philosophical Transactions of the Royal Society of London, Series B*, *368*(1622), article 20120254.

Russo, S. E., Robinson, S. K., & Terborgh, J. (2003). Size-abundance relationships in an Amazonian bird community: implications for the energetic equivalence rule. *American Naturalist*, *161*(2), 267–283.

Ruud, J. T. (1954). Vertebrates without erythrocytes and blood pigments. *Nature*, *173*(4410), 848–850.

Sabine, C. L., Feely, R. A., Gruber, N., et al. (2004). The oceanic sink for anthropogenic CO_2. *Science*, *305*(5682), 367–371.

Sadowska, E. T., Labocha, M. K., Baliga, K., et al. (2005). Genetic correlations between basal and maximal metabolic rates in a wild rodent: consequences for evolution of endothermy. *Evolution*, *59*(3), 672–681.

Sahai, R., & Nyman, L.-Å. (1997). The Boomerang Nebula: the coldest place in the universe? *The Astrophysical Journal*, *487*(2), L155–L159.

Sakai, A. (1960). Survival of the twigs of woody plants at −196 °C. *Nature*, *185*(4710), 393–394.

Sakai, A. (1970). Freezing resistance in willows from different climates. *Ecology*, *51*(3), 485–491.

Sakai, A. (1983). Comparative study on freezing resistance of conifers with special reference to cold adaptation and its evolutive aspects. *Canadian Journal of Botany*, *61*(9), 2323–2332.

Sakai, A., & Okada, S. (1971). Freezing resistance of conifers. *Silvae Genetica*, *20*(3), 91–97.

Sakai, A., & Weiser, C. J. (1973). Freezing resistance of trees in North America with reference to tree regions. *Ecology*, *54*(1), 118–126.

Salewski, V., & Watt, C. (2017). Bergmann's rule: a biophysiological rule examined in birds. *Oikos*, *126*(2), 161–172.

Salt, R. W. (1936). Studies on the freezing process in insects. *Minnesota Technical Bulletin*, *116*(1), 1–41.

Salt, R. W. (1950). Time as a factor in the freezing of undercooled insects. *Canadian Journal of Research*, *28*(5), 285–291.

Salt, R. W. (1957). Natural occurrence of glycerol in insects and its relation to their ability to survive freezing. *The Canadian Entomologist*, *89*(11), 491–494.

Salt, R. W. (1959). Survival of frozen fat body cells in an insect. *Nature*, *184*(4696), 1426.

Salt, R. W. (1961). Principles of insect cold-hardiness. *Annual Reviews of Entomology*, *6*(1), 55–74.

Salt, R. W. (1962). Intracellular freezing in insects. *Nature*, *193*(4821), 1207–1208.

Salt, R. W. (1966a). Effect of cooling rate on the freezing temperatures of supercooled insects. *Canadian Journal of Zoology*, *44*(4), 655–659.

Salt, R. W. (1966b). Factors affecting nucleation in supercooled insects. *Canadian Journal of Zoology*, *44*(1), 117–133.

Salt, R. W. (1966c). Relation between time of freezing and temperature in supercooled larvae of *Cephus cinctus* Nort. *Canadian Journal of Zoology*, *44*(5), 947–952.

Sanders, H. L. (1968). Marine benthic diversity: a comparative study. *American Naturalist*, *102*(925), 243–282.

Santelices, B., Bolton, J. J., & Meneses, I. (2009). Marine algal communities. In J. D. Witman & K. Roy (Eds.), *Marine macroecology* (pp. 153–192). Chicago: University of Chicago Press.

Santillán, M. (2003). Allometric scaling law in a simple oxygen exchanging network: possible implications on the biological allometric scaling laws. *Journal of Theoretical Biology*, *223*, 249–257.

Santos, H., Lamosa, P., Faria, T. Q., Pais, T. M., López de la Paz, M., & Serrano, L. (2008). Compatible solutes of (hyper)thermophiles and their role in protein stabilisation. In F. Robb, G. Antranikian, D. Grogan, & A. Driessen (Eds.), *Thermophiles: biology and technology at high temperatures* (pp. 9–24). Boca Raton: CRC Press.

Sarrus, P., & Rameaux, J.-F. (1839). Application des sciences accessoires et principalement des mathématiques à la physiologie générale (Rapport sur un mémoire addressé à l'Académie Royale de Médecine, séance du 23 juillet 1839). *Bulletin de l'Académie Royale de Médecine*, *3*, 1094–1100.

Sattler, B., Puxbaum, H., & Psenner, R. (2001). Bacterial growth in supercooled cloud droplets. *Geophysical Research Letters*, *28*(2), 239–242.

Savage, V. M., Deeds, E. J., & Fontana, W. (2008). Sizing up allometric scaling theory. *PLoS Computational Biology*, *4*(9), e1000171 (1000117 pp.).

Sawle, L., & Ghosh, K. (2011). How do thermophilic proteins and proteomes withstand high temperature? *Biophysical Journal, 101*(1), 217–227.

Schellman, J. A. (1997). Temperature, stability, and the hydrophobic interaction. *Biophysical Journal, 73*(6), 2960–2964.

Schleucher, E. (2002). Heterothermia in pigeons and doves reduces energetic costs. *Journal of Thermal Biology, 26,* 287–293.

Schleucher, E. (2004). Torpor in birds: taxonomy, energetics, and ecology. *Physiological and Biochemical Zoology, 77*(6), 942–949.

Schleucher, E., Prinzinger, R., & Withers, P. C. (1991). Life in extreme environments: investigations on the ecophysiology of a desert bird, the Australian Diamond Dove (*Geopelia cuneata* Latham). *Oecologia, 88*(1), 72–78.

Schleucher, E., & Withers, P. C. (2001). Re-evaluation of the allometry of wet thermal conductance for birds. *Comparative Biochemistry and Physiology, Part A, 129*(4), 821–827.

Schmid, J. (1996). Oxygen consumption and torpor in mouse lemurs (*Microcebus murinus* and *M. myoxinus*): preliminary results of a study in western Madagascar. In F. Geiser, A. J. Hulbert, & S. C. Nicol (Eds.), *Adaptations to the cold (Tenth International Hibernation Symposium)* (pp. 47–54). Armidale, Australia: University of New England Press.

Schmid, J. (2000). Daily torpor in the gray mouse lemur (*Microcebus murinus*) in Madagascar: energetic consequences and biological significance. *Oecologia, 123*(2), 175–183.

Schmid, J., Ruf, T., & Heldmaier, G. (2000). Metabolism and temperature regulation during daily torpor in the smallest primate, the pygmy mouse lemur (*Microcebus myoxinus*) in Madagascar. *Journal of Comparative Physiology B, 170*(1), 59–68.

Schmid, W. D. (1982). Survival of frogs in low temperature. *Science, 215*(4533), 697–698.

Schmidt, G. A., Ruedy, R. A., Miller, R. L., & Lacis, A. A. (2010). Attribution of the present-day total greenhouse effect. *Journal of Geophysical Research—Atmospheres, 115,* article D20106.

Schmidt-Nielsen, K. (1969). The neglected interface: the biology of water as a liquid–gas system. *Quarterly Review of Biophysics, 2*(3), 283–304.

Schmidt-Nielsen, K. (1975). *Animal physiology: adaptation and environment.* Cambridge: Cambridge University Press.

Schmidt-Nielsen, K. (1984). *Scaling: why is animal size so important?* Cambridge: Cambridge University Press.

Schmidt-Nielsen, K., Schmidt-Nielsen, B., Jarnum, S. A., & Houpt, T. R. (1956). Body temperature of the camel and it relation to water economy. *American Journal of Physiology, 188*(1), 103–112.

Scholander, P. F., Flagg, W., Walters, V., & Irving, L. (1953). Climatic adaptation in Arctic and tropical poikilotherms. *Physiological Zoology, 26*(1), 67–92.

Scholander, P. F., van Dam, L., Kanwisher, J. W., Hammel, H. T., & Gordon, M. S. (1957). Supercooling and osmoregulation in northern fishes. *Journal of Cellular and Comparative Physiology, 49*(1), 5–24.

Schrag, J. D., Cheng, C.-H. C., Panico, M., Morris, H. R., & DeVries, A. L. (1987). Primary and secondary structure of antifreeze peptides from arctic and antarctic zoarcid fishes. *Biochimica et Biophysica Acta, 915*(3), 357–370.

Schrenk, M. O., Kelley, D. S., Delaney, J. R., & Baross, J. A. (2003). Incidence and diversity of microorganisms within the walls of an active deep-sea sulfide chimney. *Applied and Environmental Microbiology, 69*(6), 3580–3592.

Schrödinger, E. (1944). *What is life? The physical aspect of the living cell.* Cambridge: Cambridge University Press.

Schroeter, B., Green, T. G. A., Kappen, L., & Seppelt, R. D. (1994). Carbon dioxide exchange at subzero temperatures: field measurements in *Umbilicaria aprina* in Antarctica. *Cryptogamic Botany, 4*(2), 233–241.

Schultz, E. T., Reynolds, K. E., & Conover, D. O. (1996). Countergradient variation in growth among newly hatched *Fundulus heteroclitus*: geographic differences revealed by common-environment experiments. *Functional Ecology, 10*(3), 366–374.

Schwartz, M. D., Ault, T. R., & Betancourt, J. L. (2013). Spring onset variations and trends in the continental United States: past and regional assessment using temperature-based indices. *International Journal of Climatology, 33*(13), 2917–2922.

Scott, C. B. (2005). Contribution of anaerobic energy expenditure to whole body thermogenesis. *Nutrition & Metabolism, 2,* article 14.

Sears, M. W., Angilletta, M. J., Schuler, M. S., et al. (2016). Configuration of the thermal landscape determines thermoregulatory performance of ectotherms. *Proceedings of the National Academy of Sciences of the United States of America, 113*(38), 10595–10600.

Secor, S. M. (2009). Specific dynamic action: a review of the post-prandial metabolic response. *Journal of Comparative Physiology, B, 179*(1), 1–56.

Seebacher, F. (2003). Dinosaur body temperatures: the occurrence of endothermy and ectothermy. *Paleobiology, 29*(1), 105–122.

Seebacher, F., & Shine, R. (2004). Evaluating thermoregulation in reptiles: the fallacy of the inappropriately applied method. *Physiological and Biochemical Zoology, 77*(4), 688–695.

Seki, S., Kleinhans, F. W., & Mazur, P. (2009). Intracellular ice formation in yeast cells vs. cooling rate: predictions from modelling vs. experimental observations by differential scanning calorimetry. *Cryobiology, 58*(2), 157–165.

Sengers, J. V., & Watson, J. T. R. (1986). Improved international formulations for the viscosity and thermal conductivity of water substance. *Journal of Physical and Chemical Reference Data*, *15*(4), 1291–1314.

Sernetz, M., Gelléri, B., & Hofmann, J. (1985). The organism as bioreactor. Interpretation of the reduction law of metabolism in terms of heterogeneous catalysis and fractal structure. *Journal of Theoretical Biology*, *117*(2), 209–230.

Sexton, J. P., McIntyre, P. J., Angert, A. L., & Rice, K. J. (2009). Evolution and ecology of species range limits. *Annual Review of Ecology, Evolution and Systematics*, *40*, 415–436.

Seymour, R. S. (2001). Biophysics and physiology of temperature regulation in thermogenic flowers. *Bioscience Reports*, *21*(2), 223–236.

Seymour, R. S., & Schultze-Motel, P. (1998). Physiological temperature regulation by flowers of the sacred lotus. *Philosophical Transactions of the Royal Society of London, Series B*, *353*(1371), 935–943.

Seymour, R. S., Silberbauer-Gottsberger, I., & Gottsberger, G. (2010). Respiration and temperature patterns in thermogenic flowers of *Magnolia ovata* under natural conditions in Brazil. *Functional Plant Biology*, *37*(9), 870–878.

Seymour, R. S., White, C. R., & Gibernau, M. (2003). Environmental biology: heat reward for insect pollinators. *Nature*, *426*(6964), 243–244.

Seymour, R. S., White, C. R., & Gibernau, M. (2009). Endothermy of dynastine scarab beetles (*Cyclocephala colasi*) associated with pollination biology of a thermogenic arum lily (*Philodendron solimoesense*). *Journal of Experimental Biology*, *212*(18), 2960–2968.

Sformo, T., Walters, K., Jeannet, K., et al. (2010). Deep supercooling, vitrification and limited survival to -100 °C in the Alaskan beetle *Cucujus clavipes puniceus* (Coleoptera: Cucujidae) larvae. *Journal of Experimental Biology*, *213*(3), 502–509.

Sgueo, C., Wells, M. E., Russell, D. E., & Schaeffer, P. J. (2012). Acclimatization of seasonal energetics in northern cardinals (*Cardinalis cardinalis*) through plasticity of metabolic rates and ceilings. *Journal of Experimental Biology*, *215*(14), 2418–2424.

Shachtman, T. (1999). *Absolute zero and the conquest of cold*. Boston: Houghton Mifflin Harcourt.

Shaklee, J. B., Christiansen, J. A., Sidell, B. D., Prosser, C. L., & Whitt, G. S. (1977). Molecular aspects of temperature acclimation in fish: contributions of changes in enzyme activities and isozyme patterns to metabolic reorganization in the green sunfish. *Journal of Experimental Zoology*, *201*(1), 1–20.

Shannon, C. E. (1948). A mathematical theory of communication. *Bell System Technical Journal*, *27*, 379–423 and 623–656.

Shannon, C. E., & Weaver, W. (1949). *The mathematical theory of communication*. Urbana: The University of Illinois Press.

Shapley, H. (1920). Thermokinetics of *Liometopum apiculatum* Mayr. *Proceedings of the National Academy of Sciences of the United States of America*, *6*(4), 204–211.

Shapley, H. (1924). Notes on the thermokinetics of dolichoderine ants. *Proceedings of the National Academy of Sciences of the United States of America*, *10*(10), 436–439.

Shapley, H. (1969). *Through rugged ways to the stars*. New York: Charles Scribner's Sons.

Sharp, C. E., Brady, A. L., Sharp, G. H., Grasby, S. E., Stott, M. B., & Dunfield, P. F. (2014). Humboldt's spa: microbial diversity is controlled by temperature in geothermal environments. *ISME Journal*, *8*(6), 1166–1174.

Sharp, K. (2001). Entropy–enthalpy compensation: fact or artifact? *Protein Science*, *10*, 661–667.

Sheridan, J. A., & Bickford, D. (2011). Shrinking body size as an ecological response to climate change. *Nature Climate Change*, *1*(8), 401–406.

Sheriff, M. J., Williams, C. T., Kenagy, G. J., Buck, C. L., & Barnes, B. M. (2012). Thermoregulatory changes anticipate hibernation onset by 45 days: data from free-living arctic ground squirrels. *Journal of Comparative Physiology, B*, *182*(6), 841–847.

Sherry, D. (2011). Thermoscopes, thermometers, and the foundations of measurement. *Studies in History and Philosophy of Science*, *42*(4), 509–524.

Shine, R. (2004). Does viviparity evolve in cold climate reptiles because pregnant females maintain stable (not high) body temperatures? *Evolution*, *58*(8), 1809–1818.

Sibley, C. G., & Ahlquist, J. E. (1991). *Phylogeny and classification of birds*. New Haven, Connecticut: Yale University Press.

Sibley, C. G., Ahlquist, J. E., & Monroe, B. L. (1988). A classification of living birds of the world based on DNA–DNA hybridization studies. *Auk*, *105*(3), 409–423.

Sibly, R. M., & Atkinson, D. (1994). How rearing temperature affects optimal adult size in ectothems. *Functional Ecology*, *8*(4), 486–493.

Sibly, R. M., Baker, J., Grady, J. M., et al. (2015). Fundamental insights into ontogenetic growth from theory and fish. *Proceedings of the National Academy of Sciences of the United States of America*, *112*(45), 13934–13939.

Sidell, B. D. (1977). Turnover of cytochrome C in skeletal muscle of green sunfish (*Lepomis cyanellus*, R.) during thermal acclimation. *Journal of Experimental Zoology*, *199*(2), 233–250.

Sidell, B. D. (1983). Cellular acclimatisation to environmental change by quantitative alterations in enzymes and organelles. In A. R. Cossins & P. Sheterline (Eds.), *Cellular acclimatisation to environmental change* (Vol. 17, pp. 102–120). Cambridge: Cambridge University Press.

Sidell, B. D., & Hazel, J. R. (1987). Temperature affects the diffusion of small molecules through the cytosol of fish muscle. *Journal of Experimental Biology, 129*(1), 191–203.

Sidell, B. D., Vayda, M. E., Small, D. J., et al. (1997). Variable expression of myoglobin among the hemoglobinless Antarctic icefishes. *Proceedings of the National Academy of Sciences of the United States of America, 94*(7), 3420–3424.

Sidell, B. D., Wilson, F. R., Hazel, J. R., & Prosser, C. L. (1973). Time course of thermal acclimation in goldfish. *Journal of Comparative Physiology, 84*(2), 119–127.

Signor, P. W. (1985). Real and apparent trends in species richness through time. In J. W. Valentine (Ed.), *Phanerozoic diversity patterns: profiles in macroevolution* (pp. 129–150). Princeton: Princeton University Press.

Siminovitch, D., & Cloutier, D. (1983). Drought and freezing tolerance and adaptation in plants: some evidence of near equivalences. *Cryobiology, 20*(4), 487–503.

Sinclair, B. J. (1999). Insect cold tolerance: how many kinds of frozen? *European Journal of Entomology, 96*(2), 157–164.

Sinclair, B. J. (2015). Linking energetics and overwintering in temperate insects. *Journal of Thermal Biology, 54*(S1), 5–11.

Sinclair, B. J., & Chown, S. L. (2005). Climatic variability and hemispheric differences in insect cold tolerance: support from southern Africa. *Functional Ecology, 19*(2), 214–221.

Singer, S. J., & Nicolson, G. L. (1972). The fluid mosaic model of the structure of cell membranes. *Science, 175*(4023), 720–731.

Singh, P., Hanada, Y., Singh, S. M., & Tsuda, S. (2014). Antifreeze protein activity in Arctic cryoconite bacteria. *FEMS Microbiology Letters, 351*(1), 14–22.

Slaughter, D., Fletcher, G. L., Ananthanarayanan, V. S., & Hew, C. L. (1981). Antifreeze proteins from the sea raven, *Hemitripterus americanus*. *Journal of Biological Chemistry, 256*(4), 2022–2026.

Small, A. M., Adey, W. H., & Spoon, D. (1998). Are current estimates of coral reef biodiversity too low? The view through the window of a microcosm. *Atoll Research Bulletin, 458*, 1–20.

Smayda, T. J. (1970). The suspension and sinking of phytoplankton in the sea. *Oceanography and Marine Biology: an Annual Review, 8*, 353–414.

Smetacek, V. (1974). On the increasing occurrence of typically plains-birds in the Kumaon Hills. *Journal of the Bombay Natural History Society, 71*(2), 299–302.

Smiles, K. A., Elizondo, R. S., & Barney, C. C. (1976). Sweating responses during changes of hypothalamus temperature in rhesus monkey. *Journal of Applied Physiology, 40*(5), 653–657.

Smit, H. (1965). Some experiments on the oxygen consumption of goldfish (*Carassius auratus* L.) in relation to swimming speed. *Canadian Journal of Zoology, 43*(4), 623–633.

Smith, A. P. (2008). Proof of the atmospheric greenhouse effect. *arXiv Preprint, arXiv: 0802.4324*.

Smith, K. R., Cadena, V., Endler, J. A., Porter, W. P., Kearney, M. R., & Stuart-Fox, D. (2016). Colour change on different body regions provides thermal and signalling advantages in bearded dragon lizards. *Proceedings of the Royal Society of London, Series B, 283*(1832), article 20160626.

Smith, P., & Smith, J. (2012). Climate change and bird migration in south-eastern Australia. *Emu, 112*(4), 333–342.

Smith, S. D., Didden-Zopfy, B., & Nobel, P. S. (1984). High-temperature responses of North American cacti. *Ecology, 65*(2), 643–651.

Snelgrove, P. V. R. (2010). *Discoveries of the census of marine life: making ocean life count*. Cambridge: Cambridge University Press.

Snow, C. P. (1959). *The two cultures*. Cambridge: Cambridge University Press.

Soffer, B. H., & Lynch, D. (1999). Some paradoxes, errors, and resolutions concerning the spectral optimization of human vision. *American Journal of Physics, 67*(11), 946–953.

Sohlenius, B., & Boström, S. (2005). The geographic distribution of metazoan microfauna on East Antarctic nunataks. *Polar Biology, 28*(6), 439–448.

Sohlenius, B., Boström, S., & Hirschfelder, A. (1995). Nematodes, rotifers and tardigrades from nunataks in Dronning Maud Land, East Antarctica. *Polar Biology, 15*(1), 51–56.

Sokolova, T., Hanel, J., Onyenwoke, R. U., et al. (2006). Novel chemolithotrophic, thermophilic, anaerobic bacteria *Thermolithobacter ferrireducens* gen. nov., sp. nov. and *Thermolithobacter carboxydivorans* sp. nov. *Extremophiles, 11*(1), 145–157.

Somero, G. N., & White, F. N. (1985). Enzymatic consequences under alphastat regulation. In H. Rahn & O. Prakash (Eds.), *Acid–base regulation and body temperature* (pp. 55–80). Berlin: Springer-Verlag.

Sømme, L. (2000). The history of cold hardiness research in terrestrial arthropods. *Cryo-Letters, 21*(5), 289–296.

Sommer, A. M., & Pörtner, H.-O. (2002). Metabolic cold adaptation in the lugworm *Arenicola marina*: comparison of a North Sea and a White Sea population. *Marine Ecology Progress Series, 240*, 171–182.

Sookias, R. B., Butler, R. J., & Benson, R. B. J. (2012). Rise of dinosaurs reveals major body-size transitions are driven by passive processes of trait evolution. *Proceedings of the Royal Society of London, Series B, 279*(1736), 2180–2187.

Sousa, T., Domingos, T., Poggiale, J.-C., & Kooijman, S. A. L. M. (2010). Dynamic energy budget theory restores coherence in biology. *Philosophical Transactions of the Royal Society of London, Series B, 365*, 3413–3428.

Sousa, T., Mota, R., Domingos, T., & Kooijman, S. A. L. M. (2006). Thermodynamics of organisms in the context of dynamic energy budget theory. *Physical Review E, 74*(5 Part 1), article 051901.

Spalding, M. D., Fox, H. E., Allen, G. R., et al. (2007). Marine ecoregions of the world: a bioregionalization of coastal and shelf areas. *Bioscience, 57*(7), 573–583.

Sparks, T. H., & Carey, P. D. (1995). The responses of species to climate over two centuries: an analysis of the Marsham phenological record, 1736–1947. *Journal of Ecology, 83*(2), 321–329.

Sparks, T. H., Jeffree, E. P., & Jeffree, C. E. (2000). An examination of the relationship between flowering times and temperature at the national scale using long-term phenological records from the UK. *International Journal of Biometeorology, 44*(2), 82–87.

Sparks, T. H., Roy, D. B., & Dennis, R. L. H. (2005). The influence of temperature on migration of Lepidoptera into Britain. *Global Change Biology, 11*(3), 507–514.

Speakman, J. R. (1997). *Doubly labelled water: theory and practice*. Berlin: Springer-Verlag.

Speakman, J. R., & Król, E. (2010). Maximal heat dissipation capacity and hyperthermia risk: neglected key factors in the ecology of endotherms. *Journal of Animal Ecology, 79*, 726–746.

Sperry, J. S., Nichols, K. L., Sullivan, J. E. M., & Eastlack, S. E. (1994). Xylem embolism in ring-porous, diffuse-porous, and coniferous trees of northern Utah and interior Alaska. *Ecology, 75*(6), 1736–1752.

Sperry, J. S., & Sullivan, J. E. M. (1992). Xylem embolism in response to freeze–thaw cycles and water stress in ring-porous, diffuse-porous, and conifer species. *Plant Physiology, 100*(2), 605–613.

Spicer, J. I., & Gaston, K. J. (1999). Amphipod gigantism driven by oxygen availability? *Ecology Letters, 2*(6), 397–401.

Spoor, W. A. (1946). A quantitative study of the relationship between the activity and oxygen consumption of the goldfish, and its application to the measurement of respiratory metabolism in fishes. *Biological Bulletin, 91*(3), 312–325.

Spotila, J. R. (1980). Constraints of body size and environment on the temperature regulation of dinosaurs. In R. D. K. Thomas & E. C. Olson (Eds.), *A cold look at the warm-blooded dinosaurs* (Vol. 28, pp. 233–252). Washington, D.C.: Westview Press, for the American Association for the Advancement of Science.

Spotila, J. R., Lommen, P. W., Bakken, G. S., & Gates, D. M. (1973). A mathematical model for body temperatures of large reptiles: implications for dinosaur ecology. *American Naturalist, 107*(955), 391–404.

Sprackling, M. (1991). *Thermal physics*. Basingstoke & London: MacMillan Education.

Stanley, S. M. (1986). Anatomy of a regional mass extinction: Plio-Pleistocene decimation of the Western Atlantic bivalve fauna. *Palaios, 1*(1), 17–36.

Stanton, T. P., Shaw, W. J., Truffer, M., et al. (2013). Channelized ice melting in the ocean boundary layer beneath Pine Island Glacier, Antarctica. *Science, 341*(6151), 1236–1239.

Staples, J. F. (2014). Metabolic suppression in mammalian hibernation: the role of mitochondria. *Journal of Experimental Biology, 217*(12), 2032–2036.

Staples, J. F., & Brown, J. C. L. (2008). Mitochondrial metabolism in hibernation and daily torpor: a review. *Journal of Comparative Physiology B, 178*(7), 811–827.

Steel, H., Verdoodt, F., Čerevková, A., et al. (2013). Survival and colonization of nematodes in a composting process. *Invertebrate Biology, 132*(2), 108–119.

Steffensen, J. F. (2002). Metabolic cold adaptation of polar fish based on measurements of aerobic oxygen consumption: fact or artefact? Artefact! *Comparative Biochemistry and Physiology, Part A, 132*, 789–795.

Steffensen, J. F., Bushnell, P. G., & Schurmann, H. (1994). Oxygen consumption in four species of teleosts from Greenland: no evidence of metabolic cold adaptation. *Polar Biology, 14*(1), 49–54.

Steffensen, J. P., Andersen, K. K., Bigler, M., et al. (2008). High-resolution Greenland ice core data show abrupt climate change happens in few years. *Science, 321*(5889), 680–684.

Stegen, J. C., Enquist, B. J., & Ferriere, R. (2009). Advancing the metabolic theory of biodiversity. *Ecology Letters, 12*(10), 1001–1015.

Stehli, F. G., Douglas, R., & Kefescegliou, I. (1972). Models for the evolution of planktonic foraminifera. In T. J. M. Schopf (Ed.), *Models in paleobiology* (pp. 116–128). San Francisco: Freeman Cooper.

Stehli, F. G., McAlester, A. L., & Helsey, C. E. (1967). Taxonomic diversity of recent bivalves and some implications for geology. *Bulletin of the Geological Society of America, 78*, 455–466.

Steig, E. J., Ding, Q., White, J. C., et al. (2013). Recent climate and ice-sheet changes in West Antarctica compared with the past 2,000 years. *Nature Geoscience, 6*(5), 372–375.

Stenvinkel, P., Fröbert, O., Anderstam, B., et al. (2013). Metabolic changes in summer active and anuric hibernating free-ranging brown bears (*Ursus arctos*). *PLoS One, 8*(9), e72934.

Stepanova, N. A. (1958). On the lowest temperatures on Earth. *Monthly Weather Review, 86*(1), 6–10.

Stephenson, N. (1998). Actual evapotranspiration and deficit: biologically meaningful correlates of vegetation distribution across spatial scales. *Journal of Biogeography, 25*(5), 855–870.

Steponkus, P. L., & Lynch, D. V. (1989). Freeze/thaw induced destabilization of the plasma membrane and the effects of cold acclimation. *Journal of Bioenergetics and Biomembranes, 21*(1), 21–41.

Stevens, S. S. (1946). On the theory of scales of measurement. *Science*, *103*(2684), 677–680.

Stevenson, T. C. E. (1864). New description of box for holding thermometers. *Journal of the Scottish Meteorological Society*, *1*, 122.

Stewart, J. R., & Lister, A. M. (2001). Cryptic northern refugia and the origins of the modern biota. *Trends in Ecology & Evolution*, *16*(11), 608–613.

Stewart, J. R., Lister, A. M., Barnes, I., & Dalen, L. (2010). Refugia revisited: individualistic responses of species in space and time. *Proceedings of the Royal Society of London, Series B*, *277*(1682), 661–671.

Stokes, G. B. (1988). Estimating the energy content of nutrients. *Trends in Biochemical Sciences*, *13*(11), 422–424.

Stomp, M., Huisman, J., Mittelbach, G. G., Litchman, E., & Klausmeier, C. A. (2011). Large-scale biodiversity patterns in freshwater phytoplankton. *Ecology*, *92*(11), 2096–2107.

Storch, D., & Šizling, A. (2007). Scaling species richness and distribution: uniting the species–area and species–energy relationships. In D. Storch, P. A. Marquet, & J. H. Brown (Eds.), *Scaling biodiversity* (pp. 300–321). Cambridge: Cambridge University Press.

Storey, K. B., & Storey, J. M. (2004). Physiology, biochemistry, and molecular biology of vertebrate freeze tolerance: the wood frog. In B. J. Fuller, N. Lane, & E. E. Benson (Eds.), *Life in the frozen state* (pp. 243–274). Boca Raton: CRC Press.

Strimbeck, G. R., Johnson, A. H., & Vaan, D. R. (1993). Midwinter needle temperature and winter injury of montane red spruce. *Tree Physiology*, *13*(2), 131–144.

Strimbeck, G. R., Kjellsen, T. D., Schaberg, P. G., & Murakami, P. F. (2007). Cold in the common garden: comparative low-temperature tolerance of boreal and temperate conifer foliage. *Trees*, *21*(5), 557–567.

Stuart, C. T., Rex, M. A., & Etter, R. J. (2003). Large-scale spatial and temporal patterns of deep-sea benthic species diversity. In P. A. Tyler (Ed.), *Ecosystems of the deep oceans* (Vol. 28, pp. 295–311). Amsterdam: Elsevier.

Studier, E. H., & Sevick, S. H. (1991). Live mass, water content, nitrogen and mineral levels in some insects from south-central lower Michigan. *Comparative Biochemistry and Physiology, Part A*, *103*(3), 579–595.

Suckling, C. C., Clark, M. S., Richard, J., et al. (2015). Adult acclimation to combined temperature and pH stressors significantly enhances reproductive outcomes compared to short-term exposures. *Journal of Animal Ecology*, *84*(3), 773–784.

Swanson, D. L., Thomas, N. E., Liknes, E. T., & Cooper, S. J. (2012). Intraspecific correlations of basal and maximal metabolic rates in birds and the aerobic capacity model for the evolution of endothermy. *PLoS ONE*, *7*(3), e34271.

Szent-Györgyi, A. (1961). Introductory comments. In W. D. McElroy & B. Glass (Eds.), *A symposium on light and life* (pp. 7–10). Baltimore: Johns Hopkins Press.

Szent-Györgyi, A. (1971). Biology and pathology of water. *Perspectives in Biology and Medicine*, *14*(2), 239–249.

Taigen, T. L. (1983). Activity metabolism of anuran amphibians: implications for the origin of endothermy. *American Naturalist*, *121*(1), 94–109.

Takai, K., Nakamura, K., Toki, T., et al. (2008). Cell proliferation at 122 °C and isotopically heavy CH_4 production by a hyperthermophilic methanogen under high-pressure cultivation. *Proceedings of the National Academy of Sciences of the United States of America*, *105*(31), 1949–10954.

Talbot, D. A., Duchamp, C., Rey, B., et al. (2004). Uncoupling protein and ATP/ADP carrier increase mitochondrial proton conductance after cold adaptation of king penguins. *Journal of Physiology*, *558*(1), 123–135.

Tanford, C. (1980). *The hydrophobic effect: formation of micelles and biological membranes* (Second ed.). New York: John Wiley & Sons.

Tansey, M. R., & Brock, T. D. (1972). The upper temperature limit for eukaryotic organisms. *Proceedings of the National Academy of Sciences of the United States of America*, *69*(9), 2426–2428.

Tarling, G., Burrows, M., Matthews, J., et al. (2000). An optimisation model of the diel vertical migration of northern krill (*Meganyctiphanes norvegica*) in the Clyde Sea and the Kattegat. *Canadian Journal of Fisheries and Aquatic Sciences*, *57*(Supplement 3), 38–50.

Taylor, C. C. W. (1999). *The atomists, Leucippus and Democritus: fragments: a text and translation with a commentary*. Toronto: University of Toronto Press.

Taylor, C. R. (1970). Strategies of temperature regulation: effect on evaporation in East African ungulates. *American Journal of Physiology*, *219*(4), 1131–1135.

Taylor, J. R. E., Rychlik, L., & Churchfield, S. (2013). Winter reduction in body mass in a very small, nonhibernating mammal: consequences for heat loss and metabolic rates. *Physiological and Biochemical Zoology*, *86*(1), 9–18.

Taylor, M. J. (1987). Physico-chemical principles in low temperature biology. In B. W. W. Grout & G. J. Morris (Eds.), *The effects of low temperatures on biological systems* (pp. 3–71). London: Edward Arnold.

Teulier, L., Rouanet, J.-L., Letexier, D., et al. (2010). Cold-acclimation-induced non-shivering thermogenesis in birds is associated with upregulation of avian UCP but not with innate uncoupling or altered ATP efficiency. *Journal of Experimental Biology*, *213*, 2476–2482.

Thatje, S., Hillenbrand, C.-D., Mackensen, A., & Larter, R. (2008). Life hung by a thread: endurance of Antarctic fauna in glacial periods. *Ecology*, *89*(3), 682–692.

Thayer, G. H. (1909). *Concealing coloration in the animal kingdom: an exposition of the laws of disguise through color* and pattern, being a summary of Abbott H. Thayer's discoveries. New York: MacMillan.

Theede, H., Schneppenheim, R., & Béress, L. (1976). Frostschutz-Glycoproteine bei *Mytilus edulis*? *Marine Biology*, 36(2), 183–189.

Thomas, C. D. (2010). Climate, climate change and range boundaries. *Diversity and Distributions*, 16(3), 488–495.

Thomas, C. D., Cameron, A., Green, R. E., et al. (2004). Extinction risk from climate change. *Nature*, 427(6960), 145–148.

Thomas, D. N. (2004). *Frozen oceans: the floating world of pack ice*. London: The Natural History Museum.

Thomas, D. N. (2012). Sea ice. In E. M. Bell (Ed.), *Life at extremes: environments, organisms and strategies for survival* (pp. 62–80). Wallingford, Oxfordshire: CAB International.

Thomas, D. N., & Dieckmann, G. S. (2002). Antarctic sea ice: a habitat for extremophiles. *Science*, 295(5555), 641–644.

Thomas, D. W., Blondel, J., Perret, P., Lambrechts, M. M., & Speakman, J. R. (2001). Energetic and fitness costs of mismatching resource supply and demand in seasonally breeding birds. *Science*, 291(5513), 2589–2600.

Thomas, E. R., Bracegirdle, T. J., Turner, J., & Wolff, E. W. (2013). A 308 year record of climate variability in West Antarctica. *Geophysical Research Letters*, 40(20), 5492–5496.

Thomas, J. A., Welch, J. J., Lanfear, R., & Bromham, L. (2010). A generation time effect on the rate of molecular evolution in invertebrates. *Molecular Biology and Evolution*, 27(5), 1173–1180.

Thomas, R. D. K., & Olson, E. C. (Eds.) (1980). *A cold look at the warm-blooded dinosaurs* (Vol. 28). Washington, D.C.: Westview Press, for the American Association for the Advancement of Science.

Thomas, T. M., & Scopes, R. K. (1998). The effects of temperature on the kinetics and stability of mesophilic and thermophilic 3-phosphoglycerate kinases. *Biochemical Journal*, 330(3), 1087–1095.

Thomashow, M. F. (2001). So what's new in the field of plant cold acclimation? Lots! *Plant Physiology*, 125(1), 89–93.

Thompson, D. A. W. (1917). *On growth and form*. Cambridge: Cambridge University Press.

Thompson, D. A. W. (1961). *On growth and form (abridged edition, edited by John Tyler Bonner)*. Cambridge: Cambridge University Press.

Thompson, D. W. J., & Solomon, S. (2002). Interpretation of recent southern hemisphere climate change. *Science*, 296, 895–899.

Thompson, M. L., Mzilikazi, N., Bennett, N. C., & McKechnie, A. E. (2015). Solar radiation during rewarming from torpor in elephant shrews: supplementation or substitution of endogenous heat production? *PLoS One*, 10(4), e0120442.

Thomson, W. (1848). On an absolute thermometric scale founded on Carnot's theory of the motive power of heat, and calculated from Regnault's observations. *Proceedings of the Cambridge Philosophical Society*, 1(1), 66–71.

Thomson, W. (1849). An account of Carnot's theory of the motive power of heat: with numerical results deduced from Regnault's experiments on steam. *Transactions of the Royal Society of Edinburgh*, 16, 541–574.

Thomson, W. (1851). On a method of discovering experimentally the relation between the mechanical work spent, and the heat produced by the compression of a gaseous fluid. *Transactions of the Royal Society of Edinburgh*, 20(2), 289–298.

Thomson, W. (1851). On the dynamical theory of heat, with numerical results deduced from Mr Joule's equivalent of a thermal unit, and M. Regnault's observations on steam. *Transactions of the Royal Society of Edinburgh*, 20(2), 261–288.

Thornley, J. H. M. (2011). Plant growth and respiration revisited: maintenance respiration defined—it is an emergent property of, not a separate process within, the system—and why the respiration:photosynthesis ratio is conservative. *Annals of Botany*, 108(7), 1365–1380.

Thorson, G. (1957). Bottom communities (sublittoral or shallow shelf). In J. W. Hedgpeth (Ed.), *Treatise on marine ecology and paleoecology* (pp. 461–534). New York: Geological Society of America.

Thuesen, E. V., & Childress, J. J. (1994). Oxygen consumption rates and metabolic enzyme activities of oceanic California medusae in relation to body size and habitat depth. *Biological Bulletin*, 187(1 (August 1994)), 84–98.

Ticehurst, C. B. (1938). *A systematic review of the genus Phylloscopus (willow-warblers or leaf-warblers)*. London: British Museum (Natural History).

Timmermann, A., Sachs, J., & Timm, O. E. (2014). Assessing divergent SST behavior during the last 21 ka derived from alkenones and *G. ruber* Mg/Ca in the equatorial Pacific. *Paleoceanography*, 29(8), 680–696.

Tjørve, E. (2003). Shapes and functions of species–area curves: a review of possible models. *Journal of Biogeography*, 30(6), 827–835.

Todgham, A. E., Hoagland, E. A., & Hofmann, G. E. (2007). Is cold the new hot? Elevated ubiquitin-conjugated protein levels in tissues of Antarctic fish as evidence for cold-denaturation of proteins *in vivo*. *Journal of Comparative Physiology B Biochemical Systematic and Environmental Physiology*, 177(8), 857–866.

Tøien, Ø., Blake, J., Edgar, D. M., Grahn, D. A., Heller, H. C., & Barnes, B. M. (2011). Hibernation in black bears: independence of metabolic suppression from body temperature. *Science*, 331(6019), 906–909.

Tolstoy, I. (1981). *James Clerk Maxwell: a biography*. Edinburgh: Canongate.

Tomashow, M. F. (2001). So what's new in the field of plant cold acclimation? Lots! *Plant Physiology, 125*(1), 89–93.

Tompa, P., & Kovacs, D. (2010). Intrinsically disordered chaperones in plants and animals. *Biochemistry and Cell Biology, 88*(2), 167–174.

Toulmin, S., & Goodfield, J. (1962). *The architecture of matter*. London: Hutchinson.

Trefil, J., Morowitz, H. J., & Smith, E. (2009). The origin of life: a case is made for the descent of electrons. *American Scientist, 97*(3), 206–213.

Trenberth, K. E., Fasullo, J. T., & Kiehl, J. (2009). Earth's global energy budget. *Bulletin of the American Meteorological Society, 311*(3), 311–323.

Trent, J. D., Chastain, R. A., & Yayanos, A. A. (1984). Possible artefactual basis for apparent bacterial growth at 250 °C. *Nature, 307*(5953), 737–740.

Truesdell, C. (1979). *The tragicomical history of thermodynamics, 1822–1854*. New York: Springer-Verlag.

Tryjanowski, P., Kuzniak, S., & Sparks, T. H. (2005). What affects the magnitude of change in first arrival dates of migrant birds? *Journal of Ornithology, 146*(3), 200–205.

Tucker, V. A. (1966). Diurnal torpor and its relation to food consumption and weight changes in the California pocket mouse *Perognathus californicus*. *Ecology, 47*(2), 245–252.

Tuoriniemi, J. T., & Knuuttila, T. A. (2000). Nuclear cooling and spin properties of rhodium down to picokelvin temperatures. *Physica B, 280*(1–4), 474–478.

Turbill, C., Bieber, C., & Ruf, T. (2011). Hibernation is associated with increased survival and the evolution of slow life histories among mammals. *Proceedings of the Royal Society of London, Series B, 278*(1723), 3355–3363.

Turner, J., Anderson, P., Lachlan-Cope, T., et al. (2009a). Record low surface air temperature at Vostok station, Antarctica. *Journal of Geophysical Research, Atmospheres, 114*, D24102.

Turner, J., Bindschadler, R., Convey, P., et al. (Eds.) (2009b). *Antarctic climate change and the environment*. Cambridge: Scientific Committee on Antarctic Research.

Turner, J., Colwell, S. R., Marshall, G. J., et al. (2005). Antarctic climate change during the last 50 years. *International Journal of Climatology, 25*(3), 279–294.

Turner, J., Lu, H., White, I., et al. (2016). Absence of 21st century warming on Antarctic Peninsula consistent with natural variability. *Nature, 535*(7612), 411–415.

Turner, J., Maksym, T., Phillips, T., Marshall, G. J., & Meredith, M. P. (2012). The impact of changes in sea ice advance on the large winter warming on the western Antarctic Peninsula. *International Journal of Climatology, 33*(4), 852–861.

Turner, J. M., Warnecke, L., Körtner, G., & Geiser, F. (2012). Opportunistic hibernation by a free-ranging marsupial. *Journal of Zoology, 286*(4), 277–284.

Turner, J. R. G., Gatehouse, C. M., & Corey, C. A. (1987). Does solar energy control organic diversity? Butterflies, moths and the British climate. *Oikos, 48*(2), 195–205.

Turner, J. R. G., Lennon, J. J., & Lawrenson, J. A. (1988). British bird species and the energy theory. *Nature, 335*(6190), 539–541.

Turner, J. S. (1988). Body size and thermal energetics: how should thermal conductance scale? *Journal of Thermal Biology, 13*(3), 103–117.

Twente, J. W., Twente, J., & Moy, R. M. (1977). Regulation of arousal from hibernation by temperature in three species of *Citellus*. *Journal of Applied Physiology, 42*(2), 191–195.

Tyndall, J. (1861a). The Bakerian Lecture: On the absorption and radiation of heat by gases and vapours, and on the physical connexion of radiation, absorption, and conduction. *Philosophical Transactions of the Royal Society of London, 151*(1861), 1–36.

Tyndall, J. (1861b). On the absorption and radiation of heat by gases and vapours, and on the physical connexion of radiation, absorption, and conduction—The Bakerian Lecture. *The London, Edinburgh, and Dublin Philosophical Magazine and Journal of Science, Series 4, 22*(1861), 169–194, 273–285.

Tzedakis, P. C., Emerson, B. C., & Hewitt, G. M. (2013). Cryptic or mystic? Glacial tree refugia in northern Europe. *Trends in Ecology & Evolution, 28*(12), 696–704.

Ugland, K. I., Gray, J. S., & Ellingsen, K. E. (2003). The species-accumulation curve and estimation of species richness. *Journal of Animal Ecology, 72*(5), 888–897.

Ultsch, G. R. (1989). Ecology and physiology of hibernation and overwintering among freshwater fishes, turtles, and snakes. *Biological Reviews, 64*(4), 435–515.

Urey, H. C. (1947). The thermodynamic properties of isotopic substances. *Journal of the Chemical Society, 1947*, 562–581.

Urrutia, M. E., Duman, J. G., & Knight, C. A. (1992). Plant thermal hysteresis proteins. *Biochimica et Biophysica Acta, 1121*(1–2), 199–206.

Ursin, E. (1967). A mathematical model of some aspects of fish growth, respiration and mortality. *Journal of the Fisheries Research Board of Canada, 24*(11), 2355–2453.

Ursin, E. (1979). Principles of growth in fishes. In P. J. Miller (Ed.), *Symposia of the Zoological Society of London, 44: Fish phenology: anabolic adaptiveness in teleosts (The proceedings of a symposium held at the Zoological Society of London on 6 and 7 April 1978)* (pp. 63–87). London: Academic Press.

Vähätalo, A. V., Rainio, K., Lehikoinen, A., & Lehikoinen, E. (2004). Spring arrival of birds depends on the North Atlantic Oscillation. *Journal of Avian Biology, 35*(3), 210–216.

Vahl, O. (1984). The relationship between specific dynamic action (SDA) and growth in the common starfish *Asterias rubens*. *Oecologia, 61*(1), 122–125.

Valdovinos, C., Navarrete, S. A., & Marquet, P. A. (2003). Mollusk species diversity in the Southeastern Pacific: why are there more species towards the pole? *Ecography*, 26, 139–144.

Valentine, J. W. (1961). Paleoecological molluscan biogeography of the Californian Pleistocene. *University of California Publications in Geological Science*, 34, 309–442.

Valentine, J. W. (1968). Climatic regulation of species diversification and extinction. *Geological Society of America Bulletin*, 79(2), 273–275.

Valentine, J. W. (1989). How good was the fossil record: clues from the Californian Pleistocene. *Paleobiology*, 15(2), 83–94.

Valentine, J. W., Foin, T. C., & Peart, D. (1978). A provincial model of Phanerozoic marine diversity. *Paleobiology*, 4(1), 55–66.

Valentine, J. W., & Jablonski, D. (1991). Biotic effects of sea level change: the Pleistocene change. *Journal of Geophysical Research*, 96(B4), 6873–6878.

van Breukelen, F., & Martin, S. L. (2015). The hibernation continuum: physiological and molecular aspects of metabolic plasticity in mammals. *Physiology & Behavior*, 30(4), 273–281.

Van Buskirk, H. A., & Thomashow, M. F. (2006). *Arabidopsis* transcription factors regulating cold acclimation. *Physiologia Plantarum*, 126(1), 72–80.

van de Ven, T. M. F. N., Mzilikazi, N., & McKechnie, A. E. (2013). Seasonal metabolic variation in two populations of an Afrotropical euplectid bird. *Physiological and Biochemical Zoology*, 86(1), 19–26.

van der Have, T. M., & de Jong, G. (1996). Adult size in ectotherms: temperature effects on growth and differentiation. *Journal of Theoretical Biology*, 183(3), 329–340.

van der Meer, J. (2006). Metabolic theories in ecology. *Trends in Ecology & Evolution*, 21(3), 136–140.

van der Meer, J., Kink, C., Kearney, M. R., & Wijsman, J. W. M. (2014). 35 years of DEB research. *Journal of Sea Research*, 94(S1), 1–4.

Van Dover, C. L. (2000). *The ecology of deep-sea hydrothermal vents*. Princeton, New Jersey: Princeton University Press.

Van Ness, H. C. (2003). H is for Enthalpy. *Journal of Chemical Education*, 80(5), 486.

Van Soest, R. W. M., Boury-Esnault, N., Vacelet, J., et al. (2012). Global diversity of sponges (Porifera). *PLoS One*, 7(4), e35105.

Vance, T. D. R., Olijve, L. L. C., Campbell, R. L., Voets, I. K., Davies, P. L., & Guo, S. (2014). Ca^{2+}-stabilized adhesin helps an Antarctic bacterium reach out and bind ice. *Bioscience Reports*, 34(4), e00121.

Vanlerberghe, G. C. (2013). Alternative oxidase: a mitochondrial respiratory pathway to maintain metabolic and signaling homeostasis during abiotic and biotic stress in plants. *International Journal of Molecular Science*, 14(4), 6805–6847.

Vanni, M. J., & McIntyre, P. B. (2016). Predicting nutrient excretion of aquatic animals with metabolic ecology and ecological stoichiometry: a global synthesis. *Ecology*, 97(12), 3460–3471.

van't Hoff, J. H. (1884). *Études de dynamique chimique*. Amsterdam: Frederik Muller.

van't Hoff, J. H. (1896). *Studies of chemical dynamics (English translation by Thomas Ewan)*. London: Williams & Norgate.

van't Hoff, J. H. (1898). *Lectures on theoretical and physical chemistry, Part 1, Chemical dynamics*. London: Edward Arnold.

Vargaftik, N. B., Volkov, B. N., & Voljak, L. D. (1983). International tables of the surface tension of water. *Journal of Physical and Chemical Reference Data*, 12(3), 817–820.

Vaughan, D. G., & Doake, C. S. M. (1996). Recent atmospheric warming and retreat of ice shelves on the Antarctic Peninsula. *Nature*, 379, 328–331.

Vaughan, D. G., Marshall, G. J., Connolley, W. M., et al. (2003). Recent rapid regional climate warming on the Antarctic Peninsula. *Climatic Change*, 60(3), 243–274.

Velleman, P. F., & Wilkinson, L. (1993). Nominal, ordinal, interval, and ratio typologies are misleading. *American Statistician*, 47(1), 65–72.

Verberk, W. C. E. P., & Bilton, D. T. (2011). Can oxygen set thermal limits in an insect and drive gigantism? *PLoS One*, 6(7), e22610.

Verberk, W. C. E. P., Bilton, D. T., Calosi, P., & Spicer, J. I. (2011). Oxygen supply in aquatic ectotherms: partial pressure and solubility together explain biodiversity and size patterns. *Ecology*, 92(8), 1565–1572.

Vercesi, A. E., Borecký, J., Maia, I. d. G., Arruda, P., Cuccovia, I. M., & Chaimovich, H. (2006). Plant uncoupling mitochondrial proteins. *Annual Review of Plant Biology*, 57(383–404).

Vermeij, G. J. (1978). *Biogeography and adaptation: patterns of marine life*. Cambridge, Massachusetts: Harvard University Press.

Vermeij, G. J. (1991). Anatomy of an invasion: the trans-Arctic interchange. *Paleobiology*, 17(3), 281–307.

Vetter, R. A. H. (1995). Ecophysiological studies on citrate-synthase: (I) enzyme regulation of selected crustaceans with regard to temperature adaptation. *Journal of Comparative Physiology B*, 165(1), 46–55.

Vézina, F., Jalvingh, K. M., Dekinga, A., & Piersma, T. (2006). Acclimation to different thermal conditions in a northerly wintering shorebird is driven by body mass-related changes in organ size. *Journal of Experimental Biology*, 209(16), 3141–3154.

Visser, M. E., Holleman, L. J. M., & Gienapp, P. (2006). Shifts in caterpillar biomass phenology due to climate change and its impact on the breeding biology of an insectivorous bird. *Oecologia, 147*(1), 164–172.

Visser, M. E., van Noordwijk, A. J., Tinbergen, J. M., & Lessells, C. M. (1998). Warmer springs lead to mistimed reproduction in great tits (*Parus major*). *Proceedings of the Royal Society of London, Series B, 265*(1408), 1867–1870.

Vogel, S. (1981). *Life in moving fluids: the physical biology of flow*. Princeton, New Jersey: Princeton University Press.

Vogel, S. (1988). *Life's devices: the physical world of animals and plants*. Princeton, New Jersey: Princeton University Press.

Vogel, S. (2012). *The life of a leaf*. Chicago: Chicago University Press.

Voituron, Y., Barre, H., Ramlov, H., & Douady, C. J. (2009). Freeze tolerance evolution among anurans: frequency and timing of appearance. *Cryobiology, 58*(3), 241–247.

Voituron, Y., Mouquet, N., de Mazancourt, C., & Clobert, J. (2002). To freeze or not to freeze? An evolutionary perspective on the cold-hardiness strategies of overwintering ectotherms. *American Naturalist, 160*(2), 255–270.

von Bertalanffy, L. (1938). A quantitative theory of organic growth (Inquiries on growth laws. II). *Human Biology, 10* (2), 181–213.

von Bertalanffy, L. (1951). Metabolic types and growth types. *American Naturalist, 85*(821), 111–117.

von Bertalanffy, L. (1957). Quantitative laws in metabolism and growth. *Quarterly Review of Biology, 32*(3), 217–231.

von Humboldt, A. (1808). *Ansichten der Natur: mit wissenschaftlichen Erläuterungen*. Tübingen, Germany: J. G. Cotta.

von Neumann, J. (1951). The general and logical theory of automata. In L. A. Jeffress (Ed.), *Cerebral mechanisms in behavior; the Hixon symposium* (pp. 1–41). New York: John Wiley & Sons.

von Neumann, J. (1966). *Theory of self-reproducing automata. Edited and completed by Arthur W. Burks*. Urbana: University of Illinois Press.

von Stockar, U., Maskow, T., Liu, J., Marison, I. W., & Patiño, R. (2006). Thermodynamics of microbial growth and metabolism: an analysis of the current situation. *Journal of Biotechnology, 121*(4), 517–533.

Vonnegut, K. (1963). *Cat's cradle*. New York: Holt, Reinhart & Winston.

Vriens, J., Nilius, B., & Voets, T. (2014). Peripheral sensation in mammals. *Nature Reviews Neuroscience, 15*(9), 573–589.

Vuarin, P., Dammhahn, M., & Henry, P.-Y. (2013). Individual flexibility in energy saving: body size and condition constrain torpor use. *Functional Ecology, 27*(3), 793–799.

Waide, R. B., Willig, M. R., Steiner, C. F., et al. (1999). The relationship between productivity and species richness. *Annual Review of Ecology and Systematics, 30*, 257–300.

Wallace, A. R. (1876). *The geographical distribution of animals: with a study of the relations of living and extinct faunas as elucidating the past changes of the Earth's surface*. New York: Harper & Brothers.

Walli, A., Teo, S. L. H., Boustany, A., et al. (2010). Seasonal movements, aggregations and diving behavior of Atlantic bluefin tuna (*Thunnus thynnus*) revealed with archival tags. *PLoS One, 4*(7), e6151.

Walsberg, G. E., & Wolf, B. O. (1996). An appraisal of operative temperature mounts as tools for studies of ecological energetics. *Physiological Zoology, 69*(3), 658–681.

Walter, D. E., & Behan-Pelletier, V. (1999). Mites in forest canopies: filling the size distribution shortfall? *Annual Review of Entomology, 44*, 1–19.

Walters, K. R., Serianni, A. S., Voituron, Y., Sformo, T., Barnes, B. M., & Duman, J. G. (2011). A thermal hysteresis-producing xylomannan glycolipid antifreeze associated with cold tolerance is found in diverse taxa. *Journal of Comparative Physiology, B, 181*(5), 631–640.

Walther, G.-R., Post, E., Convey, P., et al. (2002). Ecological responses to recent climate change. *Nature, 416*(6879), 389–395.

Wang, L. C. H. (1979). Time patterns and metabolic rates of natural torpor in the Richardson's ground squirrel. *Canadian Journal of Zoology, 57*(1), 149–155.

Wang, L. C. H., & Wolowyk, M. W. (1988). Torpor in mammals and birds. *Canadian Journal of Zoology, 66*(1), 133–137.

Wang, T., Zaar, M., Arvedsen, S., Vedel-Smith, C., & Overgaard, J. (2002). Effects of temperature on the metabolic response to feeding in *Python molurus*. *Comparative Biochemistry and Physiology Part A, 133*(3), 519–527.

Ward, D., & Seeley, M. K. (1966). Competition and habitat selection in Namib Desert tenebrionid beetles. *Evolutionary Ecology, 10*(4), 341–359.

Wasserthal, L. T. (1975). The rôle of butterfly wings in regulation of body temperature. *Journal of Insect Physiology, 21*(12), 1921–1930.

Wassmann, P., Slagstad, D., & Ellingsen, I. (2010). Primary production and climatic variability in the European sector of the Arctic Ocean prior to 2007: preliminary results. *Polar Biology, 33*(12), 1641–1650.

Wasylyk, J. M., Tice, A. R., & Baust, J. G. (1988). Partial glass formation: a novel mechanism of insect cryoprotection. *Cryobiology, 25*(5), 451–458.

Waters, J. S., & Harrison, J. F. (2012). Insect metabolic rates. In R. M. Sibly, J. H. Brown, & A. Kodric-Brown (Eds.), *Metabolic ecology: a scaling approach* (pp. 198–211). Oxford: Wiley-Blackwell.

Watson, H. C. (1835). *Remarks on the geographical distribution of British plants, chiefly in connection with latitude, elevation, and climate*. London: Longman, Rees, Orme, Brown, Green, and Longman.

Watson, J. D., & Crick, F. H. C. (1953). A structure for deoxyribose nucleic acid. *Nature, 171*(4356), 737–738.

Watson, S.-A., Peck, L. S., Tyler, P. A., et al. (2012). Marine invertebrate skeleton size varies with latitude, temperature and carbonate saturation: implications for global change and ocean acidification. *Global Change Biology*, *18*(10), 3026–3038.

Watt, C., Mitchell, S., & Salewski, V. (2009). Bergmann's rule: a concept cluster? *Oikos*, *119*(1), 89–100.

Watt, W. B. (1968). Adaptive significance of pigment polymorphisms in *Colias* butterflies. I. Variation of melanin pigment in relation to thermoregulation. *Evolution*, *22*(3), 437–458.

Watt, W. B. (1977). Adaptation at specific loci. I. Natural selection on phosphoglucose isomerase of *Colias* butterflies: biochemical and population aspects. *Genetics*, *87*(1), 177–194.

Watt, W. B. (1983). Adaptation at specific loci. II. Demographic and biochemical elements in the maintenance of the *Colias* pgi polymorphism. *Genetics*, *103*(4), 691–724.

Watt, W. B., Cassin, R. C., & Swan, M. S. (1983). Adaptation at specific loci. III. Field behavior and survivorship differences among *Colias* pgi genotypes are predictable from *in vitro* biochemistry. *Genetics*, *103*, 725–729.

Weathers, W. W. (1970). Physiological thermoregulation in the lizard *Dipsosaurus dorsalis*. *Copeia*, *1970*(3), 549–557.

Webster, M. D., & Weathers, W. W. (1990). Heat produced as a by-product of foraging activity contributes to thermoregulation by verdins, *Auriparus flaviceps*. *Physiological Zoology*, *63*(4), 777–794.

Wegner, N. C., Snodgrass, O. E., Dewar, H., & Hyde, J. R. (2015). Whole-body endothermy in a mesopelagic fish, the opah, *Lampris guttatus*. *Science*, *348*(6236), 786–789.

Wehner, R., Marsh, A. C., & Wehner, S. (1992). Desert ants on a thermal tightrope. *Nature*, *357*, 586–587.

Weibel, E. R., Bacigalupe, L. D., Schmitt, B., & Hoppeler, H. (2004). Allometric scaling of maximal metabolic rate in mammals: muscle aerobic capacity as determinant factor. *Respiration Physiology & Neurobiology*, *140*(2), 115–132.

Weibel, E. R., & Hoppeler, H. (2005). Exercise-induced maximal metabolic rate scales with muscle aerobic capacity. *Journal of Experimental Biology*, *208*, 1635–1644.

Weiser, M. D., Enquist, B. J., Boyle, B., et al. (2007). Latitudinal patterns of range size and species richness of New World woody plants. *Global Ecology and Biogeography*, *16*(5), 679–688.

Wells, R. M. G. (1987). Respiration of Antarctic fish from McMurdo Sound. *Comparative Biochemistry and Physiology, Part A*, *88*(3), 417–424.

Werning, S., Irmis, R. B., Nesbitt, S. J., Smith, N. D., Turner, A. H., & Padian, K. (2012). Early evolution of elevated growth and metabolic rates in archosaurs. *Integrative and Comparative Biology*, *52 (Supplement 1)*(S1), E189.

West, G. B., & Brown, J. H. (2004). Life's universal scaling laws. *Physics Today*, *57 (September)*(9), 36–42.

West, G. B., & Brown, J. H. (2005). The origin of allometric scaling laws in biology from genomes to ecosystems: towards a quantitative unifying theory of biological structure and organization. *Journal of Experimental Biology*, *208*(9), 1575–1592.

West, G. B., Brown, J. H., & Enquist, B. J. (1997). A general model for the origin of allometric scaling laws in biology. *Science*, *276*, 122–126.

West, G. B., Brown, J. H., & Enquist, B. J. (1999a). The fourth dimension of life: fractal geometry and allometric scaling of organisms. *Science*, *284*, 1677–1679.

West, G. B., Brown, J. H., & Enquist, B. J. (1999b). A general model for the structure and allometry of plant vascular systems. *Nature*, *400*(6745), 664–667.

West, G. B., Brown, J. H., & Enquist, B. J. (2001). A general model for ontogenetic growth. *Nature*, *413*, 628–631.

West, G. B., Woodruff, W. H., & Brown, J. H. (2002). Allometric scaling of metabolic rate from molecules and mitochondria to cells and mammals. *Proceedings of the National Academy of Sciences of the United States of America*, *99*(Supplement 1), 2473–2478.

West, G. C. (1965). Shivering and heat production in wild birds. *Physiological Zoölogy*, *38*(2), 111–120.

West, G. C. (1972). Seasonal differences in resting metabolic rate of Alaskan ptarmigan. *Comparative Biochemistry and Physiology, Part A*, *42*(4), 867–876.

Wharton, D. A. (2002). *Life at the limits: organisms in extreme environments*. Cambridge: Cambridge University Press.

Wharton, D. A., & Ferns, D. J. (1995). Survival of intracellular freezing by the Antarctic nematode *Panagrolaimus davidi*. *Journal of Experimental Biology*, *198*, 1381–1387.

Wheat, C. W., Watt, W. B., Pollock, D. D., & Schulte, P. M. (2006). From DNA to fitness differences: sequences and structures of adaptive variants of *Colias* phosphoglucose isomerase (PGI). *Molecular Biology and Evolution*, *23*(3), 499–512.

Whipp, B. J., & Wasserman, K. (1969). Efficiency of muscular work. *Journal of Applied Physiology*, *26*(5), 644–648.

White, C. R., Alton, L. A., & Frappell, P. B. (2012). Metabolic cold adaptation in fishes occurs at the level of whole animal, mitochondria and enzyme. *Proceedings of the Royal Society of London, Series B*, *279*(1734), 1740–1747.

White, C. R., Phillips, N. F., & Seymour, R. S. (2006). The scaling and temperature dependence of vertebrate metabolism. *Biology Letters*, *2*(1), 125–127.

White, C. R., & Seymour, R. S. (2004). Does basal metabolic rate contain a useful signal? Mammalian BMR allometry and correlations with a selection of physiological, ecological, and life-history variables. *Physiological and Biochemical Zoology*, *77*(6), 929–941.

White, E. P., Xiao, X., Isaac, N. J. B., & Sibly, R. M. (2012). Methodological tools. In R. M. Sibly, J. H. Brown, & A. Kodric-Brown (Eds.), *Metabolic ecology: a scaling approach* (pp. 9–20). Oxford: Wiley-Blackwell.

White, G. (1789). *The natural history and antiquities of Selborne*. London: T. Bensley, for B. White & Son.

White, R. H. (1984). Hydrolytic stability of biomolecules at high temperartures and its implication for life at 250 °C. *Nature, 310*(5976), 430–432.

Whitfield, J. (2006). *In the beat of a heart: life, energy, and the unity of Nature*. Washington, D.C.: Joseph Henry Press.

Whitfield, M. C., Smit, B., McKechnie, A. E., & Wolf, B. O. (2015). Avian thermoregulation in the heat: scaling of heat tolerance and evaporative cooling capacity in three southern African arid-zone passerines. *Journal of Experimental Biology, 218*(11), 1705–1714.

Whittaker, R. H. (1960). Vegetation of the Siskiyou Mountains, Oregon and California. *Ecological Monographs, 30*(3), 279–338.

Whittaker, R. H. (1972). Evolution of measurements of species diversity. *Taxon, 21*, 213–251.

Whittaker, R. J., Willis, K. J., & Field, R. (2001). Scale and species richness: towards a general heirarchical theory of species diversity. *Journal of Biogeography, 28*(4), 453–470.

Wicken, J. S. (1986). Entropy and evolution: ground rules for discourse. *Systematic Zoology, 35*(1), 22–36.

Wicks, L. C., & Roberts, J. M. (2012). Benthic invertebrates in a high-CO_2 world. *Oceanography and Marine Biology: an Annual Review, 50*, 127–187.

Wiersma, P., Muñoz-Garcia, A., Walker, A., & Williams, J. B. (2007). Tropical birds have a slow pace of life. *Proceedings of the National Academy of Sciences of the United States of America, 104*(22), 9340–9345.

Wiersma, P., Nowak, B., & Williams, J. B. (2012). Small organ size contributes to the slow pace of life in tropical birds. *Journal of Experimental Biology, 215*(10), 1662–1669.

Wieser, W. (1985). A new look at energy conversion in ectothermic and endothermic animals. *Oecologia, 66*(4), 506–510.

Wieser, W. (1994). Cost of growth in cells and organisms: general rules and comparative aspects. *Biological Reviews, 69*(1), 1–33.

Wilf, P., & Labandeira, C. C. (1999). Response of plant–insect associations to Paleocene–Eocene warming. *Science, 284*(5423), 2153–2156.

Wilf, P., Labandeira, C. C., Johnson, K. R., Coley, P. D., & Cutter, A. D. (2001). Insect herbivory, plant defense, and early Cenozoic climate change. *Proceedings of the National Academy of Sciences of the United States of America, 98*(11), 6221–6226.

Wilford, J. N. (1985). *The riddle of the dinosaur*. New York: Knopf.

Williams, C. B. (1964). *Patterns in the balance of nature, and related problems in quantitative ecology*. New York: Academic Press.

Williams, C. T., Sheriff, M. J., Schmutz, J. A., et al. (2011). Data logging of body temperatures provides precise information on phenology of reproductive events in a free-living arctic hibernator. *Journal of Comparative Physiology, B, 181*(9), 1101–1109.

Williams, M. J., Ausubel, J., Poiner, I., et al. (2010). Making marine life count: a new baseline for policy. *PLoS Biology, 8*(10), e1000531.

Williams, P. H., & Gaston, K. J. (1994). Measuring more of biodiversity: can higher-taxon richness predict wholesale species richness? *Biological Conservation, 67*(3), 211–217.

Williams, P. H., Humphries, C. J., & Gaston, K. J. (1994). Centers of seed plant diversity: the family way. *Proceedings of the Royal Society of London, Series B, 256*(1345), 67–70.

Willig, M. R., & Bloch, C. P. (2006). Latitudinal gradients of species richness: a test of the geographic area hypothesis at two ecological scales. *Oikos, 112*(1), 163–173.

Willis, C. K. R. (2007). An energy-based body temperature threshold between torpor and normothermia for small mammals. *Physiological and Biochemical Zoology, 80*(6), 643–651.

Willis, K. J., Rudner, E., & Sümegi, P. (2000). The full-glacial forests of central and southeastern Europe. *Quaternary Research, 53*(2), 203–213.

Willis, K. J., & Whittaker, R. J. (2002). Species diversity: scale matters. *Science, 295*(5558), 1245–1248.

Wilson, A., &Bonaparte, C. L. (1831). *American ornithology: or the natural history of the birds of the United States*. Edinburgh: Constable.

Wilson, A.-L., Brown, M., & Downs, C. T. (2011). Seasonal variation in metabolic rate of a medium-sized frugivore, the Knysna Turaco (*Tauraco corythaix*). *Journal of Thermal Biology, 36*(3), 167–172.

Wilson, E. O.,&Peter, F. M. (Eds.) (1988). *Biodiversity*. Washington, D.C.: National Academy Press.

Wilson, J. M. (1987). Chilling injury in plants. In B. W. W. Grout & G. J. Morris (Eds.), *The effects of low temperatures on biological systems* (pp. 271–292). London: Edward Arnold.

Winberg, G. G.(1960). Rate of metabolism and food requirements of fishes. *Fisheries Research Board of Canada. Translation Series, 194*, 1–202.

Winterstein, H. (1954). Der Einfluß der Körpertemperatur auf das Säure-Basen-Gleichgewicht im Blut. *Naunyn-Schmiedebergs Archiv für experimentelle Pathologie und Pharmakologie, 223*(1), 1–18.

Wisniak, J. (2010). Daniel Berthelot. Part I. Contribution to thermodynamics. *Educación Química, 21*(2), 155–162.

Wisniewski, M. E., Fuller, M., Palta, J., Carter, J., & Arona, R. (2004). Ice nucleation, propagation, and deep supercooling in woody plants. *Journal of Crop Production, 10*(19–20), 5–16.

Wittmann, A. C., & Pörtner, H.-O. (2013). Sensitivities of extant animal taxa to ocean acidification. *Nature Climate Change, 3*(11), 995–1001.

Wohlschlag, D. E. (1960). Metabolism of an Antarctic fish and the phenomenon of cold adaptation. *Ecology*, *41*(2), 287–292.

Wohlschlag, D. E. (1964). Respiratory metabolism and ecological characteristics of some fishes in McMurdo Sound, Antarctica. In M. O. Lee (Ed.), *Biology of Antarctic Seas I, Antarctic Research Series 1* (Vol. 1, pp. 33–62). Washington, D.C.: American Geophysical Union.

Wolfe, J. A. (1979). Temperature parameters of humid to mesic forests of eastern Asia and relation to forests of other regions of the Northern Hemisphere and Australasia. USGS Professional Paper 1106, Washington D.C.: US Department of the Interior.

Wolfe, J. A. (1995). Paleoclimate estimation from Tertiary leaf assemblages. *Annual Review of Earth and Planetary Sciences*, *23*, 119–142.

Wolverton, S., Huston, M. A., Kennedy, J. H., Cagle, K., & Cornelius, J. D. (2009). Conformation to Bergmann's rule in white-tailed deer can be explained by food availability. *American Midland Naturalist*, *162*(2), 403–417.

Womack, A. M., Bohannan, B. J. M., & Green, J. L. (2010). Biodiversity and biogeography of the atmosphere. *Philosophical Transactions of the Royal Society of London, Series B*, *365*(1558), 3645–3653.

Wood, R. (1999). *Reef evolution*. Oxford: Oxford University Press.

Woodd-Walker, R. S., Ward, P., & Clarke, A. (2002). Large scale patterns in diversity and community structure of surface water copepods from the Atlantic Ocean. *Marine Ecology Progress Series*, *236*, 189–203.

Woods, C. P., & Brigham, R. M. (2004). The avian enigma: 'hibernation' by common poorwills (*Phalaenoptilus nuttallii*). In B. M. Barnes & H. V. Carey (Eds.), *Life in the cold: evolution, mechanisms, adaptation, and application. Twelfth International Hibernation Symposium* (pp. 231–240). Fairbanks, Alaska: Institute of Arctic Biology.

Woodward, F. I. (1987). *Climate and plant distribution*. Cambridge: Cambridge University Press.

Wootton, R. J. (1990). *Ecology of teleost fishes*. London: Chapman & Hall.

Worland, M. R., & Block, W. (1999). Ice-nucleating bacteria from the guts of two sub-Antarctic beetles, *Hydromedion sparsutum* and *Perimylops antarcticus* (Perimylopidae). *Cryobiology*, *38*(1), 60–67.

Worland, M. R., Wharton, D. A., & Byars, S. G. (2004). Intracellular freezing and survival in the freeze tolerant alpine cockroach *Celatoblatta quinquemaculata*. *Journal of Insect Physiology*, *50*(2–3), 225–232.

Wowk, B. (2010). Thermodynamic aspects of vitrification. *Cryobiology*, *60*(1), 11–22.

Wrigglesworth, J. (1997). *Energy and life*. London: Taylor & Francis.

Wright, D. H. (1983). Species–energy theory: an extension of species–area theory. *Oikos*, *41*(3), 496–506.

Wulf, A. (2015). *The invention of nature: the adventures of Alexander von Humboldt, the lost hero of science*. London: John Murray.

Wyatt, G. R., & Kalf, G. F. (1957). The chemistry of insect hemolymph. *Journal of General Physiology*, *40*(6), 833–847.

Xiang, Q.-Y., Zhang, W. H., Ricklefs, R. E., et al. (2004). Regional differences in rates of plant speciation and molecular evolution: a comparison between eastern Asia and eastern North America. *Evolution*, *58*(10), 2175–2184.

Xiao, X., White, E. P., Hooten, M. B., & Durham, S. L. (2011). On the use of log-transformation vs. nonlinear regression for analyzing biological power laws. *Ecology*, *92*(10), 1887–1894.

Xu, H., Griffith, M., Patten, C. L., & Glick, B. R. (1998). Isolation and characterization of an antifreeze protein with ice nucleation activity from the plant growth promoting rhizobacterium *Pseudomonas putida* GR12-12. *Canadian Journal of Microbiology*, *44*(1), 64–73.

Yamashita, Y., Nakamura, N., Omiya, K., Nishikawa, J., Kawahara, H., & Obata, H. (2002). Identication of an antifreeze lipoprotein from *Moraxella* sp. of Antarctic origin. *Bioscience, Biotechnology, and Biochemistry*, *66*(2), 239–247.

Yancey, P. H. (2005). Organic osmolytes as compatible, metabolic and counteracting cytoprotectants in high osmolarity and other stresses. *Journal of Experimental Biology*, *208*(15), 2819–2830.

Yancey, P. H., & Siebenaller, J. F. (2015). Co-evolution of proteins and solutions: protein adaptation versus cytoprotective micromolecules and their roles in marine organisms. *Journal of Experimental Biology*, *218*(12), 1880–1896.

Yancey, P. H., & Somero, G. N. (1978). Temperature-dependence of intracellular pH: its role in conservation of pyruvate apparent K_m values of vertebrate lactate dehydrogenases. *Journal of Comparative Physiology*, *125*(2), 129–134.

Yesson, C., Clark, M. R., Taylor, M. L., & Rogers, A. D. (2011). The global distribution of seamounts based on 30 arc seconds bathymetry data. *Deep-Sea Research, Part I: Oceanographic Research Papers*, *58*(4), 442–453.

Yvon-Durocher, G., Allen, A. P., Cellamare, M., et al. (2015). Five years of experimental warming increases the biodiversity and productivity of phytoplankton. *PLoS Biology*, *13*(12), article 1002324.

Zachariassen, K. E., & Hammel, H. T. (1976). Nucleating agents in the haemolymph of insects tolerant to freezing. *Nature*, *262*(5566), 285–287.

Zapata, F. A., Gaston, K. J., & Chown, S. L. (2003). Mid-domain models of species richness gradients: assumptions, methods and evidence. *Journal of Animal Ecology*, *72*(4), 677–690.

Zapata, F. A., Gaston, K. J., & Chown, S. L. (2005). The mid-domain effect revisited. *American Naturalist*, *166*(5), E144–E148.

Zar, J. H. (1969). The use of the allometric model for avian standard metabolism–body weight relationships. *Comparative Biochemistry and Physiology*, *29*(1), 227–234.

Zelt, J., Courter, J., Arab, A., Johnson, R., & Droege, S. (2012). Reviving a legacy citizen science project to illuminate shifts in bird phenology. *International Journal of Zoology*, *2012*, article 710710 (710716 pages).

Zerefos, C. S., Georgiannis, V. T., Balis, D., Zerefos, S. C., & Kazantzidis, A. (2007). Atmospheric effects of volcanic eruptions as seen by famous artists and depicted in their paintings. *Atmospheric Chemistry and Physics*, *7*(15), 4027–4042.

Zeuthen, E. (1947). Body size and metabolic rate in the animal kingdom. *Comptes rendus des travaux du Laboratoire Carsberg, Serie chémique*, *26*, 17–165.

Zeuthen, E. (1953). Oxygen uptake as related to body size in organisms. *Quarterly Review of Biology*, *28*(1), 1–12.

Zhang, F.-C., Kearns, S. L., Orr, P. J., et al. (2010). Fossilized melanosomes and the colour of Cretaceous dinosaurs and birds. *Nature*, *463*(7284), 1075–1078.

Zhao, L., Wang, R., Wu, Y., Wu, W., Zheng, W., & Liu, J. (2015). Daily variation in body mass and thermoregulation in male Hwamei (*Garrulax canorus*) at different seasons. *Avian Research*, *6*, article 4.

Zheng, W.-H., Liu, J.-S., & Swanson, D. L. (2014). Seasonal phenotypic flexibility of body mass, organ masses, and tissue oxidative capacity and their relationship to resting metabolic rate in Chinese Bulbuls. *Physiological and Biochemical Zoology*, *87*(3), 432–444.

Zhou, E. H., Trepat, X., Park, C. Y., et al. (2009). Universal behavior of the osmotically compressed cell and its analogy to the colloidal glass transition. *Proceedings of the National Academy of Sciences of the United States of America*, *106*(26), 10632–10637.

Zhu, W.-L., Jia, T., Lian, X., & Wang, Z.-K. (2008). Evaporative water loss and energy metabolic in two small mammals, voles (*Eothenomys miletus*) and mice (*Apodemus chevrieri*), in Hengduan mountains region. *Journal of Thermal Biology*, *33*(6), 324–331.

Ziman, J. (1978). *Reliable knowledge: an exploration of the grounds for belief in science*. Cambridge: Cambridge University Press.

Zimov, S. A., Chuprynin, V. I., Oreshko, A. P., Chapin, F. S., Reynolds, J. F., & Chapin, M. C. (1995). Steppe-tundra transition: a herbivore-driven biome shift at the end of the Pleistocene. *American Naturalist*, *146*(5), 765–794.

Zub, K., Szafrańska, P. A., Konarzewski, M., Redman, P., & Speakman, J. R. (2009). Trade-offs between activity and thermoregulation in a small carnivore, the least weasel *Mustela nivalis*. *Proceedings of the Royal Society of London, Series B*, *276*, 1921–1927.

Index

A
2-oxyglutarate dehydrogenase 172
absolute zero 11, 23–24, 34–36, 43
absorptance 65
abyssal gigantism 304
Acrocephalus choenobenus 253
activation energy 134–138, 182, 277
activity thermogenesis.
    See thermogensis, activity
activity-thermoregulatory
    heat substitution. *See*
    thermogenesis, substitution
Adalia bipunctata 66
Aegithalos concinnus 228
aerobic respiration 165
aerobic scope hypothesis 235–236
    modification of 236–237
aerobic scope 181
alanine 97, 114
Alces americana 300
Allen, Drew 339
allometry. *See* scaling
allozyme 141
Alosa sapidissima 294
alphastat hypothesis 148–149, 161n32
alternative oxidase 208
altitude 92, 303
Alvinella pompeiana 323
Amontons, Guillaume 19, 34–35
Amphibolorus muricatus 253
Amphibolorus nuchalis. See *Ctenophorus nuchalis*
Amphibolorus vitticeps. See *Pogona vitticeps*
Anarhichas lupus 115
Anaximander 100n2
Anaximenes 100n2
Anderson, Christina 347
anemone, sea 335
Anfinsen, Christopher 101n24
Anfinsen's dogma 97
Angilletta, Michael 157
Anolis carlinensis 266n30
Anolis cristatellus 199–200

Anolis gundlachi 200
ant 131–132
Antarctic Circumpolar Current 368–369
Antarctic Peninsula 367–369
Antarctica 310–311, 361, 362, 370
Antechinus 301
anthropogenic climate change.
    See global climate change
antifreeze 114–127
    bacterial 126
    classification of 126–127
    evolution of 116–117, 125
    glycoprotein (AFGP) 114–117, 290
    protein (AFP) 115–117, 124
Antilocapra americana 179, 194n42
Aphelenchoides parientus 323
apocrine sweat glands 233, 243n41
Apodemus chevrieri 229
Aptenodytes forsteri 321
Aquifex aeolicus 322
Aradus 203
aragonite 94–95, 101n19, 367
archaea 146, 192n11, 192n10, 192n10, 295, 322, 324, 330, 334
    membrane structure 152
Archaeopteryx 238
Archimedes 44, 268
Arctic Ocean 117
Arendt, Jeff 285
Arenicola marina 148
arginine 97, 148
Aristotle 265n6
Arrhenius equation 77, 154–155, 277–278, 290, 339–341
Arrhenius, Svante 135, 160n8, 359
arrow of time 8, 22
Artemia franciscana 302
Artemia salina 317
arthropod 118
    freezing intolerance 119–121.
        See also supercooling
    freezing tolerance 121–122
Arum 208

Aschoff, Jürgen 260
asparagine 97
assimilation 75
Atkinson, David 301
Atkinson, Timothy 375
atmosphere 354–359
ATP 170, 196, 286
ATP synthesis 167, 168, 169
ATPase 143

B
Baas Becking, Louis 334
bacteria 126, 146, 165, 278, 288, 322, 324, 33,334, 340
    growth 295
    nature of species in 330
Bakker, Robert (Bob) 239
balanced energy equation 56, 74, 297
Banks, Joseph 329
Barry, Jim 375
Bartels, Heinz 271
basking 200–201, 257
Beaugrand, Grégory 375
behavioural thermoregulation.
    See thermoregulation, behavioural
Benedict, Francis 270
Bennett, Al 235
Bergmann, Carl 215, 299
Bergmann's rule 299–301
    ectotherms 301
    endotherms 299–300
Bernoulli, Daniel 16
Berthelot, Daniel 135, 160n7
Berthelot equation 135, 278, 280
beta, thermodynamic.
    See thermodynamic beta
biodiversity 329–350
    definition 329. *See also* diversity
bird 214–240
    body temperature and diet 220–221
    metabolic scaling 270–281
bivalves 335

black beetle paradox 68
Black, Joseph 28n28
black-body 43, 62, 356, 357
  black-body radiation 62–65
Bloch, Christopher 333
Block, Bill 184
blubber 59, 64
body temperature 2, 56, 196–211, 214–240, 247–250
  control 218–219
  diet 220–221
  diurnal variability in 210–211, 236, 245, 260
  endotherms 214–240
  environmental temperature 219–220
  set-point 199, 219, 245, 255
Boechera stricta 380, 384
boiling point 84, 88
bolometer 45, 54n31
Boloria charicla 383
Boltzmann factor 13, 135, 277, 280, 297
Boltzmann, Ludwig 12–14, 20, 28n29, 44
Boltzmann's constant 30, 37, 43, 136, 277, 346
Bombus 334
boreal forest 348, 363 See also taiga
Boreogadus saida 115
Bose-Einstein condensate 35, 53n19
Boyle, Robert 19
Boyle's Law 19
Boyles, Justin 260–261
brachiopods 367
bradymetabolism 215
Briffa, Keith 375
brine channels 105–106
brinicle 106
Brock, Thomas 322, 337
Brody, Samuel 175
brown adipose tissue (BAT) 226, 232
Brown, James 5n5, 267, 272
Brownian motion 12
brumation. See hibernation
bryozoans 278, 301, 335
Bubo scandiacus 68
Bufo bufo 306n10
Bufo spinulosus 203, 212n21
bumblebees 207, 334
Buxton, Patrick 68

C

calcification 94–96, 101nn21–22
calcite 94–95, 101n19, 367
calcium carbonate 367
  apparent solubility product 94–95
  See also calcite, aragonite
Calidris canutus 228
California Pleistocene molluscs 372
Calliptamus coelesyriensis 66
calorific fluid 10
calorimeter 174
calorimetry, bomb 25
Calypte anna 245
Camelus dromedarius 210
camera, thermal imaging 45–46
camouflage 68
Carassius auratus 161n26, 162n40
Carassius auratus 183
carbon dioxide 93–94, 355
  atmospheric concentration 355, 363–364
carbonic acid 93–94
carbonic anhydrase 94
cardiovascular system 177, 272–273, 298
Caribbean Pleistocene extinction 372
Carnot, Nicolas Léonard Sadi 17–18, 25n2, 35–36
carotenoid pigments 166
Cataglyphis bombycina 324
Cavendish, Henry 16
cellular environment 98–99, 147–152
  See also cytosol
Celsius, Anders 34, 53n14
Central England temperature 370, 371
Cephus cinctus 119
Ceratoserolis trilobitoides 303
Cercartetus concinnus 249
Cercartetus nanus 251
Cercartetus nanus 70
Chaenocephalus aceratus 101n17
Chaenocephalus aceratus 201
Chaetodipus californicus 251
Champsocephalus aceratus 142
Channichthyidae 92–93
Chapelle, Gauthier 303
chaperone protein 97, 125, 151, 292
characteristic energy 13
Charles, Jacques 19, 34
Charles' Law 19
Cheirogaleus medius 246
chemical potential energy 56–57, 74–79, 287, 289, 343–344, 350
chemiosmotic hypothesis 192n16
chemosynthesis 166, 315, 352n51
chilling injury 107, 118
Chlamydomonas nivalis 107–108
Chlamydosaurus kingii 203
chlorophyll 166
Chrysiptera cyanea 143
Cincindelinae 207–208

Circellium bacchus 212n33
Circumpolar Deep Water 368
citrate synthase 171, 278
Clapeyron, Émile 19
Clausius equality 30
Clausius, Rudolf 17, 20, 28n29
climate change. See global climate change
climate envelope modelling 375–379
climate sceptics 370
cline 142, 299
  latitudinal cline in gene frequency 142
clumped isotope palaeothermometry. See palaeothermometry, isotope
Clupea harengus 115
Clusella-Trullas, Susanna 66
cnidarians 278
Colaptes 374
cold-blooded 214–215
cold-hardiness 118–119
Coleoptera 337
Colias hecla 383
Colias 66, 141–142, 161n16, 295
Colobanthus quitensis 124
colour 65–68
Columba livia 233
Columbina inca 234
Comité International des Poids et Mesures (CIPM) 37
compatible solute 111, 150–151
Comte, August 2
conductance, thermal 69–70, 81n28, 254
conductivity, electrical 42–43
conductivity, thermal 42, 57–60, 88
  tissue 58–59
  water 58–59
Conférence Général des Poids et Mesures (CGPM) 37, 53n21
Conopeum reticulum 302
Conover, David 294
conservation of energy 7–8, 74
constant relative alkalinity hypothesis 148–149
convection 60–62, 69
Convention on Biological Diversity 329, 350n4
cooling rate 107
  environmental 110
  experimental 107, 119
  See also rate of temperature change
Coope, Russell 375
copepods 340
copy number 290
Coriolis effect 315

Corkrey, Ross 295–296
cosmic background radiation 35, 53n17
cost of growth 75, 288–289
  definition 288
Coturnix coturnix 234
Coulomb's Law 85
countergradient variation 294–295
Cowles, Raymond 215
Crawford, Eugene 233
Cretaceous-Palaeogene mass extinction 221, 348–349
Cricetulus triton 252
Cricetus cricetus 252
Crompton, Alfred 234–235
crustaceans 335
cryobiology 109
cryoprotectant 112–113, 120–121, 124
Cryptotis parva 253
CTD (conductivity temperature depth probe) 47, 314
Ctenophorus nuchalis 224–225
Cucujus clavipes 122
Curvularia protuberata 323
Cyanidium caldarum 322, 324
cyanobacteria 101n18, 321
Cyclocephala colasi 209
Cyprinodon julimes 323
Cyprinodon macularius 201
Cyprinodon nevadensis 162n40
Cyprinodon pachycephalus 323
cytosol 124, 151, 182

D

daily energy expenditure (DEE). *See* metabolism, routine
Dallinger, William 322
Dallingera drysdali 322
Dalton, John 12
Dansgaard-Oeschger event 362
Darwin, Charles 5, 308, 337, 350n2, 378
Dasykaluta rosamondae 253
Davy, John 204, 212n24
Decolopoda australis 303
deep time 347–349
deep-sea 332, 341
Dehnel, August 301
Dehnel's phenomenon 300–301
dehydration 109, 111
dehydrin 125
Deinonychus 238–240, 243n44
Democritus 12
Dendroides canadensis 122
denitrification 164, 192n5
Deschampsia antarctica 124
Desmodus rotundus 202

development 285
DeVries, Art 114, 120
Diamesa 321
Dichanthelium lanuginosum 323, 326
Dictyostelium discoideum 83
diffusion 182
  and temperature 152–153
diffusivity, thermal 88
diglycerol phosphate 150
dimensions 76, 100n11, 268
Dimetrodon 237
Diplodocus 239
Dipsosaurus dorsalis 156
Dipsosaurus dorsalis 61
Discovery Investigations 350n2
dispersal 373, 379
dissipation of energy in metabolism 287, 297, 392
Dissostichus mawsoni 150, 290
diversity 329–350
  and area 330–332
  and energy 342–347
  and temperature 338–342. *See also* Metabolic Theory of Biodiversity
  deep-sea 332–333
  global diversity 332–333
  sampling issues 330–331
Dorylaimus thermae 323
doubly labelled water 180, 194n43
Drebbel, Cornelis Jacobszoon 32, 52n7
Dromaeosaurus 240
Dryas octopetala 385n15
Duman, Jack 120
Dynamic Energy Budget (DEB) 77–78, 296
dynamic viscosity 88, 90–91

E

eccrine sweat glands 233, 243n41
Eckman transport 315
ecosystem engineer 346
Edaphosaurus 237
Eddington, Arthur Stanley 26n9
effective temperature 355
Eights, James 303, 307n56
Einstein, Albert 12, 26nn9–10, 44
El Niño Southern Oscillation (ENSO) 368
Electra pilosa 283n47, 302
electrical conductivity. *See* conductivity, electrical
electron transport chain (ETC) 208, 260
electron 62, 163, 165–166, 167, 168, 342
  electron acceptor 163–166

electron donor 163–166
electronegativity 84–85
elementary reaction 131–132
Elephantulus myurus 251, 257
Elephantulus rupestris 246
Emberiza pusilla 228
Embiotoca jacksoni 114
embryonic development 204
endotherm 56, 123, 196
endothermy 2, 214–240
  benefits 214, 221, 234
  definition 214–215
  evolution 223, 234–240
  facultative endothermy 215
  fossil record 237–240
  heat generation 224–228
energy flow 1
energy 6–25
  conservation of. *See* conservation of energy
  equipartition of 31, 358
  free. *See* Gibbs energy *and* Helholtz energy
  Gibbs. *See* Gibbs energy
  Helmholtz. *See* Helmholtz energy
  internal. *See* internal energy
  kinetic. *See* kinetic energy
  latent. *See* latent heat
  potential. *See* potential energy
  zero-point 23, 55n38
Enquist, Brian 267, 272
enthalpy 15
entropy 17–24, 96, 103–104, 133, 144, 146, 287–288, 292, 316–317, 391–392
  and information 22–23
  conformational entropy 145
entropy-enthalpy compensation 144
enzyme
  enzyme-catalysed reactions 138–139
  molecular adaptation to temperature 142–147
epifauna 335
Eppley, Richard 295
equilibrium freezing point 88, 103, 320
  influence of solutes 104
equilibrium mixture 132
equilibrium model of enzyme activity 155, 162n46
equilibrium, thermal 58, 79n2, 140
equilibrium, thermodynamic 31, 79n2
equipartition of energy. *See* energy, equipartition
Erithacus rubecula 253

Erwinia herbicola 111
Escherichia coli 290
Escherichia coli 83
Euclid 268
Euplectes orix 227
eurythermy 242n32, 246, 250, 324
Evans, Meredith Gwynne 137, 160n9
evaporation 69, 73, 210, 232, 245
evapotranspiration 343–347
evolution 75, 158, 273
   viviparity 204
excretion 75, 273
extinction 372–373
   bolide impact 373
   mass extinction events 373
extremophile 338
Eyring, Henry 137, 160n9
Eyring-Polanyi equation 138, 143, 154–155

F

Fahrenheit, Daniel Gabriel 32–33, 52n8, 104
Falco rusticolus 68
Farmer, Colleen 235
fat store 252
feathers 64, 228, 238
Fenchel, Tom 334
Feynman, Richard 7, 25n5, 283n32
Ficedula hypoleuca 381–382
field metabolic rate (FMR).
   See metabolism, routine
filter feeding 91
Finlay, Bland 334
fire-seeking insects 202–203
first arrival date 382
First Law of Thermodynamics 7–8, 76
Fludd, Robert 32, 52n7
fluid-mosaic model 151
flux control coefficient 192n19
food caching 246, 252
Foote, Michael 349
Fourier equation 57–58
Fourier, Jean-Baptiste Joseph 57, 79n4, 359
fractal 272, 276
Freeze, Hudson 322
freeze/thaw 109–110
freezing resistance 113–127
   bacteria 126
   insects 121–122
   intertidal invertebrates 117–118
   plants 124–125
   teleost fish 113–117
   terrestrial vertebrates 123–124
   See also supercooling, thermal hysteresis

freezing 103–127
fructose 120, 173
functions, state. *See* state functions
fundamental niche 377
Fundulus heteroclitus 142, 147, 294
fur 64, 228, 238
Furnace Creek ranch, California 312

G

Gadus morhua 115
Galilei, Galileo 13, 32, 44, 268
galinstan 40
Garrulax canorum. See Leucodioptron canorum
gastropods 335, 336, 339
Gates, David 66
Gay-Lussac, Joseph Louis 19, 34
Gazella gutturosa 210–211
Geiser, Fritz 247, 249
genetic code 31
geometric similarity 267
geothermal heat 313–314, 322, 337
Geothermobacterium ferrireducens 322
Gibbs energy 24, 28n35, 96, 98, 101n24, 132–134, 138, 143–144, 155, 168, 170, 192n18, 227, 316, 317
   standard Gibbs energy change 133
Gibbs, Josiah Willard 28n35
gigantothermy 215
Gillooly, Jamie 267
Giroud, Silvain 258–259
glacial cycles 361–363
global climate change 2, 354–384
   organism response 371–384
global climate model (GCM) 360
global warming. *See* global climate change
glucose 124, 173
glucose-6-phosphate dehydrgenase (G6PD) 142
glutamate 97
glycerol 120, 124, 150
glycine betaine 150
glycine 146
glycolysis 168, 171, 172, 287
Glyptonotus antarcticus 150, 303
Grashof number 62
Grassle, Fred 332
Graves, John 140
gravity 9, 26n12, 56, 267
Great Oxygenation Event 165
greenhouse effect 356, 357
   and real greenhouses 358–359
   discovery of 359
   enhanced greenhouse effect 364
   feedbacks in the climate system 365

   molecular processes 356
   weir as analogy 359
Greenland 361, 362
growth 75
   and latitude 294. *See also* countergradient variation
   cost of. *See* cost of growth
   determinate and indeterminate 286
   energetics of 286–289
   marine zooplankton 295
   teleost fish 295
gular fluttering 233–234
Gulick, John 337

H

Hadley cell 360
Haldane, J.B.S. 281n3
Hamilton, William 68
Hasler, Johannis 241n13
heat capacity model of enzyme activity 155, 162n46
heat capacity. *See* thermal capacity
heat increment of feeding 75
heat 6–25
   heat capacity. *See* thermal capacity
   mechanical equivalent. *See* mechanical equivalent of heat
Heinrich event 362
Heinrich, Bernd 212n32, 243n36, 245, 265n1
Heisenberg, Werner 80n7
Helmholtz energy 24
Helmholtz, Hermann Ludwig Ferdinand von 28n35
Hemitripterus americanus 115
Hemmingsen, Axel 271
Henri, Victor 161n13
Henry, William 92
Henry's Law 92, 101n15
herbivory 221
   and body temperature 221
   evolution of 221
Hermann, Jakob 16
Hertz, Paul 199
Hess, Germain 7, 25n7
Hess's Law 7, 95
Heterocephalus glaber 197, 198
heterogeneous nucleation.
   See nucleation, heterogenous
heterothermy index 247–249
heterothermy 215, 245–249, 264
   regional heterthermy 215
   regional 246
   temporal 215, 246
Heusner, Fred 271
hexactinellids 335

hibernating gland. *See* brown adipose tissue
hibernation 2, 245–264
   evolutionary aspects 263–264
   birds 263
   ectotherms 263, 266n30
Hirst, Andrew 295
Hirundo verbena 202
histidine 97, 148
HMS *Challenger* 350n2
hoar frost 107
Holeton, George 190
Holocene 362, 369
holozooplankton 336
homeothermy 215
   inertial homeothermy 215, 238
homogeneous nucleation.
      See nucleation, homogeneous
homoiothermy. *See* homeothermy
homologous gene 141
Hooker, Joseph 337
Hoplocephalus stephensii 203, 212n21
Hou, Chen 297
huddling 230
Huey, Raymond 156
Huey, Raymond 196, 199, 203
Humboldt, Alexander von 329, 338, 350n2
hummingbirds 245
Hutchinson, Evelyn 342, 352n67, 377
Huxley, Julian 268, 272, 281n7
Huxley, Thomas Henry 242n36
hydration shell 98–99
hydrogen bond 85–86, 100n8, 116, 146
   temperature and 86–87
hydrophobic interaction 96–98, 293
   membranes 96
   protein folding 97–98
   temperature 97–98, 293
hydrothermal vent 166, 315, 322–323, 335
hyperthermophile 151–152, 322
hypothalamus 218–219, 231

I
ice ages 361–363
ice core 362, 364, 369
ice nucleus 103
ice 86, 104–107
   intracellular. *See* intracellular ice
   phases 104
   ice-binding proteins 126–127
icefish. *See* Channichthyidae
ice-nucleating proteins 111
ice-scour 335
Icterus 374
ideal gas 18–19

Iduna 374
imidazole 148
Industrial Revolution 6, 25n2
infrared detection by animals 202
infrared problem 44
infrared radiation 65, 80n11, 312
insects 175, 279
   discontinuous gas exchange 175–177
   holometabolous 285, 305n2
   See also arthropod
insulation 59, 64, 228–230, 240
interbout arousal (IBA) 249
interglacial 362
intermediary metabolism 392
internal energy 14–15, 43, 56, 137
International Biological Programme (IBP) 74
International Temperature Scale (ITS) 39
intertidal 117–118, 140, 199
intracellular ice 108, 110–111
Iridomyrmex humilis 132
island biogeography theory 342
isocitrate dehydrogenase 171, 172
isoleucine 97
isotope fractionation 47–49
isotope palaeothermometry 238
isotope turnover. *See* doubly labelled water
isozyme 141
Isurus paucus 204

J
Jaeger, Edmund 264
Jetz, Walter 347
Joule, James 11
Junco hyemalis 228
Junco 374

K
Kanwisher, John 118
Keeling curve 363
Keeling, Charles 363
Kelvin, Lord 35–36, 53n20
Kemp, Tom 237
Kepler, Johannes 13
Kessel, Brina 231
kinetic energy 8–10, 43, 56, 358
   kinetic energy, translational 31
kinetic theory of gases 11–13, 134
kinetics 22, 34, 391
Kingsolver, Joel 156
Kleiber, Max 270
Kleiber's Law 270
Knutson, Roger 208
Kooijman, Sebastiaan 76
Koteja, Paweł 235

Krebs cycle. *See* tricarboxylic acid cycle
Krogh principle 195n47
Krogh, August 183, 189, 194n28
Krogh, August 269

L
lactate dehydrogenase (LDH) 140–142
lady beetle 66–67
Lagopus leucura 229
Lagopus muta 68
Lamarck, Jean-Baptiste de 208, 213n36
Lamare-Picquot, Christophe-Augustin 205, 212n28
Lamb, Horace 80n7
Lamna nasus 204
Lampris guttatus 204
Lampris guttatus 216
Land-Ocean Temperature Index (LOTI) 364–365, 386n22
Langley, Samuel Pierpont 45, 54n31
Lanthier, Clement 261
Laplace, Pierre-Simon 174, 191, 193n24
lapse rate 354
Last Glacial Maximum (LGM) 51, 348, 372–272
latent heat 19–20, 28n28, 88, 103, 232
latitude 261, 299, 313, 342
Lavoisier, Antoine-Laurent de 10, 26n13, 163, 173–174, 182, 191, 193n24
Law of Mass Action 134, 160n4
Law of Surface Area. *See* Surface Law
Lawton, John 272, 282n19, 350n17
LEA proteins 111, 125
Lepidoptera 381
Lepomis cyanellus 147
leucine 97
Leucodioptron canorum 245–246
Leuresthes tenuis 306n26
lichens 320
life 316–324
   description of 316–318
   temperature limits to 316–324
Lifson, Nathan 180, 194n43
Lindemann, Raymond 342
Linnaeus, Carolus. *See* Linné, Carl von
Linné, Carl von 34, 53n14
Liometopum apiculatum 132
Liparis gibbus 115
lipid unsaturation 226, 258
lipid 84, 96–97
Lithobates pipiens 83
Lithobates sylvaticus 123

lizard 66–67
Lobelia telekii 124
Longhurst, Alan 336
Lovelock, James 318
lower critical temperature 228
Luyet, Basile 113

M

Maciolek, Nancy 332
macroalgae 335
macroecology 3, 5, 300
Macroglossum stellatarum 207
Macrozoarces mericanus 115
Magnolia ovata 213n41
maintenance costs 75, 196, 292, 391
Maja squinado 158
mammals 214–240
    body temperature and diet 220–221
    body temperature and environmental temperature 219–220
    marsupials 216, 232
    metabolic scaling 270–281
    monotremes 216, 232
Manduca sexta 206, 302, 306n23
mannitol 120
Marinomonas primoryensis 126
Marmota breweri 254
Marmota monax 252
Marsham, Robert 379, 389n79
Martin, Andrew 339
mass action ratio 134
mass 8–9
    mass, and weight 26n12
mass-specific variables 71–72
Mauna Loa 363
Maurer, Brian 5n5
Maxwell, James Clerk 15–16
Maxwell-Boltzmann distribution 313, 0–31, 135, 277
Mayer, Julius Robert von 7, 26n7
Mayhew, Peter 349
Mayhew, Wilbur 263
Mazur, Peter 108
McNab, Brian 235, 241n5, 241n15, 271
mechanical equivalent of heat 11
Medieval Warm Period 370
Meinertzhagen, Richard 68
melanin 200–201
Melanophila acuminata 203
Mellisuga minima 253
melting point 84
    See also equilibrium freezing point
membrane 226, 258, 391
    membrane homeostasis 151–152
Menten, Maud 139, 161n13

Mercantha contracta 120
Mesenchytraeus 321
Mesocricetus auratus 258
metabolic cold adaptation (MCA) 189–191
metabolic control analysis 193n20
metabolic costs of activity 175, 177
metabolic flux 148
metabolic niche 346–347
Metabolic theory of Biodiversity 339–342, 349
Metabolic Theory of Ecology 267–281, 339
    See also ontogenetic growth model
metabolism 2, 56–57, 163–191
    active metabolism
    basal metabolism 175–178, 222–223, 227–228, 236, 256, 272, 297, 391
    control of metabolism 280
    feedback control 171
    levels of metabolism 174–181
    maintenance 256
    resting metabolism 71, 175–178, 196, 222–223
    routine metabolism 178–181, 272–273, 297
    standard metabolism 175–178
    intermediary metabolism 168–170
methanogensis 165
metapopulation dynamics 378
methane 355, 364
Methanopyrus kandleri 322
Michaelis, Leonor 139, 161n13
Michaelis-Menten equation 138, 171
Microcebus murinus, 257
Microcebus myoxinus 246, 257
microstates 20, 145
Microtus 300
Milanković cycles 361–363
Milanković, Milutin 361, 389n76
Mitchell, Peter 192n15
mitochondria 148, 165, 202, 224, 258–260, 278, 391
mitochondrial proton leak 171, 178
mitochondrial proton leak 225–226
model 57
    Gompertz function 285
    growth models 285–286, 296–298
    logistic growth model 285–286
    null. *See* null model
    reaction rate and temperature 77, 135, 154–155, 277–278, 290, 339–341
    Richards function 286

    von Bertlanffy growth model 286
Monas dallingeri 322
Moraxella 126
more individuals mechanism 343–344
Morganucodon 238
Morone saxatilis 294
Morowitz, Harold 191n2
Mortola, Jacopo 261
Morus serrator 179–180
Moseley, Henry 304
mucus 74
Mus musculus 224–225
muscle activity 196, 253
    and temperature in endotherms 253
    heat generation by 205–208
    performance and temperature 196, 253
Mustela ermine 68
Mustela nivalis 181
mutual climatic range 375–377
Myodes 300
myo-inositol 120, 150
Myotis lucifugus 252
Myoxocephalus octodecemspinosus 115
Mytilus californianus 199
Mytilus edulis 118, 307n51
Myxocephalus scorpius 114–115
Myxocephalus verrucosus 162n40
Mzilikazi, Nomakwezi 251

N

Nacella concinna 75, 118
Nagy, Kenneth 208
Nanotyrranus 240
natural history 3–4
Nelumbo nucifera 208–209
Nematoda 317, 323
Neoproterozoic Oxygenation Event 165
Nephrurus laevissimus 198
Nernst, Walter 28n34
Neuropogon acromelanus 320
Newton, Isaac 10, 13, 16, 81n33
Newton's Law of Cooling 62, 72–73, 81n33
Newton's Laws of Motion 8
niche
    fundamental. *See* fundamental niche
    metabolic. *See* metabolic niche
    realised. *See* realised niche
    thermal. *See* thermal niche
nitrification 164, 192n5
nitrogen cycle 164
Nitzschia frigida 320–321

Noether, Amalia ('Emmy') 7–8, 26n8
non-activity exercise thermogenesis (NEAT). See thermogenesis
Normalised Difference vegetation Index (NDVI) 343
normothermy 246
North Atlantic Oscillation (NAO) 368, 375, 380, 381, 383
Notothenioidea 115, 129n36
nucleation 103–104
   heterogeneous 104, 111
   homogeneous 103–104
null model 72, 348, 377, 389n72
nunatak 310
Nusselt number 62
Nyquist frequency criterion 368

O

O'Brien, Eileen 345
ocean acidification 366–367
ocean heat content 365–366
oceanic temperature 314–316
   seasonality 315
Ocymyrmex barbiger 209–210, 213n43
Ocymyrmex barbiger 324
Odocoileus hemionus 300
Odonata 381
Oncorhynchus nerka 181
Onnes, Heiki Kamerlingh 35, 53n18
ontogenetic growth model (OGM) 296–298
Onychiurus arcticus 184–185
Onymacris bicolor 68
operative temperature 199
Opsanus tau 290, 306n14
Opuntia 324
Oreotrochilus estella 245
Ornithomimus 240
orthologous gene 141, 145
Osmerus mordax 115
osmolyte 111, 150
Ostrom, John 239, 244n56, 244n59
oxycalorific coefficient 172, 173
oxygen consumption 273, 289, 392
oxygen 92–93, 172
   as final electron acceptor in aerobic metabolism 165, 172
   solubility and temperature 92–93
Oymyakon, Siberia 310–311
ozone 354, 355, 369
   Chapman cycle 369
   Montreal Protocol 369
   ozone hole 369

P

Pachycara brachycephalus 115
Pagothenia borchgrevinki 114, 116
Pagothenia borchgrevinki 150
Palaeoclimate 51
palaeothermometry 47–51
   alkenone 50
   ecological proxies 50–51
   isotope 47–50, 55n39
   Mg/Ca/Sr 50, 55n40
palmitic acid 173
Palumbi, Stephen 339
Panagrolaimus davidi 127n2
panting 233–234
paralogous gene 141
Pararge aegeria 201
Paravinella sulfincola 324
Parmesan, Camille 375
Parus major 383
Passer montanus 228
Passerina 374
Pauling, Linus 161n12
Peck, Lloyd 303
pejus temperature 157
Pelecanus occidentalis 234
peptide bond 173, 287
performance breadth 157
permittivity 85
Perognathus californicus. See *Chaetodipus californicus*
Perognathus longimembris 251
Perognathus parvus 252
Pfaundler, Leopold 135, 160n6
pH 88–90, 94, 97, 148–150, 366–367
   and temperature 89–90
   intracellular pH 148–150
Phalacrocorax auritus 234
Phalenoptilus nuttallii 253, 264
phenology 372, 379–384
phenotypic variance 380
phenylalanine 97
Pheucticus 374
Philodendron selloum 208
Philodendron solimoense 209
phlogiston 10, 193n24
Phoca vitulina 59
Phodopus sungorus 265n24
phonon 59, 79n6
phosphofructkinase (PFK) 171
phosphoglucomutase (PGM) 142
phosphoglucose isomerase (PGI) 141–142
photosynthesis 165, 320, 342, 363
Phylloscopus 3, 374
phytoplankton 106, 345
pile 206–207
pit organ 202
pit-viper 202
planarians (flatworms) 278
Planck radiation law 43–45
Planck, Max 20, 43–44, 54n28
Planck's constant 138
Planigale gilesi 253
plant hardiness zones 103
Plecoma 207
Pliny the Elder 321
Plocepasser mahali 210, 213n43
Poecile atricapillus 231
Poecile montanus 253
Pogona vitticeps 226
poikilotherm 215
Polanyi, Mihály (Michael) 137, 160n9
polar gigantism 303–304
Porter, Warren 61, 66
Pörtner, Hans-Otto 157
potential energy 8–10, 358
   bond energy 85
Power, Henry 129n49
Powers, Dennis 142
Prandtl number 61–62
pre-flight warming in insects 205–207
Priestley, Joseph 174, 193n24
primary productivity 300, 342–247
production 75
proline 146
properties, extensive 14
properties, intensive 14
protein folding 97–98, 290–293
protein synthesis 260, 391
   and temperature 289–293
proteostasis 292
proton 62, 89, 94, 342, 366–367
protonmotive force (PMF) 169, 171
Pseudantechinus macdonnellensis 257
Pseudemys scripta 148
Pseudomonas putida 126
Pseudomonas syringae 111
Pseudopleuronectes americanus 115, 116, 117
Psittacosaurus lujiatunensis 305n4
pterosaurs 240
Pterostichus brevicornis 121
Pütter equation 285, 297, 305n3
Pütter, August 285
Pycnontus sinensis 228
pyruvate dehydrogenase 171, 172
Python molurus 205

Q

Q_{10} 183–188, 255, 277–278, 295
quantum mechanics 20, 138, 145, 160n9, 166
   quantum tunnelling 182, 292

R

radiation, cosmic background. See cosmic background radiation

radiation, thermal 62–69
radiative cooling 310
radiative transfer 354
rain-beetles 207
Rameaux, Jean-François 281n5
Ramsay, James Arthur 214, 240n1
Rana esculenta 123
Rana lessonae 123
Rana sylvatica. *See Lithobates sylvaticus*
range boundary 378
range 374
Rangifer tarandus 300
Rankine, William John Macquorn 37, 54n22
rate of temperature change 392
Rattus norvegicus 83, 162n40, 178, 226
Rayleigh scattering 356
reaction centre 166
reaction norm 155–158, 295
reaction quotient 134
reaction rate 2, 131–160
  reaction rate, and temperature 140–142, 277
reactive oxygen species (ROS) 226, 339
realised niche 377
recrysallisation 121
redox couples 163–164
redox reactions 163–166, 322
reef 94, 101n18, 367, 334–335
Reeves, Robert 148
reflectance 65–68
refugia 372, 374
Regnault, Henri-Victor 35, 53n15
Reid, Clement 373
Reid's paradox 373
reserve 74
respiration 75–76
Respiratory turbinates 240
rete mirabile 62, 80n10, 204–205, 211
Retinia (Petrovia) resinella 120
Reynold's number 90
Rhabditis terrestris 323
Rhodotorula glutinis 321
Rhynchaenus fagi 120
Robertson, Charles 381
Rømer, Ole 33, 53n11
Rosenzweig, Michael 331
Ruben, John 235
RuBisCo 166
Rubner, Max 174, 194n26, 268–269
Rumford, Count 10–11, 17

S

Saccharomyces cerevisiae 108, 110
Saccharomyces cerevisiae 153
Saccharomyces cerevisiae 83

Sagredo, Giovanfrancesco 32
Sakai, Akira 124
Salamandrella keyserlingii 124
salinity 92, 105, 127n7, 315, 320
saliva spreading 234
Salt, Reginald 119
Samia cecropia 205
Sanders, Howard 336
sarcolipin 227
sarcoplasmic reticulum Ca^{2+}- ATPase (SERCA) 205, 227, 258
Sarrus, Pierre 281n5
sauropods 238
Saussure, Nicolas-Théodore de 208, 213n36
scale, interval 38
scale, metric. *See* scale, interval
scale, nominal 37–38
scale, ordinal 38
scale, ratio 38
scaling 71, 81n28, 217, 222, 252, 255, 261, 267–281
  curvature 273–275, 280, 284n51
  differences across metabolic level 274–275
  scaling exponent 268
Sceloporus jarrowi 203
Sceloporus occidentalis 78
Sceloporus undulates 78
Sceloporus undulatus 162n47
Schistocerca gregaria 66
Schmid, William 123
Schmidt-Nielsen, Knut 210
Scholander, Per 114, 190
Schrödinger, Erwin 316–317, 327n18
Scopes, Robert 154
sea-ice 106
seamount 335
sea-surface temperature (SST) 45, 364
seawater 103, 105
  freezing point 103, 105
Second Law of Thermodynamics 22, 170
Seebeck effect 42–43, 54n27
Seebeck, Thomas Johann 54n27
Séguin, Arman 193n25
Selasphorus rufus 245
Semibalanus balanoides 118
Sepia officinalis 306n18
Setophaga auduboni 374
Setophaga coronata 374
Shapley, Harlow 131, 155, 160n3
Sharp, Christine 338
shivering 205, 231, 242n35
  in incubating python 205
Siberia 367
Sidell, Bruce 147

size invariance 267
size 70–71, 206, 216, 260–263, 372, 383, 392
  and heat flow in mammals 223, 275
  and temperature 285–305.
    See also temperature-size rule
Slatkin, Montgomery 196, 203
Smetacek, Victor 354, 374
Sminthopsis crassicaudata 253
snow 107, 252
  winter snow cover 252
Snow, Charles Percy 28n30
solar irradiance 355
  spectrum 356, 357
  strength 355
Somero, George 140
sorbitol 120, 150
Sorex 301
South Pole 310, 364
Southern Ocean 117, 366
spatial scale 331–332
speciation 339, 374
species 330
species-area relationship. *See* diversity
species-energy hypothesis. *See* diversity
specific dynamic action (SDA). *See* heat increment of feeding
Spermophilus citellus 257
Spermophilus columbianus 258
Spermophilus lateralis 251
Spermophilus lateralis 258, 260
Spermophilus parryii 246, 252
Spermophilus richardsonii 257
Spermophilus tridecemlineatus 249–250, 254, 258–260
Sphyraena 140, 142, 159
sponges 335
stabilising selection 380
standard reduction potential 164
standard state 133
Stanley, Steven 373
Staples, Jim 259
state functions 15
statistical mechanics 11–13, 30
statistics 199, 220, 241n5
  importance of variability 5
  linear regression 199, 212n10, 220
  mass-specific variables 71–72
  mixing within and between species studies 5, 271
  transcendental functions 268, 282n8. *See also* dimensions
  use of residuals 283n42
Stefan, Jožef 44, 54n30

Stefan-Boltzmann law 44–45, 355
Steffensen, John 190
Steinernema feltiae 127n2
stenothermy 242n32
steppe-tundra 372, 388n60
Stevenson screen 47
Stevenson, Thomas 47, 55n33
storage. *See* reserve
Storch, David 345
stratosphere 354
Sturnus sericeus 228
sucrose 173
superconductivity 53n18
supercooling 104, 114, 119–121
surface air temperature 308–314, 364
   record high 312–313
   record low 309–311
   seasonality 313–314
surface area 270, 280
Surface Law 268–271, 281n5
surface (skin) temperature 309, 312
surface tension 88, 91
sweating 233
Sylvia atricapilla 253
Sylvia 374
symmetry 7–8
Symplocarpus foetidus 208
systems, thermodynamic.
     See thermodynamic systems
Szent-Györgyi, Albert 163, 191n3

T
tachymetabolism 215
Taeniopygia guttata 253
taiga 231, 314. *See also* boreal forest
Tamias striatus 252
Tamiasciurus hudsonicus 246
Tapinoma sessile 132
tardigrades 317–318
Tauraco corythaix 227
Teissier, George 272
temperature logger, 199 249, 257
temperature regulation 196–211
   See also thermorgulation
temperature scale 32–39
   absolute. *See* temperature scale, thermodynamic
   Celsius 34, 36, 54n21
   development of 32–37
   Fahrenheit 33
   fixed points 32–37, 53n10
   international. *See* International Temperature Scale
   thermodynamic 30, 35–37
   types of 37–38
temperature sensor 201–202
temperature 6, 29–52

air 47
   and energy flow 76
   body. *See* body temperature
   definition of 31
   measurement of 29–52
   ocean 46–47. *See also* sea-surface temperature
   See also temperature scale
temperature-size rule 301–303
Tenebrio molitor 73
Tetramitus rostratus 322
Thales of Miletus 100n2
Thamnophis sirtalis 266n30
thermal capacity 199, 221
   tissue 199
   water 11, 88, 140, 199, 221, 241n10
thermal conductance.
     See conductance, thermal
thermal conductivity.
     See conductivity, thermal
thermal denaturation of proteins 146, 154, 221–222
thermal equilibrium. *See* equilibrium, thermal
thermal hysteresis 116, 126, 129n51
thermal limits to life 2
thermal melanism 66–68
thermal niche 68
thermal performance curve 155–158, 377–379
thermal radiation. *See* radiation, thermal
thermistor 41–42, 46
thermoconformer 197
thermocouple 42–43
thermodynamic beta 13, 30, 346, 352n62
thermodynamic equilibrium.
     See equilibrium, thermodynamic
thermodynamic systems 16–17, 56
thermodynamics 1, 391–392
   classical 11, 30–31
   First Law of. *See* First law of Thermodynamics
   Second Law of. *See* Second Law of Thermodynamics
   Third Law of. *See* Third Law of Thermodynamics
   Zeroth Law of. *See* Zeroth Law of Thermodynamics
thermogenesis 224–228
   activity 227
   costs 256
   muscle 224–228
   non-activity exercise thermogenesis (NEAT) 227

   obligate 224–228
   substitution 231–232
thermogenic plants 208–209
thermogenin 226
thermohaline circulation 359
thermometer
   development of 32
   gas 34–35
   liquid-in-glass 40
   platinum resistance 39–41
   primary 38
   radiation 39, 43–45
   reversing 46–47
   secondary 39
thermometry 32–47
   in ecology 39–47
thermoneutral zone (TNZ) 70, 228–234, 242n31-33
thermoreceptors 218–219
thermoregulation 196–211, 230, 241n17
   beetles 207–208
   behavioural thermoregulation 200–204, 208–210, 232
   bumblebees 207
   butterflies 200–201, 207
   categories of 197
   effectiveness of 187–201
   lizard 199–201
   retention of metabolic heat 204.
     See also rete mirabile
thermoregulatory scope 247, 260–262
thermoscope 32, 53n9
Thermosphaeroma subequalum 323
Thermotoga maritima 322
Thermus aquaticus 322
Third Law of Thermodynamics 23, 28n34, 34
Thomas, Chris 378
Thomas, Theresa 154
Thompson, Benjamin. *See* Rumford, Count
Thompson, D'arcy Wentworth 267
Thomson, William. *See* Kelvin, Lord
Thorson, Gunnar 334–335, 351n24
three-quarter power rule. *See* Kleiber's Law
threonine 114
Thunnus 204
thyroid stimulating hormone 218
tiger-beetles. *See* Cincindelinae
Tilapia rendalli 201
Tipula trivittata 121
torpor 2
torpor 245–264
   energy use during 254–257
   evolutionary aspects 263–264

tracheae 175, 278
trade-off 142, 203, 231, 273, 303, 392
transient receptor potential (TRP) channel 201–202, 218
transition state theory 137–138
transmittance 65–66
transparency 66
trehalose 120
Trematomus bernachii 150
Trenberth, Kevin 359
tricarboxylic acid (TCA) cycle 169, 171, 192n14, 287
trimethylamine *N*-oxide (TMAO) 150
Triturus dobrogicus 204, 212n23
tropical rain forest 334–335
troposphere 354
Tryonia julimensis 323
tundra 314, 348
Turner, James 251
Twain, Mark 371, 389n80
two-factor hypothesis of low temperature injury 108
Tyndall, John 44, 354, 359, 385n1
Tyrannosaurus 239
tyrosine 97

U
ubiquitin 292
ultraviolet catastrophe 44, 54n29
Umbilicaria aprina 320–321
uncoupling proteins 225–226
undercooling. *See* supercooling
units 3
Universal Growth Model 298
Urey, Harold 48, 55n34
Ursus arctos 252
Ursus maritimus 68

V
Valenciennes, Achille 205
valine 97
vampire bat 202
van der Waals interactions 11, 85
van't Hoff factor 104
van't Hoff, Jacobus 135, 153, 160n4-5
Vitreledonella richardi 66
vitrification 112–113, 128n30-31, 130n72
vitrification 319
  arthropods 121–122
  bulk vitrification 112
  colloid glass transition 112
  plants 125
Vogel, Steven 1
von Neumann, John 317–318
Vonnegut, Kurt 127n7
Vostok, Antarctica 309–310, 326n5, 326n8
Vulpes lagopus 68

W
Wallace, Alfred Russell 329, 337, 350n2
Wang, Lawrence 257
warm-blooded 214–215
water 1, 82–100, 308, 355
  anomalous nature of 84–88
  as reactant in physiology 83–84
  as source of electrons in photosynthesis 165
  cell water content 82–83, 98–99
  cloud droplets as habitat 110
  density 86, 88
  gas solubility in 92–94
  ionisation 88–90
  phase diagram 36
  physical properties of 86–88

triple point 36, 53n21
vicinal 99, 102n28
whole body water content 83
water-splitting complex 166
Watt, James 25n2
Watt, Ward 140
wax 74, 84, 92
weather systems 360
weight 267
West, Geoffrey 267, 272
West, George 228
western Atlantic Pleistocene extinction 372
White, Charles 5
White, Gilbert 5n4, 109, 128n20, 379, 385n12
Willig, Michael 333
Wilson, Alexander 245
Wohlschlag, Donald 190
work 7, 17
World Meteorological Organisation (WMO) 309
Wright, David 342, 345, 346

X
Xanthoria candelaria 320
Xenopus 201

Y
Younger Dryas 362

Z
Zapus princeps 252
Zenaida macroura 253
Zenaida macrouri 234
zero, absolute. *See* absolute zero
Zeroth Law of Thermodynamics 29–30, 52n1
Zeuthern, Erik 271
β-galactosidase 290